FRONTIERS *of* POPULATION ECOLOGY

FRONTIERS *of* POPULATION ECOLOGY

R.B. Floyd

A.W. Sheppard

P.J. De Barro

National Library of Australia Cataloguing-in-Publication Entry

Frontiers of Population Ecology.

Bibliography.
ISBN 0 643 05781 1.

1. Nicholson, Alexander John, 1895–1969. 2. Population biology - Congresses. 3. Ecology - Congresses. I. Floyd, R.B. II. Sheppard, A.W. (Andrew Walter), 1960-. III. DeBarro, Paul J. (Paul Joseph), 1963-. IV. Nicholson, Alexander John, 1895–1969. V. CSIRO.

574.5248

This book is available from:

CSIRO Publishing
PO Box 1139 (150 Oxford Street)
Collingwood, VIC 3066
Australia

Tel. (03) 9662 7666 Int:+(613) 9662 7666
Fax (03) 9662 7555 Int:+(613) 9662 7555
Email: sales@publish.csiro.au
World Wide Web: http://www.publish.csiro.au

Melbourne, 1996.

Publisher: Kevin Jeans
Cover and layout design: Linda Kemp
Typesetting: Linda Kemp and Kylie Crane
Production Manager: Jim Quinlan

CONTENTS

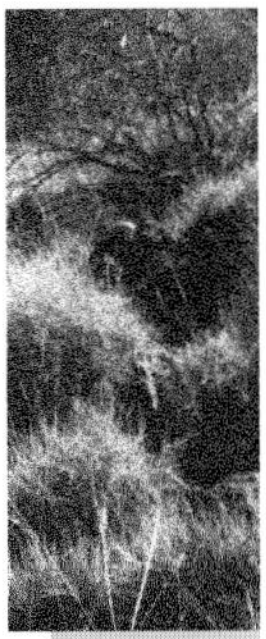

CONTENTS

CONTENTS

CONTENTS

PREFACE

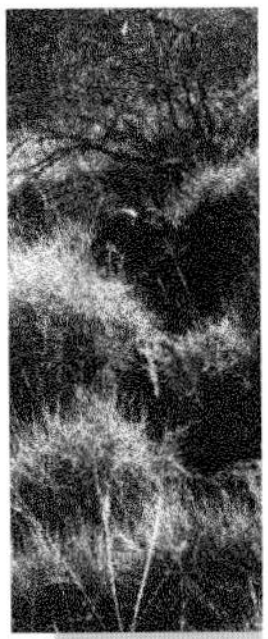

In April 1995, over 200 ecologists from a dozen countries gathered in Canberra, Australia to celebrate the centenary of the birth of the population ecologist A.J. Nicholson (1895–1969). This volume represents a selection of essays from presentations at that meeting and focuses on some of the central issues confronting modern population ecology.

John Nicholson is recognised as one of the leading ecologists of this century and is familiar to ecologists throughout the world because of his research on the factors causing population limitation and the concept of regulation through density dependence. Debate about his research on populations dominated ecology during the middle third of this century and these debates reached their zenith at the Cold Spring Harbor Symposium on Quantitative Biology in 1957. Robert MacArthur reviewed the Proceedings volume of this meeting for the *Quarterly Review of Biology* (MacArthur 1960). There he noted the significance of the symposium because it bought together the proponents of different schools of thought on the factors governing populations and was a key opportunity for a range of views to be aired. He also predicted that the debate would soon be forgotten! In fact this was not the case, although the nature of the debate did change markedly after 1957 as the focus shifted to the development of a methodology for describing density dependence (e.g. Varley and Gradwell 1960) and the consequences of non-linear relationships on population dynamics (e.g. Cook 1965). The fall out from Cold Spring Harbor was such that a great many ecologists have had the occasion to look with renewed vigour over the past four decades at the temporal and spatial trends in the abundance of individual populations and at the interactions between organisms.

Nicholson continues to shape and inform many of our contemporary ideas. His work has been consistently cited in the ecological literature over the past twenty-five years. It is now part of our common currency and lies behind a wide variety of concepts. A.J. Nicholson is remembered for three classic studies (Nicholson 1933,1954; Nicholson and Bailey 1935). For many researchers these studies underpin the debate on population regulation, competition and natural enemy-prey interactions.

It is interesting to see how influential these ideas are in determining the way in which ecologists define themselves and their interests. In 1988 a survey of British Ecological Society members was conducted in order to determine those concepts most widely used by ecologists (Cherrett 1988). This analysis identified about one-third of ecologists by their primary interests in density-dependent regulation and life-history strategies. In the space of fifty years Nicholson's ideas have become the dominant paradigm around research on populations.

The Nicholson Centenary Conference was an attempt to draw together a broad set of presentations across a range of taxa in order to have a comprehensive discussion of the current frontiers of population ecology. The meeting stimulated debate on a range of key issues and touched on major themes such as theoretical and experimental population ecology, the conservation of endangered species, the management of pests and the sustainable management of resources.

In keeping with the spirit of the Nicholson Conference, the essays presented in this volume provide a contemporary view of the major issues facing ecologists and outline some of the strengths and weaknesses of our current ecological framework and its utility in managing biological systems.

Paul Wellings
Chief
CSIRO Division of Entomology

References

Cherrett JM (1988) Key Concepts: The results of a survey of our members' opinions. In: Cherrett JM (ed) *Ecological concepts: The contribution of ecology to an understanding of the natural world.* Blackwell Scientific Publications, Oxford pp 1–16

Cook LM (1965) Oscillation in the simple logistic growth model. *Nature* **207**:316

MacArthur RH (1960) Population studies: Animal ecology and demography. *Q Rev Biol* **35**:82–83

Nicholson AJ (1933) Balance of animal populations. *J Anim Ecol* **2**:132–178

Nicholson AJ (1954) An outline of the dynamics of animal populations. *Aust J Zool* **2**:9–65

Nicholson AJ, Bailey VA (1935) The balance of animal populations. *Proc Zool Soc Lond* **3**:551–598

Varley GC, Gradwell GR (1960) Key factors in population studies. *J Anim Ecol* **29**:399–401

ALEXANDER JOHN NICHOLSON*

Ian Mackerras (Late)

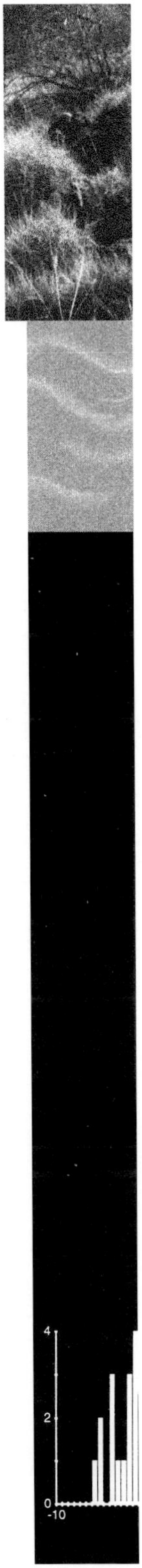

ABSTRACT

The late Ian Mackerras, a long time friend of Nicholson, reflects on Nicholsons life and achievements, particularly his contribution to the development of the sciences of population ecology and entomology and in particular, his ideas on natural selection and density dependent population regulation.

Keywords: AJ Nicholson, obituary, history, population ecology, evolution

* Originally published in 1970 (*Records of the Australian Academy of Science* **2**:66–81) and reprinted with permission of the Australian Academy of Science

Frontiers of Population Ecology, R.B. Floyd, A.W. Sheppard and P.J. De Barro (eds), CSIRO Publishing, Melbourne, 1996 pp. 1–12

Alexander John Nicholson was born at Blackall, Co. Meath, Ireland, on March 25th, 1895. His parents had also been born in Ireland, although both came from English yeoman families of, at least on his mother's side, powerfully Methodist convictions. His father was a successful designing engineer, his mother a woman of acute and independent mind, and his elder brother an inventive member of the Royal Flying Corps. Of his two younger sisters, Nan, a trained kindergarten teacher, lived in Canberra for some years during the thirties and established the first preschool centre on the northern side of the young city.

The Nicholsons moved to England, where John (he appears never to have been known as Alexander) spent the rest of his childhood, except for holidays in Ireland. He shared the family bent for mechanisms and gadgets and the exciting new motor cycles, cars and aeroplanes, but his special interests were in photography and natural history. He became a member of the Birmingham Field Naturalists' Club (annual subscription one shilling) when twelve years old and, to balance it, joined the Midland Aero Club at fifteen, although there is no record that he actually flew. He entered the Waverley Road Secondary School, then remarkable for its new science laboratories, in 1908, and, despite some concern by his masters that he was more interested in natural history than in his formal subjects, matriculated with honours and an entry prize into Birmingham University in 1912. He graduated B.Sc. in 1915 with First Class Honours in Zoology, Chemistry as a second subject, and Botany as a minor. He was not impressed with the course in chemistry, which consisted almost entirely in covering the board with structural formulae of dyestuffs, nor with what he termed the 'necrological approach' to zoology, which was concentrated on comparative anatomy and ignored physiology and the activities of living animals. Still he did get a grounding in classical evolutionary theory.

Then came the 1914–18 war. Already a member of the University Officer Training Cadets (OTC), Nicholson was commissioned 2nd Lieutenant in the Royal Field Artillery in 1915. He spent most of the next two years instructing – strenuous, because they were only allowed six weeks in which to produce gunners – and the rest of the war on active service in France and Belgium. He was promoted to 1st Lieutenant, twice mentioned in despatches, and awarded the King's Commendation for bravery. The war left him with a cardiac lesion that limited strenuous activity for the rest of his life, and it was characteristic that even his closest friends did not know of this until after his death.

He returned to Birmingham in 1919 to begin postgraduate research on two entomological problems: ecology of the natural enemies of *Chermes*, an homopteran pest of pine trees; and development of the ovaries in the mosquito *Anopheles maculipennis*. The latter was a particularly neat piece of work, which has been referred to as a source of information even in recent publications, and he received an M.Sc. for it in 1920. In March 1921 he was appointed the first McCaughey Lecturer in Entomology at the University of Sydney and given permission to spend some months studying entomological research and teaching in America, on his way to Australia. He was profoundly impressed by all that he saw, and especially by the high quality of applied entomology in America. That trip was a sound investment by the Senate of the University.

When Nicholson arrived in Sydney in October 1921, he was a slim, fresh-faced, rather shy young Englishman. He seemed at first to be rather hesitant, but nothing was further from the truth: he knew exactly what he wanted and (although he would never have thought to say so) intended to get it. He began with one great local advantage. The nominally independent McCaughey lectureship was associated with the Department of Zoology, which had recently been taken over by Launcelot Harrison who was rapidly transforming it into what was, during his too brief tenure of the Chair (he died in 1928), probably the keenest and friendliest department in the Faculty. Harrison welcomed Nicholson to his academic family. It was not difficult, because they were both first-class naturalists, they had both served in the British Army (the one in Palastine, the other in Europe), they were both equally determined not to rear their students on an undiluted diet of morphology, and, although temperamentally poles apart, they were both fundamentally friendly people. In any case, no one could remain shy or inarticulate for long in the atmosphere that Harrison was fostering, and Nicholson was very responsive to goodwill.

Up to 1921, entomology was taught as a fragment of the Arthropoda in Zoology, and (by W.W. Froggatt) as a mixture of rather primitive natural history and 'blunderbuss' control measures in Agriculture. It was Nicholson's job to give it substance in both faculties, and this was urgent, because, as his observations in America had shown him, training by apprenticeship was just not good enough any longer in either the academic or the economic field.

It is useful to pause at this point to consider the state of entomology in Australia, particularly in Sydney, fifty years ago. The professionals were all self-trained or had served some form of apprenticeship. Thus, W.W. Froggatt, Government Entomologist in New South Wales from 1896 to 1923 and a most prolific writer, got most of his training as a collector for Sir William Macleay; his successor, W.B. Gurney, joined the department in 1900, but did not obtain a B.Sc. (presumably as a part-time student) until 1925. In fact, A.R. Woodhill (who had the best of both worlds by taking Froggatt's course in 1920 and Nicholson's in 1923, and who succeeded Nicholson in the University in 1930) was, when he joined Gurney in 1924, probably the first graduate entomologist to be appointed to a Department of Agriculture in Australia. The situation was similar in the Museums, and in the Commonwealth there were only F.H. Taylor and G.F. Hill who, up to that time, had been mostly concerned with medical entomology. Almost all these people who were paid for being entomologists had built their careers on a very modest level of secondary education.

The insects attracted few students in the universities. In Brisbane, T. Harvey Johnston had two young graduates, Josephine Bancroft and O.W. Tiegs, who later became eminent, and in Sydney, Harrison had done notable work on Mallophaga (most of it, actually, while he was in Cambridge). By far the most distinguished was R.J. Tillyard, a Cambridge graduate, who must rank in part with the amateurs, for he was a school teacher for several years before resuming academic life in Sydney University in 1913 as a Research Scholar and subsequently a Linnean Macleay Fellow. He went to the Cawthron Institute in New Zealand in 1920, was elected to the Royal Society of London in 1925, produced his classical *Insects of Australia and New Zealand* in 1926, and returned finally to Australia in 1928.

In fact, productive work in entomology at that time was predominantly taxonomic and predominantly in the hands of educated amateurs, mostly professional men with rather more leisure than is common today. Some of the most productive, in terms of publication, were A.J. Turner (Lepidoptera) and A.E. Shaw (Blattodea) in Brisbane; G.A. Waterhouse (Lepidoptera), H.J. Carter (Coleoptera), E.W. Ferguson (Diptera and Coleoptera), and R.J. Tillyard from 1907 to 1913 (Odonata) in Sydney; and the Rev. Thomas Blackburn (Coleoptera) in Adelaide. Of these seven, three were medical practitioners, one a metallurgical engineer, two schoolmasters, and one a clergyman. All were active in the local scientific societies, the Sydney group particularly in the Linnean Society of New South Wales (N.S.W.) and a small but lively entomological section of the Royal Zoological Society of N.S.W.

To return to the University, Nicholson was given a detached laboratory at the western end of the Zoology School, sufficient McCaughey funds to equip it modestly, the services of Miss Gwen Burns, a very competent photographic assistant in the Zoology School, and the task of preparing two sets of classes. The more urgent, first given in 1922, was a short introductory course in first year and a full one in second year in Agriculture; the other a two-term course that could be taken by Science students as an alternative to vertebrate zoology. Nicholson described his plan, in retrospect, in the following words: '*The basis of both courses was general systematics, morphology, and general biology. Only sufficient of this was given to Agriculture students to provide a firm basis for an intelligent development of the subject of economic entomology, to which I naturally had to devote as much time as I could. With Science students I made little more than passing reference to insect pests and their control, treating entomology simply as a component of Zoology and going into greater detail with morphology, systematics and biology*'.

He had to face three problems in putting his plans into effect. He had to obtain a knowledge of the strange Australian fauna that he was seeing for the first time, to devise means to bring his courses to life, and to learn about the pest species and their control. The last was the easiest. Many of the pests were immigrants (or at least related to northern hemisphere species), he already knew something of the situation in Europe and America, and he was able to obtain State departmental help, particularly from Gurney, to fill in the local picture. The need was to foster a new approach, rather than to look for new pests.

The only solution to the first problem was to go into the field and learn for himself. His first trips were to Woy Woy and the Blue Mountains with Tillyard and A.L. Tonnoir, who were visiting Australia to fill gaps in their own research material, and Harrison took him to the Blue Mountains and Mt Kosciusko; but he received the greatest help from the band of able, enthusiastic amateurs mentioned earlier. He was quickly elected to the Linnean and Royal Zoological Societies and joined in their discussions (especially over coffee after the meetings), but, more important, somebody was always going off in the week-ends, or even during the week, to collect in some favourite spot. G.M. Goldfinch, a lepidopterist, had a roomy Essex car which he regularly filled with congenial entomologists, Waterhouse and Ferguson had cars too, Nicholson later acquired a Morris Cowley of his own, and there were frequent parties to the Zoological Society's cottage in the National Park (especially in the early spring), to many other parts of the Hawkesbury sandstone area at all seasons, and sometimes further afield. A notable later trip that he made was with P.D.F. Murray to Papua and New Guinea. On these trips Nicholson always seemed to be dawdling behind the rest of the party, but he regularly ended up with a greater number and variety of specimens than any one else. There was, too, considerable mutual stimulation, the specialists widening their interests to all orders under his influence and he learning a great deal about their own groups from them. That was not always true, for sometimes he could show them something they had not known about the biology of their own species.

The results of these activities were a remarkably wide knowledge of Australian insects (second only to Tillyard's) acquired in a remarkably short time, and a notable teaching collection of which Mrs. J.W. Evans (a daughter of Tillyard and herself an entomologist) wrote much later: '... *it became my responsibility for a year or two to care for the [Zoology] Department's entomological collections. A great deal of this had been collected by Nic across his Sydney years and I was amazed at the industry and the vastness of the field he had covered. It goes without saying that it was all, over 30 years afterwards, perfectly mounted and preserved and it brought alive most vividly AJ.N.'s early field trips in Australia*'.

There remained the problem of balancing his teaching with an adequate leavening of general biology. The special physiology of insects was little known at that time, but Nicholson taught what he could, sometimes, he thought, rather rashly. He inaugurated field excursions to show his students how insects lived and what they did, but there was not room in the timetable for enough of them. As well as pinned adults, his collections included immature stages and the results of insect activity, and these could be used in practical classes; but that was still not enough, so he resolved to supplement his efforts by photography. This could be done in the laboratory, where he was ably assisted by Miss Burns, or in the field, for which he developed an ingenious telescopic range-finder for his camera and learned to use flashlight effectively both in shade and at night. He photographed everything of interest that he saw, and thus was able to illustrate his lectures by a remarkable series of lantern-slides of larval and adult insects going about their normal business in their natural habitats. Nearly forty years later he adapted a Leica camera to do the same work in colour.

The history of entomology in Australia has been divided into four periods: Fabrician (1770–1860) when explorers sent their material to Europe for study; Macleayan (1861–1890) when the study came increasingly to be done in Australia; the period of the amateurs touched on earlier; and the period of the professionals. Nicholson's arrival marked the beginning of the transition between the last two. He put university teaching on a firm scientific foundation, and provided the first agricultural graduates to bring a new approach to economic entomology in this country. He had few students in pure science (Mary Fuller, who died young, was the most promising), and it was not until a separate honours year was instituted by the Faculty of Science after he left that the foundation he laid could be built on effectively. Nevertheless, the wind of change was blowing steadily and with increasing strength. Lectureships (and one Chair) were established in other universities, the demands of State departments grew steadily, and the Commonwealth established the Council for Scientific and Industrial Research (CSIR) in 1926 with G.A. Julius, A.C.D. Rivett and W.J. Newbigin (succeeded in 1927 by A.E.V. Richardson) as its first Executive. Young people interested in insects no longer became doctors or schoolteachers and devoted their leisure to their hobby; they took an appropriate degree in a university and made entomology their profession, with ever widening fields of research and ever increasing technical and financial resources before them. It could fairly be said that, so far as Australia is concerned, Nicholson in a sense began it all.

In 1927 he found himself with the task of preparing a presidential address for the Royal Zoological Society of N.S.W. He looked through his slides to find what subject he could illustrate best, and decided on mimicry. At about the same time a bright student had raised doubts about the generality of the rather crude concepts of population dynamics that then prevailed. This concurrence of unrelated events led to the production of a very important paper which has not received the attention it deserved. He not only gave a comprehensive review of mimicry and concealing coloration in Australian insects and demonstrated the reality of the phenomena, he also stated the basic principles and hypotheses that he was to spend much of the rest of his life in testing and amplifying. In brief, the thesis he developed was that animal populations could not survive in nature unless their densities were governed by some regulatory (feed-back) mechanism that was density-dependent in its operation; that success in searching for essential resources was the only mechanism that was truly density-dependent, all others being merely modifying; that the 'power of discovery' of natural enemies could regulate the abundance of both parasites and hosts; and that natural selection did not determine survival and abundance, it merely ensured that any form of a species that acquired an inheritable advantage, however slight, would progressively replace the previously successful form. In the immediate context, survival and abundance of the mimetic or procryptic insects studied seemed clearly to depend on density-dependent stresses operating on their immature stages, the perfection of their concealment or mimicry on quite minor selective pressures imposed by predators on the adults. The theory was well documented, it gave a credible explanation of the phenomena observed, and it was free from the objections that had been raised against previous theories, some of which, as Nicholson pointed out, were due to a failure to understand what Darwin had meant by natural selection. For this work he was awarded a D.Sc. of Sydney University in 1929.

He then concentrated on theoretical population dynamics, using arithmetical models of situations approximating to ones that he knew occurred in nature, and examining more particularly, though not exclusively, various kinds of host-parasite interactions. He soon found, to his surprise, that the equilibrium produced was not a steady one, but generally more or less violently fluctuating. V.A. Bailey began collaborating with him at this time, using more sophisticated mathematical treatment that gave generality to the results, and they found that the fluctuations were due to time displacements associated with length of life cycle. They wrote a book (finished in 1931) on the theory and its implications, but it was 'too theoretical' to find favour with the publishers' readers of the day, and their results ultimately appeared in a rather scattered series of joint and separate papers.

In the mean time, R.J. Tillyard had returned to Australia in 1928 as Chief of the newly established Division of Economic Entomology of CSIR. He gathered the nucleus of a staff about him, and they set about helping him to plan the research and the laboratories at Canberra. Tillyard proved to be an extraordinarily stimulating scientist, but an unpredictable administrator. To balance this, the Executive appointed G.A. Waterhouse (then Curator of the Division) as its Executive Officer. He remained only a year, and in 1929 Nicholson was offered the position of Deputy Chief. He accepted on the condition that he should spend most of 1930 in England as he had already planned. He undertook a number of enquiries for the Division while abroad, and arrived in Canberra towards the end of that year.

He found his position difficult, and the times were difficult too; no new developments could be planned, so he sought transfer, as a Senior Entomologist, to the sheep blowfly investigations, which had been established before the depression began. During the next year and a half, he helped to guide the entomological side of the Section's work and began two pieces of personal research. One was a study of traps and baits, for which he designed ingenious ways to reduce the action of external variables, and the other a study of the effects of temperature on the activity of the adult flies. Tillyard became ill before the trapping studies were completed, Nicholson was appointed Acting Chief in April 1933, and from then on he had little opportunity for personal research until after the war.

In 1933 he married Phyllis Heather Jarrett, M.Sc., a former CSIR research student and at the time a plant pathologist in the Division of Plant Industry, and she assisted him for some years thereafter as an unofficial technical secretary. They had two sons, Garth Alexander, who was a CSIRO scholar at Geelong Grammar School and graduated M.B., B.S. (Sydney) in 1967, and Peter John, who graduated B.Ec. (Sydney) in 1968.

The financial situation was still precarious in 1933, and the Executive had not been happy about some of the Division's work. Nicholson, aided by his senior staff (G.F. Hill, I.M. Mackerras, G.A. Currie, A.L. Tonnoir), gave it stability during this period. He planned with great care, and strove especially, as he did in all subsequent years, for a judicious balance between fundamental and applied research. He emphasised, indeed, that the need for fundamental research was probably greater at that time in economic entomology than in any other applied science. The Executive were irritated by the, to them, inordinate time he often took to develop his proposals, but they came to respect the final clarity of their presentation, their wisdom, and his unswerving scientific integrity. His appointment was confirmed in 1936. The same qualities – and Rivett's support – held the Division together during the following years, when there were moves to abandon taxonomy to the museums and distribute some of its other activities among the sister Divisions to which they were relevant.

Tillyard's main theme had been biological control of weed and insect pests. Some of his people had already found it necessary to approach their problems more broadly, and Nicholson further widened

John and Phyllis Nicholson

the approach and the range of problems under study. Biological control remained an important activity, but on a more cautious and critical basis; ecology, with emphasis on the interactions of the animals rather than the physical background, became the major theme in most investigations; biochemical studies increased; physiology and toxicology found a specific place in 1938; important protozoological and virological investigations were undertaken for the Division of Animal Health; and taxonomic research expanded. It was a fertile period, too, in the studies of life histories and general biology of the insects. In 1938 he visited research institutions abroad, and gave a paper to the 7th International Congress of Entomology in Berlin, in which he stressed the fundamental importance of detailed ecological studies of pest populations in planning the use of pesticides and assessing the results achieved. This paper was the forerunner of the modern concept of strategic pest control.

The war affected the Division in two ways: it took away nearly a third of the scientific staff on active service, and it caused a reorientation of the efforts of those that remained. In particular, a strong team, initially under F.N. Ratcliffe, was directed to studies of the pests of stored products, with results of substantial value to the general war effort; and D.F. Waterhouse did very valuable work in collaboration with the army medical services on the development of repellents and other entomological aspects of the control of malaria and scrub typhus. Ratcliffe became an Army malariologist; Waterhouse was appointed to the Reserve of Officers to be called up as needed, and so close was the collaboration that it was difficult to know at any given moment whether he was on active service or a civilian scientist. Nicholson strongly supported all these activities, served on the Medical Services Advisory Committee, and, in his spare time designed an acoustically controlled torpedo for use against submarines.

The atmosphere became more favourable after the war. The Division had made its mark, there were increasing demands for a greater range and diversity of studies, and the Executive had come to appreciate the importance of Nicholson's own field, and his reputation in it, more fully than they did before the war. Cattle tick research (begun during the war) became a major activity; the work on pests of stored grain was continued; ecological studies were extended to pasture, forest and fruit pests not previously investigated; much was to be learned about the new insecticides that had been developed during and after the war; the concept of integrated control was beginning to take shape; a distinguished school of insect physiology and biochemistry was growing up in the Division; studies of the transmission of plant viruses were begun; and a small unit of experimental population dynamics was set up for Nicholson's own work. The rate of staffing shows the overall growth most simply. In 1929, when the Canberra laboratory was occupied, the scientific staff (including graduate assistant and equivalent) was 12; in 1933, when Nicholson took over from Tillyard, it was 16, rising to 23 in 1939, and falling (in effect) to 14 during the war. It rose again to 29 in 1946, 37 in 1949 and 43 in 1959. All this expansion threw additional burdens on the Chief, and he was given progressively increasing administrative assistance to cope with them. This enabled him to do some personal research, though never as much as he could have wished.

Nicholson disliked rigid sectionalisation of research and an hierarchical arrangement of staff. His policy was 'to give each research worker full responsibility and great freedom within his allotted field of investigation', and to draw all members of a team into the planning of that team's activities. To these ends, he gradually abolished most of the pre-war sections, replacing them by three broad groups (taxonomic, ecological, physiological-biochemical) and substituting leadership resulting from recognised ability for direction. Of the three groups, taxonomy and physiology-biochemistry were relatively clearly defined, ecology more loosely linked and occupied with a less coherent range of problems and activities. There was, however, considerable flexibility. Thus, veterinary entomology and termite research retained much of their sectional identity, but with links in all three directions, and some more isolated investigations, such as the earth mite and lucerne flea studies in Western Australia, remained relatively discrete. He stressed, too, that physiological and biochemical problems could usually be defined with comparative precision and lead to the production of neat, scientifically acceptable papers, though also revealing very quickly the deficiencies of an inadequate worker; whereas ecological work generally lacked an established theoretical foundation, and was consequently likely to be relatively diffuse and productive of long, less immediately satisfying publications. The assessment of merit in these divergent fields worried him. Nevertheless, he had a keen eye for work that was intellectually demanding and likely to be intellectually rewarding, and his policy seemed, in general, to work very well. The Division was successful, which is the crucial test.

Nicholson completed his term as Chief in March 1960, and was appointed to a Senior Research Fellowship to continue his twin studies of population dynamics and natural selection. His unit was also retained for long enough to finish the experiments he had already planned, but his main task was to prepare his findings for publication. This had not been completed when he became ill, his damaged heart failed, and he died on October 28th, 1969.

It is difficult to give a brief account of his population studies because they are inherently complex and have been bedevilled by criticism and misunderstanding. It is, in fact, necessary at the outset to repeat his basic postulate in purely general terms, as all his major conclusions follow from it and it has been completely ignored by all his major critics. It is simply that no intrinsically variable phenomenon can continue to exist in the absence of a feed-back mechanism to limit its variability. In the specific context, animal populations, in places where they can exist at all, are governed fundamentally by processes that are density dependent, all other influences being merely modifying, and the actual population level at any moment being the result of interaction between the governing and modifying processes. It is to be noted, too, that the differences between the two sets of processes are qualitative, not quantitative, a fundamental point that critics have also failed to appreciate.

In the first, rather extended phase, as already noted, he developed a complex structure of theory by applying his basic premise to a variety of imagined natural situations. After the war, he was able to test many of his conclusions on experimentally manipulated populations. Initially, following H.S. Smith and his colleagues of Riverside, California, who had arrived independently at essentially similar conclusions, he used housefly puparia and the parasite *Nasonia vitripennis.* These gave trouble, so he turned to the sheep blowfly, *Lucilia cuprina*, which had proved itself to be a remarkably biddable laboratory animal, and he abandoned the examination of host-parasite interactions (temporarily, he thought) for a detailed study of intraspecific competition. Most of the results of this phase were presented, at least in outline, to various congresses and symposia, of which he attended no less than six between 1947 and 1959. It is worth noting, in view of criticisms that have been made of laboratory ecology, that the field ecology of *L. cuprina* had been studied in considerable detail by Mary Fuller, and that Nicholson's findings, though quantitatively much more precise, were completely in accord with hers.

In his mimicry paper he had developed the theme of random search and discovery as the basis of density-dependent reactions, but he soon transferred the emphasis to the competitive element that is inherent in the search, and he used the word 'competition' in this sense (sometimes to the confusion of his critics) in all his subsequent work. He distinguished between two kinds of competition: 'scramble', a continuing struggle which is wasteful of resources and leads to violent oscillations in population density; and 'contest', as in territorial animals, which conserves resources and results in steadier population levels. Basically, in his view, populations of any species are governed (not determined) by some form of intraspecific competition for essential resources (nutrients, places to live, and so on), but this can produce two distinct results. Usually (as with flies in carrion) provision of the resources is independent of the population, so their varying level in the environment serves simply as a modifying influence: only the single population is governed. When, however, the resource is itself a living organism and the association is sufficiently specific and destructive, the populations of both predator and prey (or parasite and host) are governed by the competition among the attackers for their essential living resource, each population interacting with the other. This is what has happened, for example, with *Cactoblastis* and prickly pear. On the other hand, in more usual forms of interspecific competition (as between different species of blowflies in a carcass) each species acts merely as a modifying influence on the others' populations.

Balance is the resultant of two opposing 'forces', the reproductive capacity ('power of increase') of the organisms on the one side, and the governing + modifying stresses on the other, the level of balance varying with changes in any of the three components. Thus the 'steady density' is always more or less unsteady in natural populations. Reproductive capacity is always greater, generally very much greater, than is necessary to maintain populations under favourable conditions, and Nicholson stressed that this gives the population great resilience in coping with adversity, as was abundantly demonstrated in his experiments. For example, in a stabilised population governed by competition for larval food, destruction of 99 per cent of the adults before they could lay eggs resulted (through relaxation of the larval competition) in a compensatory six-fold increase in the number of pupae maturing in the next generation. This effect could be maintained indefinitely by continuing the destruction of adults in succeeding generations, and it could be produced equally well by replacing mechanical destruction by competition between the adults for limited supplies of the protein food necessary for them to mature their eggs. That last series of experiments was particularly interesting, because it was found that larval competition could be replaced completely by adult competition; and, when mutant adults capable of maturing eggs on progressively smaller quantities of protein appeared and became selected in these cultures, balance was still maintained, but at progressively altered levels; until, finally, mutants able to mature eggs without any protein food also appeared and the population showed evident indications of beginning to explode.

These are the broadest of Nicholson's conclusions (with some illustrative examples) and they should be applicable to all animal populations, and presumably to plant populations too. However, his models were insects, in most of which there is a sharp ecological dissociation between the larval and adult populations of the species, and some of his subsidiary findings may depend upon that fact. The phenomenon of oscillation, which he examined in great detail in his later work, may be one, and so too may be some of the detailed results mentioned above. On the whole, however, it would appear that the disadvantages of using models with specialised life cycles were more than offset by the advantages of being able to define and measure their population structure with accuracy and precision.

Other aspects of intraspecific (and some of interspecific) competition could have been quoted, and the interactions between natural enemies and their hosts were studied in similar detail, though only at the theoretical level. In short, by his meticulous examination of the ways in which both governing and modifying processes could act, and of the manifold effects of altering them, Nicholson provided a system of thought which could serve as a basis for modern work on integrated control and pest management. This, indeed, was an aim that he himself had clearly in mind.

He was a slow worker (each experiment had in any case to run for at least eight months and often much longer) and an even slower writer. He was greatly concerned with the use of words and spent a great deal of time in choosing exactly the right ones to convey his meaning – and perhaps even more in eliminating redundancies and compressing what he had to say into the minimum amount of space. This made for clarity, but not always for easy reading. He was, too, sensitive to criticism, and inclined to spend more time than it warranted in answering it. The famous controversy with Andrewartha and Birch is a case in point. However, the crux of the problem lay in the complexity of his subject. Knowing, as a realist, that the time before him must steadily be growing shorter, he endeavoured to break his material into a series of papers to be completed sequentially, but at every turn he found himself faced with points that required cross-reference to other parts still not written. This grew so distracting that he reverted to his original plan to put it all into a single book. He was still organising this when he died. He left notes, graphs, and experimental records in very good order, and it is to be hoped that some one expert in the field may be found to integrate them into something resembling the treatise that he did not finish.

His studies of natural selection were more complete, though not all of his experimental evidence got into his published reports. What he did, in essence, was to clarify what Darwin and Wallace meant by natural selection and show how it worked. His thesis may be summarised in five statements.

1. Fitness, in the sense of Darwin and Wallace, is relative, like the fitness of the tall player who can manipulate a basket-ball over the heads of his shorter fellows.
2. There are two kinds of natural selection: environmental (Wallacean) in which the less fit are eliminated in periods of adversity; and competitive (Darwinian) in which the more fit displace the less fit in the struggle for existence, whether the environment varies or not. Both operate, alone or in conjunction, the Darwinian being the more powerful and the only one that can, by itself, lead to improved exploitation of the environment.
3. Natural selection operates in balanced populations which have inherent resilience in coping with variations in environmental stresses – that is, a substantial degree of homoeostasis. Consequently, any variant that possesses an inheritable quality giving it an immediate advantage, however slight, will progressively displace the previously successful members of the population, and the governing mechanisms will progressively adjust the balance to meet the new circumstances. Selection, in fact, is limited only by the capacity of the gene pool to produce advantageous variants, and hyperadaptation is the normal end result. All these occurrences were observed in the experimental populations of *Lucilia cuprina*, of which an example has been given above.
4. There is nothing purposive in natural selection. Population levels are not determined, nor necessarily much influenced, by it, nor has it necessarily anything to do with survival of the species. It can produce novelties, which may be ultimately disadvantageous like the antlers of the Irish elk, or potentially advantageous like his own strain of flies that could mature eggs without a protein meal; it does ensure 'biological improvement' of the population; and it can assist indirectly (as well as directly) in the process of speciation.
5. The post-Darwinian concept of selection producing a precise fit between organism and environment, like the exact fit of a complicated key in its lock (Fisher), is false. There is no need for this kind of 'fitness', no evidence that it exists, and no way in which either Darwinian or

Wallacean natural selection can produce it. Inferences that have been based on it should therefore be re-examined.

Looking at his publications as a whole, three stand out as having all the qualities of enduring classics: the 1927 study of mimicry, which has been too badly neglected; the 1933 statement of population theory, of which C.A. Fleschner, of Riverside, wrote (in litt.) in 1970 that it was far superior for teaching the principles of biological control to anything that has been published since; and the 1960 analysis of natural selection, which is so well documented that it should have a deep influence on modern thinking in population genetics and evolutionary theory.

Like his father and brother, Nicholson was an inventive designer of equipment. He delighted in gadgets, as he did in words and their uses, but always for the functions they could perform in solving problems that concerned him. His own were aesthetically pleasing, because they did exactly what was required of them with a minimum of complication and effort. A few have been mentioned above, the remainder were developed for his population studies. They included simple tissue grinders (published in 1951) that disintegrated all cells and most nuclei, photoelectric equipment to count blowfly eggs, a fractionator for separating the stages of *Lucilia cuprina*, a pneumatic method of capturing living flies, and an instantaneous vacuum cleaner for collecting empty puparial shells. Descriptions of the last two were left in a sufficiently advanced state to be completed as a posthumous publication.

He had another great interest, but lightly touched on so far, and that was his love of natural history. Wherever he went, he observed, photographed, and collected. After the war, he and his wife acquired a weekend cottage at Bawley Point on the coast, where they spent most of their leisure, and where he brought his colour photography to its highest level. He did not confine himself to insects, but photographed birds, frogs, plants, and indeed any living thing that caught his eye; he even developed underwater equipment to record the doings of marine animals, and used it to great effect when he visited a coral cay. He extended his records of mimicry, made new observations on pollination of orchids by male wasps, and studied the stridulatory mechanism of the whistling moth. His opportunities increased after he retired as Chief, and still more when his research fellowship ended officially in 1966, although he continued to work in the laboratory. The last nine years, when he was free from responsibility and could do what he thought most satisfying, were probably the happiest in his life.

His love of natural history was reflected also in his strong support of conservation in Australian New Zealand Association for the Advancement of Science (ANZAAS) and Academy deliberations, and in the very active part he played in the successful effort of the Royal Society of Canberra to get the Tidbinbilla Fauna Reserve established.

Nicholson had his share, too, in the gregarious activities of scientific societies. He was elected to the Linnean Society of N.S.W. in 1922, served on Council in 1928–29, and remained a member until his death. He was a member of the Royal Zoological Society of N.S.W. (President 1926–27) and of the Royal Society of Australia (President 1951, Council for several years); foundation member of the Australian Institute of Agricultural Science, the Ecological Society of Australia, and the Australian Entomological Society; foundation President of the National Parks Association of Australian Capital Territory (ACT); and a Fellow of ANZAAS (President of Section D 1947). He was a Foundation Fellow of the Academy, its first Biological Secretary in 1954–55, Vice-President in 1955–57, the first '*amicus curiae*' appointed to serve as a link between the young Council and the even younger Sectional Committees, and the initiator of the Academy's proposal to the Commonwealth Government for establishment of a research museum of Australian biology in Canberra. He made a significant contribution to the stability and steady growth of the Academy in those early difficult years.

He received several awards in addition to his higher degrees: the Clarke Medal of the Royal Society of N.S.W. in 1952; C.B.E. in 1961; Honorary Fellowship of the Royal Entomological Society of London in 1961; and Honorary Membership of the British Ecological Society in 1963, the only worker in Australia to have been so honoured. He will be long remembered for initiating the professional era in Australian entomology, for his contributions to the development of the Division of Entomology in CSIRO, for the publication of three enduring classics, and, by those who knew him, for his quiet friendliness, wise counsel, and complete integrity.

This account of John Nicholson's life and times was based on his own notes, on information provided mainly by Mrs. Phyllis Nicholson and Dr. D.F. Waterhouse, and on the memories of my friendship that lasted for forty-eight years.

References

Nicholson AJ (1921) The development of the ovary end ovarian egg of a mosquito, *Anopheles maculipennis,* Meig. *Q J Microsc Sci* **65**:395–448

Nicholson AJ (1927) A new theory of mimicry in insects. *Aust Zool* **5**:10–104

Nicholson AJ (1931) Methods of photographing living insects. *Bull Entomol Res* **22**:307–320

Nicholson AJ (1932) 'Protective' adaptations of animals. *Nature* (Lond) **130**:696

Nicholson AJ (1933) The balance of animal populations. *J Anim Ecol* **2**:132–178

Nicholson AJ (1934) The influence of temperature on the activity of sheep-blowflies. *Bull Entomol Res* **25**:85–99

Nicholson AJ, Bailey VA (1935) The balance of animal populations. *Proc Zool Soc Lond* **3**:551–598

Nicholson AJ (1935) Scientific method in the study of biology. *J Aust Inst Agric Sci* **1**:18–21

Nicholson AJ (1937) The role of competition in determining animals populations. *J CSIR (Aust)* **10**:101–106

Nicholson AJ (1939) Indirect effects of spray practice on pest populations. *Trans 7th Int Congr Entomol* (Berlin, 1938) **4**:3022–3028

Nicholson AJ (1948) Fluctuation of animal populations. *Rep Aust Assoc Adv Sci* (Perth, 1947) **26**:134–147

Nicholson AJ (1950) Progress in the control of *Hypericum* by insects. *Proc 8th Int Congr Entomol* (Stockholm 1948):96–99

Nicholson AJ (1950) Competition for food amongst *Lucilia cuprina* larvae. *Proc 8th Int Congr Entomol* (Stockholm, 1948):277–281

Nicholson AJ (1950) Population oscillations caused by competition for food. *Nature* (Lond) **165**:476–477

Nicholson AJ (1951) A simple method of disintegrating cells. *Nature* (Lond) **167**:563–564

Nicholson AJ (1954) Experimental demonstrations of balance in populations. *Nature* (Lond) **173**:862–863

Nicholson AJ (1954) Compensatory reactions of populations to stresses, and their evolutionary significance. *Aust J Zool* **2**:1–8

Nicholson AJ (1954) An outline of the dynamics of animal populations. *Aust J Zool* **2**:9–65

Nicholson AJ (1955) Density governed reaction, the counterpart of selection in evolution. *Cold Spring Harbor Symp Quant Biol* **20**:288–293

Nicholson AJ (1957) The self-adjustment of populations to change. *Cold Spring Harbor Symp Quant Biol* **22**: 153–172

Nicholson AJ (1958) Dynamics of insect populations. *Annu Rev Entomol* **3**:107–136

Nicholson AJ (1959) Density-dependent factors in ecology. *Nature* (Lond) **183**:911–912

Nicholson AJ (1960) The role of population dynamics in natural selection. In: Tax S (ed) *Evolution after Darwin: the University of Chicago centennial, vol. 1*. University of Chicago Press, Chicago, pp 477–521

Nicholson AJ (1962) Population dynamics and natural selection. In: Leeper GW (ed) *The evolution of living organisms.* Melbourne University Press, Melbourne, pp 62–73

Nicholson AJ, Bailey VA, Williams EJ (1962) Interaction between hosts and parasites when some host individuals are more difficult to find than others. *J Theor Biol* **3**:1–18

Evolutionary Theory and the Foundations of Population Ecology: The Work of A.J. Nicholson (1895–1969)

Sharon E. Kingsland

ABSTRACT

A. J. Nicholson's work in evolutionary ecology is set in the context of the modern evolutionary synthesis. From early studies of mimicry in the 1920s to experimental investigations of natural selection in laboratory populations in the 1960s, Nicholson tried to clarify Darwin's theory, emphasising the ecological rather than the genetical standpoint. His contributions to evolutionary theory can be boiled down to three ideas. First, all purposive language had to be avoided when assessing the role of natural selection. Second, in order to understand the effects of natural selection on species survival, one had to understand how populations were regulated. Third, competition within the species was the key to understanding how populations were governed. Nicholson's arguments about population regulation greatly influenced David Lack at the University of Oxford, but his concept of the population as a self-regulating entity and his ideas about competition drew criticism from entomologists, in particular W. R. Thompson, H. G. Andrewartha and L. C. Birch. Whether or not all of Nicholson's conclusions are accepted as valid, his point of view can be seen as still pertinent to modern debates about adaptation and to current concerns in conservation biology.

Key words: A.J. Nicholson, population ecology, history, evolutionary theory

Frontiers of Population Ecology, R.B. Floyd, A.W. Sheppard and P.J. De Barro (eds), CSIRO Publishing, Melbourne, 1996 pp. 13–25

In 1957 a group of ecologists gathered for a symposium at Cold Spring Harbor, New York, to discuss certain topics in population ecology that had provoked controversy for several years. Evelyn Hutchinson (1957), giving the concluding remarks to the conference, described population biologists as a 'heterogeneous unstable population', a comment on the vigorous debates that had characterised the meetings. Theodosius Dobzhansky (1957), as a member of the audience, found these controversies frankly bewildering. He concluded that they seemed to be mainly about words, that is, disagreements about definitions of basic concepts. Robert MacArthur (1960), reviewing the conference proceedings, remarked that the meeting was more notable for its 'almost religious fervor' than for any new facts.

At the centre of this controversy was A. J. Nicholson (1957), whose presentation to the conference was a restatement of an argument that he had been making for about fifteen years. The argument included both a theory of population regulation and an interpretation of how natural selection operated. Nicholson's intellectual outlook was formed at a time when there was no scientific consensus about the role of natural selection as a mechanism of evolutionary change. But he himself had always been a confirmed Darwinian, and the key to understanding his work and the controversies engendered by his work lies in appreciating the Darwinian core of his worldview. His approach to ecology was bound up with a desire to explicate and vindicate the argument Darwin had developed in the *Origin of Species.* Nicholson wove the problems of population ecology and evolutionary theory together, expounding his own 'modern synthesis' of evolutionary biology, a synthesis based not on genetics, but on an evaluation of Darwin's logic from the perspective of the population ecologist. Ironically, in trying to clear up ambiguities in the language of evolutionary biology that, he believed, impeded an understanding of the evolutionary process, Nicholson was attacked precisely for contributing to the linguistic confusion rampant in ecology by using metaphorical language that, his critics alleged, bore little relation to biological reality. I shall outline Nicholson's contributions to evolutionary ecology, review some of the issues raised in the ensuing debate, and end by looking briefly at a few scientific assessments of this controversy.

Nicholson's evolutionary logic was closely woven into his theory of population regulation. At the centre of this argument was the concept of intraspecific competition and Darwin's insight that species evolved because of competition within the species. As Mayr (1982) has argued, it was competition within the species that meant that natural selection was not merely a conservative force, a means of weeding out deviants or weak individuals and thereby preserving the form of the species, as earlier naturalists had believed. Because of the intensity of competition within the species, selection could be a genuinely creative force, enabling new varieties to replace the original populations. Without accepting the importance of competition within the species it was difficult to envision how this process of replacement might operate, and indeed many critics, not grasping Darwin's argument about competition, failed to see how natural selection could be a creative process. Nicholson's work in evolutionary biology was designed to clarify this novel aspect of Darwin's argument and to re-examine current problems in biology that had cast doubt on Darwin's theory.

Nicholson was educated at the University of Birmingham from 1912 to 1915, studying entomology, chemistry, and botany, and returned for graduate work in entomology in 1919-1920 (Mackerras 1970). Evolutionary writing at that time was commonly framed in teleological language imbued with the idea of purpose, as though nature acted with foresight to produce what was best for the species. In ecology as well, descriptions of nature as harmonious and balanced, although rooted in pre-Darwinian natural theology, were carried forward into the modern literature. One idea in particular troubled Nicholson, the idea that natural selection operated to produce a balance of nature. The assumption that selection worked to preserve the harmony of nature was common in ecological writing and possibly reflected not only religious ideas but the influence of thinkers such as Herbert Spencer, whose writings about evolution posited an equilibrium state to which natural processes tended (Kingsland 1985).

Nicholson believed that the idea that natural selection produced a balance in nature was caused partly by an erroneous concept of adaptation as the fairly close adjustment of animal to environment. The adaptationist argument was a legacy of late nineteenth-century developments in Darwinian theory and reflected the authority of A. R. Wallace, who took a strong adaptationist position and argued that all traits were adaptive and produced by natural selection. Other biologists had argued that some species differences had no adaptive significance and were not produced by natural selection. By the early 1900s the non-adaptationist position was increasingly adopted by critics of Darwinism, who voiced doubts about the presumed adaptive value of many traits. Nicholson grew up in an intellectual environment marked by scepticism about natural selection, which was particularly strong among the systematists (Richards and Robson 1926). In arguing that many differences between species were not adaptive they also suggested that the role of natural selection in producing such differences was dubious.

The anti-Darwinians took their cue from the logic of the Darwinian argument as it was expounded at that time, which implied that a trait had to confer general benefit on a species if it was to be selected (Nicholson 1932). Evidence that a trait did not benefit a species was therefore evidence against natural selection. For instance, the presence of warningly colored insects in the stomachs of birds seemed to provide evidence that protective coloration could not have been preserved by natural selection. Although mimicry had been used as strong evidence in favour of Darwin's theory by Henry Walter Bates in the 1860s, by the early twentieth century natural selection was no longer seen as the best explanation of mimicry or other forms of protective coloration (Kimler 1983). Biologists who observed that protectively colored insects or mimics were eaten in large numbers suggested that certain Darwinian explanations were merely beautiful fairy tales (Uvarov 1932).

Nicholson had also been struck by the observation that well-camouflaged species seemed to be no more successful than their relatives that lacked this adaptive trait. These observations suggested to him not that the theory of natural selection was deficient, but that one had to interpret these evolutionary processes in the light of knowledge about how populations were regulated. A new attack on the problem, one that steered around the traps caused by teleological reasoning, was in order. Nicholson, having emigrated to Australia and the University of Sydney in 1921, took up the problem of adaptation, protective coloration, and mimicry in 1927 when he had to compose an address as retiring president of the Royal Zoological Society of New South Wales (Nicholson 1927). His ideas expanded into a lengthy analysis that earned him a doctorate from the University of Sydney in 1929. Nicholson's explanation of mimicry was a compromise between extreme Darwinian gradualism and saltationist theories that had challenged Darwinian theory in the early twentieth century. E. B. Poulton had proposed a similar explanation briefly in 1912, but Nicholson's analysis was far more profound. Turner (1983) has argued that the Poulton-Nicholson interpretation of mimicry provided a more convincing alternative to R. A. Fisher's parallel explanations, which synthesised Darwinism and Mendelism in a different way. Turner pointed out that the Poulton-Nicholson explanation, which was identical to the explanation adopted later by E. B. Ford, has proven to have greater longevity in evolutionary genetics. Although Turner cannot claim that Nicholson's article directly influenced later thinkers who adopted the same explanation, he does make an important observation about Nicholson's work. Nicholson's evolutionary logic, which was drawn from natural history and population ecology, provided an alternative and a direct challenge to Fisher's evolutionary logic, drawn from population genetics, which was being developed at exactly the same time. Nicholson's continuing debate with R. A. Fisher over the problem of adaptation must be kept in mind when assessing the ecological controversies over population regulation in which Nicholson later became embroiled.

Nicholson addressed the question of how adaptations were acquired and whether these adaptations could be said to benefit a species. He argued that adaptations such as mimicry were indeed acquired by natural selection. But these adaptations did not necessarily increase the success of a species as measured by its numbers, because its numbers would be controlled by some mechanism different from the selective agent. What then accounted for the success of a species? What was needed was a theory of population regulation that would at the same time clarify Darwin's theory of natural selection. This dual purpose dominated Nicholson's research and writing for the rest of his life. Natural selection could only be understood in the light of population ecology and a theory of population control, which Nicholson proceeded to develop.

This shift in focus coincided with Nicholson's appointment as Deputy Chief to the newly formed Division of Economic Entomology of the Council for Scientific and Industrial Research. He began the job in Canberra in 1930 after spending a year doing research and observing entomological work in England and the United States. He was by this time already developing a theory of population regulation in collaboration with a physicist at the University of Sydney, Victor Albert Bailey, who helped Nicholson to convert his logical arguments into mathematical form. They intended to publish a monograph on mathematical population ecology, with a focus on predator-prey and host-parasite interactions. Oxford University Press rejected their book on the basis of a negative referee's report on a portion of the manuscript by R. A. Fisher, although Charles Elton had reportedly recommended the book to the press. Cambridge University Press also turned it down. It was deemed to be potentially too controversial in its incomplete marriage of biology and mathematics. In the end Nicholson published two articles from this unfinished book, one outlining his logical arguments in 1933 and a second adding Bailey's mathematical proofs in 1935 (Nicholson 1933; Nicholson and Bailey 1935). Both articles, titled 'The Balance of Animals Populations', developed themes rooted in Nicholson's earlier reflections on mimicry and natural selection. He started with the observation that populations were approximately in balance and argued that, this being the case, they must be regulated by causes whose severity would increase with increased population density. Competition appeared the most likely cause of density-dependent regulation. This theory of population regulation was in turn inseparably linked to Nicholson's explanation of natural selection and its effects.

The logic of Nicholson's argument, as it was repeated and elaborated over the decades, depended on drawing a subtle distinction between the effects of natural selection and the effects of competition. Selection meant preservation of favoured types, which would cause an increase in population. But population increase would, by logical inference, increase competition within the species. This competition, as it became more severe, would cause the new favoured type to replace the old, and would eventually bring the population back down to a stable level, a new equilibrium. It was competition, acting as a compensatory force to selection, which restored the balance in nature while selection was operating and afterwards (Nicholson 1937). These ideas were derived by logical argument and stated in very general terms. Having once convinced himself on logical grounds, nothing would shake his conviction that the argument was correct.

Nicholson's argument was a little convoluted at times, for he seemed to be drawing an overly fine distinction between natural selection and competition. His point was that one had to understand that populations were governed in a density-dependent manner in order to understand how selection worked and to explain apparent paradoxes, for instance that natural selection did not necessarily favour the success of the species in the sense of increasing its abundance. Nicholson was also trying to purge teleological language from evolutionary writing. In place of the language of purpose, he substituted mechanical metaphors to express his ideas of regulation. In one metaphor the population was compared to a floating balloon whose height changed depending on the ambient temperature (Nicholson 1933). Another compared the controlling function of competition to that of the governor on a steam engine (Nicholson 1937). Nicholson did not deny that such things as sudden climate changes might kill off portions of the population: rather he argued that such climatic effects were not responsible for the balance of populations.

Returning to the problem of adaptation, Nicholson argued that the prevailing view that selection produced a balance in nature was caused by the erroneous idea of adaptation as the fairly close adjustment of animal to environment. This understanding of adaptation had been a source of criticisms of natural selection in the early twentieth century. But Nicholson (1960) observed that even the architects of the neo-Darwinian Modern Synthesis were perpetuating the error by adopting a strong adaptationist position and defining adaptation in too rigid terms. He singled out for particular criticism R. A. Fisher's arguments about adaptation, for they seemed to imply that the organism was closely fitted to the environment, like a key fitting a lock (Nicholson 1955). Indeed, Fisher, alone among the founders of theoretical population genetics, was the most adamant in arguing against what he considered the 'dogma' that many specific differences were not adaptive (Kingsland 1985, pp 163-165). His ideas about adaptation had influenced E. B. Ford, whose field studies on butterflies in the 1930s were designed to demonstrate that apparently non-adaptive characters were associated genetically with adaptive changes of a more subtle nature (Ford and Ford 1930). Indeed the architects of the Modern Synthesis tended to adopt an increasingly hardened adaptationist stance as the synthesis developed (Provine 1983; Gould 1983).

Nicholson was trying to counter this increasingly rigid adaptationist position, which he did not see supported by ecological studies. True, all organisms were suited to their environments and were adapted in this general sense. But this was a loose form of adaptation, not the tight fit implied by the analogy of the lock and key. When selection operated to preserve some forms and eliminate others, the eliminated forms could not be said to be 'unfit' to live in their environment: rather they were unfit only in that they could not compete with their more effective fellows. (Nicholson 1960). Nicholson was redirecting attention to Darwin's original idea that organisms were only adapted to their environments in a relative, not an absolute sense. They were adapted relative to the level of competition they experienced from other organisms. And through genetic change and competition they evolved relative to these other organisms, not relative to the environment strictly speaking. In fact sexual selection provided many instances of adaptive change that did not serve to fit the organism to the environment, but only made the organism into a better competitor in relation to its fellows. Arguments similar to Nicholson's about adaptation were also advanced by Haldane (1932), Huxley (1942) and Dobzhansky (1955).

These ideas about evolution grounded Nicholson's theory of population regulation, with his insistence that populations were governed by density-dependent causes, especially by competition. When Nicholson first developed his ideas as logical arguments, he believed that laboratory experiments would not be appropriate tests for the many hypotheses that he and Bailey had advanced in the 1930s. His research was also thwarted by heavy administrative duties, for he had taken over as chief of the division in 1933. He changed his mind in the 1940s when he saw that two biologists in California, Paul DeBach and Harry S. Smith, had designed a set of laboratory experiments specifically to extend and test Nicholson's ideas. Moreover, they were enthusiastic about the possibilities of developing an experimental research program from Nicholson's work. Smith enthused in 1939 that Nicholson and Bailey had furnished "enough population problems to keep several laboratories busy for the next twenty years" (Smith 1939). Nicholson, encouraged by their example, then started his own experiments using sheep blowfly cultures and studied how oscillations were created in the populations (Nicholson 1950; 1954a; 1954b; 1957; 1960). These meticulous experiments followed the fluctuations of laboratory populations under different environmental conditions and examined such things as the influence of crowding on the populations. When he began to observe natural selection in the populations, he realised that these experiments would be worth pursuing as studies of evolution. He continued this line of research into the 1960s.

In all of Nicholson's work, from his earliest attempts to work through the theory of natural selection as a logical argument, through his development of a theory of population regulation, and finally to his laboratory studies, the basic argument remained unchanged. In that argument the concept of competition was inseparable from the concept of population regulation. Central to both his population ecology and his evolutionary argument was that graphic metaphor of competitive struggle that Darwin had bequeathed to evolutionary theory, with its clear echoes of the Malthusian world in which Darwin was formed. The clarity of Nicholson's logic about density-dependent regulation and the importance of competition had great impact on biologists who, like Nicholson, were examining evolutionary problems from the ecological point of view. But for those who did not accept the prevalence of competition as a given, indeed, for those who conceived of Darwinism in a fundamentally different way, this logic appeared deeply flawed, as did the strategy of model building that accompanied the argument.

By the time Nicholson was embarking on his experimental work his theories were being subjected to strong criticism from fellow entomologists. One of the earliest critics was William Robin Thompson, who in the 1930s was directing the biological control programme of the Imperial Institute of Entomology at its laboratory outside London. Thompson, although himself one of the pioneers of mathematical modeling in ecology, thought that Nicholson and Bailey were going too far in using mathematical hypotheses that could not be easily related to biological knowledge. This was not merely an intellectual issue, but involved the very survival of research programmes. With the Depression forcing major cutbacks in the funding of Thompson's research, his dream of developing a major programme in biological control to serve the British Commonwealth evaporated. These economic problems likely intensified his feeling that the kinds of mathematical approaches that were flooding the scientific marketplace, not only in ecology but also in population genetics, were threatening the long-term field studies that Thompson thought were the backbone of applied entomology.

Thompson also had strong reservations about the theory of natural selection, reservations that stemmed from his Catholicism, from which he had developed a philosophical position fundamentally opposed to the logic of Darwinism. Inasmuch as the new mathematics was being used to shore up Darwin's argument, Thompson believed that it was deeply flawed. He launched a full-scale critique of Nicholson's work, arguing that Nicholson was mistaking a mathematical or logical argument for a real cause-and-effect relationship between natural populations, that he had never demonstrated that competition had really occurred, that his writings were full of metaphorical expressions, such as the balance of nature, that were not demonstrated to exist (Kingsland 1985). Such criticisms from entomologists intensified as Nicholson's ideas began to have significant impact, not only among ecologists working on population regulation, but also among evolutionary biologists.

In England, Nicholson had several influential champions. Charles Elton discussed Nicholson's ideas with colleagues at the Bureau of Animal Population at Oxford. George Varley, also at Oxford, had designed his Ph.D. thesis to provide the first field test of some of Nicholson's predictions (Varley 1947). Varley managed to publish it only with some difficulty because the referee, W. R. Thompson, was strongly critical of it. Varley proved a consistent advocate of Nicholson's ideas, although he recognised the need to define terms more precisely and to shape the hypotheses to fit new research (Varley 1958; Varley *et al.* 1973). He in turn influenced David Lack just as he was analysing the data from his study of speciation among finches on the Galapagos Islands. Lack later drew extensively on Nicholson's work in his book *The Natural Regulation of Animal Numbers* (1954), a seminal work in evolutionary ecology with a strong Darwinian theme.

Lack did not think that Nicholson's laboratory experiments on the sheep blow-fly were of much help in interpreting how animal numbers were regulated in nature, nor did he care for Nicholson's use of the metaphor of the 'balance of nature'. But Lack was deeply influenced by Nicholson's argument for density-dependent regulation of populations and like Nicholson, he saw such regulation as a logical necessity. The three main causes of density-dependent mortality,—food shortage, predation, and disease,—operated together in a complex way, which Lack admitted was little understood. But the idea of density-dependent mortality provided a crucial step in Lack's argument that the reproductive rate of each species evolved through natural selection in such a way that the number of young surviving to independence was maximised (Lack 1954; 1966).

The success of Nicholson's logic about density-dependent regulation brought more forceful criticisms to the fore in the 1950s. Thompson's criticisms were reinforced and expanded by two younger Australian biologists, H. G. Andrewartha of the University of Adelaide and L. Charles Birch of the University of Sydney. Their collaborative treatise of 1954, *The Distribution and Abundance of Animals*, is now recognised as an ecological classic, although at the time they believed themselves to represent a minority position among ecologists. Their approach was constructed in direct opposition to Nicholson's viewpoint. Apart from a general dislike of the particular mathematical modelling strategy that Nicholson and Bailey adopted, Andrewartha and Birch (1954, pp 20-26; 404-406) had profound disagreements with Nicholson's use of the concept of competition. They objected to the broad way that Nicholson defined competition, seeming to include predator-prey relationships as forms of competition, as though the prey competed with its 'brethren' for the opportunity of being eaten by the predator. They allowed that some form of competition for limited resources existed in nature, but argued that causal explanations of population dynamics could be fashioned without introducing the concept of competition. They objected that the concept of competition did not help to resolve the ecological problem of explaining the observed densities of populations. It was preferable, they argued, to speak simply of limitations of resources. Here they concurred with Thompson.

Like Thompson they argued that the best kind of scientific explanation was one that eschewed metaphorical language. Competition suggested emotion or purpose; it was the language of the sporting arena. Nicholson (1954b) did imagine competition among animals as being like competition among humans and the terms he used, 'scramble' and 'contest', suggested these comparisons. But for Andrewartha and Birch organisms struggling for survival in the face of inadequate food supply, whether human or animal, were not comparable to a sporting event. If organisms died of starvation, the cause was limitation of food supply. Attributing the cause to competition prevented one from speaking plainly about the actual causes at work. Unlike Thompson, Andrewartha and Birch did not deny Darwin's theory of natural selection, but in discussing Darwinism they denied that Darwin had laid such emphasis on competition and argued that he had a much more sophisticated view of the evolutionary process.

Andrewartha and Birch's approach was to relate population dynamics and evolution to the daily processes of an animal's life. It was not clear to them that competition was one of the ongoing processes of daily life. In the insect populations that they studied it seemed more reasonable to attribute changes in populations to known effects, such as predation or climate. Birch's laboratory research buttressed their field work, showing that environmental conditions such as temperature influenced the rate of population growth (Birch 1948) and demonstrating that natural selection could occur in laboratory populations without competition (Birch 1955). Birch was developing an alternative interpretation of Darwinism, one that focused more on environmental effects than on competition, suggested that natural selection could occur without competition, and emphasised the element of chance in evolution. As recalled by Tom Browning (pers. comm.), who worked closely with Andrewartha at Adelaide, Andrewartha regarded evolution and ecology as different subjects in any case, so that competition would not have assumed a central role in ecological studies of distribution and abundance, no matter how relevant it might be to the study of evolutionary divergence.

However, Andrewartha and Birch criticised Lack as well for attributing differences between closely allied species to competition. Lack had assumed that competition for food was an important cause of evolutionary divergence. He used what was later called the 'principle of competitive exclusion' to argue that evolutionary divergence among species of finches on the Galapagos Islands had occurred as a means of avoiding competition (Lack 1947). But because the competition had operated in the past, there was no direct evidence of it in the present and hence no way to prove or disprove its importance. This being the case, Andrewartha and Birch argued, why invoke it? The alternative explanation, that the species had differentiated in reproductive isolation and then moved into the same territory, where they selected different places to life and different foods, was equally plausible. And the fact that the present food supply of the species appeared to be abundant indicated that there was no competition operating in the present.

Nicholson had allowed that in many populations the proximate cause of mortality might be predation, disease, or bad weather, but argued that the primary or ultimate cause was competition caused by insufficient resources to meet the needs of all individuals. For Andrewartha and Birch, this 'ultimate' cause was no explanation if a more proximate cause was available. They refused to credit the indirect evidence for competition that was already extensive in the 1950s and did not accept the validity of 'ultimate' over 'proximate' explanation that Nicholson and Lack assumed.

The strength of their reaction against competition theory suggests a more deeply rooted aversion to the idea that nature was a Malthusian competitive arena that mirrored human society. Andrewartha and Birch were not alone in rejecting this image of nature. Extending back to the nineteenth century various schools of thought had devised approaches to Darwinian theory that played down the Malthusian logic in Darwin's work (Mitman 1992; Todes 1989). Dobzhansky (1950) made a similar complaint about the emotional connotations of words such as competition, struggle, fitness and adaptive value, also arguing that such loaded words should be avoided. But Andrewartha and Birch were working against the tide of most evolutionary and ecological writing, which took competition as a given in the evolutionary process, even while admitting that direct evidence for it was scanty (Mayr 1948).

Nicholson found these disputes to be emotionally exhausting. He had hoped that some of the differences between himself and Andrewartha could be resolved prior to the Cold Spring Harbor symposium in 1957. But the meeting proved to be exceptionally tense and polarised, despite a few attempts at compromise. Nicholson (1957) reiterated his argument. Birch (1957a) emphasised the role of weather in governing the abundance and distribution of animals. He focused on the instability of natural populations and the spatial patchiness of the environment, and used stochastic models to take into account the instability of populations. The cause of mortality was not competition, but the "principle of the relative shortage of food", ultimately caused by weather. Birch agreed with Nicholson's interpretations of his "beautiful experiments with blowflies", but did not consider these a model for all species. However, he admitted that the gap between himself and Nicholson was not as great as it might appear. Andrewartha (1957), less conciliatory than Birch, adopted a more general approach in his critique of Nicholson's ideas, examining a number of basic epistemological issues that bore on this debate. These included how causal explanations should be constructed in ecology, how laboratory results could be related to the field, and what meaning could be given to theoretical models of ecological processes. Attempts by one conference participant to synthesise the theories of Thompson, Andrewartha and Birch, and Nicholson were flatly rejected by Nicholson as illogical because the compromise led to the conclusion that insect populations were controlled by density-independent factors (Milne 1957). It was clear that the opposing positions had become firmly entrenched, precluding all compromise.

A great deal of ink was spilled in various attempts to define and clarify the meaning of competition as well as other terms of population regulation. Nicholson (1954b) distinguished between two categories of competition, called 'contests' and 'scrambles', to denote competition within territorial and non-territorial species respectively. Thomas Park (1954) introduced the terms 'exploitation' and 'interference' to describe competitive interactions between species in laboratory populations. Birch (1957b) discerned four general contexts in which 'competition' was invoked, with various shades of meaning in each category, and recommended avoiding the term when another word was more precise.

Given that the argument appeared to be partly about word meanings, did efforts to impose precision on the language have any effect on the controversy? To take the case of 'competition' as the most serious instance of ambiguous terminology, it appears that debates over how to define it did not seriously inhibit biologists' use of the word. But the fact that these terminological questions were being raised did force biologists who might otherwise have ignored this controversial point to restate the evidence in favour of competition (Brown and Wilson 1956). On the other hand, important advances in population and community ecology were made by scientists who completely ignored the controversy and refused to be influenced by demands for precise definitions. Robert MacArthur took this route. Finally, attempts to make the language of ecology more precise do not appear to have influenced those whose opinions were already decided. Perhaps for this reason a more fruitful linguistic strategy at this time was Hutchinson's (1957) decision to focus on definition of the 'niche' as an "n-dimensional resource space" in response to the Cold Spring Harbor controversies. By focusing on the concept of the niche he neatly side-stepped the vexed issue of defining competition and raised a number of new questions that gave rise to a new field of research in niche theory in the 1960s. Exhorting ecologists to pay more attention to terminology to avoid confusion may not by itself be an effective means of resolving controversy, unless the linguistic precision also serves to deflect debate into another area.

However, this debate was about a great more than definitions of words: it was about the definition of ecology as a subject and control of the research agenda in ecology. It was also about anxiety that any deductive approach to ecology, especially one that used the kinds of mathematical models to create hypothetical scenarios that Nicholson and Bailey used, would completely subvert the traditional methods of natural history, methods based on long-term and laborious field studies (Birch and Browning 1993). The debate was also about how best to link evolutionary biology to ecology. At issue were competing images of Darwinism and deep disagreement over whether one should imagine nature in anthropomorphic terms, using metaphors to describe nature that were derived from human experience.

This debate operated on multiple scientific and ideological levels and has continued to the present time, although it has undergone metamorphoses as ecology has matured. Because the debate has been polarised for such a long time, it has been tempting to assess its outcome in terms of winners and losers as more evidence accumulates and views of population dynamics become more sophisticated. Andrewartha and Birch (1984, pp 258-259), reflecting on these controversies in the 1980s at a time when theories based on ecological competition were coming under widespread attack, suggested that such controversies were necessary as a way of weeding out inadequate explanations of objective facts. Because they considered their environmental explanation to be more general and more plausible than explanations based on competition theory, by implication they considered themselves the victors in this controversy. Sinclair (1989), reviewing the controversy in 1989, also saw this debate as a kind of survival of the fittest, but concluded that Nicholson's ideas had in fact survived over competing theories because they were capable of being tested and had more evidence in their favour. Thus Nicholson became a prescient forerunner of a modern and more sophisticated ecology. No matter which side was declared the winner, the image of science in both cases is similar. Controversy is seen to be beneficial because it intensifies the struggle for existence and enables the best ideas to survive, while inadequate ideas are seen as rightly falling by the wayside, or, to use current terminology, as being replaced by new 'paradigms'.

But real life is not that simple or linear. As Murdoch (1994) pointed out, such attempts at closure by declaring winners and losers also have the effect of shutting out certain lines of inquiry by shifting research priorities and hence influencing funding of research. These effects may not be desirable, depending on what goals ecologists set for their science. Murdoch considered the controversy about regulation to have been resolved by the 1990s, with both Nicholson's school and the Andrewartha and Birch school being partially correct. But he also drew attention to the fact that in the United States the study of regulation, being in conflict with the modern emphasis on non-equilibrium dynamics, had fallen well down the research agenda. He proposed to resurrect and defend the concept of regulation as central to ecology, as a problem well capable of experimental analysis, amenable to sophisticated theoretical treatment, and a powerful vehicle for uniting evolutionary ideas and the analysis of population dynamics. Murdoch also proposed that increased attention to the study of regulation was important not only to advance the discipline conceptually, but also to deal with growing environmental problems.

Murdoch's point was in part that attention to the problem of regulation would increase the authority of ecologists in evolutionary biology and in the practical sphere of environmentalism. Extending his discussion a little bit, his comments point toward an assessment of the controversy which does not entail the same irreversible process of competitive struggle and paradigm shift. His comments suggest that there are benefits to incorporating alternative approaches to science and of resurrecting ideas that have gone out of favour, if in so doing ecologists generate information that increases their authority. His comments imply that it can be prudent to retain multiple points of view, to continue the discussions and even the disagreements, and not to carry the process of competitive exclusion to the point of extinction of one of the competing points of view. Admittedly such multiple perspectives are difficult to maintain in the highly competitive scientific environment. Nicholson's history illustrates perfectly the difficulty of reaching compromise when new arguments must be framed in strong, unequivocal terms to ensure that they get a hearing.

Yet even in Nicholson's experience there were occasions on which scientists were motivated to work together despite deep disagreements on basic issues. One such instance occurred at an earlier conference at Cold Spring Harbor, convened in 1955 to discuss recent developments in population genetics. At this symposium definitions of concepts were as much at issue as they were two years later at the ecology symposium. Participants discussed ambiguities in the meaning of adaptation, fitness, and other basic concepts, all of which were seen as hindering scientific communication. Yet in reading the conference proceedings one feels a sense of unified purpose. Nicholson (1955) himself gave a paper that did not apparently raise a murmur of disagreement. Even though biologists baulked at the new mathematical methods of population genetics, and even though there were deep disagreements between the theoreticians themselves, there was open recognition that those theoreticians were contributing in an important way to the development of neo-Darwinism. Although there were agreements to disagree, the mood was described as one of progress achieved through synthesis of different viewpoints and through recognition that in evolutionary biology and genetics universal explanations were not possible. As Mayr (1955) wrote, "Oversimplifications and too much emphasis on single factors have no place in this field".

The contrast between the mood of this conference and the one on population ecology two years later could not have been more striking. No wonder Dobzhansky, who had taken a leading role in the 1955 symposium and helped to organise the ecology symposium in 1957, was perplexed at the intransigence of the ecological meetings. In reflecting on these differences, it may be relevant to consider that evolutionary biologists, especially in the United States of America, had previously confronted two serious episodes where events outside of science had threatened the nature of the scientific enterprise. The first was the fundamentalist religious attack on Darwinism in the 1920s, the second, a different kind of fundamentalism, was the challenge to western genetics by Soviet agronomist T. D. Lysenko, which deeply affected western scientists especially from the late 1940s. Both of these episodes forced scientists to react to events happening outside the domain of science and to develop strategies to combat what were seen as threats to the scientist's autonomy and integrity. These episodes changed the nature of the scientific controversies that were being conducted at these times. One of the effects of these changes was to enhance the appearance of consensus, or unified purpose, behind the scientific enterprise.

Ecologists were not involved in these episodes and perhaps were less motivated to band together in the face of this kind of external 'threat' to science. But Murdoch's comments suggest likewise that debates within science have to be conducted with a view to the larger social context of science. In particular, the need to get and maintain public support for ecology requires that ecologists emphasise their ability to address environmental problems and not close off lines of research that might help to increase their authority in these areas. Andrewartha and Birch certainly believed in the importance of ecology as an applied science, and so did Nicholson. Yet this common belief in the utility of ecology could not be translated into any fruitful theoretical synthesis, but instead precipitated a harsh competitive struggle for existence. Murdoch, in urging the synthesis that was not possible for these pioneer ecologists, nevertheless is not far removed from their spirit when he implies that ecologists should maintain research fields that speak to the public's need to have a clear demonstration of the relevance of ecological science both to broad theoretical issues and to environmental problems.

Acknowledgements

I am very grateful to Garth Nicholson for information about A. J. Nicholson's life and the Cold Spring Harbor symposium of 1957, and to Tom Browning for insights into H. G. Andrewartha's beliefs and personality.

References

Andrewartha HG (1957) The use of conceptual models in population ecology. *Cold Spring Harbor Symp Quant Biol* 22:219–232

Andrewartha HG, Birch LC (1954) *The distribution and abundance of animals.* University of Chicago Press, Chicago

Andrewartha HG, Birch LC (1984) *The ecological web: more on the distribution and abundance of animals.* University of Chicago Press, Chicago

Birch LC (1948) The intrinsic rate of natural increase of an insect population. *J Anim Ecol* 17:15–26

Birch LC (1955) Selection in *Drosophila pseudoobscura* in relation to crowding. *Evolution* 9:389–399

Birch LC (1957a) The role of weather in determining the distribution and abundance of animals. *Cold Spring Harbor Symp Quant Biol* 22:203–215

Birch LC (1957b) The meanings of competition. *Am Nat* 91:5–18

Birch LC, Browning TO (1993) Herbert George Andrewartha, 1907–1992. *Historical Record of Australian Science* 9:259–268

Brown WL, Wilson EO (1956) Character displacement. *Syst Zool* 5:49–64

Dobzhansky T (1950) Heredity, environment, and evolution. *Science* 111:161–166

Dobzhansky T (1955) A review of some fundamental concepts and problems of population genetics. *Cold Spring Harbor Symp Quant Biol* 20:1–15

Dobzhansky T (1957) Discussion following paper by HG Andrewartha. *Cold Spring Harbor Symp Quant Biol* 22:235

Ford HD, Ford EB (1930) Fluctuation in numbers and its influence on variation in *Melitaea aurinia. Trans R Entomol Soc Lond* 78:345–351

Gould SJ (1983) The hardening of the modern synthesis. In: Grene M (ed) *Dimensions of Darwinism: themes and counterthemes in twentieth-century evolutionary theory.* Cambridge University Press, Cambridge, pp 71–93

Haldane JBS (1932) *The causes of evolution.* Harper, New York

Hutchinson GE (1957) Concluding remarks. *Cold Spring Harbor Symp Quant Biol* 22:415–427

Huxley JS (1942) *Evolution: the modern synthesis.* George Allen & Unwin, London

Kimler WC (1983) Mimicry: views of naturalists and ecologists before the modern synthesis. In: Grene M (ed) *Dimensions of Darwinism: themes and counterthemes in twentieth-century evolutionary theory.* Cambridge University Press, Cambridge, pp 97–127

Kingsland SE (1985) *Modeling nature: episodes in the history of population ecology.* Chicago University Press, Chicago

Lack DL (1947) *Darwin's finches: an essay on the general biological theory of evolution.* Cambridge University Press, Cambridge

Lack DL (1954) *The natural regulation of animal numbers.* Clarendon Press, Oxford

Lack DL (1966) *Population studies of birds.* Clarendon Press, Oxford

MacArthur RH (1960) Population studies: animal ecology and demography. *Q Rev Biol* 35:82–83

Mackerras IM (1970) Alexander John Nicholson. *Rec Aust Acad Sci* 2:66–81

Mayr E (1948) The bearing of the new systematics on genetical problems: the nature of species. *Adv Genet* 2:205–237

Mayr E (1955) Integration of genotypes: synthesis. *Cold Spring Harbor Symp Quant Biol* 20:327–333

Mayr E (1982) *The growth of biological thought: diversity, evolution, and inheritance.* Harvard University Press, Cambridge Mass

Milne A (1957) Theories of natural control of insect populations. *Cold Spring Harbor Symp Quant Biol* 22:253–271

Mitman G (1992) *The state of nature: ecology, community, and American social thought, 1900–1950.* University of Chicago Press, Chicago

Murdoch WW (1994) Population regulation in theory and practice. *Ecology* 75:271–287

Nicholson AJ (1927) A new theory of mimicry in insects. *Aust Zool* 5:10–104

Nicholson AJ (1932) 'Protective' adaptations of animals. *Nature* (Lond) 130:696

Nicholson AJ (1933) The balance of animal populations. *J Anim Ecol* **2**:132–178

Nicholson AJ (1937) The role of competition in determining animals populations. *J CSIR* (Aust) **10**:101–106

Nicholson AJ (1950) Population oscillations caused by competition for food. *Nature* (Lond) **165**:476–477

Nicholson AJ (1954a) Compensatory reactions of populations to stresses, and their evolutionary significance. *Aust J Zool* **2**:1–8

Nicholson AJ (1954b) An outline of the dynamics of animal populations. *Aust J Zool* **2**:9–65

Nicholson AJ (1955) Density governed reaction, the counterpart of selection in evolution. *Cold Spring Harbor Symp Quant Biol* **20**:288–293

Nicholson AJ (1957) The self-adjustment of populations to change. *Cold Spring Harbor Symp Quant Biol* **22**:153–172

Nicholson AJ (1960) The role of population dynamics in natural selection. In: Tax S (ed) *Evolution after Darwin: the University of Chicago centennial, vol. 1*. University of Chicago Press, Chicago, pp 477–521

Nicholson AJ, Bailey VA (1935) The balance of animal populations. *Proc Zool Soc Lond* **3**:551–598

Park T (1954) Experimental studies of interspecies competition. II temperature, humidity and competition in two species of *Tribolium*. *Physiol Zool* **27**:177–238

Provine WB (1983) The development of Wright's theory of evolution: systematics, adaptation, and drift. In: Grene M (ed) *Dimensions of Darwinism: themes and counterthemes in twentieth-century evolutionary theory*. Cambridge University Press, Cambridge, pp 43–70

Richards OW, Robson GC (1926) The species problem and evolution. *Nature* (Lond) **117**:345–347

Sinclair ARE (1989) Population regulation in animals. In: Cherrett JM (ed) *Ecological concepts: the contribution of ecology to an understanding of the natural world*. Blackwell Scientific, Oxford, pp 197–241

Smith HS (1939) Insect populations in relation to biological control. *Ecol Monogr* **9**:311–320

Todes DP (1989) *Darwin without Malthus: the struggle for existence in Russian evolutionary thought*. Oxford University Press, New York

Turner JRG (1983) 'The hypothesis that explains mimetic resemblance explains evolution': the gradualist-saltationist schism. In: Grene M (ed) *Dimensions of Darwinism: themes and counterthemes in twentieth-century evolutionary theory*. Cambridge University Press, Cambridge, pp 129–169

Uvarov BP (1932) 'Protective' adaptations of animals. *Nature* (Lond) **130**:696–697

Varley GC (1947) The natural control of population balance in the knapweed gall-fly (*Urophora jaceana*). *J Anim Ecol* **16**:139–187

Varley GC (1958) Meaning of density-dependence and related terms in population dynamics. *Nature* (Lond) **181**:1780–1781

Varley GC, Gradwell GR, Hassell MP (1973) *Insect population ecology: an analytical approach*. University of California Press, Berkeley

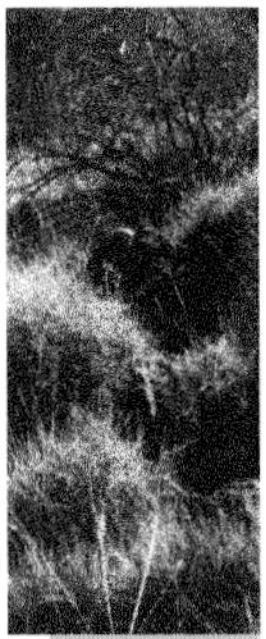

SECTION 1

POPULATION REGULATION IN THEORY AND PRACTISE

In the introductory section, Sharon Kingsland eloquently explained that the underlying motivation in A.J. Nicholson's theory of population regulation was to provide a mechanism on which to rest his neo-Darwinian beliefs concerning natural selection. In this context, Nicholson considered population regulation through density dependent competition from a theoreticians perspective with a belief that the underlying mechanism and its presence was unquestionable, as without it there was no theory. When viewed however by field-based empiricists such as naturalists or ecologists with experience of the full complexity in natural systems, Nicholson's theory appeared overly simplistic and highly dogmatic. This dichotomy of viewpoint between theorists and empiricists is still widespread among ecologists and in ecological thought, but may be an important feature in the progressive questioning process of ecological deduction. Chapters in this book were selected with this diversity of ecological reasoning in mind and a balance was sought between theoretical and empirical contributions. Practicality is also a theme with particular emphasis being placed on the application of ecology to management and conservation of organisms in their environment.

As a firm starting point, Bill Murdoch and Roger Nisbet offer a synthesis of the theme of this book and the conference it follows, from their perspective as strong proponents of population regulation. They provide an informative account of the state and complexity of modelling techniques aimed at describing one or more species populations and the interactions. They focus on the key theoretical approach of individual-based models, where populations are treated as a group of independent entities interacting in space, and they highlight key studies, such as their own; the red scale/*Aphytis* biological control system, where empirical and theoretical approaches are working hand in hand.

Population regulation is almost synonymous with density dependence, the latter now being the prerequisite for proposing the former, however demonstrating density dependence remains a contentious issue either by statistical analysis or by field experiment. David Fox and James Ridsdill-Smith provide a statistical critique of existing statistical tests, suggesting that perhaps we have progressed little in this area in 20 years since the older tests are still the most generally applicable. Naomi Cappuccino and Susan Harrison review density dependent experiments and find that the results do not clearly support historical theoretical arguments (including those of Nicholson), especially with regard to insect herbivores. There is a clear message from their synthesis, including some of their own work, that dispersal is a frequent behaviour that may generate density dependence in field populations, however for many studies density dependent dispersal is one of the hardest features to substantiate.

Rick McGarvey explores another avenue of evidence for density dependence, also examined by Nicholson (1957); that of harvesting. The removal of one life stage can cause increases in the population density of other stages due to the effect of reduced competition on survival. If populations are not regulated by density dependent competition, then harvested populations should smoothly decline to zero. Ample evidence against this is seen from the extremely rapid demise of the many celebrated fisheries following many years of consistently high catches. McGarvey uses two analytical methods on field data that suggest harvested scallop populations from the Canadian Atlantic show massively increased spawn survival even when one half or more of the adult biomass is being removed.

Contributions from both Mark Ritchie and Richard Fleming concern the actual and likely effects of temperature variation on population control. Ritchie's paper illustrates the specific case, and uses experiments to test model-generated hypotheses that density dependent mortality increases with temperature in grasshoppers, while density independent mortality remains constant. Fleming discusses the more general scenario of global warming on North American forest insect pests. He predicts that warming is likely to increase not only the destructive outbreaks of existing pests, but possibly lead to the appearance of new pest species from the existing fauna as the longer-lived host trees are poorly adapted to the changed climate. As in Ritchie's example, this can come about through increasing density dependence at higher temperatures leading to greater population fluctuation in insects, because of the destabilising effects of overcompensatory regulation.

Tony Dixon and colleagues have been modelling within season population change in various arboreal aphids. For the Turkey oak aphid, the model they have developed suggests that increases in population size result from intraspecific competition for resources driven by resource quality and are predictable in all seasons except summer. Summer population decline may be under the influences of external factors, such as seasonal variation in temperature, and is largely caused by migration of winged forms. The drive to migrate from the oak trees is certainly not population density and yet it is a high-risk strategy. Why it happens, remains the enigma of aphid population dynamics.

Mike Begon and colleagues outline some of their recent work on three species interactions between host, parasitoid and pathogen. These studies provide explicit examples of how theoretical and empirical studies can be complimentary in the extension of predator-prey theory beyond simple two species interactions. This certainly represents an ecological frontier leading towards the mechanistic understanding of population ecology within community dynamics. As with Nicholson's blowflies, the context of such studies requires development and demonstration beyond lab-based systems into the multi-stranded web of natural ecosystems. The indirect effects in multi-species interactions represent the forces of the silent majority and are the mortar that bonds community structure.

Emphasis turns to mammals with an invited essay by Tony Sinclair. He provides a clear and perhaps general explanation of how population regulation is related to individual size. Average size is critical to the type of regulation in populations because of its inverse relationship with the intrinsic rate of increase (r_{max}) and risk of predation, both of which contribute to the strength of any density dependent dynamics. Populations of large mammals resist food shortage through decreased fecundity, and mortality tends to affect mainly juveniles. Sinclair argues that these expressions of food shortage effects early in life, lead to population stabilising density dependence and thus gentle population fluctuations, and that this shows an intrinsic adaptation in large mammals determining the rate of population decline. In small mammals, food shortage tends to cause increased mortality to all ages, and small animals also tend to exhibit stronger behavioural responses to intraspecific competition such as dispersal. Combined with a faster potential population recovery rates due to a higher r_{max}, small mammals, like the insect populations of Ritchie and Fleming, tend to experience overcompensating density dependence. Add the higher predation (the Nicholsonian delayed density dependent response) experienced by small animal populations and demographic fluctuation increases to include cyclic behaviour.

As a classic example of the factors important in small mammal population dynamics, Charlie Krebs and colleagues describe the state of play in the largest ecological experiment designed to understand the regulatory forces in the snow-shoe hare population cycle. The cyclic dynamics of some northern temperate small mammal populations have been a feature that has fascinated Krebs and his colleagues for many years and yet the relative importance of the responsible underlying factors has remained enigmatic. In this "all or nothing" attempt to untangle the ecological interactions, both food limitation and predation appear to interact as the cause of population decline and appear to be linked to the predation-avoidance behavioural capacity of hares under food-related stress. Clearly the classic description of this cycle as an interaction between two trophic levels (predator and prey) was too simplistic.

The dispersal behaviour of small marsupials has been studied and modelled by Michael McCarthy. He shows that during the seasonal dispersal of young kangaroo rats from multiple nest sites, the degree of competition, determined by density and availability of suitable territories, is an important, but poorly understood feature in the theoretical prediction of median dispersal distances and dispersal success of spatially structured populations.

John Harwood and Pejman Rohani review of evidence for population regulation in marine mammals, and provide a fascinating account of what is known about the population ecology of these popular creatures. Their discussion interestingly parallels Sinclair's contribution, as marine mammals all weigh in excess of the 30 kg threshold used to define large mammals. Harwood and Rohani first stress that the speed with which exploited populations of many marine mammal species have become endangered is a clear indication that these creatures do not exhibit dynamics comparable to the sustainable yield scenarios of e.g. McGarvey's scallops. This results from their typical large-mammal intrinsic demographic characteristics. Intrinsic features of large mammal populations also constrain population recovery potential following disturbance. Harwood and Rohani highlight, therefore, the innate risk to such animals, through either direct exploitation or the degradation of the marine environment by man, and show how recent research has used this understanding of marine mammal dynamics to generate conservation and management procedures.

This section concludes with a comprehensive look at how the Nicholsonian view holds for plant populations. Michael Crawley and Mark Rees use a carefully selected series of case studies of single and multi-species systems to show how the deterministic approach of Nicholson has also contributed to a clearer understanding of plant competition and population dynamics. Using single species, they show how simple models incorporating intraspecific competition for space and the law of constant yield for biomass, can adequately describe the range of possible population behaviours and determine those conditions (e.g. seed limitation) requiring careful assessement in the field to define how self-replacement dynamics regulates population size. Lastly they illustrate, using multispecies plant systems, how Nicholson's understanding of the relative importance of intraspecific over interspecific competition in the maintenance of coexistence holds true in plant communities. It is interesting that the possibilities for species coexistence, despite only minimal differences in their ecology, are now largely accepted for plants even though field work and theory are still only in the process of providing mechanisms.

Frontiers of Population Ecology

William W. Murdoch and Roger M. Nisbet

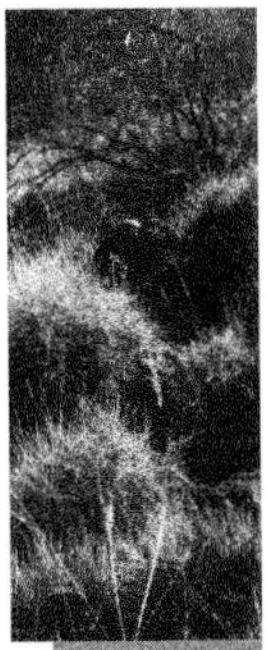

ABSTRACT

Population ecology has seen rapid advances on many fronts in recent years, especially on the theoretical side. Great improvements have been made, for example, in the modelling of the dynamics of disease, the spread of genetically altered organisms, the probability of persistence of rare and endangered species, the potential response of population and communities to large-scale environmental change, and the interaction of species in several trophic levels. We have chosen to emphasise two areas from among the many that could be discussed: modelling of population dynamics based directly on the properties of individual organisms, and spatially distributed populations, based again on interactions among individual organisms. The advantage of individual-based models is their potential for providing direct insight into real systems and for facilitating experimental tests of ideas about processes affecting population dynamics. We illustrate this potential with stage-structured models of parasitoid–host interactions in insects, and discuss advances needed to connect spatial models with real systems. The advances in modelling need to be matched with equivalent progress in experimental and observational analysis of real systems. This is perhaps more difficult to achieve. We discuss some promising examples and outline possible future research.

Key words: Population ecology, individual-based models, stage-structured models, spatial models, model hierarchies

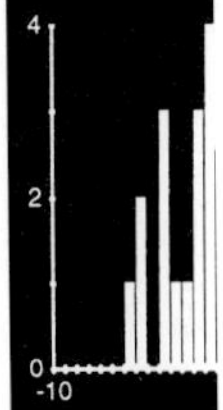

Frontiers of Population Ecology, R.B. Floyd, A.W. Sheppard and P.J. De Barro (eds), CSIRO Publishing, Melbourne, 1996, pp. 31–43

Introduction

Population ecology is rich in challenging problems. It aims to understand, *inter alia*, the dynamics of natural, exploited and managed populations, including persistence, regulation, dynamic patterns (stability, cycles, chaos, periods, amplitudes) and long-term average abundances, and to predict the effects on these population characteristics of changes in the biotic and physical environment, including natural and human-induced alterations. It aims to do this for particular populations, but also to establish general patterns so that probable population properties might be related to life history features, trophic status, community structure, etc. To the extent that a community's properties are but the manifestation of the joint properties of its constituent populations, population ecology also needs to explain the structure and dynamics of communities.

Some recent advances in population ecology

To obtain an adequate perspective on population ecology's frontiers, we need to take a brief look back at the distance we have travelled. The above definition of population ecology's problems is close to, if a little more explicit than what it was when Nicholson (1933) published his first major opus on the topic. But today, we are richer in our understanding of these problems and in approaches for solving them. Much of this development has taken place only in the last decade or so. We now have a crisper delineation of the links among extinction, persistence and regulation, of the connections between local and global dynamics, and of the different forms that population dynamics might take. Perhaps less happily, we have also learned that population dynamics also have the potential to be more complex than we might have hoped (e.g. May 1976; Hastings and Higgins 1994).

We have in addition expanded the scale of our questions. In part, this has arisen from analyses of the potentially crucial roles of spatial heterogeneity and linkages among spatially dispersed organisms and populations (a view we owe more to Andrewartha than to Nicholson), and in part from the pressing need to solve problems with a large spatial extent – e.g. the effects of global warming or conservation of particular species or of biodiversity in general (e.g. Kareiva *et al.* 1993).

Recent advances in attacking old questions have come in a variety of forms. There is now a firmer and richer conceptual and theoretical framework for population dynamics, and our major emphasis below will be to pursue aspects of this framework. Also encouraging is the start that has been made on a closer integration of population and ecosystem ecology, as ecosystem ecologists have recognised the important role that species differences can sometimes play in ecosystem processes (Vitousek 1990). Better tools also exist for empirical work; we mention just three.

First, if environmental problems have a larger spatial reach than before, remote sensing and methods for handling large and spatially extended data sets (such as geographical information systems (GIS)) have developed to analyse them (Walker this volume). Observations in the field have also grown in scope, comparative work on trophic structure and dynamics in Swedish lakes being one example (Persson *et al.* 1992). A start has also been made on large-scale experiments on the determinants of community structure, again in lakes (Carpenter and Kitchell 1993), though the problem of replication here is daunting. Second, where once ecologists argued over the meaning of the few long-term data sets in existence, there are now much richer troves of such data, for example, for hundreds of species of insects and birds, many from a wide geographic area, and there are improved (if not yet entirely satisfactory) statistical methods for analysing them (e.g. Hanski 1990; Turchin and Taylor 1992; Sugihara 1994). Finally, new molecular genetic techniques are still stronger in promise than performance, but that situation seems likely to improve soon (Avise 1993).

In addition, some earlier questions have been more or less answered. For example, hundreds of field experiments have now been done on central problems defined in the 1950s and 1960s: Is competition between species widespread? Does predation set limits to distribution, affect community diversity and suppress the density of some populations? Indeed, the experimental path blazed by Connell (1961) and Paine (1966) has become a broad highway. In their simplest form these are the easy questions of population ecology, answerable by fairly short-term ANOVA-type experiments; by contrast much remains to be done when they are asked of multi-species interactions or are taken to include indirect interactions. Experiments about dynamics are more difficult, which is why the analyses of long term observations mentioned above have been so important. Some field experiments on dynamics have been done, for example on metapopulation and refuge dynamics (Walde 1994; Murdoch *et al.* 1996), but they are typically more difficult, and this is an area that sorely needs attention.

From before the days of Nicholson's early work, mathematical models have played a central role in formulating and analysing population problems. In spite of the frequent (and frequently justified) empiricist's complaints about ecological models, we are convinced that models and theory are essential to the discipline's advancement. Theoretical advances have been made on a broad front, for example on the dynamics of disease (Anderson and May 1991), the detection of chaotic dynamics (Sugihara 1994; Ellner 1991), techniques for modelling persistence of rare and endangered species (Tuljapurkar 1990; Lande 1993), the potential response of population and communities to large-scale environmental change (Kareiva *et al.* 1993), and the development of individual-based models (DeAngelis and Gross 1992), to name a few.

The frontiers of population ecology are thus broad, with strong advances in all of the areas mentioned. We cannot here do justice to this wide-ranging research. But perhaps its central success, and the activity we think should be a major focus of future research, has been adding ecological realism to ecological theory. We believe this necessarily implies a close interplay between theory, and experiment and observation; it is this interaction that has distinguished the most promising recent research. Theory for competing species, for example, has been enriched by Chesson's (1994) stochastic approaches, which incorporate the effects of random variation in the environment on coexistence, and from Tilman's (1994) explicit incorporation of resource dynamics and spatial interactions. And in both cases modelling has been accompanied by a complementary experimental program.

We next emphasise research that we believe is especially suited to incorporating realism and that combines naturally with experiment, namely the individual-based approach. We look first at models emphasising differences among individuals, and then at spatial models.

We shall see that such efforts to incorporate realism come with a cost: models become more complex. Such complexity tends to be antithetical to the other major aim of any science: generality. Our proposed solution is the development of clusters, or hierarchies, of models that explicitly include simple models as tools for understanding and giving insight into both the ecology and the more complex models. We mention connections between simple and complex models in the next three sections, which deal with different approaches to incorporating realism, and give it more weight in the final section of the paper.

Models based on individual physiology, behaviour and life history

The individual-based approach recognises that population dynamics is the expression of the sum of individual properties: life history, physiology and behaviour. Since the individual is also the unit of evolution, and the focus of much evolutionary ecology, individual-based models also provide a framework for bringing together evolutionary and ecological studies (Metz *et al.* 1992; Murdoch and Briggs in press).

It might be argued that population models have always been individual-based: the Lotka-Volterra predator–prey model assumes individuals encounter each other at random and all individuals in the population have the same properties. But recent work is based on the explicit recognition that properties vary among individuals and that such variation may have dynamic consequences (DeAngelis and Gross 1992). We discuss here two approaches: stage-structured models that follow the dynamics of groups of similar individuals, and computer simulations of the ensemble of every individual in the population. The former exemplify 'i-state distribution' models, the latter are 'i state-configuration' models.

Stage- and state-structured models of insect parasitoids and hosts

Insects and many other organisms pass through a series of stages, all the members of which have roughly the same properties. The stages might be egg, larva and pupa, but one could also distinguish between large and small larvae, or among different instars. Gurney and Nisbet (1983) introduced a general formalism using delay-differential equations for describing populations in which dynamics are more or less continuous in time so that generations potentially can overlap, and used it in a model of Nicholson's blowflies. Murdoch *et al.* (1987) applied these techniques to the interaction between the parasitoid *Aphytis* and its host California red scale, which is a world wide pest of citrus and is widely controlled by *Aphytis.* A body of theory is now developing for such parasitoid-host interactions (Murdoch and Briggs in press). In addition, the approach has been applied to the within-generation dynamics of univoltine species with non-overlapping generations (Godfray *et al.* 1994; Rohani *et al.* 1994), though the models become less tractable with the need to use different formalisms for within- and between-generation dynamics.

Applications to the *Aphytis*–red scale interaction exemplify the key advantage of this approach: the structure of the models can be made to fit closely the individual life histories and behaviours of the species under study. Scale insects grow through a series of instars distinguished mainly by their size; they also grow within each instar. The parasitoid appears to use red scale size in 'deciding' how to respond to the various vulnerable instars. Since scale growth is temperature dependent at the low densities prevailing when control is successful, scale in the model can simply be divided into 'stages' (i.e. size classes) defined by how they are treated by the parasitoid, each stage having a fixed duration in units of physiological time (degree-days).

Figure 1 shows that our most recent models include most of the known size-dependent behaviour of the parasitoid. The adult scale, both reproductive and pre-reproductive, are invulnerable; the smallest scale are only eaten, to provide nutrients to mature parasitoid eggs; somewhat larger scale receive a single male egg; yet larger scale almost always receive a single female egg; the largest scale may receive 2 female eggs (occasionally more, and sometimes also a male egg). These size-dependent parasitoid responses have potentially important dynamical effects. For example, the invulnerable adult stages are stabilising (Murdoch *et al.* 1987; Murdoch *et al.* 1996), and the size-dependent sex-allocation and host-feeding induce a kind of delayed density dependence in the parasitoid recruitment rate and hence, depending on the strength of the density dependence and the length of the developmental delay, can either increase stability or induce a different kind of instability (Murdoch *et al.* 1992b; Briggs *et al.* 1995).

These stage-structured models can serve several purposes. First, as indicated above, they can yield general insight into how various life history features might affect population dynamics. Godfray and Hassell's (1989) demonstration of host stage-structure inducing the almost discrete generations seen in some tropical parasitoid–host interactions is a further illustration. This type of analysis has only begun, and we believe this is a potentially rich vein that can be mined for additional broad insight into the dynamics of different types of insects and other organisms. The models can be adapted to mimic different life histories and life styles, so future theory can be aimed at ecologically and

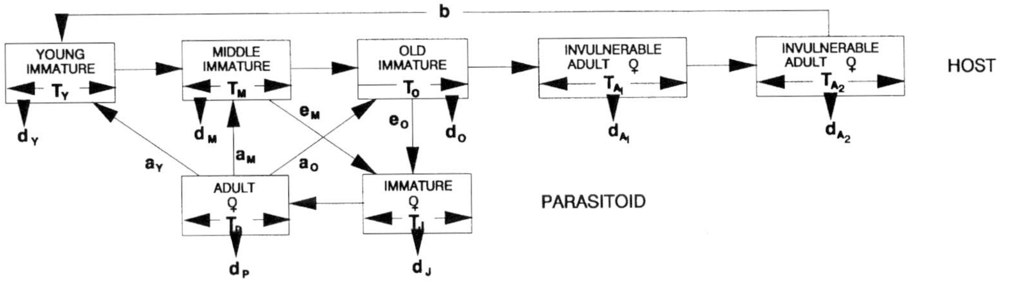

Fig. 1. Size- and state-dependent interactions between the parasitoid *Aphytis* and its host, red scale. The figure also depicts the components of a sequence of models describing different combinations of these interactions and a model containing all components. Only females are modelled. Scale and parasitoid stages suffer background death rates, d_i. The adult scale, both pre-reproductive and reproductive stages, is invulnerable to parasitoid attack. Three vulnerable immature instars can be distinguished (attacked at rate a_i). The smallest is only host-fed thus producing nutrients for egg-maturation. The fraction of larger hosts that are host-fed or receive a male egg, or that receive a female egg changes continuously with host size, but two main stages can be distinguished. Intermediate-sized hosts are mainly host-fed or receive a male egg, and the largest stage receives mainly female eggs. Hosts may also yield different numbers of female eggs (e_i). Whether a host is parasitised or receives an egg also depends on the number of mature eggs carried by the parasitoid (egg load), and the probability of laying an egg, at a given egg load, increases with host size.

behaviourally defined classes of organisms. In this respect it will be less general than early simple models in ecology hoped to be, but it will have a greater chance of being relevant to the real world.

A particular comparative use of the models has already provided insight into those features that distinguish more- from less-successful biological control agents of pest species. While the classical approach to this problem was to try to define the features that make the best control agents, stage-structured models have been used to predict which of two parasitoid species is likely to be a better bet, or to explain why one parasitoid species has been more successful than another (Waage 1990; Godfray and Waage 1991; Gutierrez *et al.* 1993).

Second, by matching real life histories, physiologies and behaviours, the models raise the possibility of testing hypotheses in real systems, through prediction of field dynamics using models whose structure, functions and parameter values have been derived from independent observations and experiments of individual organisms. By stimulating this goal, however, they inevitably lead to the uncovering of more and more details of particular systems, which leads, in turn, to the need for new techniques for modelling these details, to reduction in tractability of the models, and to the need to devise ways of making correct judgments about which details can be omitted.

Finally, since the behaviour of parasitoids in the models is 'state-dependent,' they are a natural vehicle for incorporating into population dynamics insights gained from evolution-inspired behavioural investigations based on considerations of optimality. For example, experiments in the last few years have shown that the 'decisions' of a searching female parasitoid are typically influenced by her egg load, i.e. the number of mature eggs in her ovaries. This is especially true in species that develop additional mature eggs using nutrients obtained from feeding on the body fluids of some host individuals. The decisions include time-in-patch, clutch size, sex-allocation, and whether to parasitise or host-feed on a host of a given size (Minkenberg *et al.* 1992). For example, in *Aphytis* (Fig. 1) the probability of host-feeding on a host of a given size decreases rapidly with egg load (Collier *et al.* 1994), as is predicted (approximately) by state-variable models optimising lifetime fitness (Collier *et al.* 1994; Collier 1995).

Incorporation of realism, of course, comes at a price—potentially increased complexity. Models that incorporate parasitoid behaviour based on egg load would seem to need to distinguish among not only different host stages but also groups of female parasitoids with 0, 1, 2... mature eggs. Surprisingly, however, Briggs *et al* (1995) have shown that, provided death rate does not depend on egg load, egg-load dependent behaviour can be faithfully portrayed in several situations by keeping the parasitoid population unstructured and letting each parasitoid respond to the mean population egg load. Our hope must be that we can establish such simplifying solutions in other situations, though the mathematical issues involved are daunting (Val and Metz 1991).

With a combination of stage- and state-dependent behaviour, we are now approaching a model of the *Aphytis*–scale interaction that is quite close to the details of the real system in the field and that should yield quantitative predictions that can be tested in the field. But the realism of the model raises the issue of its tractability. The model in Fig. 1 is cumbersome. We can obtain semi-analytic results using numerical approaches that draw stability boundaries in parameter space, but much of the analysis depends on simulations of the interaction with particular parameter values. Since the model has many parameters, sensitivity analysis over realistic ranges is a potentially enormous task. The challenge is therefore to develop techniques (substituting dependence of individual behaviour on mean state, where possible, is a powerful example) that help us understand the model's behaviour and that bound its potential behaviour. Such difficulties raised by model complexity also motivate the development of theory as clusters of models, discussed in the final section below.

Simulations of collections of individuals

Many organisms, such as trees and fish, grow continuously and indeterminately. A natural way to model such populations is to characterise each individual in the population by its 'state,' and follow the dynamics of the entire ensemble, including changes in the state of individuals as they grow and mature.

This 'i-state configuration' approach has recently been applied with success to understanding factors influencing annual variation in recruitment to the year 1 class of fish populations. A good example is provided by a series of studies of striped bass in the Potomac river (van Winkle *et al.* 1993). These studies show, for example, that the observed 150-fold between-year variation in recruitment can be explained by the combined effects of just two factors such as size-distribution of females and the density of zooplankton prey (Cowan *et al.* 1993), and that the model predictions of variables such as juvenile densities, mortality and growth rates are close to those seen in the field (Rose and Cowan 1993). These results, however, also illustrate the current limits of this approach, since the predictors are themselves population variables that one would like to predict.

A crucial feature of many of these models is stochastic variation in initial size and growth rate among individual larvae and juvenile fish. Since survival of immature fish can be strongly size-dependent, especially under predation pressure, the stochastic variation can induce substantial changes in the number and size-distribution of surviving fish (Rice *et al.* 1993). While some types of random variation among individuals can be incorporated in i-state distribution models (e.g. partial differential equations), DeAngelis *et al.* (1993) point out that serially correlated randomness, for example in daily growth rate, is best captured by the i-state configuration models (i.e. simulations of ensembles of individuals, see Efford this volume), and that such randomness can have a major effect on survival and ultimate size-distributions of recruiting fish.

As this example illustrates, these models have been most useful in answering quite narrowly defined questions about intra-generational dynamics in particular systems. They have yet to be extended to long-term multi-generation dynamics, which may also require that they become truly multi-species models, to take into account the dynamics of, for example, the prey species, as we noted above.

As the models thus become ineluctably even more complex, it will also be useful to connect them to simpler models with known properties (e.g. DeAngelis *et al.* 1993).

The simulation approach, however, still faces difficulties. Two are perhaps paramount. First, the very complexity of the models makes them difficult to understand, in the sense that it may not be clear what elements of the model are responsible for which dynamic features. Even with enormously increased capacity for testing the models' sensitivities, it may be impossible in practice to explore fully the implications of various generic and particular assumptions, and their interactions. The models that appear to have avoided this problem, such as the fish models discussed above, are much simplified. They do not follow in realistic detail the dynamics of any species other than the central one of interest, and they do not look at multiple-year dynamics. Second, although their testability is in principle their primary advantage, they are so data-hungry that testing predictions independently of data used to parameterise the model is in practice very difficult. The need for clusters of simpler models whose properties are more transparent, as discussed below, is therefore especially great in this type of modelling (e.g. Gurney *et al.* in press).

Adding realism: models incorporating space

Studying spatial effects is a growth industry in population ecology (Kareiva and Wennergren 1995; see Part 3 this volume) and again we can do little more than comment on a few approaches we have found particularly interesting. As with non-spatial dynamics, there is a marked contrast in approach between those who seek to model specific systems and those whose target is generality. The former start with computational tools such as geographic information systems (GIS) which permit incorporation of much geographical (e.g. topographical) detail, the latter incorporate space into traditionally formulated population models. An important 'frontier' involves linking these approaches in such a way that the roles of space and geography can be distinguished. This is particularly important in conservation contexts, and interesting studies are in progress, for example on the spotted owl (McKelvey *et al.* 1993). Progress will be facilitated by improved understanding of simple, spatially explicit, non-geographical models.

Much traditional spatial theory focuses on two types of model. **Metapopulation models** describe populations made up of subpopulations on distinct patches. Each subpopulation maintains its integrity over time scales longer than a generation, thereby permitting patch-specific vital rates. Recent research on metapopulation models has been reviewed by Hastings and Harrison (1994). The key insight from this work has been that population persistence may be greatly increased by metapopulation structure provided the combined effect of interactions between and within patches is to produce a 'blinking lights' situation where the various subpopulations fluctuate asynchronously. **Reaction-diffusion-advection models** describe populations where vital rates and flows between neighbouring points in space are functions of local density and environment. These models take a variety of forms, the simplest mathematically being couched in terms of partial differential equations (Okubo 1980), but they include multi-compartment models of real systems (Ross *et al.* 1993).

There is a wide body of literature on these approaches, and a key job for the immediate future is to develop seamless connections between the different approaches. Here again we believe that the key to progress is careful incorporation of ecological mechanisms and that the individual-based approach has much to offer (e.g. Durett and Levin 1994). An example of the theoretical problems associated with relating individual-based and metapopulation models occurs in our own research on the zooplankter *Daphnia*. That research has yielded experimentally-based rules about reproduction, mortality, and movement of *Daphnia* (Gurney *et al.* 1990; McCauley *et al.* 1990; Cuddington and McCauley 1994). However, there are many problems in moving from this broad insight to viable theory applicable to larger and/or more complex systems. Typical *Daphnia* densities are 10^3 m^{-3};

Daphnia eat microalgae (typical natural density is 10^9 edible algal cells m^{-3}) whose growth rate depends on the distribution of inorganic substrates over a spatial scale of 10^{-6} m . A small lake might have a volume of 10^6 m^3. Thus there is no possibility of modelling explicitly the state or environment of every individual in a lake (an exercise in futility which would demand unattainable computer power and yield no insight); instead we need to establish a hierarchy of simplifications that enable us to move up in spatial scale and across levels of biological organisation, but still relating properties of individuals to population dynamics. Workers with individual-based models have tackled the computational issue by working with 'super-individuals' and by relating the individual-based model to i-state distribution models for which much theory exists (see previous section and DeAngelis *et al.* 1993) but many unresolved issues remain. In essence the problem is one of averaging and there are appealing analogies with scaling issues in the theory of fluid turbulence (Levin 1992), but due to the long time scales of some biological interactions that theory becomes more complex.

One important insight, which relates individual behaviour to metapopulation dynamics, comes from recent work by deRoos *et al.* (1991; see also McCauley *et al.* 1993) who studied models in which space is represented as a lattice of small cells, each a potential location for only one individual. Mathematically, the models are stochastic cell automata; however the emphasis on individuals distinguishes them from other applications of cell automata in metapopulation studies (e.g. Hassell *et al.* 1991). The main result was that when movement was restricted, the interactions within the system determined a 'correlation length' such that population fluctuations at points separated by greater distances are largely uncorrelated. Thus 'patches' are emergent properties caused by interactions among individuals, and population persistence is promoted by the familiar blinking lights scenario. Similar dynamics have been found using spatially-explicit, individual-based simulations in continuous space (R.M. Nisbet unpublished), thereby establishing that the emergence of the metapopulation dynamics is not an artefact due to the discretisation of space in the lattice models.

We do not yet have adequate theory relating 'correlation length' to the interactions among individuals, and the development of such theory is a high priority. However, as metapopulation-like dynamics can occur in simulations of only a few thousand individuals, we conclude that they may be of importance in quite small systems as well as in the large, global scale systems discussed by Kareiva and Wennergren (1995).

A related, inverse, problem is encountered in attempts to model data obtained by remote sensing. Here, the need for linking interactions among individuals to 'natural' spatial scales such as the correlation length described above arises because remote sensing data are determined on fixed grids whose size is set by technology, not ecology. Thus analysis requires assumptions about sub-grid-scale processes, reflecting how local transportation, growth and death events affect the measured population distribution. Modelling will be facilitated by improved understanding of the principles involved in scaling up biological information obtained from studies on individuals and small populations.

There is again great need to confront theory with observation and experiment from real systems, and this has nowhere proved more difficult than in spatial processes. Harrison (1994) suggests there is virtually no convincing evidence for the 'blinking light' process of metapopulation dynamics, and there has been very little experimentation (Walde 1994). Bringing spatial dynamics into the experimental arena of ecology is surely one of the major areas in need of the frontier spirit.

Connecting models and field systems

Our emphasis on models is decidedly not meant to imply a lesser role for observation and experiment. On the contrary, arguably the best of population ecology in recent years has combined theory and experiment, and we argued that the great advantage of the individual-based approach is its potential for incorporating information on individual physiology and behaviour and for facilitating experimental tests. But challenging theory with experiment is not easy.

Experimental analyses in the field, and particularly field tests of individual-based models, have not kept up with advances in modelling. This is to be expected. Since the models are more complex, more work is needed to establish the functional relationships and estimate the parameter values; since the predictions are more detailed, more effort is needed to compare real and expected outcomes. But the problem is deeper. There are inevitably limits (which may grow less constraining with time) to the degree of detail that can be compared between model and reality, and the problem is to find ways of nevertheless testing the models. For example, we know a great deal about the behaviour of individual parasitoids such as *Aphytis*, as a function of their own state (egg load) and that of the environment (e.g. host sizes available). But one possible variable to be predicted, pest abundance in the field, depends heavily on *Aphytis* attack rate, which in the field is affected by weather, daylength, microclimate that changes rapidly in time and space, the microstructure of the habitat, the physical distribution of hosts within this structure, host size-distribution, parasitoid egg loads, the size-distribution of adult female *Aphytis*, and so on. Even some basic variables are very difficult to estimate accurately, for example the density of searching adult parasitoids, their search rate, and their instantaneous survival. These factors may make it difficult to predict even time-averaged densities. Instead, tests may need to focus on only qualitative dynamics, or variables measuring relative quantities such as age distributions, or numbers and qualities of individuals at crucial periods of the year, or patterns across space. This also argues for the need to choose empirical systems where the theory in question can be most easily tested.

The difficulty of usefully applying reductionist and mechanistic individual-based models to field conditions in which individuals are subjected to multiple driving forces that change rapidly in space and time is broader than population ecology. Recently oceanographers, for example, have begun to try to replace essentially curve-fitting phenomenological models of marine primary productivity with the equivalent of individual-based models, i.e. models of collections of algal cells controlled by the underlying physiological and biochemical mechanisms of photosynthesis and their responses, in the individual cell, to changes in the physical and chemical environment. From this mechanistic basis it is hoped to gain a better understanding of production on the large scale, and a sounder basis for predicting the effects of, for example, global climate change (e.g. Lande and Lewis 1989; Pahl-Wostl and Imboden 1990).

In both examples, we know that ignoring individual differences, and mere extrapolation, will frequently mislead us. Yet equally large problems may arise from the propagation of the errors made in predicting the time course of collections of individuals. Developing solutions to this generic problem is one of the major tasks facing the individual-based approach. Yet again, we should not ignore simple models. For example, though we have developed with our colleagues detailed individual-based models of the interaction between the freshwater zooplankter *Daphnia* and its algal prey, we found a highly simplified but individual-based and rigorously parameterised model of this system, served well to test against field data hypotheses concerning the effect of enrichment on the abundance of plankton populations (Nisbet *et al.* in press). Other hypotheses, notably those relating to spatial heterogeneity and individual variability, can only be tested with the aid of individual-based models. We see a continuing need for clusters of models, with insight coming both from relationships among models and from the connections between the models and real-world dynamics.

Realism, complexity and generality: hierarchies of models

There is an inherent tension between making a model more realistic, which usually makes it more particular, limited in application and more complex, and developing general theory (Murdoch *et al.* 1992a). We believe the solution to this dilemma lies in development of clusters, or hierarchies of models, a practice we recommended above as a way of understanding more complex models. When the behaviour of a complex, realistic model can be understood in terms of simpler models, the former can be viewed as derivatives of the latter in a hierarchy of models.

The simpler models need not be realistic, though they may be, but rather should allow us to understand the workings of the more realistic and more complicated model. The simpler models may have the additional advantage of yielding broad insight into how particular dynamical mechanisms work. We mention three examples. First, Murdoch *et al.* (in prep.) have shown that the range of dynamical behaviour seen in the complex stage-structured host–parasitoid model in Fig. 1, and in Godfray and Hassell's (1989) model with density-dependence in the parasitoid attack rate, are already present or foreshadowed in the simplest model of this system, which contains only an adult invulnerable host stage and an immature vulnerable stage (Murdoch *et al.* 1987).

The second example comes from the individual-based spatial model of deRoos *et al.* (1991) discussed earlier. The rules in that model were chosen so that if the system was well mixed, the dynamics were similar to those in the Rosenzweig-MacArthur prey–predator model: large amplitude cycles of both prey and predator populations. Thus a well-understood, simple, differential equation model emerged as a limiting case of the more complex model, and understanding the role of space in stabilising the population dynamics was facilitated through detailed comparison of the dynamics in the simple and the individual-based model.

Finally, Chesson (1994) has developed a general theory covering the three classes of mechanisms discovered to date that allow coexistence of competing species. The first, classical 'Lotka-Volterra' mechanism, is unaffected by temporal variation in the environment. The other two require such variation. One is non-linear responses (such as functional responses) to temporal changes in the abundance of the shared limiting factor, coupled with different responses in the competing species (e.g. Armstrong and McGehee 1980). The other is the collection of mechanisms that exemplify Chesson's 'storage effect,' whereby each species is able to 'store' the population gains made during favourable periods in a life history stage that declines in abundance only slowly during unfavourable periods (examples are resistant seeds or potentially long-lived adults).

Chesson's achievement is to have developed a hierarchy of models that together spans the range from general to specific and testable models. At the top of this hierarchy is a truly general model, with generic functions describing the underlying dynamics, that incorporates all three classes of mechanisms. On an intermediate level of generality and specificity are three models, one for each of the three classes of coexistence mechanisms, again with generic rather than specific functions. At the lowest level of generality, but the highest of specificity, are three models each describing a particular 'storage' mechanism (the 'lottery', a seedbank, and recruitment variation with resource partitioning), each of which can be found in a particular ecological system. A version of the seedbank model is now being tested in the field.

Such collections of models may seem less manageable than earlier visions of a general theory wholly contained in the Lotka-Volterra or Nicholson-Bailey models and a few of their variants. But their greater complexity actually represents an overall gain. Our modelling tools now allow us to look at more interesting questions than were dreamed of by any of these founders of ecological theory, and the resulting models at the more particular end of the spectrum can be both a stimulant to the experimentalist and a tool for understanding important problems in the management of ecological systems.

Acknowledgments

We are grateful to Sue Swarbrick, Steve Gaines and Sebastian Diehl for useful comments on the manuscript. The research was supported by NSF grant 93-06354 and USDA grant 94373020616.

References

Anderson RM, May RM (1991) *Infectious diseases of humans: dynamics and control.* Oxford Univ Press, Oxford, UK

Armstrong RA, McGehee R (1980) Competitive exclusion. *Am Nat* **115**:151–170

Avise JC (1993) *Molecular markers, natural history and evolution.* Chapman & Hall, New York, USA

Briggs CJ, Nisbet RM, Murdoch WW, Collier TR, Metz JAJ (1995) Dynamical effects of host-feeding in parasitoids. *J Anim Ecol* **64**:403–416

Carpenter SR, Kitchell JF (1993) *The trophic cascade in lakes.* Cambridge Univ Press, Cambridge

Chesson P (1994) Multispecies competition in variable environments. *Theor Pop Biol* **45**:227–276

Collier TR (1995) Physiological realism and dynamic state variable models of parasitoid host-feeding. *Evol Ecol* **9**:217–235

Collier TR, Murdoch WW, Nisbet RM (1994) Egg load and the decision to host-feed in the parasitoid, *Aphytis melinus*. *J Anim Ecol* **63**:299–306

Connell JH (1961) Effects of competition, predation by *Thais lapillus*, and other factors on natural populations of the barnacle *Balanus balanoides*. *Ecol Monogr* **31**:61–104

Cowan JH, Rose KA, Rutherford ES, Houde ED (1993) Individual-based model of young-of-the-year striped bass population dynamics. II. Factors affecting recruitment in the Potomac River, Maryland. *Trans Am Fish Soc* **122**:439–458

Cuddington KM, McCauley E (1994) Food-dependent aggregation and mobility of the water fleas *Ceriodaphnia dubia* and *Daphnia pulex*. *Can J Zool – Rev Can de Zool* **72**:1217–1226

DeAngelis DL, Gross LJ (1992) *Individual-based models and approaches in ecology.* Routledge, Chapman and Hall, New York, USA

DeAngelis DL, Rose KA, Crowder LB, Marschall EA (1993) Fish cohort dynamics: Application of complementary modeling approaches. *Am Nat* **142**:604–622

deRoos AM, McCauley E, Wilson WG (1991) Mobility versus density-limited predator-prey dynamics on different spatial scales. *Proc R Soc Lond B* **246**:117–122

Durrett R, Levin SA (1994) Stochastic spatial models – A users guide to ecological applications. *Phil Trans R Soc Lond B* **343**:329–350

Ellner S (1991) Detecting low-dimensional chaos in population dynamics data, a critical review. In: Logan JA, Hain FP (eds) *Chaos and insect ecology.* Virginia Agricultural Experiment Station, Blacksburg, Virginia Polytech Inst State Univ, USA, pp 65–92

Godfray HCJ, Hassell MP (1989) Discrete and continuous insect populations in tropical environments. *J Anim Ecol* **58**:153–174

Godfray HCJ, Waage JK (1991) Predictive modelling in biological control: The mango mealy bug (*Rastrococcus invadens*) and its parasitoids. *J Appl Ecol* **28**:434–453

Godfray HCJ, Hassell MP, Holt RD (1994) The population dynamic consequences of phenological asynchrony between parasitoids and their hosts. *J Anim Ecol* **63**:1–10

Gurney WSC, Nisbet RM (1983) The systematic formulation of delay-differential models of age and size structured populations. In: Freedman HI, Strobeck C (eds) *Population Biology.* Springer-Verlag, Germany, pp 163–172

Gurney WSC, McCauley E, Nisbet RM, Murdoch WW (1990) The physiological ecology of *Daphnia*: A dynamic model of growth and reproduction. *Ecology* **71**:716–732

Gurney WSC, Middleton DAJ, Nisbet RM, McCauley E, Murdoch WW, deRoos AM (1996) Modelling predators and their impacts. *Proc R Soc Edinburgh* (in press)

Gutierrez AP, Neuenschwander P, van Alphen JJM (1993) Factors affecting biological control of casava mealybug by exotic parasitoids: a ratio-dependent supply-demand driven model. *J Appl Ecol* **30**:706–721

Hanski I (1990) Density dependence, regulation and variability in animal populations. *Phil Trans R Soc Lond B* **330**:141–150
Harrison S (1994) Metapopulations and conservation. In: Edwards PJ, May RM, Webb NR (eds) *Large-scale ecology and conservation biology*. Blackwell, Oxford, UK, pp 111–128
Hassell MP, Comins H, May RM (1991) Spatial structure and chaos in insect population dynamics. *Nature* **353**:255–258
Hastings A, Harrison S (1994) Metapopulation dynamics and genetics. *Annu Rev Ecol Syst* **25**:167–188
Hastings A, Higgins K (1994) Persistence of transients in spatially structured ecological models. *Science* **263**:1133–1136
Kareiva P, Wennergren U (1995) Connecting landscape patterns to ecosystem and population processes. *Nature* **373**:299–302
Kareiva PM, Kingsolver JG, Huey RB (1993) *Biotic interactions and global change*. Sinauer Associates, Sunderland, MA, USA
Lande R (1993) Risks of population extinction from demographic and environmental stochasticity and random catastrophes. *Am Nat* **142**:911–927
Lande R, Lewis MR (1989) Models of photoadaptation and photosynthesis by algal cells in a turbulent mixed layer. *Deep-Sea Res* **36**:1161–1175
Levin SA (1992) The problem of pattern and scale in ecology. *Ecology* **73**:1943–1967
May RM (1976) Simple mathematical models with very complicated dynamics. *Nature* **261**:459–467
McCauley E, Murdoch WW, Nisbet RM, Gurney WSC (1990) The physiological ecology of *Daphnia*: Development of a model of growth and reproduction. *Ecology* **71**:703–715
McCauley E, Wilson WG, deRoos AM (1993) Dynamics of age-structured and spatially structured predator-prey interactions: Individual-based models and population-level formulations. *Am Nat* **142**:412–442
McKelvey K, Noon BR, Lamberson RH (1993) Conservation planning for species occupying fragmented landscapes: the case of the northern spotted owl. In: Kareiva PM, Kingsolver JG, Huey RB (eds) *Biotic interactions and global change*. Sinauer Associates, Sunderland, MA, USA, pp 424–450
Metz JAJ, Nisbet RM, Geritz SAH (1992) How should we define 'fitness' for general ecological scenarios? *Trends Ecol & Evol* **7**:198–202
Minkenberg OPJM, Tatar M, Rosenheim JA (1992) Egg load as a major source of variability in insect foraging and oviposition. *Oikos* **65**:134–142
Murdoch WW, Briggs CJ Theory for biological control: recent developments. *Ecology* (in press)
Murdoch WW, Nisbet RM, Blythe SP, Gurney WSC, Reeve JD (1987) An invulnerable age class and stability in delay-differential parasitoid-host models. *Am Nat* **129**:263–282
Murdoch WW, McCauley E, Nisbet RM, Gurney WSC, deRoos AM (1992a) Individual-based models: combining testability and generality. In: DL DeAngelis and LJ Gross (eds) *Individual-based models in ecology*. Chapman and Hall, New York, USA, pp 18–35
Murdoch WW, Nisbet RM, Luck RF, Godfray HCJ, Gurney WSC (1992b) Size-selective sex-allocation and host feeding in a parasitoid-host model. *J Anim Ecol* **61**:533–541
Murdoch WW, Swarbrick SL, Luck RF, Walde S, Yu DS (1996) Refuge dynamics and metapopulation dynamics: Experimental test. *Am Nat* **147**:424–444
Nicholson AJ (1933) The balance of animal populations. *J Anim Ecol* **2**:132–178
Nisbet RM, McCauley E, Gurney WSC, Murdoch WW, de Roos, AM (in press) Simple representations of biomass dynamics in structured populations. In: Othmer HG, Adler FR, Lewis MA, Dallon J (eds) *Case studies of mathematical modeling in biology*. Prentice-Hall, in press
Okubo A (1980) *Diffusion and ecological problems: mathematical models*. Springer-Verlag, Heidelberg, Germany
Pahl-Wostl C, Imboden DM (1990) DYPHORA: A dynamic model for the rate of photosynthesis of algae. *J Plankt Res* **12**:1207–1221
Paine RT (1966) Food web complexity and species diversity. *Am Nat* **100**:65–75
Persson L, Diehl S, Johansson L, Andersson G, Hamrin SF (1992) Trophic interactions in temperate lake ecosystems: A test of food chain theory. *Am Nat* **140**:59–84

Rice JA, Miller TJ, Rose KA, Crowder LB, Marschall EA, Trebitz AS, DeAngelis DL (1993) Growth rate variation and larval survival: Inferences from an individual-based size-dependent predation model. *Can.J Fish Aquat Sci* **50**:133–142
Rohani P, Godfray HCJ, Hassell MP (1994) Aggregation and the dynamics of host–parasitoid systems: A discrete generation model with within-generation redistribution. *Am Nat* **144**:491–509
Rose KA, Cowan JH (1993) Individual-based model of young-of-the-year striped bass population dynamics. I. Model description and baseline simulations. *Trans Am Fish Soc* **122**:415–438
Ross AH, Gurney WSC, Heath MR (1993) Ecosystem models of Scottish sea lochs for assessing the impact of nutrient enrichment. *ICES J Mar Sci* **50**:359–367
Sugihara G (1994) Nonlinear forecasting for the classification of natural time series. *Phil Trans R Soc Lond A* **348**:477–495
Tilman D (1994) Competition and biodiversity in spatially structured habitats. *Ecology* **75**:2–16
Tuljapurkar S (1990) *Population dynamics in variable environments.* Springer-Verlag, Germany
Turchin P, Taylor AD (1992) Complex dynamics in ecological time series. *Ecology* **73**:289–305
Val J, Metz, AJ (1991) Asymptotic exact finite dimensional representation of models for physiologically structured populations: the concepts of weak and strong linear chain trickery. In: Arino O, Kimmel M (eds) *Proc. 3rd Int Congr Math Pop Dyn*, S Wuertz, Winnipeg.
van Winkle W, Rose KA, Chambers RC (1993) Individual-based approach to fish population dynamics: An overview. *Trans Am Fish Soc* **122**:397–403
Vitousek PM (1990) Biological invasions and ecosystem processes: Towards an integration of population biology and ecosystem studies. *Oikos* **57**:7–13
Waage J (1990) Ecological theory and the selection of biological control agents. In: Mackauer M, Ehler LE, Roland J (eds) *Critical issues in biological control.* Intercept, Andover, UK pp 135–157
Walde S (1994) Immigration and the dynamics of a predator-prey interaction in biological control. *J Anim Ecol* **63**:337–346

Detecting density dependence

David R. Fox and T.J. Ridsdill-Smith

Abstract

A number of papers have appeared in the literature recently in which the performance of tests of density dependence in population time series data were compared. We believe that the search for an 'optimal' test for density dependence is unproductive and the practice of assessing competing tests *via* computer simulation flawed. It is unlikely that a single test will have uniformly superior power characteristics under all conditions. The difficulty with simulation studies is the lack of communality between them. We have observed differences in simulation results that have as much to do with differences in model parameterization, simulation run length, characterisation of the stochastic component, and methods of analysis of the results as they do with any inherent difference in the performance of the tests for density dependence being compared. We advocate the use of Bulmer's first test as a *de facto* standard because it makes minimal assumptions about the response-generating mechanism, has reasonable power characteristics, and is easy to implement. It is a test of serial correlation in the log abundance data and consequently is affected by the presence of autocorrelated errors. However, this should not be interpreted as a failing of the test, but rather a caveat to its use.

Key words: Density dependence, statistical tests, test selection, time series

Frontiers of Population Ecology, R.B. Floyd, A.W. Sheppard and P.J. De Barro (eds), CSIRO Publishing, Melbourne, 1996 pp. 45–51

Introduction

Although it is widely considered that regulation occurs in biological populations, the detection and measurement of this phenomenon *via* an analysis of time series data has proved difficult (Murdoch 1994). Population regulation is an important concept for the management and control of pests where we wish to retard population growth, and in conservation where it is desired to maintain populations or at least to prevent extinction.

In an unregulated population, fluctuations in population size are considered to represent a random walk, while regulation is characterised by bounded fluctuations in abundance with a propensity to return to an equilibrium value (Murdoch 1994). Both behaviours are encapsulated in the simple model: $X_{t+1} = r + \beta X_t + e_{t+1}$ where X_t is the logarithm of population size at time t, e_t is a zero mean, finite variance random variable, and r and β are model parameters. When $\beta = 1$ the population follows a random walk with drift ($r \neq 0$) or without drift ($r = 0$). Stationarity conditions require $|\beta| < 1$ with the usual notion of regulation corresponding to $-1 < \beta < 0$. This model represents the situation of a stabilising density dependent factor in a local population and is the density dependence described by Nicholson (1954) based on his studies of intraspecific competition in blowflies for a limited food resource. It is important to realise that there are other stabilising mechanisms involving, for example, spatial heterogeneity in the environment and the dispersal and movement of organisms. This is the type of control described by Andrewartha and Birch (1954), and these spatial processes have also been widely considered in the literature. We will, however, concentrate on tests for stabilising density dependent factors in local populations. Regulation in a population is thus characterised by a measurable feedback process which can be quantified by examining a series of census observations from a population. A number of procedures have been developed to test for density dependence, although useful recommendations to practising biologists have been obscured by on-going debates over what these are actually detecting (Wolda and Dennis 1993; Holyoak and Lawton 1993; Wolda *et al.* 1994) coupled with inconclusive results from comparative simulation studies (Holyoak 1993).

In this paper we look at several tests of density dependence and some of the difficulties in designing comparative studies. We also discuss the selection of an appropriate statistical test, particularly in relation to the effects of non-independent errors and delayed density dependence.

Tests for density dependence

The development of a new test for density dependence is usually motivated by perceived or known shortcomings of existing tests. Each new test introduces its own set of limitations and restrictions and so the cycle continues (Fox and Ridsdill-Smith 1995). The increasing availability of affordable computing power is accelerating this trend, which seems to have increased the computational complexity rather than the universal applicability of recent tests. These points are illustrated in relation to five of the more widely used tests.

Regression techniques

Key factor analysis was developed to investigate stage specific mortality within populations (Morris 1959; Varley and Gradwell 1960). This has been adapted to examine density dependence in time series of population abundance. The slope b of the regression of X_{t+1} on X_t is used to test the null hypothesis of density independence (H_0: $\beta = 1$) against an alternative of direct density dependence (H_0: $\beta < 1$). Tests of the regression method using artificial data indicate that density dependence is present in many situations where it is not present (Maelzer 1970).

Bulmer's first test

Bulmer (1975) suggested that it was inappropriate to treat the slope b of the regression test (see above) as if it was an ordinary regression coefficient, since β is equivalent to the first-order serial correlation coefficient between the X's, which cannot exceed unity. Bulmer's test is based on the statistic $R=V/U$, where

$$U=\sum_{t=1}^{n-1}(X_{t+1}-X_t)^2,\ V=\sum_{t=1}^{n}(X_{t+1}-\overline{X})^2 \text{ and } \overline{X}=\frac{1}{n}\sum_{t=1}^{n}X_t$$

This is an indirect method of inferring density dependence via an examination of serial correlation between the X's and the test rejects the hypothesis of density independence for small values of R.

Randomisation test of Pollard et al.

Pollard *et al.* (1987) devised a Randomisation test which they claim does not suffer from the lack of power observed for Bulmer's test in the presence of a temporal trend. Their procedure is based on the proportion of times some test statistic T obtained from random permutations of the data is exceeded by the sample value for the observed series. If the proportion of values of the test statistic that are less than or equal to the observed value is less than the level of significance, the hypothesis of density independence is rejected.

Test for limitation of Reddinguis and den Boer

Reddinguis and den Boer (1989) also criticised earlier tests and proposed a new test statistic. It is a non-parametric test based on a comparison of the rank of the log-range of the field data within the collection of log-ranges obtained from all possible permutations of the coefficients of net reproduction (growth rates) that were derived from the field data. When density is stabilised, the actual sequence of net reproductive values is expected to have a lower log-range than in a significant proportion of permuted series. In practice it is very similar to the test of Pollard *et al.*, but uses the logarithmic range ($X_{max} - X_{min}$) as the basis of inference rather than a correlation coefficient. The lack of computing power in desktop PCs at the time the test was developed (1970s) posed a serious limitation to the effectiveness of this test, although today this is no longer a problem.

Parametric bootstrap likelihood ratio (PBLR) test of Dennis and Taper

As a motivation for the development of their own procedure, Dennis and Taper (1994) suggest that the randomisation test of Pollard *et al.* (1987) has low power. They report up to a 50% increase in power over the Pollard *et al.* test when the data are generated from the stochastic logistic model: $N_{t+1} = N_t\, exp(a + bN_t + Z_t)$, where a, b and are constants and Z_t is a unit normal random variable. The PBLR method uses a resampling method such as the Jacknife or Bootstrap (Efron and Tibshirani 1993) to estimate the sampling distribution of a likelihood ratio test statistic under the assumption that the data have been generated by the stochastic logistic model (hence the reference to parametric). An appropriate percentile of this distribution is then used as a critical value in the test procedure.

Comparative studies

The five examples given above serve to illustrate the pattern of development of statistical tests for density dependence. Each new test is usually accompanied by simulation results attesting to the claimed superiority (usually measured in terms of improved power), although unfortunately these simulations are often quite limited in scope and therefore conclusions can not be readily generalised. The number and diversity of statistical test procedures has provided fertile ground for authors of comparative studies in which attempts are made to rate the tests (Maelzer 1970; Itô 1972; Holyoak 1993 and many others). While this would appear to be an entirely reasonable objective, a definitive ranking is unlikely to emerge. Instead, the outcome is usually a highly qualified recommendation of the sort 'test A is preferred to test B when the data conform to model x, except when the effects of y are present, otherwise test C should be used'. The use of computer simulations to evaluate competing tests is fraught with difficulties. Issues which require careful forethought include: i) choice of model parameters and the treatment of 'infeasible' parameter combinations; ii) length of simulation run; iii) representation of a model's stochastic component; iv) handling of 'start-up' effects; and v) choice of statistical methods to analyse the output. We believe that the different approaches to these issues are in part responsible for the confusion generated by different comparative studies.

Notwithstanding our reservations about the utility of comparative studies, a number of general observations emerge: i) inconsistent results arise from different studies; ii) different test results are to be expected in part by virtue of the different assumptions built into the tests; and iii) the power of many tests is affected by non-independent error terms, by spatial and temporal trends, by short series length, and different models used to synthesise the data; iv) regression tests generally perform poorly (Maelzer 1970; Holyoak 1993).

Selecting a test of density dependence

When investigating density dependence, it is important to distinguish between the statistical hypothesis of density dependence and the ecological hypothesis (i.e. the biological mechanism of negative feedback on growth rate; Dennis and Taper 1994). Statistical analyses of time series data have a limited capacity to reveal mechanisms of regulation (Reddinguis and den Boer 1989; Murdoch 1994). Wolda *et al.* (1994) noted that statistical density dependence (SDD) is often indicated in time series data for which the notion of density dependence is inappropriate (for example, rainfall data) and cautioned against inferring biological density dependence (BDD) using test results for SDD. The key issue here is that any stationary process can be interpreted as a regulated process, but the interpretation of the regulatory mechanism is entirely subjective.

The researcher needs to be quite clear about the objectives of his or her study. These will usually involve a requirement to: *describe* an observed series; *infer* something from the observed series, or to *model* the mechanism responsible for the observed series. A purely descriptive analysis of the data would be uncommon, with most studies attempting to infer and/or model biological density dependence. We have suggested (Fox and Ridsdill-Smith 1995) a staged approach to modelling in a spirit similar to *post-hoc* analyses in ANOVA models where additional testing is not advised in the absence of a significant ANOVA result. Likewise, we do not believe the fitting of complex models in order to better understand the mechanisms of regulation is warranted when the hypothesis of density independence has not been rejected. Admittedly, the hypothesis of density independence can be couched and hence tested in terms of the parameters of some arbitrarily chosen model, but we see two difficulties with this approach: i) there would seem to be no *a priori* grounds for preferring one functional form over another and ii) conflicting test results may arise from different models and the way in which their parameters are estimated (Holyoak 1993; Dennis and Taper 1994).

For a test of density dependence to be useful it should: i) make minimal assumptions about the response-generating mechanism; ii) be simple to implement; and iii) be unbiased and have good power characteristics over a wide range of alternative hypotheses. Tests 3 to 5 above may satisfy the first and third of these criteria, but we are not convinced that the heavy computational burden outweighs the claimed lack of power of simpler alternatives. We suggest the adoption of Bulmer's first test (Bulmer 1975) as a *de facto* test since it satisfies all three criteria. In making this recommendation, we acknowledge the limitations and weaknesses of Bulmer's test. For example, it has poor power in the presence of temporal trends (Pollard *et al.* 1987) and in short time series an 'aberrant' initial value may have a large leverage on the outcome of the test. We argue that atypical values should be left out of the analysis in this situation, although further investigations into cross-validation methods are required. The correct application of any parametric test is predicated on certain assumptions having been reasonably met. In this sense, tests of density dependence are no different and there is an onus of responsibility on the user to ensure that the application of Bulmer's test is correct.

Statistical issues to be considered

Autocorrelation

Bulmer's test is a test of serial correlation in (log) abundance data and is known to be affected by autocorrelated errors (Reddingius 1990, Solow 1990). To illustrate, consider a lagged dependent variable model with an AR(1) error structure:

$$x_{t+1} = r + \beta x_t + e_{t+1} \qquad |\beta| < 1$$
$$e_{t+1} = \phi e_t + u_{t+1} \qquad |\phi| < 1$$

where β is a measure of the density dependent effect; ϕ a measure of the autoregressive effect; and u_t is a mean zero random variable with finite variance σ^2. Furthermore, we let ρ denote the first-order serial correlation between the x's.

A test of (statistical) density dependence is equivalent to testing the pair of hypotheses H_0: $\beta = 1$ versus H_1: $\beta < 1$. Bulmer's test was developed under the explicit assumption that $\phi = 0$ and in this case the hypotheses in terms of β can be equivalently re-expressed in terms of ρ. When this is not the case, Bulmer's test (or indeed any test) cannot distinguish between an autoregressive effect in the errors and a true density dependent effect in the data. This indeterminacy is revealed by writing the pair of equations above in terms of the backward shift operator B where $BX_t = X_{t-1}$ (Box and Jenkins 1970) as $(1 - \phi B)(1 - \beta B)x_t = u_t$.

Delayed density dependence

Bulmer's test cannot distinguish whether density dependence is direct or delayed for the simple reason that it is essentially a test of the first-order serial correlation between the x's. Problems arise because this quantity is simultaneously affected by both direct and delayed density dependence. To see this, consider a simple extension to the previous model by introducing a delayed density dependent effect:

$$x_{t+1} = \beta_0 x_t + \beta_1 x_{t-1} + e_{t+1}$$

with

$$\beta_0 + \beta_1 < 1;\ \beta_1 - \beta_0 < 1 \text{ and } |\beta| < 1$$

It can be shown that in this case the first-order serial correlation between the x's is:

$$\rho = \frac{\beta_0}{1 - \beta_1}$$

Thus, if the density dependence is entirely delayed ($\beta_0 = 0$; $\beta_1 < 0$) Bulmer's test will (within the limits of sampling variability) suggest no density dependent effect is present. If on the other hand, both forms of density dependence are present ($\beta_0 < 0$; $\beta_1 < 0$) the power of Bulmer's test is reduced since ρ is increased (less negative). Holyoak (1994) observed that Bulmer's test rejected density independence 'too frequently' when the data contained only delayed density dependence. However, the immediately preceding model provides a counter-example. The reason Holyoak (1994) observed 'over-detection of non-delayed density dependence' with Bulmer's test was due to non-zero ρ values that were an artefact of the particular model used to synthesise the data. To examine the effects of delayed density dependence, Holyoak (1994) generated data from the model:

$$N_{t+2} = N_{t+1}\, exp\,[r_t(1 - \alpha N_t)]$$

where α is a fixed parameter and the parameter r_t was normally distributed. One can show either mathematically or empirically, that this procedure will result in data having non-zero first and second-order serial correlations in the x's. It therefore comes as no surprise that Bulmer's test will reject an hypothesis of density independence more frequently than the nominal level of significance in this situation.

Conclusions

The widespread availability of cheap and powerful computers has accelerated the development of computationally intensive methods of testing for density dependence. Following the development of a new test is the inevitable set of performance comparisons, and while some of the short-comings of a previous test may have been overcome, each new test is not without its own limitations and drawbacks. Furthermore, the methods by which results from comparative simulations are analysed further complicates the issue (Fox and Ridsdill-Smith 1995). The statistical behaviour of tests is not well studied, so that they tend to produce unreliable results (Murdoch 1994). We are not necessarily suggesting that the development of new tests and the refinement of existing ones should stop altogether, however we feel that researchers should be provided with clear and unambiguous recommendations. As a start we suggest the use of Bulmer's (first) test (Bulmer 1975) as a *de facto* standard, insisting that caveats to its use be well understood and appreciated. We make this recommendation because: i) the test is easy to conduct (Bulmer's test statistic and critical values can be computed on a calculator) and ii) the test has reasonable power characteristics over a wide range of alternative hypotheses (Fox and Ridsdill-Smith 1995). Finally, we believe important discussion on biological density dependence has been sidetracked by the continual stream of claims and counter-claims about the superiority of different statistical procedures to detect this phenomenon. The sooner biologists realise that a globally optimal test (i.e. over all response generating mechanisms and error structures) does not exist, the sooner they can get off the statistical treadmill and focus on the important biological issues.

Acknowledgments

We wish to thank Richard Morton and Barry Longstaff for their helpful comments. Financial assistance was provided by Australian wool growers through the International Wool Secretariat.

References

Andrewartha HG, Birch LC (1954) *The distribution and abundance of animals.* Univ Chicago Press, Chicago USA

Box GEP, Jenkins GM (1970) *Time series analysis: Forecasting and control.* Holden-Day, CA USA

Bulmer MG (1975) The statistical analysis of density dependence. *Biometrics* **31**:901–911

Dennis B, Taper ML (1994) Density dependence in time series observations of natural populations: Estimation and Testing. *Ecol Monogr* **64**:205–224

Efron B, Tibshirani RJ (1993) *An introduction to the bootstrap.* Chapman and Hall. New York USA

Fox DR, Ridsdill-Smith TJ (1995) Test for density dependence revisited. *Oecologia* **103**:435–443

Holyoak M (1993) New insights into testing for density dependence. *Oecologia* **93**:435–444

Holyoak M (1994) Identifying delayed density dependence in time-series data. *Oikos* **70**:296–304

Holyoak M, Lawton JH (1993) Comment arising from a paper by Wolda and Dennis: Using and interpreting the results of tests for density dependence. *Oecologia* **95**:592–594

Itô Y (1972) On the methods for determining density-dependence by means of regression. *Oecologia* **10**:347–372

Maelzer DA (1970) The regression of Log N_{n+1} on Log N_n as a test of density dependence: An exercise with computer-constructed density-dependent populations. *Ecology* **51**:810–822

Morris RF (1959) Single-factor analysis in population dynamics. *Ecology* **40**:580–588

Murdoch WW (1994) Population regulation in theory and practice. *Ecology* **75**:271–287

Nicholson AJ (1954) Experimental demonstrations of balance in populations. *Nature* **173**:862

Pollard E, Lakhani KH, Rothery P (1987) The detection of density dependence from a series of annual censuses. *Ecology* **68**:2046–2055

Reddingius J (1990) Models for testing: A secondary note. *Oecologia* **83**:50–52

Reddinguis J, den Boer PJ (1989) On the stabilization of animal numbers. Problems of testing. 1. Power estimates and estimation errors. *Oecologia* **78**:1–8

Solow AR (1990) Testing for density dependence: A cautionary note. *Oecologia* **83**:47–49

Varley GC, Gradwell GR (1960) Key factors in population studies. *J Anim Ecol* **29**:399–401

Wolda H, Dennis B (1993) Density dependence tests, are they? *Oecologia* **95**:581–591

Wolda H, Dennis B, Taper ML (1994) Density dependent tests, and largely futile comments: Answers to Holyoak and Lawton (1993) and Hanski, Woiwood and Perry (1993). *Oecologia* **98**:229–234

Density-perturbation experiments for understanding population regulation

Naomi Cappuccino and Susan Harrison

ABSTRACT

We review the use of experimental manipulations of population density to test for the existence and causes of regulation. Relatively few such studies have been performed, but they typically show clearer evidence for regulation than do the far more abundant census and life-table studies. Potentially regulating density dependence is nearly always exerted by the lower trophic level (food quality as well as quantity), rather than by predators or parasitoids. Such bottom-up effects were prevalent even among herbivorous insects, albeit at a lower frequency than in other taxa. Unfortunately, many studies finding density dependence lack critical information that would allow us to infer regulation. We discuss the 'ideal' experimental design for detecting regulation and examine the difficult issue of when temporal density dependence may be inferred from the results of spatial manipulations of density.

Key words: Density dependence, perturbation experiments, plant-herbivore interactions, dispersal

Frontiers of Population Ecology, R.B. Floyd, A.W. Sheppard and P.J. De Barro (eds), CSIRO Publishing, Melbourne, 1996, pp. 53–64

Introduction

Among A.J. Nicholson's many contributions to the field of population ecology was to suggest a powerful method for detecting regulation in natural populations – the density-perturbation experiment (Nicholson 1957). Despite the passion that the issue of population regulation has ignited over the last five decades, and despite the well argued reiteration of the need for perturbation experiments by Murdoch (1970), Nicholson's call for perturbation experiments has gone largely unheeded. Most of the debate over whether or not populations are regulated has relied upon phenomenological evidence in the form of time-series and k-factor analyses. The debates have dragged on for several decades, largely due to the inconclusiveness of these analyses (e.g. Gaston and Lawton 1987; Murdoch and Walde 1989; Holyoak and Lawton 1993; Hanski *et al.* 1993; Wolda and Denis 1993). Although newer, more robust analyses and longer time-series are now allowing us to address the question of population regulation with more powerful methods than ever before (Turchin 1995), the density-perturbation experiment remains an effective and under-utilised tool for understanding population regulation.

The density-perturbation experiment is a useful method for many reasons. First, experiments can reveal the causal mechanisms of regulation more conclusively than non-experimental methods; thus, they are the best way to resolve the current debate over the prevalence of top-down and bottom-up regulation in different types of systems. Second, they are an efficient way to measure the form and strength of density dependence, although this has rarely been done. Third, they can be used to assess spatial variation in the causes, form and strength of population regulation, which recent theory suggests may be important.

Here, we first discuss some striking findings from a recent review of density-perturbation experiments (Harrison and Cappuccino 1995). Then, we will illustrate density-perturbation experiments by describing two studies of insect herbivores that are regulated by their resources. Finally, we will discuss how density-perturbation experiments may be elaborated to address the many different forms that regulation may take.

Review of density perturbation studies

In an earlier paper (Harrison and Cappuccino 1995), we reviewed density-perturbation studies published between 1970 and 1993. These studies were found by searching the journals *Ecology* and the *Journal of Animal Ecology* and all papers having cited Murdoch's (1970) call for perturbation experiments. We retained all papers reporting the results of density manipulations, regardless of whether these had been performed in the laboratory or in the field. In all, we found 60 studies including a wide range of taxa and habitats. We divided potential regulatory agents into the following categories: 'bottom-up' (regulation by food resources), 'top-down' (predation, parasitism and disease) and 'lateral' (interference competition, territoriality, dispersal or cannibalism when these could not be directly linked to resource availability).

Direct density dependence was strikingly common in these experimental studies (Fig. 1). Of the 84 tests for density dependence (some studies reported more than one experiment or species), 79% showed direct density dependence. As is true of all literature reviews, there is always the possibility that a bias toward reporting only 'positive' results may have inflated our assessment of the preponderance of density dependence. On the other hand, since so-called 'non-equilibrium' theory is currently in vogue in North America (Koetsier *et al.* 1990), we might have expected that experiments failing to find density dependence would be rapidly published as evidence for this new world view, at least in the journal *Ecology*.

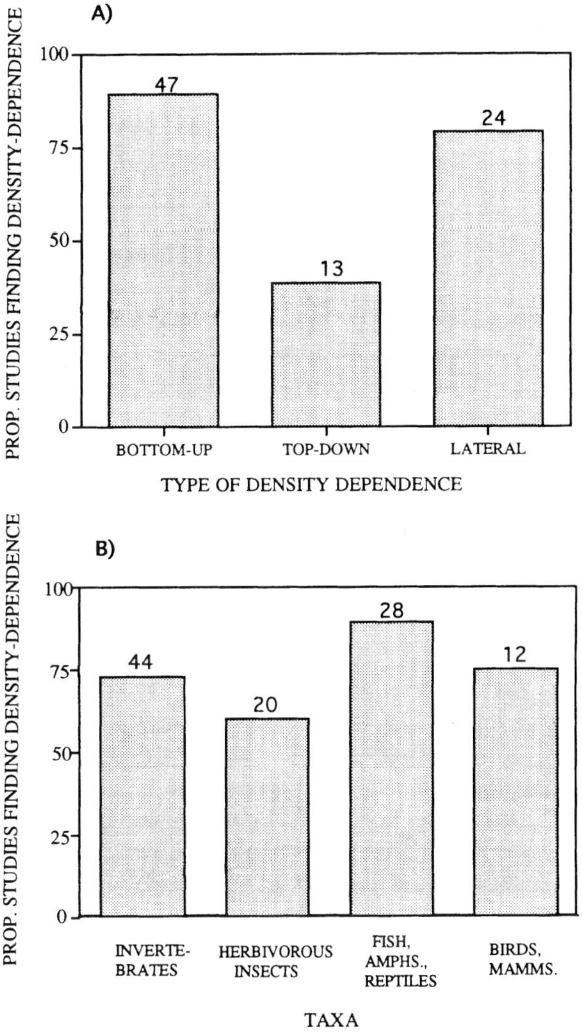

Fig. 1. Proportion of tests looking for density dependence that found it, categorised by (a) type of factor studied (see text for explanation) and (b) taxa of study organisms (in addition to being depicted as a separate category, herbivorous insects are included in 'inverts'. Numbers above bars represent total number of studies looking for a particular effect or at a particular taxonomic grouping. For references used in constructing this figure, see Harrison and Cappuccino 1995.

Evidence for 'bottom-up' regulation appears to be much more common than evidence for 'top-down' regulation (Fig. 1a). Although the prevalence of bottom-up forces in natural systems has recently been discussed (e.g. Hunter and Price 1992), it has not previously been clear if these forces were potentially regulatory (i.e. density-dependent) or large but density-independent. That top-down effects are studied less often than bottom-up effects is not surprising since the spatial scale needed to detect a numerical response by natural enemies is often prohibitively large.

Density-dependent 'lateral' factors were found in 79.2% of the studies that looked for them. Territoriality and density-dependent dispersal were common lateral forces. Again, the regulatory potential of these factors is not always clear. When some individuals disperse or fail to win a territory as a result of high density, it is important to know the fate of these individuals – whether they survive and reproduce – before these lateral forces may be considered regulatory (e.g. Smith and Arcese 1989). It is also unclear how often lateral forces are underlain by resource limitation; such cases would be more appropriately considered as bottom-up density dependence.

One final point that emerged from this review is that the preponderance of density dependence applies to herbivorous insects as well as to other taxa (Fig. 1b). Especially interesting is that most density dependence in insect herbivores was bottom-up (66.7%). This runs counter to two opposing views regarding the population dynamics of insect herbivores: that their populations are unregulated and that they are regulated by interactions with their natural enemies. Although life-table studies of herbivorous insects have stressed the importance of natural enemies (e.g. Cornell and Hawkins 1995), whether predation and parasitism are temporally density dependent and, thus, potentially regulatory, has rarely been determined in such studies. In the debate over the relative importance of top-down versus bottom-up forces, it is important to differentiate between forces that are potentially regulatory and forces that, although strong, are density independent, and thus not regulatory.

Case studies of herbivorous insects regulated by their resources

'Early larval death' in Eurosta solidaginis

Eurosta solidaginis (Diptera: Tephritidae) is a non-outbreak herbivore that forms galls in the stems of goldenrod *Solidago altissima*. Over the last 15 years, this species and several others have been censused in over 20 old fields in south-central New York State, USA (analyses on the first six years of this census have been published by Root and Cappuccino 1992). During this time, *E. solidaginis* has been present at moderate densities in most fields and, like other gall-makers on goldenrod, its populations have fluctuated little (Root and Cappuccino 1992).

Cappuccino (1992) altered *E. solidaginis* gall density in 20 large patches of goldenrod in old fields near Ottawa, Ontario, Canada, by collecting all galls in the patches, letting the occupants emerge and varying the number of adult flies released in each patch. Natural enemies emerging from collected galls were returned to their patch of origin, so only *E. solidaginis* density, and not enemy density, was perturbed. The year following the release, galls were collected once again from each goldenrod patch, and the occupants allowed to emerge to assess mortality rates.

Galls in populations whose densities had been more severely reduced showed a reduction in mortality in the post-perturbation generation. This density-dependent change in mortality from one year to the next could not be ascribed to enemies. The only mortality agent whose strength diminished in density-reduced patches was 'early larval death', a form of mortality in which the gall-maker larva dies early in its development and the resultant gall is small or shrivelled. 'Early larval death' diminished slightly but did not disappear when galls were protected by mesh bags, indicating that although it may sometimes result from an enemy that attacks the gall early without leaving a trace (failed parasitism, for example), much early larval death is likely to be caused by a defensive reaction of the host plant. Resistance in goldenrod to the gall-maker is common (McCrea and Abrahamson 1987; Maddox and Root 1987) and resistance mechanisms include a hypersensitive response of the plant to the burrowing larva (Anderson *et al.* 1989). Although this particular response occurs before gall formation, other plant defensive reactions, or simply the unsuitability of certain stems, may result in larval death at later stages of development. *E. solidaginis* densities are thus likely to be limited by the number of suitable stems in a patch of goldenrod (*cf.* Sheppard and Woodburn this volume pp. 277–290).

Starvation and reduced fecundity in Orgyia vetusta

Outbreaks of the western tussock moth (*Orgyia vetusta*) may cause host plants (bush lupine, *Lupinus arboreus*) on the California coast to be completely defoliated and may last a remarkably long time (>10 years at one site). However, outbreaks appear to remain spatially localised within much larger areas of the host plant (Fig. 2). Harrison (1994) conducted a series of experiments at Bodega Marine Reserve to determine how the tussock moth can remain at such high densities for long periods and why outbreaks do not necessarily spread to adjacent stands of the host plant.

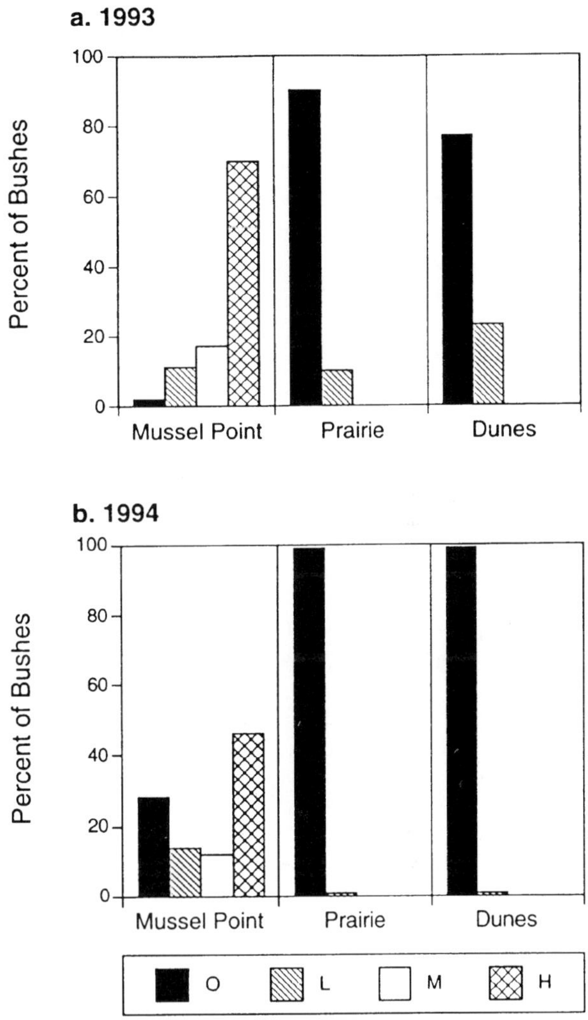

Fig. 2. Percentage of bushes in different categories of Orgyia larval abundance in three areas at Bodega Marine Reserve: MP = Mussel Point (outbreak area); PR = Prairie and DU = dunes (non-outbreak areas). O = none; L = 1–9; M = 10–100; H >100.

Lupine bushes were caged and inoculated with a realistic range of numbers of tussock moth eggs, and larvae then developed to pupation on these bushes, causing from negligible to complete defoliation. Larval survival, pupal sex ratio and adult fecundity (number of eggs per female) all showed direct density dependence (Table 1a), which was attributable to resource availability since cages prevented predation and emigration. From these density-dependent relationships, it was possible to predict the numbers of larvae per bush and eggs per egg mass in a population at its resource-determined equilibrium. These predictions agreed well with data from the natural population (Table 1b), suggesting that the population was near the density at which the resource supply constrained its further growth (Harrison 1994).

The failure of the outbreak to spread was all the more remarkable in that the population was food-limited while being surrounded by large areas of available host plants. Larvae reared in the field performed equally well in outbreak and non-outbreak areas, suggesting that differences in host quality did not explain the failure to spread (Harrison 1994). Slow dispersal appeared to be an important factor; adult females of this moth are flightless, and two experimental outbreaks created

Table 1. Results of density-perturbation experiment on Orgyia vetusta, 1992

a. Regressions of per capita survival, sex ratio and fecundity on the initial number of tussock moth eggs or larvae per bush.

	Mean	SD	slope	intercept	r^2	P
Early larval survival	0.420	.14	–0.12	–0.34	0.13	<0.05
Late larval survival	0.720	.45	–0.42	1.16	0.51	<0.001
Larvae to pupae survival	0.860	.22	–0.24	0.65	0.29	<0.001
Pupal sex ratio	0.500	.19	–0.11	0.88	0.19	<0.05
Egg mass size	194	104	–96.3	531	0.55	<0.001

b. Population parameters observed on 15 bushes in the outbreak area in 1992, and predicted from the experimental results (above).

	Observed	Predicted
Larvae per bush:		
Young bushes	85±25	92
Mature bushes	591±288	736
Eggs per egg mass	117±48	155

by placing 30 000 tussock moth eggs in uninfested lupine stands spread only 2 m per generation (median) (Harrison 1994). Predation also appeared to play a role; larvae and pupae experienced much higher predation when moved from the outbreak area to non-outbreak areas (Harrison and Wilcox 1995; Fig. 3). This was apparently not because of differences in predator abundance, but because an important generalist predator (the ant *Formica lasioides*) was 'satiated' by abundant prey within the outbreak area (Harrison and Wilcox 1995; Fig. 4).

Another unusual attribute of the tussock moth population was its relatively constant density over time. In other insects, fluctuations may occur because insect attack reduces the quality of host plants for the next generation (Haukioja and Neuvonen 1987; Karban and Myers 1989), or because insects that develop at high population densities produce poor-quality offspring (Rossiter 1992). Tussock moths were reared on bushes defoliated to varying degrees in the previous year, with the assay insects obtained from mothers reared at varying population densities. This experiment revealed no effects of either prior foliar damage or maternal crowding on insect growth or survival (Harrison 1995). Moreover, bushes that survived defoliation grew substantially, and their biomass one year later was unaffected by previous defoliation (Harrison and Maron 1995).

These experiments illustrate that mechanisms of population regulation may vary with spatial scale. The size of the tussock moth population is regulated at a local level by resource abundance, but at a larger spatial scale by a limited dispersal and by predation. This study also reinforces the important distinction between immediate negative feedbacks on population growth such as resource depletion, which may be stabilising, versus delayed feedbacks, which appear to be absent in this system, such as long-term induced defences or maternal effects, which are likely to be destabilising (cf. Berryman 1987, 1990; Turchin 1990).

Perturbation experiments for understanding the many facets of population regulation

The changing concept of regulation

Density perturbation experiments such as those we have reviewed are mainly useful for revealing evidence for regulation in its simplest form – that brought about by direct, local density dependence with no time lags. Most of the studies we reviewed were not designed to allow determination of the

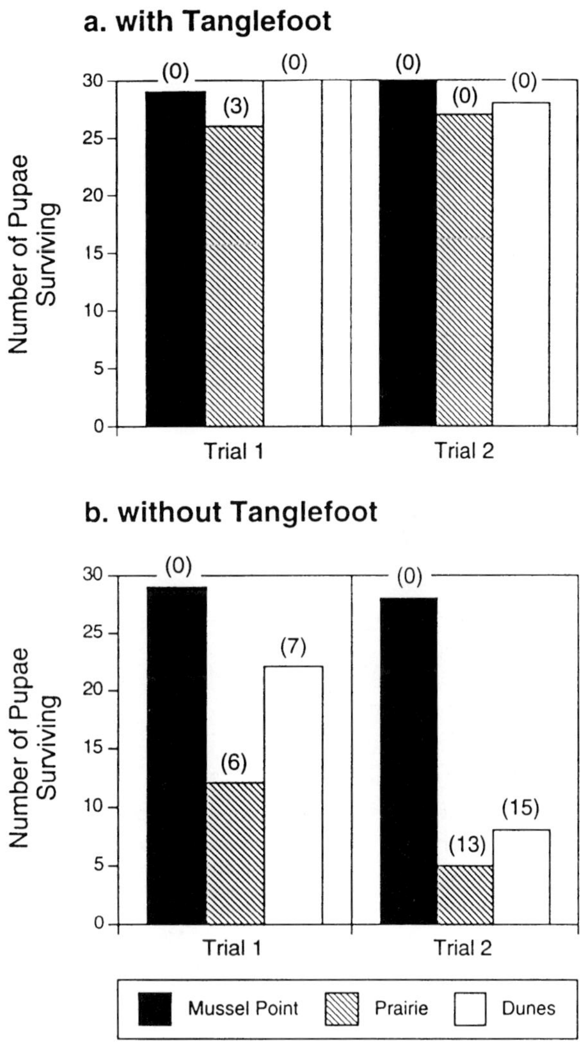

Fig. 3. Survival of Orgyia pupae on plants (a) with Tanglefoot to exclude predators and (b) without Tanglefoot, in three areas at Bodega Marine Reserve: MP = Mussel Point (outbreak area); PR = Prairie and DU = dunes (non-outbreak areas). Number of pupae (out of 30 initial) surviving after one week is shown. Trial 1 was conducted in 1993; trial 2 in 1994. In parentheses, number of observed instances of ants feeding on pupae.

form or strength of density dependence. Few studies included a spatial component or lasted long enough to allow the detection of delayed density dependence. Future manipulations of density should be designed to provide both more rigorous tests of local regulation, as well as to distinguish between the many other processes by which we now agree populations might be regulated.

Throughout much of this century, debate as to whether or not populations are regulated became bogged down by a very narrow view of how regulation could work. Much of the debate was based on the commonness, or rarity, of regulation *via* direct and immediate density-dependent feedback around a point equilibrium. However, regulation may take other forms, as Nicholson (1954) was one of the first to note. Recent, more inclusive definitions of regulation require that populations be bounded (Murdoch 1994) or have a stationary probability distribution of densities (Dennis and Taper 1994; Turchin 1995). This definition includes populations that experience only weak and sporadic density dependence (density-vague dynamics *sensu* Strong 1983), as well as those with cyclic or chaotic dynamics.

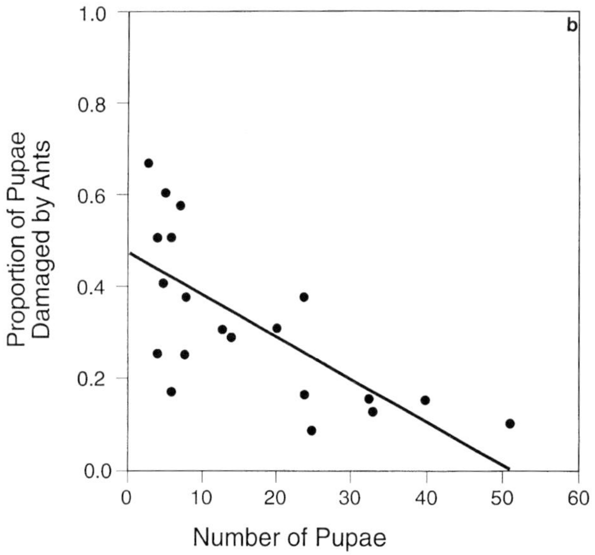

Fig. 4. Predator satiation experiment, 1994: proportion of pupae damaged by ants versus number of pupae per bush, after 6 days.

A very popular idea arising from the regulation debates was that species subject to local environmental fluctuations or catastrophes may persist through so-called metapopulation dynamics, in which local populations go extinct but are recolonised by dispersers from other populations (Gilpin and Hanski 1991). In a milder version of this theory, local populations do not necessarily go extinct, but are stabilised by dispersal among numerous asynchronously fluctuating populations (Taylor 1991). Although the origin of the metapopulation concept is typically traced to Levins (1969), this form of regulation was anticipated by many early authors. For example, Nicholson's (1954) explanation for the persistence of locally unstable host–parasitoid interactions – that they form a network of 'fragmented oscillations' in which local populations undergo extinction and recolonization – was essentially metapopulation regulation.

We have now come full circle to the Nicholsonian, pluralist view of the balance of nature. By adopting a broad definition of regulation we may avoid the polemics of the past but we also risk weakening the utility of the concept. This becomes obvious if we consider some of the organisms whose populations meet our criteria for regulation in the broad sense. Regulated (bounded) populations include such outbreak species as the spruce budworm, *Choristoneura fumiferana* (Murdoch and Walde 1989), a forest pest species that fluctuates between outbreak levels at which it destroys large stands of balsam fir, *Abies balsamea,* and levels at which it is barely detectable in censuses (Fleming this volume). Surely the dynamics of such forest pests are radically different from those of, say, *E. solidaginis.* If we accept a broad definition of regulation, and, consequently, the assertion that most populations are regulated, we must turn our attention to the new question of *how* populations are regulated. Evidence for different types of regulation may come from time-series analysis, but experiments will be necessary to nail down regulatory mechanisms. Below, we discuss some issues involving the mechanisms and processes of regulation and how they might be resolved using density-perturbation experiments.

Regulation through density-dependent dispersal

In herbivorous insects, dispersal may often be density-dependent at the scale of individual plants, such as Ohgushi and Sawada (1985) found for female lady beetles ovipositing on thistles and Crawley and Gilman (1989) found for cinnabar moth caterpillars that defoliated ragwort plants.

Although dispersal may equalise the densities of insects per plant, it can only regulate the size of the population over time if it leads to increased mortality or decreased fecundity. The demographic consequences of dispersal have seldom been investigated, but their severity may be expected to vary with the distances between plants and with characteristics of the 'matrix', such as vegetation type and predator abundance. In turn, if mortality caused by density-dependent dispersal is important, the equilibrium abundances of insects per plant should be highest in places where plants are dense and the habitat matrix is favourable to successful dispersal (e.g. predators are sparse). The effects of plant spacing and matrix type on herbivore loads have been investigated in the context of agroecology (reviewed by Kareiva 1983), but since agricultural systems are usually 'reset' every year, such studies generally last only one plant generation and do not address long-term dynamics.

Perturbation experiments can be used to test the role of density-dependent dispersal in regulating natural insect populations. The first step would be to show that density-dependent dispersal occurs, by manipulating densities per plant and either observing dispersal, measuring it indirectly or assessing the production of dispersive forms (e.g. Denno and Roderick 1992). The second and more difficult step is to measure the resultant mortality or loss of fecundity. Assuming that it would be impossible to track the fates of dispersers directly, this could perhaps be done indirectly by crossing a manipulation of per-plant herbivore density with a manipulation of plant spacing, background vegetation (e.g. clipped vs. natural) or ground-dwelling predators (excluded or not). The prediction would be that per capita change in the herbivore population is affected by an interaction between its own density and the second factor. For example, population decline with respect to density might be disproportionately greater in plots with widely spaced plants than in plots with dense plants.

Populations with complex dynamics

Many insect species undergo dramatic fluctuations in number, varying from levels at which they are barely detectable in the habitat to levels at which they defoliate large stands of their host plants. Recent analysis of time-series data for forest pest species has suggested that complex dynamics (oscillatory behaviour such as limit cycles and chaos) is a common form of regulation in such populations (Turchin and Taylor 1992). Although such analyses are useful for identifying the type of regulation acting on these populations, they do not inform us of the mechanisms involved; for this, we must fall back on experiments (Myers and Rothman 1995).

Large-scale perturbations of forest pest species with apparently complex dynamics are more numerous than those assessing density dependence in non-outbreak, non-pest species (reviewed above). This is probably because the interest in doing perturbations of cyclic populations is the dynamics, and experiments must thus be done on a spatial and temporal scale appropriate to this question. Experiments on density dependence in non-pest species may often be often done for reasons other than understanding dynamics.

Myers and Rothman (1995) identify several types of perturbation experiments that can be performed to understand cyclic dynamics. Some authors have suppressed the population peak by spraying or cropping the population (e.g. Myers 1993). Others have augmented the population during the non-outbreak or increase phase (e.g. Liebhold and Elkinton 1989; Gould *et al.* 1990; Myers 1990). The results of these manipulations allow the general conclusion that regional dynamics of fluctuating species are quite resilient to perturbation (Myers and Rothman 1995).

Metapopulation regulation

Evidence for metapopulation regulation has come primarily from the observation that populations correspond to four postulates (Gilpin and Hanski 1991): 1) their distribution is spatially fragmented, 2) the dynamics of individual subpopulations is not stable, 3) subpopulation dynamics are asynchronous and 4) dispersal between subpopulations is sufficient to allow for the recolonization of vacant habitat patches. Many populations, especially of butterflies, possess these characteristics (Hanski and Kuussaari 1995), suggesting that they are indeed regulated at the metapopulation level. However, since the key to metapopulation regulation lies in the dispersal rate – which must be sufficient to allow recolonization but not so great as to result in a synchronisation of subpopulation dynamics – the most convincing evidence for this form of regulation would come from the manipulation of the dispersal (Reeve 1990). This might be done by erecting barriers to dispersal (e.g. Kareiva 1990) or by manipulating the spatial structure of the habitat in order to increase the distance between patches. Although some experimental studies have shown that the spatial subdivision of habitats affects population dynamics and predator-prey interactions (e.g. Fahrig and Paloheimo 1988; Kareiva 1987, 1990; Walde 1991; Roland 1993), none have fully borne out the premises of the metapopulation model (Harrison 1991; Taylor 1991; Hastings and Harrison 1994). Moreover, very few empirical studies have attempted to examine the roles of both local and metapopulation processes in regulation (Murdoch 1994). One way to do this would be to perturb densities of subpopulations that vary, either naturally or experimentally, in their degree of isolation from other subpopulations. If metapopulation dynamics were important to regulation, then less-isolated populations should converge faster than more-isolated ones.

Regulation: the next generation

Now that the question of population regulation has been resolved (Turchin 1995), the challenge facing the next generation of ecologists is to develop ways of predicting when to expect which kind of dynamics (Lawton 1992). In this endeavour, too, Nicholson paved the way. While others during Nicholson's era were concerned with classifying the *factors* that influence populations (e.g. Andrewartha 1957; Milne 1957), Nicholson (1954) had already developed a classification of *dynamics* that includes most of the dynamical possibilities mentioned above, including both direct and delayed density dependence and his precursor to metapopulation dynamics, fragmented oscillations. He also outlined the conditions under which we should expect different dynamical outcomes depending on the nature of the regulating factor and the rapidity of the reaction to density. This, like his call for experiments, has remained largely ignored and deserves more attention.

If we wish to understand the frequency of different types of regulation in nature, we sorely need data on non-outbreak, non-pest species. We must also avoid choosing study species based on how well they seem *a priori* to fit certain preconceived views of how nature works. For example, given the present 'non-equilibrial' trend in ecological thinking in North America, it is tempting for new graduate students to chose study organisms that fit the postulates of metapopulation regulation. Other types of spatio-temporal population structures, for example species that are sparsely distributed throughout their range instead of clumped into relatively discrete subpopulations, have received very little attention. Finally, we wish to stress that the least 'trendy' form of regulation i.e. simple local regulation, remains a viable alternative to the more fashionable forms of dynamics receiving most of the attention in recent years.

References

Anderson SS, McCrea KD, Abrahamson WG, Hartzel LM (1989) Host genotype choice by the ball gall-maker *Eurosta solidaginis* (Diptera: Tephritidae). *Ecology* **70**:1048–1054

Andrewartha HG (1957) The use of conceptual models in population ecology. *Cold Spr Harbor Symp Quant Biol* **22**:219–236

Berryman AA (1987) The theory and classification of outbreaks. In: Barbosa P, Schultz JC (eds.) *Insect outbreaks.* Academic Press, San Diego, USA pp 3–30

Berryman AA (ed) (1990) *Dynamics of forest insect populations: Patterns, causes and implications.* Plenum Press, New York, USA

Cappuccino N (1992) The nature of population stability in *Eurosta solidaginis*, a nonoutbreaking herbivore of goldenrod. *Ecology* **73**:1792–1801

Cornell HV, Hawkins BA (1995) Survival patterns and mortality sources of herbivorous insects: some demographic trends. *Am Nat* **145**:563–593

Crawley MJ, Gillman MP (1989) Population dynamics of cinnabar moth and ragwort in grassland. *J Anim Ecol* **58**:1035–1050

Dennis B, Taper B (1994) Density dependence in time series observations of natural populations: estimation and testing. *Ecol Monogr* **64**:205–224

Denno RF, Roderick GK (1992) Density-related dispersal in planthoppers: effects of interspecific crowding. *Ecology* **73**:1323–1334

Fahrig L, Paloheimo J (1988) Effect of spatial arrangement of habitat patches on local population size. *Ecology* **69**:468–475

Gaston KJ, Lawton JH (1987) A test of statistical techniques for detecting density dependence in sequential censuses of animal populations. *Oecologia* **74**:404–410

Gilpin M, Hanski I (1991) *Metapopulation dynamics: Empirical and theoretical investigations*. Academic Press, London, UK

Gould JR, Elkinton JS, Wallner WE (1990) Density-dependent suppression of experimentally created gypsy moth *Lymantria dispar* (Lepidoptera: Lymantriidae) populations by natural enemies. *J Anim Ecol* **59**:213–233

Hanski I, Kuussaari M (1995) Butterfly metapopulation dynamics. In: Cappuccino N, Price PW (eds) *Population dynamics: novel approaches and synthesis*. Academic Press, San Diego, USA, pp 49–171

Hanski I, Woiwod I, Perry J (1993) Density dependence, population persistence and largely futile arguments. *Oecologia* **95**:595–598

Harrison S (1991) Local extinction and metapopulation persistence: an empirical evaluation. In: Gilpin ME, Hanski I (eds) *Metapopulation dynamics: Empirical and theoretical investigations*. Academic Press, London, UK, pp 73–88

Harrison S (1994) Resources and dispersal as factors limiting a population of the tussock moth (*Orgyia vetusta*), a flightless defoliator. *Oecologia* **99**:27–34

Harrison S (1995) Lack of strong induced or maternal effects in tussock moths (*Orgyia vetusta*) on bush lupine (*Lupinus arboreus*). *Oecologia* **101**:309–316

Harrison S, Maron JL (1995) Impacts of defoliation by tussock moths (*Orgyia vetusta*) on bush lupine (*Lupinus arboreus*). *Ecol Ent* **20**:223–229

Harrison S, Cappuccino N (1995) Using density-manipulation experiments to study population regulation. In: Cappuccino N, Price PW (eds) *Population dynamics: novel approaches and synthesis*. Academic Press, San Diego, USA, pp 131–147

Harrison S, Wilcox C (1995) Evidence that predator satiation may restrict the spatial spread of a tussock moth (*Orgyia vetusta*) outbreak. *Oecologia* **101**:309–316

Hastings A, Harrison S (1994) Metapopulation dynamics and genetics. *Ann Rev Ecol Syst* **25**:167–188

Haukioja E, Neuvonen S (1987) Insect population dynamics and the induction of plant resistance: the testing of hypotheses. In: Barbosa P, Schultz JC (eds) *Insect outbreaks*. Academic Press, San Diego, USA, pp 411–432

Holyoak M, Lawton JH (1993) Comments arising from a paper by Wolda and Dennis: using and interpreting the results of tests for density dependence. *Oecologia* **95**:592–594

Hunter MD, Price PW (1992) Playing chutes and ladders: heterogeneity and the relative roles of bottom-up and top-down forces in natural communities. *Ecology* **73**:724–732

Karban R, Myers JH (1989) Induced plant responses to herbivory. *Ann Rev Ecol Syst* **20**:331–348

Kareiva PM (1983) Influence of vegetation texture on herbivore populations: resource concentration and herbivore movement. In: Denno RF, McClure MS (eds) *Variable plants and herbivores in natural and managed ecosystems*. Academic Press, New York, USA, pp 259–289

Kareiva PM (1987) Habitat fragmentation and the stability of predator-prey interactions. *Nature* **326**:388–390

Kareiva PM (1990) Population dynamics in spatially complex environments: theory and data. *Phil Trans R Soc Lond* B **330**:175–190

Koetsier P, Dey P, Mladenka G, Check J (1990). Rejecting equilibrium theory – a cautionary note. *Bull Ecol Soc Am* **71**:229–230

Lawton JH (1992) There are not 10 million kinds of population dynamics. *Oikos* **63**: 337–338

Levins R (1969) Some demographic and genetic consequences of environmental heterogeneity for biological control. *Bull Ent Soc Am* **15**:237–240

Liebhold AM, Elkinton JS (1989) Elevated parasitism in artificially augmented populations of *Lymantria dispar* (Lepidoptera: Lymantriidae) *Env Ent* **18**:986–995

Maddox GD, Root RB (1987) Resistance to 16 diverse species of herbivorous insects within a population of goldenrod, *Solidago altissima*: genetic variation and heritability. *Oecologia* **72**:8–14

McCrea KD, Abrahamson WG (1987) Variation in herbivore infestation: historical vs. genetic factors. *Ecology* **68**:822–827

Milne A (1957) Theories of natural control of insect populations. *Cold Spr Harbor Symp Quant Biol* **22**:253–271

Murdoch WW (1970) Population regulation and population inertia. *Ecology* **51**:497–502

Murdoch WW (1994) Population regulation in theory and practice. *Ecology* **75**:271–287

Murdoch WW, Walde SJ (1989) Analysis of insect population dynamics. In: Grubb P, Whittaker JB (eds) *Toward a more exact ecology*. Blackwell, Oxford, UK, pp 113–140

Myers JH (1990) Population cycles of western tent caterpillars: experimental introductions and synchrony of populations. *Ecology* **71**:986–995

Myers JH (1993) Population outbreaks in forest Lepidoptera. *Am Scient* **81**:240–251

Myers JH, Rothman L (1995) Field experiments to study regulation of fluctuating populations of forest Lepidoptera. In: Cappuccino N, Price PW (eds) *Population dynamics: novel approaches and synthesis*. Academic Press, San Diego, USA, pp 229–250

Nicholson AJ (1954) An outline of the dynamics of animal populations. *Aust J Zool* **2**:1–8

Nicholson AJ (1957) Comments on paper by T. B. Reynoldson. *Cold Spr Harbor Symp Quant Biol* **22**:326

Ohgushi T, Sawada H (1985) Population equilibrium with respect to available food resource and its behavioral basis in an herbivorous lady beetle, *Henosepilachna niponica*. *J Anim Ecol* **54**:781–796

Reeve JD (1990) Stability, variability and persistence in host–parasitoid systems. *Ecology* **71**:422–425

Roland J (1993) Large-scale forest fragmentation increases the duration of tent caterpillar outbreak. *Oecologia* **93**:25–30

Root RB, Cappuccino N (1992) Patterns in population change and the organization of the insect community associated with goldenrod. *Ecol Monogr* **62**:393–420

Rossiter MC (1992) The impact of resource variation on population quality in herbivorous insects: a critical aspect of population dynamics. In: Hunter MD, Ohgushi T, Price PW (eds) *Effects of resource distribution on animal–plant interactions*, Academic Press, San Diego, USA, pp 13–42

Smith JNM, Arcese P (1989) How fit are floaters? Consequences of alternative territorial behavior in a non-migratory sparrow. *Am Nat* **133**:830–845

Strong DR (1983) Density-vague ecology and liberal population regulation in insects. In: Price PW, Slobodchikoff CN, Gaud WS (eds) *A new ecology: Novel approaches to interactive systems*. John Wiley and Sons, New York, USA pp 313–327

Taylor AD (1991) Studying metapopulation effects in predator-prey systems. In: Gilpin ME, Hanski I (eds) *Metapopulation dynamics: Empirical and theoretical investigations*. Academic Press, London, UK, pp 305–323

Turchin P (1990) Rarity of density dependence or population regulation with lags? *Nature* **344**:660–663

Turchin P (1995) Population regulation: old arguments and a new synthesis. In: Cappuccino N, Price PW (eds) *Population dynamics: novel approaches and synthesis*. Academic Press, San Diego, USA, pp 19–40

Turchin P, Taylor AD (1992) Complex dynamics in ecological time series. *Ecology* **73**:289–305

Walde SJ (1991) Patch dynamics of a phytophagous mite population: effect of number of subpopulations. *Ecology* **72**:1591–1598

IS LONG-TERM PERSISTENCE OF HARVESTED POPULATIONS EVIDENCE OF DENSITY DEPENDENCE?

Richard McGarvey

ABSTRACT

Most harvested marine animal populations persist over many generations after the arrival of fishing, attaining a new age-structure where the life span and lifetime egg production (egg per recruit) of an average female are truncated. In this paper, the average survival of Georges Bank sea scallops (*Placopecten magellanicus*) from egg to age of first reproduction is estimated for the population under both natural and harvested regimes. The hypothesis of density dependence is validated if a higher average survival rate of spawn is observed under exploitation. Average survival rate of eggs to age of recruitment is estimated by two methods: 1) For the present exploited population, it is estimated empirically by survey as the ratio of the average annual numbers of age-3 recruits divided by average eggs released per year. 2) For both unexploited and exploited cases, the average survival rate from egg to age of first reproduction is inferred under the assumption of stationarity. This method relies on a relation, derived as an extension of Lotka's theory of demographics to stationary populations, that the average survival rate from egg to age of first reproduction equals the reciprocal of average lifetime egg production. Stationary age distributions are derived using the coefficients of natural and natural plus fishing mortality, respectively. Average female lifetime egg production is obtained by multiplying population numbers-at-age by fecundities-at-age and summing over a full lifespan. Early life history survival is estimated as the reciprocal of lifetime egg production. Employing Methods (1) and (2), the estimates of average survival rate from egg to age 3 are 32 and 63 times greater under harvesting.

Key words: density dependence, harvesting theory, scallop, *Placopecten magellanicus*, exploitation

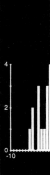

Frontiers of Population Ecology, R.B. Floyd, A.W. Sheppard and P.J. De Barro (eds) CSIRO Publishing, Melbourne, 1996 pp. 65–77

Introduction

Both before and after Nicholson (1933) elaborated the hypothesis of intraspecific competition as the means by which the densities of populations are regulated, many studies have been carried out to test for this response in the field and laboratory, including the early work of Gause (1931) and Pearl (1927). It remains a subject of empirical investigation in modern population ecology (e.g. Sinclair; Lonsdale; Cappuccino and Harrism; this volume). This study investigates density dependence in a harvested sea scallop (*Placopecten magellanicus*) stock on Georges Bank off Nova Scotia. Considerable scientific effort is devoted to understanding the dynamics of fish stocks because of their economic value. Choosing harvested populations for study affords the advantage of abundant information, from both literature and raw data. One or several estimates for basic population life history parameters of *P. magellanicus* have been reported, including growth, mortality and reproduction.

After passing from a natural to an intensively exploited regime, often in a generation or two, the vast majority of harvested marine populations persist under much greater than natural rates of adult removal. Often half or more ($F \geqslant 0.7$) of the adult (fishable-sized) biomass is taken annually. Generally, only local and global depletion to unprofitable levels of density or aggregation avert local extinction. Fisheries thus offer a useful test of density dependence and may be considered as inadvertent mortality manipulation experiments without controls on whole populations over long times.

The two variables chosen for quantifying density dependence in this study are the survival rates of eggs to age of first reproduction in the harvested and unharvested populations. The hypothesis of density dependence in reproductive success predicts that the survival rate of spawn will be higher in the population subject to rapid removal of adults. To undertake this comparison, two independent methods of calculating early life history survival are applied. The first method is by survey measurement of average recruitment and egg production. The second method is by inference from the reconstructed age-structure of the populations before and after harvesting. The latter method depends on the assumption of stationarity, in particular, that the time-averaged numbers in each age (or size) class approximate to the steady-state values.

In a population exhibiting no long-term trend of increase or decrease in average total size, there must, on average, be one-to-one replacement of mothers by daughters that reach the age of reproduction in the next generation (recruits). When, because of (often substantially) shorter lifespan, fewer eggs are produced, the survival rate of those eggs must increase proportionately in order for the population to persist. If the average survival rate of eggs remained unchanged under harvesting, less than one-to-one replacement would result, and the population would decline exponentially.

Data and methods

Two independent sources of information were employed to calculate egg-to-recruit survival rates: 1) random stratified sample size-frequency survey data from the Northern Edge and Northeast Peak of Georges Bank, and 2) life-history parameters from published field studies of *P. magellanicus* populations. This northeastern subpopulation on Georges Bank is the largest and densest, from which is taken approximately half the total annual harvest of this scallop species found in shallow temperate coastal waters of the Northwest Atlantic. These two independent sources of information served as the basis of the two methods for estimating egg-to-recruit survival in the present exploited population of Georges Bank. For the natural population, survival was deduced from life-history parameters alone, no direct observation or measurements being available from the years before exploitation.

Survey data

The survey estimates for egg-to-recruit survival employed the sample size frequencies from the Northern Edge and Northeast Peak population. Recruitment was quantified as the numbers under the youngest age class peak, age 2. The derivation of the time averages of age-2 population numbers and total female egg production from these size-specific survey numbers over the years 1977–1988, and the method of deriving the confidence intervals of these estimates were described by McGarvey *et al.* (1992). Estimated average annual numbers of age-2 scallops (30–60 mm), were

$$[\bar{N}_{2-}, \bar{N}_{2}, \bar{N}_{2+}] = [5180, 5784, 18730] \times 10^6$$

and estimates for egg production by all adult sizes, $\bar{S}$, were

$$[\bar{S}_{-}, \bar{S}, \bar{S}_{+}] = [38.43, 45.75, 59.93] \times 10^{15} \text{ eggs}$$

where $\bar{N}_{2-}$ and $\bar{S}_{-}$ are lower and $\bar{N}_{2+}$ and $\bar{S}_{+}$ upper 95% confidence limits on $\bar{N}_2$ and $\bar{S}$.

Life history parameters

Parameters taken from the literature are 1) the average age-specific fecundities, and 2) the average instantaneous mortalities, both from natural causes, *M*, and due to fishing, *F*.

The age-specific fecundity vector employed in calculating total population egg production was derived from field measurements of MacDonald and Thompson (1985), for the Sunnyside Newfoundland 10 m study subpopulation. MacDonald and Thompson measured the weight change of female gonads during the month of spawning. The conversion of spent gonad biomass to total eggs released follows the method of Langton *et al.* (1987) assuming a neutrally buoyant spherical egg of diameter 67 mm. This conversion assumes that all gonad weight change is converted to egg biomass. Thus, these fecundity values represent maximum numbers of (possibly unfertilised) eggs spawned. Histological examination corroborated the assumption that most of the change in gonad biomass during spawning consists of eggs released (MacDonald and Thompson 1985; Langton *et al.* 1987). The resulting fecundity vector as a function of age, *a*, is denoted $\{m_a; a = a_r, \ldots a_{\max}\}$, a_r being the age of first reproduction.

Transforming the independent variable from age to size, fecundity was additionally obtained as a function of scallop shell height, $\{m_h; h = h_r, \ldots, h_{\max}\}$. Details of this derivation and a graphical comparison of the size-based fecundity vector, showing good agreement with Langton *et al.* (1987), were presented previously (McGarvey *et al.* 1992) .

The age-structure of the unexploited population was inferred from *M*, the instantaneous natural mortality. For Northwest Atlantic sea scallops, estimates of *M* were reported from three field studies. Dickie (1955) obtained a value of $M = 0.10$ from the ratio of empty to live shells in the Bay of Fundy population. Merrill and Posgay (1964) applied this method to the Georges Bank population and obtained an estimate of 0.10 ± 0.02. MacDonald and Thompson (1986) directly measured mortality in the same inshore Newfoundland population from which the fecundity vector is derived, divers monitoring 4 m × 4 m quadrats over a 1.5 year study period. They reported natural mortality as a function of size, confirming the approximate average value of 0.1, with evidence of slightly lower values at mid-sizes of 95–135 mm (ages 5–10) and higher natural mortalities at younger and older ages. Considerable variability in natural mortality due to episodic temperature and starfish predation events has been observed (Dickie and Medcof 1963). While bearing this variability in mind, the consistency with which this average estimate of 0.1 was found by two methods of measurement augments confidence in its accuracy, in particular, for adult scallops in three subpopulations, including the one under study here. A constant $M = 0.1$ is therefore employed for adult scallops, of ages 3 and older. To simulate senescence at the maximum age of 20 reported in previous estimates

Table 1. Age-specific rates of instantaneous natural and fishing mortality assuming a lifespan of 20 years.

Age a	Natural mortality a to $a^{+}1$	Fishing mortality a to $a^{+}1$
2	0.4	0.0
3	0.1	0.403
4	0.1	0.918
5	0.1	0.966
6	0.1	0.966
7	0.1	0.966
8	0.1	0.966
9	0.1	0.966
10	0.1	0.966
11	0.1	0.966
12	0.1	0.966
13	0.1	0.966
14	0.1	0.966
15	0.1	0.966
16	0.1	0.966
17	0.2	0.966
18	0.4	0.966
19	0.6	0.966
20	infinite	

of lifespan (MacDonald 1986), mortality was set to rise with ages 17–20 to 1.0 (Table 1). An estimate of $M = 0.4$ for 2-year old juveniles was taken from MacDonald and Thompson (1986).

The average fishing mortality, $F = 0.966$, was derived by Robert *et al.* (1991) from length-to-age converted cohort analysis (Pope 1972) of Canadian commercial harvest samples. Dickie (1955) and Caddy (1972) tabulated the increase in scallop dredge capture efficiency over ages 3–5 from field measurements.

Adult survival rates

Population numbers include only females. Recruitment is defined as the numbers of females reaching the age of first reproduction, $a_r = 3$. The derivation of average age-specific survival rates from age of recruitment to older ages $\{l_{a_r,a} = a_r,\ldots,a_{max}\}$ was carried out in the direct manner by iteration:

$$N^u{}_4 = e^{-M_3} \cdot N^u{}_3$$
$$N^u{}_5 = e^{-M_4} \cdot N^u{}_4$$
$$\vdots$$

where $N^u{}_a$ equals the average numbers of female scallops of age a in the Northern Edge and Northeast Peak for the unexploited case, and where $\{M_{a_r}; a = a_r,\ldots,a_{max}\}$ are the instantaneous natural mortalities during the year from age a to age $a+1$.

Similarly, for exploited scallops,

$$N^e{}_4 = e^{-(M_3+F_3)} \cdot N^e{}_3$$
$$N^e{}_5 = e^{-(M_4+F_4)} \cdot N^e{}_4$$
$$\vdots$$

where $N^e{}_a$ are average numbers of age a female scallops in the exploited case.

The survivals from age of recruitment to all older age classes, a, were then obtained for both the unexploited,

$$l^u{}_{a_r,a} = \frac{N^u{}_a}{N^u{}_{a_r}} \qquad a = a_r, \ldots, a_{\max}$$

and exploited

$$l^e{}_{a_r,a} = \frac{N^e{}_a}{N^e{}_{a_r}} \qquad a = a_r, \ldots, a_{\max}$$

populations.

Estimation of egg-to-recruit survival

The two methods for estimating egg-to-recruit survival are described below.

Method 1

The first method, direct measurement by survey, used the annual size-frequency samples summarised above to derive the estimates of average annual egg production, $\overline{S}$, and average annual recruitment, $\overline{R}$.

To obtain $\overline{R}$, the average numbers reaching the age of first reproduction, age 3, from age-2 numbers, $\overline{N}_2$, the value of $M_2 = 0.4$ for age-2 juveniles was employed. The age-2 estimates of McGarvey *et al.* (1992) included both males and females. A sex-ratio factor of one-half (MacKenzie *et al.* 1978) was therefore introduced to count only females. With

$$\overline{R} = \frac{1}{2} e^{-M_2} \overline{N}_2 \qquad \text{we obtain}$$
$$\left[\overline{R}_-, \overline{R}, \overline{R}_+\right] = [1736, 1938.5, 6277.5] \times 10^6 \text{ age} - 3 \text{ female recruits.}$$

The estimates for average egg production, $\overline{S}$, may be taken without modification from McGarvey *et al.* (1992), where the factor of one-half is already explicit.

The survey method estimate of average egg-to-recruit survival is then simply their ratio, $\overline{R}/\overline{S}$.

Method 2

The second method employed life history parameters derived from the literature, in particular, the instantaneous natural and fishing mortality coefficients, M and F. These allowed the calculation of average survival rates from age of recruitment to older ages which, in turn, yielded the steady-state age structure. Multiplying these numbers-at-age by fecundities-at-age and summing, yielded the 'average lifetime egg production' (ALEP).

In a steady-state, survival from egg to age of first reproduction equals one divided by lifetime egg production of an average female. Intuitive and mathematical reasoning for this equality and how it may be simply derived from numbers-at-age and fecundities-at-age under the assumption of steady-state are presented below and its relationship to Lotka's (1907a) equation for exponential population growth is discussed. This equality of steady-state population reproduction may be written

$$\begin{pmatrix} \text{average} \\ \text{egg}-\text{to}-\text{recruit} \\ \text{survival} \end{pmatrix} = \frac{\overline{R}}{\overline{S}} = \frac{1}{ALEP}$$

The intuitive basis for this relation lies in the stationary state requirement of one-to-one replacement; each mother at the age of first reproduction will, in her lifetime, on average, produce only one daughter that will survive to that same age and succeed her.

The mathematical basis for this equality is simple and is presented here for the first time in application to the question of density dependence. Beginning with the steady-state age-specific expressions for annual population egg production,

$$\sum_{a=a_r}^{a_{\max}} N_a \, m_a$$

and annual recruitment to age of first reproduction, N_{a_r}, we have for their ratio

$$\frac{\overline{S}}{\overline{R}} = \frac{\sum_{a=a_r}^{a_{max}} N_a \, m_a}{N_{a_r}} = \sum_{a=a_r}^{a_{max}} \frac{N_a}{N_{a_r}} m_a = \sum_{a=a_r}^{a_{max}} l_{a_r,a} \, m_a . \qquad (1)$$

The left-hand side is the inverse of the average survival probability of eggs to age of recruitment. The right-hand side is the average lifetime egg production of a female that has reached age. Inverting both sides of the equation we obtain

$$\frac{\overline{R}}{\overline{S}} = \frac{1}{ALEP}$$

as asserted above. The adult population numbers-at-age, $\{N_a;\ a = a_r,\ldots,a_{\max}\}$ are assumed to be steady-state values, or averages when the approximation of steady states by means is valid. The question of how well this equality holds as an approximation when the strict assumption of steady-state is relaxed for non-growing but dynamically varying populations will determine how useful it is in practice.

Following the seminal work of Lotka on population demographics (1907a, 1907b, 1922), much of the theory and applications have described populations undergoing constant exponential growth (Cole 1954; Caughley 1967; Laughlin 1965; Leverich and Levin 1979). A fundamental equality of this theory is the equation, expressed in discrete-time form by Caughley (1967) of which Eqn 1 above is the analogue for steady populations:

$$\sum_{a=0}^{\infty} e^{-ra} \, l_{0,a} \, m_a = 1 \qquad (2)$$

The equality of Eqn 1 for steady-state populations differs from Eqn 2 in three ways: i) Insofar as Eqn 1 describes a stationary population, $r \to 0$, on average, over observable time scales. ii) The sum in Eqn 1 is taken over 'adult' ages starting from a_r rather than 0. iii) Because the sum in Eqn 1 is taken from the age of first reproduction, the right-hand side no longer equals 1, but instead equals the egg-to-recruit survival we seek to quantify in exploited and unexploited populations.

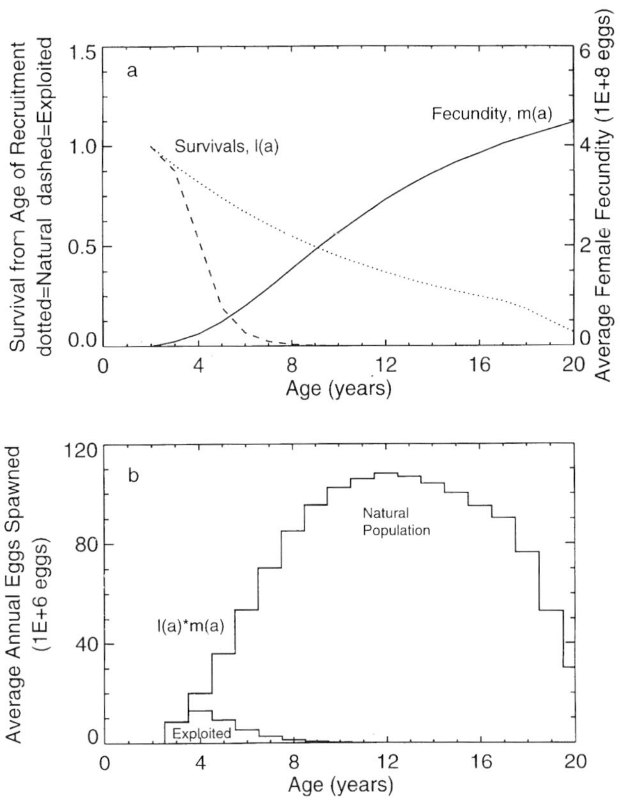

Fig. 1. Age-specific breakdown of the results of the calculation of average lifetime egg production. The fecundity vector and the two ogives for survival from age 3 for the two cases of natural and exploited populations in panel (a), are multiplied in each age to obtain the average annual contribution to total egg production from each cohort in panel (b). The sums under the two curves in panel (b) are the lifetime egg production estimates for the exploited and unexploited populations.

Results

From survey Method 1, using size as the independent variable,

$$\frac{\bar{R}}{\bar{S}} = \frac{1}{23.6 \ \textit{million eggs}}$$

Using Method 2, with $M = 0.1$ and $F = 0.966$,

$$\left(\frac{1}{ALEP}\right) = \frac{1}{46.2 \ \textit{million eggs}}$$

These two independent estimates of egg-to-recruit survival are obtained for the exploited population. Confidence intervals on these estimates are derived in the Appendix. The factor of 2 difference in the two estimates reflect comparable uncertainties in each, notably, in absolute average recruitment numbers, $\bar{R}$, for Method 1, and in average fishing mortality, F, and average natural mortality, M, for Method 2.

For the natural population, using $M = 0.1$,

$$\left(\frac{1}{ALEP}\right) = \frac{1}{1481 \ \textit{million eggs}}$$

The two independent estimates, Methods 1 and 2, therefore yield survival rates of an average egg to age 3, of 63 and 32 times greater, respectively, in the exploited compared with the natural population. This outcome of at least an order of magnitude increase is robust with respect to the uncertainties in the underlying estimates, in particular of $\bar{R}$, F, and M, as detailed in the Appendix.

The numbers of eggs produced by an average female at various ages in the two cases of exploited and natural populations are plotted in Fig. 1b. The sums under the two curves are the average lifetime egg productions in the two mortality regimes.

Discussion

To test for density dependence, a standard and direct approach is to measure instantaneous rates of growth, r, in a population at different levels of density (cf. Sinclair; Lonsdale; Fox and Ridsdill-Smith; this volume). In practice, this is often only possible with populations undergoing relatively rapid change in population size over the time scale of observation, in order that the population under study be in a different level of density at each measurement value. This approach thus applies to populations in transition.

The approach adopted here is different insofar as it compares two stationary states in both of which r is approximately zero on average. Density is not measured directly. Rather, two flow regimes of the population are compared, one in which adult removal is an order of magnitude greater. Effectively, it is the *turnover* that is quantified in the two cases under study rather than the absolute levels of stock size or density. This reasoning should hold for populations that are not in a true steady-state, but persist while fluctuating about some rough average values, since even small long-term average values of r above or below zero will yield exponential change in population size. Phrased in the classical notation of $r = b - d$ (Pearl 1927), if r is roughly zero in the unexploited population, then when d is increased by an order of magnitude under harvesting, b must also be increasing by a comparable amount, or the population would decline exponentially. The analysis above adds two aspects to this reasoning by Pearl: 1) Age structure is explicit, extrapolating from the classical demographics of Lotka to non-exponentially growing populations. 2) By applying Eqn 1 describing steady-state population reproduction we infer, under assumptions of non-growth, the change in survival rate of spawn needed to maintain the population under harvesting.

'Egg-to-recruit survival' is taken in a general sense, potentially manifested through a number of density-dependent response mechanisms. Reduced lifetime egg production can be compensated by increased fecundity of females, faster growth, lower adult natural mortality, as well as by enhanced survival of eggs to age of reproduction. The quantities of survival derived above cannot distinguish among them. It can only be inferred that some population response must occur which allows an average egg to 'survive' or be produced by mothers of a given age with one to two orders of magnitude greater probability. The very large size of this inferred change, together with the results of sensitivity analysis on adult natural mortality, leaves an approximate order of magnitude or more of survival increase to explain under harvesting.

Begon *et al.* (1986), note that the characteristic equation of Lotka and Caughley, Eqn 2, does not offer a self-evident biological meaning. Comparing Eqn 2 with its analogue, Eqn 1, derived here for steady-state populations, allows one interpretation. If the right-hand side of Eqn 2 is the reciprocal of the average survival of an egg to the lower bound of age over which the sum is taken, a_r, as we have proven for stationary populations, then we can understand why it takes the value of 1, namely because the survival rate from an egg of age 0 to age $a_r = 0$ is, by definition, 1. By letting the sum in the steady-state formulation start from age of first reproduction, rather than 0, the right-hand side becomes more intuitively meaningful and allows its use in quantifying the rate of early life history survival. If a similar derivation is carried out for the case of exponentially growing populations, assuming, as Lotka did, constant overall intrinsic rate r, constant inter-age survivals, and stationary age structure, the result is that the survival rate of eggs spawned in any year to age of first reproduction, a_r years later, is also constant, yielding,

$$\frac{S(t)}{R(t)} = \sum_{a=a_r}^{\infty} e^{-ra}\, l_{a_r,a}\, m_a$$

for all t where recruitment, $R(t) = N_{a_r}(t + a_r)$, are numbers reaching age a_r that were spawned in year t.

The approximation of steady states by averages is a fundamental assumption in this analysis. In general, natural populations of marine organisms fluctuate subject to a wide number of dynamic or stochastic influences. But while few populations resemble a true steady-state, many tend, over varying time scales, to fluctuate about a relatively constant level. Cyclical, seasonal, or simply randomly fluctuating populations would fall into this category. The analysis presented here applies to populations that do not exhibit a tendency to either rise or fall over long times, in particular, populations undergoing exponential growth or decay are excluded. Constant exponential change is the subject of Lotka's original demographic theory, and has been extensively and effectively advanced by Cole (1954), Laughlin (1965), Caughley (1967) and others. It is a natural and simple step to extend this demographic theory to stationary (though fluctuating) populations.

Chesson (1978), considering five widely occurring forms of population variability, presented a comprehensive review of the long-term stability properties, i.e. the conditions under which predator–prey systems will persist, through an analysis of the long-term tendency of these model systems to remain in a bounded neighbourhood. For the method presented here, it is necessary not only that the population trajectories be bounded above and below as Chesson has shown, but also that over long times the population means converge to the steady states. This is a stricter assumption and will not always apply, varying with the form of dynamic behaviour.

The theoretical literature of exponentially growing populations and predator–prey systems has generally tended to support the validity of the approximation of steady states by means. Lotka showed that the stationary age-structure was stable in populations undergoing constant exponential growth, tending to return to the stationary age distribution when fluctuations impinge. Volterra (1927) calculated the relationship between means and steady states in the original 2-variable Lotka-Volterra model, and showed that time averages of predator and prey variables over one cycle exactly equal their analytic steady states. This approximation has been shown to extend to a more elaborate predator–prey model of the scallop population studied here. The age-structured prey variable was subject to lognormal stochastic environmental variability and a range of assumed dependences on adult density of egg-to-recruit survival. In this open-access fishery model, which is mathematically a form of predator–prey model, the steady states fell well within one standard error of the variable time averages over nearly two cycle periods (McGarvey 1994).

The assumption that density dependence acts in exploited populations has long been accepted in fisheries science. Ricker (1954) published a seminal review applying Nicholson's ideas to fish stocks and graphed dynamic trajectories of single-species populations, yielding steady states, cycles, and what were later shown by May (1974) to be a new form of behaviour, chaos. Fishery models which consider population reproduction normally assume some form of density dependence, whether severe (Ricker) or moderate (Beverton and Holt 1957). Average lifetime egg production (known as 'egg per recruit') is often calculated in fisheries management applications to quantify the fractional decline in egg production per female under harvesting.

Surplus fishery production models, first proposed by Graham (1935) and formalised mathematically by Schaefer (1954, 1957) depend critically on the assumption of density dependence acting at high (near natural) levels of stock abundance to slow production of new biomass each year. At high exploitation rates, and thus low stock abundance, lower egg production is the limiting factor. In between lies the optimal level of exploitation known as maximum sustainable yield for the case of the original Schaefer model with purely logistic biomass growth. Thus the comparison of production models to data from real fisheries is a broadly applied, if indirect, test for density dependence in fish stocks.

Bayley (1988) considered 59 tropical fisheries and found good agreement with a single production model. The data diverged from the model curve in the way the fisheries declined in annual catch at high levels of exploitation. This implies that these fish stocks are more resilient – responding with more rapid growth when their populations are reduced by some external factor such as fishing – than pure logistic growth would have implied. Pella and Tomlinson (1969) found the same trend for Pacific yellowfin tuna, finding a better fit by shifting the production curve to allow greater harvests at higher levels of fishing effort than predicted by the Schaefer model. Thus fitting production models indicates a greater ability of fish populations to increase biomass production under harvesting than a pure logistic model of population biomass growth would imply.

In the fisheries literature, controlled experimental fishing to directly test for density dependence in an exploited population has been rare, and only one example is known to me, namely the controlled removal of freshwater crayfish in an Ontario lake by Momot (1990, 1993). The experimental outcome yielding good agreement with the Schaefer production model, pre-recruit survival in the population increasing with the number of trap days. Regulation appeared to be mediated by maturing males insofar as their higher abundance was observed to diminish the survival of hatching and juvenile stages. Terrestial mammal studies of the effect of reproductive success and population growth under removal by hunting also have yielded evidence of density dependence, notably including the studies of Chaetum and Severinghaus (1950), Clutton-Brock *et al.* (1985), and Skogland (1985).

This approach, of measuring density dependence in populations subject to continuous removal, has thus tended to yield significant measurable results with consistency. This is in agreement with the more theoretical approach presented here. Since essentially all fish stocks experience a decrease in average lifetime egg production under exploitation, the implications for exploited fish stocks and the implied increase in egg to recruit survival is consistent with the more direct emperical studies cited above, providing two forms of evidence for density dependence in populations subject to harvesting.

References

Bayley PB (1988) Accounting for effort when comparing tropical fisheries in lakes, river-floodplains, and lagoons. *Limnol Oceanogr* **33**:963–972

Begon M, Harper JL, Townsend CR (1986) *Ecology: individuals, populations and communities.* Blackwell, Oxford UK

Beverton RJH, Holt SJ (1957) *On the dynamics of exploited fish populations.* Fisheries Investment Series 2, Vol 19, UK Ministry of Agriculture and Fisheries, London UK

Caddy JF (1972) Size selectivity of the Georges Bank offshore dredge and mortality estimate for scallops from the northern edge of Georges in the period June 1970 to 1971. *Int Comm Northw Atl Fish Redbook* **3**:79–86

Caughley G (1967) Parameters for seasonally breeding populations. *Ecology* **48**:834–839

Chaetum EL, Severinghaus CW (1950) Variations in fertility of white-tailed deer related to range conditions. *Trans North Amer Wildl Conf* **15**:170–189

Chesson P (1978) predator–prey theory and variability. *Ann Rev Ecol Syst* **9**:323–347

Clutton-Brock TH, Major M, Guinness FE (1985) Population regulation in male and female red deer. *J Anim Ecol* **54**:831–846

Cole LC (1954) The population consequences of life history phenomena. *Quart Rev Biol* **29**:103–137

Dickie LM (1955) Fluctuations in abundance of the giant scallop, *Placopecten magellanicus* (Gmelin), in the Digby area of the Bay of Fundy. *J Fish Res Board Can* **12**:797–857

Dickie LM, Medcof JC (1963) Causes of mass mortalities of scallops (*Placopecten magellanicus*) in the southwestern Gulf of St. Lawrence. *J Fish Res Board Can* **20**:451–482

Gause GF (1931) The influence of ecological factors on the size of population. *Am Nat* **65**:70–76

Graham M (1935) Modern theory of exploiting a fishery and application to North Sea trawling. *J Cons Int Explor Mer* **10**:264–274

Langton RW, Robinson WE, Schick D (1987) Fecundity and reproductive effort of sea scallops *Placopecten magellanicus* from the Gulf of Maine. *Mar Ecol Prog Ser* **37**:19–25

Laughlin R (1965) Capacity for increase: a useful population statistic. *J Anim Ecol* **34**:77–91

Leverich WJ, Levin DA (1979) Age-specific survivorship and reproduction in *Phlox drummondii*. *Am Nat* **113**:881–903

Lotka AJ (1907a) Relationship between birth rates and death rates. *Science* **26**:21–22

Lotka AJ (1907b) Studies on the mode of growth of material aggregates. *Am J Sci 4*, **24**:199–216

Lotka AJ (1922) The stability of the normal age distribution. *Proc Natl Acad Sci* **8**:339–345

MacDonald BA (1986) Production and resource partitioning in the giant scallop *Placopecten magellanicus* grown on the bottom and in suspended culture. *Mar Ecol Prog Ser* **34**:79–86

MacDonald BA, Thompson RJ (1985) Influence of temperature and food availability on the ecological energetics of the giant scallop *Placopecten magellanicus*. II. Reproductive output and total production. *Mar Ecol Prog Ser* **25**:295–303

MacDonald BA, Thompson RJ (1986) Production, dynamics and energy partitioning in two populations of the giant scallop *Placopecten magellanicus* (Gmelin). *J Exp Mar Biol Ecol* **101**:285–299

MacKenzie CL Jr, Merrill AS, Serchuk FM (1978) Sea scallop resources off the northeastern U.S. coast, 1975. *Mar Fish Rev* **40**:19–23

May RM (1974) Biological populations with nonoverlapping generations: stable points, stable cycles, and chaos. *Science* **186**:645

McGarvey R (1994) An age-structured open-access fishery model. *Can J Fish Aquat Sci* **51**:900–912

McGarvey R, Serchuk FM, McLaren IA (1992) Statistics of reproduction and early life history survival of the Georges Bank sea scallop (*Placopecten magellanicus*) population. *J Northw Atl Fish Sci* **13**:83–99

Merrill AS, Posgay JA (1964) Estimating the natural mortality rate of the sea scallop (*Placopecten magellanicus*). *Int Comm Northw Atl Fish Res Bull* **1**:88–106

Momot WT (1990) Yield estimates for the virile crayfish, *Orconectes virilis* (Hagen, 1970) employing the Schaefer logistic model. *J Shellfish Res* **9**:373–381

Momot WT (1993) The role of exploitation in altering the processes regulating crayfish populations. In: Holdich DM, Warner GF (eds), *Freshwater Crayfish IX*. University of Southwestern Louisiana, Lafayette, LA USA

Nicholson AJ (1933) The balance of animal populations. *J Anim Ecol* **2**:131–178

Pearl R (1927) The growth of populations. *Quart Rev Biol* **2**:532–548

Pella JJ, Tomlinson PK (1969) A generalized stock production model. *Inter-Am Trop Tuna Comm Bull* **13**:419–496.

Pope JG (1972) An investigation of the accuracy of virtual population analysis using cohort analysis. *Int Comm Northw Atl Fish Res Bull* **9**:65–74

Ricker WE (1954) Stock and recruitment. *J Fish Res Board Can* **11**:559–623

Robert G, Black GAP, Butler MAE (1991) Georges Bank Scallop Stock Assessment – 1990. *Can Atl Fish Sci Advi Comm Res* Doc 91/69

Schaefer MB (1954) Some aspects of the dynamics of populations important to the management of commercial marine fisheries. *Bull Inter-Am Trop Tuna Comm* **1**:27–56

Schaefer MB (1957) A study of the dynamics of fishery for yellowfin tuna in the eastern tropical Pacific Ocean. *Bull Inter-Am Trop Tuna Comm* **2**:247–285

Skogland T (1985) The effect of density-dependent resource limitations on the demography of wild reindeer. *J Anim Ecol* **54**:359–374

Taylor JR (1982) *An introduction to error analysis*. University Science Books, Mill Valley CA, USA

Volterra V (1927) Variations and fluctuations in the numbers of coexisting animal species (Translators: Scudo FM, Ziegler JR). In: *The golden age of theoretical ecology: 1923–1940.* Springer-Verlag Lecture Notes in Biomathematics, Vol 22, 1978, Berlin Germany

Appendix

Confidence intervals for the estimate of egg-to-recruit survival are calculated in three ways: 1) from the confidence intervals of the survey means from which the survival is calculated for the exploited population, 2) by sensitivity analysis of the calculated values of (1/*ALEP*) using M and F, for both exploited and unexploited populations, and 3) by the variance in the two independent estimates, 1) and 2), obtained for the exploited population.

Confidence intervals from survey

The confidence intervals for the estimate of $\bar{R}/\bar{S}$ from the survey numbers-at-height (Method 1) may be derived by applying the classical theory of error propagation (Taylor 1982). The formula for $\bar{R}/\bar{S}$ in terms of the independent variables from which it is calculated is:

$$\frac{\bar{R}}{\bar{S}} = \frac{\frac{1}{2}e^{-M_2}\bar{N}_2}{\bar{S}}$$

The following errors were assigned to each dependent variable: 1) The uncertainty in the assumed sex ratio of 1:1 (MacKenzie *et al.* 1978) expressed by the factor of (1/2) is negligible compared to the three variables to follow. 2) Formal estimates for the uncertainty of natural mortality of age 2 scallops, M_2, were not reported by MacDonald and Thompson (1986). 95% confidence intervals of $(M_2) = \pm 50\%\cdot M_2$, i.e. from $M_2 = 0.2$ to $M_2 = 0.6$, were arbitrarily adopted. 95% confidence intervals for $\bar{N}_2$ and $\bar{S}$ were reported by McGarvey *et al.* (1992). These are 3) $\delta_+(\bar{N}_2) = +250\%\cdot\bar{N}_2$ and $\delta_-(\bar{N}_2) = -27\%\cdot\bar{N}_2$ and 4) $\delta_+(\bar{S}) = +31.5\%\cdot\bar{S}$ and $\delta_-(\bar{S}) = -16\%\cdot\bar{S}$, for the upper and lower values. The overall uncertainty in $\bar{R}/\bar{S}$ is calculated from the formula (Taylor 1982, p. 177) assuming errors in these variables, 1) – 4), are independent:

$$\pm\delta(\bar{R}/\bar{S}) = \sqrt{\left[\frac{d(\bar{R}/\bar{S})}{dM_2}\cdot\pm\delta(M_2)\right]^2 + \left[\frac{d(\bar{R}/\bar{S})}{d\bar{N}_2}\cdot\pm\delta(\bar{N}_2)\right]^2 + \left[\frac{d(\bar{R}/\bar{S})}{d\bar{S}}\cdot\pm\delta(\bar{S})\right]^2}$$

The resulting estimates for upper and lower errors are $\delta_+(\bar{R}/\bar{S}) = +250\%\cdot\bar{R}/\bar{S}$ and $\delta_-(\bar{R}/\bar{S}) = -37\%\cdot\bar{R}/\bar{S}$ respectively. The dominant contribution to the error is in the estimation of absolute recruit numbers. In particular, the large upper limit for average age-2 numbers from the Georges Bank survey is due to uncertainty about variation with size in the sample gear capture efficiency, particularly for those smaller sizes at which scallops first become susceptible to capture.

Table 2. Sensitivity analysis of average lifetime egg production (in millions of eggs) for exploited and unexploited populations, assuming the given range of estimates of natural mortality M in both mortality regimes, and of fishing mortality, F, in the exploited regime. Egg to age 3 survivals, by Method 2, are the reciprocals of the tabulated lifetime egg productions. Method 1 yielded an estimate of 23.6 million eggs for the exploited population.

M	*F*(exploited)			*F*(unexploited)
	0.7	0.966	1.2	0
0.05	77	50	39	2377
0.1	69	46	36	1481
0.2	55	39	32	642

Sensitivity analysis of

Because of the iterative method of calculating the steady-state age-specific population numbers in Method 2, a closed analytic form for *1/ALEP* in terms of M and F would be cumbersome. A formal calculation of 95% confidence intervals by standard error analysis is therefore impractical. A form of sensitivity analysis was instead undertaken, recalculating *1/ALEP* with chosen likely upper and lower limits for M and F. Ranges of

$$M = [0.05, 0.1, 0.2]$$

and

$$F = [0.7, 0.966, 1.2]$$

conservatively estimate 95% confidence intervals. In particular, for M, Merrill and Posgay (1964) reported 0.1 ± 0.02. The range of estimates for *ALEP* is given in Table 2. The variation in the factor of increase in egg-to-recruit survival between unexploited and exploited populations depends on the assumed choice for M, varying from factors of 11 to 20 for $M = 0.2$, and from 30 to 61 for $M = 0.5$.

Confidence intervals from two independent estimates

The two methods of estimating egg-to-recruit survival, by survey and by calculation of *1/ALEP*, allow a third method of estimating confidence intervals for the exploited population. Thinking of these two estimates as a sample of means, of sample size 2, their standard deviation is, by definition, the standard error of their means. This approach yields an estimate for the 95% confidence interval of ± 63% above and below their average of (1/34.9 million eggs). This is of similar magnitude to the confidence intervals for each estimate independently calculated.

Interaction of temperature and resources in population dynamics: an experimental test of theory

Mark E. Ritchie

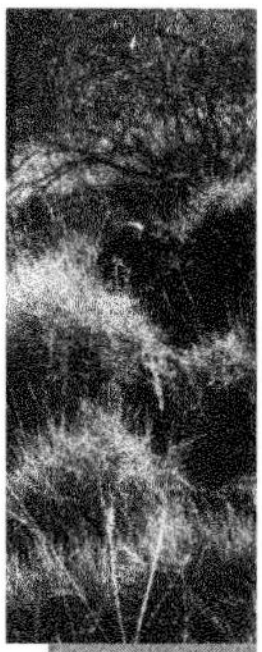

ABSTRACT

Much debate in ecology has focused on the roles of density-independent factors (e.g. temperature) and density-dependent factors (e.g. resources) in regulating populations of animals, particularly ectotherms. I incorporate parameters expressed as general functions of temperature into a simple mechanistic model of resource-limited population growth. This model predicts that increased temperature should lead to greater density-dependent per capita mortality rates and lower equilibrium densities. I tested these predictions in a field experiment by monitoring mortality in grasshopper (Orthoptera: Acrididae) populations stocked at different densities in replicate cages and subjected to one of three different thermal treatments (shaded, control, or greenhouse). With increased temperature, per capita mortality rates were significantly higher and grasshopper densities at the end of the experiment were significantly lower. Increased temperature was associated with increased density-dependent mortality but not density-independent mortality. These results support the predictions of the model. In addition, strong interactions between temperature and resources may make species' populations more sensitive to thermal environmental change than would be expected from species' thermal tolerances.

Key words: temperature, density dependence, resource-limited population model, Acrididae, environmental change

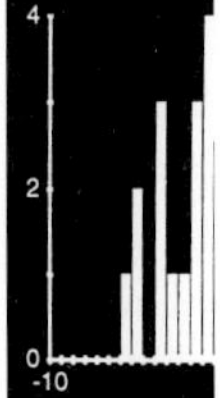

Frontiers of Population Ecology, R.B. Floyd, A.W. Sheppard and P.J. De Barro (eds), CSIRO Publishing, Melbourne, 1996, pp. 79–91

Introduction

Ecologists have strongly debated the role of biotic versus abiotic factors in controlling the dynamics of populations. These factors have been strongly associated with the types of mechanisms limiting population growth: density dependence (Nicholson and Bailey 1935; Lack 1954) or density independence (Davidson and Andrewartha 1948; Andrewartha and Birch 1954). Proponents of density dependence have argued that per capita population growth rates decline with population density and that populations are controlled by competition or predation interactions (Lack 1954; Hutchinson 1978; Hanski 1990). Proponents of density independence have argued instead that per capita population growth rates are largely controlled by abiotic factors such as weather (Andrewartha and Birch 1954; Lawton and Strong 1981; Strong *et al.* 1984). Despite early ideas and experiments that linked abiotic factors and biotic interactions (Lotka 1925; Park 1954), the traditional view has been that the two types of factors are mutually exclusive.

Studies during the past 20 years strongly suggest that abiotic and biotic factors are not mutually exclusive and are likely to interact (Magnuson *et al.* 1979; Kingsolver 1989; Tracy 1992). Many physiological and behavioural processes that directly determine organisms' ability to gather and compete for limiting resources or to exploit prey depend on an abiotic factor, environmental thermal conditions (Magnuson *et al.* 1979; Gates 1980; Porter 1989). Thus, thermal conditions probably have strong effects on the outcome of density dependent interactions (e.g. competition, predation) for many species (Dunham 1993). Chesson (1990, 1994) and Pacala and Tilman (1994) use these relationships to show that interaction between abiotic factors and biotic interactions may help explain population persistence and species coexistence in competitive environments. Despite these ideas, the interaction between temperature and density dependent interactions has generally not been explored in theoretical studies (Tracy 1992; *cf.* Ives and Gilchrist 1993), and few specific hypotheses exist.

Thermal conditions and biotic interactions are most likely to be linked for ectotherms, because their metabolism, locomotion, and physiological processing of resources are sensitive to their thermal environment (Kingsolver 1989; Dunham 1993). Resource acquisition, metabolism, and conversion of resources into growth and reproduction all change predictably with temperature for ectotherms (Porter 1989). These typical functions are based on temperatures measured in black-body laboratory environments, but the effect of the complete array of thermal factors characteristic of realistic environments, including radiation and convection, can be incorporated by considering temperature to be T_e, standard operative temperature (Porter and Gates 1969; Bakken 1976; Gates 1980). General expressions of resource intake, metabolism and conversion for ectotherms as a function of temperature (Porter 1989) can be incorporated in simple models of population dynamics (Schoener 1973; Lomnicki 1988) to reveal some general predictions about the interaction between thermal conditions and resources in limiting population dynamics.

One case where temperature is likely to interact with resources in affecting population dynamics is for strictly resource-limited populations (sensu Schoener 1973), i.e. where a fixed supply of resource, S, must be shared by all individuals in the population. These resources must be spent on requirements for maintenance at a rate $m(T)$ and remaining resources are then converted into growth or reproduction at a rate $c(T)$. These rates are expressed as functions of temperature, T, to reflect their potential temperature-dependence. In continuous time, the dynamics of such a population can be expressed as

$$\frac{dN}{dt} = Nc(T)\left[\frac{S}{N} - m(T)\right] \quad (1)$$

where N is population size or density. In this simple model, the per capita rate of increase, r, is

$$r = c(T)S / N - c(T)m(T) \tag{2}$$

and the carrying capacity or equilibrium density is

$$K = \frac{S}{m(T)} \tag{3}$$

The change in these characteristics with temperature depend on the typical shapes of the functions $c(T)$ and $m(T)$. For the majority of ectotherm species and their environments, $c(T)$ is a constant (Porter 1989). This follows because conversion efficiency is not a rate; it depends more on the biochemical forms involved during growth and reproduction rather than on rate-dependent reactions (Calder 1984). Requirements, $m(T)$, should increase with temperature at an increasing rate (the familiar Q_{10} effect; Porter 1989). Consequently,

$$\frac{\partial c(T)}{\partial T} = 0, \ \frac{\partial m(T)}{\partial T} > 0 \tag{4}$$

From these assumptions, per capita mortality and equilibrium density should change with temperature in the following way:

$$\frac{\partial r}{\partial T} < 0, \ \frac{\partial K}{\partial T} < 0 \tag{5}$$

An explicit assumption of these predictions is that $\partial S/\partial T = 0$, i.e. that resource supply rate does not change with temperature. In addition, these predictions apply only if the population is resource limited; if per capita population growth is limited by time for resource consumption or digestive capacity, changes in r and K with temperature may be different (Schoener 1973, M. Ritchie unpublished).

In this paper, I test these predictions when applied to experimental populations of a common North American grasshopper *Melanoplus sanguinipes* (Orthoptera: Acrididae). This insect, a herbivorous univoltine ectotherm, typically hatches at high densities, declines throughout the summer, and reaches an equilibrium density in late summer (Belovsky 1986; Schmitz 1993; Joern and Klucas 1993; Ritchie and Tilman 1993). They are easily established in temporary 'microcosm' field cages, in which their initial densities are known and their subsequent dynamics can be easily measured. In most cases, these grasshopper populations appear to be limited by the supply of digestible plant material. For the grasshoppers in this experimental system, mortality occurs daily (i.e. continuously), there are no births, and resource requirements depend on grasshoppers' allocation of resources to survival versus growth or reproduction. Under these assumptions, resource requirements include resources spent on maintenance, growth, and reproduction, and mortality results when resource requirements exceed per capita resource availability. Furthermore, $r < 0$ because there are only deaths and no births, and increased temperature is likely to increase mortality rate μ (where $\mu = -r$). Equilibrium densities (K) should result when densities decline so that per capita resource availability equals demand and mortality rate is zero.

These assumptions were applied to the basic resource-limited population dynamics model (Eqn 1) and its associated predictions (Eqn 5) to generate three hypotheses which could be tested by this grasshopper system;

1. Resource supply rate does not change with temperature.
2. Per capita mortality rate is greater at higher temperatures.
3. Equilibrium densities are lower at higher temperatures.

These hypotheses were tested using a field cage experiment with differing densities of grasshoppers and three thermal environment treatments designed to determine if increased temperature affected population characteristics in the manner predicted by Eqn 5.

Methods

The study was conducted from June – August 1994 at the Millville Animal Research facility of Utah State University, located 10 km south of Logan, Utah. Experimental cages (see below) were placed on a dryland pasture (elevation 1393 m) with gravelly soils at the foot of the Bear River mountain range. The pasture was grazed in spring, but cattle had been removed by 15 June. Vegetation was non-native grassland dominated by the grasses *Bromus inermis* and *Agropyron cristatum*, bindweed (*Convolvulus arvense*), and alfalfa (*Medicago sativa*). The grasshoppers *M. sanguinipes, Melanoplus femur-rubrum, Melanoplus confusus,* and *Camnula pellucida* were the dominant large insect herbivores at the site.

The effects of temperature on population dynamics were tested using a randomised two-factor design experiment, with grasshopper density and thermal environment the main effects on grasshopper mortality and final density. Sixty-three windowscreen cages (40 × 40 cm base, 1m tall, inserted 5 cm into the soil) were placed on the pasture in a 7 × 9 grid, with cages separated by 5 m (cf. Belovsky 1986; Ritchie and Tilman 1992; Schmitz 1993). After clearing the inside of cages of all large arthropods, cages were sealed at the top with binder clips. On 30 June, blue polyester tarpaulin covers (1.8 × 1.2 m) were erected on wooden frames mounted on steel posts over 21 of the 63 cages, such that they shaded cages from 1000–1400 hr. Over another 21 cages, 1.8 × 1.2 m wooden frames covered with clear polyethylene sheets were mounted on steel posts to create miniature greenhouses over cages. The remaining 21 cages were left as controls.

Standard operative temperature (T_e), a measure of the actual thermal environment experienced by organisms, is defined as the temperature in a black-body environment (Gates 1980; Vispo and Bakken 1993) that produces the same heat flux or body temperature of a grasshopper experiencing solar radiation, radiation from surrounding objects, and convection in a real environment.

Operative temperatures in dried mounted grasshopper nymph models were measured at half hourly intervals from 0600–2100 hr each day during the first four days of the experiment in three randomly selected cages from each thermal treatment (a total of 9 cages). This indirect method was used because body temperatures of live grasshoppers are extremely difficult to measure in the field (Bakken 1976; Chappell 1982). A needle thermistor probe was inserted into the grasshopper nymph model and connected the probe to a BHR® E-5 platinum digital thermometer, to obtain four separate, stable (± 0.1°C over 1 min) readings over 5 mins. At each reading, the model was positioned 2 cm above the ground, touching plant leaves or stems, at a different random location.

Prior to stocking cages, total nymphal densities of all species were measured as 32.4 ± 8.3 m^{-2} (S.E.) on 3 July from ten 0.5 m diameter ring counts (Capinera 1987) at random locations throughout the experimental grid. On 5 July, seven densities of 2, 3, 4, 6, 9, 12, and 16 third to fourth instar nymphs of *M. sanguinipes* were released per cage with three replicate cages for each thermal treatment. This provided initial stocking rates of 12.5, 18.7, 25, 37.5, 56.2, 75, and 100 nymphs m^{-2}. Grasshoppers for each cage were weighed collectively prior to release.

The number of grasshoppers in each cage was counted at 1, 3, 5, 8, 13, 18, 21, 25 and 31 days after stocking. Final densities were achieved when numbers of grasshoppers had remained constant in each cage for at least 10 days. After final counts, the remaining grasshoppers were removed from the cages and all green plant material inside each cage was clipped, dried at 45°C, and weighed. This material was then ground in a Wiley Mill through a 0.6 mm mesh screen, redried, and digested *in vitro* in 2 g L^{-1} pepsin and 0.1 N HCl for 48 h at 38°C (Terry and Tilley 1964; Belovsky and Slade 1995). The percent of material digested estimates the dry matter digestibility of green plant material to herbivores. An index of food abundance for grasshoppers in each cage was obtained by multiplying green biomass and *in vitro* dry matter digestibility for each cage.

The per capita mortality rates (μ, where $\mu = -r$, from Eqn 2), were estimated by fitting the grasshopper declines in each cage to the following equation, obtained by solving Eqn 1:

$$N(t) = N_f - (N_f - N_0) exp(-\beta t) \quad (6)$$

where N_f is final density observed in the cage, N_0 is initial density in the cage, t is time since the beginning of the experiment and β is an estimated regression parameter that corresponds to the product $cm(T)$ in Eqn 2 (c is assumed to be constant with temperature) for the per capita mortality rate. The final density was assumed to represent a within-season equilibrium within cages (Fig. 2), so $K = S/(m(T)) = N_f$, and as the initial mortality rate was being estimated, so $N = N_0$. After making these substitutions in Eqn 2 and simplifying,

$$\mu = \beta\ (1 - N_f/N_0) \quad (7)$$

All statistical analysis was performed with SPSS 6.1 for Macintosh® statistical package. The food-limited population growth model was fitted to the grasshopper dynamics within cages and β was estimated with non-linear regression. Differences in operative temperature among thermal treatments were analysed with repeated measures ANOVA. Differences in the parameter β, per capita mortality rates, and final densities were tested with two-way ANCOVA, with the final food abundance index as a covariate. Density-independent and density-dependent mortality rates were estimated for each thermal treatment from the intercept and slope, respectively, of linear regressions of per capita mortality rate against initial grasshopper density, with individual cages for each thermal treatment as sample points. Effects of initial grasshopper density and thermal treatment on green biomass, digestibility, and final food abundance were analysed with two-way ANOVA. All comparisons of means following ANOVA or ANCOVA were performed with Fisher's LSD test ($\alpha = 0.05$).

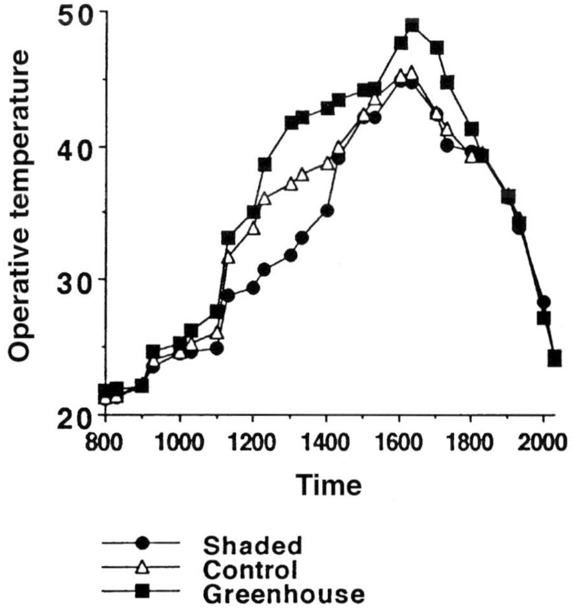

Fig. 1. Mean standard operative temperatures (°C) measured inside grasshopper cages subjected to different thermal treatments (shaded, control, and greenhouse) during daylight hours over three days during the experiment.

Results

Operative temperatures differed significantly among thermal treatments (P = 0.005) especially during the period 1100–1500 hr (Fig. 1). Daytime (0700–2100 hr) average operative temperatures in each treatment were 33.0°C in shaded cages, 34.8°C in controls, and 36.2°C in greenhouse cages. The windowscreen cages used in this experiment were cooler by an average of 1°C of operative temperature than similar locations for grasshopper models outside cages.

Green plant biomass at the end of the experiment averaged 110 ± 20 g m^{-2} (S.E.) and did not differ among treatments (P > 0.74). *In vitro* digestibility of green plant material averaged 39.7 ± 1.2 % and also did not differ among treatments (P > 0.22). Consequently, the product of green biomass and digestibility in each cage at the end of the experiment (final food abundance) also did not differ among treatments (P > 0.74) and averaged 42.4 ± 7.9 g m^{-2}.

Numbers of grasshoppers typically declined rapidly for the first 5 days but varied little over the final 18 days of the experiment (Fig. 2). Declines of grasshoppers within each cage fit the non-linear function in Eqn 6 very well ($0.85 < r^2 < 0.99$). The parameter β (Fig. 3a), which corresponded to an estimate of the product of conversion efficiency and resource requirements, correlated positively with final food abundance (P = 0.05) and negatively with final grasshopper density (P = 0.02). After controlling for these effects, β differed significantly with thermal treatment (P = 0.04) and initial grasshopper density (P = 0.004), but the interaction term was not significant. As expected from the results for β, initial per capita mortality rate also correlated positively with final food abundance (P = 0.02). It also differed significantly with thermal treatment (Fig. 3b, P < 0.001) and initial grasshopper densities (P < 0.001), but the interaction term was not significant. Overall, β and mortality rate increased with average temperature across the thermal treatments, which increased significantly between shaded and greenhouse cages (Fig. 3a,b). Densities of grasshoppers over the final 10 days of the experiment did not change in most cages (Fig. 2) and thus, final density reflected this stable density. Final density correlated negatively with final food abundance (P = 0.05), varied significantly among thermal treatments (Fig. 3c, P = < 0.001) and increased with initial grasshopper densities (P = < 0.001). The interaction was not significant.

For each thermal treatment, per capita mortality rate regressed significantly and positively with initial grasshopper density for all three thermal treatments (Fig. 4a) (shaded: r^2 = 0.37, N = 21,

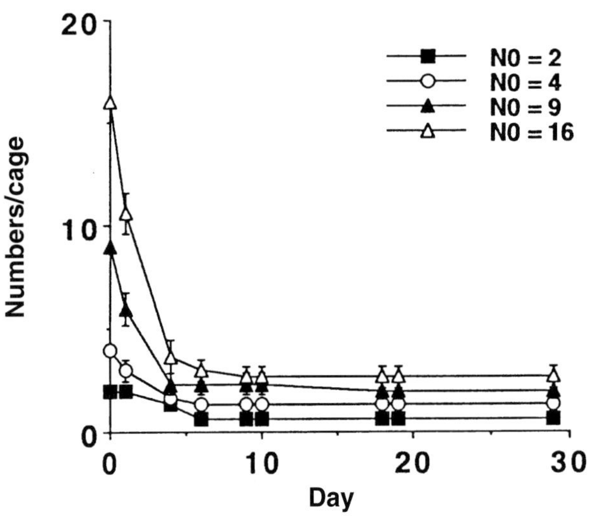

Fig. 2. Example trajectories of grasshopper numbers per cage over time after the beginning of the experiment (Day). Data are means (± S.E., n = 3) for control cages at four different initial stocking rates (NO: 2, 4, 9, and 16 per cage).

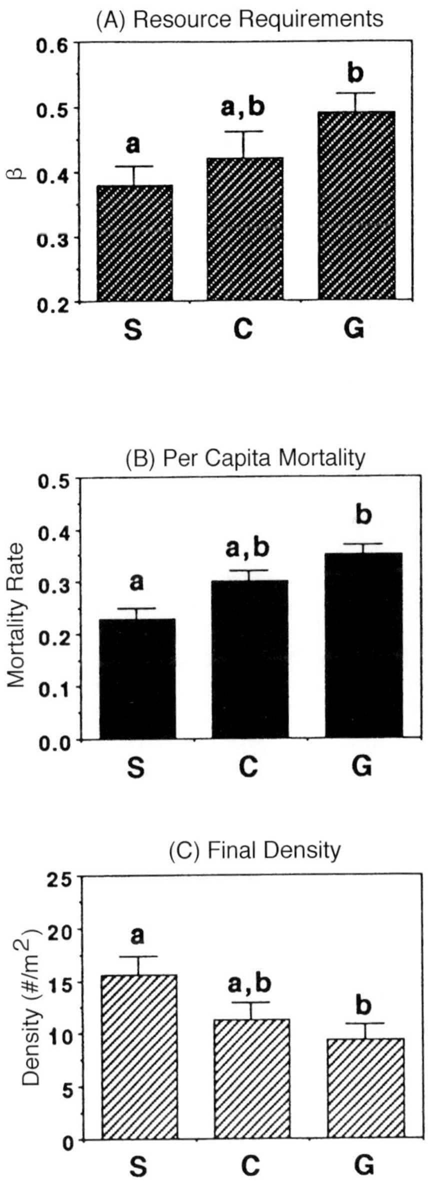

Fig. 3. Effects of thermal treatment (S, shade; C, control; G, greenhouse) on (A) the parameter β (rate of decline) which was estimated by fitting the resource limited population model (Eqn 1) to the trajectory of grasshoppers/cage over time, (B) calculated per capita mortality rate, and (C) final density. Values are means (± S.E., n = 21) pooled over all initial stocking densities for each treatment. Differences in lower case letters indicate significant contrasts.

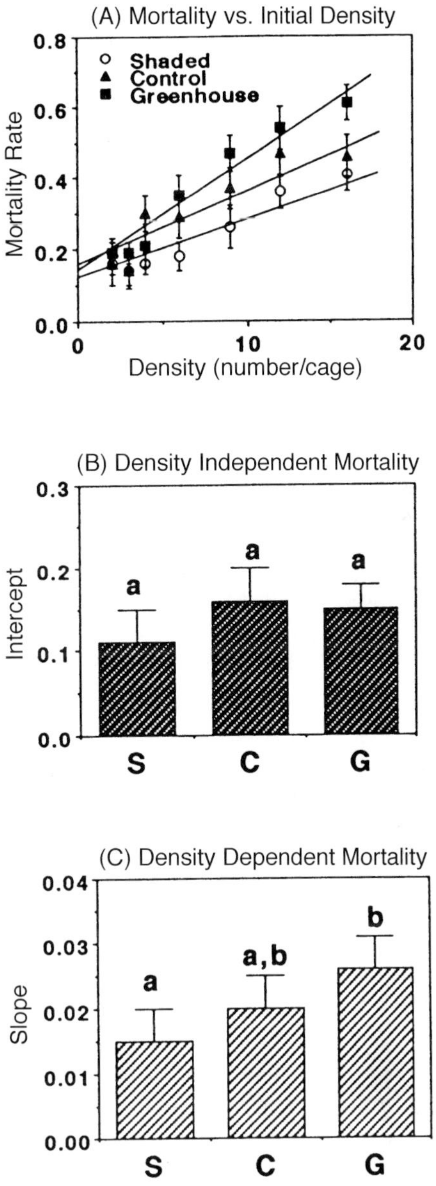

Fig. 4. (A) Relationship of calculated initial daily per capita mortality rate vs initial stocking density (number/cage) for each thermal treatment (S, shade; C, control; G, greenhouse). Data are presented as means for each treatment × density combination (± S.E.) to reduce clutter, but regression lines (see text for r^2 values) were estimated with cages as sample points (n = 3 for each treatment × density combination). (B) Estimated y-intercepts of regression lines (±S.E.). (C) Estimated slopes of regression lines (± S.E.). Differences in lower case letters indicate significant contrasts.

$P = 0.004$; control: $r^2 = 0.45$, $N = 20$, $P < 0.001$; greenhouse: $r^2 = 0.47$, $N = 21$, $P < 0.001$). The intercepts of these regressions did not differ significantly (Fig. 4b), but the slope of the regression for greenhouse cages was significantly greater than that for shaded cages (Fig. 4c).

Discussion

The results suggest that density dependent components of population dynamics in resource limited invertebrate populations may be affected by temperature. Specifically, the simple model of food limited population growth (Eqn1) successfully predicted responses of per capita mortality rate and population density to temperature (Eqn 5). However, density independent mortality exhibited little change with temperature, which suggests that temperature had stronger effects on density dependent mechanisms than on density independent ones.

Temperature effects on mortality and density

The pattern of grasshopper decline in each cage (Fig. 2) fits well with that expected (Eqn 6) for a purely resource limited population (Schoener 1973). The basic model (Eqn 1) however, predicts that final grasshopper density should be constant with respect to initial density, in this experiment final density increased with initial density (Fig. 2). Although explanations other than the simple resource-limited population model are needed to fully describe the observed dynamics (see below), grasshoppers still appeared to be resource-limited. For a simple food limited system in which resource and consumer dynamics are explicitly modelled (see Appendix), equilibrium resource density, R^*, should be inversely related to equilibrium consumer density, N^*, because $R^* = S/(aFN^*)$, where a is the consumption rate of resource per unit time per unit resource and F is time for feeding each day. Since $N^* = N_f = S/m(T) = cS/\beta$, then $R^* = \beta/aFc$. These predictions suggest the following expected patterns: i) a positive correlation between estimates of β and final food abundance, ii) a negative correlation between final density and final food abundance, and iii) a negative correlation between β and final density. All these patterns were observed in the data, so the simple resource-limited population growth model (Eqn 1) to predict the response of grasshopper populations to temperature described both grasshopper and plant dynamics under the controlled conditions within cages. This conclusion might not apply outside the cages as predators or habitat heterogeneity might reduce the influence of food limitation for grasshoppers.

As expected from the typical relationship between metabolism and temperature (Porter 1989), β increased significantly with operative temperature across the thermal treatments. Per capita mortality rate in the resource limited model is predicted to be directly proportional to b (Eqns 2 and 7), and, as expected from the results for β, also increased significantly with operative temperature. These patterns strongly support the prediction of Eqns 2 and 5, that per capita mortality rate should increase with temperature because resource requirements increase with temperature.

Densities in most cages were virtually constant during the final 10 days of the experiment, and probably reflected a within-season equilibrium in the absence of other factors that affect uncaged grasshoppers. Equations 3 and 5 predict that equilibrium density should decrease with an increase in temperature because resource requirements increase and fewer individuals can be supported by available resources. The fact that final density declined with increased temperature across thermal treatments (Fig. 3c) strongly supports this prediction. The increase in β across thermal treatments suggests that the pattern of final density may have been due to increased resource requirements with temperature.

An alternative hypothesis to explain these results is that the thermal treatments affected resource supply, S, or resource availability (final food abundance) and therefore grasshopper mortality rate and final density (Eqns 2 and 3). However, final plant biomass, *in vitro* digestibility, and food abundance in the experiment did not differ among thermal treatments. According to the resource-limited model, final density (N_f) = $c\ S/\beta$ (see Appendix). Thus, the product ($N_f \cdot \beta$) for each cage should be proportional to S for that cage, since c is assumed to be a constant. The product ($N_f \cdot \beta$) did not differ significantly between thermal treatments, again suggesting that resource supply was constant across thermal treatments. Therefore, the effects of temperature on resource requirements (β), and its consequences for per capita mortality rate and final grasshopper density, seems to be the most likely mechanism driving the experimental results.

For a given thermal treatment, both β and final density increased significantly with initial grasshopper density, even though the model predicts them to be constant. The results for β suggest that resource requirements increased with initial density. Individual grasshoppers may have increased their allocation of resources to growth and or reproduction in response to density, i.e. exhibited plastic life history shifts in response to a higher expected mortality rate (Monk 1985; Sanchez *et al.* 1988). Alternatively, greater interference among individuals with greater densities of grasshoppers could have caused individual grasshoppers to increase their activity and therefore their resource requirements. The pattern for final density is perhaps best explained by initial density independent mortality within the experiment (see below, Fig. 4a). With density independent mortality, even cages with initial densities near equilibrium will experience some mortality, thereby yielding a final density below equilibrium. Thus, only cages that begin at densities well above equilibrium should actually achieve the predicted equilibrium density. These discrepancies suggest that the overall population dynamics of these grasshoppers are more complex than those predicted by the simple resource-limited model. However, they do not detract from the patterns in mortality rates and final densities that support the simple model's predictions.

The experiment suggests that temperature and density effects interact. Although no significant interaction between the effects of thermal treatment and initial density was detected for any response variable with ANOVA, more detailed analysis shows that higher temperatures intensified density dependent effects. Per capita mortality rate was clearly density dependent (Fig. 4a), and the per capita density dependent mortality rate (slope of the regression between mortality and density) increased significantly with temperature (Fig. 4c). However, density independent mortality (the intercept of the regression) did not differ with temperature (Fig. 4b). These results suggest that temperature actually had its greatest effect on a density dependent component of population dynamics, which contradicts the traditional view that abiotic factors primarily affect populations in a density independent manner (Andrewartha and Birch 1954; Lawton and Strong 1981; Strong *et al.* 1984). Effects of temperature on density-dependent components may apply particularly within the upper and lower lethal temperature limits of organisms by affecting resource requirements and acquisition rates. The stronger the density-dependent effects in controlling population dynamics, the stronger the response to temperature is likely to be. Temperature may affect populations in a density independent manner when temperatures exceed lethal limits.

Other Implications

Numerous ecological models predict that temperature can affect population stability. For ectotherms, increased temperature should increase per capita mortality rate and/or resource requirements, either of which may decrease the likelihood of population stability (May 1974; Lomnicki 1988). These predictions could not be tested explicitly in this short-term study. Nevertheless, increased temperature is known to destabilise realistic simulation models of ectotherm population dynamics (Logan and Hilbert 1983; Wollkind *et al.* 1988; Doveri *et al.* 1993), and numerous insect populations exhibit dynamics that approach limit cycles and chaos in the absence of other factors (Turchin and Taylor 1992; Ellner and Turchin 1995). Given that a 40% increase in per capita mortality rate was associated with a 3°C increase in average daytime operative temperature in my experiment, it seems reasonable to expect that small temperature changes might have large effects on the stability of ectotherm populations.

Results from this and other studies suggest several implications of temperature–resource interactions for applied problems. Anthropogenic influences on ecosystems (e.g. agriculture, greenhouse gas emission, deforestation, urbanisation, etc.) often produce large, rapid changes in the thermal environment. Such changes are usually thought to impact species by producing conditions that exceed species' tolerance levels (Peters 1992). Species' persistence may be much more sensitive to changes in the thermal environment than expected from these tolerances; small changes in temperature within non-lethal bounds may influence extinction rates through the demographic or deterministic mechanisms described. Strong temperature–resource interactions may increase this sensitivity. No studies known to me have predicted an association between sensitivity of population stability and strong temperature × resource interactions. Ives and Gilchrist (1993) and Ives (1995) argue that density dependence should ameliorate the effects of environmental changes such as temperature, but, they assume that density dependence and temperature effects act on different components of population growth and do not interact.

The experiment presented here supports the predictions (Eqn 5) of a simple model of population dynamics in which resource utilisation is temperature dependent (Eqn 1), and suggests that temperature may affect density dependent components of population dynamics more strongly than density independent ones. Support of these predictions occurred largely because the experimental system met most of the assumptions of the model, i.e. that caged grasshopper populations were resource limited, resource supply rate and availability was not affected by temperature, and no other factors, e.g. predation, affected the population. Therefore, this study does not show whether such interactions are important for real populations. It suggests, however, that considering the interaction of temperature and resources may greatly improve our understanding of the sensitivity of population dynamics to environmental change.

Acknowledgments

I thank my wife Estelle and Sharon Casapulla for help with the field experiment. The manuscript benefited from comments by P. Chesson, E.W. Evans, and discussions with J. Chase, A. Joern, C. Luecke, and G.E. Belovsky. The study was funded by the Utah Agricultural Experiment Station (Project 816) and the National Science Foundation (BSR 9007125).

References

Andrewartha HG, Birch LC (1954) *The distribution and abundance of animals.* Univ Chicago Press, Chicago, USA

Bakken GS (1976) A heat transfer analysis of animals: unifying concepts and the application of metabolism data to field ecology. *J Theor Biol* **60**:337–384

Belovsky GE (1986) Generalist herbivore foraging and its role in competitive interactions. *Am Zool* **26**:51–69

Belovsky GE, Slade JB (1995) Dynamics of two Montana grasshopper populations: relationships among weather, food abundance, and intraspecific competition. *Oecologia* **101**:383–396

Calder WA III (1984) *Size, function, and life history.* Cambridge Univ Press, Cambridge, UK

Capinera, JL (ed) (1987) *Integrated pest management on rangeland: a shortgrass prairie perspective.* Westview Press, Boulder, USA

Chappell MA (1982) Metabolism and thermoregulation in desert and montane grasshoppers. *Oecologia* **56**:126–131

Chesson P (1990) Geometry, heterogeneity and competition in variable environments. *Phil Trans R Soc Lond B* **330**:165–173

Chesson P (1994) Multispecies competition in variable environments. *Theor Pop Biol* **45**:227–276

Davidson J, Andrewartha HG (1948) The influence of rainfall, evaporation and atmospheric temperature on fluctuations in the size of a natural population of *Thrips imaginis* (Thysanoptera). *J Anim Ecol* **17**:200–222

Doveri F, Scheffer M, Rinaldi S, Muratori S, Kuznetsov Y (1993) Seasonality and chaos in a plankton-fish model. *Theor Pop Biol* **43**:159–183

Dunham AE (1993) Physiology-based modeling in assessing global climate change. In: Kareiva PM, Kingsolver JG, Huey RB (eds) *Biotic interactions and global change*. Sinauer Associates, Sunderland, MA, USA, pp 98–119

Ellner S, Turchin P (1995) Chaos in a noisy world: evidence from time series analysis. *Am Nat* **145**:343–375

Gates DM (1980) *Biophysical ecology*. Springer-Verlag, New York, USA

Hanski I (1990) Density dependence, regulation and variability in animal populations. *Phil Trans R Soc Lond B* **330**:141–150

Hutchinson GE (1978) *An introduction to population ecology*. Yale Univ Press, New Haven, USA

Ives AR (1995) Predicting the response of populations to environmental change. *Ecology* **76**:926–941

Ives AR, Gilchrist G (1993) Climate change and ecological interactions. In: Kareiva PM, Kingsolver JG, Huey RB (eds) *Biotic interactions and global change*. Sinauer Associates, Sunderland, USA, pp 120–146

Joern A, Klucas G (1993) Intra- and interspecific competition in adults of two abundant grasshoppers (Orthoptera: Acrididae) from a sandhills grassland. *Envir Ent* **22**:352–361

Kingsolver JG (1989) Weather and the population dynamics of insects: integrating physiological and population ecology. *Physiol Zool* **62**:314–334

Lack DL (1954) *The natural regulation of animal numbers*. Oxford Univ Press, Oxford, UK

Lawton JH, Strong DR (1981) Community patterns and competition in folivorous insects. *Am Nat* **118**:317–338

Logan JA, Hilbert DW (1983) Modeling the effect of temperature on arthropod population systems. In: Lauenroth WK, Skogerboe GV, Flug M (eds) *Analysis of ecological systems: state-of-the-art in ecological modelling*. Elsevier, Amsterdam, Netherlands, pp 113–122

Lomnicki A (1988) *Population ecology of individuals*. Princeton Univ Press, Princeton, USA

Lotka, AJ (1925) *Elements of physical biology*. Williams and Wilkins, Baltimore, USA

Magnuson JJ, Crowder LB, Medvick PA (1979) Temperature as an ecological resource. *Am Zool* **19**:331–343

May RM (1974) Biological populations with nonoverlapping generations: stable points, stable cycles and chaos. *Science* **186**:645–647

Monk KA (1985) Effect of habitat on the life history strategies of some British grasshoppers. *J Anim Ecol* **54**:163–177

Nicholson AJ, Bailey VA (1935) The balance of animal populations, Part I. *Proc Zool Soc London* **3**:551–598

Pacala SW, Tilman D (1994) Limiting similarity in mechanistic and spatial models of plant competition in heterogeneous environments. *Am Nat* **143**:222–257

Park T (1954) Experimental studies of interspecies competition II, temperature, humidity, and competition in two species of *Tribolium*. *Physiol Zool* **27**:177–229

Peters RL (1992) Conservation of biological diversity in the face of climate change. In: Peters RL, Lovejoy T (eds) *Global warming and biological diversity*. Yale Univ Press, New Haven, USA, pp 15–30

Porter WP (1989) New animal models and experiments for calculating growth potential at different elevations. *Physiol Zool* **62**:286–313

Porter WP, Gates DM (1969) Thermodynamic equilibria of animals with environment. *Ecol Monogr* **39**:245–270

Ritchie ME, Tilman D (1992) Interspecific competition among grasshoppers and their effects on plant abundance in experimental field environments. *Oecologia* **89**:524–532

Ritchie ME, Tilman D (1993) Predictions of species interactions from consumer-resource theory: experimental tests with grasshoppers and plants. *Oecologia* **94**:516–527

Sanchez NE, Onsager JA, Kemp WP (1988) Fecundity of *Melanoplus sanguinipes* (F.) under natural conditions. *Can Ent* **10**:29–37

Schmitz OJ (1993) Trophic exploitation in grassland food chains: simple models and a field experiment. *Oecologia* **93**:327–335

Schoener TW (1973) Population growth regulated by intraspecific competition for energy or time: some simple representations. *Theor Pop Biol* **6**:265–307

Strong DR, Lawton JK, Southwood R (1984) *Insects on plants: community patterns and mechanisms.* Harvard Univ Press, Cambridge, USA
Terry RA, Tilley JMA (1964) The digestibility of the leaves and stems of perennial ryegrass, cocksfoot, timothy, tall fescue, lucerne, and sainfoin, as measured by an *in vitro* procedure. *J Brit Grassld Soc* **19**:363–372
Tracy CR (1992) Ecological responses of animals to climate. In: Peters RL, Lovejoy T (eds) *Global warming and biological diversity.* Yale Univ Press, New Haven, USA, pp 171–179
Turchin P, Taylor AD (1992) Complex dynamics in ecological time series. *Ecology* **73**:289–305
Vispo CR, Bakken GS (1993) The influence of thermal conditions on the surface activity of the thirteen-lined ground squirrel. *Ecology* **74**:377–389
Wollkind DF, Collings JB, Logan JA (1988) Meta-stability in a temperature-dependent model system for predator-prey outbreak interactions on fruit trees. *Bull Math Biol* **50**:379–409

Appendix

The simple resource-limited model for growth of a single consumer population (Eqn 1 in the text) actually arises from a trophic model of the dynamics of both the consumer and its resource. In its simplest form, this model assumes that the consumer, with density N, has a simple functional response, $f(R)$ and that the resource, with density R, is supplied at a constant rate S. Thus:

$$\frac{dR}{dt} = S - f(R)N$$

$$\frac{dN}{dt} = cN\left[f(R) - m\right] \qquad \text{(A1)}$$

where c is conversion efficiency and m is resource requirements for the consumer. The simplest functional response (Type I) is $f(R) = aFR$, where a is a constant that represents the search rate (area/time) of the consumer and F is time available for search within the period dt (e.g., time per day). We can also define the product cm as the parameter β, which was estimated in the field experiment. By making these two substitutions, we arrive at

$$\frac{dR}{dt} = S - aFRN$$

$$\frac{dN}{dt} = N\left[caFR - \beta\right] \qquad \text{(A2)}$$

A population is only strictly resource-limited (i.e., individuals share a fixed supply of resources) when the resource is at equilibrium (Schoener 1973). Solving for R at equilibrium yields $R^* = S/(a\,F\,N)$. Substituting R^* in the equation for consumer dynamics yields, the following equation, which is equivalent to Eqn 1 in the text.

$$\frac{dN}{dt} = N\left[\frac{cS}{N} - \beta\right] \qquad \text{(A3)}$$

When both consumer and resource are at equilibrium, the expected relationships between final grasshopper density (N^*), food abundance (R^*), resource requirements (included in β), search rate ($a\,F$), and resource supply (S) are:

$$R^* = \frac{S}{aFN^*} = \frac{\beta}{aF}$$

$$N^* = \frac{cS}{\beta} \qquad \text{(A4)}$$

Note that R^* and N^* should be inversely related to each other, and that correlation of R^* with β should be positive, while correlation of N^* with β should be negative. Note also that search rate and resource supply, which are unmeasured, can influence these relationships and the response of R^* and N^* to treatments.

Forest-insect defoliator interaction in Canada's forests in a warming climate

Richard A. Fleming

Abstract

Many expect climate warming to have a greater influence on Canada's future forests by altering disturbance factors rather than through direct effects on the trees. Insect defoliator populations are one of the dominating disturbance factors in Canada's forests and during outbreaks trees are often killed over vast areas. If the predicted climate shifts occur, the damage patterns caused by insects may be substantially changed, particularly those of insects whose temporal and spatial distributions are very dependent on climatic factors. Recent research is referred to which indicates that climate warming may already be influencing some insect life-cycles. Available information is examined and scenarios developed describing how forest insect defoliators might respond to climate warming. The spruce budworm, *Choristoneura fumiferana*, is used as an illustrative case study. The importance of threshold effects, natural selection, rare but extreme events, phenological synchrony, and historical factors are emphasised.

Key words: Choristoneura fumiferana, climate warming, phenological synchrony, plant quality, natural selection, *Abies balsamea*

Frontiers of Population Ecology, R.B. Floyd, A.W. Sheppard and P.J. De Barro (eds), CSIRO Publishing, Melbourne, 1996, pp. 93–101

Introduction

Although the world's climate has sometimes changed very quickly in the past (Dansgaard *et al.* 1989), the general consensus among climatologists and atmospheric scientists is that the rate of change of the world's climate is quicker now than ever experienced previously (Bolin *et al.* 1986; Houghton *et al.* 1990). The abruptness of this change may make it impossible for many forest ecosystems to react as integral units and for many areas to experience smooth transitions from one assemblage of species to another (Perry *et al.* 1990). Instead, system components such as some ecological guilds or species or age classes will probably respond quicker than others, resulting in new combinations, both at the borders and at the centre of the current ranges of many ecosystems.

The range of the spruce budworm, *Choristoneura fumiferana* Clem. (Lepidoptera: Tortricidae), mirrors that of white spruce, *Picea glauca* (Moench) Voss., in Canada's boreal forest. Records of severe defoliation extend to within 150 km of the Arctic Circle (Volney and Cerezke 1992). During outbreaks, which typically last 5–15 years, population densities can reach 10^8 fourth instar larvae ha^{-1}; between outbreaks the budworm can remain rare (10^5 fourth instar larvae ha^{-1}) for 20–60 years. In Canada, the spruce budworm accounts for the majority of insect-caused losses to forest productivity. This native insect attacks spruce, *Picea* spp., and balsam fir, *Abies balsamea* (L.) Mill., and often kills most trees in dense, mature fir stands when outbreaks are not controlled.

Life histories provide a basis for exploring possible reactions of insect defoliator populations to climate warming. The spruce budworm has one generation a year, overwintering as a tiny second instar larva, resuming activity in early May, and then passing through four feeding instars before pupating in late June. Mating, moth dispersal, and oviposition occur in July. The non-feeding first instar larvae hatch and move to overwintering sites on the branches in mid-August.

There are at least two aspects to anticipating the possible effects of climate warming on insect populations. The first concerns the distinction between the direct effects of climate on insect per capita growth rates and the indirect effects mediated through interactions and feedbacks between the insect species of concern and other species and abiotic components of the environment. The second aspect concerns the distinction between the effects on insect populations of the predicted changes in average climatic conditions and the predicted changes in the variability about those average conditions.

Although climate directly affects budworm survival (Lucuik 1984), and to some extent, fecundity (Harvey 1983b), self-regulation brought about by competitive interactions among individuals (cf. Nicholson 1955) is expected to modify the extent to which climate-induced gains in survival and fecundity produce gains in the per capita growth rate. Competition among spruce budworm in the same generation, however, seems to have little impact except among feeding larvae at very high densities (e.g. Fleming and van Frankenhuyzen 1992; Sanders 1991). Therefore, limitations on climate-induced changes in the spruce budworm's rate of population growth must come mainly through indirect effects such as those associated with trophic interactions. Wellington (1948) has shown that in the past the net result of these indirect effects has been a tendency for per capita growth rates to increase during warm, dry years.

Geographical migration of ecosystems

Integrated migration

Since the possible indirect effects of climate warming include those mediated through interactions with other species (Kingsolver 1989), it is useful to consider how ecosystems as a whole might react to climate warming. The simplest assumption is that as current climatic zones migrate towards the poles (and to higher altitudes), ecosystems will track suitable environmental conditions from one geographic region to another as complete integrated units (e.g. Farrow 1991). This implies that although the geographic distribution of pests may shift in response to climate warming, their impact (per unit area) should change little because they will remain embedded within the same ecosystems and hence subject to the same feedback structure as before. To varying degrees this assumption underpins the inductive approaches (reviewed by Sutherst *et al.* 1995) which infer a species' (or ecosystem's) climatic regime from its observed geographical distribution, and then predict responses to climate warming by matching projected future climates with this regime.

Piece-by-piece migration

An alternative assumption is that ecosystems may not migrate as integrated units, but rather, move piece-by-piece. Such a scenario is predicted by the dynamic models of trophic interactions of forest-forest pest systems derived by Antonovsky *et al.* (1990), including one model whose parameter values were specifically estimated for the spruce budworm-forest system. Key to this line of reasoning is the expectation that, near the southern boundary of the spruce budworm's host trees, tree senescence and regeneration rates will increase and decrease, respectively, in response to climate warming (Rizzo and Wicken 1992). But because seedling mortality is particularly high for these tree species in hot, dry weather (e.g. Sims *et al.* 1990), established trees are expected to survive, albeit in less than ideal environmental conditions, long after almost all regeneration has died (e.g. balsam fir, black spruce, and white spruce can live up to 150–200 years). In effect, the model predicts that the youngest age classes of these tree species, and all species dependent on these age classes for habitat, microenvironment, etc., will retreat north before established trees and before the spruce budworm which can survive on older host trees. Thus the spruce budworm-forest system is expected to retreat north piece-by-piece from its southern boundary in response to climate warming.

An important aspect of piece-by-piece migration is that historical factors may determine the nature of the eventual prevailing condition of an area. For instance, in analysing their simple forest–forest pest models, Antonovsky *et al.* (1990) showed that threshold effects, which have been documented in North American forest ecosystems (e.g. Perry *et al.* 1989), were implicit in the mathematics. In the model representing the spruce budworm–forest interaction, the existence of thresholds depended on the 'per capita' rates of tree senescence and regeneration. Both these rates are climate dependent (Rizzo and Wicken 1992).

Specifically, the model implied that for identical intermediate values of each of these rates (i.e. for a particular range of climates), two different steady states were possible for the spruce budworm–forest system. The steady state imposed depended on whether the density of tree species hosting the spruce budworm was above or below a threshold density when the local climate moved into this particular range of climates. If the host tree density in the model was low when such a new climate began to persist, then regeneration could not overcome weed competition for the site so neither the host trees nor the spruce budworm could become established. This steady state involves local extinction for both the host tree species and spruce budworm.

Alternatively, if the spruce budworm's host tree populations were already dense and well established when such a new climate began to persist, then the host tree–budworm assemblage could hold the site indefinitely because the trees' canopy would shade out the weeds and allow the more shade-tolerant regeneration of the host tree species to succeed. In this case, the model system would eventually settle at a stable equilibrium at which host trees species and budworm coexist. Weed competition would not become a factor unless changes in the rates of host tree senescence and regeneration eventually resulted in opening the overstory to the point where weeds invaded and began to inhibit regeneration.

Thus, theoretically, two sites experiencing the same climate, and hence, similar rates of senescence and regeneration of the spruce budworm's host trees, may eventually be occupied by quite different ecosystems purely because of different initial host tree densities when that climate arrived. Under these circumstances, climate matching could be misleading: the difference in steady state ecosystem composition between the two sites might well be ascribed to differences in the values of irrelevant climatic variables rather than to the historical levels of key populations.

Phenological synchrony

The focus will now be on particular processes and aspects of spruce budworm ecology that may be most sensitive to climate warming. This shift of focus is done in the context that the spruce budworm–forest ecosystem may migrate (and react) in either an integrated, or more likely, a piece-by-piece manner, in response to climate warming.

With host plants

The spruce budworm, like many other herbivores and about 50% of forest insect pests (Martineau 1984), has a life history in which it has synchronised the peak of its nutritional demand with the time when developing, rather then mature, host plant foliage is most accessible. For instance, as balsam fir foliage matures, leaf growth slows but photosynthesis continues, and this can result in increased leaf carbohydrate content relative to nitrogen concentration (Little 1970). The excess carbon is often stored as secondary metabolites (Bryant *et al.* 1983) which either purposefully (e.g. Rhoades and Cates 1976; Haukioja 1980), or incidentally (Tuomi *et al.* 1988), can impede herbivory. In addition, because animals need tissue concentrations of 7–14% nitrogen, while plants typically provide only 0.5–4%, Mattson (1980) and White (1984) have suggested that dietary nitrogen can limit herbivore growth. Thus, developing plant tissue may offer the advantage to the herbivore of being low in secondary metabolites (defensive chemicals), high in nitrogen (nutritional value) and low in fibre (which can limit digestibility).

In a warmer climate, however, foliage will probably develop faster, thus decreasing the 'window of opportunity' when herbivores can consume immature foliage. This may reduce breeding success in certain homeothermic herbivores (Sedinger and Flint 1991), but the effect on poikilothermic herbivores, which also develop faster at higher temperatures, is less certain. For instance, simulation of the phenological development of the spruce budworm and its host tree's foliage suggest that the insect is well synchronised with white spruce (Volney and Cerezke 1992). Another simulation study (Régnière and You 1991) found that the budworm is so tightly synchronised with balsam fir that changes in seasonal weather patterns have little influence on defoliation.

With natural enemies

According to the most recent current theories of budworm population dynamics (Royama 1992), a 'complex' of numerically responding natural enemies drives the budworm outbreak cycle, almost as though it were a classic predator–prey relationship. Royama (1992) specifically identifies two univoltine parasitoids as key members of this 'complex'.

Nicholson and Bailey (1935) were among the first to recognise that simple models provide a general tactical approach to studying host–parasitoid dynamics. To see how climate change might affect host–parasitoid dynamics, Hassell *et al.* (1993) developed a model in which both the host insect and its parasitoid have four stages in their life-cycle: egg, larva, pupa, and adult. Only the host larva is susceptible to attack and only the adult parasitoid is capable of attack. Hence, the rate of attack depends on the synchrony between the parasitoid's adult stage and the host's larval stage. It was assumed that, because of differences in the biological constraints on host and parasitoid physiology, as the climate warms, the host larvae emerges progressively earlier relative to the emergence of the adult parasitoids. Simulations showed that this effectively provided a 'temporal refuge' whereby many hosts escaped any possibility of attack by passing completely through their susceptible larval stage and pupating before the first adult parasitoid emerged. The earlier the host's phenological development relative to the parasitoid's, the greater the host's density at the stable equilibrium which characterises the model's long term dynamics. As the asynchrony between host and parasitoid development increased further, a point was eventually reached at which the 'temporal refuge' provided complete protection and the host population escaped parasitoid regulation altogether.

Drought stress on the host plant

Global circulation models (GCMs) predict generally drier conditions with an increased probability of droughts and heat waves for much of Canada's boreal forest, particularly in the south-west (Hengeveld 1991). There are at least two reasons to expect that some insect pests would do well in such a climate. First, drought stress may alter host plant physiology to the benefit of insect herbivores such as the spruce budworm (but see Larsson 1989). One purported example of such a benefit is that late-instar spruce budworm larvae are very sensitive to sucrose concentration which acts as a feeding stimulant, and thus helps the insect attain greater body size (Harvey 1974), and possibly also helps the insect find and accept suitable host material. Mattson and Haack (1987) report that balsam fir sucrose concentrations are typically near 0.004 M, but are almost three times this level in moisture-stressed trees. Consequently, the level in moisture-stressed fir falls in the range of the peak feeding response of sixth-instar spruce budworm larvae, which occurs at sucrose concentrations of between 0.01 and 0.05 M (Albert *et al.* 1982). This may help explain the increased growth of spruce budworm larvae on host plants suffering low to moderate moisture stress (Mattson *et al.* 1983).

Second, the thermal environment of drought-stressed plants is often very suitable for insect pests. This is partly because stomatal closure reduces transpirational cooling so that plants suffering drought stress are typically 2–4°C warmer than well-watered plants (Mattson and Haack 1987), and partly because of the higher air temperatures and lower humidities generally associated with drought. For instance, as the microenvironment of the spruce budworm warms towards its optimum, the insect usually develops more quickly (e.g. Lysyk 1989), suffers less early instar mortality due to cool wet springs (e.g. Lucuik 1984), and has greater fecundity (e.g. Sanders *et al.* 1978). Moreover, budworm mortality due to natural enemies may decline during drought conditions because the optimum temperature for budworm development of about 26.6°C (Hudes and Shoemaker 1988) exceeds that of many of its natural enemies [e.g. the microsporidian parasite *Nosema fumiferanae* (Wilson 1974), the solitary endoparasitoid *Apanteles fumiferanae* (Nealis and Fraser 1988), and the entomophthoralean fungal pathogen *Erynia (Zoophthora) radicans* (Perry and Fleming 1988)].

Thus, some increase in spruce budworm survival and fecundity can be expected if the climate becomes generally warmer, drier, and more drought-prone. This may have little effect on the insect's population dynamics other than to produce some increase in average densities over the long term. At the other extreme, however, synergistic interaction between drought-induced changes in host plant quality and drought-induced changes in the insect's thermal environment may allow the budworm to improve its performance to the point that it can escape natural enemy regulation more easily. More frequent (Mattson and Haack 1987) and more severe outbreaks can be expected under these circumstances. Studies at the southern limit of budworm distribution may throw light on this.

Natural selection

Many insect species already have a few genotypes pre-adapted to climate change as a consequence of their sizeable populations. For example, Crawford and Jennings (1989) estimate a 'low density' spruce budworm population at 100000 fourth instar larvae ha^{-1}. Since outbreaks involve densities 1000 times greater (Royama 1992) and may extend over as much as 72 million ha (Hardy *et al.* 1986), this represents approximately 7.2×10^{15} insects! Given such large populations, and mutation rates of 10^{-5} to 10^{-4} (Sager and Ryan 1961), it follows that many rare genotypes will gain representation in every generation and at least a few of these genotypes will likely be 'preadapted' to a warmer environment.

Natural selection can also operate on more common genotypes. In species with wide latitudinal distributions, prevalent genotypes in the warmer part of the range may enter cooler parts as the climate warms. For example, Harvey (1983 a,b) found a genetically based cline in spruce budworm egg weights which appears to be an adaptation to winter conditions. Northern females tend to put their egg-laying resources into fewer (150 versus 250 per 100 mg moth) but larger (0.22 versus 0.16 mg) eggs than females from the southern part of the range. The larger egg is necessary to give the young insect enough food reserves to survive the long northern winters (Harvey 1985). If climate warming shortens northern winters, however, one can expect selection in the north to favour the 67% fecundity advantage of the southern genotypes, all else being equal. Thus, natural selection need not operate through catastrophic episodes of mortality in responding to climate change. Rather, many of the expected shifts in the distribution and abundance of populations will probably be the result of subtle, non-fatal effects.

Generally, the GCMs forecast a long trend toward warmer climates (Houghton *et al.* 1990). This represents many generations of directional selection, especially for insects, since many have at least one generation per year. Under these circumstances, genotypes better adapted to a warmer climate can be expected to represent an ever-increasing proportion of their populations, and consequently, their populations can be expected to become progressively better adapted to warmer climates. This has implications for closed environment experiments in which insects (or other organisms) are exposed to conditions simulating climates projected for the distant future. Because any opportunity for evolutionary change is effectively excluded by such experiments, the results may substantially underestimate a species' ability to deal with those projected climates.

In addition, some insects may have already responded to increased temperatures. For example, Jones and Wigley (1990) report that the mean temperature in the northern hemisphere has risen by about 0.4°C since 1964. Accelerated phenological development represents one possible indicator of recent insect response (though not necessarily a genotypic one) to climate warming. This realisation motivated Fleming and Tatchell (1994, 1995) to examine the records of daily suction trap catches from the Rothamsted Insect Survey. The records for five of the most thoroughly sampled aphid species at each of the eight longest running trap sites were examined to see if flight periods were occurring earlier in more recent years. These authors found that the flight periods had moved ahead by an average of about 3–7 days over the last 25 years. As anticipated under a climate warming scenario, they found that although individually most of the trends lacked statistical significance, when all species-site combinations were considered together, trend direction was significantly consistent ($P < 0.0001$).

Conclusions

At least five aspects of forest–insect interactions may greatly influence how the boreal forest's insect-based disturbance regimes will react to climate warming. First, there is tremendous variation among insect species in their trophic relationships, in their life histories, and in the characteristics of their 'outbreak' patterns. Second, because of natural selection, future insect populations may evolve a genotypic makeup which progressively diverges from the population's present composition. Partly because of this, what were otherwise innocuous insect species may become forest pests in a warming climate. Third, multiple and complex feedback loops which often characterise insect population dynamics and trophic interactions can be extremely sensitive to weather (e.g. Kingsolver 1989). Fourth, exceptional weather may be more influential than long term averages in determining the impact of climate warming (see also Fleming and Volney 1995; Solbreck 1991). Finally, the possible existence of historical, threshold, and transient effects complicates attempts to forecast the future behaviour of forest-insect population systems in a warming climate.

Our success in integrating the effects of climate change on some of the processes underlying boreal forest–defoliator interactions will go a long way to determining the reliability of our forecasts. Ideally, the large-scale models resulting from such integration would then provide a sound basis for projecting the dynamics of the boreal forest's insect-based disturbance regimes into the future. To be practical, however, it will be an overwhelming task to model all of these processes and traits explicitly for each important boreal forest-defoliator interaction. Consequently, we must resort to an insightful interdisciplinary mix of available approaches and results to develop sound forecasting procedures for boreal forest–defoliator interactions.

Acknowledgments

The Canadian Forest Service Green Plan provided financial support. I thank Roger Farrow, Vince Nealis, and Bob Sutherst for very helpful reviews of an earlier draft.

References

Albert PJ, Clearley C, Hanson F, Parisella S (1982) Feeding responses of eastern spruce budworm larvae to sucrose and other carbohydrates. *J Chem Ecol* **8**:233–239

Antonovsky MY, Fleming RA, Kuznetsov YA, Clark WC (1990) Forest-pest interaction dynamics: the simplest mathematical models. *Theor Popul Biol* **37**:343–367

Bolin B, Doos BR, Jagger J, Warwick RA (eds) (1986) *The greenhouse effect, climate change and ecosystems.* SCOPE 29. John Wiley, Chichester, UK

Bryant JP, Chapin FS III, Klein DR (1983) Carbon nutrient balance of boreal plants in relation to vertebrate herbivory. *Oikos* **40**:357–368

Crawford HS, Jennings DT (1989) Predation by birds on spruce budworm *Choristoneura fumiferana*: functional, numerical, and total responses. *Ecology* **70**:152–163

Dansgaard W, White JWC, Johnsen SJ (1989) The abrupt termination of the Younger Dryas event. *Nature* **339**:532–533

Farrow, RA (1991) Implications of potential global warming on agricultural pests in Australia. *EPPO Bulletin* **21**:683–696

Fleming RA, Tatchell GM (1994) Long term trends in aphid flight phenology consistent with global warming: Methods and some preliminary results. In: Leather SR, Watt AD, Mills NJ, Walters KAF (eds) *Individuals, populations and patterns in ecology.* Intercept, Andover, UK, pp 63–71

Fleming RA, Tatchell GM (1995) Shifts in the flight periods of British aphids: a response to climate warming? In: Harrington R, Stork NE (eds) *Insects in a changing environment.* Academic Press, London, UK, pp 479–482

Fleming RA, van Frankenhuyzen K (1992) Forecasting the efficacy of operational *Bacillus thuringiensis* Berliner applications against spruce budworm, *Choristoneura fumiferana* (Lepidoptera: Tortricidae), using dose ingestion data: Initial models. *Can Entomol* **124**:1101–1113

Fleming RA, Volney WJA (1995) Effects of climate change on insect defoliator population processes in Canada's boreal forest: some plausible scenarios. *Water, Air, and Soil Pollution* **82**:445–454

Hardy Y, Mainville M, Schmitt DM (1986) *An atlas of spruce budworm defoliation in eastern North America, 1938–1980.* USDA For Serv, Cooperative State Research Service. Miscellaneous Publ No 1449. USDA, Washington, USA

Harvey GT (1974) Nutritional studies of eastern spruce budworm (Lepidoptera: Tortricidae). I. Soluble sugars. *Can Entomol* **106**:353–365

Harvey GT (1983a) A geographic cline in egg weights in *Choristoneura fumiferana* (Lepidoptera: Tortricidae) and its significance in population dynamics. *Can Entomol* **115**:1103–1108

Harvey GT (1983b) Environmental and genetic effects on mean egg weight in spruce budworm (Lepidoptera: Tortricidae). *Can Entomol* **115**:1109–1117

Harvey GT, (1985) Egg weight as a factor in the overwintering survival of spruce budworm (Lepidoptera: Tortricidae) larvae. *Can Entomol* **117**:1451–1461

Hassell MP, Godfray HCJ, Comins HN (1993) Effects of global change on the dynamics of insect host–parasitoid interactions. In: Kareiva PM, Kingsolver JG, Huey RB (eds) *Biotic interactions and global change*. Sinauer Associates, Sunderland, MA, USA, pp 402–423

Haukioja E (1980) On the role of plant defences in the fluctuation of herbivore populations. *Oikos* **35**:202–213

Hengeveld H (1991) *Understanding atmospheric change: A survey of the background science and implications of climate change and ozone depletion*. SOE Report 91–2, Atmos Environ Serv, Environment Canada, Ottawa

Houghton JT, Jenkins GJ, Ephraums JJ (eds) (1990) *Climate change: The IPCC scientific assessment*. Cambridge Univ Press, Cambridge, UK

Hudes ES, Shoemaker CA (1988) Inferential method for modeling insect phenology and its application to the spruce budworm (Lepidoptera: Tortricidae). *Environ Entomol* **17**:97–108

Jones PD, Wigley TML (1990) Global warming trends. *Sci Am* **263**:66–73

Kingsolver JG (1989) Weather and the population dynamics of insects: integrating physiological and population ecology. *Physiol Zool* **62**:314–334.

Larsson S (1989) Stressful times for the plant stress – insect performance hypothesis. *Oikos* **56**:277–283

Little CHA (1970) Seasonal changes in carbohydrate and moisture content in needles of balsam fir (*Abies balsamea*). *Can J Bot* **48**:2021–2028

Lucuik GS (1984) Effect of climatic factors on post-diapause emergence and survival of spruce budworm (*Choristoneura fumiferana*) larvae (Lepidoptera: Tortricidae). *Can Entomol* **116**:1077–1084

Lysyk TJ (1989) Stochastic model of eastern spruce budworm (Lepidoptera: Tortricidae) phenology on white spruce and balsam fir. *J Econ Entomol* **82**:1161–1168

Martineau R (1984) *Insects harmful to forest trees*. Can For Serv, Govt Publ Centre, Ottawa

Mattson WJ (1980) Herbivory in relation to plant nitrogen content. *Oecologia* **81**:186–191

Mattson WJ, Haack RA (1987) The role of drought in outbreaks of plant-eating insects. *BioScience* **37**:110–118

Mattson WJ, Slocum SS, Koller CN (1983) Spruce budworm performance in relation to foliar chemistry of its host plants. USDA For Serv Gen Tech Rep NE-85:55–56.

Nealis VG, Fraser S (1988) Rate of development, reproduction, and mass-rearing of *Apanteles fumiferanae* Vier. (Hymenoptera: Braconidae) under controlled conditions. *Can Entomol* **120**:197–204

Nicholson AJ (1955) An outline of the dynamics of animal populations. *Aust J Zool* **2**:9–65

Nicholson AJ, Bailey VA (1935) The balance of animal populations. *Proc Zool Soc Lond* **3**:551–598.

Perry DA, Amaranthus MP, Borchers JG, Borchers SL, Brainerd RE (1989) Bootstrapping in ecosystems. *BioScience* **39**:230–237

Perry DA, Borchers JG, Borchers SL, Amaranthus MP (1990) Species migrations and ecosystem stability during climate change: the belowground connection. *Conserv Biol* **4**:266–274

Perry DF, Fleming RA (1988) The timing of *Erynia radicans* resting spore germination in relation to mycosis of *Choristoneura fumiferana*. *Can J Bot* **67**:1657–1663

Régnière J, You M (1991) A simulation model of spruce budworm (Lepidoptera: Tortricidae) feeding on balsam fir and white spruce. *Ecol Model* **54**:277–297

Rhoades DF, Cates RG (1976) Toward a general theory of plant antiherbivore chemistry. *Rec Adv Phytochem* **10**:168–213

Rizzo B, Wicken E (1992) Assessing the sensitivity of Canada's ecosystems to climatic change. *Clim Change* **21**:37–56

Royama T (1992) *Analytical population dynamics*. Routledge, Chapman and Hall, New York, USA

Sager R, Ryan FJ (1961) *Cell heredity*. John Wiley & Sons, New York, USA

Sanders CJ (1991) Biology of North American spruce budworms. In: van der Geest LPS, Evenhuis HH (eds) *Tortricid pests, their biology, natural enemies and control*. Elsevier, Amsterdam, Netherlands, pp 579–620

Sanders CJ, Wallace DR, Lucuik GS (1978) Flight activity of female eastern spruce budworm (Lepidoptera: Tortricidae) at constant temperatures in the laboratory. *Can Entomol* **107**:1289–1299

Sedinger JS, Flint PL (1991) Growth rate is negatively correlated with hatch date in black brant. *Ecology* **72**:496–502

Sims RA, Kershaw HM, Wickware GM (1990) *The autecology of major tree species in the north central region of Ontario*. Ontario Ministry of Natural Resources, Northwestern Ontario Forest Technology Development Unit. OMNR Publication 5310. OMNR, Thunder Bay, Canada

Solbreck C (1991) Unusual weather and insect population dynamics: *Lygaeus equestris* during an extinction and recovery period. *Oikos* **60**:343–350

Sutherst RW, Maywald GF, Skarratt DB (1995) Predicting insect distributions in a changing climate. In: Harrington R, Stork NE (eds) *Insects in a changing environment*. Academic Press, London, UK, pp 479–482

Tuomi J, Niemelii P, Chapin FS III, Bryant JP, Siren S (1988) Defensive responses of trees in relation to their carbon/nutrient balance. In: Mattson WJ, LeVieux J, Bernard-Dagenm C, (eds) *Mechanisms of woody plant defenses against insects: Search for patterns*. Springer-Verlag, New York, USA, pp 57–72

Volney WJA, Cerezke HF (1992) The phenology of white spruce and the spruce budworm in northern Alberta. *Can J For Res* **22**:198–205

Wellington WG (1948) The light reactions of the spruce budworm, *Choristoneura fumiferana* Clemens (Lepidoptera: Tortricidae). *Can Entomol* **80**:56–82

White TCR (1984) The abundance of invertebrate herbivores in relation to the availability of nitrogen in stressed foods. *Oecologia* **63**:90–105

Wilson GF (1974) The effects of temperature and UV radiation on the infection of *Choristoneura fumiferana* and *Malacosoma pluviale* by a microsporidian parasite, *Nosema* (Perzia) *fumiferanae* (Thom.). *Can J Zool* **52**:59–63

POPULATION REGULATION IN APHIDS

A.F.G. Dixon, P. Kindlmann and R. Sequeira

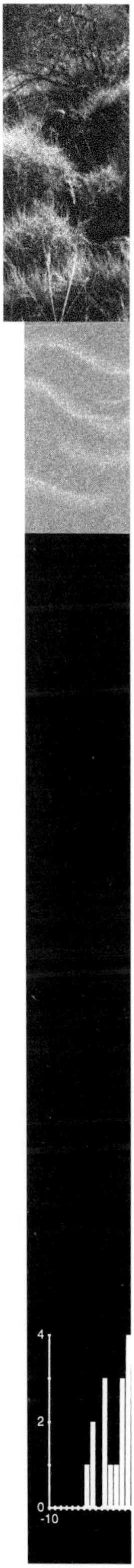

ABSTRACT

Recent analyses of aphid population dynamics have used yearly totals of the suction trap catches of aphids, but it is questionable whether such analyses reveal anything more than that aerial numbers are regulated in some way. The numbers of the host specific Turkey-oak aphid, *Myzocallis boerneri*, on two *Quercus cerris* trees have been monitored weekly for 18 years. This census data was analysed to determine the nature of the density dependence and the time scale over which it operated. In addition, an individual based simulation model was developed for the aphid and used to identify possible mechanisms of regulation. The partial autocorrelation function (PACF) plots of seasonally adjusted data revealed that aphid abundance is regulated by first-order density dependent processes acting over a time frame of a month or less. A correlation matrix also confirmed that the spring and autumn increases and the numbers overwintering are predictable, but not the extent of the summer decline. The model was used to predict seasonal population trends. The output very closely matched the empirical data, suggesting that the regulatory process is intraspecific competition for resources, causing an immediate size related reduction in recruitment and increase in the tendency to migrate. The correlation matrix indicated a negative correlation between the numbers in July and those next spring. The negative correlation between the numbers of first generation in spring and the last generation in autumn, observed in other tree-dwelling aphids, is weak or not present in this species.

Key words: Population regulation, time series analysis, individual based model, aphids, *Myzocallis boerneri*

Frontiers of Population Ecology, R.B. Floyd, A.W. Sheppard and P.J. De Barro (eds), CSIRO Publishing, Melbourne, 1996 pp. 103–114

Introduction

During the course of a season aphids complete several generations and show definite patterns of fluctuation in abundance, which reflect seasonal changes in the quality of their host plant (Dixon 1970a). The seasonal fluctuations in abundance confound the search for the mechanisms of population regulation. In an attempt to overcome this Dixon (1970b, 1971) analysed the relationship between the numbers of aphids present at the beginning and end of a season, that is, the numbers entering and leaving the annual life cycle. More recent analysis (Turchin 1990; Turchin and Taylor 1992; Woiwod and Hanski 1992) have used yearly totals, which also avoids the complexity generated by the seasonal changes in host quality. Both methods have identified density dependence in natural populations of aphids.

These analyses, however, reveal nothing of the mechanisms of regulation. This is particularly true where yearly totals have been used, as this obscures the nature of the density dependence and the time scale over which it acts. In fact it is questionable whether such analyses reveal anything more than can be gleaned from the observation that populations persist for many years – the numbers are regulated in some way (Royama 1977). The earlier method, however, did indicate that regulation occurred within rather than between years.

The objective of this study was to determine the mechanisms that regulate the numbers of tree dwelling aphids. An analysis of census data was used to determine the nature of the density dependence and the time scale over which it operated, correlation analysis tested the predictability of the dynamics and an individual based simulation model identified possible mechanisms of population regulation in the Turkey-oak aphid, *Myzocallis boerneri* Stroyan.

Material and methods

Life cycle of M. boerneri

In the United Kingdom, *M. boerneri* occurs mainly on Turkey-oak (*Quercus cerris* L.) where it is not known to have specific natural enemies such as parasitoids. A previous analysis has also indicated that insect predators are not important in regulating the abundance of this aphid (Dixon 1990). The aphid can complete its entire life cycle on Turkey-oak. The cycle begins at the end of April when eggs laid the previous autumn hatch and produce the first spring generation, also referred to as fundatrices. The fundatrix and subsequent generations are winged and asexual. There are several such generations throughout spring, summer and early autumn, each 2–3 weeks in duration. Sexual morphs begin to appear by the middle of October, to coincide with the beginning of leaf fall. Males are winged but females are apterous. After mating, eggs are laid on the twigs. The end of November marks the end of leaf fall and the completion of the aphid's 7-month life cycle.

The data set

The temporal pattern of density fluctuations of *M. boerneri* populations on two mature Turkey Oak trees in Norwich, England, from 1975 to 1992, were analysed. The populations, labelled A and B, were sampled at weekly intervals from the beginning of May to the end of November each year. Each week, 10 leaves were selected at random at each of 8 fixed sampling points around the circumference of the trees and the numbers of young aphids (instars I–III), instar IV individuals and adults (including oviparous females and males) were counted. Totals of all life stages, N_t, were transformed to $\mathrm{Ln}(N_t+1)$ prior to analysis.

Time series analysis

The data were analysed with the aid of autoregressive methodology, including autocorrelation function (ACF) and partial autocorrelation function (PACF) plots (Box and Jenkins 1976; Brockwell and Davis 1987; Turchin 1990; Turchin *et al.* 1991). The ACF estimates the average correlation between the population sizes, *N*, at time *t* and *t-i*, i.e. now and *i* months ago in our case.

ACFs provide information about periodicity and stationarity of the population dynamics. The ACF of periodic fluctuations, i.e. repeated at regular time intervals, shows a significant positive autocorrelation with a certain time lag. The ACF of stationary dynamics, i.e. those with relatively constant means and variances, decays either in a monotonic or oscillatory fashion, to zero (Box and Jenkins 1976). PACFs estimate the same average correlations as ACF but adjust for the intervening correlations. The PACF indicates the order of density dependence, i.e. the direct influence of, for example, past density on current population size. Three different autocorrelation analyses were done.

In the first analysis, weekly census data were analysed on a year-by-year basis. The objective of this analysis was to determine the nature of and variability in the autocorrelation structure of density within years.

In the second analysis, the original series of weekly totals for populations A and B were transformed to series of monthly means per 80 leaves per week. The computation of monthly mean densities provided 7 census points each year. This served three main purposes: 1) To determine whether or not the annual pattern of spring increase followed by summer decline in density is periodic – repeated year after year – it was necessary to standardise the number of census points in a year. This was not possible on the weekly scale because the length of the season in weeks differed between years. 2) To facilitate the statistical removal of seasonality (see third analysis below). 3) To encompass the aphid's generation time (2–3 weeks), which is longer than the weekly sampling interval; individuals from one generation are usually represented in the following 1 or 2 weekly estimates of population density.

Seasonality is the most important ecological cause of 'periodic nonstationarity', giving rise to fluctuations in mean density at regular intervals (Nisbet and Gurney 1982; Turchin and Taylor 1992). Periodic nonstationarity that is statistically detectable as a significant autocorrelation at the periodic time lag can be eliminated by data adjustment (de-seasoning) using a transformation available in SPSS (1988) of the form $X_t = m + (N_t - S_t)$, where X is seasonally adjusted density, m is the grand mean of the series, N is unadjusted density and S the corresponding seasonal average density at time t (Box and Jenkins 1976; Makridakis and Wheelwright 1978; Wei 1990). As the first two analyses have shown a seasonal effect, the monthly data were adjusted for regular seasonal changes in density in the third analysis using the procedure described above, to identify possible deterministic relationships in the data in the absence of seasonal effects.

Correlation analysis

To find out, when in the season the dynamics are predictable and when extrinsic phenomena have to be considered, a correlation matrix of the monthly data was calculated.

The individual based simulation model

To determine whether in the absence of natural enemies, the responses of the aphids to it's own abundance could give the dynamics observed in the field, an individual based simulation model was developed. The details of the model are published in Kindlmann and Dixon (1996). Here only a brief summary and a flow diagram (Fig. 1) are presented:

The sizes of the soma and gonads of larvae and the size of the gonads and age of adults were monitored in a matrix, which contained one row for each individual. The changes in the size of the gonads and soma of larvae were assumed to follow the Kindlmann and Dixon (1989) model:

$$\frac{ds}{dt} = as^{\alpha} - Rg, s(0) = s_0$$
$$\frac{dg}{dt} = Rg, g(0) = g_0 \qquad (1)$$

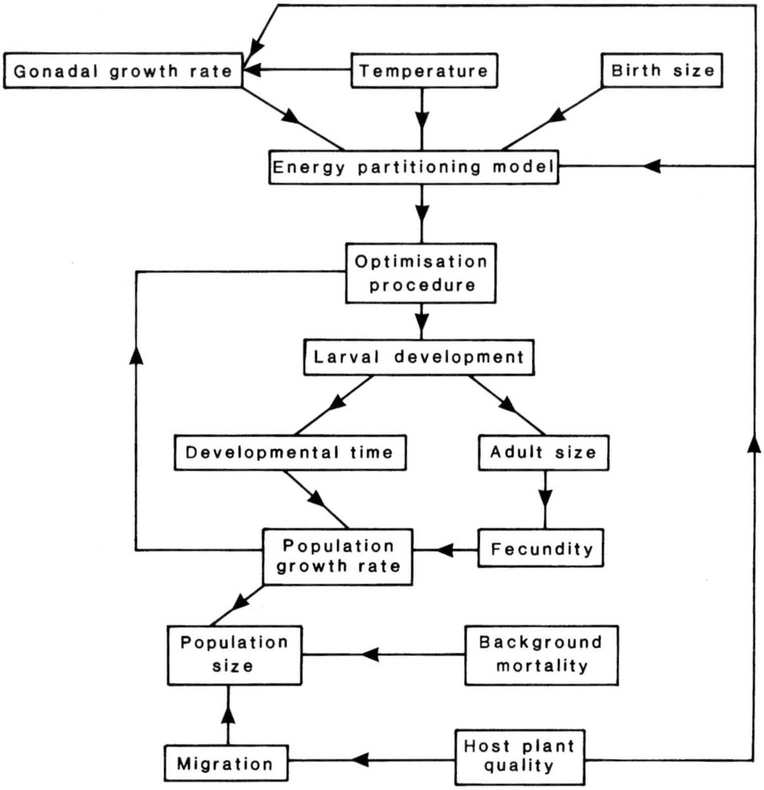

Fig. 1. A flow diagram of the simulation model.

where *s* and *g* are sizes of soma and gonads at time *t*, respectively, *R* is the gonadal growth rate, *a* is the assimilation rate, $a = 2/3$ is a constant (surface to volume ratio) and s_0 and g_0 are birth sizes of soma and gonads. The size of gonads at birth was assumed to be

$$g_0 = (1 + .01 s_0) a^2 \quad (2)$$

The gonadal growth rate, *R*, was assumed to be equal to 0.5 mg mg^{-1} day^{-1}, as this is the value common in aphids (Kindlmann and Dixon 1992). The host quality dependent assimilation rate, *a*, was estimated as

$$a(t) = \left((t - t_{\min}) / t_{\min} \right)^2 \cdot (A_{\max} - A_{\min}) / (d / C + 1) \quad (3)$$

where *d* is the cumulative aphid population density (calculated from the model), *C* a scaling constant, A_{max} the maximum value of *a* at the beginning of the season ($t = 0$), A_{min} the minimum value of *a* at time t_{min} (on an uninfested tree, i.e. $d = 0$ throughout the season). Temperature is included in the model through the effect of seasonal changes in temperature on *a*. A detailed analysis of the effect of between year variation in temperature on $a + R$ is currently being carried out. If the season begins in April, the value $t_{min} = 135$ is realistic for the time of minimum host plant quality, i.e. August and was used in most of the simulations. The values of A_{max}, A_{min} and *C* used in the simulations were varied over the range observed in aphids. The simulations of Eqn 1 were stopped, when *ds/dt* became negative (at a time referred to as *D*, developmental time), i.e. when the animal became adult.

For adults, the size of the gonads at maturity, *g(D)*, determines fecundity. The shape of the reproduction curve (number of offspring born per mother per day), *F(t)*, was assumed to be positive only for *2D>t>D* and triangular:

$$F(t) = 2Rg(D) / (s_0 + g_0)(2 - t / D) \tag{4}$$

where *t* is the age of the individual (Kindlmann *et al.* 1992). As only integer increments in the number of individuals can be simulated in the individual based model, a step length 100 times longer than in the simulation of larval growth was used. Fractions of *F(t)* were rounded randomly up or down with a probability determined by the value of the fraction.

The mortality due to predators (cf. Dixon 1990), aphids falling from the leaves etc., was assumed to be constant. The proportion of aphids that fly away from a colony, *m*, was assumed to be a linear function of the number of aphids (larvae + adults) present in the colony, *x*, with positive slope *S* and positive intercept *i* (Barlow and Dixon 1980), times a linear function depending on *a*:

$$m = (Sx + i) \cdot (A_{max} - a) / (A_{max} - A_{min}) \tag{5}$$

The second term in Eqn 5 simulates the reluctance of aphids to migrate, when living on a good host plant, and their willingness when on a poor quality plant.

The success of an aphid in finding another host plant is assumed to be in the range of 0.01 (Taylor 1977). Therefore, the immigration from other colonies was assumed to be equal to 0.01 *m*. For the same reason as stated for calculating fecundity, the time step used was 100 times longer than for larval development.

Because the output of the model was dependent on, e.g. fecundity being randomly rounded up and down, each simulation was repeated 10 times. The calculations were performed on an HP Apollo computer using FORTRAN language.

Results

Weekly data

The weekly data series (Fig. 2) show rapid fluctuations in density and a regular seasonal pattern of change within years. Density climbs to a spring peak and then declines to a summer trough. In some years there is a small recovery in density in the autumn (cf. Fig. 4).

The ACF plots in Fig. 2 indicate that both data series are oscillatory and nonstationary, i.e. the oscillations in density are perpetuated indefinitely. The negative half of the wave in the ACF plot for Population B (inset, Fig. 2) differs in shape and length from its counterpart for Population A. This indicates that the declining phase of density in summer in Population B is longer and subject to more stochastic disturbances than in Population A.

Monthly data

The transformed data (Fig. 3), consisting of monthly mean density, retain the main features of the original data, i.e. the series are oscillatory and nonstationary. The ACF plot of the monthly data series for Population A (inset, Fig. 3) indicates that density fluctuations are homogenous and periodic at time lags of 7 months (length of the season). This means that the data must be corrected for periodic fluctuations prior to any analysis of density dependence.

Seasonally adjusted data

The average seasonal pattern of density changes for population A is shown in Fig. 4. Figure 5 shows the monthly data series for population A after adjustment for seasonality. The ACF plot (inset, Fig. 5) reveals that the de-seasonalised data retain an oscillatory tendency but the oscillations are aperiodic and heterogeneous in frequency and amplitude. The PACF plot (inset, Fig. 5) shows a

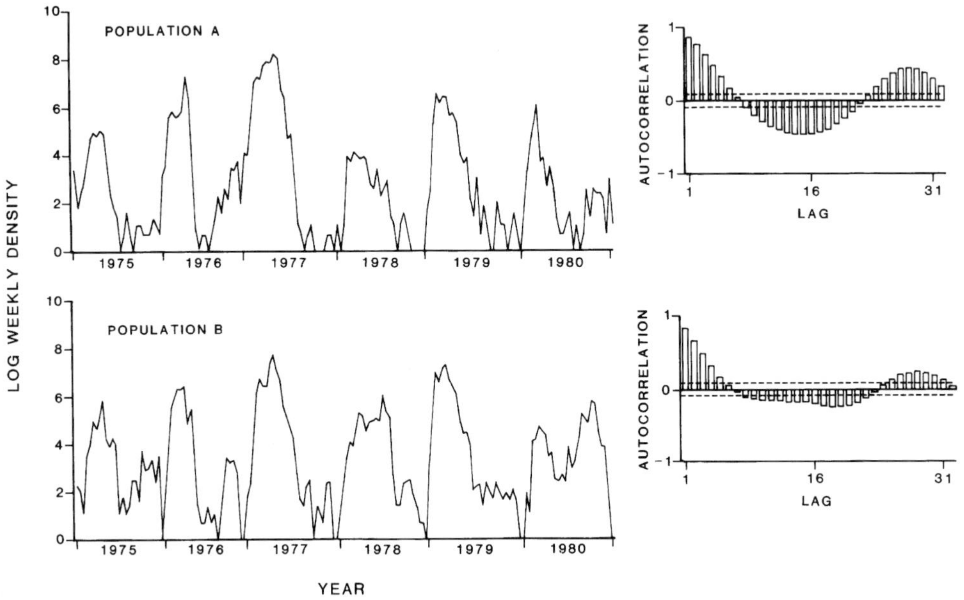

Fig. 2. Weekly log density (total + 1) of M. boerneri populations A and B, from 1975 to 1980. Data for 1981-1992 are not shown. Each abscissa tick marks the beginning of a year. The inset ACF plots indicate sinusoidal nonstationarity in the corresponding data series; dashed lines represent twice the standard error of the autocorrelation.

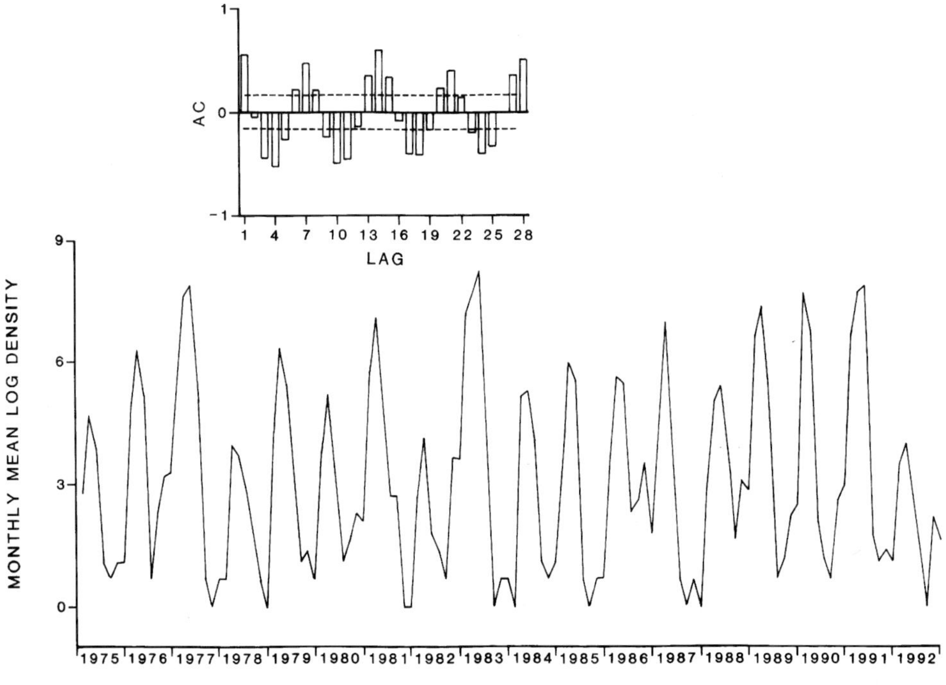

Fig. 3. Monthly mean log density/80 leaves/week for population A, from 1975 to 1992. Each year has 7 monthly totals, from May to November. The ACF plot (inset) shows sinusoidal nonstationarity and periodicity of 7 months; dashed lines represent twice the standard error of the autocorrelation.

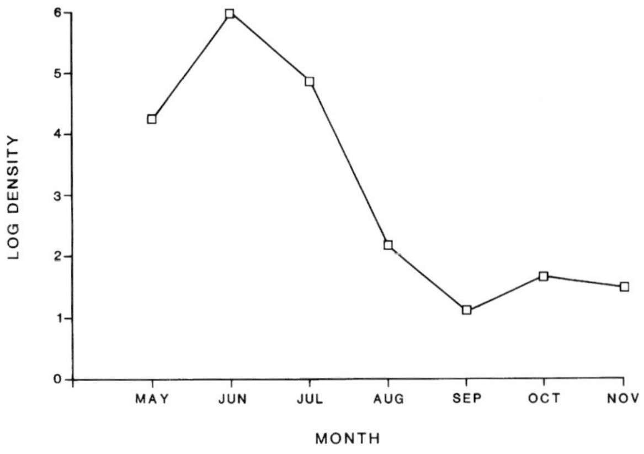

Fig. 4. The average seasonal pattern of density changes, from May to November, for the period 1975-1992.

single significant spike at lag 1 month and small statistically insignificant PACs at all higher lags. However, the lags at 4, 5 and 7 months are very close to being significant at $P < 0.05$. This pattern is consistent with the theoretical PACF of a first-order (direct or non-delayed) density-dependent process (see Fig. 3.2 in Royama 1992). In mathematical terms a first-order density-dependent process could be described by the equation $N_t = \mu + \emptyset N_{t-1}$, where N is density at time t, μ is average value or mean of the series, and $-1 < \emptyset < 1$. Depending on the value of $\emptyset$, a graph of this function shows monotonic or oscillatory decay to the average value, μ.

Correlation analysis

The correlation coefficients between population sizes in individual months and one month later, and those between population sizes in July and seven consecutive months are shown in Fig. 6, together with their significance levels. The population density in the next month can only be predicted from the present one in May (for June), October (for November) and in November (for the next May). Therefore, the pattern of spring population increase and overwintering are predictable but that of the summer decline is either stochastic, or depends on external factors like temperature, wind speed etc.

October densities are positively correlated also with those in the next May and June (R = 0.83 and R = 0.82 respectively – not depicted). This again confirms that overwintering is predictable. Interestingly, July densities are *negatively* correlated with densities the following spring (May and June, Fig. 6b). This suggests the possibility of a 'see-saw effect' – with large summer densities followed by low densities the next spring and *vice versa.*

Simulation results

The common feature of almost all the simulations was the sharp increase in total numbers at first, followed by a decline and then another increase in abundance. The peak number was reached very early each year in all cases – during the first 15–40 days. The more or less broad trough in population numbers never preceded the instant when host quality and therefore a was lowest but occurred some 15–30 days later.

The average seasonal pattern of density changes of the Turkey-oak aphid, from May to November, for the period 1975–1992, predicted by the model is given in Fig. 7a. The model's ability to mimic the dynamics in three years with very different patterns of density changes (1975 – the spring peak followed by a decline and a long plateau in autumn; 1981 – a high spring peak followed by a monotonous decline, second peak not present; 1982 – both spring and autumn peaks relatively low and of the same size) is illustrated in Figs. 7b-d.

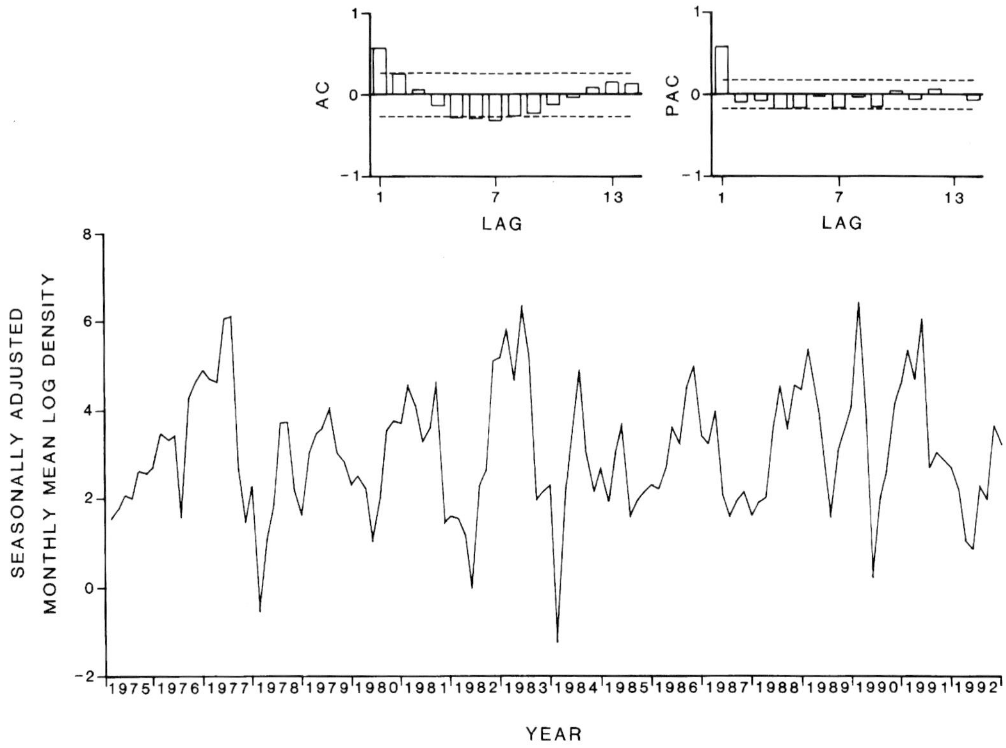

Fig. 5. Monthly mean log density/80 leaves/week for population A, adjusted for seasonality. The ACF plot (inset) indicates that the adjusted data series is stationary (i.e. autocorrelations become statistically insignificant over time) and aperiodic. The PACF plot indicates a strong central tendency in the data; dashed lines represent twice the standard error of the autocorrelation.

A global view of the model's behaviour obtained from 10 runs of the model (Kindlmann and Dixon 1996) indicated a strong positive correlation between spring and summer numbers, a weak positive one between summer and autumn densities and no correlation between the peak and summer densities. This is similar to the outcome of correlation analysis of real data shown in Fig. 6a. Moreover, this indicates that regulation occurs mainly in the spring to summer period, when large perturbations in spring densities are nearly damped out, yielding only small differences in resulting summer densities. The plot of peak to autumn densities revealed a slight tendency to a negative correlation – the 'see-saw effect'. This, however, was only present if cumulative density dependence was included in the model (Kindlmann and Dixon 1996).

Discussion

This study of the Turkey-oak aphid has again revealed the importance of seasonality in the population dynamics of aphids. The seasonal trend in abundance is similar each year irrespective of population density (Fig. 4). This seasonal cyclical fluctuation in abundance is a consequence of population processes driven by changes in the quality of the host plant.

Analysis of the seasonally adjusted census data reveals that the aphid is regulated by a first-order (direct) density dependent process acting over a period of a month or less. Time-delayed effects of density do not appear to be important in the regulation of Turkey-oak aphid populations. Density is strongly regulated around a seasonal mean or equilibrium value. In the field, the population

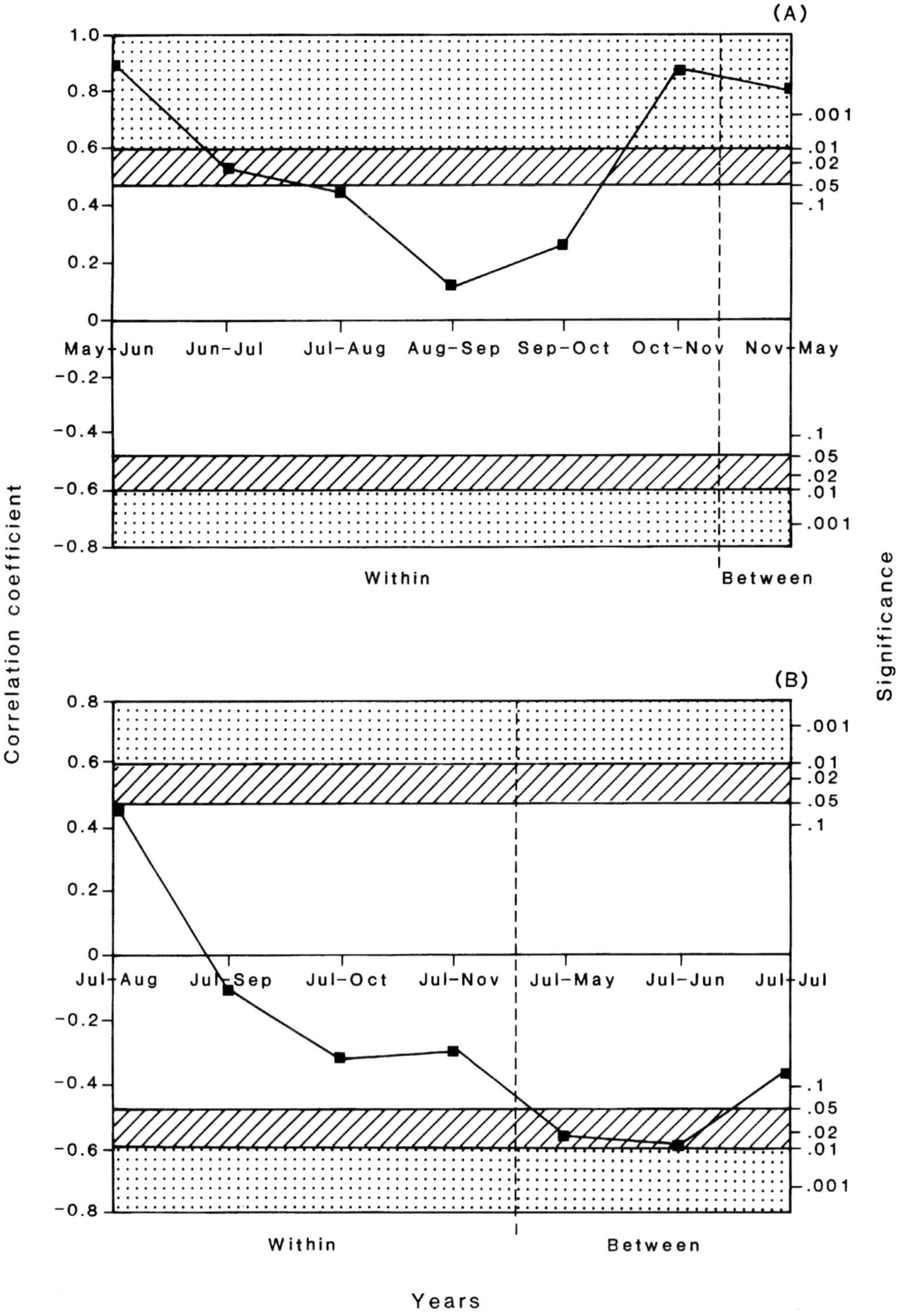

Fig. 6. Correlation coefficients between densities in (A) two consecutive months and (B) July and other months. Shaded areas are critical values for correlation coefficients at significance levels indicated.

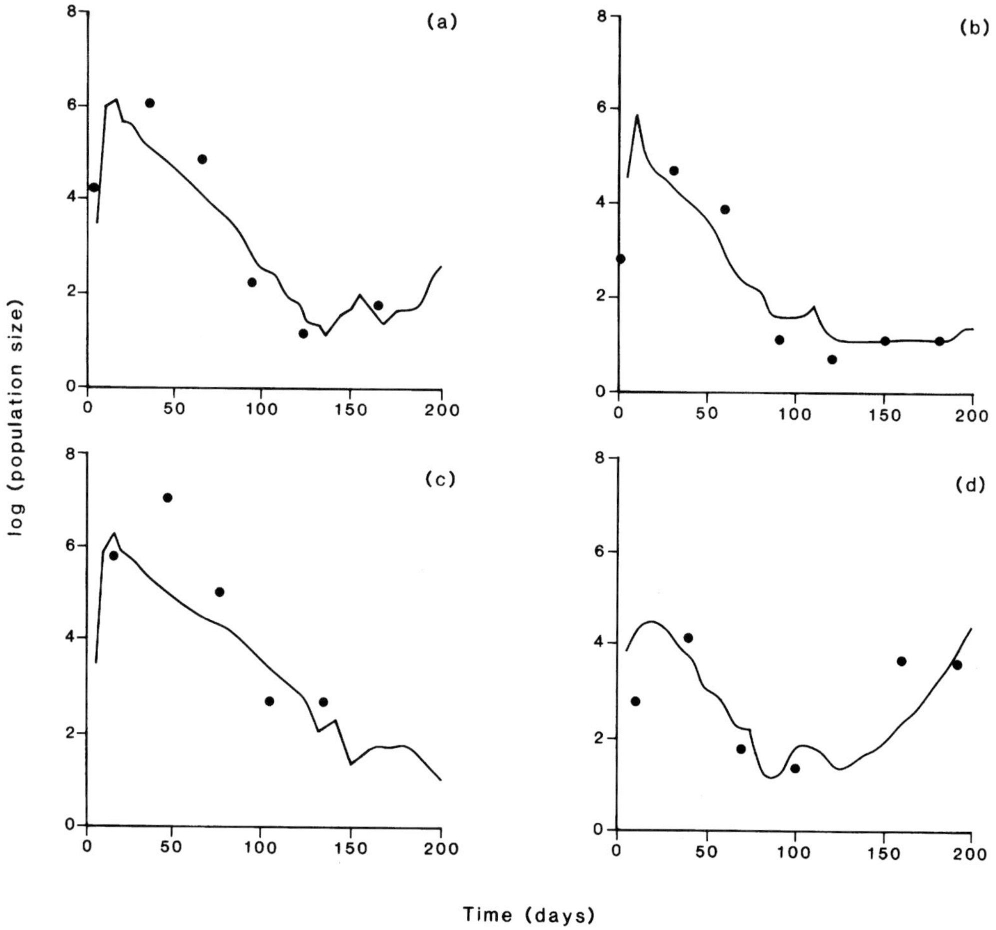

Fig. 7. (a) The average seasonal pattern of density changes of *M. boerneri* populations on two Turkey-oak trees in Norwich, England, from 1975 to 1992, fitted by the model. Parameters: A_{max} = 2.5, A_{min} = .01, R =.5, s_0 = 25, t_{min} = 135, S = .005, i = .1, C = 2.10^6. Log of initial number of aphids: 1.5. Empirical data on density changes of *M. boerneri* for 1975 (b), 1981 (c) and 1982 (d) and simulation results showing the same trends. Parameters are the same as in Fig. 2., with the following exceptions: 1975 - S = .01, 1981 - t_{min} = 185, 1982 - t_{min} = 100, A_{max} = 1. Points: empirical data, lines: model results.

density is subjected to frequent perturbations following which direct density dependent processes come into operation and bring density back to the seasonal equilibrium.

The output of an individual based simulation model very closely mimics the population trends observed in the field (Fig. 7) and supports the contention that regulation is occurring within rather than between years. This model includes the effect of intraspecific competition between aphids for resources and the cumulative effect this drain on resources has on the quality of the plant for aphids but not the effect of natural enemies, which have previously been shown to be unimportant (Dixon 1990). The close match between the model's output and reality suggests that aphid abundance is mainly regulated by competition for resources through its effects on adult size and recruitment, and the migratory tendency of aphids. Interestingly the dominant seasonal trend, the dramatic collapse of the aphid population to very low levels of abundance each summer, is driven by migration.

This exodus, which occurs at the same time each year, often in the absence of natural enemies, is a common feature of many tree dwelling aphids. As the population fluctuations of a particular species of aphid tend to be synchronised on all the trees in an area this exodus raises some very interesting questions: Where are the aphids going and what is the adaptive significance of this high risk strategy for an individual aphid? These questions remain unanswered and constitute one of the most intriguing conundrums of aphid population ecology.

Another feature of the population dynamics of the tree dwelling aphids is the 'see-saw effect'. This is the negative correlation between the number of first generation aphids (fundatrices) and the last generation in autumn (sexuals), which is marked in the lime, pecan and sycamore aphids (Dixon 1970b, 1971; Liao and Harris 1985) and is either weak or not present in the Turkey-oak aphid (this study). This see-saw effect could be driven by a cumulative density effect operating in one of several ways: through its effect on the quality of the host plant at the end of the year as in the model presented here, through the aphid or both through the aphid and the host plant. High numbers of the pecan aphid early in a year adversely affect the quality of pecan food for the aphid later in the year (Wood *et al.* 1985; Bumroongsook and Harris 1991). In the sycamore aphid, *Drepanosiphum platanoidis* Schr., the effect appears to operate through the aphid with high aphid abundance early in the year resulting in the production of very small aphids, which are late in coming into reproduction in autumn and have a low reproductive rate (Dixon 1975). That is, the system has a 'memory', which transfers information on abundance from spring to autumn. In species where this effect is marked there is likely to be a pronounced see-saw in abundance. In the Turkey-oak aphid the partial autocorrelation functions of seasonally adjusted data at time lags of 4, 5 and 7 months (Fig. 5) and the results of the correlation matrix indicate the operation of a cumulative density effect (Fig. 4). Whether weak or strong this effect is a consequence of a delayed response to aphid abundance in spring.

The processes regulating the abundance of the Turkey-oak aphid appear to be driven mainly by intraspecific competition for resources. This interpretation conforms with that of previous studies on the lime, pecan and sycamore aphids (Chambers *et al.* 1985; Liao and Harris 1985; Dixon 1990; Dixon *et al.* 1993). As the time scale over which most of this regulation occurs is considerably less than a year the analysis of data sets consisting of yearly totals of aerial populations can make little or no contribution to our understanding of what regulates aphid abundance.

Interestingly, the closely related insects such as coccids appear to differ from aphids in being regulated by top down processes. This has been very elegantly demonstrated by Murdoch and Nisbet (this volume) for the red scale, also using an individual based simulation model. This raises the question why some phytophagous insects are regulated by top down and other by bottom up processes. The answer to this question, however, is dependent on knowing more about population regulation in other groups of phytophagous insects.

Acknowledgments

This work was supported by the NERC grant GR3/8026A, NATO research grant CRG920553 and ESF Programme for Population Biology travel grant POB/9459.

References

Barlow ND, Dixon AFG (1980) *Simulation of lime aphid population dynamics.* Pudoc, Wageningen, Netherlands

Box GEP, Jenkins GM (1976) *Time series analysis: forecasting and control.* Holden Day, Oakland, CA, USA

Brockwell PJ, Davis RA (1987) *Time series: Theory and methods.* Springer-Verlag, New York, USA

Bumroongsook S, Harris MK (1991) Nature of the conditioning effect on pecan by blackmargined aphid. *Southwest Ent* **16**:267–275
Chambers RJ, Wellings PW, Dixon AFG (1985) Sycamore aphid numbers and population density II. Some processes. *J Anim Ecol* **54**:425–442
Dixon, AFG (1970a) Quality and availability of food for a sycamore aphid population. In: Watson A (ed) *Animal populations in relation to their food resources.* BES Symposium X, Blackwell, Oxford, UK pp 271–287
Dixon AFG (1970b) Stabilization of aphid populations by an aphid induced plant factor. *Nature* **227**:1368–1369
Dixon AFG (1971) The role of intra-specific mechanisms and predation in regulating the numbers of the lime aphid, *Eucallipterus tiliae* L. *Oecologia* **8**:179–193
Dixon AFG (1975) Effect of population density and food quality on autumnal reproductive activity in the sycamore aphid, *Drepanosiphum platanoides* (Schr). *J Anim Ecol* **44**:279–304
Dixon AFG (1990) Population dynamics and abundance of deciduous tree-dwelling aphids. In: Hunter M, Kidd N, Leather SR, Watt A (eds.) *Population dynamics of forest insects.* Intercept, Andover, UK, pp. 11–23
Dixon AFG, Wellings PW, Carter C, Nichols JFA (1993) The role of food quality and competition in shaping the seasonal cycle in the reproductive activity of the sycamore aphid. *Oecologia* **95**:89–92.
Kindlmann P, Dixon AFG (1989) Developmental constraints in the evolution of reproductive strategies: telescoping of generations in parthenogenetic aphids. *Funct Ecol* **3**:531–537
Kindlmann P, Dixon AFG (1992) Optimum body size: effects of food quality and temperature, when reproductive growth rate is restricted. *J Evol Biol* **5**:677–690
Kindlmann P, Dixon AFG (1996) Population dynamics of a tree-dwelling aphid: individuals to populations. *Ecol Modelling*, in press
Kindlmann P, Dixon AFG, Gross LJ (1992) The relationship between individual and population growth rates in multicellular organisms. *J Theor Biol* **157**:535–542
Liao HT, Harris MK (1985) Population growth of the blackmargined aphid on pecan in the field. *Agric Ecosys Environ* **12**:253–261
Makridakis S, Wheelwright SC (1978) *Forecasting: methods and applications.* John Wiley, New York, USA
Nisbet RM, Gurney WSC (1982) *Modelling fluctuating populations.* John Wiley and Sons, Chichester, UK
Royama T (1977) Population persistence and density dependence. *Ecol Monogr* **47**:1–35
Royama T (1992) *Analytical population dynamics.* Chapman and Hall, London, UK
SPSS (1988) SPSS-X *Trends Manual.* SPSS Inc, Chicago, USA
Taylor LR (1977) Migration and spatial dynamics of an aphid, *Myzus persicae. J Anim Ecol* **46**:411–423
Turchin P (1990) Rarity of density dependence or population regulation with lags? *Nature* **344**:660–663
Turchin P, Taylor AD (1992) Complex dynamics in ecological time series. *Ecology* **73**:289–305
Turchin P, Lorio PL, Taylor AD, Billings RF (1991) Why do populations of southern pine beetles Coleoptera: Scolytidae fluctuate? *Envir Ent* **20**:401–409
Wei WWS (1990) *Time series analysis: univariate and multivariate methods.* Addison-Wesley Publishing, Redwood City, CA, USA
Woiwod IP, Hanski I (1992) Patterns of density dependence in moths and aphids. *J Anim Ecol* **61**:619–629
Wood BW, Tedders WL, Thompson JM (1985) Feeding influence of three pecan aphid species on carbon exchange and phloem integrity of seedling pecan foliage. *J Amer Soc Hort Sci* **110**:393–397

POPULATION DYNAMICS BEYOND TWO SPECIES: HOSTS, PARASITOIDS AND PATHOGENS

Michael Begon, Roger G. Bowers, Steven M. Sait
and David J. Thompson

ABSTRACT

The importance of extending the frontiers of population ecology to encompass three-species population dynamics is stressed, and three-species studies in general, both theoretical and empirical, are reviewed briefly. More specifically, theoretical studies of host–host–pathogen systems are examined in order to contrast their implications, with and without the inclusion of host self-regulation, first for the role of shared pathogens in community structure, and second for the microbial control of pests. Also, long-term replicated population dynamics data are used to argue for the potential importance of undetected host infections (with pathogens) in hindering a proper understanding of host–parasitoid population data, and for emphasising the possible complexities of relationships between two-species and three-species dynamics. Finally, the role is illustrated of host–pathogen–parasitoid studies within individual hosts in providing mechanisms underlying the dynamics of their populations.

Key words: Three-species interactions, pathogens, parasitoids, community structure, population theory

Frontiers of Population Ecology, R.B. Floyd, A.W. Sheppard and P.J. De Barro (eds), CSIRO Publishing, Melbourne, 1996, pp. 115–126

Introduction

Population ecology (both empirical and theoretical) has been preoccupied very largely with studies of single species and interactions between species pairs. This has been so, not because of a widespread belief that species or species pairs exist in isolation, but rather in recognition of the practical difficulties of extending frontiers to encompass three or more species, and in the belief, or hope, that the narrower focus nonetheless captures satisfactorily the essence of population ecology. The advantages, however, of extending attention even to three species are clear. First, an incremental step is taken towards the reality of a web of interactions between species, holding out the hope of progress in our understanding of the ecology of populations. Second, the various three-species combinations (what Robert Holt has called 'community modules'; Fig. 1) may themselves be seen as simple but realistic building blocks from which ecological communities are constructed, forging new links between population and community ecology. And third, a deeper understanding can be developed of two-species interactions, since such interactions often occur not directly between the two species concerned but indirectly, through a third species.

Progress in the ecology of populations beyond two species has gained increasing momentum over the past two decades. Theoretical studies have examined dynamics when single 'prey' species are attacked by two or more predators, parasites or pathogens (e.g. May and Hassell 1981; Dobson 1985; Hassell and May 1986; Hochberg *et al.* 1990; Hochberg and Holt 1990) and when a predator, parasite or pathogen attacks two or more species of prey (e.g. Roughgarden and Feldman 1975; Comins and Hassell 1976; Holt 1977; Holt and Pickering 1985; Bowers and Begon 1991; Begon *et al.* 1992; Begon and Bowers 1994; Holt and Lawton 1993; Norman *et al.* 1995). Further studies, especially related to plants, have examined the dynamics of interspecific competition when the resource that is being competed for also has explicit dynamics (see Tilman 1990). These can be viewed as exercises in introducing 'mechanism' into the interactions between exploitative competitors (Tilman 1990), but they can also be seen as part of a more general movement into three-species ('predator–predator–prey') dynamics.

This in turn emphasises that the traditional sharp distinction within population ecology between competition and predation stems simply from the pragmatic tendency to study species pairs in isolation. Three-species community modules, by contrast (Fig. 1), present exploitative competition in terms of two linked predator–prey interactions, and portray two-prey species being attacked by a predator as the basis for 'apparent competition' between them (Holt 1977). Thus, competition and predation may not survive as separate chapter headings in ecological texts as the frontiers of population ecology are extended beyond two species.

Empirical studies of population ecology beyond two species have been rare, especially if one excludes investigations which, for example, monitor the co-occurrences of large numbers of potential competitors, but take little or no interest in the dynamics of their shared resources, nor of the competitors themselves, and hence fail to address questions at the heart of population ecology – those concerning the dynamics of populations. Examples include the early work of Park (1948), who examined the dynamics when a shared sporozoan parasite, *Adelina triboli*, altered the outcome of competition between two species of flour beetle, *Tribolium confusum* and *T. castaneum.* Similar results were obtained by Sibma *et al.* (1964) using oats, barley and a root-feeding nematode, *Heterodera avenae*, and by Bentley and Whittaker (1979) and Cottam *et al.* (1986) using dock plants and the chrysomelid beetle *Gastrophysa viridula* – work extended by Hatcher *et al.* (1994) to include a rust fungus pathogen of the plants. Bess *et al.* (1961) examined the successive displacement amongst three species of braconid parasitoid from the genus *Opius* on Hawaii and the percentage of hosts (*Dacus dorsalis*) parasitised by each. The importance of apparent competition has been supported by Schmitt (1987), with the co-occurrence of gastropods and bivalves being determined by pressure from their shared invertebrate predators, and by Grosholz (1992), with the co-occurrence

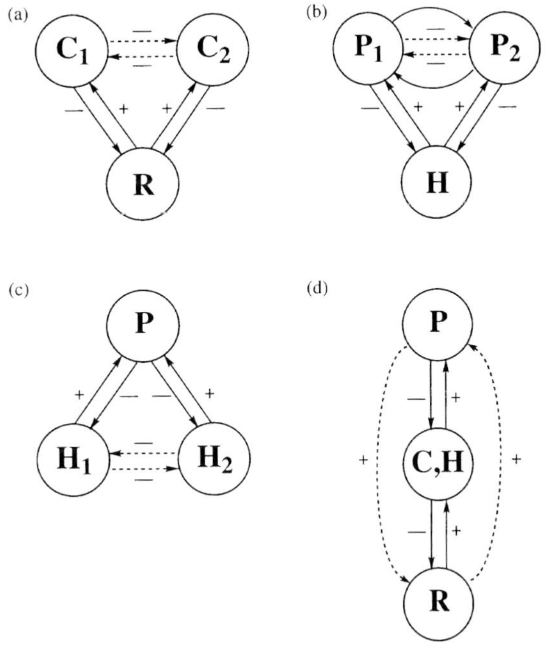

Fig. 1. Examples of community modules comprising three species. Solid lines indicate direct interactions, dotted lines indirect interactions. (a) Exploitative competition between two consumers, C_1, C_2, for a shared resource, R. (b) A combination of interference and exploitative competition between two predators, pathogens, parasites or parasitoids, P_1, P_2, for a shared prey or host, H. (c) Two hosts or prey, H_1, H_2, sharing a common predator, P, and hence subject to apparent competition. (d) A resource, R, supporting a consumer, C or H, supporting a predator, P.

of two species of terrestrial isopod being determined by a shared viral infection. Tilman and co-workers have been able to support the predictions of consumer-consumer-resource models both in diatoms (Tilman *et al.* 1981) and higher plants (Tilman and Wedin 1991a,b).

Here, two recent three-species investigations are examined. The first is theoretical and highlights the way in which host-host-pathogen dynamics may both influence community structure and guide procedures for the release of pathogens in biological control. The second is empirical and highlights the dangers of focusing too narrowly on two-species interactions and the subtleties and complexities that emerge once a third species is included.

Host–host–pathogen models

There has been a series of theoretical investigations of host-host-pathogen interactions. The first was by Holt and Pickering (1985) who modelled hosts with neither immunity nor self-regulation, and pathogens with direct transmission. This was followed by Bowers and Begon (1991) who included free-living stages of the pathogen, by Begon *et al.* (1992) who included host self-regulation but with direct transmission, by Begon and Bowers (1994) who included both host self-regulation and free-living pathogens, and by Norman *et al.* (1995b) who included immunity with and without host self-regulation. Here we contrast the implications of these models, with and without the inclusion of host self-regulation, first for the role of shared pathogens in community structure, and second for the microbial control of pests.

Pathogens, host coexistence and community structure

In the absence of host self-regulation, Holt and Pickering (1985) found a marked similarity between the behaviour of their host-host-pathogen system and that of a conventional Lotka-Volterra system of interspecific competition. Assuming that the pathogen could regulate each host on its own (the equivalent, in Lotka-Volterra competition, of assuming that each consumer could be supported alone by the shared resource) four outcomes were possible: two involved the predictable elimination of one host by the other, a third involved elimination of one or other host contingent on initial host densities, and the last was the coexistence of both hosts with their shared pathogen. Just as the outcomes in Lotka-Volterra competition depend on the relative strengths of intra- and interspecific competition, here they depend on the relative strengths of intra- and interspecific infection: coexistence, for example, is only possible when both species are more affected by intra- than interspecific infection, that is, by the partitioning of enemy-free space in the process of apparent competition.

Including host self-regulation, however, increases the number of paths to coexistence (Begon *et al.* 1992). Host self-regulation can still occur through the partitioning of enemy-free space, but it can also occur through (i) 'resource-mediation', that is, as a result of a ready availability of host resources overcoming the adverse effects on the hosts of apparent competition, or, more specifically, high r levels (intrinsic resources) and high K levels (resource-use efficiency) in the hosts relative to the effects of competition; this immediately invites comparisons with 'predator-mediated coexistence' in conventional competition; (ii) a combination of elements of the partitioning of enemy-free space and resource-mediation when neither alone would be sufficient for coexistence; and (iii) interspecific transmission between hosts, neither of which would be able to support the pathogen alone, inviting comparisons with a mutualistic relationship between two consumers that also share a resource.

Clearly, the potential role of pathogens in community structure, as evidenced by the dynamics of these community modules, is profound. More generally, it is apparent that the analysis of three-species systems holds out the prospect of the conceptual unification of apparently disparate areas of population ecology.

Microbial pest control

One specific example of a host-host-pathogen system is a pest host, a pathogen introduced to control that pest, and a non-target host that it is undesirable to harm and may even be of conservation interest. Models of the dynamics of host-host-pathogen systems may therefore be used to provide a framework within which the releases of microbial pest control agents may be planned. Two questions are likely to be of particular concern: Do the pest and pathogen between them threaten the non-target host? And might the non-target either dilute or enhance the effect of the pathogen on the pest?

The overall contrast between the findings of Begon and Bowers (1994), who included host self-regulation, and those of Bowers and Begon (1991), who did not, may be summarised by noting that in the absence of self-regulation, the threat to the non-target was apparently much less and the chances of the non-target enhancing pest control apparently much greater. This contrast can be illustrated by focusing on just one of the possible combinations of two hosts – that in which both would be regulated by the pathogen to a stable equilibrium or stable limit cycle if they interacted with the pathogen alone (Table 1). Conclusions are arrived at in each case by selecting, from the whole possible range of outcomes for these types of hosts, those that are appropriate for the parameter values of a target, pest host and a non-target host. Crudely, this means assuming that the pathogen is more pathogenic to (and/or has a better chance of regulating) the target than the non-target, and that the target (a pest) has a high intrinsic rate of increase.

Without host self-regulation, the pest is not merely regulated but eliminated by the combination of the pathogen and the non-target. The latter is likely to equilibrate at a density of free-living pathogens greater than the pest can tolerate – the converse of competitive elimination in consumer-

Table 1. Predicted outcomes from models investigating the behaviour of a pest host, a pathogen released to control that pest, and a non-target host, when both hosts would be regulated by the pathogen if they interacted with it alone.

Host	No host self-regulation (Bowers and Begon 1991)	Host self- regulation (Begon & Bowers 1994)
	Diseased non-targets produce many pathogens and/or pathogen particles have high survival rate	Diseased non-targets produce few pathogens and/or pathogen particles have low survival rate
Pest	Eliminated	Coexistence at a lower density than when alone with the pathogen
Non-target	Regulated as when alone with pathogen	Coexistence at a lower density than when alone with the pathogen

consumer-resource interactions (see Tilman 1990). With host self-regulation, on the other hand, while pest elimination remains a possibility, coexistence of both hosts with the pathogen seems more likely, with both pest and non-target at lower abundances than they would be if they interacted with the pathogen alone. The lowered abundance of the non-target, while typically slight in itself, may assume much greater importance once viewed in the light of the effects of further natural enemies (see below). The models suggest, therefore, that in the planning of microbial releases, the strength of host self-regulation would need to be assessed alongside values for other relevant parameters. Of course, what emerges can only be an uncertain prediction playing its own part in the overall balancing of risks and benefits of microbial release. Nonetheless, such balancing exercises may be seriously weakened by the neglect of available planning frameworks.

Host–parasitoid–pathogen dynamics

This section first describes aspects of the population dynamics of the Indian meal moth, *Plodia interpunctella*, its granulosis virus (GV), and the parasitoid *Venturia canescens* maintained in various combinations in population cages. Methodological details of the experimental system may be found in Sait *et al.* (1994a) and Begon *et al.* (1995). This is followed by a brief examination of interactions between parasitoids and viruses within individual hosts.

Host–parasitoid population dynamics

Figure 2 illustrates data for the population dynamics of three replicate host–parasitoid interactions: the tendency for the populations to undergo coupled cycles in abundance is clear. These are 'generation' cycles, with a period just exceeding the length of a host generation – a pattern predicted for host–parasitoid interactions (Godfray and Hassell 1989) but with little empirical support. However, since host populations alone follow cycles with a similar period (Begon *et al.* 1995; see also Fig. 4), it is likely that those in Fig. 2 are not simply host–parasitoid cycles, but owe something, at least, to the underlying host cycles.

Figure 3, on the other hand, illustrates population data from a further series of host–parasitoid interactions maintained under the same conditions, in which fluctuations are erratic within replicates and show no clear pattern to link replicates. The difference is that the hosts in the second series were also supporting a GV infection – this was a three-species rather than a two-species interaction. But which series is more representative of field observations on host–parasitoid (or other prey–predator) dynamics? How often is the host or prey species in such interactions also affected by pathogens, which nonetheless remain undetected simply because the hosts are not subjected to post-mortems?

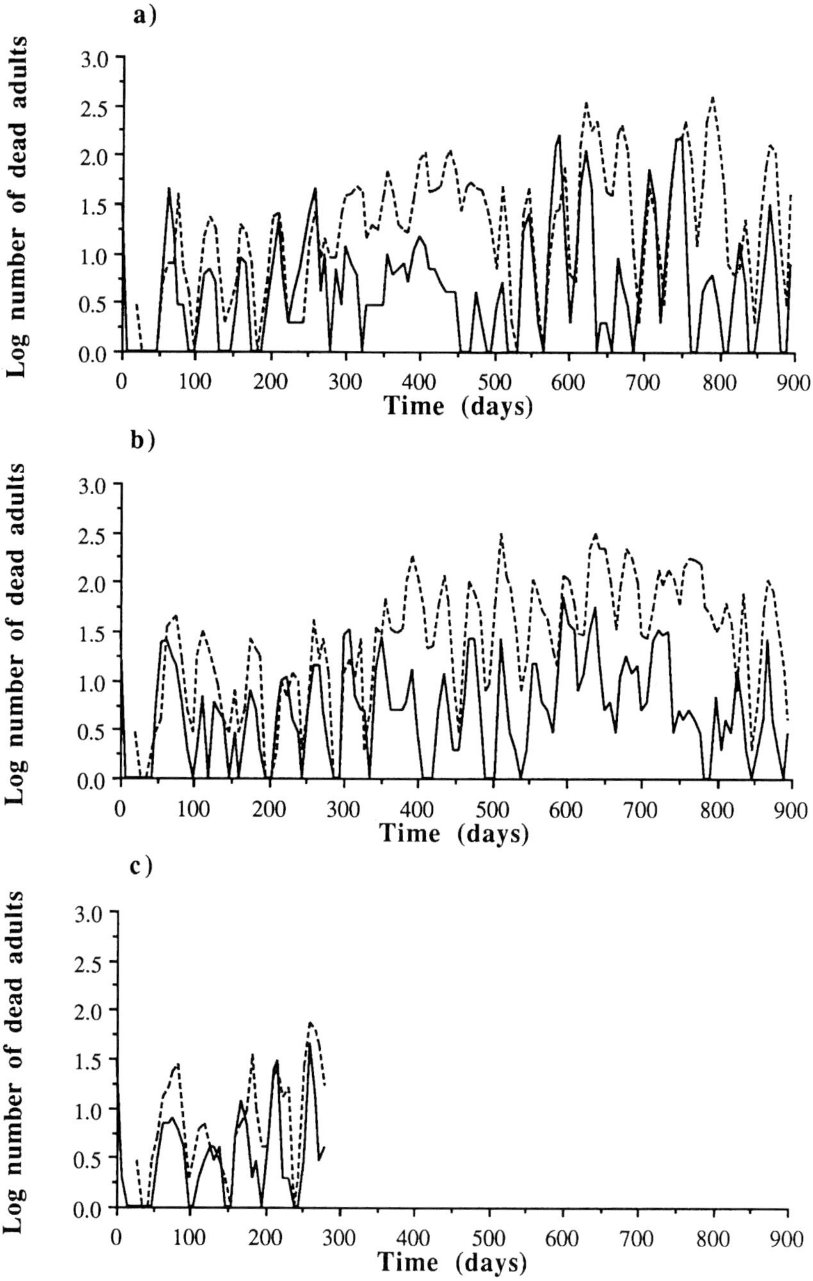

Fig. 2. The population dynamics, in three replicate population cages, of the interaction between a host, *Plodia interpunctella*, solid line, and its parasitoid, *Venturia canescens*, dotted line, where additional replicates showing comparable patterns have been omitted for clarity. Numbers refer to dead adults removed from the populations each week. Autocorrelation functions (AFCs) were calculated to establish the period, strength and consistency of coupled host–parasitoid population cycles. For each host (H) and each parasitoid (P) population, the cycle period (the lag, in weeks, at which the ACF reached its peak) and the ACF of that period (a measure of the strength of the periodicity) are as follows: (a) H: 6, 0.372; P: 6, 0.413; (b) H: 6, 0.377; P: 6, 0.567; (c) H: 7, 0.394; P: 7, 0.481.

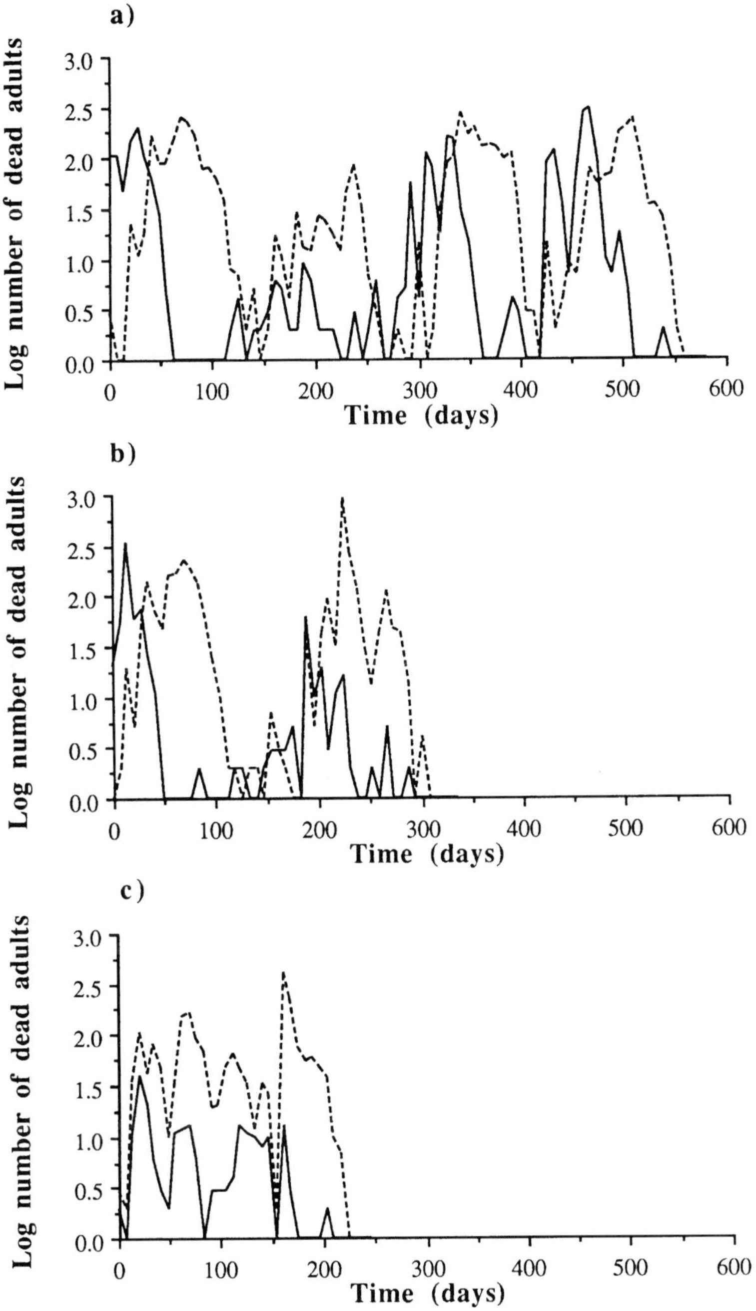

Fig. 3. The population dynamics, in another three replicate population cages, (a) (b) (c) of the interaction between *P. interpunctella*, solid line, and *V. canescens*, dotted line, where additional replicates showing comparable patterns have been omitted for clarity. In all these cases, however, the *P. interpunctella* population was also infected with a granulosis virus. For each host (H) and each parasitoid (P) population, the cycle period (the lag, in weeks, at which the ACF reached its peak) and the ACF of that period (a measure of the strength of the periodicity) are as follows: (a) H: 20, 0.279; P: 20, 0.507; (b) H: 27, 0.315; P: 25, 0.342; (c) H: 15, 0.204; P: 32, 0.223.

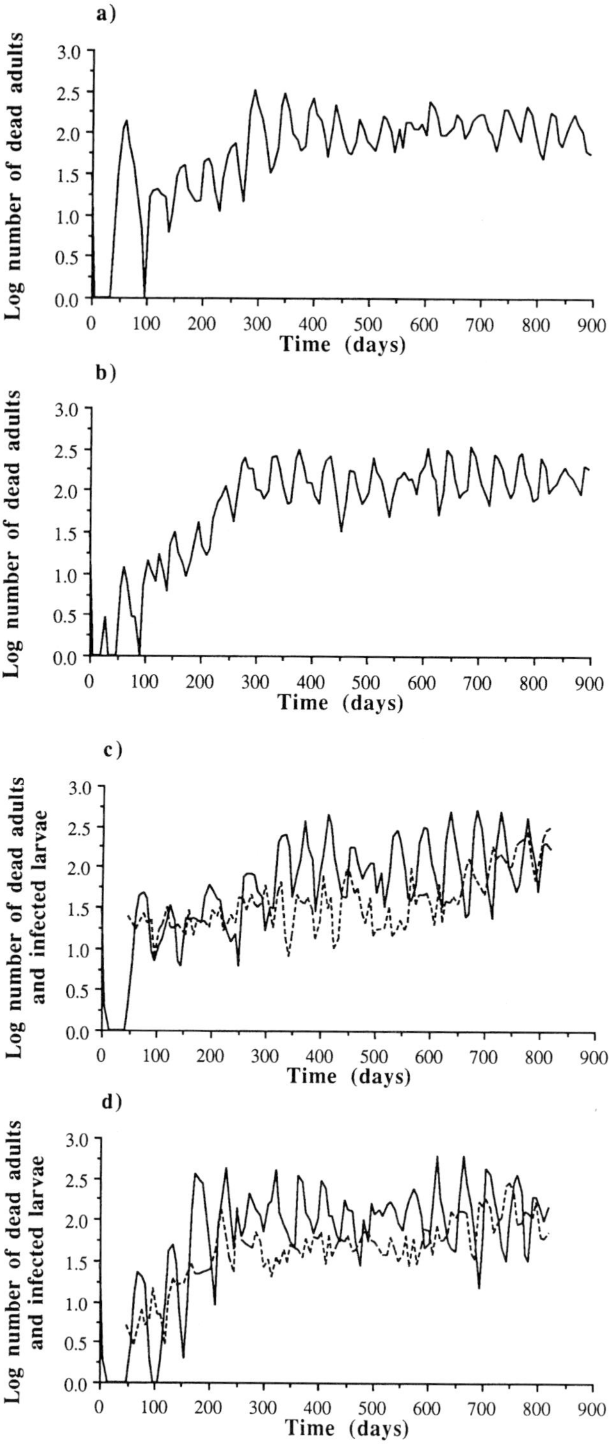

Fig. 4. The population dynamics of *P. interpunctella* alone (a) and (b), and the *P. interpunctella*-GV interaction (c) and (d) in replicated population cages. Additional replicates showing comparable patterns have been omitted for clarity. The solid lines refer to uninfected adult hosts, while the dotted lines refer to numbers of infected host larvae in the one-sixth of the host food replaced each week (larvae being returned to the population). For each host population, the cycle period (the lag, in weeks, at which the autocorrelation function reached its peak) and the ACF of that period (a measure of the strength of the periodicity) are as follows: (a) 6, 0.512; (b) 6, 0.666; (c) 6, 0.505; (d) 7, 0.594.

We contend that such undetected infections are likely to be common, especially when the prevalence of overtly infected hosts is low, and when host-alone dynamics are only weakly affected by the presence or absence of infection. In fact, this was the case for *P. interpunctella* and its GV (Fig. 4). The population dynamics of the host with and without the pathogen are effectively indistinguishable; and whereas the Figure shows the number of diseased larvae to be comparable with the number of healthy adults produced each week, a typical host fecundity of around 200 eggs (Sait *et al.* 1994b) suggests a prevalence of infection of around 1% and certainly less than 5%. Overall, therefore, these data emphasise that (i) complex patterns can emerge in three-species population dynamics (Fig. 3) that could not readily be predicted from observations on the component two-species population dynamics alone (Figs 2 and 4), (ii) interactions may commonly be classified as 'two-species' when they are actually more complex, and (iii) whereas such simplification may often be justified on grounds of pragmatism or because interactions with further species are only weak, it may also often provide a barrier to the proper understanding of the dynamics of the populations concerned.

Interactions in individual hosts

The three-species perspective, aside from being more realistic, holds out the prospect of understanding population dynamics (Fig. 3) in terms of the mechanisms of interaction amongst individuals of the three species concerned, and especially, here, how a pathogen with so little effect on host dynamics in its own right can nonetheless have such a profound, qualitative impact when the parasitoid is present. Here, a taste of these mechanisms is provided by the range of possible outcomes when individual *P. interpunctella* are attacked ('parasitised') by the parasitoid, *V. canescens*, in their fourth instar but then acquire a GV infection later in the fourth instar. Note that the parasitoid in such a case will not start significant development until the fifth host instar (Harvey *et al.* 1994).

Figure 5 provides a summary of these outcomes. At a given dose of the pathogen (1.95×10^7 infectious units), parasitised larvae are less likely (57%) than unparasitised larvae (71%) to become diseased – even before significant parasitoid development has been initiated, the parasitoid shows a strong tendency to exclude the virus from their shared resource. Moreover, amongst those parasitised hosts that do become overtly diseased, three outcomes are possible. First, in only a small proportion (22%) does the GV effectively prevent parasitoid development so that a virus-filled host cadaver is the outcome. Virus productivity is as high here as in unparasitized hosts, and the GV is thus the clear 'winner', but these cases represent only 12% of the original cohort compared to 71% of unparasitised hosts. Second, in 39% of hosts, a parasitoid larva – not an adult – emerges from the host, having been prevented by the virus from completing its development. However, the quantities of infectious virus particles present in the doomed parasitoid larva and the remains of the host are substantially smaller than those in a normally infected host. Thus, neither virus nor parasitoid can truly be said to have won, though of the two, the virus fares slightly better. Finally, from a further 39% of diseased hosts, an adult parasitoid emerges and thus 'wins' – compared to 90% adult parasitoids from non-diseased hosts. However, the adults emerge at both a significantly smaller size and after a significantly longer period of development than parasitoids developing in uninfected hosts (Fig. 5).

Overall, therefore, for this sequence of parasitisation and infection, the parasitoid may have marginally more of an effect on the virus than the virus has on the parasitoid, but both clearly suffer in the interaction. A similar picture emerges at other virus doses and when, for example, the host is parasitised in the third instar and infected in the fourth. On the other hand, the virus can readily infect early host instars I and II (Sait *et al.* 1994c) that the parasitoid is effectively incapable of parasitising (Sait *et al.* 1995). Moreover, when infection precedes parasitisation, the virus usually wins, since *V. canescens* has no chance of developing unless the infected host survives to the fifth instar.

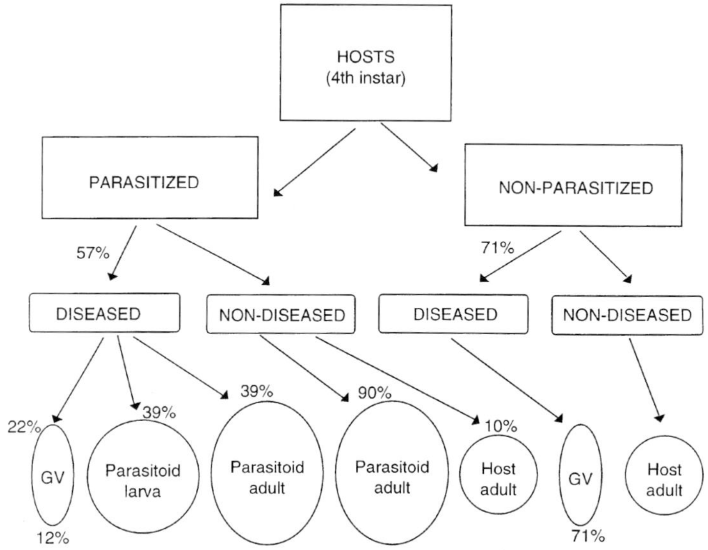

Fig. 5. A flow diagram documenting the fates of individual *P. interpunctella* (as percentages) when either parasitised by *V. canescens* and subsequently infected with granulosis virus in the fourth instar, or when only infected with virus. The two percentage values at the bottom of the diagram refer to the original cohorts succumbing to GV infection alone when parasitised or not parasitised, respectively. The full figures for parasitoid adults emerging from diseased and non-diseased hosts, respectively, are: weights 0.74 ± 0.07 mg, 1.31 ± 0.03 mg (mean ± S.E.), $t = 7.5$, $P < 0.01$; development times 26.2 ± 1.3 days, 24.7 ± 2.1 days (mean ± S.E.), $t = 2.4$, $P < 0.01$.

Hochberg *et al.* (1990) modelled host-pathogen-parasitoid interactions, but in a highly seasonal environment that is not applicable to the present system. Nonetheless, it is interesting to note that they found that coexistence of parasitoid and pathogen could occur at constant, cyclic or chaotic densities (see Fig. 3). They also found that coexistence required one enemy to be superior at exploiting healthy hosts and the other to be competitively superior within jointly infected and parasitized individual hosts. In the present case, comparing Figs 2 and 4, it seems likely that *V. canescens* was a more effective exploiter of healthy hosts. However, taking all five host instars together, it is much less easy to suggest whether *V. canescens* or the GV was competitively superior within hosts. What is especially clear, perhaps, is that there is no simple answer to that question.

As to why the combination of pathogen and parasitoid have so much more of an effect on host dynamics together than either (and especially the pathogen) do alone, the explanation may lie in the partitioning of preferred larval instars between the two 'predators'. Far from being a niche-partitioning favouring coexistence, because this separation is sequential, it is likely to exert a squeeze on the host. With either predator alone, an age-class of larvae escape – with both together there is no escape.

Conclusion

Extending both the theoretical and empirical frontiers of population ecology to embrace three-species interactions is both possible and extremely valuable. Complexities in the underlying interactions are revealed, but the challenge of unravelling these complexities is surely one that is worthy of ecologists' attentions if population ecology is to make a concerted move beyond the monitoring and quantitative analysis of population dynamics into an understanding of underlying mechanisms.

References

Begon M, Bowers RG (1994) Host-host-pathogen models and microbial pest control: the effect of host self-regulation. *J Theor Biol* **169**:275–287

Begon M, Bowers RG, Kadianakis N, Hodgkinson DE (1992) Disease and community structure: the importance of host self-regulation in a host-host-pathogen model. *Am Nat* **139**:1131–1150

Begon M, Sait SM, Thompson DJ (1995) Persistence of a predator-prey system: refuges and generation cycles? *Proc R Soc Lond B* **337**:131–137

Bentley S, Whittaker JB (1979) Effects of grazing by a chrysomelid beetle, *Gastrophysa viridula*, on competition between *Rumex obtusifolius* and *Rumex crispus*. *J Ecol* **67**:79–90

Bess HA, van den Bosch R, Haramoto FH (1961) Fruit fly parasites and their activity in Hawaii. *Proc Hawaii Ent Soc* **17**:367–378

Bowers RG, Begon M (1991) A host-host-pathogen model with free-living infective stages, applicable to microbial pest control. *J Theor Biol* **148**:305–329

Comins H, Hassell MP (1976) Predation in multi-prey communities. *J Theor Biol* **62**:93–114

Cottam DA, Whittaker JB, Malloch AJC (1986) The effects of chrysomelid beetle grazing and plant competition on the growth of *Rumex obtusifolius*. *Oecologia* **70**:452–456

Dobson AP (1985) The population dynamics of competition between parasites. *Parasitology* **91**:317–347

Godfray HCJ, Hassell MP (1989) Discrete and continuous insect populations in tropical environments. *J Anim Ecol* **58**:153–174

Grosholz ED (1992) Interactions of intraspecific, interspecific, and apparent competition with host-pathogen population dynamics. *Ecology* **73**:507–514

Hassell MP, May RM (1986) Generalist and specialist natural enemies in insect predator-prey interactions. *J Anim Ecol* **55**:923–940

Harvey JA, Harvey IF, Thompson DJ (1994) Flexible larval growth allows use of a range of host sizes by a parasitoid wasp. *Ecology* **75**:1420–1428

Hatcher NE, Paul ND, Ayres PG, Whittaker JB (1994) The effect of an insect herbivore and a rust fungus individually, and combined in sequence, on the growth of two *Rumex* species. *New Phytol* **128**:71–78

Hochberg ME, Hassell MP, May RM (1990) The dynamics of host–parasitoid–pathogen interactions. *Am Nat* **135**:74–94

Hochberg ME, Holt RD (1990) The coexistence of competing parasites. I. The role of cross-species infection. *Am Nat* **136**:517–541

Holt RD (1977) Predation, apparent competition and the structure of prey communities. *Theor Pop Biol* **12**:197–229

Holt RD, Lawton JL (1993) Apparent competition and enemy-free space in insect host–parasitoid communities. *Am Nat* **142**:623–645

Holt RD, Pickering J (1985) Infectious disease and species coexistence: a model in Lotka-Volterra form. *Am Nat* **126**:196–211

May RM, Hassell MP (1981) The dynamics of multiparasitoid-host interactions. *Am Nat* **117**:234–261

Norman R, Bowers RG, Begon M (1996) The effects of recovery and immunity on the population dynamics of two hosts and a shared pathogen. *Theor Pop Biol* (in press)

Park T (1948) Experimental studies of interspecific competition. I. Competition between populations of the flour beetles *Tribolium confusum* Duval and *Tribolium castaneum* Herbst. *Ecol Monogr* **18**:267–307

Roughgarden J, Feldman M (1975) Species packing and predation pressure. *Ecology* **56**:489–492

Sait SM, Begon M, Thompson DJ (1994a) Long-term population dynamics of the Indian meal moth *Plodia interpunctella* and its granulosis virus. *J Anim Ecol* **63**:861–870

Sait SM, Begon M, Thompson DJ (1994b) The effects of a sublethal baculovirus infection in the Indian meal moth, *Plodia interpunctella*. *J Anim Ecol* **63**:541–550

Sait SM, Begon M, Thompson DJ (1994c) The influence of larval age on the response of *Plodia interpunctella* to a granulosis virus. *J Invert Pathol* **63**:107–110

Sait SM, Andreev RA, Begon M, Thompson DJ, Harvey JA, Swain RD (1995) *Venturia canescens* parasitizing *Plodia interpunctella*: host vulnerability – a matter of degree. *Ecol Ent* **20**:199–201

Schmitt RJ (1987) Indirect interactions between prey: apparent competition, predator aggregation, and habitat segregation. *Ecology* **68**:1887–1897

Sibma L, Kort J, de Wit CT (1964) Experiments on competition as a means of detecting possible damage by nematodes. *Jaarboek, Inst voor Biol Scheikundig Onderzoek van Landbouwgewassen.* **1964**:119–124

Tilman D (1990) Mechanisms of plant competition for nutrients: the elements of a predictive theory of competition. In: Grace JB, Tilman D (eds) *Perspectives on plant competition.* Academic Press, New York, USA, pp 117–141

Tilman D, Mattson M, Langer S (1981) Competition and nutrient kinetics along a temperature gradient: an experimental test of a mechanistic approach to niche theory. *Limnol Oceanog* **26**:1020–1033

Tilman D, Wedin D (1991a) Plant traits and resource reduction for five grasses growing on a nitrogen gradient. *Ecology* **72**:685–700

Tilman D, Wedin D (1991b) Dynamics of nitrogen competition between successional grasses. *Ecology* **72**:1038–1049

Mammal populations: fluctuation, regulation, life history theory and their implications for conservation

A.R.E. Sinclair

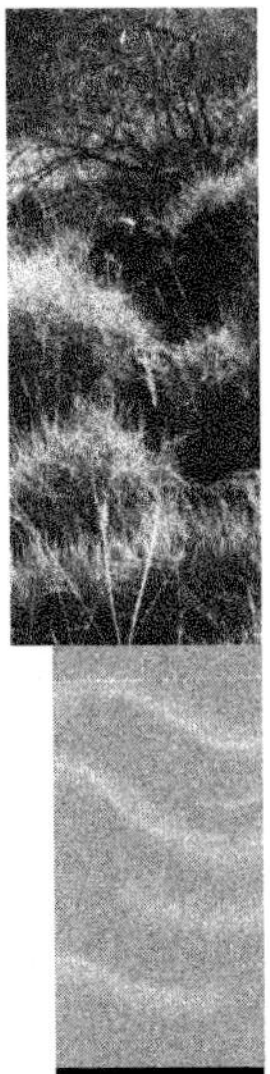

ABSTRACT

Mammal populations exhibit a range of variability inversely related to body size when considered over absolute time. However, there is no relationship between population variability and body size over the length of a generation, so all species show the same intrinsic degree of variability. Small species are not subject to more severe extrinsic perturbations than larger species. Therefore, the variability seen in small species is due to their high intrinsic rates of increase (r_m) which allows them to recover faster from extrinsic perturbations. At the same time, the magnitude of decrease in a population must also be determined intrinsically, since the populations (in this analysis) were stationary in the longterm. Thus, in a given environment, both large and small species experience the same negative environmental effects, and the degree to which the species are buffered from these effects is inversely related to r_m and positively related to body size. This implies that species are not simply passive responders to a stochastic environment, but are adapted to tolerate decline, or resist it, as a function of r_m, as predicted by life history theory.

Since this tolerance or resistance to decline is a measure of density dependence in the population, it follows that small species must have high overcompensating density dependence while large species have weaker stabilising density dependence. This conclusion, which is in agreement with empirical studies, provides a generalisation on both the prevalence and intensity of density dependence in mammal populations. Thus, the evidence supports the hypothesis that population fluctuations are explained more by strong, overcompensating density dependence than by weak density dependence and high extrinsic stochasticity.

The intrinsic (density dependent) response to fluctuations appears to occur earlier in life (through fecundity and early juvenile mortality) in larger species, later (through juvenile and adult mortality) in smaller species. This trend may explain the inverse relationship of the strength of density dependence with body size seen above, because mortality responds faster to environmental change than does reproduction. Behaviour, particularly, social organisation and dispersal, is part of the specific intrinsic response or adaptation to perturbation, and contributes to the overcompensating density dependence in small species.

A strong overcompensating behavioural response leads to population cycles or chaotic fluctuations (which appear cyclic) and their frequency is inversely related to body size.

Causes of mortality involving food shortage affect all species but especially large ones. Predation causing regulation appears more frequently in small species, and

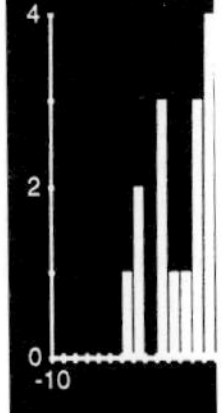

Frontiers of Population Ecology, R.B. Floyd, A.W. Sheppard and P.J. De Barro (eds), CSIRO Publishing, Melbourne, 1996 pp. 127–154

because it produces delayed effects, it contributes to population cycles. The role of disease remains obscure.

Since the strength of density dependence is inversely related to body size, it is the larger species which are the most prone to extinction when at low density. Thus, large species must be conserved at relatively higher population size. In contrast, smaller species should be conserved by allowing dispersal between sub-populations.

Key words: Body size, population variability, intrinsic rate of increase, density dependence, viable population size

Introduction

The natural regulation of populations is one of the central themes in ecology. It underlies our thinking in intraspecific competition theory, predator-prey dynamics, life history theory and the action of natural selection. It is also fundamental to how we will conserve rare and endangered species. Nicholson's (1933) classic paper on 'The balance of animal populations‘ highlighted the concept that populations persisted in nature through a negative feedback mechanism that later came to be called 'density dependence'. In essence, the per capita rate of increase (r) of a population is negatively related to population density, as seen for example (Fig. 1) in the Serengeti wildebeest (*Connochaetes taurinus*). The decrease in r occurs through either an increase in per capita mortality or a decrease in per capita natality. Populations experiencing such a negative feedback mechanism are 'regulated'. (The term 'centripetality' (Caughley 1987) is synonymous with regulation).

In the sixty years since Nicholson proposed this idea of regulation there has been much debate on whether it occurs in nature. Some of the debate is founded on misunderstanding of the terms, and this is reviewed elsewhere (Sinclair 1989; Murdoch 1994; Sinclair and Pech 1996; Krebs 1995). Other issues concern the evidence for regulation and how it works. One of the most persistent misconceptions is that there should be some measurable evidence from census data of an equilibrium point, or points, if regulation occurs: however, regulation merely requires a negative feedback process with density, and an equilibrium, though potentially existing, may never be achieved. Thus, the distinction between 'equilibrium' and 'non-equilibrium' is unnecessary and confusing if both terms refer to regulation. I define 'non-equilibrium' as implying random drift.

It is now recognised that populations which do not experience density dependence will show a random drift to extinction (Reddingius 1971; Murdoch 1970, 1994). Thus, for populations to remain extant they must experience some form of negative feedback related to population density. This predicts that regulated populations exist for longer periods than do unregulated populations. However, because it is impossible to compare the times to extinction of two such populations the prediction cannot be tested. Therefore, we are confined to testing the subsidiary prediction that if populations are regulated we should be able to detect density dependence in them.

Some populations fluctuate more than others. It has been suggested that those with greater variability experience weaker density dependence and thus are more prone to environmental perturbations than are more stable populations (Fowler 1981; Strong 1986; Sinclair 1989). The alternative hypothesis suggests that population variability derives from very strong, overcompensating density dependence producing cyclicity and chaos (May 1975; Stubbs 1977; Bellows 1981; Turchin and Taylor 1992). If smaller species exhibit greater population variability than larger species, then the two hypotheses may be distinguished by whether or not smaller species show more variability than that predicted by size:

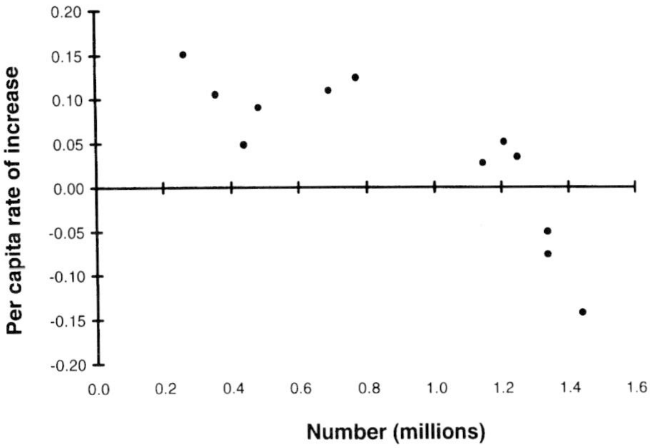

Fig. 1. Wildebeest per capita rate of increase (r) as a function of population size N (data from Sinclair 1995a) ($r = 0.186 - 0.166N$, n = 12, $P < .002$).

if they do not then variability can be accounted for by the intrinsic properties of the species through rate of increase (supporting the 'strong DD hypothesis'); while if they do then additional stochasticity must derive from the environment, supporting the 'weak DD hypothesis'.

Here I examine our knowledge of mammal populations and ask four questions. First, what is the nature of population variability? Second, what evidence is there for regulation? Third, what are the causes of regulation? and fourth, how does regulation affect the conservation of small populations?

Stability of mammal populations and life history theory

Mammal species range in size over seven orders of magnitude from the 2 g shrews to the 150 tonne blue whales. This is a range greater than any other vertebrate group (including the known extinct reptiles) and probably all invertebrate groups too. Mammals being warm-blooded, are constrained by the physical problems of heat loss at smaller sizes and heat load at larger sizes due to their surface area – volume ratios. Thus, small species with their high metabolic requirements (Brody 1945; Kleiber 1961) are sensitive to environmental changes in food supply and temperature to a greater extent than are large species. Such physiological and nutritional sensitivity with size is reflected in the stability of populations and their response to environmental change.

Over a period of time populations will tend to fluctuate around a mean rate of increase of zero. When conditions improve (e.g. through a milder climate, more food, fewer predators) the population will increase either through an increase in fecundity or a decrease in mortality. The rate of increase is determined by two things, extrinsic factors such as the amount of food available, and intrinsic biological characteristics which convert the extra food into increased reproduction and lower mortality. When external factors are not limiting, then the potential maximum rate of increase, (the intrinsic rate of increase, r_m, (Fisher 1930)) is determined by species specific biological features. Using data in Appendix 2, Fig. 2 illustrates for mammals that r_m is related to body weight by

$$r_m = 1.375\, W^{-.315} \qquad (1)$$

where W is mean adult live weight of females in kilograms, and r_m is the maximum instantaneous rate over a year. Since these values of r_m were measured under field conditions, they are approximate and show variability in Fig. 2. However, they approach the expected scaling of –0.25 for body weight (Charnov 1993), and are similar to other data for mammals (Caughley and Krebs 1983) and invertebrates and vertebrates combined (Blueweiss *et al.* 1978).

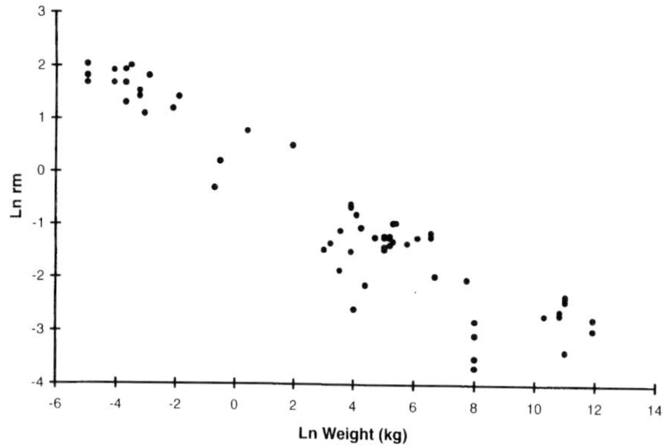

Fig. 2. Natural log of the intrinsic rate of increase (r_m) as a function of natural log body weight of species given in Appendix 2.(Ln r_m = 0.319 - 0.315 Ln*W*, n = 61, *P*<.000).

Since r_m is a biological feature of a species, it follows that smaller species should respond to environmental change faster than large species. Thus, smaller species should show a greater degree of population variability over time, and superficially this would seem to be so. Rodent populations appear to fluctuate to a greater extent than do those of elephants. Demonstrating this relationship, however, has had its problems (Anderson *et al.* 1982; Glazier 1986; Ostfeld 1988; Pimm and Redfearn 1988; McArdle *et al.* 1990; McArdle and Gaston 1993; Xia and Boonstra 1992; Hanski *et al.* 1994). Previous work on population variability has used the Coefficient of Variation (CV) from a series of censuses. This measure, however, can be distorted by both bias errors (e.g. the census does not represent the population) and sampling errors. Some of these problems do not apply to mammals – sample censuses, even of small mammals, usually do represent the population on a regional level. Others do apply – it is difficult to compare data from true counts with those using indices of density, activity etc. In addition, CV does not account for differences in time intervals between censuses – one cannot compare ten counts of a rodent population at six month intervals with ten censuses of a large ungulate counted irregularly between one and five years. What we are interested in is the rate of change in population, and this can be measured by the multiplication rate per unit time. In order to have a symmetrical value for decrease as well as increase, the instantaneous rate of change between censuses adjusted to an interval of one year (r_t) is given by

$$r_t = [\text{Ln}\ (N_{n+t} / N_n)] / t \qquad (2)$$

where N_n and N_{n+t} are consecutive counts, and t is the interval in years between them.

A further issue that arises when comparing species is the difference in generation time. Again some rodents can have two, even three generations per year, elephants one every quarter century. Thus, the instantaneous rate of change (r_t) should be multiplied by generation time T. This is the mean age in years of reproductive females. In data sets where there is little or no long term trend in population, r_t is approximately zero and variability can be measured directly by the standard deviation, SD (r_t). This allows comparisons between different types of data.

Appendix 3 presents estimates of r_t, SD (r_t) and T for a range of different sized mammal species, calculated from published data sets. Figure 3 shows the relationship between Ln [SD (r_t)] and Ln body weight (W) in kg. There is a clear negative relationship between population variability over a year and body weight such that

$$\text{SD}\ (r_t) = 0.805\ W^{-.316} \qquad (3)$$

Population variability weighted by generation time shows no relationship with body weight (Fig. 4). This is because generation time is also related to body weight (Millar and Zammuto 1983) by

$$T = 1.74\ W^{0.27} \qquad (4)$$

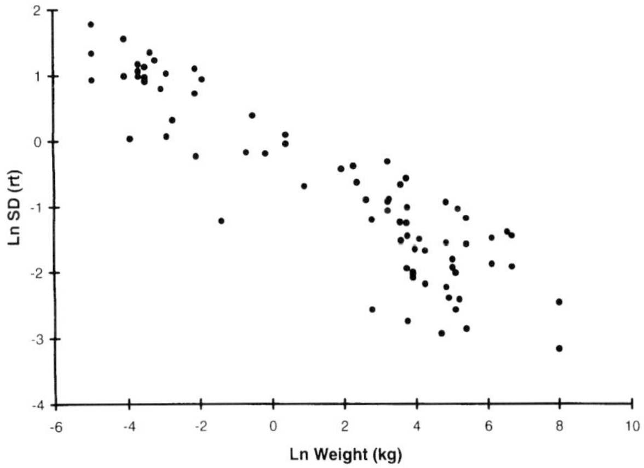

Fig. 3. Natural log of the Standard Deviation of r_t as a function of natural log body weight of mammal species given in Appendix 3. The instantaneous rate of increase (r_t) is calculated over one year. (Ln SD (r_t) = -0.207 - 0.316 Ln*W*, n = 79, *P*<.000).

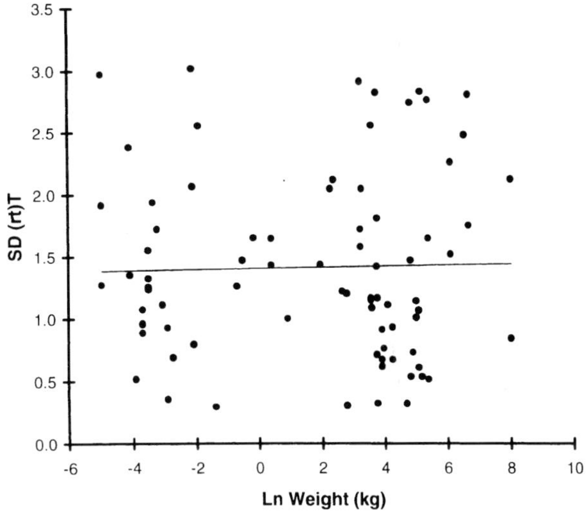

Fig. 4. The product of Standard Deviation of r_t and generation time (*T*) is a constant with respect to body weight. Data from Appendix 3. (SD (r_t) *T* = 1.407 + 0.004 Ln*W*, n = 79, *P*>.8).

Because r_m and *T* are inversely related the exponents in Eqns 1 and 4 have opposite sign and they tend to cancel each other out (Caughley 1977; Fowler 1988). Thus, both $r_m T$ and SD $(r_t) T$ are independent of body weight.

The significance of this result is that the degree of variability in a population is the same for all species, irrespective of size, relative to their generation time. Thus, small species are not intrinsically more variable than larger ones. This result is predicted by life history theory. Species can adapt to perturbations in their environment either by improving reproduction (*r* species) or by improving survivorship (*K* species). However, species cannot do both simultaneously, one parameter is improved only at the cost of the other. Due to physical constraints of body size small species adapt by increasing reproduction, so survivorship is low. These adaptations result in increased population variability in absolute time (in this case one year) as these species respond to short-term seasonal fluctuations in the environment (Fig. 3). In contrast, large species cannot increase reproduction but can improve survivorship through physiological adaptations. These species have a buffer against environmental changes. Thus, the degree of population decrease that a population shows is as much an adaptation, through survivorship, as is population increase through reproduction. Furthermore, what ever approach a species takes in adapting to environmental variations, the optimal solution (which is the product of reproduction and survivorship) should be about the same across all species. This prediction is confirmed in Fig. 4. The mathematical derivation of this relationship is given by M. Ritchie in Appendix 1.

Population cycles

Over the animal kingdom regular periodic fluctuation in population size (cycles) is extremely rare. Mammals are unusual in having several species which exhibit population cycles. Cyclicity is best demonstrated by autocorrelation, spectral, or phase-shift analysis (Finerty 1980; Garsd and Howard 1982; Henttonen *et al.* 1985; Oksanen 1990; Sinclair *et al.* 1993), but this requires very long data sets, as for some microtine rodents and the snowshoe hare (*Lepus americanus*). Usually such data are unavailable and to avoid this problem an index of cyclicity, *S*, has been calculated as

$$S = \sqrt{\frac{\left[\sum\left(\log N_i - \overline{\log N_i}\right)^2\right]}{n-1}} \tag{5}$$

where N_i is an annual index of density and n is the number of years in the sample (Stenseth and Flamstad, 1980; Hansson and Henttonen 1985). Amongst small rodents and insectivores there is a continuum in the degree of cyclicity ranging from no periodicity to strong periodicity (Krebs 1979; Taitt and Krebs 1985; Hansson and Henttonen 1985; Sandell *et al.* 1991). The value of *S* is spuriously increased in short data sets. This aside, S is perhaps a better reflection of amplitude than of the regularity of the cycle period (Sandell *et al.* 1991).

True population cycles are unrecorded in tropical latitudes and are a feature of northern rather than southern latitudes. For the vole *Clethrionomys glareolus* in Fennoscandia, *S* increases with latitude (Fig. 5; Hansson and Henttonen 1985). Turchin (1993) found that temperate microtine populations showed stability while northern populations showed intrinsic fluctuations that were amplified by environmental noise. Most small mammals show cycles in the northern holarctic (Finerty 1980; Henttonen *et al.* 1985) with the characteristic periodicity being 3–4 years for voles, shrews and their predators (Marcstrom *et al.* 1990; Korpimaki and Norrdahl 1991; Korpimaki *et al.* 1991; Hornfeldt 1994) and 10 years for the snowshoe hare and its predators (Keith 1963; Sinclair *et al.*1993). Peterson *et al.* (1984), following Calder (1983), have suggested that cycle period (C) is related to body weight in kilograms (*W*) by

$$C = 8.15\, W^{.26} \tag{6}$$

Equation 6 suggests that frequency of the cycle ($1/C$) is directly related to r_m and is an intrinsic biological property of the species (Calder 1983; Cockburn 1988). Again, from Eqns 4 and 6, the ratio T/C is a constant relative to body weight. Thus, small mammals are not intrinsically more cyclic than large mammals, the data for the former are simply more abundant and more visible.

Chaos

The degree of variability in a population could be related to the strength of negative feedback, or density dependence, acting on the population. This has been modelled by May (1975) and Bellows (1981) for simple systems with discrete breeding seasons. A population *N* at time $t + 1$ is given by

$$N_{t+1} = N_t R \,/\, [1 + (aN_t)^b] \tag{7}$$

where $R = N_{t+1} / N_t$, the net reproductive rate (or finite rate of increase), $a = (R - 1)/K$, and b is a constant. K is the carrying capacity in the logistic equation. Mortality, k, is given by

$$k = \mathrm{Log}\, [1 + (aN_t)^b] \tag{8}$$

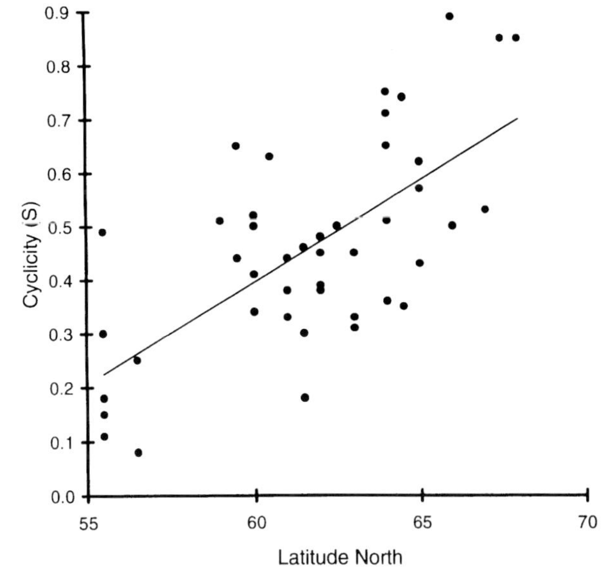

Fig. 5. A measure of cyclicity in population numbers of small mammal species (*S*) increases with latitude in Fennoscandia (from Hansson and Henttonen 1985).

When *k* is plotted against Log N_t, *b* is the slope and a measure of the strength of density dependence. In Eqn 7 when $b = 1$, density dependence is exactly compensating, the population equilibrates monotonically at all values of *R*, and there is no fluctuation. With higher values of *b* and *R*, populations first show damped oscillations, then stable limit cycles, and finally at high values of *b* (>4) and *R* (>10) chaotic behaviour. Chaos results in cyclic fluctuations which are predictable in the short term but not over longer timescales (Hastings and Powell 1991) because both the periodicity and amplitude change. Chaos is produced when density dependence is strongly overcompensating ($b >> 1$). Hanski *et al.* (1993) and Turchin (1993) modelled microtine rodent fluctuations and those of their mustelid predators and produced chaotic oscillations similar to field data. The Soay sheep population on St. Kilda, UK (Grenfell *et al.* 1992) also exhibits strong overcompensatory density dependent mortality and fluctuations which appear to be chaotic.

In general, one would expect to see chaotic fluctuations in species with a finite rate of increase $R > 4$ or $r_m > 1.4$. This value in Eqn 1 predicts that such fluctuations should be seen in species with body weights less than 950 g. Similarly, stable limit cycles should be found in species with $r_m > 0.7$ and body weights less than about 8.5 kg. Empirical data support these predictions. For mammals larger than this one should expect to see damped oscillations after perturbations in the smaller species and monotonic dampening for the largest species. Thus, European beavers (*Castor fiber*) appear to have exhibited an eruption and subsequent oscillation following their reintroduction to Sweden (Hartman 1994), and similar behaviour is seen in muskrats, *Ondatra zibethicus* (Hengeveld 1989), and North American beaver, *Castor canadensis* (Johnston and Naiman 1990). In contrast, large ungulates erupting where dispersal is not impeded show a monotonic levelling out, as in wildebeest and African buffalo (Sinclair 1977, 1995a), wood bison (Larter 1994) and Arabian oryx (Stanley-Price 1989). Where large mammal populations are constrained, as reindeer were on St. Matthew island of Alaska (Klein 1968), or introduced as exotics where indigenous prey are not adapted to the invader (e.g. thar in New Zealand; Caughley 1970), then oscillations can occur, sometimes leading to extinction.

In summary, the review of mammal population fluctuations suggests that variability can be accounted for largely by intrinsic biological properties of the species, in particular r_m, *T* and *W*.

These biological properties lead both to differential sensitivity and responsiveness to environmental stochasticity, (i.e. small species respond more rapidly to environmental fluctuations), and to inherent oscillations (cycles, chaos) due to overcompensating density dependence.

Density dependence and regulation

In general, the most robust evidence for density dependence and regulation comes from experimental perturbations of a population (Sinclair 1989). As the population rebounds density dependent mortality can be detected, as seen for example in the Yellowstone elk (Fig. 6a) (Bester and Lenglart 1982; Houston 1982), southern elephant seal (*Mirounga leonina*) (Fig. 6b), red deer (Clutton-Brock *et al.* 1985), reindeer (Skogland 1985), wildebeest (Fig. 1) and African buffalo

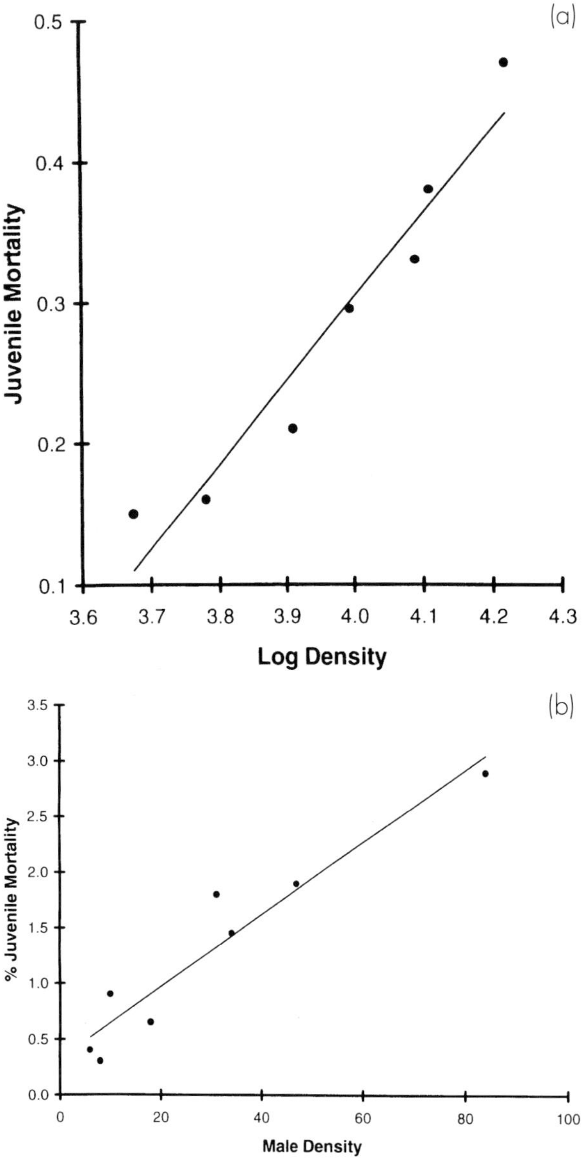

Fig. 6. Density dependence in juvenile life stages. (a) Juvenile mortality (*y*) as a function of log initial population size (*x*) in Yellowstone elk ($y = 0.595\ x - 2.078$, $P<.005$)(Houston 1982), (b) pup mortality (*y*) is a function of male density per km of colony (*x*) in southern elephant seals ($y = 0.03\ x + 0.32$, $P<.0002$)(Bester and Lenglart 1982).

(Sinclair 1977; Sinclair *et al.* 1985). There is a paucity of similar analyses for carnivores and small mammals. Detection of density dependence in census data from populations that have not been experimentally perturbed is obscured by sampling error and environmental stochastic events. However, some evidence in small mammals has now appeared from analytical and simulation models. These approaches suggest that in small mammals density dependence is delayed, as in voles (Turchin 1990; Hanski *et al.* 1993) and snowshoe hares (Fig. 7; Keith *et al.* 1977; Trostel *et al.* 1987), and such delays are the underlying demographic causes of cycles. The direct density dependence observed for large mammals results in damped oscillations and more stable population sizes (reviews in Fowler 1981, 1987; Fowler and Smith 1981).

Table 1 gives the frequency distribution of studies which record the life stage where density dependence occurs for various mammal groups. In large herbivores the majority of studies considered that regulation occurred either through reproductive loss and delayed maturation, for example in moose (Mech *et al.* 1987; McRoberts *et al.* 1995), red deer (Albon *et al.* 1983) and elephant (Laws *et al.* 1975), or through calf mortality as in elk (Fig. 6a) and reindeer (Skogland 1985) in the first stress season (winter or dry period). About 10% of the studies considered adult mortality was regulating as in African buffalo (Sinclair 1977) or caribou (Fancy *et al.* 1994). No studies found mortality of subadults (late juvenile) important. In contrast, the few studies of small mammals where life stage for regulation was recorded, indicated that subadult expulsion was the primary stage, followed by overwinter adult mortality. This latter mortality is implied in many of the studies of microtine cycles but is not explicitly stated as such, so I have omitted them here.

Amongst terrestrial carnivores there are too few records to draw conclusions. In marine mammals, however, by far the majority of studies consider reproduction to be density dependent, for example in killer whales (Olesiuk *et al.* 1990). The remaining studies reported that mortality of pups on the breeding grounds was regulating (Fig. 6b; cf. Harwood and Rohani this volume).

One must be cautious with interpreting these numbers because there are distortions in recording some of the life stages – early stages of small mammals and later stages in marine mammals are both difficult to observe. Despite this, it appears that a trend exists relating the life stage for regulation to body size: the largest mammals are regulated at the earliest stage (fecundity), intermediate sizes largely at the juvenile stage, and the smallest mammals at the late juvenile or adult stage.

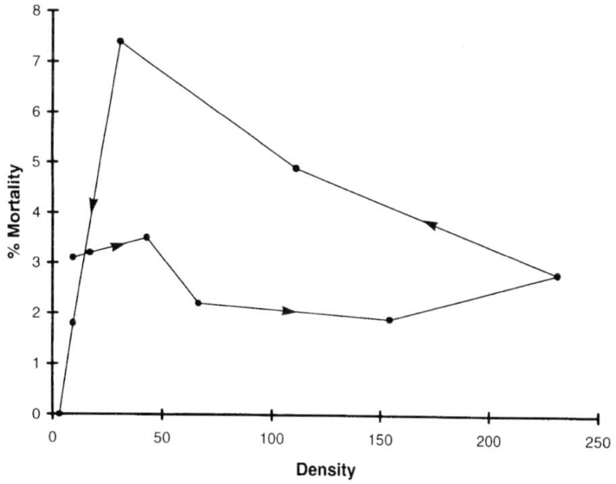

Fig. 7. Percent of adult snowshoe hares killed by great horned owls (in sequence from 1966 to 1975) shows an anticlockwise spiral with density (animals km^{-2}) indicating delayed density dependence (redrawn from Keith et al. 1977).

Table 1. Number (%) of reports of separate populations demonstrating density dependence at different life stages.

Group	Fecundity	Early Juvenile	Late Juvenile	Adult	Total
Small mammals	2 (10)	1 (5)	13 (62)	5 (24)	21
Carnivores	1 (14)	1 (14)	3 (43)	2 (29)	7
Large terrestrial herbivores	52 (48)	43 (39)	0	14 (13)	109
Large marine mammals	36 (71)	14 (27)	0	1 (2)	51

Sources: Those given in Sinclair (1989) as well as Armbruster and Lande 1993; Bester and Lenglart 1982; Choquenot 1991; Clutton-Brock *et al.* 1991; Crete and Huot 1993; Fancy *et al.* 1994; Fowler 1981, 1987, 1990; Fowler and Smith 1981; Frylestam 1979; Ginsberg *et al.* 1995; Hanski *et al.* 1993; Harwood and Prime 1978; Henttonen *et al.* 1989; Hornfeldt 1994; Kunkel and Mech 1994; Laurenson 1995; Lindstrom 1989; Mech *et al.* 1987; McRoberts *et al.* 1995; Oksanen and Oksanen 1992; Olesiuk *et al.* 1990; Ostfeld *et al.* 1993; Robertson *et al.* 1992; Sinclair 1995b; C. Packer *pers. comm.*

Table 2. Number (%) of reports of separate populations recording cause of density dependence. Population totals differ from those in Table 1 because not all studies record both life stage and cause. Disease includes both micro and macro parasites.

Group	Food	Predators	Disease	Social	Total
Small mammals	17 (27)	27 (42)	1 (2)	19 (29)	64
Terrestrial carnivores	5 (38)	1 (8)	4 (31)	3 (23)	13
Large terrestrial herbivores	77 (91)	6 (7)	2 (2)	0	85
Large marine mammals	8 (50)	4 (25)	1 (6)	3 (19)	16

Sources: as for Table 1 but in addition Abramsky and Tracy 1979; Anderson *et al.* 1981; Anderson and Trewhella 1985; Englund 1970; Erlinge *et al.* 1984; Henttonen *et al.* 1987; Krebs 1979; Messier 1994; Owen-Smith 1991; Peterson 1994; Trout *et al.* 1992; van Ballenberghe and Ballard 1994.

Causes of regulation

In an earlier review of this aspect (Sinclair 1989) I noted that most studies merely recorded the causes of mortality that limit populations. This is relatively uninformative unless one also knows whether the mortality is density dependent. The k - factor analysis (see Eqn 8; Varley and Gradwell 1968; Podoler and Rogers 1975) can tell us which mortalities account for most of the total population change annually. These are the 'key factors' and they can be either density independent or density dependent. For mammals such analyses are rare. Nevertheless, data on which causes produce density dependence are accumulating, and Table 2 presents the frequency of case histories where the cause of density dependent mortality was identified. Many but not all of the studies are those referenced in Table 1.

Food Supply, Predation and other Extrinsic Factors

For large mammals, both terrestrial and marine, the overwhelming cause of regulation is food supply. In marine mammals it is assumed that lack of food affects fertility, and growth and survival of juveniles. This has been suggested as the cause of decline in the northern fur seal, *Callorhinus ursinus*, but remains to be substantiated (Trites 1992). In terrestrial ungulates it is the food supply during the limiting season which is critical. For example, winter food supply regulates numbers of Soay sheep, red deer (Clutton-Brock *et al.* 1985, 1991) and reindeer (Skogland 1985) in Europe, and some deer in North America (Klein and Olsen 1960); while dry season food supply regulates donkeys and kangaroos in Australia (Choquenot 1991; Caughley *et al.* 1987) and buffalo, wildebeest and greater kudu in Africa (Sinclair 1975, 1977; Owen-Smith 1991). Although, food supply is also implicated in some microtine populations in Scandinavia (Oksanen and Oksanen 1992; Hornfeldt 1994), food does not seem to be the predominant cause of small mammal regulation. However, resources for such small species are difficult to measure and so food, as a cause of density dependence, is likely to be under represented. Food addition experiments (Taitt and Krebs 1981; Ims 1987; Sinclair *et al.* 1988; Krebs *et al.* this volume) show that microtines and leporids are responding to changes in food supply. For carnivores there are very few studies which specify the density dependent process and its cause. In Serengeti lions food shortage causes cub mortality (Schaller 1972; Bertram 1979). The wolves of Isle Royale appear to be tracking their moose food supply (Peterson 1994).

In contrast to food supply, predation is the predominant cause of regulation in small mammals. This has been particularly well studied in Fennoscandia (e.g. Erlinge *et al.* 1983; Hannson 1987; Korpimaki 1986, 1994; Korpimaki and Norrdahl 1991; Henttonen *et al.* 1987, 1989) for microtines and shrews. A predator exclusion experiment on the arctic coast of Canada (Reid 1995) points to predator regulation of collared lemmings (*Dicrostanyx groenlandicus*) in that non-cycling population. The snowshoe hare in North America is possibly regulated by predation (Trostel *et al.* 1987) or by an interaction of food and predation (Keith 1974; Krebs *et al.* 1986, this volume; Sinclair *et al.* 1988) or by food, stress, and predation combined (R. Boonstra *pers. comm.*). Foxes regulate European rabbits in Australia (Newsome *et al.* 1989; Pech *et al.* 1992)

In large African mammals, it appears that the smaller species are or may be regulated by predation while the larger species are regulated by food (Sinclair 1975, 1985, 1995b). In North America, although it has been proposed that ungulates are generally regulated by predators (Keith 1974; Connolly 1978; Bergerud 1980) the evidence is equivocal. Some small caribou herds may be regulated by wolves (Seip 1992) but the large migratory herds are not (Fancy *et al.* 1994) just as the migratory herds of wildebeest and white-eared kob are not (Fryxell and Sinclair 1988). Some moose populations are regulated by wolves and grizzly bears, *Ursos arctos* (Ballard *et al.* 1987; Gasaway *et al.* 1992; Messier 1994; van Ballenberghe and Ballard 1994), but the moose on Isle Royale are not (Messier 1991). Good information on other ungulates, small carnivores and marsupials is almost entirely lacking. However, the general hypothesis, at this stage, is that the relative roles of food supply and predation in regulating mammal populations switch over as body size goes from large to small. Even if food supply is the predominant factor, predation could influence mortality through avoidance of predation risk (McNamara and Houston 1987) as in snowshoe hares (Hik 1995) and wildebeest (Sinclair and Arcese 1995). In contrast to food supply, predators can act alone by keeping prey populations well below the limit set by food supply.

Disease and parasites usually act synergistically with food supply by accelerating the mortality rate (Sinclair 1977) and this could cause either greater stability or fluctuations through overcompensation (cf. Harwood and Rohani this volume). However, there are some recorded cases where disease acts alone as a density dependent factor (Hone 1994). Regulation occurred from tuberculosis in European badgers (Anderson and Trewella 1985), myxomatosis in European rabbits (Trout *et al.* 1992), and rabies in foxes (Anderson *et al.* 1981). The viral disease, rinderpest in African ruminants (Sinclair 1979; Dobson 1995) and canine distemper virus in both lions (C. Packer pers. comm.) and wild dogs, together with rabies in the latter (Ginsberg *et al.* 1995), may have acted this way but the evidence is circumstantial.

Social behaviour and intrinsic factors

Intrinsic factors, such as genes, physiology, and behaviour, are not alternatives on the same hierarchical level as the extrinsic factors reviewed above. Intrinsic and extrinsic factors have been viewed as alternatives in the past (Wynne-Edwards 1962; Lack 1966; Chitty 1967; Krebs 1978) and this has lead to unnecessary confusion. Perhaps the best way to unravel the misunderstanding is to start by considering that firstly, all biological characteristics of a species (the intrinsic factors) are adaptations that improve individual fitness. Secondly, such adaptations may be responding to the extrinsic regulatory factor directly to mitigate its effects and improve fitness. For example, red squirrels hoard food for winter use (behavioural adaptation) and ungulates put on fat for the same reason (physiological adaptation). Adaptations may also be responding to other selection pressures and thus indirectly affect an animal's response to the extrinsic regulating factor. For example, avoidance of predation increases starvation mortality mentioned above, and conversely feeding behaviour reduces vigilance and increases predation mortality. Territorial behaviour in ungulate males (for mating and reproductive success) increases predation rate relative to non-territory holding females (pers. obs.). These arguments apply to all adaptations (intrinsic factors) no matter how remotely related to the extrinsic regulating factor, because there is always a cost to each adaptation. Thirdly, such costs or consequences of adaptation may result in either an increase or decrease in mean population level relative to populations without such adaptations. Of course we usually cannot measure such differences because the comparison is not available.

Adaptations affecting the mean population level cannot be selected for through natural selection acting on individuals or their relatives. Therefore, it is not possible to select for a mean population size below the level determined by food, through natural selection acting only on individual behaviour, physiology or other intrinsic traits. This is because to achieve such a population result would require the evolution of 'spite'. This can be illustrated by considering the example of territoriality: if territories were initially an adaptation to sequester food and maximise individual reproductive success, then 'spite' would occur if a gene produced behaviour enlarging a territory beyond that necessary for maximum reproductive success. The spiteful gene would incur a cost of the extra defence, but a relative benefit by excluding other individuals who would not breed at all. However, unless all others were excluded from breeding, the remaining non-spiteful territory holders would have the double benefit accruing from those excluded by the spiteful territory holder plus having no cost. Thus, the non-spiteful gene would have the higher fitness and displace the spiteful gene. If all other territory holders are excluded then the argument reduces to a single spiteful individual in its territory with lower fitness, and a single non-spiteful individual elsewhere with higher fitness. The end result is the same. So far there is no conclusive evidence for spite in nature (Gadaghar 1993; Keller *et al.* 1994).

There are two important conclusions from this discussion. First, all behaviour and physiology will affect an individual's response to extrinsic regulating factors such as food supply, predation and disease, and so intrinsic factors are always involved in regulation – studies which reach only this conclusion are true but trivial. Secondly, intrinsic factors are complimentary in action to extrinsic factors, not alternatives. Intrinsic factors respond to the action of the extrinsic, and it is the combined result which determines the population outcome. Caughley and Krebs (1983) suggested that small species (< 30 kg) were regulated more by intrinsic factors, larger species more by extrinsic ones. I suggest that this dichotomy is not as distinct as originally proposed because all species are influenced by both types of factors. Rather I would amend their conclusion by suggesting that smaller species show intrinsic responses through behaviour while larger species respond through physiology.

I have previously discussed (Sinclair 1989) the old argument of group selection: a population with a gene for reduced reproduction at fixation should reach a level below that set by the food supply, and thereby should avoid damaging the supply and last longer than populations without such a gene (Wynne-Edwards 1962). Group selection is both unnecessary and unsupported, because in theory, natural selection should more than counteract group selection (Bell 1987), and in general observations do not support the two predictions above (Sinclair 1989).

The types of behaviour which have been cited as contributory to regulation (Table 2) are territoriality (carnivores, ungulates, some rodents such as squirrels) and dispersal in most mammal groups (Stenseth and Lidicker 1992). These mechanisms respond to increasing densities by excluding an increasing proportion of the population from sources of good food (or shelter and protection) and so produce proportionally higher mortality from starvation and predation in the population as a whole.

Population cycles

Because population cycles are a special characteristic of mammals I review the hypotheses for their cause. Early suggestions for disease or stress have not been substantiated. Chitty (1960) suggested that the decline phase of vole (*Microtus agrestis*) cycles was the outcome of intraspecific aggression during the previous peak phase. This suggestion lead to the 'Genetic hypothesis' (Chitty 1967, 1987; Krebs 1978) which proposes that the population is genetically polymorphic for aggression, with low reproducing but highly aggressive morphs at high density alternating with high reproducing, non-aggressive types at low density. The selection pressure for aggression is the degree of social interaction, having become emancipated from extrinsic factors in the past. Conceptually it is not clear how such a mechanism can evolve without resorting to the arguments of group selection or spite which are of dubious validity. Despite considerable research in the past thirty years there is no compelling empirical evidence to support the genetic component of the hypothesis (Krebs *et al.* 1973; Krebs and Myers 1974; Tamarin 1978; Boonstra and Boag 1987; Boonstra *et al.* 1994).

This research, however, has highlighted the important role of dispersal in population cycles (Krebs *et al.* 1973; Krebs and Myers 1974; Abramsky and Tracy 1979; Lidicker 1975; Stenseth and Lidicker 1992) and the distinction between species with dispersal in the increase phase versus those with dispersal in the peak phase. The dispersal is density dependent but how extrinsic factors are connected with the timing of dispersal remains speculative. However, Stenseth and Oksanen (1987) propose that cycles in microtines require an interaction of predators, herbivores and food supply. In addition, social behaviour causing dispersal in the herbivores is necessary to keep fluctuations in bounds that allow predators to persist. Hestbeck (1982, 1987) proposes another model involving different interactions of social behaviour and food supply depending on density and habitat type. Boonstra (1994) has proposed that delayed density dependent inhibition of maturation in juveniles causes a shift in age structure. In turn this reduces survival and reproduction and so numbers drop and remain low until age structure changes again, thus producing a cycle.

Turning to more extrinsic causes of cycles Haukioja (1980) and Haukioja *et al.* (1983) proposed that plant availability is delayed through secondary chemical defence and so cycles develop in the herbivore. However, subsequent work shows that phenolic secondary compounds do not produce herbivore cycles (Oksanen *et al.* 1987), and instead plant production cycles are a consequence of herbivore feeding (Oksanen and Erikson 1987). The plant defence hypothesis has been proposed for the snowshoe hare cycle (Bryant *et al.* 1991) but the data show, as in microtines, that plant cycles are a consequence of and not a cause of the hare cycle (Sinclair *et al.* 1988).

In Fennoscandia, evidence is accumulating that predation is the cause of small mammal cycles. Southern parts of this area have non-cyclic small mammals regulated by generalist predators (Erlinge *et al.* 1983). In northern areas predators may also be regulating together with food supply (Hansson and Henttonen 1985; Hornfeldt 1994). Specialist predators produce the cycles through delayed density dependence (Hanski *et al.* 1993) in both microtines (Hansson and Henttonen 1988) and shrews (Henttonen *et al.* 1989), a conclusion also reached for North American voles (Ostfeld *et al.* 1993). In Europe cycles are closely synchronised not only between various vole species and shrews (Angelstam *et al.* 1985; Marcstrom *et al.* 1990; Korpimaki and Norrdahl 1991) but also between voles and their specialist predator, the least weasel (*Mustela nivalis*) (Korpimaki *et al.* 1991), raptors and owls (Korpimaki and Norrdahl 1989, 1991). Synchrony does not have to be caused by a continuously present factor such as predation. It could be caused by some aperiodic environmental event (Leslie 1959). However, in this case one would see cycles in different species and in different areas drift out of phase only to become reset, not necessarily at the same phase at the next climatic event. Since this is not seen, it is highly improbable that random climatic events could continuously keep many species (with different r_m) in phase. Such interspecific synchrony is also strong evidence that social behaviour is not the sole cause of the cycle for there is no reason that different species with different intrinsic dynamics (r_m and cycle period are related to body size, see above) would end up with their cycles in phase.

In general, predation is implicated in synchronising the cycles of microtines. Predation may be interacting with food supply and social behaviour through dispersal (see Krebs *et al.* this volume).

Fluctuations and conservation

Caughley (1994) points out that small populations which are stable in the face of perturbations over a long period of time, and hence have evolved in such an environment, are less of a conservation problem than those species populations which have been reduced recently and which continue to have negative rates of increase. The issue is what causes this negative rate of increase. Theoretically, the existence of a density dependent response in a population should provide a buffer against extinction, because the lower the population the greater the per capita rate of increase (Fig. 1). However, some models show that even exponentially growing populations, when at low density, have a high probability of extinction in the presence of density independent catastrophic declines (Mangel and Tier 1994; Foley 1994).

For large mammals in particular Young (1994) has demonstrated that catastrophes causing 25%–70% mortality are relatively common (through starvation in herbivores, disease in carnivores). Thus, for conservation, if large mammal populations are reduced to low levels, then i) the above models may have validity, and ii) population viability estimates would be overly optimistic.

Apart from catastrophes, there are other processes which affect rates of increase at low population densities. There are many examples of introduced or remnant small populations, particularly in vertebrates, which do not increase (for example the North Atlantic right whale, *Eubalaena glacialis*; Kraus 1990) or do so only slowly (e.g. the Pacific sea otter) until they reach some higher number, after which they increase exponentially. Such population behaviour was exhibited by the reintroduced wood bison in northern Canada (Gates and Larter 1990; Larter 1994), and the artificially reduced plains bison and pronghorn antelope (Fig. 8a,b) in Yellowstone Park (Singer and Norland 1994) – the per capita rate of increase actually showed inverse density dependence at low population size (P = .034 for regression of r on pronghorn population size 165–500). This suggests that there is some 'threshold population' size below which the rate of increase is depressed. The reason that a depressed r_t was not exhibited by wildebeest in Fig. 1, is that the population was already well established and above the threshold size.

Reasons for low r_t at low population density are both environmental and intrinsic. Thus, even a small number of predators eating a constant number produce a depensatory (inverse density dependent) mortality resulting in a 'predator pit' (Peterman *et al.* 1979; Sinclair and Pech 1995). At very low population sizes the 'Allee effect' may occur where the sexes have difficulty in finding mates. This was modelled for the sperm whale (Botkin *et al.* 1980). In addition, even small random environmental mortalities can overwhelm the intrinsic rate of increase, and large mammals with low r_m are particularly prone to this (Mangel and Tier 1994).

Small mammals are also prone to extinctions because of their high population variability (Fig. 3).

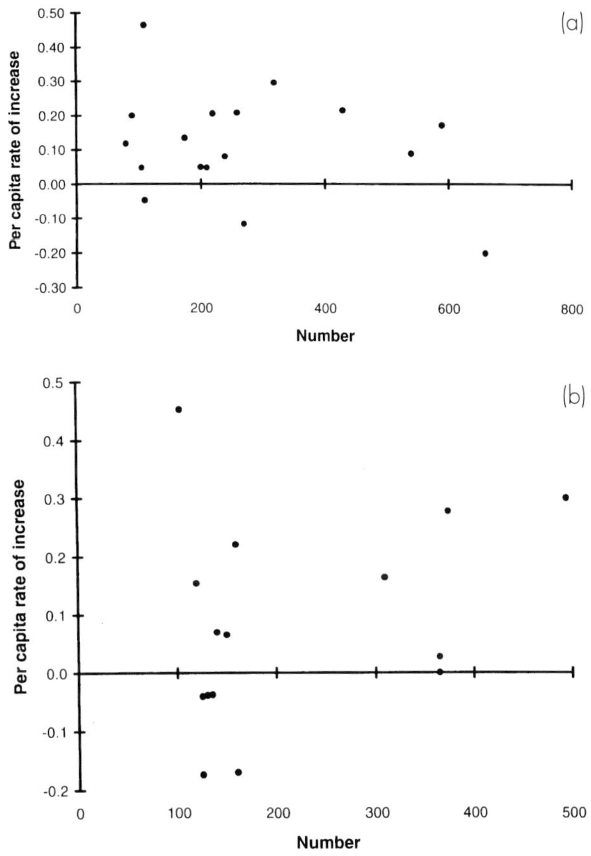

Fig. 8. Per capita rates of increase (r) are depressed at low densities in large mammal populations: (a) Yellowstone plains bison, (b) Yellowstone pronghorn antelope (data from Singer and Norland 1994).

Thus, on islands in Finland, extinction rates are inversely related to body weight in shrews, and this is ascribed to greater susceptibility to environmental fluctuations (Peltonen and Hanski 1991). Silva and Downing (1994) calculate that minimum viable densities (MVD) decrease as the –0.68 power of body weight. Consequently small mammal MVD are 1000 times higher than those of larger species. Although, small mammals are not intrinsically more variable (Fig. 4) they do exhibit a greater frequency of population fluctuation, and thus a greater probability of extinction in absolute time.

Since small populations have a depressed r_t and high probability of extinction, we either need to maintain large populations, or rely on several small interconnected populations. Thus, minimum reserve size for elephants must be 2600 km^2 for a 1% probability of extinction in 1000 years (Armbruster and Lande 1993). Although not all reserves need to be this size for other species, the figures suggest that the strategy of several smaller reserves with dispersal between them may be a more viable option. Small local populations may fluctuate, show chaotic behaviour, or even go extinct, but through dispersal with other subpopulations the whole 'metapopulation' remains extant. Density dependence is still necessary in the whole system, but metapopulation analysis suggests that linkage between subpopulations is either stabilising (Allen *et al.* 1993; Barcompte and Sole 1994) or causes cyclic and chaotic fluctuations depending on the number of subpopulations and the degree of linkage (Ruxton 1993, 1994). We now need more empirical data to test these predictions.

In conclusion, our present knowledge of population fluctuations and regulation suggests that populations reduced to small sizes but high density (those living in small habitat fragments of the original range) are likely to experience stochastic events and local extinction. This applies to all mammal species and is best overcome through dispersal. In many cases this will require deliberate human intervention. Populations reduced to low density within their range are likely to exhibit reduced r_t through both biotic and abiotic factors and so are also prone to extinction. Intervention in these cases should aim at achieving a higher density. This applies more to the larger species.

Conclusion

Populations of small mammals are neither intrinsically more variable nor subject to greater environmental perturbation relative to generation time, an observation predicted by life history theory. The greater population variability of small species in absolute time is a consequence of their greater r_m, an intrinsic property of species, responding to the same environmental perturbations experienced by all species. The greater r_m of small species results in overcompensating density dependence, and this can produce cycles or chaotic oscillations. Density dependence appears to act in later life stages of small mammals, but in progressively earlier life stages as species become larger. The intrinsic properties of variability affect a species vulnerability to extinction, and hence our response to their conservation.

Acknowledgments

I thank Rudy Boonstra, John Harwood, David Hik, Wes Hochachka, Jim Hone, Erkki Korpimaki and Roger Pech for constructive critiques of the manuscript. I also thank Andrea Byrom, John Harwood and Tim Karels for information they provided. Tina Raudzus helped with literature search and Simon Mduma with computing. The Vertebrate Biocontrol Centre, Gungahlin, Canberra, funded my airfare to the Nicholson Centenary meeting.

References

Abramsky Z, Tracy CR (1979) Population biology of a 'noncycling' population of prairie voles and a hypothesis on the role of migration in regulating microtine cycles. *Ecology* **60**:349–361

Albon SD, Mitchell B, Staines BW (1983) Fertility and body weight in female red deer: A density-dependent relationship. *J Anim Ecol* **52**:969–980

Allen JC, Schaffer WM, Rosko D (1993) Chaos reduces species extinction by amplifying local population noise. *Nature* **364**:229–232

Anderson RM, Gordon DM, Crawley, MJ, Hassell MP (1982) Variability in the abundance of animal and plant species. *Nature* **296**:245–248

Anderson RM, Jackson HC, May RM, Smith AM (1981) Population dynamics of fox rabies in Europe. *Nature* **289**:765–771

Anderson RM, Trewhella W (1985) Population dynamics of the badger (*Meles meles*) and the epidemiology of bovine tuberculosis (*Myobacterium bovis*). *Phil Trans R Soc Lond B* **310**:327–381

Angelstam P, Lindstrom E, Widen P (1985) Synchronous short-term population fluctuations of some birds and mammals in Fennoscandia -occurrence and distribution. *Holarctic Ecol* **8**:285–298

Armbruster P, Lande R (1993) A population viability analysis for African elephant (*Loxodonta africana*): How big should reserves be? *Conserv Biol* **7**:602–610

Ballard WB, Whitman JS, Gardner CL (1987) Ecology of an exploited wolf population in south-central Alaska. *Wildl Monogr* **98**:1–54

Bascompte J, Sole RV (1994) Spatially induced bifurcations in single species population dynamics. *J Anim Ecol* **63**:256–264

Baskin Y (1993) Blue whale population may be increasing off California. *Science* **260**:287

Batzli GO, Pitelka FA (1971) Condition and diet of cycling populations of the California vole, *Microtus californicus*. *J Mammal* **52**:141–163

Bayliss P (1987) Kangaroos dynamics. In: Caughley G, Shepherd N, Short J (eds) *Kangaroos: their ecology and management in the sheep rangelands of Australia.* Cambridge Univ Press, Cambridge, UK, pp 119–134

Bell G (1987) (Review of) Evolution through group selection. V.C. Wynne-Edwards. *Heredity* **59**:145–147

Bellows TS (1981) The descriptive properties of some models for density dependence. *J Anim Ecol* **50**:139–156

Bergerud AT (1980) Review of the population dynamics of caribou and wild reindeer in North America. In: Reimers E, Gaare E, Skjenneberg S (eds) *Proc 2nd Int Reindeer/Caribou Symp.* Direktoratet for vilt og ferskvannsfisk, Trondheim, pp 556–581

Bergerud AT, Nolan MJ, Curnew K, Mercer WE (1983) Growth of the Avalon Penninsula, Newfoundland, caribou herd. *J Wildl Manage* **47**:989–998

Bertram BCR (1979) Serengeti predators and their social systems. In: Sinclair ARE, Norton-Griffiths M (eds) *Serengeti: Dynamics of an ecosystem.* Univ Chicago Press, Chicago, USA, pp 221–248

Best PB (1993) Increase rates in severely depleted stocks of baleen whales. *ICES J Mar Sci* **50**:169–186

Bester MN, Lenglart P-Y (1982) An analysis of the southern elephant seal *Mirounga leonina* breeding population at Kerguelen. *S Afr J Antarct Res* **12**:11–16

Blueweiss L, Fox H, Kudzma V, Nakashima D, Peters R, Sams S (1978) Relationships between body size and some life history parameters. *Oecologia* **37**:257–272

Boonstra R (1994) Population cycles in microtines: the senescence hypothesis. *Evol Ecol* **8**:196–219

Boonstra R, Boag PT (1987) A test of the Chitty hypothesis: inheritance of life history traits in meadow voles, *Microtus pennsylvanicus. Evolution* **41**:929–947

Boonstra R, Hochachka WM, Pavone L (1994) Heterozygosity, aggression, and population fluctuations in meadow voles (*Microtus pennsylvanicus*). *Evolution* **48**:1350–1363

Botkin DB, Schimel DS, Wu LS, Little WS (1980) Some comments on density dependent factors in sperm whale populations. *Rep Int Whal Commn* (Special Issue 2), pp 83–88

Boutin S, Krebs CJ, Boonstra R, Dale MRT, Hannon SJ, Martin K, Sinclair ARE, Smith JNM, Turkington R, Blower M, Byrom A, Doyle FI, Doyle C, Hik D, Hofer L, Hubbs A, Karels T, Murray DL, Nams V, O'Donoghue M, Rohner C, Schweiger S (1995) Population changes of the vertebrate community during a snowshoe hare cycle in Canada's boreal forest. *Oikos* **74**:69–80

Brody S (1945) *Bioenergetics and growth.* Reinhold, New York, USA

Broten MD, Said M (1995) Population trends of ungulates in and around Kenya's Masai Mara Reserve In: Sinclair ARE, Arcese P (eds) *Serengeti II: Dynamics, management and conservation of an ecosystem.* Univ Chicago Press, Chicago, USA, pp 169–193

Bryant JP, Kuropat PJ, Reichardt PB, Clausen TP (1991) Controls over the allocation of resources by woody plants to chemical antiherbivore defense. In: Palo RT, Robbins CT (eds) *Plant defenses against mammalian herbivory,* CRC Press, Boca Raton, USA, pp 83–102

Calder WA (1983) An allometric approach to population cycles of mammals. *J Theor Biol* **100**:275–282

Calder WA (1984) *Size, function and life history.* Harvard Univ Press, Cambridge, MA, USA

Caughley G (1970) Eruption of ungulate populations, with emphasis on Himalayan thar in New Zealand. *Ecology* **51**:53–72

Caughley G (1977) *Analysis of vertebrate populations.* John Wiley and Sons, London, UK

Caughley G (1987) Ecological relationships. In: Caughley G, Shepherd N, Short J (eds) *Kangaroos: their ecology and management in the sheep rangelands of Australia.* Cambridge Univ Press, Cambridge, UK, pp 159–187

Caughley G (1994) Directions in conservation biology. *J Anim Ecol* **63**:215–244

Caughley G, Gunn A (1993) Dynamics of large herbivores in deserts: kangaroos and caribou. *Oikos* **67**:47–55

Caughley G, Krebs CJ (1983) Are big mammals simply little mammals writ large? *Oecologia* **59**:7–17

Caughley G, Shepherd N, Short J (eds) (1987) *Kangaroos: their ecology and management in the sheep rangelands of Australia.* Cambridge Univ Press, Cambridge, UK

Charnov EL (1993) *Life history invariants.* Oxford Univ Press, Oxford, UK

Chitty D (1960) Population processes in the vole and their relevance to general theory. *Can J Zool* **38**:99–113
Chitty D (1967) The natural selection of self-regulation behaviour in animal populations. *Proc Ecol Soc Aust* **2**:51–78
Chitty D (1987) Social and local environment of the vole *Microtus townsendii*. *Can J Zool* **65**:2555–2566
Choquenot D (1991) Density-dependent growth, body condition, and demography in feral donkeys: testing the food hypothesis. *Ecology* **72**:805–813
Clutton-Brock TH, Lonergan ME (1994) Culling regimes and sex ratio biases in highland red deer. *J Appl Ecol* **31**:521–527
Clutton-Brock TH, Major M, Guiness FE (1985) Population regulation in male and female red deer. *J Anim Ecol* **54**:831–846
Clutton-Brock TH, Price OF, Albon SD, Jewell PA (1991) Persistent instability and population regulation in Soay sheep. *J Anim Ecol* **60**:593–608
Cockburn A (1981) Population processes of the silky desert mouse *Pseudomys apodemoides* (Rodentia), in mature heathlands. *Aus J Zool* **8**:499–514
Cockburn A (1988) *Social behaviour in fluctuating populations*. Croon Helm, London, UK
Commonwealth of Australia (1984) *Kangaroo management programs of the Australian states*. Commonwealth of Australia, Canberra
Connolly GE (1978) Predation and predator control. In: Schmidt JL, Gilbert DL (eds) *Big game of North America: Ecology and management*. Stackpole Books, Harrisburg, PA, USA pp 369–394
Couturier S, Brunelle J, Vandal D, St-Martin G (1990) Changes in the population dynamics of the George River caribou herd, 1976–87. *Arctic* **43**:9–20
Crete M, Huot J (1993) Regulation of a large herd of migratory caribou: summer nutrition affects calf growth and body reserves of dams. *Can J Zool* **71**:2291–2296
Cumming DHM (1983) The decision-making framework with regard to the culling of large mammals in Zimbabwe. In: Owen-Smith RN (ed) *Management of large mammals in African conservation areas*. Haum, Pretoria, South Africa, pp173–186
Dobson A (1995) The ecology and epidemiology of rinderpest virus in Serengeti and Ngorongoro Conservation Area. In: Sinclair ARE, Arcese P (eds) *Serengeti II: Dynamics, management and conservation of an ecosystem*. Univ Chicago Press, Chicago, USA, pp 485–505
Damuth J (1991) Of size and abundance. *Nature* **351**:268–269
Englund J (1970) *Some aspects of reproduction and mortality rates in Swedish red foxes (Vulpes vulpes), 1961–63 and 1966–69*. Viltrevy : 1–82
Erlinge S, Goransson G, Hansson L, Hogstedt G, Liberg O, Nilsson T, von Schantz T, Sylven M (1983) Predation as a regulating factor in small rodent populations in southern Sweden. *Oikos* **40**:36–52
Erlinge S, Frylestam B, Goransson G, Hogstedt G, Liberg O, Loman J, Nilsson IN, von Schantz T, Sylven M (1984) Predation on brown hare and ring-necked pheasant populations in southern Sweden. *Holarctic Ecol* **7**:300–304
Fancy SG, Whitten KR, Russell DE (1994) Demography of the Porcupine caribou herd, 1983–1992. *Can J Zool* **72**:840–846
Finerty JP (1980) *The population ecology of cycles in small mammals*. Yale Univ Press, New Haven, CN, USA
Fisher RA (1930) *The genetical theory of national selection*. Clarendon Press, Oxford, UK
Foley P (1994) Predicting extinction times from environmental stochasticity and carrying capacity. *Conserv Biol* **8**:124–137
Foster JB, Kearney D (1967) Nairobi National Park census, 1966. *E Afr Wildl J* **5**:112–120
Foster JB, McLaughlin R (1968) Nairobi National Park game census, 1967. *E Afr Wildl J* **6**:152–154
Fowler CW (1981) Density dependence as related to life history strategy. *Ecology* **62**:602–610
Fowler CW (1987) A review of density dependence in populations of large mammals. In: Genoways H (ed) *Current mammalogy*. Plenum Press, New York, USA, pp 401–441
Fowler CW (1988) Population dynamics as related to rate of increase per generation. *Evol Ecol* **2**:197–204
Fowler CW (1990) Density dependence in northern fur seals (*Callorhinus ursinus*). *Marine Mamm Sci* **6**:171–195

Fowler CW, Smith TD (eds) (1981) *Dynamics of large mammal populations.* John Wiley and Sons, New York, USA

Frylestam B (1979) Structure, size, and dynamics of three European hare populations in southern Sweden. *Acta Theriol* **33**:449–464

Fryxell JM, Sinclair ARE (1988) Causes and consequences of migration by large herbivores. *Trends Ecol & Evol* **3**:237–241

Gadagkar R (1993) Can animals be spiteful? *Trends Ecol & Evol* **8**:232–234

Garsd A, Howard WE (1982) Microtine population fluctuations: An ecosystem approach based on time series analysis. *J Anim Ecol* **51**:225–234

Gasaway WC, Boertje RD, Grangaard DV, Kelleyhouse DG, Stephenson RO, Larsen DG (1992) The role of predation in limiting moose at low densities in Alaska and Yukon and implications for conservation. *Wildl Monogr* **120**:1–59

Gates CC, Larter NC (1990) Growth and dispersal of an erupting large herbivore population in northern Canada: The Mackenzie wood bison (*Bison bison athabascae*). *Arctic* **43**:231–238

Ginsberg JR, Alexander KA, Creel S, Kat PW, McNutt JW, Mills MGL (1995) Handling and survivorship in the African wild dog (*Lycaon pictus*): A survey of five ecosystems. *Conserv Biol* **9**:665–674

Glazier DS (1986) Temporal variability of abundance and the distribution of species. *Oikos* **47**:309–314

Grenfell BT, Price OF, Albon SD, Clutton-Brock TH (1992) Overcompensation and population cycles in an ungulate. *Nature* **355**:823–826

Halle S, Lehmann U (1992) Cycle-correlated changes in the activity behaviour of field voles, *Microtus agrestis*. *Oikos* **64**:489–497

Hanley JP, Bygott JD, Packer C (1995) Ecology, demography, and behaviour of lions in two contrasting habitats: Ngorongoro Crater and the Serengeti plains. In: Sinclair ARE, Arcese P (eds) *Serengeti II: Dynamics, management and conservation of an ecosystem* Univ Chicago Press, Chicago, USA, pp 315–331

Hanski I, Turchin P, Korpimaki E, Henttonen H (1993) Population oscillations of boreal rodents: regulation by mustelid predators leads to chaos. *Nature* **364**:232–235

Hanski I, Henttonen H, Hansson L (1994) Temporal variability and geographical patterns in the population density of microtine rodents: A reply to Xia and Boonstra. *Am Nat* **144**:329–342

Hansson L (1987) An interpretation of rodent dynamics as due to trophic interactions. *Oikos* **50**:308–318

Hansson L, Henttonen H (1988) Rodent dynamics as community processes. *Trends Ecol & Evol* **3**:195–200

Hansson L, Henttonen H (1985) Gradients in density variations of small rodents: the importance of latitude and snow cover. *Oecologia* **67**:394–402

Hartman G (1994) Long-term populaton development of a reintroduced beaver (*Castor fiber*) population in Sweden. *Conserv Biol* **8**:713–717

Harwood J, Prime JH (1978) Some factors affecting the size of British grey seal populations. *J Appl Ecol* **15**:401–411

Hastings A, Powell T (1991) Chaos in a three-species food chain. *Ecology* **72**:896–903

Hatter IW, Janz DW (1994) Apparent demographic changes in black-tailed deer associated with wolf control on Northern Vancouver Island. *Can J Zool* **72**:878–884

Haukioja E (1980) On the role of plant defences in the fluctuation of herbivore populations. *Oikos* **35**:202–213

Haukioja E, Kapiainen K, Niemela P, Tuomi J (1983) Plant availability hypothesis and other explanations of herbivore cycles: Complimentary or exclusive alternatives? *Oikos* **40**:419–432

Heard DC, Ouellet J-C (1994) Dynamics of an introduced caribou population. *Arctic* **47**:88–95

Hengeveld R (1989) *Dynamics of biological invasions.* Chapman and Hall, London, UK

Henttonen H, Haukisalmi V, Kaikusalo A, Korpimaki E, Norrdahl K, Skaren UAP (1989) Longterm population dynamics of the common shrew *Sorex araneus* in Finland. *Ann Zool Fennici* **26**:349–356

Henttonen H, McGuire AD, Hansson L (1985) Comparisons of amplitudes and frequencies (spectral analysis) of density variations in long-term data sets of *Clethrionomys* species. *Ann Zool Fennici* **22**:221–227

Henttonen H, Oksanen T, Jortikka A, Haukisalmi V (1987) How much do weasels shape microtine cycles in the northern Fennoscandian Taiga? *Oikos* **50**:353–365

Heske EJ, Brown JH, Mistry S (1994) Long-term experimental study of a chihuahuan desert rodent community: 13 years of competition. *Ecology* **75**:438–445

Hestbeck JB (1982) Population regulation of cyclic mammals: The social fence hypothesis *Oikos* **39**:157–163

Hestbeck JB (1987) Multiple regulation states in populations of small mammals: A state transition model. *Am Nat* **129**:520–532

Hik D (1995) Does risk of predation influence population dynamics? Evidence from the cyclic decline of snowshoe hares. *Wildl Res* **22**:115–129

Hofer H, East M (1995) Population dynamics, population size, and the community system of Serengeti spotted hyenas. In: Sinclair ARE, Arcese P (eds) *Serengeti II: Dynamics, management and conservation of an ecosystem* Univ Chicago Press, Chicago, USA, pp 332–363

Hone J (1994) *Analysis of vertebrate pest control*. Cambridge Univ Press, Cambridge, UK

Hornfeldt B (1994) Delayed density dependence as a determinant of vole cycles. *Ecology* **75**:791–806

Houston DB (1982) *The northern Yellowstone elk*. Macmillan, New York, USA

Ims RA (1987) Responses in spatial organization and behaviour to manipulations of the food resource in the vole, *Clethrionomys rufocanus*. *J Anim Ecol* **56**:585–596

Jimenez JE, Feinsinger P, Jaksic FM (1992) Spatiotemporal patterns of an irruption and decline of small mammals in northcentral Chile. *J Mamm* **73**:356–364

Johnston CA, Naiman RJ (1990) Aquatic patch creation in relation to beaver population trends. *Ecology* **71**:1617–1621

Keith LB (1963) *Wildlife's ten-year cycle*. Univ Wisconsin Press, Madison, USA

Keith LB (1974) Some features of population dynamics in mammals. *Proc Int Congr Game Biol, Stockholm* **11**:17–58

Keith LB, Todd AW, Brand CJ, Adamcik RS, Rusch DH (1977) An analysis of predation during a cyclic fluctuation of snowshoe hares. *Proc Int Congr Game Biol* **13**:151–175

Keller L, Milinski M, Frischknecht M, Perrin N, Richner H, Tripet F (1994) Spiteful animals still to be discovered. *Trends Ecol & Evol* **9**:103

Kleiber M (1961) *The fire of life*. John Wiley and Sons, New York, USA

Klein DR (1968) The introduction, increase, and crash of reindeer on St. Matthew Island. *J Wildl Manage* **32**:350–367

Klein DR, Olsen ST (1960) Natural mortality patterns of deer in southeast Alaska. *J Wildl Manage* **24**:80–88

Korpimaki E (1986) Predation causing syndronous decline phases in microtine and shrew populations in western Finland. *Oikos* **46**:124–127

Korpimaki E (1994) Rapid or delayed tracking of multiannual vole cycles by avian predators? *J Anim Ecol* **63**:619–628

Korpimaki E, Norrdahl K (1989) Avian and mammalian predators of shrews of Europe: regional differences, between-year and seasonal variation, and mortality due to predation. *Ann Zool Fennici* **26**:389–400

Korpimaki E, Norrdahl K (1991) Numerical and functional responses of kestrels, short-eared owls, and long-eared owls to vole densities. *Ecology* **72**:814–826

Korpimaki E, Norrdahl K, Rinta-Jaskari T (1991) Responses of stoats and least weasels to fluctuating food abundances: Is the low phase of the vole cycle due to mustelid predation? *Oecologia* **88**:552–561

Kraus SD (1990) Rates and potential causes of mortality in North Atlantic right whales (*Eubalaena glacialis*). *Marine Mamm Sci* **6**:278–291

Krebs CJ (1978) A review of the Chitty hypothesis of population regulation. *Can J Zool* **56**:2463–2480

Krebs CJ (1979) Dispersal, spacing behaviour, and genetics in relation to population fluctuations in the vole *Microtus townsendii*. *Fortschr Zool* **25**:61–77

Krebs CJ (1995) Two paradigms of population regulation. *Wildl Res* **22**:1–10

Krebs CJ, Gaines MS, Keller BL, Myers JH, Tamarin RH (1973) Population cycles in small rodents. *Science* **179**:35–41

Krebs CJ, Gilbert BS, Boutin S, Sinclair ARE, Smith JNM (1986) Population biology of snowshoe hares. I. Demography of food supplemented populations in the southern Yukon, 1976–84. *J Anim Ecol* **55**:963–982

Krebs CJ, Myers J (1974) Population cycles in small mammals. *Adv Ecol Res* **8**:267–399

Kunkel KE, Mech LD (1994) Wolf and bear predation on white-tailed deer fawns in northeastern Minnesota. *Can J Zool* **72**:1557–1565

Lack D (1966) *Population studies of birds*. Clarendon Press, Oxford, UK

Larter NC (1994) *Plant-herbivore dynamics associated with an erupting ungulate population: A test of hypotheses*. PhD dissertation, Univ British Columbia, Vancouver, Canada

Laurenson MK (1995) Implications of high offspring mortality for cheetah population dynamics. In: Sinclair ARE, Arcese P (eds) *Serengeti II: Dynamics, management and conservation of an ecosystem*. Univ Chicago Press, Chicago, USA, pp 385–399

Laws RM, Parker ISC, Johnstone RCB (1975) *Elephants and their habitats*. Clarendon Press, Oxford, UK

Le Henaff D, Crete M (1989) Introduction of muskoxen in northern Quebec: The demographic explosion of a colonizing herbivore. *Can J Zool* **67**:1102–1105

Leslie PH (1959) The properties of a certain lag type of population growth and the influence of an external random factor on a number of such populations. *Physiol Zool* **32**:151–159

Lidicker W (1975) The role of dispersal in the demography of small mammals. In: Petrusewicz K, Golley FB, Ryszkowski I (eds) *Small mammals: Productivity and population dynamics*. Cambridge Univ Press, Cambridge, UK, pp 103–128

Lindstrom E (1989) Food limitation and social regulation in a red fox population. *Holarctic Ecol* **12**:70–79

Lindzey FG, van Sickle WD, Ackerman BB, Barnhurst D, Hemker TP, Laing SP (1994) Cougar population dynamics in southern Utah. *J Wildl Manage* **58**:619–624

MacLulich DA (1937) Fluctuations in the numbers of the varying hare (*Lepus americanus*). Univ Toronto Press, Toronto, Canada

Mangel M, Tier C (1994) Four facts every conservation biologist should know about persistence. *Ecology* **75**:607–614

Marcstrom V, Hoglund N, Krebs CJ (1990) Periodic fluctuations in small mammals at Boda, Sweden from 1961 to 1968. *J Anim Ecol* **59**:753–761

May RM (1975) Biological populations obeying difference equations: stable points, stable cycles and chaos. *J Theor Biol* **47**:511–524

McArdle BH, Gaston KJ (1993) The temporal variability of populations. *Oikos* **67**:187–191

McArdle BH, Gaston KJ, Lawton JH (1990) Variation in the size of animal populations: Patterns, problems and artefacts. *J Anim Ecol* **59**:439–454

McCullough DR (1983) Rate of increase of white-tailed deer on the George Reserve: a response. *J Wildl Manage* **47**:1248–1250

McCullough DR (1992) Concepts of herbivore population dynamics. In: McCullough DR, Barrett RH (eds) *Wildlife 2001: Populations*. Elsevier, New York, USA, pp 967–984

McNamara JM, Houston AI (1987) Starvation and predation as factors limiting population size. *Ecology* **68**:1515–1519

McRoberts RE, Mech LD, Peterson RD (1995) The cumulative effect of consecutive winter snow depth on moose and deer populations: A defence. *J Anim Ecol* **64**:131–135

Mech LD (1977) Productivity, mortality, and population trends of wolves in northwestern Minnesota. *J Mamm* **58**:559–574

Mech LD, McRoberts RE, Peterson RD, Page RE (1987) Relationship of deer and moose populations to previous winters' snow. *J Anim Ecol* **56**:615–627

Messier F (1991) The significance of limiting and regulating factors on the demography of moose and white-tailed deer. *J Anim Ecol* **60**:377–393

Messier F (1994) Ungulate population models with predation: A case study with the North American moose. *Ecology* **75**:478–488

Miller JS, Zammuto RM (1983) Life histories of mammals: An analysis of life tables. *Ecology* **64**:631–635
Murdoch WW (1970) Population regulation and population inertia. *Ecology* **51**:497–502
Murdoch WW (1994) Population regulation in theory and practice. *Ecology* **75**:271–287
Newsome AE, Catling PC, Corbett LK (1983) The feeding ecology of the dingo II. Dietary and numerical relationships with fluctuating prey populations in south-eastern Australia. *Aus J Ecol* **8**:345–366
Newsome AE, Parer I, Catling PC (1989) Prolonged prey suppression by carnivores – predator-removal experiments. *Oecologia* **78**:458–467
Nicholson AJ (1933) The balance of animal populations. *J Anim Ecol* **2**:132–178
Oksanen L (1990) Exploitation systems in seasonal environments. *Oikos* **57**:14–24
Oksanen L, Ericson L (1987) Dynamics of tundra and taiga populations and herbaceous plants in relation to the Tihomirov-Fretwell and Kalela – Tast hypotheses. *Oikos* **50**:381–388
Oksanen L, Oksanen T (1992) Long-term microtine dynamics in north Fennoscandian Tundra: The vole cycle and the lemming chaos. *Ecography* **15**:226–236
Oksanen L, Oksanen T, Lukkari A, Siren S (1987) The role of phenol-based inducible defense in the interaction between tundra populations of the vole *Clethrionomys myrtillus*. *Oikos* **50**:371–380
Olesiuk PF, Bigg MA, Ellis GM (1990) Life history and population dynamics of resident killer whales (*Orcinus orca*) in the coastal waters of British Columbia and Washington State. *Rep Int Whal Commn* (special issue) **12**:209–243
Ostfeld RS (1988) Fluctuations and constancy in populations of small rodents. *Am Nat* **131**:445–452
Ostfeld RS, Canham CD, Pugh SR (1993) Intrinsic density-dependent regulation of vole populations. *Nature* **366**:259–261
Owen-Smith N (1990) Demography of a large herbivore, the greater kudu *Tragelaphus strepsiceros*, in relation to rainfall. *J Anim Ecol* **59**:893–913
Pech RP, Sinclair ARE, Newsome AE, Catling PC (1992) Limits to predator regulation of rabbits in Australia: Evidence from predator removal experiments. *Oecologia* **89**:102–112
Peltonen A, Hanski I (1991) Patterns of island occupancy explained by colonization and extinction rates in shrews. *Ecology* **72**:1698–1708
Peterman RM, Clark WC, Holling CS (1979) The dynamics of resilience: shifting stability domains in fish and insect systems. In: Anderson RM, Turner BD, Taylor LR (eds) *Population Dynamics, Symp Brit Ecol Soc* **20**: 321–341, Blackwell, Oxford, UK
Peterson RO (1994) *Ecological studies of wolves on Isle Royale, annual report – 1993–1994.* Isle Royale Nat Hist Assoc, Houghton, MI, USA
Peterson RO, Page RE, Dodge KM (1984) Wolves, moose, and allometry of population cycles. *Science* **224**:1350–1352
Pimm SL, Redfearn A (1988) The variability of population densities. *Nature* **334**:613–614
Podoler H, Rogers D (1975) A new method for the identification of key factors from life-table data. *J Anim Ecol* **44**:85–114
Reddingius J (1971) Gambling for existence. *Acta Biotheroretica*, suppl 1, **20**:1–208
Reid DG (1995) *Factors limiting population growth of non-cyclic collared lemmings at low densities.* PhD dissertation, Univ British Columbia, Vancouver, Canada
Robertson A, Hiraiwa-Hasegawa M, Albon SD, Clutton-Brock TH (1992) Early growth and suckling behaviour of Soay sheep in a fluctuating population. *J Zool* **227**:661–671
Runyoro VA, Hofer H, Chausi EB, Moehlman PD (1995) Long-term trends in the herbivore populations of the Ngorongoro Crater, Tanzania. In: Sinclair ARE, Arcese P (eds) *Serengeti II: Dynamics, management and conservation of an ecosystem* Univ Chicago Press, Chicago, USA, pp 146–168
Ruxton CD (1993) Linked populations can still be chaotic. *Oikos* **68**:347–348
Ruxton CD (1994) Local and ensemble dynamics of linked populations. *J Anim Ecol* **63**:1002
Sandell M, Astrom M, Atlegrim O, Danell K, Edenius L, Hjalten J, Lundberg P, Palo T, Pettersson R, Sjoberg G (1991) 'Cyclic' and 'non-cyclic' small mammal populations: An artificial dichotomy. *Oikos* **61**:281–284
Schaller GB (1972) *The Serengeti lion.* Univ Chicago Press, Chicago, USA

Seip DR (1992) Factors limiting woodland caribou populations and their interrelationships with wolves and moose in southeastern British Columbia. *Can J Zool* **70**:1494–1503

Silva M, Downing JA (1994) Allometric scaling of minimal mammal densities. *Conserv Biol* **8**:732–743

Sinclair ARE (1975) The resource limitation of trophic levels in tropical grassland ecosystems. *J Anim Ecol* **44**:497–520

Sinclair ARE (1977) *The African buffalo.* Univ Chicago Press, Chicago, USA

Sinclair ARE (1979) The eruption of the ruminants. In: Sinclair ARE, Norton-Griffiths M (eds) *Serengeti: Dynamics of an ecosystem*, Univ Chicago Press, Chicago, USA, pp 82–103

Sinclair ARE (1985) Does interspecific competition or predation shape the African ungulate community? *J Anim Ecol* **54**:899–918

Sinclair ARE (1989) Population regulation in animals. In: Cherrett JM (ed) *Ecological concepts.* Blackwell, Oxford, UK, pp 197–241

Sinclair ARE (1995a) Serengeti past and present. In: Sinclair ARE, Arcese P (eds) *Serengeti II: Dynamics, management and conservation of an ecosystem.* Univ Chicago Press, Chicago, USA, pp 3–30

Sinclair ARE (1995b) Population limitation of resident herbivores. In: Sinclair ARE, Arcese P (eds) *Serengeti II: Dynamics, management and conservation of an ecosystem.* Univ Chicago Press, Chicago, USA, pp 194–219

Sinclair ARE, Arcese P (1995) Population consequences of predation-sensitive foraging: The Serengeti wildebeest. *Ecology* **76**:882–891

Sinclair ARE, Dublin H, Borner M (1985) Population regulation of Serengeti wildebeest: A test of the food hypothesis. *Oecologia* **65**:266–268

Sinclair ARE, Gosline JM, Holdsworth G, Krebs CJ, Boutin S, Smith JNM, Boonstra R, Dale M (1993) Can the solar cycle and climate synchronize the snowshoe hare cycle in Canada? Evidence from tree rings and ice cores. *Am Nat* **141**:173–198

Sinclair ARE, Krebs CJ, Smith JNM, Boutin S (1988) Population biology of snowshoe hares. III Nutrition, plant secondary compounds and food limitation. *J Anim Ecol* **57**:787–806

Sinclair ARE, Pech RP (1996) Density dependence, stochasticity, compensation and predator regulation. *Oikos* **75**:164–173

Singer FJ, Norland JE (1994) Niche relationships within a guild of ungulate species in Yellowstone National Park, Wyoming, following release from artificial controls. *Can J Zool* **72**:1383–1394

Skinner JD, van Aarde RJ (1983) Observations on the trend of the breeding population of southern elephant seals, *Mirounga leonina*, at Marion Island. *J Appl Ecol* **20**:707–712

Skogland T (1985) The effects of density-dependent resource limitations on the demography of wild reindeer. *J Anim Ecol* **54**:359–374

Spinage CA (1970) Population dynamics of the Uganda defassa waterbuck (*Kobus defassa ugandae* Neumann) in the Queen Elizabeth Park, Uganda. *J Anim Ecol* **39**:51–78

Stanley-Price MR (1989) *Animal re-introductions: The Arabian oryx in Oman.* Cambridge Univ Press, Cambridge, UK

Stenseth NC, Flamstad E (1980) Reproductive effort and optimal reproductive rates in small rodents. *Oikos* **34**:23–34

Stenseth NC, Lidicker WZ (eds) (1992) *Animal dispersal: Small mammals as a model.* Chapman and Hall, London, UK

Stenseth NC, Oksanen L (1987) Small rodents with social and trophic interactions in a seasonally varying environment. *Oikos* **50**:319–326

Strong DR (1986) Density-vague population change. *Trends Ecol & Evol* **1**:39–42

Stubbs M (1977) Density dependence in the life cycles of animals and its importance in K- and r- strategies. *J Anim Ecol* **46**:677–688

Taitt MJ, Krebs CJ (1981) The effect of extra food on small rodent populations. II Voles (*Microtus townsendii*). *J Anim Ecol* **50**:125–137

Taitt MJ, Krebs CJ (1985) Population dynamics and cycles. In: Tamarin R (ed) *Biology of New World* Microtus. Am Soc Spec Publ No 8, USA, pp 567–620

Tamarin RH (1978) Dispersal, population regulation, and k-selection in field mice. *Am Nat* 112:545–555
Trites AW (1992) Northern fur seals: why have they declined? *Aquat Mamm* 18:3–18
Trostel K, Sinclair ARE, Walters CJ, Krebs CJ (1987) Can predation cause the 10-year hare cycle? *Oecologia* 74:185–192
Trout RC, Ross J, Tittensor AM, Fox AP (1992) The effect on a British wild rabbit population (*Oryctolagus cuniculus*) of manipulating myxomatosis. *J Appl Ecol* 29:679–686
Turchin P (1990) Rarity of density dependence on population regulation with lags? *Nature* 344:660–663
Turchin P (1993) Chaos and stability in rodent population dynamics: Evidence from non-linear time-series analysis. *Oikos* 68:167–172
Turchin P, Taylor AD (1992) Complex dynamics in ecological time series. *Ecology* 73:289–305
van Ballenberghe V, Ballard WB (1994) Limitations and regulation of moose populations: The role of predation. *Can J Zool* 72:2071–2077
Varley GC, Gradwell GR (1968) Population models for the winter moth. In: Southwood TRE (ed) *Insect abundance.* Oxford Univ Press, London, UK, pp 132–142
Wynne-Edwards VC (1962) *Animal dispersion in relation to social behaviour.* Oliver and Boyd, Edinburgh, UK
Xia X, Boonstra R (1992) Measuring temporal variability of population density: a critique. *Am Nat* 140:883–892
Young TP (1994) Natural die-offs of large mammals: Implications for conservation. *Conserv Biol* 8:410–418

Appendix I

Variability in mammalian population growth rate scaled by generation time: some theoretical expectations.

M. Ritchie

Figure 4 in the text suggests that, for mammals, variability in per capita population growth rate scaled by generation time does not vary with body size. Here it is mathematically demonstrated that such size invariance is expected from the way that life history characteristics of mammals scale with body size. If the population dynamics for a species are described with a logistic equation

$$r = r_{max}(K–N)/K \qquad \text{(A1)}$$

where r is per capita instantaneous rate of increase, r_{max} is a physiological maximum intrinsic rate of increase that is proportional to r_m, the actual intrinsic rate of increase (as measured for real animals), N is population density, and K is the carrying capacity of the population as set by the species' environment. The per capita rate of population change scaled to generation time is $\mathrm{Ln}(l)$, where l is the finite rate of increase or net reproductive rate, such that

$$\mathrm{Ln}(l) = rT \qquad \text{(A2)}$$

where T is generation time. Per capita growth rate of the population scaled to generation time is therefore

$$rT = r_{max}\, T\,(K - N)/K \qquad \text{(A3)}$$

Data from this paper show that the scaling relationships for the key life history characteristics are $r_{max} = aW^{-0.31}$ and $T = b\, W^{0.27}$, where W is mass and a and b are proportionality constants. This yields the product is $r_{max}\ T = a\ b\ W^{-0.04}$. Given the standard errors normally associated with mammal allometric relationships (Calder 1984), the exponent –0.04 is unlikely to be significantly different from zero, so that $r_{max}T$ is approximately ab, or a constant k. Thus, Eqn A3 can be rewritten

$$rT = k((K–N)/K) \qquad \text{(A4)}$$

The variability in rT is then

$$\mathrm{Var}(rT) = k^2\,\mathrm{Var}((K–N)(1/K)) \qquad \text{(A5)}$$

which, given that $1/K$ is very small and T is a constant that reflects life history traits, simplifies to

$$T\,\mathrm{SD}(r) = k\,\mathrm{SD}(K - N) \qquad \text{(A6)}$$

where SD is standard deviation. Because a linear transformation does not affect the variance of a random variable, Eqn A6 shows that variability in rT or $\mathrm{Ln}(l)$ is directly proportional to either extrinsic variation (i.e., K is a random variable on the interval $(0, \infty)$ and N is a constant long-term average density) or intrinsic variation (i.e. N is a random variable on the interval $(0, \infty)$ and K is a constant), as they are defined in the text. Thus, Eqn A6 predicts the relationship obtained in Fig. 4, namely that there is no *a priori* expectation that variability in population growth rate should correlate with body size when growth rate is scaled to generation time. This conclusion appears to be robust to the choice of the model, as other simple models, including those with density-independent or mechanistic parameters, provide the same conclusions.

Size invariance of variability in population growth rate is expected even if one assumes that *average* N or K scale with body size (e.g. Damuth 1991). Average N may scale with size without changing $\mathrm{Var}(K - N)$ when K is a random variable, and the same is true for average K when N is a random variable. Variability should scale with body size only if factors controlling population growth differ inherently in *their* variability for different sized mammals, e.g., carrying capacity varies more for smaller mammals than for large ones. There may indeed be constraints imposed by body size on the magnitude of variation in N or K as random variables (Damuth 1991), but such constraints are not suggested by the data in Fig. 4.

Appendix 2

Data for generation time (T years), intrinsic rate of increase over one year (rmt) and body weight (W kg) used in Fig. 2.

Species	T	r_{mt}	W	Source
Macropus fuliginosus	4.00	0.260	25	Bayliss 1987
Macropus rufus	4.00	0.330	35	Bayliss 1987
Microtus townsendii	0.33	6.182	0.055	Krebs 1979
Microtus californicus	0.33	4.620	0.04	Batzli and Pitelka 1971
Microtus oeconomus	0.33	4.158	0.04	Henttonen *et al.* 1987
Microtus agrestis	0.33	5.348	0.025	Henttonen *et al.* 1987
Microtus agrestis	0.33	3.692	0.025	Hannson 1987
Microtus agrestis	0.33	5.416	0.025	Halle and Lehmann 1992
Microtus agrestis	0.33	6.920	0.025	Marcstrom *et al.* 1990
Sorex araneus	0.50	7.700	0.007	Hannson 1987
Sorex araneus	0.50	5.400	0.007	Marcstrom *et al.* 1990
Sorex araneus	0.50	6.182	0.007	Henttonen *et al.* 1989
Clethrionomys glariolus	0.50	5.416	0.017	Henttonen *et al.* 1987
Clethrionomys glariolus	0.50	6.846	0.017	Hannson 1987
Phyllotis darwini	0.50	3.000	0.047	Jimenez *et al.* 1992
Octodon degus	1.00	3.308	0.123	Jimenez *et al.* 1992
Spermophilus parryii	1.20	0.740	0.5	A.Byrom and T.Karels pers.com.
Chinchilla lanigera	1.00	1.228	0.6	Jimenez *et al.* 1992
Lepus americanus	1.50	2.190	1.5	MacLulich 1937
Mustela nivalis	0.50	7.480	0.03	Korpimaki *et al.* 1991
Mustela erminea	1.00	4.160	0.15	Korpimaki *et al.* 1991
Vulpes vulpes	2.20	1.660	7	Peterson 1994
Ursus arctos	8.00	0.300	180	McCullough 1992

Continued over page

Species	T	r_{mt}	W	Source
Capriolus capriolus	4.00	0.230	20	Caughley 1977
Odocoileus hemionus	5.00	0.223	50	Hatter and Janz 1994
Odocoileus virginianus	5.00	0.518	50	McCullough 1983
Odocoileus virginianus	5.00	0.550	50	Caughley 1977
Antilocapra americana	5.00	0.450	60	Singer and Norland 1994
Ovis canadensis	5.00	0.350	70	McCullough 1992
Damaliscus korrigum	6.00	0.292	110	Sinclair 1995b
Rangifer tarandus	7.00	0.300	150	Skogland 1985
Rangifer tarandus	7.00	0.290	150	Bergerud *et al.* 1983
Rangifer tarandus	7.00	0.231	150	Caughley and Gunn 1993
Rangifer tarandus	7.00	0.244	150	Heard and Ouellet 1994
Kobus ellipsiprymnus	8.00	0.290	175	Foster and Kearney 1967
Oryx leucoryx	8.00	0.252	180	Stanley-Price 1989
Cervus elephas	6.00	0.383	200	Houston 1982
Cervus elephas	6.00	0.270	200	Caughley 1977
Tragelaphus strepsiceros	8.00	0.385	220	Owen-Smith 1990
Ovibos moschatus	9.00	0.260	320	La Henaff and Crete 1989
Syncerus caffer	10.00	0.290	450	pers. obs.
Bison bison	10.00	0.320	700	Larter 1994
Bison bison	10.00	0.295	700	Singer and Norland 1994
Giraffa camelopardalis	14.00	0.140	800	Foster and Kearney 1967
Loxodonta africana	25.00	0.030	3000	Armbruster and Lande 1993
Loxodonta africana	25.00	0.025	3000	Cumming 1983
Loxodonta africana	25.00	0.060	3000	Caughley and Krebs 1983
Arctocephalus gazella	10.00	0.155	34	Skinner and van Aarde 1983
Arctocephalus tropicalis	10.00	0.075	55	Skinner and van Aarde 1983
Phoca vitulina	15.00	0.118	80	Harwood and Rohani this volume
Mirounga angustirostris	25.00	0.131	2300	Harwood and Rohani this volume
Orcinus orca	20.00	0.046	3000	Olesiuk *et al.* 1990
Eschrichtius robustus	40.00	0.067	30000	Best 1993
Balaena glacialis	50.00	0.068	50000	Best 1993
Balaena glacialis	50.00	0.073	50000	Best 1993
Balaena mysticetus	50.00	0.034	60000	Best 1993
Megaptera novaeangliae	60.00	0.094	60000	Best 1993
Megaptera novaeangliae	60.00	0.088	60000	Best 1993
Megaptera novaeangliae	60.00	0.097	60000	Best 1993
Balaenaptera musculus	80.00	0.063	150000	Baskin 1993
Balaenaptera musculus	80.00	0.051	150000	Best 1993

Appendix 3

Data for SD (r_t) used in Figs. 3 and 4.

Species	n	T	r_t	SD (r_t)	W	Source
Perameles nasuta	9	2.00	–0.019	0.826	0.85	Newsome *et al.* 1983
Trichosurus vulpecula	9	2.00	–0.169	0.500	2.5	Newsome *et al.* 1983
Macropus rufogriseus	9	3.00	0.127	0.406	13.8	Newsome *et al.* 1983
Macropus fuliginosus	12	4.00	0.000	0.395	25	Bayliss 1987
Macropus giganteus	9	4.00	0.151	0.728	25	Newsome *et al.* 1983
Macropus rufus	12	4.00	0.037	0.287	35	Bayliss 1987
Macropus rufus	8	4.00	0.002	0.291	35	Commonwealth of Aus. 1984

Continued on next page

Species	n	T	r_t	SD (r_t)	W	Source
Vombatus ursinus	9	5.00	0.039	0.409	26	Newsome *et al.* 1983
Microtus agrestis	55	0.33	0.010	2.890	0.025	Marcstrom *et al.* 1990
Microtus agrestis	7	0.33	0.000	2.683	0.025	Hannson 1987
Microtus agrestis	19	0.33	0.242	3.249	0.025	Halle and Lehmann 1992
Mictotus agrestis	31	0.33	−0.287	2.914	0.025	Henttonen *et al.* 1987
Microtus townsendii	10	0.33	0.067	2.804	0.055	Krebs 1979
Microtus townsendii	11	0.33	0.200	1.074	0.055	Krebs 1979
Akodon olivaceus	13	0.50	−0.254	2.504	0.03	Jimenez *et al.* 1992
Apodemus flavicollis	55	0.50	−0.224	3.880	0.035	Marcstrom *et al.* 1990
Clethrionomys glariolus	31	0.50	0.144	2.700	0.017	Henttonen *et al.* 1987
Clethrionomys glariolus	7	0.50	0.231	4.766	0.017	Hannson 1987
Clethrionomys ruficanus	31	0.50	0.071	2.644	0.03	Henttonen *et al.* 1987
Clethrionomys rutilus	31	0.50	0.113	2.476	0.03	Henttonen *et al.* 1987
Lemmus lemmus	31	0.50	−0.324	1.378	0.065	Henttonen *et al.* 1987
Marmosa elegans	13	0.50	−0.100	3.444	0.04	Jimenez *et al.* 1992
Phyllotis darwini	13	0.50	−0.166	2.224	0.047	Jimenez *et al.* 1992
Pseudomys apodemoides	13	0.50	−0.013	1.036	0.02	Cockburn 1981
Sorex araneus	7	0.50	−0.184	5.944	0.007	Hannson 1987
Sorex araneus	55	0.50	−0.030	3.834	0.007	Marcstrom *et al.* 1990
Sorex araneus	46	0.50	0.060	2.540	0.007	Henttonen *et al.* 1989
Chinchilla lanigera	13	1.00	−0.134	1.472	0.6	Jimenez *et al.* 1992
Octodon degus	13	1.00	−0.400	3.020	0.123	Jimenez *et al.* 1992
Rattus fuscipes	9	1.00	−0.057	0.794	0.125	Newsome *et al.* 1983
Rattus lutreolus	9	1.00	0.132	2.067	0.122	Newsome *et al.* 1983
Tamiasciurus hudsonicus	8	1.00	0.013	0.294	0.25	Boutin *et al.* 1995
Lepus americanus	91	1.50	−0.004	0.956	1.5	MacLulich 1937
Lepus americanus	18	1.50	−0.017	1.098	1.5	Boutin *et al.* 1995
Spermophilus parryii	5	1.50	−0.060	0.841	0.5	A.Byrom and T.Karels pers.comm.
Mustela nivalis	17	0.50	−0.006	3.104	0.03	Hanski *et al.* 1993
Mustela erminea	13	1.00	−0.138	2.560	0.15	Korpimaki *et al.* 1991
Vulpes vulpes	23	2.20	−0.033	0.652	7	Peterson 1994
Lynx canadensis	7	3.00	−0.032	0.682	9.7	Boutin *et al.* 1995
Canis latrans	7	4.00	−0.041	0.530	10.7	Boutin *et al.* 1995
Canis lupus	10	5.00	−0.023	0.217	36	Mech 1977
Canis lupus	12	5.00	−0.013	0.512	36	Foley 1994
Crocuta crocuta	11	6.00	−0.013	0.112	70	Hofer and East 1995
Felis concolor	14	7.00	−0.007	0.134	50	Foley 1994
Felis concolor	8	7.00	0.069	0.130	50	Lindzey *et al.* 1994
Panthera leo	20	8.00	0.022	0.091	135	Hanley *et al.* 1995
Gazella thomsoni	5	4.00	−0.016	0.076	16	Foster and Kearney 1967, 1968
Gazella thomsoni	13	4.00	−0.040	0.300	16	Runyoro *et al.* 1995
Phacochoerus aethiopicus	5	4.00	−0.117	0.190	53	Foster and Kearney 1967, 1968
Aepyceros melampus	5	5.00	−0.056	0.142	42	Foster and Kearney 1967, 1968
Aepyceros melampus	5	5.00	0.055	0.565	42	Sinclair 1995b
Aepyceros melampus	7	5.00	−0.075	0.284	42	Broten and Said 1995
Alcelaphus buselaphus	7	5.00	−0.009	0.107	126	Broten and Said 1995
Antilocapra americana	12	5.00	0.012	0.222	60	Singer and Norland 1994
Gazella granti	5	5.00	0.033	0.064	43	Foster and Kearney 1967, 1968

Continued over page

Species	n	T	r_t	SD (r_t)	W	Source
Gazella granti	13	5.00	0.005	0.362	43	Runyoro *et al.* 1995
Gazella granti	7	5.00	−0.052	0.233	43	Broten and Said 1995
Odocoileus hemionus	8	5.00	0.011	0.123	50	Hatter and Janz 1994
Ovis aries	37	5.00	0.000	0.344	25	Clutton-Brock *et al.* 1991
Ovis canadensis	20	5.00	−0.062	0.186	70	Singer and Norland 1994
Cervus elephas	16	6.00	0.011	0.089	180	Clutton-Brock and Lonegan 1994
Damaliscus korrigum	5	6.00	0.002	0.053	110	Sinclair 1995b
Alcelaphus buselaphus	5	7.00	0.071	0.210	126	Sinclair 1995b
Alcelaphus buselaphus	15	7.00	−0.004	0.392	126	Runyoro *et al.* 1995
Rangifer tarandus	7	7.00	0.203	0.163	150	Courturier *et al.* 1990
Rangifer tarandus	17	7.00	0.059	0.144	150	Caughley and Gunn 1993
Connochaetes taurinus	7	8.00	−0.026	0.076	163	Sinclair 1995a
Connochaetes taurinus	10	8.00	−0.003	0.133	163	Runyoro *et al.* 1995
Kobus ellipsiprymnus	6	8.00	0.133	0.354	175	Spinage 1970
Tragelaphus strepsiceros	11	8.00	0.033	0.206	220	Owen-Smith 1990
Equus burchelli	7	9.00	−0.008	0.057	220	Sinclair 1995a
Equus burchelli	18	9.00	−0.005	0.307	220	Runyoro *et al.* 1995
Bison bison	8	10.00	0.040	0.248	700	Singer and Norland 1994
Syncerus caffer	8	10.00	−0.019	0.226	450	pers obs.
Syncerus caffer	8	10.00	−0.049	0.152	450	Broten and Said 1995
Giraffa camelopardalis	5	12.00	−0.107	0.146	800	Foster and Kearney 1967, 1968
Giraffa camelopardalis	5	12.00	0.149	0.234	800	Sinclair 1995b
Loxodonta africana	7	25.00	−0.031	0.085	3000	Broten and Said 1995
Orcinus orca	15	20.00	0.009	0.042	3000	Olesiuk *et al.* 1990

VERTEBRATE COMMUNITY DYNAMICS IN THE BOREAL FOREST OF NORTH-WESTERN CANADA

C.J. Krebs, A.R.E. Sinclair and S. Boutin

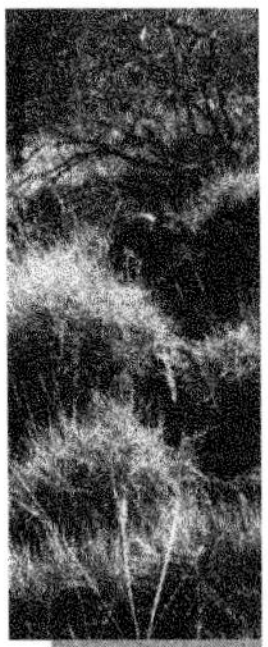

ABSTRACT

The vertebrate community of the boreal forests of northern Canada fluctuates dramatically in a ten year cycle focused on the snowshoe hare (*Lepus americanus*). Since 1986 we have been analysing the structure of this vertebrate community to understand the impact of the hare cycle on all the other species. We have conducted large-scale experiments in a factorial design by adding food and reducing predation pressure. We will complete these experiments in 1996. The major finding to date for snowshoe hares is of a large interaction between food and predation, so that the combined treatment of reduced predation plus added food delayed the cyclic decline and maintained high numbers. We do not know the mechanism by which the interaction between food and predation occurs. Risk-sensitive foraging and condition-dependent stress responses are possibilities that require further work.

Key words: Snowshoe hare, *Lepus americanus*, population cycles, predation, food limitation

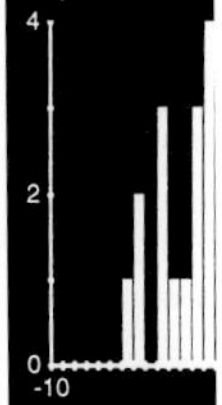

Frontiers of Population Ecology, R.B. Floyd, A.W. Sheppard and P.J. De Barro (eds), CSIRO Publishing, Melbourne, 1996, pp. 155–161

Introduction

The snowshoe hare cycle is one of the classic events of the boreal spruce forests that occupy about one-third of Canada (Keith 1963, 1990). Ten-year cycles occur throughout most of this range in habitats unfragmented by human activities. Because both generalist and specialist predators eat snowshoe hares, the hare cycle has repercussions for a broad variety of species in the boreal forest. In order to gain some understanding of how this system works, we began in 1986 a series of large-scale manipulations in undisturbed boreal forests of the southwestern Yukon. We manipulated nutrient levels, food supplies and predation pressure directly, and observed their consequences on hares and other species. We report some preliminary results of these manipulations. We made the operational decision to manipulate large areas of 1 km^2 at the onset of this research with the realisation that we might not be able to replicate all treatments at this scale. Our focus has been on the major processes operating in this community. In this paper we discuss two issues: 1) What is the effect of food and predation on snowshoe hare numbers? 2) How are survival rates affected by these treatments?

Methods

We have utilised a series of sampling methods to estimate the abundance of the major bird and mammal species in this community. Most of these are described in detail elsewhere (Boutin *et al.* 1995; Krebs *et al.* 1992) and will not be given here. Live-trapping and radio-telemetry are our key methods for census of the more abundant species, and sighting indices are used for the less abundant animals.

The experimental design was to manipulate food supplies and predation mortality in a factorial design. Our experimental units were 1 km^2 blocks of boreal forest separated at least 1 km from one another. We provided supplemental food (commercial rabbit chow, 16% protein) year round *ad lib* to two areas, and we constructed an electric fence around another area to keep out mammalian predators (see Krebs *et al.* 1995). We also combined these treatments on one area to produce a predator exclusion + food addition area. We could not replicate these predator exclosure treatments because of cost. The electric fence successfully deterred mammalian predators (lynx, coyotes, foxes). It did not prevent predation by great-horned owls or other raptors, and consequently these treatments must be thought of as predator-reduction areas rather than predator elimination areas. In our study area about two-thirds of the predation losses are due to mammal predators and one-third to birds, on average. We carried out these manipulations from 1986 to the present, but the electric fences were not fully operational until the winter of 1988–89 when hares had reached peak densities.

Results

We have the most detailed snowshoe hare information from control area No. 1, which was studied by Hik (1994, 1995). Figure 1 shows the population changes from 1986 to 1994. The population increased from 1986 to spring 1989 after a prolonged low phase from 1983 to 1986. The annual rate of increase was 3–4 fold in both 1986–87 and 1987–88 and was reduced to a 1.5 fold increase from 1988–89. These population estimates were highly consistent between open population Jolly-Seber estimates and closed population CAPTURE estimates (the robust design of Pollock *et al.* 1989). The peak phase occurred during 1989 and 1990, and the decline began in the autumn of 1990 and continued through 1991 and 1992 to a low in 1993. During each of the two decline years the population fell about 70% from spring to spring.

The largest impact on hare numbers has come from the combined predator exclosure + added food treatment (Krebs *et al.* 1995). Figure 2 contrasts the spring densities of the control areas with that of the combined treatment area. The combined treatment population increased about 70% from spring 1989 to spring 1990, while the control populations were nearly constant. During the cyclic decline

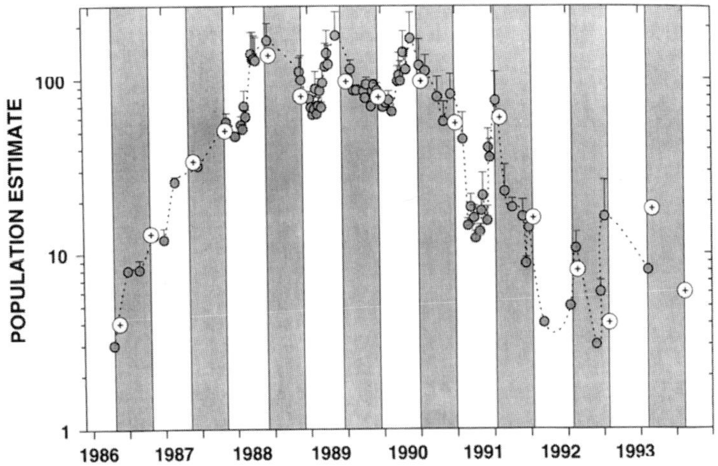

Fig. 1. Snowshoe hare population estimates for control area No. 1, 1986–1994. Jolly-Seber open population estimates (shaded) with 95% upper confidence limits and program CAPTURE closed population estimates (circles with crosses) from each April and October live trapping. Winter months are shaded. The effective trapping area is approximately 60 ha. There is in general good agreement between the open and the closed population estimates.

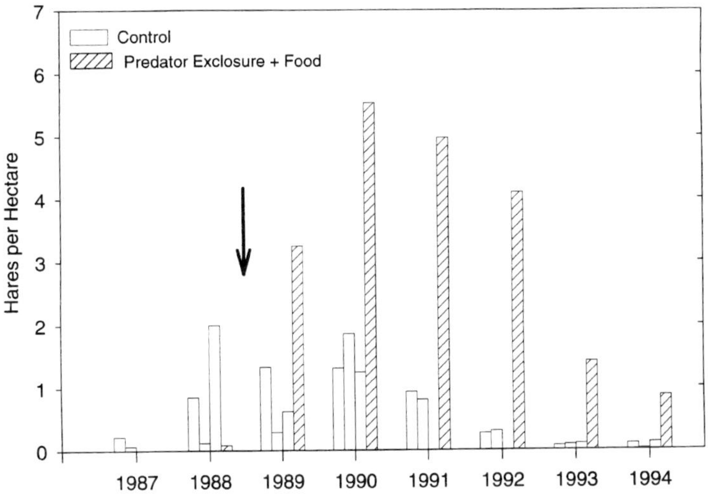

Fig. 2. Snowshoe hare closed population density estimates from each spring for three control populations and the combined treatment of predator exclosure + food addition. The arrow marks the completion of the electric fence on the experimental area, which was not replicated.

from 1990 to 1992 the control population declined at a substantially higher rate than the combined treatment population (Fig. 3). In 1992–93 the two populations declined about the same amount (65–75%) and the combined treatment population declined further from 1993 to 1994.

These population changes are a joint product of reproduction, survival, and movements. We know that the reproductive rate of hares was higher on the combined treatment area during the decline phase than it was on the control areas (Krebs *et al.* 1995). We have detailed survival data from radio-

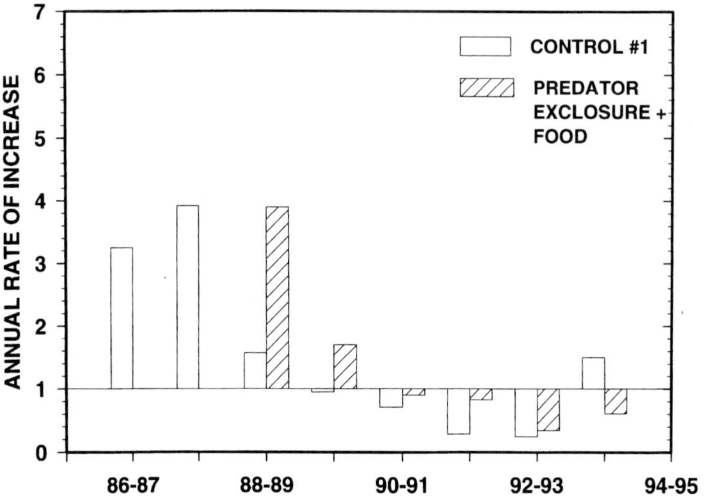

Fig. 3. The finite annual rate of population increase for the control No. 1 population and the combined treatment area. Finite rates (N_{t+1}/N_t) estimated from closed population estimates from spring to spring for each year.

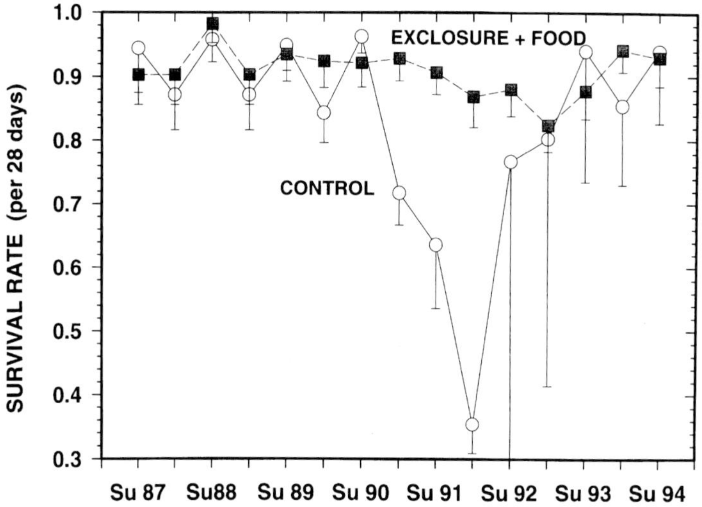

Fig. 4. Survival rates (per 28 days) estimated from radio-telemetry of snowshoe hares on the control No. 1 area and on the combined treatment area. The lower 95% confidence limits are shown. The year is subdivided into two equal periods: summer (April to September) and winter (October to March).

collared hares on all areas and these show a striking contrast between the control and combined treatment plots (Fig. 4). During the increase and peak phases until the summer of 1990 survival was good and there were no systematic differences in survival on the two areas. Beginning in the winter of 1990–91 and continuing until the winter of 1992–93 survival remained high on the combined treatment area and dropped precipitously on the control area. It is these survival differences (Fig. 4) that preserve the large density differences (Fig. 3) on the combined treatment area.

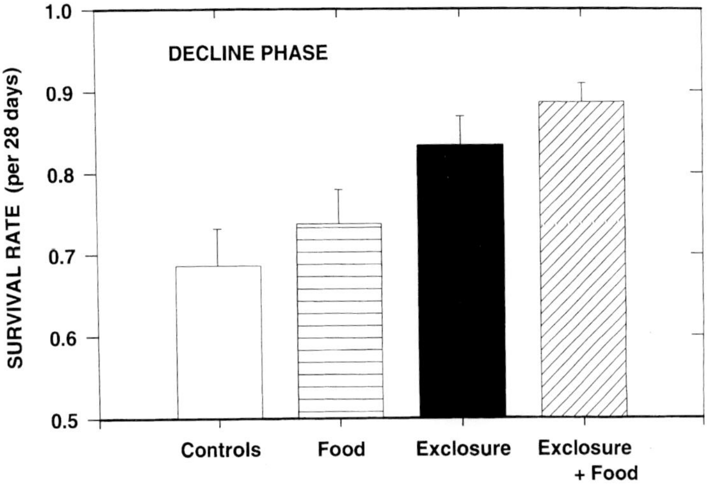

Fig. 5. Average survival rates (per 28 days) estimated from radio-telemetry for snowshoe hares from the factorial treatments during the cyclic decline from autumn 1990 to spring 1993. The upper 95% confidence limit is shown for each treatment.

We can factor out the relative effects of food and predator exclusion on survival rates by comparing survival rates on areas in which only food was added and the area on which only predators were excluded (Krebs *et al.* 1992). Figure 5 shows the average survival rate of radio-collared hares during the two years of population decline from autumn 1990 to autumn 1992. The effect of excluding predators by itself appears to exert a stronger effect on survival rates than does the provisioning of food by itself. The combined effect of predator exclusion and food addition appears to be an additive combination of the separate effects. Since the majority of the losses of radio-collared hares is caused by direct predation, as opposed to starvation (Boutin *et al.* 1986), these results confirm our conclusion that the major effects on hares are due to predators.

Discussion

In an ideal world we would identify the factors necessary for generating cycles by eliminating them one by one and then determining if a cycle in numbers still occurred. There are several problems with this approach. If a factor interaction is necessary for causing cycles, then we will have to manipulate both factors simultaneously. This is the major conclusion of our research – that the snowshoe hare cycle is caused by an interaction between predation and food. But if this is correct we have to explain why the combined treatment population also declined (Fig. 2). We should have expected that the experimental area remained at a high density indefinitely, if our conclusions are correct. Can we not interpret Fig. 2 as showing that the treatments merely delayed the cycle and caused it to occur at a higher carrying capacity?

The key to understanding this dilemma is spatial scale. The treated area is a small patch in a more or less continuous spruce forest. In the spring of 1992 there was a 10-fold difference in hare density inside and outside the combined treatment area. By spring 1993 this difference had increased to over 20-fold. Hares in this type of landscape diffuse from high to low density areas. The combined treatment area declined from 1992–94 under the joint pressure of avian predation (which could not be controlled) and emigration or diffusion. Juvenile dispersal is particularly prevalent in snowshoe hares (Boutin *et al.* 1985) and in spite of good reproductive rates in the combined treatment area (Krebs *et al.* 1995), there was little gain in population during summer 1992 or summer 1993. From summer 1990 through summer 1993 the combined treatment population of hares increased only 10% on average during the summer reproductive period. During periods of population growth, summer reproduction typically increased hare population from 300–500% by the autumn (Keith 1990; Keith and Windberg 1978; Krebs *et al.* 1986). Clearly a large fraction of the juvenile hares were either dying or emigrating from the high density combined treatment area. However, we do not know if it is death or emigration that prevents further increase. There are no known direct causes of social mortality (like infanticide) in hares (Krebs *et al.* 1992), and we doubt that avian predation could have been intensive enough to take all these missing juveniles.

Figure 4 shows clearly the large difference in mortality rates between control and treated areas during the cyclic decline. Because both reproduction and survival remained high on the combined treatment area from 1990–93, we conclude that our experiment did indeed stop the hare cycle on this local area until spatial processes of diffusion produced a decline.

We do not think that snowshoe hares are ever short of food in an absolute sense (Krebs *et al.* 1986; Smith *et al.* 1988; Sinclair *et al.* 1988), and if this is accepted then we need to determine what causes hare reproduction to change during the cycle (Keith 1990) and what produces the interaction between predation and food. Hik (1994, 1995) has argued that predation risk operating through risk-sensitive foraging constrains reproduction in snowshoe hares. Boonstra (1994) has argued that physiological stress may be the key process producing a reproductive effect from predation-risk-avoidance behaviour. It is important to test these ideas on snowshoe hares and to determine how individual hares perceive predation risk so that it can be manipulated in field populations.

The explanation of the hare cycle that comes from our experiments is similar to that proposed by Wolff (1980) from studies in central Alaska and by Keith (1990) from his extensive studies on hares in central Alberta (see Sinclair this volume). We differ primarily in that we emphasise that food is not always in short supply for hares, and that feeding experiments by themselves cannot alter the demography of the decline phase in the same way that predation manipulations can. While both Wolff and Keith emphasised a model of the snowshoe hare cycle based on starvation and absolute food shortage in the peak phase followed by predation mortality in the decline phase, we propose a model emphasising predation mortality, which causes most hare deaths, and a relative shortage of food caused by risk-sensitive foraging. We do not know which of these models applies to large areas of the boreal forest without further study, particularly in eastern Canada.

Acknowledgments

We thank the Natural Sciences and Engineering Research Council of Canada for their financial support of this research, and the Arctic Institute of North America for the use of their Kluane Base. We thank all the postgraduate students, undergraduate assistants, and technicians who have contributed to this effort, particularly V. Nams, S. Schweiger, and M. O'Donoghue. This is publication # 83 from the Kluane Ecosystem Project.

References

Boonstra R (1994) Population cycles in microtines: the senescence hypothesis. *Evol Ecol* 8:196–219

Boutin S, Gilbert BS, Krebs CJ, Sinclair ARE, Smith JNM (1985) The role of dispersal in the population dynamics of snowshoe hares. *Can J Zool* 63:106–115

Boutin SA, Krebs CJ, Sinclair ARE, Smith JNM (1986) Proximate causes of losses in a snowshoe hare population. *Can J Zool* 64:606–610

Boutin S, Krebs CJ, Boonstra R, Dale MRT, Hannon SJ, Martin K, Sinclair ARE, Smith JNM, Turkington R, Blower M, Byrom A, Doyle FI, Doyle C, Hik D, Hofer L, Hubbs A, Karels T, Murray DL, Nams V, O'Donoghue M, Rohner C, Schweiger S (1995) Population changes of the vertebrate community during a snowshoe hare cycle in Canada's boreal forest. *Oikos* 74:69–80

Hik DS (1994) Predation risk and the 10-year snowshoe hare cycle. Ph.D. thesis, Dept Zoology, Univ British Columbia, Vancouver, Canada 140 pp

Hik DS (1995) Does predation risk influence population dynamics? Evidence from the cyclic decline of snowshoe hares. *Wildl Res* **22**:115–129

Keith LB (1963) *Wildlife's ten-year cycle.* Univ Wisconsin Press, Madison, USA

Keith LB (1990) Dynamics of snowshoe hare populations. *Current Mammal* **2**:119–195

Keith LB, Windberg LA (1978) A demographic analysis of the snowshoe hare cycle. *Wildl Monogr* **58**:1–70

Krebs CJ, Gilbert BS, Boutin S, Sinclair ARE, Smith JNM (1986) Population biology of snowshoe hares. I. Demography of food-supplemented populations in the southern Yukon, 1976–84. *J Anim Ecol* **55**:963–982

Krebs CJ, Boonstra R, Boutin S, Hannon S, Martin K, Sinclair ARE, Smith JNM, Turkington R (1992) What drives the snowshoe hare cycle in Canada's Yukon? In: McCullough DR, Barrett RE (eds) *Wildlife 2001: Populations.* Elsevier, London, UK pp 886–896

Krebs CJ, Boutin S, Boonstra R, Sinclair ARE, Smith JNM, Dale MRT, Martin K, Turkington R. (1995) Impact of food and predation on the snowshoe hare cycle. *Science* **269**:1112–1115

Pollock KH, Winterstein SR, Bunick CM, Curtis PD (1989) Survival analysis in telemetry studies: the staggered entry design. *J Wildl Manag* **53**:7–15

Sinclair ARE, Krebs CJ, Smith JNM, Boutin S (1988) Population biology of snowshoe hares III. Nutrition, plant secondary compounds and food limitation. *J Anim Ecol* **57**:787–806

Smith JNM, Krebs CJ, Sinclair ARE, Boonstra R (1988) Population biology of snowshoe hares II. Interactions with winter food plants. *J Anim Ecol* **57**:269–286

Wolff JO (1980) The role of habitat patchiness in the population dynamics of snowshoe hares. *Ecol Monogr* **50**:111–130

Natal dispersal distance under the influence of competition

Michael A. McCarthy

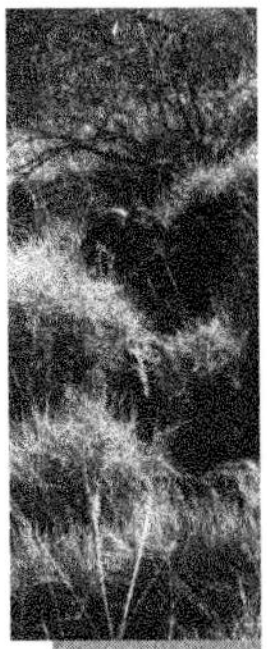

ABSTRACT

Previous attempts to model effects of competition on natal dispersal distance have considered effects of competition between territorial residents and dispersers, and effects of competition among nest-mates. Such models may be expressed in terms of an exponential model of dispersal in which an individual moves in a straight line to the first encountered vacancy, where vacancies are arranged randomly in space. These models suggest that competition will increase dispersal distances, but they do not include competition among dispersers from multiple nests. In this paper, the exponential model of dispersal is modified to incorporate the latter. Analytical solutions of the model demonstrate that competition among identical dispersers may have a significant influence on dispersal patterns and that competition alone may cause large variation in dispersal distances. A result is that competition among individuals may increase or decrease dispersal distances depending on the nature and magnitude of competition. When individuals disperse simultaneously from multiple nests and the number of dispersers exceeds the number of vacant territories, individuals that are close to vacancies are more likely to disperse successfully, and dispersal distances will decline. Competition among dispersers can help to explain observed patterns of dispersal by banner-tailed kangaroo rats. Such competition reduces dispersal success, and has consequences for models of spatially structured population dynamics.

Key words: Natal dispersal, competition, dispersal models, habitat fragmentation, *Dipodomys spectabilis*

Frontiers of Population Ecology, R.B. Floyd, A.W. Sheppard and P.J. De Barro (eds), CSIRO Publishing, Melbourne, 1996, pp. 163–171

Introduction

Dispersal of individuals influences the dynamics, genetic structure, evolution and social behaviour of populations (Chepko-Sade and Halpin 1987). There is often considerable variation in the dispersal distances of individuals within a population, and sex-biased dispersal is common in birds and mammals (Greenwood 1980). A number of causes of variation in dispersal patterns have been proposed such as inbreeding avoidance (Howard 1960; Greenwood 1980), declining habitat (Howard 1960), social aggregation (Stamps 1991), helping at nests (Woolfenden and Fitzpatrick 1984, 1986), the chance location of vacant territory (Murray 1967; Waser 1985; Efford this volume) and competition among dispersers for those vacancies (Murray 1967; Tonkyn and Plissner 1991). Models may be useful to help understand how these various factors influence dispersal (see also Efford this volume). For example, when variation in dispersal distance is greater than is expected from chance alone, it is often proposed that there is a genetic basis for dispersal (Howard 1960; Lidicker 1962; Bunnel and Harestad 1983; Waser 1985).

Inherent differences between individuals may contribute to observed variation in natal dispersal distance (Rees 1993). However, it is useful to consider how variation in dispersal distance may arise in a population of identical competing individuals (Murray 1967; Waser 1985; Tonkyn and Plissner 1991; Caley 1991). Previous models of the effect of competition on natal dispersal have considered the movement of individuals from their natal site to the first encountered territorial vacancy, with vacancies distributed randomly in space. Competition for vacancies may be incorporated into these models by assuming site pre-emption, with the first individual that arrives at a vacancy excluding subsequent dispersers. Murray (1967) used simulation to show that the chance location of vacant territories and competition among nest-mates and other dispersers could cause considerable variation in dispersal distance. Waser (1985) obtained analytical solutions for the cases in which an individual disperses to the nearest vacancy or to the first vacancy encountered in a straight line. Buechner (1987) showed that Waser's straight-line dispersal model provided a reasonable fit to observed dispersal distributions, but in several cases data need to be omitted to obtain the good fit. Recently, Porter and Dooley (1993) argued that short-distance dispersers were more likely to be observed than long-distance dispersers, and that the good fits obtained by Buechner (1987) may be due to these biases in measuring dispersal rather than to the validity of the model.

Previous models of the effect of competition on dispersal suggest that increased competition for vacancies leads to increased dispersal distances (Murray 1967; Waser 1985; Tonkyn and Plissner 1991). However, these models only consider dispersal from an isolated nest, and simulations have shown that they may not predict dispersal distances reliably when individuals disperse from multiple nests (Rodgers and Klenner 1990). Furthermore, Jones *et al.* (1988) demonstrated that increased competition for vacancies was associated with shorter dispersal distances by banner-tailed kangaroo rats (*Dipodomys spectabilis*), which is opposite to predictions of the models.

In this paper, competition among dispersers from multiple natal sites is included in a model of dispersal. Analytical solutions are obtained and conditions are identified under which increased competition may increase or decrease dispersal distances. Predictions of the model are compared to data on dispersal distances provided by Jones *et al.* (1988). The significance of the results are discussed in relation to models of spatially structured population dynamics.

Dispersal from a single nest

Before dispersal from multiple nests is modelled, it is useful to consider a solitary individual that disperses from a nest in a straight-line to the first encountered territorial vacancy (Fig. 1a). Assuming that vacancies are distributed randomly in space, the probability that there are no vacancies along a line of length x may be derived from a Poisson distribution. Thus, the probability that the distance

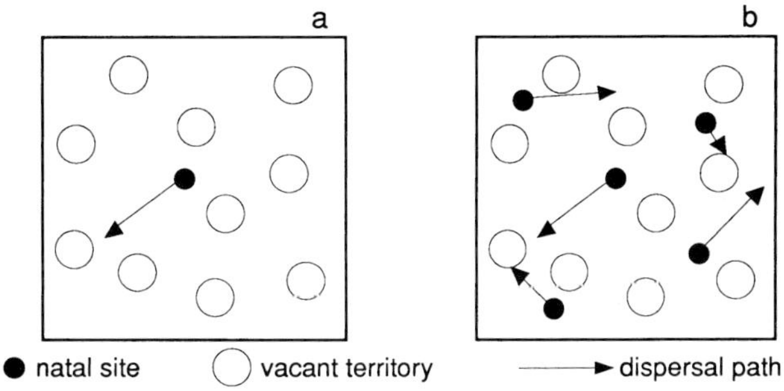

Fig. 1. Schematic representation of dispersal by a solitary individual (a) and of simultaneous dispersal by several individuals from multiple nests (b). In the latter case, vacant territories are acquired by the first individual to reach the site at which point the individual stops dispersing.

dispersed (X_s) is greater than x may be expressed as

$$Pr(X_s > x) = e^{-vx},$$

where v is the density of vacant territories. Therefore, the cumulative probability function of the distance dispersed is

$$F_s(x) = 1 - Pr(X_s > x) = 1 - e^{-vx}.$$

This is an exponential distribution, which is the continuous space analogue of Waser's (1985) geometric distribution (Rees 1993). The equation gives the probability that the dispersal distance is less than x given that an individual moves in a straight line from its natal site to the first encountered vacancy. The median dispersal distance by individuals in this model is

$$M_s = \ln 2 / v. \tag{1}$$

As density of vacancies declines (i.e. as the density of residents increases), individuals will tend to be forced to move further to obtain a territory. In this sense, the model accounts for competition between the disperser and the established residents, but it does not account for the reduction in vacancies that would occur as other individuals disperse.

Dispersal from multiple nests

Previously, movement of competing individuals from multiple nests has been modelled in two different ways. Murray (1967) simulated individuals sequentially, with each disperser leaving their nest only when the previous disperser had settled. For this case, the probability of finding a vacancy varies among individuals but remains constant in space. Under sequential dispersal, dispersal distances increase with increasing competition (Murray 1967; McCarthy 1995). Rodgers and Klenner (1990) simulated simultaneous dispersal, with individuals leaving their nest at the same time and moving at a constant speed to the nearest vacancy. Simultaneous dispersal is different from sequential dispersal because the probability of finding a vacancy is the same for all individuals when they leave the nest, but the probability declines with distance from the nest as competing individuals settle. This type of dispersal is analysed below.

Consider an individual that is dispersing from its natal site to the first encountered vacancy, where vacancies are distributed randomly in space. If the density of vacancies at distance x from the natal site is defined as $l(x)$, then the probability of an individual settling within the infinitesimally small interval $(x, x+dx)$ is equal to the probability that the individual has not settled previously $(1-F_m(x))$ multiplied by $l(x)dx$, where $F_m(x)$ is the cumulative probability function of the dispersal distance. This probability is equivalent to the value of the probability density function of the dispersal distance ($f_m(x)$) multiplied by dx. Thus:

$$f_m(x)dx = [1 - F_m(m)]l(x)dx.$$

Prior to dispersal the density of vacancies will equal v. As an individual moves from its natal site, other dispersers will settle and the density of vacancies ($l(x)$) will decline with the distance moved. If dispersers move simultaneously and at the same rate and if other dispersers are distributed randomly in space with density n (Fig. 1b), then the density of vacancies will decline linearly with the proportion of individuals that settle such that $l(x) = v - nF_m(x)$. Thus:

$$f_m(x) = [1 - F_m(x)][v - nF_m(x)].$$

Note that $f_m(x)$ is the derivative of $F_m(x)$, so by integrating and using the equality $F_m(0) = 0$ it is possible to obtain the cumulative probability function

$$F_m(x) = \frac{v\left[e^{(v-n)x} - 1\right]}{ve^{(v-n)x} - n}, \qquad v > n. \tag{2}$$

The median dispersal distance for this probability function is

$$M_m = \frac{1}{v-n}\ln\left(\frac{2v-n}{v}\right), \qquad v > n. \tag{3}$$

These solutions (Eqns 2 and 3) are valid when the density of vacancies is greater than the density of dispersers ($v > n$). For the case $v = n$, the cumulative probability function is

$$F_m(x) = \frac{vx}{vx+1},$$

and the median is equal to $1/v$.

When there are insufficient vacancies for all dispersers ($v < n$), only a proportion (v/n) of the individuals will settle. In this case, the cumulative probability function given above (Eqn 2) is still strictly accurate for describing the proportion of dispersers that settle within distance x of their natal sites. However, a proportion ($1 - v/n$) of the dispersers will never settle, and it would be useful to know the distribution of dispersal distance for those individuals that actually settle. This may be obtained with the assistance of conditional probability theory. The probability of settling within distance x given that the individual eventually settles is equal to the probability of settling within distance x (for all dispersers) divided by the probability of actually settling. When competition for vacancies is saturated ($v < n$), it is necessary to divide the probability function (Eqn 2) by v/n to describe the distribution of dispersal distance of those individuals that actually settle. Therefore, the cumulative probability function when $v < n$ is

$$F_m(x) = \frac{n\left[e^{(v-n)x} - 1\right]}{ve^{(v-n)x} - n}, \qquad v < n,$$

and the median dispersal distance in this case is

$$M_m = \frac{1}{n-v}\ln\left(\frac{2n-v}{n}\right) \qquad v < n. \tag{4}$$

The influence of competition among dispersers is illustrated by plotting the median dispersal distance against different densities of dispersers and vacancies (Fig. 2). In the case of simultaneous dispersal, the effect of competition on dispersal distances will depend on the level of competition. Increasing competition will increase dispersal distances when there is an excess of vacant habitat ($v > n$). In contrast, if competition for vacancies is saturated ($v < n$), the individuals that settle will tend to be those that are closer to vacancies, and the individuals that are far from vacancies will tend

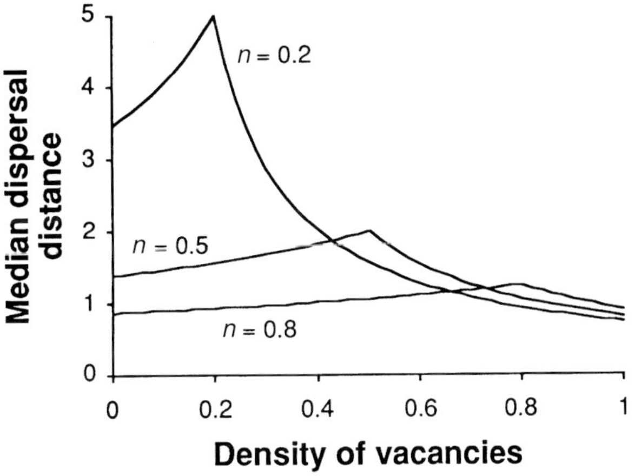

Fig. 2. Median dispersal distance versus the density of vacancies (v) and for different densities of dispersers (n = 0.2, 0.5, 0.8) for the case where individuals disperse simultaneously in a straight-line from their nests (Eqns 3 and 4). When the number of dispersers exceeds the number of vacancies ($n > v$), the median dispersal distance declines with increasing competition among dispersers.

to remain unrecorded. When $v < n$, dispersal distances will decline with increasing competition. This is a more complex pattern than that obtained from previous models of the effect of competition on dispersal, which showed a monotonic increase in median dispersal distance with increasing competition (Waser 1985; Tonkyn and Plissner 1991).

Testing models with data

The simultaneous dispersal model was tested by using the data presented by Jones *et al.* (1988, their Fig. 1) to compare the observed dispersal distance of banner-tailed kangaroo rats with the predicted median dispersal distance. Median dispersal distance in each year was predicted from the observed density of juveniles and adults. It was assumed that the density of juveniles in May represented the density of dispersers (n). The greatest density of adults observed during the seven year study was assumed to represent the adult carrying capacity (the maximum density of mounds that could be occupied). The density of vacancies in each year (v) was obtained by subtracting the density of adults in May from the adult carrying capacity. The density of vacancies and dispersers were used to predict median dispersal distance in each year for the simultaneous dispersal model (Eqns 3 and 4) and the exponential dispersal model (Eqn 1). Observed median dispersal distance in each year was based on those individuals that dispersed from their natal mound because the simultaneous dispersal model does not accommodate dispersal of zero distance.

Observed median dispersal distances were plotted against median dispersal distances predicted by the exponential model and the simultaneous dispersal model (Fig. 3). The density of dispersers was greater than the density of vacancies in two of the seven years. The negative correlation between the predictions of the exponential model and the observed data indicates that the exponential model is unsuitable for predicting effects of competition on median dispersal distance of banner-tailed kangaroo rats ($r_s = -0.464$). For each year, differences between observed and predicted median dispersal distances were tested with the sign test (Daniel 1990). The two largest dispersal distances predicted by the exponential model (185 and 246 metres) were significantly different from the observed median dispersal distances ($P < 0.05$; sign test).

The predictions of the simultaneous dispersal model were substantially better than the exponential model which ignores competition among dispersers. There was a positive correlation between the predictions of the simultaneous dispersal model and the observed data, but this was not significantly greater than zero ($r_s = 0.571$; $P = 0.1$). The dispersal distances predicted by the simultaneous dispersal model were in relatively close agreement with the expected relationship (Fig. 3). None of

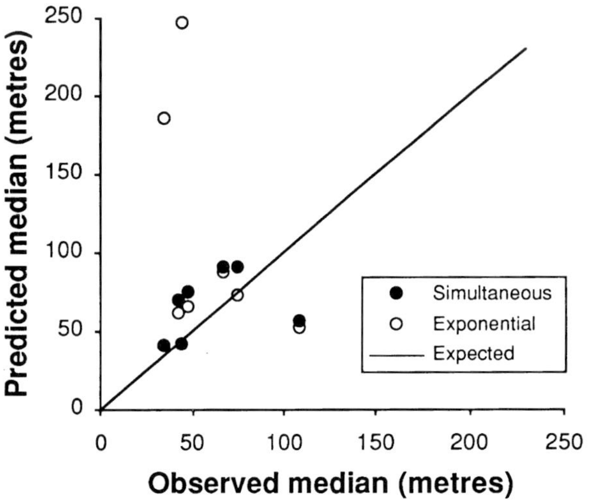

Fig. 3. Observed versus predicted median dispersal distance for banner-tailed kangaroo rats that leave their natal site (data from Jones *et al.*, 1988). Open circles indicate predictions of the exponential model, which ignores competition among dispersers. Closed circles correspond to predictions of the simultaneous dispersal model, which includes competition among dispersers. The line indicates what would be a perfect relationship between observed and predicted dispersal distances.

the median dispersal distances predicted by the simultaneous model was significantly different from the observed median ($P > 0.05$; sign test).

The median dispersal distances of banner-tailed kangaroo rats were not predicted accurately if competition among dispersers was ignored. The analysis, however, should be treated with some caution because it is based on only seven data points (years of data) and the confidence limits around the observed median dispersal distance for each year are relatively large. In 1984 and 1986, the number of observed dispersers was particularly small (sample sizes of four and nine), making estimates of median dispersal distance subject to substantial error (Daniel 1990). The correlation between the observed median dispersal distance and dispersal distance predicted by the simultaneous model was strengthened when these two data points were excluded from the analysis ($r_s = 0.9$; $P = 0.05$).

Habitat fragmentation

The dispersal models used above may be linked to other models that explore the consequences of dispersal strategies. For example, Lande (1987) generalised Levins' (1970) metapopulation model to consider how fragmentation may affect the persistence of territorial species. Lande (1987) used a model of dispersal that ignored competition among dispersers to predict the probability that an individual will find a vacant territory before it dies. Alternatively, Eqn 2 could be used to account for competition among dispersers, thereby generalising the results of Lande (1987). Eqn 2 gives the proportion of dispersers that find a vacant territory within a given distance when individuals disperse simultaneously prior to reproduction. Thus, taking the approach of Lande (1987), an equilibrium (stable population size) is given by

$$\frac{v\left[e^{(v-n)m}-1\right]}{ve^{(v-n)m}-n}R_0' = 1, \tag{5}$$

where m is the maximum distance of dispersal after which individuals die, and R_0' is the number of offspring produced per lifetime for each individual that obtains a territorial vacancy.

At extinction, the density of vacancies (v) will equal the maximum density of individuals that could be supported in the landscape (h), and the density of dispersers (n) will equal zero because no offspring will be produced. It follows that a population can only persist when

$$h > \ln\left(\frac{R_0'}{R_0'-1}\right)/m. \tag{6}$$

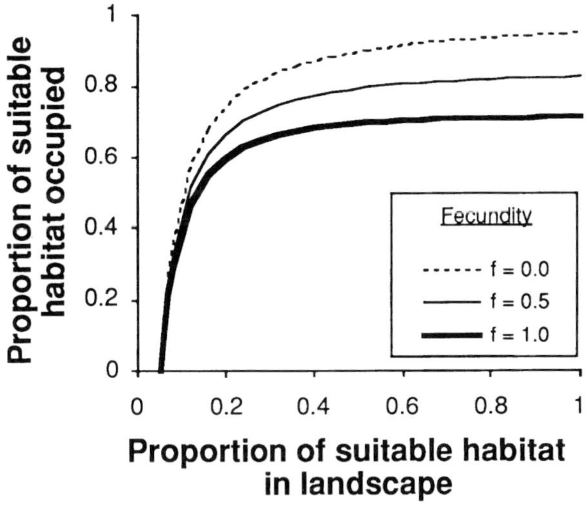

Fig. 4. The proportion of suitable habitat occupied at equilibrium ($[h-v]/h$) as a function of the proportion of habitat in the landscape that is suitable (h), for different levels of competition among dispersers (Eqn 5 assuming $R_0' = 2.5$, $m = 10$). As fecundity of individuals (f) increases, the density of dispersers ($n = f[h-v]$) increases and the proportion of suitable habitat occupied at equilibrium decreases. The curve ignoring the effect of competition among dispersers ($f = 0.0$) is the same as that given by Lande (1987).

Extinction will occur when the amount of suitable habitat (h) declines below this threshold (e.g. due to fragmentation or competition).

The influence of fragmentation on the equilibrium density of a population was also considered. It was assumed that each surviving individual produced, on average, a constant number of dispersing juveniles per year. Thus, the density of nest sites is given by $n = f[h-v]$, where f is the mean number of dispersers produced per individual per year. Following the approach taken by Lande (1987) Eqn 5 was solved for v at different values of h and f (Fig. 4), demonstrating that the equilibrium density of a population may be overestimated by Lande's (1987) model, which does not incorporate competition among dispersers. The extinction threshold, however, remains the same. If competition among dispersers is ignored, metapopulation models may overestimate dispersal success and occupancy rates.

Discussion

Previous attempts at modelling effects of competition on dispersal have considered dispersal in discrete space (Murray 1967; Waser 1985; Tonkyn and Plissner 1991). Whilst this may be a more realistic representation of territoriality for certain species (e.g. some birds), differences between the continuous space models presented here and the discrete models are mostly trivial. It may be possible to convert the continuous models to discrete models by integration and appropriate parameterisation (Rees 1993). Furthermore, both types of models include a limited amount of suitable habitat that is distributed randomly in space, and variation in the amount of suitable habitat has the same qualitative effect on dispersal distances.

When individuals move simultaneously to obtain vacancies, those individuals closest to vacancies will be favoured. Whilst it is unlikely that all individuals will disperse at exactly the same time, dispersal may be limited to particular periods. Such a seasonal pattern of dispersal is observed in banner-tailed kangaroo rats, which die over winter if they are unable to obtain a territorial mound (Jones *et al.* 1988).

Local familiarity may play a role in the establishment and defence of territory. When the competitive ability of an individual declines with distance from its natal site, the model of competition among simultaneous dispersers will be a better representation of dispersal than the exponential model. The simultaneous dispersal model predicts changes in dispersal distance in response to changes in the density of dispersers and the density of territorial residents. As formulated in this paper, these changes are described in terms of changes in competition for vacant territory. Other mechanisms, such as changes in social behaviour in response to population density (Jones *et al.* 1988), might predict similar patterns.

Dispersal models should be used cautiously for predicting the distance moved by juveniles that are in search of vacant territory. These models are theoretical simplifications that have numerous restrictive assumptions. Factors such as competition, habitat heterogeneity and sociality can all affect dispersal patterns. The value of the models developed in this paper is that they demonstrate that competition among identical dispersers can have significant effects on the dispersal patterns of territorial animals. The models do not necessarily provide reliable predictions of dispersal distances in wild populations. Nevertheless, a model that included competition among simultaneous dispersers provided better predictions of the median distance dispersed by banner-tailed kangaroo rats than a model that ignored such competition. The qualitative results have consequences for models of competition, dispersal and spatially structured population dynamics.

Acknowledgments

I am grateful to Mark Burgman, Peter Chesson and Hugh Possingham, whose comments improved earlier versions of this manuscript. This study was supported by an Australian Postgraduate Award.

References

Buechner M (1987) A geometric model of vertebrate dispersal: tests and implications. *Ecology* 68:310–318

Bunnel FL, Harestad AS (1983) Dispersal and dispersion of black-tailed deer: models and observations. *J Mammal* 64:201–209

Caley MJ (1991) A null model for testing distributions of dispersal distances. *Am Nat* 138:524–532

Chepko-Sade BD, Halpin ZT (eds) (1987) *Mammalian dispersal patterns.* Univ Chicago Press, Chicago, USA

Daniel WW (1990) *Applied nonparametric statistics.* PWS-KENT, Boston, MA, USA

Greenwood PJ (1980) Mating systems, philopatry, and dispersal in birds and mammals. *Anim Behav* 28:1140–1162

Howard WE (1960) Innate and environmental dispersal of individual vertebrates. *Am Midl Nat* 63:152–161

Jones WT, Waser PM, Elliot LF, Link NE, Bush BB (1988) Philopatry, dispersal, and habitat saturation in the banner-tailed kangaroo rat, *Dipodomys spectabilis. Ecology* 69:1466–1473

Lande R (1987) Extinction thresholds in demographic models of territorial populations. *Am Nat* 130:624–635

Levins R (1970) Extinction. In: Gerstenhaber M (ed) *Some Mathematical Questions in Biology.* American Mathematical Society, Providence, RI, USA, pp 75–108

Lidicker WZ Jr (1962) Emigration as a possible mechanism permitting the regulation of population density below carrying capacity. *Am Nat* 96:29–33

McCarthy MA (1995) Stochastic population models for wildlife management. Ph.D. thesis, Univ Melbourne, Australia

Murray BG Jr. (1967) Dispersal in vertebrates. *Ecology* 48:975–978

Porter JH, Dooley JL Jr (1993) Animal dispersal patterns: a reassessment of simple mathematical models. *Ecology* 74:2436–2443

Rees M (1993) Null models and dispersal distributions: a comment on an article by Caley. *Am Nat* 141:812–815

Rodgers AR, Klenner WE (1990) Competition and the geometric model of dispersal in vertebrates. *Ecology* 71:818–822

Stamps JA (1991) The effect of conspecifics on habitat selection in territorial species. *Behav Ecol Sociobiol* 28:29–36

Tonkyn DW, Plissner JH (1991) Models of multiple dispersers from the nest: predictions and inference. *Ecology* 72:1721–1730

Waser PM (1985) Does competition drive dispersal? *Ecology* **66**:1170–1175

Woolfenden GE, Fitzpatrick JW (1984) *The Florida Scrub Jay: demography of a cooperative-breeding bird.* Princeton Univ Press, Princeton, USA

Woolfenden GE, Fitzpatrick JW (1986) Sexual asymmetries in the life history of the Florida Scrub Jay. In: Rubenstein DI, Wrangham RW (eds) *Ecological aspects of social evolution: Birds and mammals.* Princeton Univ Press, Princeton, USA, pp 87–107

THE POPULATION BIOLOGY OF MARINE MAMMALS

John Harwood and Pejman Rohani

ABSTRACT

A number of features make the population dynamics of the 113 extant species of pinnipeds and cetaceans unusual. Most pinnipeds and many large cetaceans have been subject to extensive commercial exploitation in the past. Their potential rate of increase varies from 15% per annum in some fur seal species to 3% for some large whales. In the few cases where density-dependent processes have been identified, these have only a weak effect on the dynamics of their populations. Recent events have suggested that rare, episodic mass mortalities (caused by disease or failure in the food supply) may have an important effect on population dynamics. Some species are hovering on the verge of extinction, despite strenuous conservation attempts, whereas others have increased from a handful of individuals to population sizes of hundreds of thousands following the cessation of exploitation. Finally, there is growing public concern about the potential effect of pollution (in all its forms), over-fishing and climate change on the conservation status of marine mammals. In this paper, we describe the insights which theoretical population biology and detailed studies of the reproductive histories of individual whales and seals provide into the importance of these effects.

Key words: Population dynamics, marine mammals, seals, whales, conservation, management

Frontiers of Population Ecology, R.B. Floyd, A.W. Sheppard and P.J. De Barro (eds), CSIRO Publishing, Melbourne, 1996, pp. 173–190

Introduction

Marine mammals have been extensively exploited by man for millennia (see reviews in Bonner 1982; Tønnessen and Johnsen 1982; Busch 1985). In the 18th, 19th and early 20th centuries this exploitation caused massive declines in the abundance of the most commercially important species. Species which were easily accessible to European or North American sealers and whalers were driven to the edge of extinction in the early phases of commercial exploitation. More elusive and remote species were progressively over-exploited as technological innovations made them accessible to commercial operations. In the second half of this century, a number of attempts have been made to develop management policies to place this exploitation on a sustainable basis. In general, attempts to apply these policies have exposed the level of scientific ignorance of the factors which determine the abundance of marine mammals. For economic reasons, few marine mammals are now exploited commercially, but human activities remain the most serious threat to the persistence of the 113 surviving species of cetaceans and pinniped (scientific names for all species mentioned in the text can be found in Table 1, for pinnipeds, and Appendix 1 for cetaceans). For example, up to half a million dolphins were killed annually in the purse seine fishery for tuna in the eastern tropical Pacific during the 1960s and early 1970s (Joseph 1994).

Current interest in the population biology of marine mammals is concerned with two rather different issues. The first involves those species which are abundant or which are recovering from former exploitation. Here the problem is to determine what level of harvest (either as a result of deliberate or accidental killing) these populations can sustain and to develop a management regime that does not compromise their long-term persistence. The second issue relates to populations which are at low levels or are declining. Here, the problem is to identify the factors which may be responsible for the decline, decide which are most influential, and develop a management approach that not only minimises the risks from these factors but maximises the chances of recovery. The first issue is in the domain of resource management, the second in the domain of conservation biology. In this paper we will consider the contribution which theoretical population biology can make to these issues. In particular, we will focus on the work of A J Nicholson, but we will also draw heavily on the work of the late Graeme Caughley.

In order to understand both of the above issues, we need to identify the factors which may be important in determining population size in marine mammals and understand how these factors operate. Marine mammals are long-lived animals which are difficult to study; their populations are not amenable to conventional experimentation, although they have frequently been subjected to uncontrolled perturbation in the form of exploitation by man. We therefore need to look to the results of studies of more amenable species to help identify important factors. Nicholson (1957) recognised the importance of both extrinsic and intrinsic factors in determining population dynamics, but he considered that intrinsic factors were likely to be most important in determining the stability of natural populations. However, Caughley and Krebs (1983) concluded that the dynamics of mammals larger than 30 kg were more likely to be regulated by extrinsic than intrinsic factors, whereas the reverse was likely to be the case for those smaller than 30 kg. There are few marine mammals which weigh less than 30 kg.

Nicholson (1958) was a firm believer in the stabilising effect of density dependence, which he referred to as 'density-governed reaction'. He used the term 'stability' to indicate the absence of indefinite growth or decline in population numbers (Nicholson 1957) rather than the presence of a point equilibrium; nowadays we refer to this as persistence. As we shall see, there is little empirical evidence to indicate that marine mammal populations fluctuate around some equilibrium point, but they are undoubtedly persistent and this is a feature of their dynamics which theoretical models must replicate.

Table 1. Current status and rates of increase for living pinniped. Values are taken from Reijnders *et al.* (1993).

Common Name	Species	Status	Rate of Change
Northern sea lion	*Eumetopias jubatus*	Decreasing	82% decline since 1956
Californian sea lion	*Zalophus californianus*	Increasing	?
Southern sea lion	*Otaria byronia*	Decreasing	Decline of at least 80% since 1930s in parts of range
Australian sea lion	*Neophoca cinerea*	Rare but apparently stable	
Hooker's sea lion	*Phocarctos hookeri*	Rare but apparently stable	
Guadalupe fur seal	*Arctocephalus townsendi*	Rare but apparently increasing	
Galapagos fur seal	*A. galapagoensis*	Apparently stable	
Juan Fernandez fur seal	*A. philippii*	Rare but apparently increasing	
South American fur seal	*A. australis*	Apparently stable	
Subantarctic fur seal	*A. tropicalis*	Increasing	Up to 15% p.a.
Antarctic fur seal	*A. gazella*	Increasing	9–10% p.a.
Cape fur seal/Australian fur seal	*A. pusillus*	Increasing in southern Africa	3% p.a.
New Zealand fur seal	*A. forsteri*	Apparently increasing	
Northern fur seal	*Callorhinus ursinus*	Decreasing	–7% p.a.
Walrus	*Odobenus rosmarus*	Unknown	
Grey seal	*Halichoerus grypus*	Increasing	Up to 7% p.a
Harbour seal	*Phoca vitulina*	Increasing in many parts of range	Up to 12.5%
Largha seal	*Phoca largha*	Unknown	
Ringed seal	*P. hispida*	Some local populations decreasing	
Baikal seal	*P. sibirica*	Unknown	
Caspian seal	*P. caspica*	Unknown	
Harp seal	*P. groenlandica*	May be increasing in NW Atlantic	
Ribbon seal	*P. fasciata*	Unknown	
Hooded seal	*Cystophora cristata*	Unknown	
Bearded seal	*Erignathus barbatus*	Unknown	
Mediterranean monk seal	*Monachus monachus*	Decreasing	
Hawaiian monk seal	*M. schauinslandi*	Decreasing	
Weddell seal	*Leptonychotes weddellii*	Unknown	
Ross seal	*Ommatophoca rossii*	Unknown	
Crabeater seal	*Lobodon carcinophagus*	Unknown	
Leopard seal	*Hydrurga leptonyx*	Unknown	
Southern elephant seal	*Mirounga leonina*	Decreasing	–2–6% p.a.
Northern elephant seal	*M. angustirostris*	Increasing	14% p.a.

Nicholson (1957) provided one of the first experimental examples of the way in which harvesting at one stage in an animal's life cycle could increase population density at a subsequent stage, by increasing the growth rate of the surviving animals. This practical demonstration was undoubtedly influential in the development of the theories of sustainable yield which were applied in the early 1960s. During this period, approaches developed for fisheries management were applied to large whales, and those developed for the management of terrestrial wildlife were applied to seals. In the

1970s and 1980s the application of techniques from theoretical population biology showed why those approaches were unsuitable for the management of long-lived species with low intrinsic rates of increase (e.g. May *et al.* 1979; Harwood 1981).

In this paper, we first review some recent developments in theoretical population biology. We then consider how these may be relevant to understanding the population biology of marine mammals. In a highly influential paper, Caughley (1994) reviewed the current state of conservation biology and identified two paradigms: the small population paradigm which is concerned with theoretical risks of extinction for small populations; and the declining population paradigm which is concerned with the practical reasons why particular species or populations decline. He concluded that the small population paradigm needs more practical application and that the declining population paradigm needs more theory. In the final sections of this paper, we try to use information from marine mammals to assist in the development of both paradigms. First, we review the factors that have been implicated in the current and potential decline of some marine mammal populations and search for generalities. Finally, we consider how theoretical and practical concerns may be integrated into a common conceptual framework along the lines developed by the International Whaling Commission's (IWC) Scientific Committee as part of its Revised Management Procedure.

Lessons from theory

Nicholson's Work

Density dependence

Even now, there appears to be some confusion in the ecological literature about what is meant by density dependence (e.g. Murray 1994). For the purposes of this review we consider a process to be density dependent if survival or fecundity rates (or some factor which is highly correlated with these demographic variables) change monotonically with density. Some of the strongest experimental evidence for density dependence was provided by Nicholson's series of classic experiments on laboratory cultures of the Australian sheep-blowfly, *Lucilia cuprina* Wied., supplied with a constant supply of food and water. The cages were maintained under predetermined conditions and the numbers of individuals at the various developmental stages were noted on a daily basis (Nicholson 1957). In these experiments population density had a marked influence on dynamics. When the cages contained constant supplies of food for the larvae and ample resources for the adults, very pronounced cyclic population dynamics were observed. These were explained by Nicholson in terms of 'scramble' competition: when the number of adults in the population was high, a large number of eggs were laid but, due to the intense competition for resources, the resultant larvae were too small to pupate successfully and few adults emerged. The larvae from these adults experienced little competition and most managed to pupate, resulting in unchecked population growth. Eventually, high densities of adults developed and the cycle was repeated (Nicholson 1957). Precisely the same factors were at work when adults were presented with a limited protein supply. This time, the scramble competition was for the minimum amount of protein required for egg production.

Nicholson (1957) provided a verbal argument why 'contest' competition (where only a finite number of individuals have access to sufficient resources, thus depriving the inferior competitors of vital resources) should result in less violent oscillations. Two decades later, Hassell (1975) used a rigorous mathematical approach to confirm that contest competition is more stabilising than scramble competition. His model is given by a single difference equation describing the population densities of successive generations (N' and N)

$$N' = \lambda N(1 + aN)^{-b} \qquad (1)$$

where λ is the finite rate of increase and the parameters a and b define the strength of density dependence. More specifically, $b = 1$ reflects the situation where contest competition is in operation; scramble competition is observed as b approaches infinity. Hassell (1975) showed that the stability properties of this model depend solely on λ and b. As Nicholson correctly argued, contest is more stabilising when λ is large. However, when λ is small, the system is stable irrespective of the mechanism of competition.

Natural enemies

Nicholson (1958) also argued that mortality caused by natural enemies could act to stabilise population densities. There is a large body of experimental and theoretical work on the effects of natural enemies, much of which has focussed on the interaction between insect parasitoids and their hosts. The adult female parasitoid lays her eggs on, in or near the bodies of its host, usually another insect, which is eventually killed by the feeding parasitoid larva. This simple relationship between host attack and parasitoid recruitment means that host-parasitoids are extremely useful subjects for the study of predator–prey interactions. Simple models of these interactions, however, give rise to highly unstable and non-persistent dynamics in the form of diverging oscillations (Nicholson and Bailey 1935; Hassell 1978). Since natural host–parasitoid systems obviously persist, this discovery sparked off a search for mechanisms that could stabilise this interaction at the local population level.

Theoretical models have demonstrated that numerous factors, such as behavioural interference between searching adult parasitoids (Hassell and Varley 1969), heterogeneity in host susceptibility to parasitism (Bailey *et al.* 1962; Hassell and Anderson 1984), classical density dependence (Murdoch and Stewart-Oaten 1989; Ives 1992) and spatial heterogeneity in the risk of parasitism (Hassell and May 1973, 1974; May 1978; Chesson and Murdoch 1986; Pacala *et al.* 1990; Rohani *et al.* 1994) may have a stabilising influence on host–parasitoid systems.

Andrewartha and Birch (1954) and Nicholson (1957) recognised the importance of the spatial structure of populations for persistence, but a theoretical framework (based on the concept of the metapopulation: a collection of populations linked by dispersal) for evaluating the effects of spatial structure only emerged in the 1970s. It has had a profound influence on the way in which we view issues regarding persistence.

Metapopulation approaches

The development of the metapopulation concept has been succinctly reviewed by Hanski (1991). Metapopulation theory, in its classical form, is concerned with the balance between extinction and recolonization in these local populations (Levins 1969). More recent approaches, however, have explicitly concentrated on the population dynamics of metapopulations (Reeve 1988; Doebelli 1995; Rohani *et al.* in press). Some studies have demonstrated how local dispersal between a sufficiently large number of non-persistent populations can result in metapopulation persistence (Hassell *et al.* 1991; Comins *et al.* 1992; Rohani and Miramontes 1995). Spatial structure also enhances the coexistence of competing species by creating stable spatial segregation that may, at extremes, give rise to small 'islands' within the habitat (Hanski 1983; Tilman 1994; Hassell *et al.* 1994). These findings are important since they have added a new dimension to they way in which real-world persistence is viewed. Rather than assuming that regional persistence arises solely as a result of stable local dynamics, it is now understood that metapopulations can persist in spite of non-persistence at the local population scale.

Other models derived from this theory have been primarily concerned with the proportion of suitable habitat patches which are occupied (Gotelli and Kelley 1993; Tilman *et al.* 1994; Kareiva and Wennergren 1995). In these models, it is important to determine the geographical scale on which colonisation and extinction take place. If movement between local populations is frequent, they should be considered together as a single population; if dispersal rates are very low, each local population will have its own, unique dynamics. Hanski and Gilpin (1991) distinguish between the local scale (at which individuals move and interact with each other in the course of their routine feeding and breeding activities), and the metapopulation scale (at which individuals move infrequently from one population to another, typically only one or a few times per generation). Clearly, for mobile animals, like marine mammals, the local scale can cover a wide geographical area.

Metapopulation theory suggests that local populations exist on a knife edge of persistence. Many metapopulation models have two stable states, one of which corresponds to metapopulation extinction (Hanski 1991). Even a large population may become extinct if, by chance, it falls below the threshold where the rate of recolonisation is less than the rate of extinction. Increasing habitat fragmentation can accentuate this process because it reduces population size, decreases the average size of habitat patches, and increases predation by increasing the average distance between patches (Burgman *et al.* 1993)

It is not surprising that the metapopulation model has been seen as a paradigm for the management of rare and endangered species, which are often confined to virtually isolated patches of suitable habitat. However, Harrison (1991) found few examples which fitted the classical metapopulation model, where local populations periodically become extinct and the resulting unoccupied patches are recolonised after a few generations. She concluded that local extinction is more an incidental rather than a central feature for most so-called 'metapopulations'. In a later paper (Harrison 1994), she suggested that "any set of conspecific populations, possibly but not necessarily interconnected" can be considered as a metapopulation. By this definition, the significance of metapopulation structure for a species' dynamics must be evaluated on a case-by-case basis.

Individual-based models

Some ecologists have long argued that conventional population models are too simplistic. These models often ignore individual variations in size, age or genotype that are likely to affect population dynamics (Caswell and John 1992). The recognition that a population is not composed of a uniform collection of individuals, with identical average demographic parameters and environmental responses, highlights the need for a modelling approach in which individuals are the fundamental objects of study. Individual-Based Models (IBMs), treat the population as a collection of individuals, incorporating explicit rules for individual biology and their interaction with the environment (Driessen and Visser 1993; Judson 1994). The widespread use of IBMs has been largely facilitated by recent advances in computer technology, which makes the tracking of large numbers of individuals feasible.

The emergence of IBMs in ecology is a relatively recent development and, although many (e.g. Huston *et al.* 1988; Judson 1994) argue that IBMs point to the direction in which modelling should turn, the number of studies that have used this approach is still quite low. In a series of pioneering studies (McCauley *et al.* 1993; Wilson *et al.* 1993), McCauley and his colleagues showed that models which accounted explicitly for individual differences in behaviour could lead to the kinds of dynamics predicted by conventional population-oriented models. A number of other studies (Trebitz 1991; Madenjian *et al.* 1993; Karev 1994; Dunham and Overall 1994) have examined the predictive capabilities of IBMs. In general, the predictions of IBMs seem to be sensitive to the way in which individual behaviour is modelled (Bjornstad and Hansen 1994), and it is therefore difficult to draw general conclusions. As a result, the extent to which the use of IBMs will contribute to our understanding of population dynamics remains to be seen.

The population dynamics of marine mammals

Rates of increase and density dependence

Rates of increase

The pelagic life style of most marine mammals, their long lives and the fact that many species have been heavily exploited by man have all made it difficult to collect meaningful time series of abundance. A recent review of the conservation status of all dolphins, porpoises and whales (Reeves and Leatherwood 1994) has revealed the paucity of information on their population dynamics.

Marine mammals are long-lived and usually breed seasonally. As a result, their dynamics are best modelled by age-structured difference equation models. For these models, λ in Eqn 1 can be estimated approximately as exp – $\{(\mathrm{Ln}\ R_0)/T\}$, where R_0 is the expected number of offspring produced by the average individual in its lifetime and T is average generation time (May 1981).

Documented rates of increase in cetacean populations are typically rather low. Olesiuk *et al.* (1990) estimated an annual rate of increase of 2.9% for a killer whale population in the coastal waters of British Columbia and Washington State. Other estimates are 3.3% for California gray whales (Buckland *et al.* 1993), 4.8% for bottlenose dolphins (Wells and Scott 1990), 6.8% for southern right whales off the South African Coast (Best and Underhill 1990) and 7.7% for long-finned pilot whales around the Faroe islands (Martin and Rothery 1993). Although Best (1993) found that the recovery rates of some seriously depleted stocks of large whales were as high as 14.4%, these values may provide a poor indication of the long-term potential for recovery of these populations because the contribution of immigration and unbalanced age-structures to the observed rate of increase cannot be evaluated.

Most pinniped species breed colonially at a small number of traditional sites and it has therefore been easier to monitor changes in abundance than has been the case for cetaceans. Table 1 summarises the available information for the 33 extant species. Overall, 11 species are known to be increasing in abundance, four are stable and seven are known to be decreasing. The status of 11 is unknown, but in most of these cases populations are relatively large. Most of the increasing species are otariids in the southern hemisphere, which were heavily exploited in the 19th and early 20th century but which now have extensive legislative protection. Some of these populations are believed to be increasing at rates of up to 15% per annum, although the average is around 8%. Declines in numbers appear to be largely the result of increased mortality, usually anthropogenic in origin.

Density dependent processes

Because of their pelagic life-styles, it has been extremely difficult to obtain accurate estimates of demographic parameters for cetaceans. On the few occasions where a density-dependent relationship between parameter values and population size has been detected, its potential to regulate the population has been weak (Horwood 1987).

The best evidence for the existence of density-dependent processes in cetacean populations comes from observations made after the massive reduction of large whale biomass in the Southern Ocean which occurred during the first half of this century. Laws (1977) has reviewed this evidence. He calculated that the reduction in large whale numbers created a 'surplus' of some 150 million tonnes of krill (*Euphausia supurba*, the principle prey species of all the large whales in the Southern Ocean) each year. Gambell (1973) showed that fertility rates in mature blue and fin whales increased from 25% to 50–60% in the years before the Second World War and that pregnancy rates for sei whales increased from 25% to 60% between 1946–48 and 1969–70. Lockyer (1972) has presented evidence that the age of sexual maturity in fin whales decreased from about 10 years in the period 1910–30 to 5–6 years in the 1970s. Size at maturity did not change during this time (Laws 1962), suggesting that growth rates had increased. May *et al.* (1979) used these results to illustrate the multispecies consequences of a commercial fishery for krill in the Southern Ocean. However, the evidence marshalled by Laws (1977) must be interpreted with caution, since it is based entirely on samples collected from whales killed in the course of commercial exploitation. These animals were certainly not a representative sample of the populations involved, and the whaling industry's selectivity may have changed over the 60 years covered by these studies.

In contrast to cetaceans, density-dependent changes in survival and fecundity have often been demonstrated in pinniped populations, particularly on the breeding grounds. For example, Fowler (1990) recorded density-dependent changes in eight different population processes in the northern fur seal. Harwood and Prime (1978), and Doidge *et al.* (1984) showed how pup survival decreased with increased population density in grey seals and Antarctic fur seals, respectively. Bowen *et al.* (1981) showed that fertility rate increased and age at sexual maturity decreased in a harp seal population during a period of declining numbers. However, the increase in the population fertility rate was entirely explained by the change in age at sexual maturity. In all of these cases the documented density-dependent processes had a very weak effect on the population growth rate (Smith and Polachek 1981; Harwood 1981; Doidge *et al.* 1984) and were insufficient to limit population numbers within a reasonable range.

The population growth rate of pinnipeds is much more sensitive to changes in adult survival than it is to changes in any other demographic parameter (e.g. Harwood and Prime 1978). It remains possible, therefore, that adult survival may vary in a density-dependent fashion and regulate population numbers. However, given the practical difficulties involved in obtaining a reliable point estimate of this parameter, it seems unlikely that sufficient data will ever be available to allow the nature of density-dependent changes in adult survival to be quantified. In addition, pinnipeds now have few natural enemies and predation is unlikely to generate density-dependent changes in survival. Pathogens may act in this way, but the available evidence (see below) suggests that most diseases of marine mammals cause density-independent mortality. Changes in per capita food availability should affect female investment in reproduction long before they affect survival. Food-related changes in female investment have been documented, but these are usually linked to episodic collapses in local prey populations, which are only weakly linked to population density.

We know little about the foraging behaviour of marine mammals and the competition mechanisms which may influence them. However, their prey are typically heterogeneously distributed over a wide area and may occur in dense aggregations. Ecological theory suggests that spatial heterogeneity in prey distributions is likely to be an important stabilising factor (Hassell and May 1973; Rohani *et al.* 1994). The large stores of fat found in most cetaceans and pinnipeds provide a substantial buffer against temporary shortages of food, making it unclear whether competition between marine mammals for food should be in the form of scramble or contest. In the only example where Eqn 1 has been fitted to data from a marine mammal (Harwood and Prime 1978) the value of b was not significantly different from unity, implying contest competition. However, the significance of this is largely academic since the values of λ documented above place all marine mammals in a region of parameter space where their dynamics are likely to involve point equilibria for all values of b (Hassell 1975).

A major difference between marine mammals and many other animals is their long reproductive cycle. Once they have reached puberty, which may take 20 years for bowhead whales (Schell *et al.* 1989), female cetaceans come into oestrus at apparently irregular intervals; although this may be seasonal to an extent. In migratory cetaceans (for example, baleen whales), ovulations within a population are synchronised such that births occur at a specific stage in the migration cycle – typically in the warmer waters of the wintering grounds (Martin 1990). Pregnancy can last from 10–16 months; lactation may extend from four months to more than 3.5 years (Martin and Rothery 1993; Martin 1990).

This aspect of the biology of marine mammals introduce time delays into their dynamics, because the density-dependent processes which have been identified so far act on fertility or juvenile survival. Time delays have been demonstrated to be potentially destabilising (Beddington and May 1975; Murray 1989), depending on the size of λ and the length of the time delay. However, May (1980) has shown that such destabilising effects are unlikely given the combinations of demographic parameters which are observed in marine mammal populations.

Extrinsic factors

A number of large predators, including sharks (particularly the great white shark *Carcharhinus carcharias*), killer whales and leopard seals, are known to prey on marine mammals (for a review of predators of pinnipeds see Riedman 1990). However, it seems unlikely that these can cause sufficiently high levels of mortality to have a significant effect on the dynamics of marine mammals.

The mass mortalities of seals and dolphins in the North Sea, the Mediterranean Sea and the east coast of the USA since 1987 (reviewed by Harwood and Hall 1990) have indicated that disease can cause significant additional mortality in some marine mammal populations. However, the rapid recovery of harbour seal populations around the North Sea since 1988 (Anon. 1994), and evidence from other marine mammal populations which have been affected by disease outbreaks (Geraci *et al.* in press) suggests that the effects of disease may be only temporary. The tendency of pinniped to aggregate for breeding or resting out of water, and the tendency of many cetacean species to form large social groups provide ideal opportunities for the transmission of infectious diseases. This reduces the potential for mortality induced by such diseases to act in a density dependent way because the probability that a pathogen will be transmitted from an infected to a susceptible individual is virtually independent of population size. However, large-scale mortality induced by disease agents can increase the risks of extinction for species or local populations which are already at very low levels. The potential for the introduction of novel disease agents into marine mammal populations has increased in recent years because long-distance movements of domestic animals and humans have become much more frequent. Such introductions into naive populations may cause significant mortality.

Changes in food availability may affect marine mammal populations in a variety of different ways. In general, a decrease in availability below some threshold will reduce the efficiency of foraging, and thus net energy intake per day. Species may respond by emigrating, by slowing their growth rate, or by economising on some aspect of reproduction. The effect of El Niño events on seal populations in the Pacific has clearly demonstrated the results of large scale changes in food availability (Trillmich and Ono 1991). However, the population consequences of these events appear to be similar to those caused by disease. Although they may result in the loss of most of a cohort in the year in which they occur, this has only a temporary effect on the species' long-term dynamics.

Metapopulations and marine mammals

The metapopulation concept is probably applicable to colonially breeding pinnipeds and to cetacean species which form long-term stable social groups. Recent evidence obtained using techniques from molecular genetics (e.g. Allen *et al.* 1995; Amos *et al.* 1993; Hoelzel and Dover 1991) indicates that gene flow between colonies and groups is relatively restricted, despite the fact that marine mammals are physically capable of moving very long distances. On the evolutionary scale, as has been shown in theoretical models, such a population structure probably leads to greater population persistence because the risks of synchronised outbreaks of disease and of local failures in prey availability are reduced (Allen *et al.* 1993; Ruxton 1994); when one local population becomes extinct, the site is recolonised by dispersing individuals from the other populations. However, the effects of human exploitation, which are often directed at colonies or social groups, can result in a species having a highly fragmented distribution, as is the case for the Mediterranean monk seal (Harwood *et al.* in press).

Individual-based models

Individual marine mammals show a wide variation in body size (adult elephant seals and grey seals can differ in size by a factor of two) and in foraging behaviour. Since both of these features can have a significant effect on an individual's reproductive success, and since the proportion of animals showing different characteristics may vary during a population's history, it is important to examine their potential effects on the population dynamics of marine mammals.

Durant and Harwood (1992), and Durant *et al.* (1992) constructed an IBM of the dynamics of the highly endangered Mediterranean monk seal, primarily to identify the relationship between extinction risk and population structure and to develop management approaches aimed at minimising these risks. Brault and Caswell (1993) developed a matrix model of the population dynamics of the killer whale, using stable social groups ('pods') as the population unit. They found that most of the inter-pod variance in growth rate was due to variance in the reproductive output of individual females, but that this variance was not significantly greater than expected on the basis of the observed variation within the population. Chivers and DeMaster (1993) developed an IBM for dolphin populations to investigate the way in which the proportion of sexually mature females and the average age at sexual maturity might vary with population density. They concluded that population growth in dolphins is likely to be regulated by combination of density-dependent parameters.

Thus, to date, the results from the application of individual-based models to marine mammal populations are similar to those discussed earlier: they tend to confirm the results of more general models but do not provide any unique general insight.

Actual and potential causes of decline in marine mammal populations

A large number of extrinsic factors have been implicated in changes in the survival and reproduction of marine mammals. International bodies have classified these 'threats' in different ways. The International Council for the Exploration of the Sea (Anon. 1994) has divided them into factors which act directly on survival and reproduction (ie where the factors immediately result in death, abortion or sterility), and those that have an indirect effect (for example by increasing vulnerability to fatal infection, or by reducing foraging efficiency). Reijnders *et al.* (1993) divided them into immediate threats (deliberate killing and incidental catches by fishermen), intermediate threats (episodic mass mortalities, habitat degradation), and longer term threats (climate change and loss of genetic diversity). These differences are presentational rather than conceptual, and we have adopted the terminology of Reijnders *et al.*

Caughley (1994) has stressed the need to develop a conceptual framework for the 'declining population paradigm' and to formulate tight hypotheses for the potential impact of the factors which are believed to be important in a population's decline. However, it has often proved difficult to follow the latter recommendation for marine mammals. This is, in part, because intermediate and longer term threats may act synergistically. For example, the effects of pollution may only be evident when an animal is in poor condition as a result of changes in food availability or when it is challenged by a disease agent. The effects of disturbance may only be important when food availability is low. Because of this, it is often not possible to determine the relative importance of particular threats in any general way. Nevertheless, one generality does emerge clearly from the following review of perceived threats: all are anthropogenic in origin.

Immediate threats

Deliberate killing

Deliberate killing, either for commercial purposes or to reduce perceived damage to fishing gear or catches, will obviously increase the mortality rate for a population. Since adult mortality rates are low for many marine mammal species, even apparently small amounts of additional mortality can have significant effects on population dynamics. The tendency of marine mammals to aggregate in particular areas, and the vulnerability of seals when they are on land, means that high mortality rates can be imposed even when a species is relatively rare. Deliberate killing can make animals wary, and therefore more vulnerable to the effects of disturbance.

The world market for seal products is severely depressed (apart from demand in the Far East for certain items because of their reputed aphrodisiac properties) and very few seals are killed for commercial purposes (Reijnders *et al.* 1993). However, some seals are still killed by the indigenous peoples of Alaska, northern Canada, Greenland and Siberia. In general, reported catches appear to be small in proportion to the size of the population from which they are taken. However, in some cases the size of the population is poorly known. In the case of the Caspian seal, reported catches are large in proportion to estimated population size, and the latter is large is relation to the area of suitable habitat.

At present, the members of the International Whaling Commission have agreed to a pause in commercial whaling, although Norway lodged a formal objection to this agreement and has recently, and quite legitimately, resumed whaling for minke whales. If commercial whaling under the aegis of the IWC resumes, it will be under the tight restrictions of the Revised Management Procedure (RMP). Commercial exploitation of small cetaceans, either for meat or as bait in fishing, continues in a number of countries (Reeves and Leatherwood 1994).

Incidental catch

Incidental catches have exactly the same effects on population dynamics as deliberate killing. Shifts in fishing practice to the use of passively fished gear over the last 30 years have led to a significant increase in the numbers of marine mammals which are caught in fishing gear in some areas.

At least 20 species of seal are known to be caught incidentally during commercial fishing operations (Woodley and Lavigne 1991). Such incidental catches have been implicated in the decline of a number of species (harbour seals around Japan and in Alaska and the Baltic Sea, harp seal in the Barents Sea, Mediterranean and Hawaiian monk seals, northern fur seals, Steller sea lion, Hooker's sea lion). At least 57 of the 80 species of cetacean are known to be killed incidentally in different fisheries worldwide (Perrin *et al.* 1994). In some cases many thousands of individuals are involved.

It has proved difficult and expensive to document the size of incidental catches accurately. However, it has proved even more difficult to determine the ecological impact of these catches on the target population. In most cases, there are no estimates of the size of this population. In a few cases, estimates have been made of the number of marine mammals in the area where a particular fishery is operating. However, simply dividing the incidental catch by this number does not provide a reliable estimate of the mortality rate imposed by the fishery. The mammals in the area where the fishery operates may be only a part of a much larger population or, conversely, there may be a number of smaller, but discrete, populations within that area. At present, there is no simple way of resolving these problems: an integrated approach for determining the population structure of the marine mammals within the geographical area where a fishery operates is required.

Although the ecological impact of most incidental catches cannot be determined, public concern about these is so great that some operational response by the fishery is often required. The vulnerability of marine mammals to incidental capture in fishing gear appears to be related to their age and feeding strategies, and the types of gear used within a species' range. It may therefore be possible to identify those local geographical areas where incidental catches are likely to be a serious problem by bringing together available information on the foraging range of the most vulnerable marine mammals, the distribution of different types of fishing gear, and the likelihood of capture if a particular gear type is encountered. These data could then be used to modify fisheries practice to reduce the risks of incidental capture in a cost-effective way.

Intermediate and longer term threats

Disturbance

The potential effects of disturbance on seal populations have been reviewed by Reijnders *et al.* (1993). The most dramatic evidence of the effects which disturbance can have comes from changes in the numbers and distribution of Hawaiian monk seals on certain islands (Gerodette and Gilmartin 1990). The number of seals using Kuril Atoll declined from over 100 to less than 20 following the establishment of a US Coast Guard station in 1960. Conversely, the numbers of seals using Tern Island at French Frigate Shoals increased sharply after the Coast Guard station there was vacated in 1979.

Disturbance may have other long-term effects apart from redistribution. Disturbed animals may forage less efficiently, and be more vulnerable to the effects of pollution and disease through increased stress. However, these effects have never been quantified.

Pollution

Marine mammals, as top predators, are vulnerable to the effects of bioaccumulating pollutants, particularly those (like the organohalogens), which are fat-soluble. Organochlorines and some heavy metals are known, from experimental studies of other mammals, to have an effect on reproduction and the immune system, but sensitivity to these effects varies widely between species. The basic pathways through which these compounds act is well documented in other mammal species, but the situation is complicated in marine mammals because many compounds are only mobilised under certain conditions. Their effects may therefore only become obvious when the animals are stressed (for example, due to changes in food availability, because of disturbance, or after exposure to disease).

Changes in the physical environment

Physical factors can be as important as biological ones in determining the suitability of an area for marine mammals. However, although there is good information for many species on the distribution of sites used by seals for hauling out or pupping, little is known about what makes these sites suitable. Locations which are used for hauling out are often very specific and traditional, suggesting that there are certain factors which make them preferred. For example, walruses prefer haul out sites which have easy access to deep water because the animals find it difficult to move across extensive shallow areas.

Human activities can change the physical attributes of an area and make it much less suitable for seals. For example, the disappearance of the Mediterranean monk seal from many parts of its range is closely correlated with the development of coastal resorts. Icebreakers and the presence of offshore structures may alter ice characteristics and result in changes in the distribution of ringed seals (Stirling 1988). While there may be some benefit from improved access to certain feeding areas, this may also make the seals more vulnerable to predation by polar bears.

Global environmental change

One exogenous factor that could have a profound effect on the distribution, abundance and productivity of marine mammal populations is environmental change. The two changes in global conditions which have caused most concern in recent years are the dramatic rise in atmospheric CO_2 levels and increased exposure to ultraviolet radiation (UV-B) as a result of chlorine-induced depletion of the stratospheric ozone layer. The impact of these factors on marine ecosystems has not been studied to the same extent as for terrestrial systems and, in most instances, their effects can only be inferred.

Although a doubling of the atmospheric CO_2 level is expected to result in a 2.5°C increase in mean global surface air temperature, the sea surface temperature may respond more slowly (Tynan and DeMaster 1994). The changes in sea surface temperature and the consequences for the Antarctic ice edge could have strong ecological impacts since higher values of primary productivity are associated with the physical processes of ice retreat. In spring time, meltwater from receding ice edges increases the stability and stratification of the water column, which permits the growth of phytoplankton in a well-illuminated region (Smith and Nelson 1986). This is particularly important for the distribution of marine mammals since krill (mostly *E. supurba*), which are a major cetacean prey item, graze on epipontic algae during the winter and hence their abundance tends to be strongly correlated with the stable regions of the ice edge (Laws 1992). This effect was demonstrated for some cetaceans by Hammond and Zheng (1994) who showed that the density of minke whales declines away from the ice edge.

The reported seasonal reduction in ozone concentration (called the 'ozone hole') also has important consequences for primary productivity. The depletion of the ozone layer is accompanied by enhanced UV-B irradiance which depresses the rate of carbon fixation to depths of 25 m, as well as altering the pigment content of phytoplankton. These, combined with UV-B-induced photoinhibition in phytoplankton blooms in the Antarctic marginal ice zone, result in a minimum reduction in primary production between 6–12%. Recently, Roemmich and McGowan (1995) have shown that the biomass of macrozooplankton off southern California has declined by 80% since 1951, while the surface layer has warmed by up to 1.5°C in some places.

The consequences of climate change for marine mammals are hard to predict given the paucity of information currently available on what constitutes critical habitat for these species and their likely response to changes in prey distribution and availability.

Models for management

Following its agreement on a pause in commercial whaling, the IWC asked its Scientific Committee to develop a safe strategy for the sustainable exploitation of whales. The Scientific Committee was set three specific objectives: catch limits should be as stable as possible; catches should not be allowed on stocks at levels below 54% of the estimated carrying capacity; and the highest possible continuing yield should be obtained from the stock. The outcome of this research was the Revised Management Procedure (RMP) which, although accepted by the IWC, has not yet been implemented.

The RMP employs a simple discrete time model that requires a time-series of catches, estimates of absolute population size and specified value of productivity. The model then uses this information to estimate nominal catch limits (IWC 1994). An important feature of the RMP is that it has undergone extensive and rigorous simulation trials to examine its robustness to a number of factors including:

- incorrect assumptions regarding the underlying dynamics of the stock;
- bias and variable collection frequency in abundance data;
- errors in previous catch data;
- deteriorating environment;
- errors in assumptions about the stock identity.

The development of the RMP demonstrated that, in principle, it should be possible to manage whaling safely, with the price that catches (if allowed) will be very much lower than in the past. However, whether or not the RMP will be put into practice depends on a number of technical and political considerations.

The approach exemplified in the RMP is similar to that advocated by Sainsbury (1988) for the management of complex multispecies fisheries in tropical Australia, and to that proposed by a working group of the United Nations Environment Programme (Anon. 1992) for evaluating the potential effects of culls of marine mammals designed to benefit fisheries. In each case, the basis for management is a simple control model which contains little biological information. However, the robustness of this model to violations in its assumptions is tested by a range of 'deeper' simulation models which attempt to capture biological reality. There is no requirement for these models to provide precise predictions, and they are therefore much easier to specify than the more formal models which are actually used to formulate advice. If the management model is not robust to the hypotheses which are encapsulated in these simulations two options are available: either the form of the model can be modified to make it more robust, or the simulation model can be used to identify what environmental characteristics need to be monitored as part of the management procedure to avoid problems.

This approach could be applied effectively to many of the problems identified in the preceding sections. As the review of threats indicated, only in the case of deliberate or accidental killing has it been possible to actually quantify their impact on population dynamics. At present, concerns which have been expressed about 'intermediate' threats have not provided any clear formulation of their likely effects on marine mammal populations, rather than individuals. The approach developed for the RMP provides exactly the kind of conceptual framework advocated by Caughley (1994) for assessing these threats. Scientists who propose that a particular threat may be important should also suggest ways in which these threats can be incorporated into population models and how their potential consequences could be monitored. We do not think this is an unreasonable requirement. This approach does not require precise predictions, only the suggestion of a range of possible effects and how they could act. In other words, the way in which the threat is likely to affect the population should be formulated as a scientific hypothesis.

References

Allen JC, Schaffer WM, Rosko D (1993) Chaos reduces species extinction by amplifying local population noise. *Nature* **364**:229–232.

Allen PJ, Amos WA, Pomeroy PP, Twiss SD (1995) Evidence of strong genetic isolation between two British grey seal breeding colonies. *Molecular Ecol* **4**:653–662

Amos W, Twiss S, Pomeroy PP, Anderson SS (1993) Male mating success and paternity in the grey seal (*Halichoerus grypus*): a study using DNA fingerprinting. *Proc R Soc Lond B* **252**:199–207

Andrewartha HG, Birch LC (1954) *The distribution and abundance of animals.* Univ Chicago Press, Chicago, USA

Anon (1992) Report of the Scientific Advisory Committee of the Marine Mammal Action Plan. UN Envir Prog

Anon (1994) Report of the ICES workshop on the distribution and sources of pathogens in marine mammals. ICES CM 1994/N:2

Bailey VA, Nicholson AJ, Williams EJ (1962) Interaction between hosts and parasites when some hosts are more difficult to find than others. *J Theor Biol* **3**:1–18

Beddington JR, May RM (1975) Time delays are not necessarily destabilizing. *Math Biosci* **27**:109–117

Best PB (1993) Increase rates in severely depleted stocks of baleen whales. *ICES J Mar Sci* **50**:169–186

Best PB, Underhill LG (1990) Estimating population size in southern right whales *(Eubalaena australis)* using naturally marked animals. *Rep Int Whal Commn* special issue **12**:183–189

Bjornstad ON, Hansen TF (1994) Individual variation and population dynamics. *Oikos* **69**:167–171

Bonner WN (1982) *Seals and man: a study of interactions.* Univ Washington Press, Seattle, USA

Bowen WD, Capstick CK, Sergeant DE (1981) Temporal changes in the reproductive potential of female harp seals (*Pagophilus groenlandicus*). *Can J Fish Aquat Sci* **38**:495–503

Brault S, Caswell H (1993) Pod-specific demography of killer whales (*Orcinus orca*). *Ecology* **74**:1444–1454

Buckland ST, Breiwick JM, Cattanach KL, Laake JL (1993) Estimated population size of the California gray whale. *Mar Mam Sci* **9**:235–249

Burgman MA, Ferson S, Akcakaya HR (1993) *Risk assessment in conservation biology*. Chapman and Hall, London, UK

Busch BC (1985) *The war against the seals: A history of the North American seal fishery*. McGill-Queen's Univ Press, Kingston and Montreal, Canada

Caswell H, John AM (1992) From the individual to the population in demographic models. In: DeAngelis DL, Gross LJ (eds) *Individual-based models and approaches in ecology: populations, communities and ecosystems*. Chapman and Hall, New York, USA, pp 36–61

Caughley G (1994) Directions in conservation biology. *J Anim Ecol* **63**:215–244

Caughley G, Krebs CJ (1983) Are big mammals simply little mammals writ large? *Oecologia* **59**:7–17

Chesson PL, Murdoch WW (1986) Aggregation of risk: relationships among host–parasitoid models. *Am Nat* **127**:696–715

Chivers SJ, DeMaster DP (1993) An individual-based model to evaluate life history parameters as indicators of density dependence in delphinid populations. Paper SC/45/SM10 presented to the Sci Comm Int Whal Commn

Comins HN, Hassell MP, May RM (1992) The spatial dynamics of host–parasitoid systems. *J Anim Ecol* **61**:735–748

Doebelli M (1995) Dispersal and dynamics. *Theor Pop Biol* **47**:82–106

Doidge DW, Croxall JP, Baker JR (1984) Density dependent pup mortality in the Antarctic fur seal, *Arctocephalus gazella*, at South Georgia. *J Zool* **202**:449–460

Driessen G, Visser ME (1993) The influence of adaptive foraging decisions on spatial heterogeneity parasitoid population efficiency. *Oikos* **67**:209–218

Dunham AE, Overall KL (1994) Population response to environmental change – life-history variation, individual-based models, and the population dynamics of short-lived organisms. *Am Zool* **34**:382–396

Durant SM, Harwood J (1992) Assessment of monitoring and management strategies for local populations of the Mediterranean monk seal *Monachus monachus*. *Biol Conserv* **61**:81–92

Durant SM, Harwood J, Beudels RC (1992) Monitoring and management strategies for endangered populations of marine mammals and ungulates. In: McCullough DR, Barrett RH (eds) *Wildlife 2001: Populations*. Elsevier, London, UK, pp 252–261

Fowler CW (1990) Density dependence in northern fur seals (*Callorhinus ursinus*). *Mar Mamm Sci* **6**:171–195

Gambell R (1973) Some effects of exploitation on reproduction in whales. *J Reprod Fert* (suppl) **19**:533–555

Geraci JR, Harwood J, Lounsbury VJ (in press) The ecological importance of marine mammal die-offs. In: Reynolds JE, Twiss JR (eds) *Marine Mammals*. Smithsonian Institute Press, Washington, USA

Gerrodette T, Gilmartin WG (1990) Demographic consequences of changed pupping and hauling sites of the Hawaiian monk seal. *Conserv Biol* **4**:423–430

Gotelli NJ, Kelley WG (1993) A general model of metapopulation dynamics. *Oikos* **68**:36–44

Hammond PS, Zheng Y (1994) Towards a framework for investigating the effects of environmental change on whale populations. Paper SC/46/O28 submitted to 46th Meeting of the Sci Comm Int Whal Commn

Hanski I (1983) Coexistence of competitors in patchy environment. *Ecology* **64**:493–500

Hanski I (1991) Single-species metapopulation dynamics: concepts, models and observations. In: Gilpin M, Hanski I (eds) *Metapopulation dynamics: Empirical and theoretical investigations*. Academic Press, London, UK, pp 17–38

Hanski I, Gilpin M (1991) Metapopulation dynamics: brief history and conceptual domain. In: Gilpin M, Hanski I (eds) *Metapopulation dynamics: Empirical and theoretical investigations*. Academic Press, London, UK, pp 3–16

Harrison S (1991) Local extinctions in a metapopulation context: an empirical evaluation. In: Gilpin M, Hanski I (eds) *Metapopulation dynamics: Empirical and theoretical investigations.* Academic Press, London, UK, pp 73–88
Harrison S (1994) Metapopulations and conservation. In: Edwards PJ, May RM, Webb NR (eds) *Large-scale ecology and conservation biology.* Blackwells, Oxford, UK, pp 111–128
Harwood J (1981) Managing grey seal populations for optimum stability. In: Fowler CW, Smith TD (eds) *Dynamics of large mammal populations.* John Wiley and Sons, New York, USA, pp 159–172
Harwood J, Hall AJ (1990) Mass mortality in marine mammals: its implications for population dynamics and genetics. *Trends Ecol & Evol* **5**:254–257
Harwood J, Prime JH (1978) Some factors affecting the size of British grey seal populations. *J Appl Ecol* **15**:401–411
Harwood J, Stanley H, Vanderlinden C, Beudels M-O (in press) Metapopulation dynamics of the Mediterranean monk seal. In: McCullough DR (ed) *Metapopulations and wildlife conservation and management.* Elsevier, NewYork, USA
Hassell MP (1975) Density dependence in single-species populations. *J Anim Ecol* **44**:283–295
Hassell MP (1978) The dynamics of arthropod predator-prey systems. Princeton Univ Press, Princeton, USA
Hassell MP, Anderson RM (1984) Host susceptibility as a component in host–parasitoid systems. *J Anim Ecol* **53**:611–621
Hassell MP, Comins HN, May RM (1991) Spatial structure and chaos in insect population dynamics. *Nature* **353**:255–258
Hassell MP, Comins HN, May RM (1994) Species coexistence and self-organising spatial dynamics. *Nature* **370**:290–292
Hassell MP, May RM (1973) Stability in insect host-parasite models. *J Anim Ecol* **42**:693–726
Hassell MP, May RM (1974) Aggregation in predators and insect parasites and its effects on stability. *J Anim Ecol* **43**:567–594
Hassell MP, May RM, Pacala SW, Chesson PL (1991) The persistence of host–parasitoid associations in patchy environments. *Am Nat* **138**:568–583
Hassell MP, Varley GC (1969) New inductive population model for insect parasites and its bearing on biological control. *Nature* **223**:1133–1136
Hoelzel AR, Dover GA (1991) Genetic differentiation between sympatric killer whale populations. *Heredity* **66**:191–195
Horwood J (1987) *The sei whale: population biology, ecology and management.* Croom Helm, London, UK
Huston M, DeAngelis DL, Post W (1988) New computer-models unify ecological theory – computer simulations show that many ecological patterns can be explained by interactions among individual organisms. *Bioscience* **38**:682–691
Ives AR (1992) Density-dependent and density-independent parasitoid aggregation in model host–parasitoid systems. *Am Nat* **140**:912–937
IWC (1994) Report of the Scientific Committee. *Ann Rep Int Whal Commn* **44**:41–73
Joseph J (1994) The tuna-dolphin controversy in the Eastern Pacific Ocean; biological, economic and political impacts. *Ocean Development and International Law* **25**:1–30
Judson OP (1994) The rise of the individual-based model in ecology. *Trends Ecol & Evol* **9**:9–14
Kareiva P, Wennergren U (1995) Connecting landscape patterns to ecosystem and population processes. *Nature* **373**:299–302
Karev GP (1994) Individual-based models of forest vegetation dynamics. *Doklady Academii Nauk* **337**:273–275
Laws RM (1962) Some effects of whaling on the southern stocks of baleen whales. In: LeCren ED, Holdgate MW (eds) *The exploitation of natural animal populations.* Blackwell, Oxford, UK, pp 137–158
Laws RM (1977) The significance of vertebrates in the Antarctic marine ecosystem. In: Llano G (ed) *Adaptations within Antarctic ecosystems.* Smithsonian Institute Press, Washington, USA, pp 411–438
Laws RM (1992) Antarctica and environmental change – closing remarks. *Phil Trans R Soc Lond B* **338**:329–334

Levins R (1969). Some demographic and genetic consequences of environmental heterogeneity for biological control. *Bull Ent Soc Am* **15**:237–240

Lockyer CH (1972) The age of sexual maturity in the southern fin whale (*Balaenoptera physalus*) using annual layer counts in the ear plug. *J Cons Int Explor Mer* **34**:276–294

Madenjian CP, Carpenter SR, Eck GW, Miller MA (1993) Accumulation of PCBs by lake trout (*Salvenius-Namaycush*) – an individual-based model approach. *Can J Fish Aquat Sci* **50**:97–109

Martin AR (1990) *Whales and dolphins*. Salamander Books, London, UK

Martin AR, Rothery P (1993) Reproductive parameters of female long-finned pilot whales (*Globicephala melas*) around the Faroe Islands. *Rep Int Whal Commn* special issue **14**:263–304

May RM (1978) Host–parasitoid systems in patchy environments: a phenomenological model. *J Anim Ecol* **47**:833–843

May RM (1980) Mathematical models in whaling and fisheries management. In: Oster G (ed) *Some mathematical questions in biology, Vol. 13*. The American Mathematical Society pp 1–64

May RM (1981) Models for single populations. In: May RM (ed) *Theoretical ecology: principles and applications*. Blackwell, Oxford, UK, pp 5–29

May RM, Beddington JR, Clark CW, Holt SJ, Laws RM (1979) Management of multispecies fisheries. *Science* **205**:267–277

McCauley E, Wilson WG, de Roos AM (1993) Dynamics of age-structured and spatially structured predator-prey interactions: individual-based models and population-level formulations. *Am Nat* **142**:412–442

Murdoch WW, Stewart-Oaten A (1989) Aggregation by parasitoids and predators: effects on equilibrium and stability. *Am Nat* **134**:288–310

Murray BG (1994) On density dependence. *Oikos* **69**:520–523

Murray JD (1989) *Mathematical Biology*. Springer-Verlag, Berlin, Germany

Nicholson AJ (1957) The self-adjustment of populations to change. *Cold Spr Harbor Symp Quant Biol* **22**:153–173

Nicholson AJ (1958) Dynamics of insect populations. *Ann Rev Entomol* **3**:107–136

Nicholson AJ, Bailey VA (1935) The balance of animal populations Part I. *Proc Zool Soc Lond* **3**:551–598

Olesiuk PF, Bigg MA, Ellis GM (1990) Life history and population dynamics of resident killer whales (*Orcinus orca*) in the coastal waters of British Columbia and Washington State. *Rep Int Whal Commn* special issue **12**:209–43

Pacala SW, Hassell MP, May RM (1990) Host–parasitoid associations in patchy environments. *Nature* **344**:150–153

Perrin WF, Donovan GP, Barlow J (eds) (1994) Cetaceans and gillnets. *Rep Int Whal Commn* special issue **15**, Cambridge, UK

Reeve JD (1988) Environmental variability, migration, and persistence in host–parasitoid systems. *Am Nat* **132**:810–836

Reeves RR, Leatherwood S (1994) *Dolphins, porpoises, and whales*. International Union for the Conservation of Nature and Natural Resources. Gland, Switzerland

Reijnders P, Brasseur S, van der Toorn J, van der Wolf P, Boyd I, Harwood J, Lavigne D, Lowry L (1993) *Seals, fur seals, sea lions and walrus*. International Union for the Conservation of Nature, Gland, Switzerland

Riedman M (1990) *The pinnipeds: seals, sea lions, and walruses*, Univ California Press, Berkley and Los Angeles, USA

Roemmich D, McGowan J (1995) Climatic warming and the decline of zooplankton in the California current. *Science* **267**:1324–1326

Rohani P, Miramontes O (1995) Host–parasitoid metapopulations: the consequences of parasitoid aggregation on spatial dynamics and searching efficiency. *Proc R Soc Lond B* **260**:335–342

Rohani P, Godfray HCJ, Hassell MP (1994) Aggregation and the dynamics of host–parasitoid systems: A discrete-generation model with within-generation redistribution. *Am Nat* **144**:491–509

Rohani P, May RM, Hassell MP (1996) Metapopulations and local stability: The effects of spatial structure. *Theor Biol* (in press)

Ruxton GD (1994) Low levels of immigration between chaotic populations can reduce system extinctions by inducing asynchronous regular cycles. *Proc R Soc Lond B* **256**:189–193
Sainsbury KJ (1988) The ecological basis of multispecies fisheries, and management of a demersal fishery in tropical Australia. In: Gulland JA (ed) *Fish population dynamics*. John Wiley and Sons, London, UK, pp 349–382
Schell DM, Saupe SM, Haubenstock N (1989) Bowhead (*Balaena mysticetus*) growth and feeding as estimated by sigma 13C techniques. *Mar Biol* **103**:433–443
Smith WO, Nelson DM (1986) Importance of ice-edge phytoplankton in the Southern Ocean. *Bioscience* **36**:251–257
Smith TD, Polachek T (1981) Reexamination of the life table for northern fur seals with implications about population regulatory mechanisms. In: Fowler CW, Smith TD (eds) *Dynamics of large mammal populations*. John Wiley and Sons, New York, USA, pp 99–120
Stirling I (1988) Attraction of polar bears, *Ursus maritimus*, to offshore drilling sites in the eastern Beaufort sea. *Polar Record* **24**:1–8
Tilman D (1994) Competition and biodiversity in spatially structured habitats. *Ecology* **75**:2–16
Tilman D, May RM, Lehman CL, Nowak MA (1994) Habitat destruction and the extinction debt. *Nature* **371**:65–66
Tønnessen JN, Johnsen AO (1982) *The history of modern whaling*. Australian National Univ Press, Canberra, Australia
Trebitz AS (1991) Timing of spawning in largemouth bass – implications of an individual-based model. *Ecol Model* **59**:203–227
Trillmich F, Ono KA (eds) (1991) *Pinnipeds and El Niño*. Springer-Verlag, Berlin, Germany
Tynan CT, DeMaster DP (1994) Predictions of Antarctic climatic and ecological response to increase CO_2 and decreasing ozone. Paper SC/46/O3 submitted to 46th Meeting of the Sci Comm Int Whal Commn
Wells RS, Scott MD (1990) Estimating bottlenose dolphin population parameters from individual identification and capture-recapture techniques. *Rep Int Whal Commn* special issue **12**:407–415
Wilson WG, de Roos AM, McCauley E (1993) Spatial instabilities within the diffusive Lotka-Volterra system: individual-based simulation results. *Theor Pop Biol* **43**:91–127
Woodley TH, Lavigne DM (1991) Incidental capture of pinnipeds in commercial fishing gear. International Marine Mammal Association. Technical Report 91–101

Appendix 1

Scientific names of cetacean species referred to in the text

Bowhead whale	*Balaena mysticetus*
Gray whale	*Eschrichtus robustus*
Blue whale	*Balaenoptera musculus*
Fin whale	*Balaenoptera physalus*
Sei whale	*Balaenoptera borealis*
Minke whale	*Balaenoptera acutorostrata*
Killer whale	*Orcinus orca*
Long-finned pilot whale	*Globicephala melas*
Bottlenose dolphin	*Tursiops truncatus*

THE BALANCE OF PLANT POPULATIONS

Michael J. Crawley and Mark Rees

ABSTRACT

A.J. Nicholson made fundamental and lasting contributions to theoretical ecology, and our purpose is to show how his approach has improved our understanding of plant population dynamics. Drawing on case histories from single-species (feral oilseed rape and naturally regenerating English oak) we illustrate the circumstances under which plant recruitment might be expected to be seed-limited, and hence to define the conditions under which simple theoretical models of self-replacing dynamics might be appropriate. Next, we investigate the mechanisms responsible for promoting coexistence in multi-species plant systems, drawing on case histories involving annual (sand dune) and perennial plants (mesic grasslands). We emphasise the importance of distinguishing between transient (successional) dynamics and equilibrium (self-replacing) dynamics, and show that there is evidence from different ecosystems to support a variety of different mechanisms for coexistence (e.g. trade offs between dispersal and competitive ability in dune annuals; differences between perennial grassland plants in their pH and biomass tolerances). Nicholson's contribution to each of these fields was substantial, yet many of the misconceptions with which he wrestled 50 years ago (e.g. the role of density dependent processes in population regulation, the necessary and sufficient conditions for coexistence, the dynamic consequences of different kinds of competition) are still widespread amongst plant ecologists today.

Key words: Plant populations, transient dynamics, equilibrium dynamics, seed limitation, coexistence.

Frontiers of Population Ecology, R.B. Floyd, A.W. Sheppard and P.J. De Barro (eds), CSIRO Publishing, Melbourne, 1996 pp. 191–212

Introduction

A.J. Nicholson held what would nowadays be regarded as very up-to-date views on the way that ecological science ought to be carried out. He stressed that the interplay of observation, theory and experiment was most likely to produce important insights, and he was forthright in his criticism of those ecologists who he felt were anti-theory (see Table 1). With hindsight, it is clear that their prejudice was then, as now, based more on a phobia of mathematics than on any reasoned critique of the practice of theoretical ecology. Arguably, Nicholson's main contribution was to champion the argument that population size is not (and can not be) determined by a single factor (e.g. by climate or competition) but by the *interaction* of density independent and density dependent processes. He argued that the consequences of a given ecological factor on population dynamics could only be understood by incorporating the relevant process in a theoretical model. He pointed out that intuition was seldom a good guide to the behaviour of ecological systems, especially in trying to assess the relative impact of different processes on population dynamics. It is clear from his 1930's papers (see Table 1) that the Lotka Volterra competition model and, in particular, the coexistence criterion (that intraspecific competition must be more important than interspecific competition) had a major impact on his thinking. Nicholson's realisation that systematic search by individuals led to random search by populations (what he referred to as the area of discovery), and that this led to the creation of refuges from competitors or natural enemies, had a profound influence on subsequent theoretical models of predation and competition.

In addition to his fundamental contribution to insect host–parasitoid dynamics, his work anticipated many of the most important recent developments in theoretical ecology, including coexistence through temporal heterogeneity (Nicholson 1954, p45; Grubb 1977; Chesson 1985), coexistence in metapopulations (Nicholson 1960, p441; Atkinson and Shorrocks 1981), the ideal free distribution (Nicholson 1957, p169; Fretwell and Lucas 1970), enemy regulation of herbivore populations (Nicholson 1954, p50; Hairston *et al.* 1960), coexistence and threshold levels of resource depletion (Nicholson 1954, p57; Tilman 1982), and so on (see Table 1).

Dynamics of single species populations

The essential properties of a population of a single species of annual plant can be understood by constructing so called 'Ricker curves' which relate seed density in year t to that in year $t+1$ (May and Oster 1976). This type of model structure is appropriate only for those habitats where seed-limited recruitment is the norm (i.e. where there are ample microsites available for colonisation in every year, such as in sand dunes, salt marshes and the margins of arable fields; Symonides 1988; Watkinson and Davy 1985). The slope of the relationship between seed densities in successive years, evaluated at the equilibrium, determines whether the population will be characterised by a stable equilibrium point, population cycles or chaos. If the slope is between 0 and 1 the equilibrium point is approached smoothly, if it is between 0 and –1 there will be damped oscillations and finally when the slope is less than –1 the population will exhibit cycles and, if the slope is sufficiently steep, chaos (Fig. 1). In many plant populations the relationship is relatively flat topped, so that the slope at intersection is much shallower than –1, which results in stable dynamics.

Consider the simplest case of a population of annual plants with no seed dormancy. In order to predict the number of seeds next year (and then from next year to the year after that, and hence the long-term dynamics of the population) we need to know four things: 1) how total biomass production is related to initial seed density; 2) the distribution of biomass between individuals; 3) the relationship between individual biomass and fecundity; and 4) the fraction of seeds that survive over-winter to initiate the subsequent generation.

Table 1. Quotations from A.J. Nicholson on a variety of topics that have subsequently become major foci of work in theoretical ecology. Parentheses show the date and page numbers of the quotations (see References).

On the scientific practice of ecology

The common and reiterated insistence upon the paramount importance of observation and experiment, and the deprecation of 'theorizing' (which seems to be the fashionable word for any deliberate and sustained thought) indicates a gross misunderstanding of the scientific method. There appears to be a widespread idea that the facts of nature can be revealed by observation and experiment alone, so avoiding both the pitfalls and labour of thought (1954, p. 54)

On plant population dynamics and lottery models

When a plant species invades a new and favourable area, population growth is at first exponential, the population consisting dominantly of young individuals. By the time that crowding significantly impedes further population increase, the number of individuals greatly exceeds the number of mature plants the space can accommodate, for the space requirements of young plants are small. Subsequently, there is a fall in density as the more powerful, or more fortunate, maturing individuals displace their neighbours, until equilibrium is reached when the only space available to young individuals is that vacated by the dead (1954, p. 40-41)

On estimating the importance of different processes in population dynamics

If an attempt be made to assess the relative importance of the various factors known to influence a population, no reliance whatever must be placed upon the proportion of animals destroyed by each. Instead, we must find which of the factors are influenced, and how readily they are influenced, by changes in the density of animals (1933, p. 136)

It is the interaction of the insects themselves that produces balance and so limits density, while physical factors, by modifying this interaction, influence the position at which the insects themselves limit their population (1933, p. 134)

Non-reactive factors cannot determine population densities for, if sufficiently favourable, they permit indefinite multiplication or, if not, they cause populations to dwindle to extinction. On the other hand, they inevitably limit distribution to those areas within which they are favourable (1954, p. 59)

On coexistence and temporal heterogeneity

If the fluctuating conditions are definitely unfavourable at times, a population progressively decreases during those times, but at the intermittent favourable periods the population tends to adjust itself to the prevailing conditions. This determines the density at the beginning of each unfavourable period and therefore also the general level from which the population falls during these periods. Density governance is merely relaxed from time to time and subsequently resumed, and it remains the influence which adjusts population density in relation to environmental favourability (1954, p. 45)

On coexistence and spatial heterogeneity

For the steady state to exist, each species must possess some advantage over all other species with respect to some one, or group, of the control factors to which it is subject (1933, p. 147)

In nature different species competing for the same requisite are seldom, if ever, completely coextensive both spatially and ecologically. Each of several competing species with overlapping distributions can maintain itself in those places where it possesses an advantage over all the others; and it can also spread into neighbouring habitats of the others and remain there indefinitely. The tendency for the locally inferior species to be displaced being offset by continued invasion from the areas in which it enjoys advantage (1960, p. 499)

On natural selection and population dynamics

The function of natural selection is to select - not to produce balance (1933, p. 137)

Natural selection continually tends to disturb population balance by improving the properties of competing species, instead of producing balance, as it is so often supposed to do (1954, p. 53)

It should be noted that the selection of advantageous properties in one species may cause it to deplete some requisite it shares with another to below the threshold density for the competing species, which it therefore displaces (1954, p. 53)

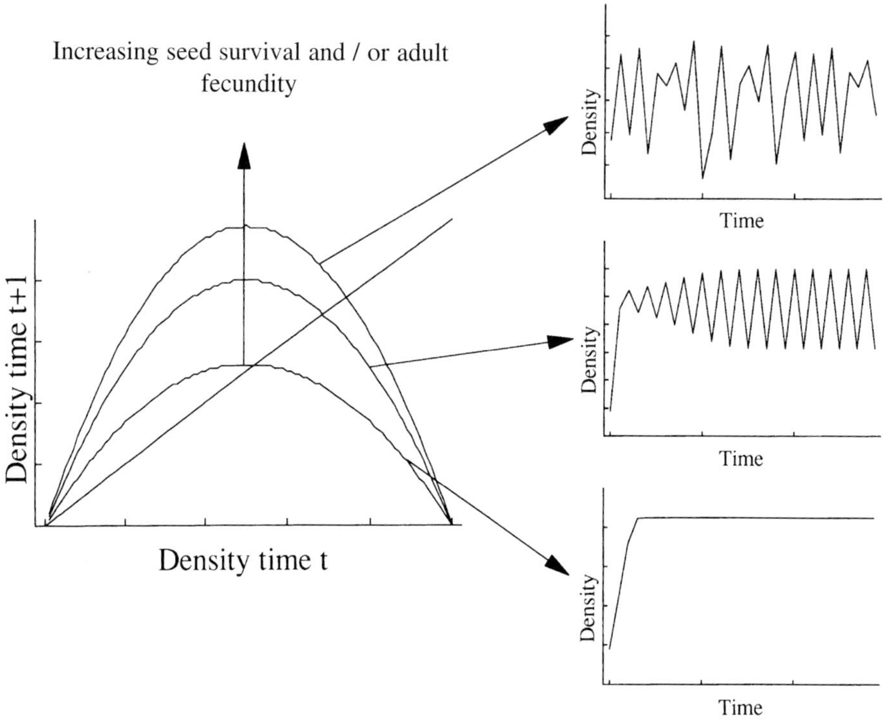

Fig. 1. Relationship between successive population densities and the resulting population dynamics. The slope of the 'Ricker' curve, evaluated at the equilibrium point (where it cuts the 45 degree line) determines the type of dynamics observed (see text for details).

Biomass production and initial seed density

The relationship between total biomass and initial seed density is described by an empirical law – the law of constant yield – which states that total biomass production rapidly becomes independent of the density of seeds sown (Kira *et al.* 1953; Donald 1951; Palmblad 1968). Coupled with realistic assumptions about the asymmetry of plant competition (e.g. the Nicholsonian contest for light where large plants have a disproportionately adverse effect on small plants, on a weight for weight basis; Weiner 1990), the law of constant yield arises naturally for most mathematical functions describing the effects of neighbours on individual biomass (Pacala and Weiner 1991).

Distribution of biomass among individuals

The distribution of biomass among individuals in a population depends on many factors such as differences in germination date, genetic makeup, abiotic environment, herbivore pressure and competitive interactions. Empirical studies demonstrate that most plant populations consist of many small individuals and relatively few large ones (Obeid *et al.* 1967; Ogden 1970) and the mechanism causing this asymmetry is thought to be competition for light. Tall individuals shade short ones but not vice versa and hence tall plants become taller, so generating positive feedback which accentuates the differences between small and large individuals; asymmetric competitive interactions generate positively skewed patterns in the distribution of biomass among individuals.

Size-fecundity schedules

Plant size-fecundity schedules have received considerable attention in the literature (Samson and Werk 1986; Rees and Crawley 1989; Klinkhamer *et al.* 1992). To a good approximation, the majority of plant species show a linear relationships between reproductive and vegetative biomass (Samson and Werk 1986; Rees and Crawley 1989), although some non-linear relationships have also been found (Klinkhamer *et al.* 1992; Rees and Brown 1992). If it is generally true that plants have small threshold sizes for reproduction, so that seed production is directly proportional to shoot biomass, then total seed set by a plant population will be independent of the size distribution. Hence, we need only to know the total biomass produced, and not the distribution of biomass between individuals within the population, in order to predict the total seed set. It follows, therefore, that the law of constant yield for biomass implies constant seed set at high plant densities.

If, however, plants exhibit a substantial threshold size for reproduction, then the proportion of total biomass allocated to reproduction by individual plants will be an increasing function of their biomass. In this situation, both the spatial arrangement of neighbours and the symmetry of competition between individuals could have important effects on total seed set and hence on long-term dynamics. Whether or not size differences between individuals are important depends, therefore, on the existence of a threshold size for reproduction. Nicholson's idea of scramble competition involved a wastage of resources, which, in the case of plants implies the existence of mature plant biomass that produces no seeds at all. In our view, this is uncommon and relatively unimportant in the dynamics of most plant populations. Typical plant dynamics appear to be characterised by a direct proportionality between size and fecundity and to be dominated by strongly asymmetric, highly stabilising contest competition.

The proportion of seeds that survives over winter

There are two cases to consider. First, the yield-density relationship might be flat topped, resulting in stable population dynamics regardless of the probability of over-winter seed mortality; the equilibrium population size simply increases as over-winter seed survival increases. Alternatively, the yield-density relationship might be humped, so that as the fraction of seeds surviving becomes greater, so the absolute value of the slope of the yield-density curve (evaluated at equilibrium) becomes steeper, and the population dynamics become less and less stable. Likewise, increasing the fertility of a habitat would result in an increase in average plant fecundity and this, in turn, would steepen the slope of the yield-density relationship at equilibrium producing a similar, destabilising effect on population dynamics.

These results are easily extended to include the effects of dormancy and the formation of a seed bank. Inclusion of a seed bank is stabilising because even following a high density reproductive crash there are plenty of seeds to recruit in the following year (Pacala 1986). A good example of the biology described above is given by *Abutilon theophrasti* (Thrall *et al.* 1989) which has a reproductive threshold of 0.41 g . Calibrated models predict that if *A. theophrasti* had no seed bank then its population would exhibit persistent cycles. Inclusion of a seed bank stabilises the population model, resulting in damped oscillations which correspond closely to the dynamics observed in the field.

These simple models make the counter-intuitive prediction that the effects of stress on plant population dynamics will tend to *increase* population stability, whilst lowering equilibrium population size. Stability is enhanced by increasing the size differential between dominant and subordinate plants, by reducing fecundity, and by increasing seedling mortality, any or all of which might be expected in stressed plant populations. Some effects of stress may, however, lead to a decrease in population stability, for example by reducing or eliminating the seed bank. In other words, intuition is a poor guide for predicting the effects of particular stresses on plant population dynamics. As Nicholson repeatedly emphasised, the only way to make sensible predictions about the impact of particular processes on population dynamics is to incorporate their effects into a model.

Analysis

Do analyses of plant census data support the prediction that many plant populations are characterised by stable dynamics? A common form of analysis is to regress population size in year *t* on density in year *t*–1 and/or *t*–2 (Ricker curves). If no significant relationship is found, then some botanists have assumed that population sizes are de-coupled, with the implication that there are no density dependent processes operating to stabilise the population. As Nicholson understood, both of these conclusions are incorrect. Suppose we have a simple population with two important features: 1) the parameters are such that the equilibrium point is stable and approached smoothly (see above); and 2) the per capita rate of increase is a random variable that changes from year to year because of, e.g. variations in the intensity of drought. A typical population trajectory produced from the model is shown in Fig. 2a. A regression of density at time t on density at time *t*–1 shows no significant relationship (Fig. 2b). However, there is a strong, highly significant relationship between the logarithmic growth rate $ln(N_{t+1}/N_t)$ and density. The correct analysis (Fig. 2c) demonstrates that density dependent processes *are* operating and that the population *is* regulated. In fact we can go further because the relationship in Fig. 2c describes a population with a stable equilibrium point that is approached by a smooth trajectory. Hence, using an appropriate form of analysis we can discover a great deal about the underlying population dynamics (Turchin and Taylor 1992). The incorrect analysis (Fig. 2b) yields incorrect and misleading information.

Population persistence of plants in variable environments

Populations inhabit spatially and temporally variable environments and this has obvious and important implications for population persistence. Here, we develop a simple model for an annual

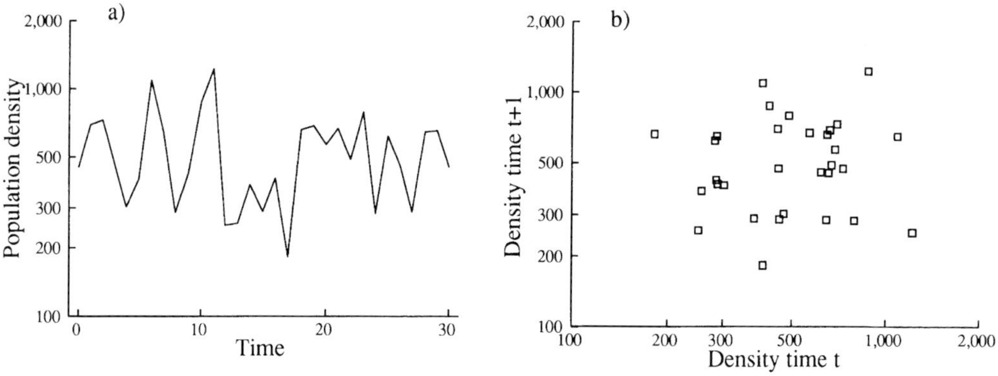

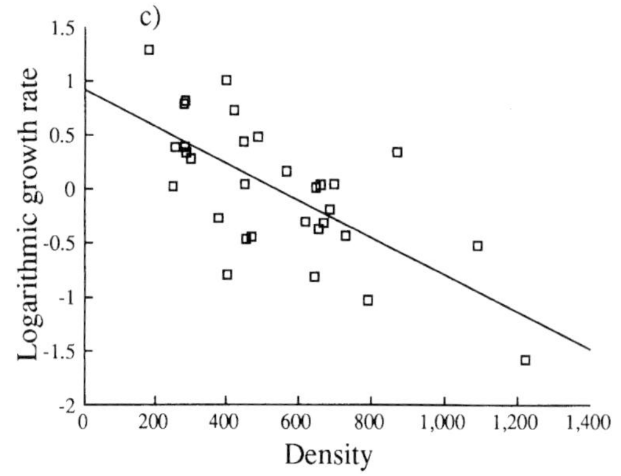

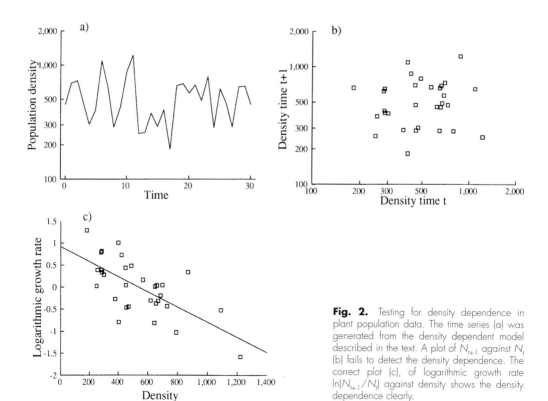

Fig. 2. Testing for density dependence in plant population data. The time series (a) was generated from the density dependent model described in the text. A plot of N_{t+1} against N_t (b) fails to detect the density dependence. The correct plot (c), of logarithmic growth rate $\ln(N_{t+1}/N_t)$ against density shows the density dependence clearly.

plant inhabiting an environment where the factors that limit growth vary in both time and space. We begin by considering spatial variation alone. The condition for population persistence is simply stated; we require that on average each seed replaces itself with more than one seed when population size is small (this is the invasion criterion;

$$\frac{dN}{dt} = 0$$

for small *N*). If this is the case, then the population will increase when rare and eventually reach some density dependent equilibrium. In contrast, if each seed replaces itself with less than a single seed then population size will decrease and the population will become extinct. By applying these intuitively obvious criteria it is relatively easy to explore how changes in the environment alter the likelihood of population persistence. Before we do so, however, note two things. First, population persistence is not the same as population stability; a population can persist without necessarily being stable. Second, although the models are derived with annual plants in mind, the insights also apply to perennials. For perennials we need to construct more complicated, stage-structured models in order to predict persistence (e.g. Caswell 1989), but although the calculations are algebraically more complex, the predictions are qualitatively similar.

Assume that the habitat consists of a large number of microsites. Some fraction (1 – *E*) of these contain perennial plants and any seeds from the annual that germinate at a site occupied by a perennial die before reproduction (Crawley and May 1987). Recruitment from seed fails often completely in the presence of perennial vegetation (Fenner 1985; Rees and Brown 1991; Rees and Long 1992). We assume that some fraction *d* of the seeds die over winter and that in spring a fraction *g* seeds germinate. Those that germinate in empty microsites produce *F* seeds and those that germinate in microsites occupied by perennials produce none. Hence, the number of seeds next year, S_{t+1}, produced from the S_t seeds this year is

$$S_{t+1} = (1-d)(1-g)S_t + (1-d)gEFS_t \qquad (1)$$

The first term represents the seeds that survive but do not germinate (the seed bank), while the second predicts the number of seeds produced by those seeds that survive and germinate in empty microsites. Dividing S_{t+1} by S_t gives the finite rate of increase (λ) and for persistence we require l > 1. Because we are interested in the behaviour of the population only when population size is small, we have not included any density dependence in the model. Intuitively, the formation of a seed bank (*g* < 1) might be expected to foster persistence of the annual. This is not necessarily the case – the formation of a seed bank can drive the population extinct (Rees and Long 1992). This result occurs because delaying germination does not increase the probability of a seed germinating in an empty microsite and so cannot increase the finite rate of increase. There is a cost to forming a seed bank that results from the death of seeds before germination, which is why forming a seed bank makes persistence more difficult. The consideration of simple models focuses attention on the entire life-cycle, rather than on just, e.g. adult fecundity, a further point on which Nicholson was adamant. If the aim is to understand the effects of a particular factor on dynamics then how that factor affects seed mortality, germination, the fraction of sites suitable for colonisation *and* average adult fecundity needs to be known.

This simple model assumes that germination biology can be summarised by a single parameter (*g*). However, experimental work has demonstrated that the probability of a seed germinating is reduced in the presence of established plants (Gorski *et al.* 1977; King 1975; Rice 1985; VanTooren and Pons 1988; Rees and Brown 1991). This may occur through changes in the red – far red ratio of the incident radiation, alteration of the range of temperature fluctuations, or a reduction in soil nitrogen levels. It is therefore necessary to modify the model to allow for different germination probabilities in empty microsites and in those occupied by perennials. If we assume that the probability of germinating in an unoccupied microsite is g_u, and that the probability of germination in a microsite occupied by an established perennial plant is *lower* (g_0), we obtain the following model

$$S_{t+1} = (1-d)\{1-\Omega\}S_t + (1-d)g_u EFS_t \tag{2}$$

where $\Omega = g_u E + g_0(1-E)$. For persistence we require

$$(1-d)\{1-g_0(1-E)+g_u E(F-1)\} > 1 \tag{3}$$

Comparison with the first model is easier if we set

$$g = g_0(1-E) + g_u E$$

so that the fraction of seeds that germinate in each time interval is equal (this ensures that the cost of forming a seed bank, i.e. mortality of buried seeds, is the same in both models). Persistence then becomes easier whenever $g_u > g_0$. In other words, inhibition of germination by perennial plants promotes persistence. It is worth re-emphasising that it is not delaying germination *per se* that promotes persistence, but the seed's germination response to the presence of established plants (we might call this the 'clever seeds' versus 'dumb seeds' distinction).

These simple models assume that all microsites not containing established perennials are equally suitable for growth. However, we would expect the presence of perennial neighbours might reduce microsite quality by shading or nutrient uptake. Several studies have demonstrated that the relationship between plant fecundity F_c, and the weight or number of neighbours is a non-linear function (Goldberg 1987; McConnaughay and Bazzaz 1987; Miller and Werner 1987; Pacala and Silander 1985) often well described by the simple exponential:

$$F_c = F.e^{-\alpha i} \tag{4}$$

where F is the fecundity of a plant with no neighbours; i is the number of perennial neighbours and α is a decay parameter. In order to determine the condition for persistence of an annual plant in such an environment, we must calculate its average fecundity. This may be approximated by expanding F_c about the mean number of perennial neighbours and taking expectations giving

$$\approx F\exp(-\alpha\bar{i}) + \frac{F\alpha^2\sigma^2}{2}\exp(-\alpha\bar{i}) \tag{5}$$

The first term on the right-hand side is the fecundity of a plant with the average number of perennial neighbours ($\bar{i}$), the second term is positive and proportional to the variance (σ^2) in the number of neighbours, demonstrating that variance in microsite quality promotes persistence relative to the average environment. Thus, the spatial arrangement of perennial plants can be important in determining persistence of annuals. If perennials are spatially aggregated (resulting in a large σ^2) then this will promote persistence relative to a more even spatial distribution because of the provision of refuges for the annual from competition by the perennials.

Temporal variation

So far we have assumed that the fraction of sites available for colonisation is constant from year to year. However, many annuals live in successional environments where the fraction of sites available for colonisation (E) varies through time. In the simplest successional environment virtually all microsites will be available for colonisation after a large-scale disturbance ($E \approx 1$), whereas if there is no disturbance (or the last disturbance occurred several years ago), then virtually all sites will be occupied by perennial plants ($E \approx 0$). Consider the case where a constant proportion of the seeds germinate. In a constant environment, we have seen that the formation of a seed bank made persistence more difficult, but in randomly varying environments this simple result no longer holds. If the probability of germination is high, the population rapidly declines in years when there is no large-scale disturbance, resulting in extinction. A lower germination rate results in slower decay of the seed bank between disturbances resulting in persistence. However, the germination rate cannot be too low, otherwise most seeds die before they have a chance to germinate and this drives the population to extinction. As we have already seen, seeds which can alter their germination probabilities in relation to the suitability of the microsite for seedling establishment, strongly promotes persistence. This is because seeds are not wasted by germinating in the perennial-dominated environments that occur during the years between the large-scale disturbances, and so seeds are available in the seed bank when disturbance eventually destroys the perennial cover and opens up potential microsites for the annuals (Rees and Long 1992).

Thus we see that changing the assumptions of population models alters our predictions about population behaviour. It is exactly this property of models that so exercised Nicholson's contemporary scientific adversaries (e.g. Thompson who distrusted their simplicity and Fisher who disliked their determinism, cf. Kingsland this volume). But modelling shows what can happen, given certain conditions and certain assumptions; models define what is possible. It is the job of observation and experiment to separate the actual from the possible.

Case Study 1: Oilseed rape, an alien annual of ephemeral habitats

While populations of annual plants are extremely convenient to observe, to experiment with and to build models of, they tend often to inhabit ephemeral, early successional conditions and hence can never exhibit year-to-year patterns of population dynamics driven by the classic, text book paradigm of

$$N_{t+1} = \lambda N_t . f(N_t) \qquad (6)$$

where λ is the per capita rate of increase and $f(N_t)$ is some density dependent function.

Vigorous growth of perennial plants generally means that the above ground populations of annual plants are locally extinct within 2 to 4 years (Crawley *et al.* 1993), so the pattern of dynamics is more a reflection of the nature of the disturbance regime, than an unfolding of a seed-limited, purely density-dependent process. There are habitats like deserts and sand dunes where annual plants do have long-term, self replacing dynamics, and where simple competition models borrowed from animal ecology can be employed: the most widespread model embodies Nicholson's concepts of scramble and contest competition

$$N_{t+1} = \frac{\lambda N_t}{(1+aN_t)^b} \qquad (7)$$

where $b = 1$ represents contest competition and $b = \infty$ scramble (Hassell 1975). The equilibrium population size

$$N^* = \frac{\lambda^{1/b} - 1}{a} \qquad (8)$$

increases with λ, the rate of increase, and declines with b, the nature of density dependence and with the scaling factor, a. The stability properties of the equilibrium are well known and depend upon the slope of the recruitment curve evaluated at equilibrium (i.e. on both λ and b, but not on the scaling parameter, a; see above); cycles and chaos are associated with values of b larger than those exhibited by typical plant populations (Hassell *et al.* 1976; Watkinson 1980).

Feral populations of oilseed rape present something of a paradox. Yellow drifts of flowers of oilseed rape (*Brassica napus* subsp. *oleifera*) are a characteristic feature of British motorway verges in early spring. On the other hand, it is extremely difficult to establish oilseed rape populations by sowing seeds into established vegetation, and populations which are experimentally established on disturbed ground tend to go extinct after only a few years (Crawley *et al.* 1993). Two pieces of work carried out by Crawley and Brown (1995) on roadside populations of oilseed rape set out to address 4 related questions: 1) under what circumstances is recruitment in oilseed rape seed-limited; 2) what is the relative importance of imported seed, the seed bank and seed produced by resident plants in determining recruitment; 3) what is the annual turnover in patch occupancy by oilseed rape; and 4) how long do established oilseed rape populations persist in the absence of soil disturbance?

The apparent permanence of motorway verge rape populations could result from self-replacement of local populations by recruitment from seed produced by resident plants. Alternatively, the impression of permanence might be an illusion, with local extinction of rape patches compensated by recruitment following soil disturbance in other places. Patches of rape might establish from a seed bank, from seeds dispersed from nearby fields or feral populations, or from imported seed (e.g. from rape seed spilled from lorries).

Feral oilseed rape populations in 3658 100m-quadrats on the verges of London's orbital motorway (the M25) have been studied annually since 1993. There is substantial turnover in site occupancy between years; for example, 55% of quadrats had a different population density in 1994 than in 1993; 53% of the quadrats occupied in 1993 were locally extinct in 1994, and 20% of the empty quadrats in 1993 were occupied by oilseed rape in 1994. Verges next to the carriageway carrying traffic towards the main rape-seed crushing plant at Erith in Kent had significantly more plants than the opposite verge carrying traffic away from Erith (Fig. 3). Mean rape densities were also higher in

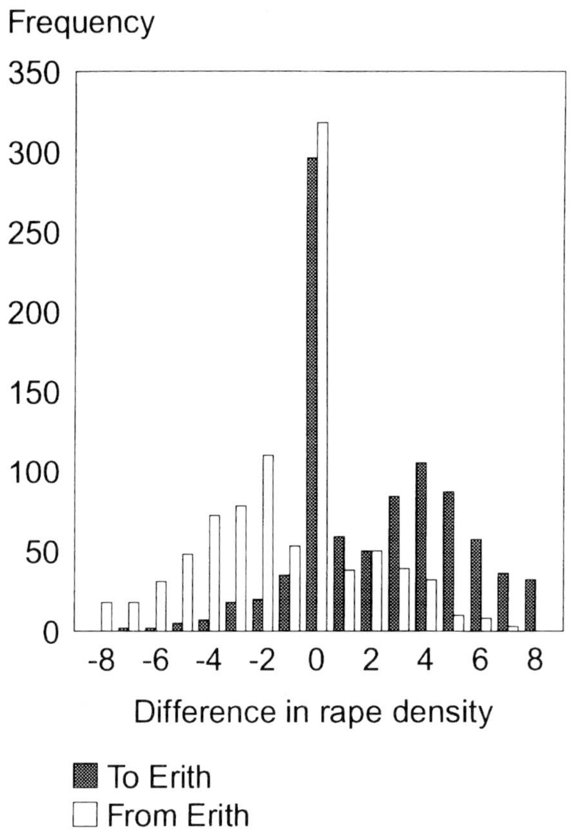

Fig. 3. Frequency of differences in density scores for the number of flowering shoots of feral oilseed rape per 100 m strip of London's orbital motorway verge (0=none, 1=1, 2=2, 3=3 to 4, 4 = 5 to 8, 5 = 8 to 16, 6 = 17 to 32, 7 = 33 to 64, 8 = more than 65) for both verges in April 1993, where the score for each strip on one verge is subtracted from the score for the matching strip on the opposite verge. The excess of quadrats with more oilseed rape plants on the side of the road heading towards the seed processing plant (solid bars), and the excess of quadrats containing fewer plants on the verge going away from the plant (open bars) are significant, $G = 1435.0$, $d.f. = 15$, $p << 0.0001$. From Crawley & Brown (1995).

the vicinity of exit and entry slip roads than on sections of verge between motorway junctions, but densities were not affected by the presence of rape crops or rape volunteers in adjacent fields. The apparent permanence of oilseed rape populations on motorway verges clearly belies substantial turnover in patch occupancy. In the absence of soil disturbance, rapid secondary succession (principally the growth of perennial grasses) tends to lead to local extinction within 3 years. Given sufficient soil disturbance, however, rape population density appears to be seed limited, and seed spillage can cause a 2- to 5-fold increase in mean population density.

We conclude that recruitment in oilseed rape is seed-limited on those sections of motorway verge where the disturbance regime is sufficient to prevent the development of a closed grass sward, but not seed limited in other sections. The relative importance of the seed bank, seed produced by resident plants and imported (spilled) seed in regeneration is difficult to assess from an observational study, but the probability of extinction was lower in quadrats where local seed production was high in the previous year. In the absence of disturbance, however, our observations suggest that local extinction within 2 to 4 years is the typical fate of a feral oilseed rape population.

For this system it is clear that a stochastic model is needed which takes account of frequent colonisation and local extinction. The classic single-species models (Eqn 6) can deal with neither of these processes because we require that $N_{t+1} > 0$ for $N_t = 0$ (establishment) and that $N_{t+1} = 0$ for $N_t > 0$ (local extinction). The model would work as a two stage process; the first stage determines whether the site will be occupied or not, for sites which were and which were not occupied in the previous year. Next, for the occupied patches, the model predicts the likely population size. The probability of extinction was lower, and the probability of a high population was greater, for populations with high rape densities in the previous year (Crawley and Brown 1995), but extreme density-independent transitions were noted (e.g. some vacant sites in 1993 became high density sites in 1994 and some high density sites in 1993 were extinct in 1994). The model would need to take explicit account of the probability of soil disturbance and the rate of introduction of seed. The system highlights the need to study the 'empty quadrats' as well as the occupied quadrats if we are to understand establishment and hence the pattern of larger-scale population dynamics (Crawley 1990).

Case Study 2: Natural regeneration in English oak

The question of whether or not recruitment is seed-limited is a fundamental issue in plant population dynamics, and in many cases we do not know whether an increase in seed production would lead to an increase in the number of seedlings let alone to an increase in the density of adult plants. Recruitment might not be seed-limited because of microsite-limitation (a shortage of suitable places for recruitment) or predator-limitation (too many seed-feeding or seedling-feeding animals; Crawley 1990). An oak tree might begin producing acorns after 25 years or so, and continue to produce an average of 10,000 acorns per year for the next 200 years. The dynamics of an oak population can be portrayed as follows:

$$N_{t+1} = sN_t + FsN_t\, f[K - sN_t] \tag{9}$$

where survivorship, s, is high (often close to 1 for established plants) and $K - sN_t$ is the amount of open space available for oak regeneration (*Quercus robur* can not regenerate beneath a closed oak canopy, partly because it is shade intolerant and partly because its seedlings can not withstand the incessant rain of invertebrate herbivores from a mature canopy overhead). F is the annual per capita replacement of mature oaks, and is microsite limited to the extent that sN_t is close to K. The system is highly resilient to repeated recruitment failure (either $F = 0$ because of herbivory or bad weather, or because N is close to K- microsite limitation; $f[0]=0$) because adult plant survivorship is so high. Note the similarity of this model to the seed-bank models introduced earlier; in this case the storage effect (Chesson 1985, this volume) comes through high adult tree survivorship (s) while in the case of the annual plant models, the storage effect came through protracted seed dormancy (low germination rate, *g*).

Over a 17 year period, *Q. robur* has shown a pattern of alternate bearing at several sites in south east England (Fig. 4), with significant (but not complete) synchrony between individual trees (Crawley and Long 1995). Seedling recruitment was assessed in three ways: by annual destructive sampling, by monitoring permanent quadrats, and by ageing destructive samples of saplings using basal ring counts. A continuum of responses was observed, and it appears that seed density, microsites and predators can all affect oak recruitment, but to differing degrees in different places. In Sunningdale, recruitment was seed-limited and seedling density was correlated with acorn production over the full range of acorn densities. This pattern probably occurred because vertebrate herbivores were scarce as a result of high numbers of cats and foxes, and there was a continuous supply of freshly disturbed ground (herbaceous borders) in which squirrels and jays could bury acorns. In contrast, oak recruitment in nearby Silwood Park was predator-limited; high rabbit densities ensured that most acorns were eaten before they could be buried, and such oak seedlings as did appear were repeatedly browsed. Sowing extra acorns in Silwood Park did not lead to increased recruitment in any of the 7 years in which the experiment was attempted. Windsor Great Park and rabbit-fenced areas of Silwood Park were intermediate between Sunningdale and Silwood; oak recruits were produced in substantial numbers but only in certain peak years of acorn production (e.g. 1987 and 1989). Recruitment was herbivore-limited in the low years of the acorn cycle (because of seed predation by insects, woodmice and birds), but seed-limited in the peak years; burial of acorns was the key factor determining the probability of seedling recruitment inside the rabbit fences.

Peak acorn crops caused predator satiation in some years, and for some acorn-feeding animals but not others. Attack by the principal invertebrate acorn-feeders (the gall wasp *Andricus quercuscalicis* and the weevil *Curculio glandium*) was inversely density dependent in most years, with relatively

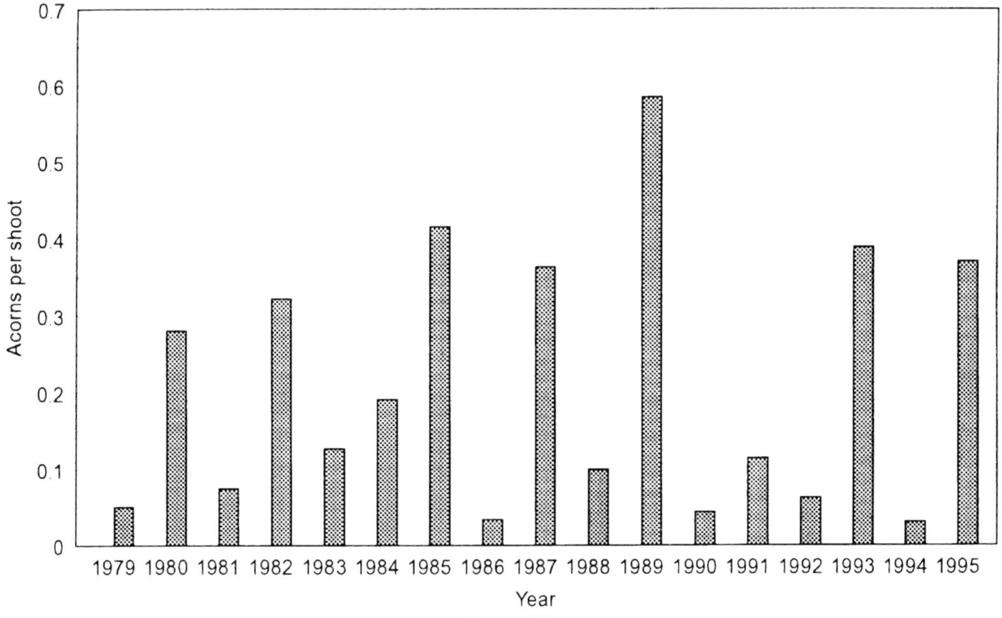

Fig. 4. Alternate acorn bearing in Quercus robur over the period 1979 to 1995 in Silwood Park, Berkshire. From Crawley and Long (1995) with new data for 1994 and 1995.

high percentage acorn loss during the low years of the cycle. These invertebrates, particularly *A quercuscalicis,* can reduce the size of the effective acorn crop below the level necessary for satiation of vertebrate herbivores (e.g. there was no seedling recruitment after the peak crop of 1993 because of > 85% loss to gall-insects). Satiation of post-dispersal vertebrate seed-predators like rabbits, woodmice, grey squirrels, wood pigeons and jays occurred after the highest peaks of acorn production in some habitats (Crawley and Long 1995).

Thus, while it is clear that predator satiation can work, especially where vertebrate herbivores are kept at relatively low densities by natural enemies and/or hunting, it is equally clear that predator satiation does not work in all peak years or in all habitats. Oak recruitment was seed-limited in open sites where vertebrate herbivores were scarce, but was microsite-limited beneath a dense oak forest canopy. In other places, especially those where rabbits were abundant, oak recruitment was herbivore-limited, and it appears that invertebrate herbivores can affect the rate of tree recruitment in some years by reducing the availability of sound acorns below the threshold necessary for satiation of the vertebrate acorn-feeders. For this system, there is simply no general way of predicting whether more seed will mean more plant recruitment (compare with Eqns 1 and 6, above). Huge changes in the rate of tree recruitment can occur when there are dramatic changes in the abundance of 'keystone' herbivores. For example, there was mass recruitment of oak in grassland habitats in Silwood Park following the eradication of rabbits by the myxoma virus during the 1950's and these trees now form even-aged, reproductively mature, closed canopy woodland (Dobson and Crawley 1995).

Multispecies dynamics

Our interest here is in the question of coexistence. Gause's principle of competitive exclusion predicts dominance by the single, best adapted plant species in environments that are spatially and temporally uniform. It emphasises that interspecific competition is typically destabilising and tends to lead to reduced species richness. Nicholson understood this, but it is clear that he had a keen appreciation of the (then very modern) findings of the Lotka-Volterra competition model, that coexistence requires intraspecific competition to be more important than interspecific competition. Along with other of his contemporaries (e.g. Skellam 1951), Nicholson could see that there were many ways that the Lotka-Volterra criterion might be realised, and he speculated that partial niche overlap, with spatially separated refugia for each of the species might be an important mechanism of coexistence (thus anticipating much of what is nowadays known as metapopulation dynamics). Other coexistence mechanisms might involve different regeneration niches (Grubb 1977; Chesson 1985) or trade-offs between competitive ability and other aspects of life history (e.g. fecundity, dispersal ability or herbivore tolerance). It is possible, of course, that a local population might be thrown together by chance alone, and might persist for long periods in ecological time (c. 500 yrs) without any coexistence mechanisms at all, especially if the individuals were long-lived.

Case Study 3: Coexistence in Dune Annuals

Dune annuals provide an excellent model system for the study of coexistence; they are small and have little carryover of seed from one year to the next, so their biology closely matches the assumptions of the single-species models discussed above. The study was carried out in collaboration with Peter Grubb and Dave Kelly and a full description of the work is given in Rees *et al.* (1995) and Rees (1995). The study site was Holkham on the North Norfolk coast, and every year from 1979 to 1988 the number of flowering individuals of four species of winter annual were counted in 1000 permanent quadrats each 10 cm by 10 cm (Fig. 5). The species were *Erophila verna* (Brassicaceae), *Cerastium semidecandrum* (Caryophyllaceae), *Myosotis ramosissima* (Boraginaceae) and *Valerianella locusta* (Valerianaceae).

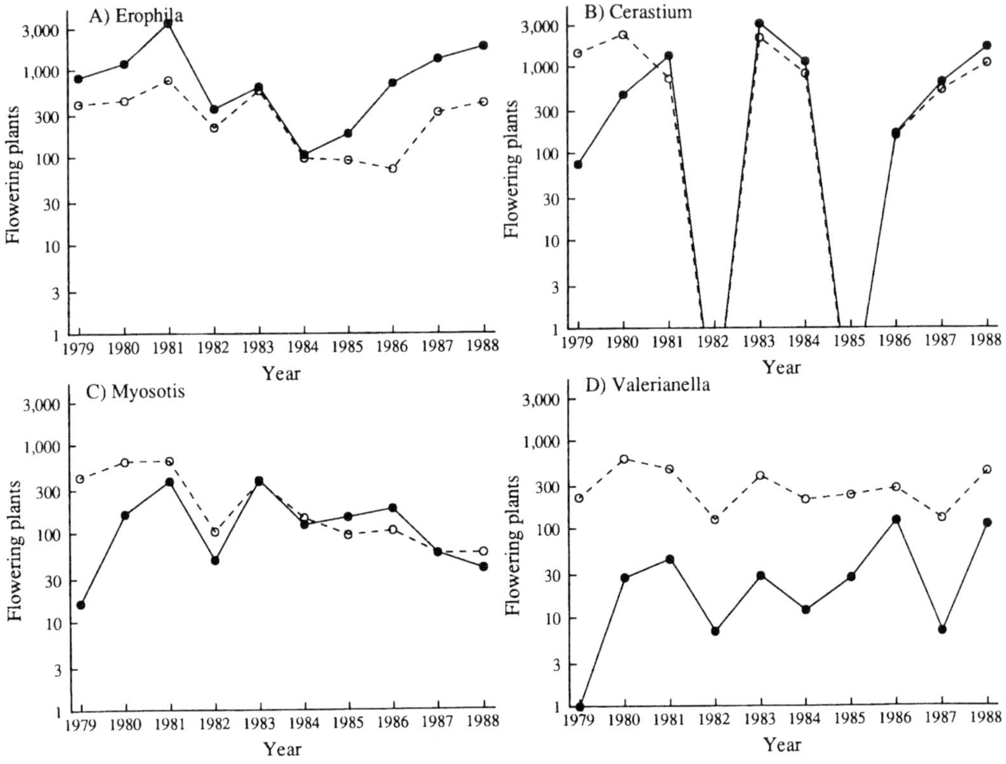

Fig. 5. Population trajectories for four annuals in two dune study areas at Holkham, Norfolk. A) *Erophila verna*, B) *Cerastium semidecandrum*, C) *Myosotis ramosissima* and D) *Valerianella locusta*. Open circles = area A, closed circles = area B.

The analysis of this spatially structured population census data uses spatial variation in density to determine how the density of individuals in the 10 cm by 10 cm quadrats affects the population growth of each species. So if we know the density of individuals in each quadrat in two consecutive years we can design customised maximum likelihood estimators in order to fit specific population models to the data (Rees *et al.* 1995). Data analysis demonstrates that population growth is significantly density dependent in almost all cases. Most importantly, the analysis shows that intraspecific density effects are common, whereas interspecific density effects are both uncommon and weak. We can also study the dynamics of theoretical models, using the parameter values estimated from the data. General population dynamic theory (Watkinson 1980; Thrall *et al.* 1989; Rees and Crawley 1989, 1991) predicts that populations should show stable dynamics, and this implies that the eigenvalues all lie in the unit interval. The agreement between general theory and the fitted models is excellent, with all the eigenvalues lying in the unit interval and most of them exhibiting values greater then zero (predicting a smooth approach to the equilibrium rather than cyclic behaviour, see Fig. 6). Numerical simulation demonstrated that even when damped oscillations were predicted the maximum amplitude of oscillation was small and the time to reach the equilibrium was normally no more than 5 years.

But does the observed level of density dependence matter? Does it influence the population sizes of these generally rather sparse, tiny plants that we observe in nature? Comparing the predicted mean population density using the fitted model, which includes both intra and interspecific effects, with

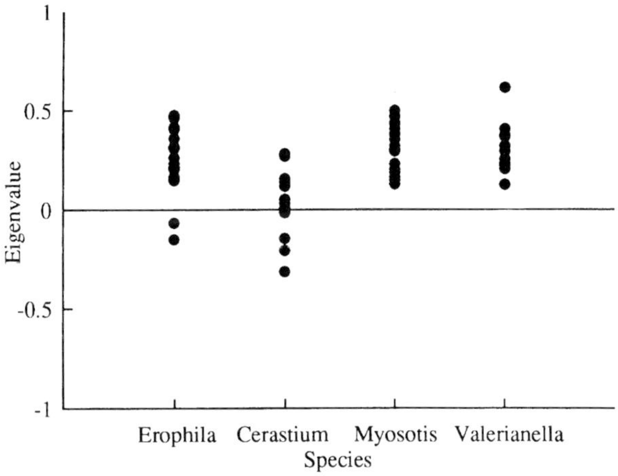

Fig. 6. Eigenvalues of the fitted models of the dynamics of four dune annuals indicating population stability (eigenvalues > 0) to damped oscillations about a stable equilibrium point (eigenvalues between 0 and -1). As predicted (see text), cyclic and chaotic dynamics (eigenvalues < -1) were not found.

the density independent form of the model shows that on average, population density would have been greater by 3-fold for *Cerastium*, 2-fold for *Erophila*, 1.5 fold for *Myosotis*, and 1.9 fold for *Valerianella* if density dependence was eliminated. Clearly, spatial density dependent processes do have important effects on both regulation and population size. Further analysis demonstrates the extremely weak effects of interspecific interactions amongst these four annual species. This is consistent with Law and Watkinson (1989) who found no significant pairwise interactions in 73% of the experimental studies of competition they reviewed.

Of the various mechanisms that might be invoked to explain the coexistence of these 4 annuals, temporal storage effects can be eliminated, since none of the species possesses a seed bank. The species do, however have different seed weights and different average fecundities. The mean seed weights (mg) at Holkham were *Valerianella* 0.80, *Myosotis* 0.17, *Cerastium* 0.08 and *Erophila* 0.03 (Kelly 1982). Moreover, seed weight appears to be linked with the probability of seedling survival, with *Valerianella* and *Myosotis* having the highest probability of survival (41% and 27% respectively), whereas in *Erophila* and *Cerastium* survival was about 20% (difficulties in separating these species at the seedling stage prevent more accurate estimation; Grubb *et al.* 1982; Kelly 1982). The differences in seed weight influence the numbers of seeds produced per plant, with *Erophila* and *Cerastium* producing more seeds than *Myosotis,* the average seed production per plant for *Erophila*, *Cerastium* and *Myosotis* was 15.5, 20.3, and 7.5 respectively (Kelly 1982). These differences also appear to be linked with the observed population densities (*Erophila* and *Cerastium* > *Myosotis* > *Valerianella*). These patterns generate a type of competition-colonisation trade-off; large seeded species have a greater probability of surviving in a microsite but because fewer seeds are produced the larger-seeded species have a lower probability of colonising any given microsite (and vice versa for small seeded species). This particular trade-off can strongly promote coexistence (Skellam 1951; Levins 1970; Tilman 1994). Recent comparative work using data from eight dune systems has demonstrated that seed weight is negatively correlated with several measures of abundance suggesting that the patterns observed at Holkham may be quite general (Rees 1995).

The idea of a competition-colonisation trade-off is entirely consistent with the observed weak interactions between species. In this case the large seeded species appear to have a competitive advantage over smaller seeded species (as observed in experimental studies of dune annuals reviewed by Rees 1995). As a result we would expect that when individuals of a large seeded species are close to those of a smaller seeded species, negative effects on performance of individuals of the smaller seeded species would be detected but there would be little or no effect of the smaller seeded species on the larger one. This is exactly the effect documented by Kelly (1982). These strong interactions between individuals of different species do not translate into strong population level effects for two reasons: 1) the larger-seeded species are rare and so they interact infrequently with the smaller-seeded species; and 2) the species are spatially aggregated, and this lowers the frequency of interspecific contacts even further.

In summary, the weakness of measured interspecific interactions probably arises as a result of three aspects of the ecology of sand dune system: 1) the soil is extremely nutrient poor which means that the individual plants are small, and this sets the scale of environmental heterogeneity, both biotic and abiotic, that the plants can exploit; 2) the positive links between seed size and competitive ability and the negative correlation between seed size and seed production mean that the strong competitors are generally rare, and this reduces their impact on the weaker competitors at the population level; and 3) the species are aggregated/segregated in space which reduces the frequency of interspecific contacts. These spatial patterns could reflect local dispersal and/or niche differentiation to spatial variation in the environment.

Case Study 4: Coexistence in perennials of mesic grassland

In considering the coexistence of long-lived, clonal perennial plants, where storage effects are potentially strong and where interspecific competition is likely to be important most of the time (and highly asymmetric, at least between adult plants and recruiting seedlings) different kinds of processes need to be considered. The model of Crawley and May (1987) can be extended to include competition between many perennial plant species. The potential colonists are ranked from most to least competitive (a competitive species is one which is capable of invading a monoculture of a less competitive species, but the latter would not be able to invade a monoculture of the former). Then, starting with the most competitive species, we have an equilibrium proportion of free (invasible) space, E. As shown by Crawley and May (1987) this system is invasible by a second species so long as the net rate of increase of the invader

$$\lambda > \frac{1}{E}$$

(and see Skellam 1951). Coexistence is possible even though the interspecific competition is completely one-sided (species 1 always occupies space at the expense of species 2). The coexistence is possible because of the Nicholsonian competitor-free-space (the botanical analogue of his area of discovery); because of mortality of perennial ramets, coupled with limited fecundity and/or dispersal, the superior competitor is not able to form a monoculture of 100% cover. The third species then attempts to invade the two-species community made up by the most-competitive and second-most-competitive species, and so on. Then the number of potentially coexisting perennial plant species is given by i in

$$\lambda_{i+1} < \frac{1}{1-\Sigma p_i} \tag{10}$$

where p_i is the proportion of the total space occupied by the i th species. Species are accumulated until the rate of increase of the next most competitive species (the i+1 th) is less than the reciprocal of the total unoccupied space. Monocultures are observed where $p_1 = 1$; these communities are uninvasible no matter how high the fecundity of the next best competitor. This argument is close in spirit to the models presented by Tilman (1994) and May and Nowak (1994) for cases where there is a competitive-ability/rate-of-increase trade-off (and is not dissimilar to the r/K dichotomy).

The object of this section is to contrast the dynamics of an equilibrium community of perennials (Park Grass, Rothamsted) and the transient dynamics of recently perturbed communities (Nash's Field, Silwood Park). Equilibrium plant communities are few and far between, and it is reasonable to suppose that, in the absence of a long run of data to demonstrate the contrary, it is a good working hypothesis that the plant community one studies exhibits transient rather than equilibrium dynamics. The mesic grasslands of Park Grass, Rothamsted, is a good example of a British plant community that is likely to be in equilibrium (Tilman *et al.* 1994). Set up by Lawes and Gilbert in 1856, their floristic composition has remained reasonably stable since about 1900. It took more than 40 years for the transient dynamics caused by the new experimental treatments to damp away (nutrients were applied at various rates, and sheep were no longer allowed to graze the aftermath of the first (June) hay crop; instead, a second cut was taken in autumn).

Species richness varies from monocultures of *Holcus lanatus* on the unlimed plots receiving 150 kg ha^{-1} N in the form of (acidifying) ammonium sulphate, to over 40 species on the limed, unfertilised control plots. These trends in species richness reflect two potent, but essentially artefactual processes; 1) more extreme abiotic conditions (very low soil pH in this case) mean that there is a smaller pool of adapted species capable of coexisting on a given plot; and 2) higher productivity leads to larger individual plants and hence to lower total plant population density – the smaller the number of individuals, the lower will be the species richness per unit area. Explanations of differences in the species richness which might give insights to the mechanisms allowing coexistence on plots with a similar species-pool size and similar mean plant shoot weights, must await further experimental and theoretical study.

There are beautifully clear examples of niche differentiation between species on different plots on Park Grass, and 2-dimensional ordinations of relative abundance (proportion of total biomass made up by the species in question) using peak biomass and soil pH as the axes defining the niche-space, show an almost jigsaw-like set of differences in niche location. The niche maps of the Park Grass plots also suggest that there may be within-species niche differentiation (as represented by isolated 'islands' in this 2-dimensional niche space). It remains to test whether the individuals inhabiting these different regions of niche space are genetically distinct, but reciprocal transplant experiments with the early-flowering grass *Anthoxanthum odoratum* have already shown that that genetic differentiation can occur rapidly on Park Grass (Davies and Snaydon 1976). It is still not clear, however, whether the high diversity Park Grass plots retain their species richness through stasis (the coexistence of essentially immortal clones) or through equilibrium dynamics (rapid turnover with recruitment and mortality, with each species exhibiting the ability to increase when rare). Work is starting on seed-limitation in Park Grass this year in an attempt to answer some of these questions.

Because manipulative experiments can not be carried out on Park Grass (to preserve its condition) a replica of the experiment was initiated at Nash's Field, Silwood Park. This has smaller plots than Park Grass (4 m^2 compared with 200 m^2), but it has the advantage of being properly replicated and randomised. Extra experimental treatments include manipulated levels of plant competition (minus grass and minus herb herbicide treatments with an untreated control) and factorial combinations of herbivore exclusion treatments (rabbits with fences, insects with insecticide, molluscs with slug pellets). The nutrients N, P, K and Mg are applied singly and as 'all-but' treatments; thus we have N and everything but N (i.e. P, K and Mg), P and everything but P (N, K and Mg), and so on. There are untreated controls and plots that get the full set of nutrients. One of the principal motivations for setting up the Nash's Field experiment was that the early, transient dynamics of the Park Grass experiment were not documented. There are no data on the changes in floristic composition and species richness that occurred during the first 7 to 10 years, as the originally homogeneous grassland reacted to the imposition of a mosaic of different nutrient treatments. Those species which chanced to be in the right place (e.g. *H. lanatus* on plots which happened to receive nitrogen fertiliser, legumes on plots that received potassium but not nitrogen, and so on) would have prospered, while those which found themselves in the wrong place (e.g. legumes on N-fertilised plots) would have suffered a more or less rapid demise. The Nash's Field experiment affords an opportunity to document these effects and to see the extent to which the responses are modified by the kind and intensity of interspecific plant competition. For example, many of the herbaceous species show pronounced competitor release on the minus-grass, herbicided plots. Some species only show higher peak biomass following competitor release when certain nutrients are added; some only show competitor release on the unfertilised control plots; still others show competitor release under all the nutrient treatments.

There are effects of vertebrate herbivory, as with the interaction between rabbit grazing and nitrogen fertilisation. Inside the fences, the plots receiving nitrogen produce substantially more biomass, as expected. Outside the fences, however, rabbit grazing is so concentrated on the N-fertilised plots that their biomass is significantly *lower* than the non-nitrogen fertilised plots. The grazed, plus-nitrogen plots are billiard-table-like lawns, compared with the grazed non-nitrogen plots which are tussocky with conspicuous standing dead organic matter and abundant bryophytes.

A series of companion plots were also set up, where 4 of the dominant grasses from Park Grass were grown in monocultures and in 2- and 4-species combinations. The soil was fumigated with methyl bromide following cultivation to destroy the soil seed bank and to reduce the weeding effort involved in establishing the monocultures. The methyl bromide also killed the plant pathogen and mycorrhizal fungal populations. The biomass yields during the first year after sowing were much higher than the estimated monoculture yields from Park Grass (Fig. 7; 2-fold for *Festuca rubra*, 2-fold for *A. odoratum*, 1.5-fold for *Alopecurus pratensis* and 1.3-fold for *Arrhenatherum elatius*), under the same regime of nutrient input (150 kg ha^{-1} yr^{-1} N, 225 kg ha^{-1} yr^{-1} K and 35 kg ha^{-1} yr^{-1} P). As the years have passed, the yields of the sown plots have declined towards the Park Grass levels, presumably as pathogen and mycorrhizal populations built up (Fig. 7). These transient dynamics were probably caused directly by the elimination of soil organisms, rather than an indirect result

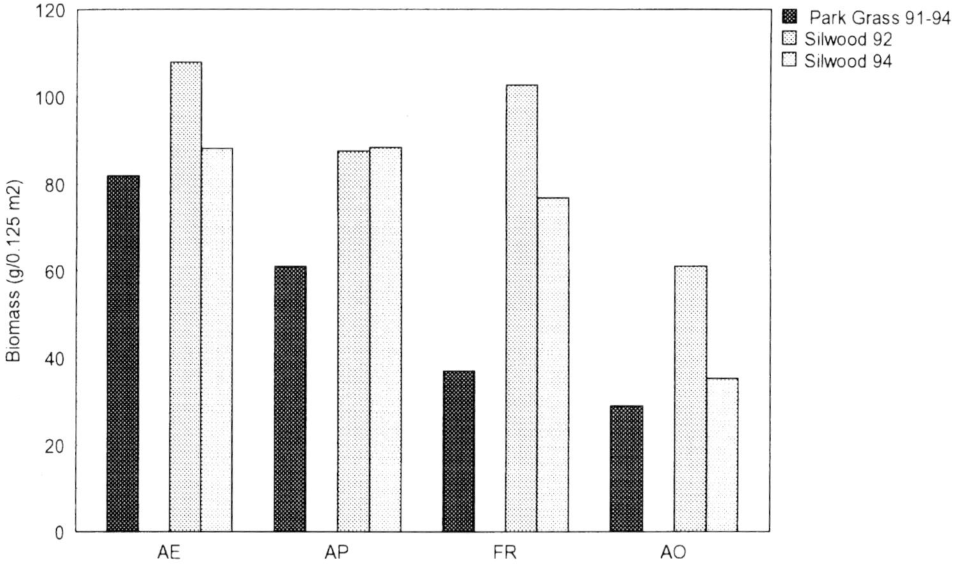

Fig. 7. Transient dynamics of four dominant perennial grasses in Nash's Field, Silwood Park compared with Park Grass, Rothamsted shown as peak biomass of samples of monocultures from different plots, reflecting the ranked mean size of individuals of these grasses. Nash's Field plots, established from seed following methyl bromide soil fumigation are shown in the first and third growing season. AE = *Arrhenatherum elatius*, AP = *Alopecurus pratensis*, FR = *Festuca rubra*, AO = *Anthoxanthum odoratum*.

through extra fertiliser input from decomposition of roots etc. killed by the fumigation (M.J. Crawley and D. Tilman, unpublished). The drain placed on above ground plant productivity by subterranean fungal populations appears to be very great, and acts in addition to any direct, pathological effects of soil micro-organisms (cf. van der Putten *et al.* 1993).

The Nash's Field work predominantly shows the importance of animal activity in providing the disturbance to produce the competition-free microsites required for seedling recruitment. Molehills, worm casts, rabbit scrapes and ant middens all form locally important seed beds. By excluding these animals in various combinations, we hope to be able to measure reductions in seedling recruitment. In the absence of such disturbance, seed-sowing experiments have shown that recruitment of few, if any, of the resident species are seed-limited in Nash's Field.

Discussion

Nicholson emphasised repeatedly that the only way to understand the population-level consequences of changes in demographic parameters was to incorporate the changes in a theoretical model for the dynamics of the entire life cycle. Many apparently important mortality factors which act early in the life cycle (e.g. at the seedling stage), can turn out to have negligible long-term impact on dynamics because of compensating density dependence which acts later in the life cycle. We support Nicholson's view, but stress the caveat that the consequences of a given parameter change often depend critically upon the detailed structure of the model in which the parameter is assessed. Thus, for example, in our model of plant dynamics (Eqn 1) the incorporation of dormancy, and the consequent development of a seed-bank, *reduces* the likelihood of persistence in a model with seeds which germinate irrespective of the perennial cover in the microsite ('dumb seeds'), whereas dormancy can *increase* persistence in a model with seeds where the probability of germination is reduced in microsites which are beneath a leafy canopy, because the red to far-red ratio has been shifted ('clever seeds' Eqn 3).

In general, it will be impossible to distinguish processes that are statistically significant from those that are ecologically important on the basis of within-generation studies of plant demography. For instance, the impact of slugs on seedling mortality might be assessed by exclusion using chemical molluscicide. Such studies typically show substantial and highly significant effects of herbivory on seedling density within the first few weeks following germination. It would be wrong, however, to conclude that slugs influence between-generation plant dynamics on this evidence alone. It is plausible that by the following growing season, the number of plants is no greater in the molluscicide treated plots than on the untreated controls (e.g. Crawley *et al.* 1993; Rees and Long 1992).

One of the principal barriers to furthering our understanding of plant dynamics is the inability of the present generation of theoretical models (variations on the theme of Eqn 6) to cater adequately for the processes of establishment and local extinction. Similarly, we need to develop new statistical models for the analysis of plant population data which allow for extinction and colonisation (the issue of what to do with the 'empty quadrats'; Crawley 1990). At present, quadrats exhibiting establishment or local extinction have to be excluded from conventional analyses because they lead to 'zero divide' or 'log(zero)' errors in computing the response variable $y=ln(N_{t+1}/N_t)$. New methods (Rees *et al.* 1995) allow for local extinction but do not deal with colonisation. A further difficulty is that existing models like Eqn 6 explicitly assume that seed-limitation is universal; as we saw from the case studies above, this is simply not what happens in the field. Recruitment is seed limited some of the time and in some places, but certainly not all of the time or in all places. Herbivore limitation when plant density is low, or microsite limitation when density is relatively high, appear to be quite common under field conditions. These questions about seed limitation arise as a direct result of having excluded spatial structure from the formulation of the model; including spatial structure allows different processes to act in different places, with potentially profound effects on model predictions (Chesson 1985; Pacala and Crawley 1992).

On the issue of coexistence, Nicholson's views on competitor-free-space continue to be highly influential. In recent years, conventional wisdom has swung away from an emphasis on competitive exclusion towards potentially unlimited numbers of coexisting species. We can see plenty of ways in which large numbers of species could coexist in equilibrium (niche differences, like adaptations to underlying spatial heterogeneity, or trade-offs between competitive ability and dispersal; Skellam 1951; Crawley and May 1987; Tilman 1994) and even more ways in non-equilibrium systems (temporal variations in relative recruitment success coupled with storage effects (Grubb 1977; Chesson 1985). Broadly speaking, coexistence will occur when species obtain some advantage or other when they become rare (this might be an absolute, density dependent advantage, or a frequency dependent advantage). The problem lies in devising unequivocal tests which can distinguish between equally plausible mechanisms whereby this 'rare species advantage' might come about. Nicholson tackled questions like this using a mix of theory and laboratory experiments. We favour a mix of theory and field work, both observational and experimental. It is only when we have produced parameterized community models that correctly predict both the number of coexisting species and their patterns of relative abundance, for several different community types, that we can claim to have understood the processes that maintain plant diversity. For plants at least, current progress suggests that this goal is attainable.

References

Atkinson WD, Shorrocks B (1981) Competition on a divided and ephemeral resource: a simulation model. *J Anim Ecol* **50**:461–471

Caswell H (1989) *Matrix population models.* Sinauer Associates, Sunderland, MA, USA

Chesson PL (1985) Coexistence of competitors in spatially and temporally varying environments: a look at the combined effects of different sorts of variability. *Theor Pop Biol* **28**:263–287

Crawley MJ (1990) The population dynamics of plants. *Phil Trans R Soc Lond B,* **330**:125–140

Crawley MJ, Brown SL (1995) Seed limitation and the dynamics of feral oilseed rape on the M25 motorway. *Phil Trans R Soc Lond B* **259**:49–54

Crawley MJ, Hails RS, Rees M, Kohn D, Buxton J (1993) Ecology of transgenic oilseed rape in natural habitats. *Nature* **363**:620–623

Crawley MJ, Long CR (1995) Alternate bearing, predator satiation and seedling recruitment in *Quercus robur L. J Ecol* **83**:683–696

Crawley MJ, May RM (1987) Population dynamics and plant community structure: competition between annuals and perennials. *J Theor Biol* **125**:475–489

Davies MS, Snaydon RW (1976) Rapid population differentiation in a mosaic environment. III. Measures of selection pressures. *Heredity* **36**:59–66

Dobson A, Crawley MJ (1994) Pathogens and the structure of plant communities. *Trends Ecol & Evol* **9**:393–398

Donald CM (1951) Competition among pasture plants. I. Intra-specific competition among annual pasture plants. *Aust J Ag Res* **2**:355–376

Fenner M (1985) *Seed ecology.* Chapman and Hall, London, UK

Fretwell SD, Lucas HL (1970) On territorial behaviour and other factors influencing habitat distribution in birds. *Acta Biotheor* **19**:16–36

Goldberg DE (1987) Neighbourhood competition in an old-field plant community. *Ecology* **68**:1211–1223

Gorski T, Gorska K, Nowicki J (1977) Germination of seeds of various herbaceous species under leaf canopy. *Flora Batava* **166**:249–259

Grubb PJ (1977) The maintenance of species-richness in plant communities: the importance of the regeneration niche. *Biol Revs* **52**:107–145

Grubb PJ, Kelly D, Mitchley J (1982) The control of relative abundance in communities of herbaceous plants. In: Newman EI (ed) *The plant community as a working mechanism*. Blackwell, Oxford, UK, pp 79–97
Hairston NG, Smith FE, Slobodkin LB (1960) Community structure, population control and competition. *Am Nat* **94**:421–425
Hassell MP (1975) Density dependence in single species populations. *J Anim Ecol* **44**:282–295
Hassell MP, Lawton JH, May RM (1976) Patterns of dynamical behaviour in single species populations. *J Anim Ecol* **45**:471–485
Kelly D (1982) *Demography, population control and stability in short-lived plants of chalk grassland*. Unpublished Ph.D. thesis, Univ Cambridge, UK
King TJ (1975) Inhibition of seed germination under leaf canopies in *Arenaria serpyllifolia, Veronica arvensis* and *Cerastium holosteoides*. *New Phytol* **75**:87–90
Kira T, Ogawa H, Shinozaki K (1953) Intraspecific competition among higher plants. I. Competition-density-yield inter-relationship in regularly dispersed populations. *J Inst Polytech Osaka City Uni* **4**:1–16
Klinkhamer PGL, Meelis E, de Jong TJ, Weiner J (1992) On the analysis of size-dependent reproductive output in plants. *Funct Ecol* **6**:308–316
Law R, Watkinson AR (1989) Competition. In: Cherrett JM (ed) *Ecological concepts*. Blackwell, Oxford, UK, pp 243–285
Levins R (1970) Extinction. In: Gerstenhaber M (ed) *Some mathematical problems in biology*. Mathematical Society, Providence, RI, USA, pp 77–107
May RM, Oster GF (1976) Bifurcations and dynamic complexity in simple ecological models. *Am Nat* **110**:573–599
May RM, Nowak MA (1994) Superinfection, metapopulation dynamics and the evolution of diversity. *J Theor Biol* **170**:95–114
McConnaughay KDM, Bazzaz FA (1987) The relationship between gap size and performance of several colonizing annuals. *Ecology* **68**:411–416
Miller TE, Werner PA (1987) Competitive effects and responses between plant species in a first-year old-field community. *Ecology* **68**:1201–1210
Nicholson AJ (1933) The balance of animal populations. *J Anim Ecol* **2**:131–178
Nicholson AJ (1954) An outline of the dynamics of animal populations. *Aust J Zool* **2**:9–65
Nicholson AJ (1957) The self adjustment of population change. *Cold Spr Habor Symp Quant Biol* **22**:153–173.
Nicholson AJ (1960) The role of population dynamics in natural selection. In: Tax S, Callender C (eds) *Evolution after Darwin*. Univ Chicago Press, USA, pp 477–521
Obeid M, Machin D, Harper JL (1967) Influence of density on plant to plant variations in fiber flax, *Linum usitatissimum*. *Crop Sci* **7**:471–473
Ogden J (1970) Plant population structure and productivity. *Proc NZ Ecol Soc* **17**:1–9
Pacala SW (1986) Neighbourhood models of plant population dynamics. 4. Single-species and multispecies models of annuals with dormant seeds. *Am Nat* **128**:859–878
Pacala SW, Crawley MJ (1992) Herbivores and plant diversity. *Am Nat* **140**:243–260
Pacala SW, Silander JA Jr (1985) Neighbourhood models of plant population dynamics. I. Single-species models of annuals. *Am Nat* **125**:385–411
Pacala SW, Silander JA Jr (1990) Field tests of neighborhood population dynamic models of two annual weed species. *Ecol Monogr* **60**:113–134
Pacala SW, Weiner J (1991) Effects of competitive asymmetry on a local density model of plant interference. *J Theor Biol* **149**:165–179
Palmblad IG (1968) Competition in experimental populations of weeds with emphasis on the regulation of population size. *Ecology* **49**:26–34
Rees M (1995) Community structure in sand dune annuals: Is seed weight a key quantity? *J Ecol* **83**:857–863
Rees M, Grubb PJ, Kelly D (1995) Quantifying the impact of competition and spatial heterogeneity on the structure and dynamics of a four-species guild of winter annuals. *Am Nat* **147**:1–32
Rees M, Brown VK (1991) The effects of established plants on recruitment in the annual forb *Sinapis arvensis*. *Oecologia* **87**:58–62

Rees M, Brown VK (1992) Interactions between invertebrate herbivores and plant competition. *J Ecol* **80**:353–360

Rees M, Crawley MJ (1989) Growth, reproduction and population dynamics. *Funct Ecol* **3**:645–653

Rees M, Crawley MJ (1991) Do plant populations cycle? *Funct Ecol* **5**:580–582

Rees M, Long MJ (1992) Germination biology and the ecology of annual plants. *Am Nat* **139**:484–508

Rice KJ (1985) Responses of *Erodium* to varying microsites: the role of germination cueing. *Ecology* **66**:1651–1657

Samson DA, Werk KS (1986) Size-dependent effects in the analysis of reproductive effort in plants. *Am Nat* **126**:667–680

Skellam JG (1951) Random dispersal in theoretical populations. *Biometrika* **38**:196–218

Symonides E (1988) Population dynamics of annual plants. In: Davy AJ, Hutchings MJ, Watkinson AR (eds) *Plant population ecology*. Blackwell, Oxford, UK, pp 221–248

Thrall PH, Pacala SW, Silander JA (1989) Oscillatory dynamics in populations of an annual weed species *Abutilon theophrasti*. *J Ecol* **77**:1135–1149

Tilman D (1982) *Resource competition and community structure*. Princeton Univ Press, Princeton, USA

Tilman D (1994) Competition and biodiversity in spatially structured habitats. *Ecology* **75**:2–16

Tilman D, Dodd ME, Silvertown J, Poulton PR, Johnston AE, Crawley MJ (1994) The Park Grass Experiment: insights from the most long-term ecological study. In: Leigh RA, Johnston AE (eds) *Long-term experiments in agricultural and ecological sciences*. CABI, Wallingford, UK, pp 287–303

Turchin P, Taylor AD (1992) Complex dynamics in ecological time series. *Ecology* **73**:289–305

van der Putten WH, van Dijk C, Peters BAM (1993) Plant-specific soil-borne diseases contribute to succession in foredune vegetation. *Nature* **362**:53–56

Van Tooren BF, Pons TL (1988) Effects of temperature and light on the germination in chalk grassland species. *Funct Ecol* **2**:303–311

Watkinson AR (1980) Density-dependence in single-species populations of plants. *J Theor Biol* **83**:345–357

Watkinson AR, Davy AJ (1985) Population biology of salt marsh and sand dune annuals. *Vegetatio* **62**:487–497

Weiner J (1990) Asymmetric competition in plant populations. *Trends Ecol & Evol* **5**:60–364

SECTION 2

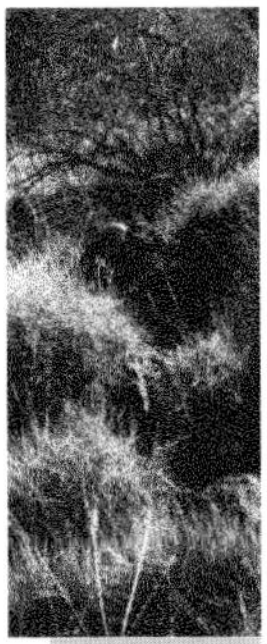

Two species interactions

The papers in this section focus on interactions between species rather than population regulation (see Section 1) and begins with a review by Nigel Barlow and Steve Wratten of predator/prey and parasite/host interactions. Firstly they considered the Nicholson and Bailey contribution to two species interactions in the light of the contributions of Thompson, and Lotka and Voltera, and concluded that Nicholson and Bailey had a more lasting impact. Their impact on the subsequent development of the theory of predator/prey and parasite/host interactions is documented in some detail. The review focuses on the development of mathematical models of functional response, host density-dependence, and stability and persistence. Finally, they consider how the theory has been translated into the practical context of biological control and integrated pest management. They conclude that the lack of application of Nicholson and Bailey models and their modern derivatives to the development of classical biological control is a contributing factor to the high failure rate of such programs. Further, they consider that the increasing pressures from environmental agencies and the public demanding greater predictability of outcomes will necessitate a greater use of these models to provide reassurance that releases will be both successful and safe. Finally, they conclude that theory based on steady accumulation of empirical generalities rather than on global models will have the greatest impact upon advances in the body of theory associated with classical biological control.

Louise Vet then describes how behaviour of the individual, particularly parasitoid foraging, may affect population dynamics. The paper discusses the current state of the theory of parasite/host interactions and suggests that theory of these interactions need to be extended to include the effects of food plant on the interaction. The role of learning as a source of behavioural variability and the influence of the host plant are seen as key factors in the parasite/host interaction. The third topic to be considered is the spatial aspects of parasitoid searching strategies. The strategies adopted by different species are shown to be greatly influenced by host distribution, and knowledge of parasitoid behaviour in response to host distribution is considered necessary for a better understanding of the population dynamics and the evolutionary consequences of spatial variation in parasitism rates. The paper concludes that the effect of complex individual decision making processes on population dynamics may well be best approached using individual-based models. These models may improve our understanding of local instability. Furthermore, Vet suggests that linking of evolutionary models with population models may enable researchers to understand the evolution of particular host selection and searching strategies.

Sources of stability which balance the seemingly intrinsic instability of host-parasitoid systems and an understanding of how these stabalising mechanisms interact are the basis of Andrew Taylor's contribution. Taylor uses models of aggregation of risk, superparasitism and within-host competition to demonstrate that stabilising mechanisms often act in concert where the influence of one mechanism is modified by the effect of another. Taylor concludes that the study of a single factor will seldom, if ever, produce a satisfactory explanation of observed stability and consequently, the real role of theory in studies of parasite/host interactions is the identification of factors which may be useful in under-standing empirical systems. Taylor further suggests that a return to the observational approach to understanding temporal dynamics may be the way ahead for increasing our ability to explain the maintenance of stability.

The paper by Robert Knell, Mike Begon and David Thompson considers the influence of transmissibility of an insect pathogen on the long term dynamics and age structure of the host population. By modelling transmis-sibility and testing the model empirically they demonstrated that the rate of infection, i.e. the transmission coefficient, was a useful predictor of long-term dynamics of parasite/host systems.

The strong link between Nicholsonian theory and biological control is the subject of the contribution by Andy Sheppard and Tim Woodburn. In biological control of an exotic pest, control is achieved through either resource-limited agents exerting control on an abundant exotic host or a natural enemy-regulated agent being utilised in an natural enemy-free environment to bring about density-dependent control. The study of the biological control of weeds belonging to the Cardueae, demonstrated that resource limitation and not freedom from natural enemies was the main regulatory mechanism of control agent populations, suggesting that control agents of weeds which are rare in their native range, my not be regulated by natural enemies.

The dynamics of disease in natural plant populations are discussed by Jeremy Burdon. From the framework of pathogen/host interactions, it is concluded that a population may be a convenient unit to study short-term evolutionary change, but less appropriate for investigation of changes in pathogen virulence and the coevolution of pathogen/host systems. Rather, it may be better to consider a pathogen population as a mosaic of genetic types or demes which are relatively isolated from each other in terms of the effects of genetic drift, migration and selection. As a consequence, the effect of pressures exerted on the demes by the host and vice versa will be independent and therefore, a population when considered at the species level, may have the appearance of being unperturbed when at the genetic level it is in a constant state of flux.

Andrew Watkinson, Kevin Newsham and Alastair Fitter summarise the dynamics of individuals in plant populations as being dependent upon the intrinsic rate of increase of the plant, intraspecific competition for resources, interspecific competition, natural enemies, mutualism and refuge effects. Population theorists have addressed each of these with the exception of the role of mutualism and as a consequence there is no theoretical framework with which to treat this factor. They investigated the role of mutualism in population dynamics using arbuscular mycorrhizal fungi and winter annual grass, and concluded that the mutualistic relationship between the species may have a significant effect on the abundance of individuals and therefore on the dynamics of populations.

In the insect world, the interaction between herbivores and plants reaches its pinnacle in the interaction between sap-feeding insects and their hosts. In the paper by Roger Farrow and Rob Floyd, the interaction between insect genotypes and host phenotypes is discussed using various example of sap-feeding insects. They review the various responses of host trees to sap-feeding and conclude that rather than single plant compounds influencing herbivore population dynamics, it is the complex interaction of phenological, morphological and biochemical variability of the host population that has the greatest influence on population structure.

The previous papers in this section have dealt mostly with the direct effects that different types of interactions have on the dynamics of plant and animal populations. In the final paper, Marti Anderson investigates the role that indirect effects have on the dynamics of populations. Using the example of grazing marine gastropods, Anderson demonstrates that the direct interaction between two species, in this case grazing gastropods and algae, can have a significant indirect benefit for a third species, oysters. Algae and oysters were shown to be competing for the same resource, i.e. a bare rock surface, and consequently the action of gastropods to remove the algae also removed the competition.

Ecology of predator-prey and parasitoid-host systems: Progress since Nicholson

Nigel D. Barlow and Stephen D. Wratten

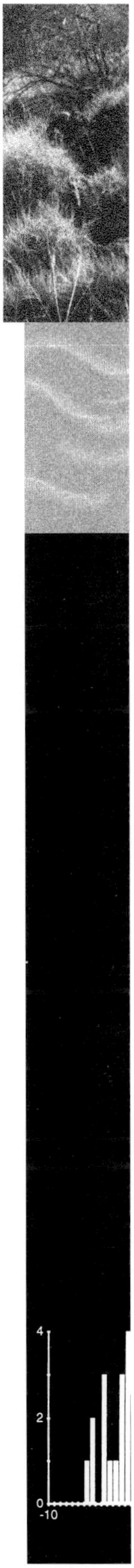

ABSTRACT

This review has three aims. First, we consider Nicholson's contribution to the understanding of predator-prey and parasitoid-host ecology, including his impact on the density-dependence debate in the context of predator-prey dynamics. Secondly and somewhat selectively, we critically review subsequent developments in understanding with a Nicholsonian focus on theory. Finally, we assess the relevance of theory and Nicholsonian ideas in the practical context of biological control.

Key words: density dependence, biological control, theory, predator-prey models, tri-trophic interactions

Frontiers of Population Ecology, R.B. Floyd, A.W. Sheppard and P.J. De Barro (eds), CSIRO Publishing, Melbourne, 1996 pp. 217–243

Introduction

Nicholson made two major contributions to population ecology. The first of these was the concept of the balance of nature, and the necessity for density-dependence and competition to maintain this balance. Perhaps surprisingly, Nicholson never considered the interaction between host density-dependence (i.e. regulation due to factors other than natural enemies, such that the per capita rate of increase declines with density), and predation or parasitism. Yet his advocacy for density-dependence, population regulation and the balance of nature, and the stimulating debate which developed around this and conflicting points of view, is probably an even more significant contribution to ecology than his parasitoid-host models. Indeed, it is interesting to see that the debate persists today, in terms of the significance of density-dependence in particular populations and whether or not it is realistic to consider populations in terms of carrying capacities or whether it is more fruitful to focus on responses to environmental perturbations. Smith (1935) coined the terms 'density-dependent' and 'density-independent', and Howard and Fiske (1911), working on gypsy moth in the USA, had earlier recognised the significance and distinction between the factors, which they termed 'facultative' and 'catastrophic' respectively. However, Nicholson's contribution to the debate was distinguished by its rigour, in terms of experimentation, theory and thinking. An example of the clarity of this thinking was the observation (Nicholson 1933) that population densities for any one species often differ consistently from place to place, the differences frequently associated with observable differences in environment. If the populations were not in a state of balance, how could these consistent differences persist? In his experimental work on population regulation, Nicholson (1954) identified two different types of competition which he referred to as 'scramble' and 'contest'. This established for the first time characteristic extremes of intra-specific competitive behaviour which later single-species models had to be able to accommodate (e.g. Hassell 1975). Perhaps the most interesting development in the 'balance of nature' debate involves recent work on the stabilising effect at a metapopulation level of spatial density-dependence, resulting from heterogeneity and differential movement between subpopulations governed by unstable models (e.g. Hassell *et al.* 1991). This, and the contribution of host density-dependence to predator-prey models, are considered further below.

Nicholson's second major contribution was an extension of his theory of competition and random search. With physicist collaborator V.A. Bailey, he provided an explicit model for host-parasitoid interactions which, alone among the several models already proposed, led to the development of most subsequent theory. Some of these developments were initiated by Nicholson himself, with various co-workers. For example, he initiated the quest for stability in parasitoid-host models, a quest which dominated subsequent theoretical work. He also demonstrated that heterogeneity of risk, associated with variations in the area of discovery between host individuals, could provide this stability. Nicholson was well aware that his original parasitoid-host model was unstable, referring in later papers to the difference between 'steady states' and 'stable steady states'. Mathematical analysis of the stability of his models was another pioneering contribution, anticipating an explosion of interest in identifying and understanding stability in model ecological systems. Finally, Nicholson's models led to the concept of 'Nicholsonian searching efficiency', which is still used as a valuable, if variable, measure of parasitoid performance under field conditions. It is defined as:

$$\frac{\ln\left(N_t / N_{t+\tau}\right)}{P_t}$$

where N_t and $N_{t+\tau}$ are the host densities before and after parasitism has acted and P_t is the initial density of searching parasitoids.

The Nicholson-Bailey model

The explicit model proposed by Nicholson and Bailey (1935) arose from the application of Nicholson's (1933) 'competition curve', which he conceived to explain how the simple process of random searching imposed a limit on the sequestering of any kind of desired resource. Random searching means that some areas are covered more than once while others are missed; the 'area covered' per individual is therefore less than the 'area traversed', and the greater the number of individuals the greater the difference between the two and the greater the effect of competition. When applied to parasitoids searching for hosts, such competition would clearly lead to a stable population of parasitoids so long as the supply of hosts in each generation was fixed. Thus, random search does imply density-dependence in the parasitoid equation. Unfortunately for Nicholson and his quest for explicit mechanisms which would produce the balance of nature he believed in, the host-parasitoid model only provided half the answer. In Nicholson's own terminology, it led to a 'steady state' but not a 'stable steady state'. The other predator-prey models available at the time (Thompson 1924, Lotka-Volterra, cited in May 1974) also could not provide this stability. The ability of the competition curve to confer the desired state of balance on the parasitoid population was negated by the interactive nature of the model and the resulting delayed density-dependence embodied within it. Only if the initial number of hosts in each generation was constant or otherwise unaffected by previous parasitism, would random searching then stabilise the parasitoid population. Perhaps surprisingly, Nicholson himself did not consider the obvious possibility of forcing the host to be more constant and more independent of the parasitoid by granting it other independent density-dependent mechanisms, which would then allow the competition curve to stabilise the parasitoid. However, Nicholson and co-workers did identify other realistic modifications to the basic model which would provide the elusive balance he sought, notably variation in attack rates between hosts (Bailey *et al.* 1962).

The Nicholson-Bailey (N-B) host-parasitoid model embodies five key assumptions: 1) parasitoids and hosts have discrete, synchronised generations; 2) one host found leads to one new parasitoid; 3) parasitoids have unlimited potential fecundity; 4) parasitoids have a constant, characteristic 'area of discovery' (Nicholson 1933) or searching efficiency, implying that the number of encounters each parasitoid makes with hosts is directly proportional to host density, in other words a linear 'functional response' to host density; 5) parasitoids search randomly, so the chance of a host being missed is given by the zero term of a Poisson distribution. If the total number of encounters with hosts is N_e and the number actually attacked is N_a, then:

from assumption 4: $$N_e = aN_tP_t \tag{1}$$

and from assumption 5: $$N_a / N_t = 1 - e^{-N_e/N_t} \tag{2}$$

This gives the N-B model:

$$N_{t+1} = \lambda N_t e^{-aP_t} \tag{3a}$$

$$P_{t+1} = N_t\left[1 - e^{-aP_t}\right] \tag{3b}$$

where N and P are host and parasitoid densities in generations t and t+1, λ is the net rate of increase of hosts and a the parasitoid's area of discovery or searching efficiency. The 'area traversed' is aP_t, the 'area covered' is

$$\left[1 - e^{-aP_t}\right]$$

and the relationship between the two is Nicholson's 'competition curve'. As Hassell (1978) points out, the N-B model is equally applicable to predators given the addition of a suitable reproduction function.

The N-B model was conceived at a time when ecological theory was sparse, the only other existing predator-prey models being those of Thompson (1924) and Lotka-Volterra (cited in May 1974). Thompson also used the assumption of random search but with the number of encounters related to the parasitoid's fecundity, F, rather than the number of hosts it was capable of finding. Thus the number of hosts encountered is constant and independent of host density. The model is:

$$N_e = FP_t$$

Hence:

$$N_{t+1} = \lambda N_t e^{-FP_t/N_t}$$

$$P_{t+1} = N_t\left[1 - e^{-FP_t/N_t}\right]$$

The Lotka-Volterra model was formulated as differential equations:

$$dN_t / dt = N_t\left(r - bP_t\right)$$

$$dP_t / dt = P_t\left(-d + cN_t\right)$$

where r is the intrinsic rate of increase of the host, d the death rate of parasitoids in the absence of hosts and b and c are interaction terms between the populations. This model gave results superficially different to that of N-B, namely stable limit cycles with an amplitude dependent on the starting values, compared with divergent oscillations. However, May (1974) and Hassell (1978) show that this is simply a consequence of the lag inherent in the difference equations of the N-B model; re-casting the Lotka-Volterra model in difference equation form, in which the proportion of hosts missed is $[1 - aP_t]$ rather than e^{-aP_t}, gives behaviour qualitatively similar to the N-B model, especially for small values of aP_t (Hassell 1978).

Development of theory post-Nicholson

The N-B model had a greater impact on subsequent work than either Thompson's or those of Lotka and Volterra. In the case of Thompson's this is probably because the Nicholsonian assumption of constant searching efficiency is marginally more realistic than Thompson's assumption of a constant number of encounters per parasitoid, particularly in terms of the host-parasitoid dynamics it generates. However, N-B models incorporating a functional response to host density (see below) are essentially a hybrid of those of N-B and Thompson. Why the Lotka-Volterra model did not lead to an evolution of predator-prey and parasitoid-host theory based on differential equations, like the theory for disease-host dynamics, is less clear. There seems no obvious reason for the difference in approaches, since there are probably as many or more parasitoid-host or predator-prey interactions with continuous generations as there are with discrete ones, and equally, many disease-host interactions with discrete generations or cycles of infection. One suspects that the answer has more to do with the overwhelming influence of a few key workers in the respective fields than with the dictates of biology. Continuous-time predator-prey models do exist but they have maintained a consistently lower profile in the literature than discrete-time ones. This is changing slightly with a recent revival of interest in so-called ratio-dependent predation, and continuous-time models are discussed briefly below.

Developments in the N-B model itself addressed some of its most important and least realistic assumptions. Assumption 1 relating to discrete generations was not easily modified, although Beddington *et al.* (1978) included within the structure of an N-B type model two host generations per parasitoid generation and vice versa, and a single parasitoid generation attacking a continuously reproducing host. In general, though, the assumption of discrete, synchronised generations in the basic N-B model is probably one of the main constraints to its practical application in the field of biological control (see below).

Assumption 2, that one host parasitised leads to one new parasitoid was easily rendered more realistic by adding a mortality parameter to the parasitoid equation 3b. Given that many parasitoids undergo a free-living pupal stage (e.g. Hassell 1980), it is highly likely that mortality exists in such cases. Yet it is more often omitted than included in post-Nicholsonian theoretical studies. This may be because it does not affect stability of the model (e.g. Hassell and May 1973), but it will affect the extent of host suppression by a parasitoid and stability relative to the degree of suppression. Hence it is of some importance in the context of biological control. For example, Hassell (1980) found that *Cyzenis* larvae and puparia in the soil not only suffered heavy mortality, presumably from generalist predators which also attacked the host, but that the mortality was density-dependent.

Modifying assumption 2 to allow for predators was less simple, involving the derivation of a numerical response relating the predator rate of increase, combining fecundity, survival and rate of development, to the number of prey ingested or to prey density (reviewed in Hassell 1978). The choice of ingestion rate or prey density as the independent variable is of some biological significance, and is discussed further below in the context of continuous-time models and ratio-dependence. As pointed out by Hassell (1978), the crucial difference between models for the rates of increase of parasitoids and predators is that the former allow parasitoid reproduction at any host density greater than zero, whereas predators can only reproduce at prey densities exceeding some threshold. Furthermore, the effect is to make the predator models less stable than their parasitoid equivalents and to introduce equilibria which are only locally stable, making the outcome more dependent on initial predator-prey densities (Hassell 1978).

The remaining assumptions of the N-B model were the principal targets of post-Nicholsonian theory. Its major themes were functional responses in the 1950s and 60s (assumptions 3 and 4), and stabilising effects, including non-random searching and relatively simple dispersal models in the 1970s and explicit spatial effects in the 1980s and early 90s (assumption 5). Additional key areas of post-Nicholsonian theory included the recognition and incorporation into models of more realistic and complex predator or parasitoid behaviour, and the evolution of continuous-time and ratio-dependent theory paralleling the discrete-generation N-B paradigm. These major themes are considered in turn, all but the functional response relating to the quest for stability.

The functional response

Nicholson assumed that the number of hosts attacked per parasitoid increased without limit in proportion to host density (assumptions 3 and 4). This is clearly impossible, since the rate of parasitism must ultimately be limited either by parasitoid fecundity or by the fact that each encounter takes a finite time. Work on the functional response was dominated by the rigorous experimental and modelling studies of Holling (1959, 1965), an approach he called experimental components analysis. Using the praying mantis as a model animal, Holling derived a relationship between number of prey attacked and prey density which has an asymptotic form which Holling designated a 'Type 2' response. It is represented in his studies and in most subsequent models by a Michaelis-Menten function, which is readily derived from the assumption that there is a handling time associated with each prey item caught or host parasitised. This in turn reduces the time available for catching more prey. Thus Eq. 1 for the number of encounters per unit time changes to:

for parasitism $$N_e = a\left(1-\frac{HN_e}{P_t}\right)N_tP_t$$

and for predation $$N_e = a\left(1-\frac{HN_a}{P_t}\right)N_tP_t \quad (4)$$

where H is handling time as a proportion of total time. The difference between the two equations arises because parasitoids waste time re-encountering already parasitised prey (assuming the absence of superparasitism-avoiding behaviour such as host marking), whereas for predators there are no multiple encounters and handling time is associated with prey actually attacked. Re-arranging the parasitoid equation gives:

$$N_e = \frac{aN_tP_t}{1+aHN_t} \quad (5)$$

and if searching is systematic rather than random, $N_a = N_e$ and the same equation applies to predation. This is the asymptotic Type 2 functional response relating prey attacked to prey density, also called the 'disc equation' after an experimental validation using blindfolded human subjects searching for sandpaper discs (Holling 1959). In the case of predation, handling time includes all time-consuming activities related to prey capture, such as eating, digestion and time taken for hunger levels to increase. For parasitism it is the time taken to oviposit once a host is encountered, which includes any waiting time for the host to be susceptible. For example, the Braconid *Microctonus hyperodae* cannot oviposit in a stationary Argentine stem weevil host (*Listronotus bonariensis*), because of the need to penetrate between the tergites (CB Phillips, unpublished). Strictly, handling time for a parasitoid should include the effect of egg depletion, since this features in the attack equation in exactly the same way and is likely to be much more significant (see below).

Combining the functional response with the Nicholsonian assumption of random searching by the P_t parasitoids or predators (assumption 5) gives the 'random predator' and 'random parasitoid' equations (e.g. Rogers 1972). Thus, N_e from Eqs. 4 and 5 is substituted into Eq. 2 to give:

for predators $$N_a = N_t\left[1-e^{-aP_t\left(1-\frac{HN_a}{P_t}\right)}\right] \quad (6)$$

for parasitoids $$N_a = N_t\left[1-e^{\frac{-aP_t}{1+aHN_t}}\right] \quad (7)$$

Note that Eq. 7 effectively combines the Nicholson-Bailey and Thompson models, with (handling time/total time) re-interpreted as 1/fecundity (i.e. H in Eq. 7 = $1/F$ in Thompson's model); it behaves like the Nicholson-Bailey when N_t is small, and like Thompson's model when N_t is large.

Ivlev (1961) developed an alternative version of the asymptotic Type 2 response, also based on detailed experimental analysis, but in this case of fish feeding behaviour:

$$N_a = N_{amax} \cdot P_t\left[1-e^{-cN_t}\right]$$

where N_{amax} is the maximum number of prey attacked per predator per unit time (= total time/handling time in Eq. 4) and c is a constant equivalent to the area of discovery/N_{amax}. In addition, Ivlev elaborated the functional response to include the effects of i) prey aggregation, such that an increase in the degree of aggregation was equivalent to an increase in the mean density, ii) the preference of a feeding species for different types of food, and iii) competition between predators for prey.

Watt (1959) also derived an equation for the functional response of parasitoids to host density, of similar form to Ivlev's, but including explicit competition between the parasitoids represented by a term P_t^{1-b}:

$$N_a = N_{amax} \cdot P_t \left[1 - e^{-cN_t P_t^{1-b}} \right]$$

The predator-prey equation of Frazer and Gilbert (1976) was rather more complex, and was based on empirical fitting of a function to detailed experimental data on coccinellids feeding on aphids. The function was used by Gutierrez and Baumgaertner (1984) as a general description of trophic interactions with units in biomass rather than numbers, and its form was as follows:

$$N_a = N_t \left[1 - e^{\left(-N_{amax} \cdot \frac{P_t}{N_t} \right) \left(1 - e^{\frac{-aN_t}{N_{amax}}} \right)} \right]$$

It collapses to an N-B model when N_t tends to 0 and to a Lotka-Volterra model when N_t and P_t both tend to 0 (Gutierrez and Baumgaertner 1984).

The other functional responses characterised by Holling were Types 1 and 3. The first represented a linear relationship between number attacked and prey density, up to an upper limit at which the number attacked abruptly levelled off. This is effectively a hybrid between the functional responses implicit in the N-B and Thompson parasitism models, and has rarely been invoked since. The Type 3 response is a sigmoid relationship between number attacked and prey density, attributable to increases in the area of discovery with prey density. This is often represented by an equation of the form:

$$N_a = \frac{N_{amax} \cdot N_t^2 P_t}{b^2 + N_t^2}$$

where b is the prey density at which the attack rate is half the maximum (e.g. May 1974).

Overall, functional responses have featured less in subsequent theoretical developments and practical applications of predator-prey theory than have extensions and modifications to the N-B model. In other words, such developments have focused on the relationship between proportion attacked and predator/parasitoid rather than prey/host density. This is partly because Type 2 responses are invariably de-stabilising, with Type 3 responses de-stabilising in discrete-generation models (Hassell 1978), and the whole emphasis of post-Nicholsonian predator-prey theory has been a quest for factors stabilising the interaction. However, stability of a whole model, expressed as the proportion of parameter space occupied by the stable region, is not necessarily a useful criterion. Of more practical importance is the model's stability relative to particular parameter combinations or equilibrium points. For example, adding a functional response and handling time to an N-B host-parasitoid model with host density-dependence can stabilise an otherwise unstable interaction, for particular values of searching efficiency (a), intrinsic rate of host increase (r) and host parasitoid-free

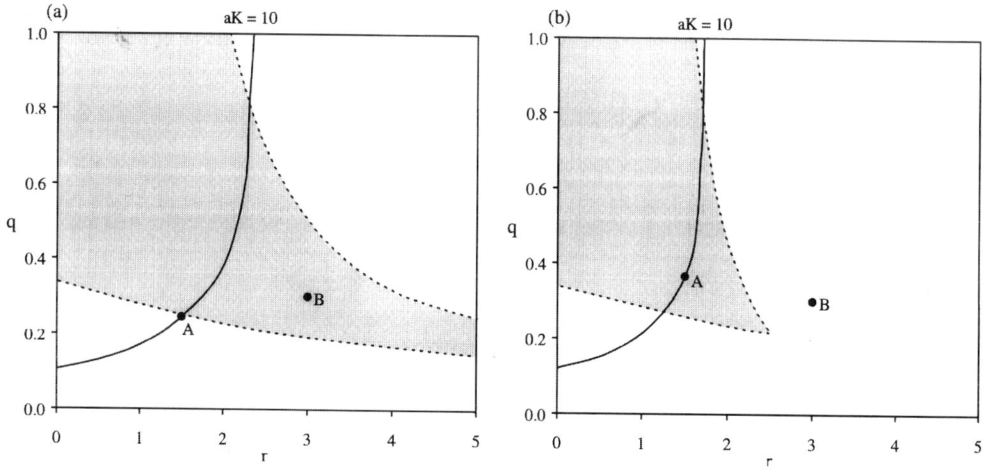

Fig. 1. Stability boundaries (shaded areas indicate stability) and host suppression (vertical axes, q = host density/K), as functions of host intrinsic rate of increase (bottom axis, r) and for a particular value of parasitoid searching efficiency x host carrying capacity (curve on the graph, aK = 10), based on a Nicholson-Bailey model with host density-dependence (Model 2 in May *et al.* (1981) Fig. 5a with $k \to \infty$). (a) without handling time, (b) with handling time (handling time/total time = 0.1). Point A represents a constant parameter combination (r = 1.5 and aK = 10). Point B represents a constant level of host suppression and a given r value but varying aK (q = 0.3, r = 3).

equilibrium (K), but it will be destabilising for any given value of r and host suppression q. The effect is shown in Fig. 1 by the positions of point A relative to the stability boundaries in Fig. 1a and 1b, using Model 2 in May *et al.* (1981) with $k \to \infty$. Fig. 1a is equivalent to Fig. 5a in May *et al.* (1981). The stabilising effect of handling time is due to an effective reduction in mean searching efficiency. However, it does result in a reduced level of host suppression, conveniently measured by the parameter q which is the ratio of host equilibrium density to K. If stability of the two models is compared for the same values of r and q (though not necessarily the same equilibrium parasitoid density), then it is true that a model achieving a q value of 0.3 for r = 3 is less stable with handling time (point B in Fig. 1b) than without it (point B in Fig. 1a). In the latter case, the absence of handling time must be compensated for by a reduced searching efficiency to maintain the same q value.

Another reason for neglect of the functional response is because handling time/total time has been regarded as generally small and therefore of little practical significance (e.g. Table 3.1 in Hassell 1978). However, this may reflect the way handling times have been interpreted and measured: in the case of parasitoids, considering handling time/total time as the inverse of fecundity can lead to a significant term in the random parasitoid equation (Eq. 7) if the parasitoid has a low fecundity (Yamada 1987) or the host density is high. Thus the neglect of the functional response in recent theoretical developments may be unwarranted, and it contrasts notably with Hassell's (1978) assertion that 'the functional response to prey density is a crucial component of any predator-prey model'.

Type 3 responses represent a partial exception to this neglect and have enjoyed a stable, if still rather low-profile, life in the literature. As suggested above, they are stabilising in continuous-time predator-prey models, over certain ranges of prey density. Moreover, they offer one of the few biologically realistic explanations for population systems exhibiting multiple stable states (e.g. Ludwig *et al.* 1978; Hestbeck 1987) and outbreak behaviour. As such they are of considerable theoretical and practical interest, since outbreak behaviour is a poorly understood ecological phenomenon, and a 'predator pit' associated with complex density-dependence arising from a sigmoid functional response may be responsible for cases of biological control (but see Beddington *et al.* 1978). The term 'predator pit' reflects the fact that under the above circumstances, instead of increasing steadily as prey density declines, the per capita rate of increase of the prey may drop sharply and become negative at low densities because of predation, before rising and become positive again as density declines further towards zero (e.g. Pech *et al.* 1992).

The quest for stability

The possible stabilising mechanisms in Nicholson's model involve density-dependent host population growth, density-dependent parasitism of the host arising from mutual interference between parasitoids or aggregation to high host densities, or heterogeneity of risk of parasitism among hosts.

Host density-dependence

The stabilising effect of density-dependent host population growth was considered by Hassell and Varley (1969), and in more detail by Beddington *et al.* (1978) and May *et al.* (1981). One of the most widely used models of this kind, and the one analysed so thoroughly by Beddington *et al.* (1978), is:

$$N_{t+1} = N_t e^{r(1-N_t/K)} e^{-aP_t}$$

$$P_{t+1} = N_t \left[1 - e^{-aP_t}\right]$$

As pointed out by May *et al.* (1981) this model is of questionable biological validity since it implies that host mortality from parasitism follows that from density-dependence, whereas the parasitoids search for hosts before the density-dependence has acted. At best this is unlikely. May *et al.* (1981) further show that the dynamics of models in which the order of parasitism and density-dependence changes are qualitatively different. In particular, if parasitism precedes overcompensating density-dependence it can increase rather than decrease the host equilibrium density, a finding of obvious significance for biological control.

Beddington *et al.* (1978) effectively discounted the role of host density-dependence as a stabilising factor in biological control, showing that this mechanism alone could not give the degree of host suppression exhibited by six examples of successful biological control programmes in the field. Whether for this reason or not, host density-dependence rarely featured in subsequent predator-prey theory, the implicit but unstated assumption being that density-dependence is unlikely to be significant at host densities associated with parasitoid- or predator-induced equilibria. On the other hand, it seems intuitively reasonable that host density-dependence will be a significant component of stability and equilibrium densities in some systems, particularly where their behaviour is otherwise oscillatory. Conversely, although the effect has not been explicitly demonstrated it is likely that temporal density-dependence in subpopulations may exert a destabilising effect on a metapopulation by reducing the spatial asynchrony and heterogeneity within it (see below). It would appear that more empirical evidence is needed to identify the relative roles of predation or parasitism and other sources of host density-dependence in the dynamics of real populations, in order to justify or refute the assumption that one factor may effectively be ignored in considering the nature of each equilibrium.

Mutual interference and pseudo-interference

Mutual interference between parasitoids was another early candidate for a stabilising mechanism (Hassell and Varley 1969), invoked to explain an observed negative effect of increasing parasitoid densities on searching efficiency and parasitism rates. The effect was included in the N-B model by raising parasitoid density to a power, in a similar way to Watt (1959), giving a function for parasitism (i.e. proportion of hosts surviving parasitism):

$$f(N_t, P_t) = e^{-QP_t^{1-m}}$$

where Q is the quest constant and m the interference constant. Free *et al.* (1977) showed that the same effect could theoretically arise from aggregation of parasitoids to clumped distributions of hosts, in the absence of actual interference. They termed the effect 'pseudo-interference' because it led to exactly the same model.

Spatial effects: implicit spatial models

In the early seventies there was a surge of interest in the role of spatial effects in non-random search and consequent stability of N-B type models. The treatment of spatial behaviour and heterogeneity in these models ranged from implicit to partially explicit and explicit. The former retained an N-B type format of only two equations within which spatial effects were simplified and subsumed through an appropriate choice of parasitism function. Thus they were described by May (1978) and Perry (1987) as 'models of intermediate complexity'. Nicholson himself, in collaboration with colleagues V.A. Bailey and E.J. Williams, was an early pioneer in the development of models incorporating non-random search, and well understood the potential stabilising influence of heterogeneity of risk. In his case this took the form of variations in the area of discovery, with some hosts harder to find than others (Bailey *et al.* 1962). This generalised earlier work by Varley (1947) which showed that the presence of a host refuge could stabilise the N-B model. At the same time, these authors also produced one of the first stability analyses of parasitoid-host models, anticipating a theme which later became a central part of predator-prey theory. Temporal refuges involving parasitoid-host asynchrony or an invulnerable host age-class have the same effect as spatial refuges (e.g. May 1974). The refuge effect in these discrete generation models (e.g. Hassell and May 1973) is a net outcome over the lifespan of the parasitoid. In practice, within-generation host movement may largely negate the stabilising effect of a physical refuge even if at any one time a high proportion of hosts are invulnerable. Rather different was the model of Hassell and May (1973), which was partially explicit in a spatial sense since it applied an N-B model separately to a number of patches. However, rather than model migration of parasitoids and hosts explicitly, their densities were assigned to the patches in each generation according to specified distributions. Analysis showed that stability was enhanced by low prey increase rates, a high degree of clumping of the prey and a high level of predator aggregation in response to prey density; the more clumped the prey, the less predator aggregation required for stability.

May's (1978) phenomenological model was particularly interesting and captured the spatial effect implicitly. May used the negative binomial distribution rather than the Poisson to characterise the distribution of parasitoid attacks among hosts, which neatly by-passed the need for specification of the actual spatial distributions of parasitoid and host, but relied instead on the availability of a measure of the distribution of attacks between hosts. May's model is:

$$N_{t+1} = \lambda N_t \left(1 + \frac{aP_t}{k}\right)^{-k}$$

$$P_{t+1} = N_t \left[1 - \left(1 + \frac{aP_t}{k}\right)^{-k}\right]$$

where k is the negative binomial clumping parameter, low values representing increasing aggregation and $k \to \infty$ giving a Poisson distribution of attacks and the N-B model. As Hassell (1980) demonstrated, the distribution of attacks between hosts (k) may itself relate to host density, giving the model:

$$N_{t+1} = \lambda N_t \left(1 + \frac{aP_t}{bN_t}\right)^{-bN_t}$$

$$P_{t+1} = N_t \left[1 - \left(1 + \frac{aP_t}{bN_t} \right)^{-bN_t} \right]$$

where b is a constant relating the negative binomial k value to N_t. With a functional response the model becomes:

$$N_{t+1} = \lambda N_t \left[1 + \frac{aP_t}{bN_t(1+aHN_t)} \right]^{-bN_t}$$

$$P_{t+1} = N_t \left[1 - \left(1 + \frac{aP_t}{bN_t(1+aHN_t)} \right)^{-bN_t} \right]$$

In addition to describing and fitting these significant variants of May's model, Hassell's (1980) paper represented one of very few attempts to apply a theoretical, N-B type model to a real biological control system, involving the winter moth and its parasitoid *Cyzenis albicans*. In this respect it was a landmark. Later work by Chesson and Murdoch (1986) showed that May's model really represents parasitoid aggregation independent of host density, or for parasitoid densities constant between patches but with searching efficiency varying. It can represent parasitoid aggregation to high host densities but only if searching efficiency or the negative binomial parameter k is related to average host density.

Spatial effects: metapopulation models

In contrast to recent work in the early nineties, the earliest spatially explicit models developed in the seventies focused on between-patch dynamics and stability, largely ignoring the dynamics of predators and prey within patches. Rather they considered persistence of prey or predators in terms of persistence of patches which contained one or the other (e.g. Maynard-Smith 1974; Hastings 1977). Patches were typically characterised by a limited number of states, such as empty, prey only and prey plus predator. Within-patch dynamics generally took the form of a deterministic succession through these states, returning to empty as predators exterminated prey, while between-patch dynamics were typically simulated as stochastic movements of prey and predators. Maynard-Smith (1974) distinguished between 'island models' in which dispersal occurred from each patch to every other patch, and 'stepping stone models' in which dispersal was restricted to neighbouring patches. Conclusions from his 'island' model, which were representative of those from other similar work, were that persistence was easily obtained and that it was enhanced by large numbers of patches in the models, high prey migration rates, limited predator migration rates and the presence of some cells as prey refuges. Of some relevance for biological control, substantial prey suppression could be achieved if prey migration rate was low and predator migration rate high, but this required an island model and many cells. A deterministic equivalent to this model also gave a persistent equilibrium with low prey density and most cells empty. Hastings (1977) analysed the stability of an analytical version of a similar model, with the time for a patch to change states from 'prey + predator' to 'empty' taking the form of a variable finite lag. Other transitions, from 'empty' to 'prey' and 'prey' to 'prey + predator', were determined by the respective migration rates. Again, Hastings found a large region of stability, which depended on the migration rates, and that if both rates were high the prey could be highly predator-limited even if the predator density was extremely low.

Most of these early spatial models were stochastic, at least in terms of the movement of predators and prey, and by including more realistic within-patch dynamics, Chesson (1981) showed that such stochastic models gave predictions of overall metapopulation densities consistently different from those of their deterministic equivalents. This was because of the interaction between stochasticity and the non-linear within-patch dynamics, and while it may be possible to mimic the overall metapopulation behaviour using a deterministic model, Chesson suggested that such a treatment would be purely descriptive rather than explanatory. Thus, stochasticity was held to be necessary to maintain spatial and temporal variability and the resulting stability and persistence. If within-patch stochasticity is identified with Andrewartha's and Birch's (1954) view of populations influenced primarily by environmental factors, then Chesson's conclusions rather ironically suggest that an assemblage of populations behaving according to Andrewartha and Birch rules gives a stable Nicholsonian metapopulation.

In a highly significant paper, Hassell *et al.* (1991) later showed that stochasticity is not necessary to maintain spatial heterogeneity and metapopulation stability, at least for discrete-time subpopulation models. Indeed, they used the original N-B model for within-patch dynamics, showing that the metapopulation could be stabilised by local movement of parasitoids and hosts. Unlike many of the earlier models, this was a 'stepping stone' one rather than an 'island' one, with dispersal occurring only to neighbouring patches. The stability results from the inherent spatial density-dependence, which is itself a consequence of diffusive movement between patches of unequal density, of the parasitoid density-dependence and overall delayed density-dependence in the N-B patch models, and of the maintenance of heterogeneity due presumably to the lags inherent in a discrete-time model and the consequent within-patch delayed density-dependence. The model showed that local dispersal combined with inherently unstable, oscillatory local dynamics can lead to self-organised spatial structures described as spiral waves, crystal lattices or chaos, depending on the relative migration rates. Echoing Chesson's earlier work, this model also imparted a conceptually Nicholsonian stability to the metapopulation but suggested that the dynamics of the latter were an emergent property and could not necessarily be described by a Nicholsonian model for the entire metapopulation. Unlike Chesson, though, Hassell *et al.* (1991) considered that such deterministic models could be applied at the subpopulation level. The same authors (Hassell *et al.* 1994) later showed that the self-organised patterns allow three or more species to coexist, often spatially separated but in an otherwise homogenous environment. This suggests that it may not be necessary to seek deterministic, environmental causes for spatial variations in local abundance, a finding analogous to the discovery of chaos in single-species dynamics.

The issue of whether spatial heterogeneity is stabilising in corresponding continuous-time models has been the subject of some recent debate, with Murdoch and Oaten (1989) suggesting that it is and Godfray and Pacala (1992) holding the contrary view. However, the difference appears to arise from the rather different models used by the respective authors, that of Murdoch and Oaten being a whole-population model rather than an explicit metapopulation one (Murdoch *et al.* 1992). However, it does appear that, unlike the discrete-time model of Hassell *et al.* (1991), a continuous time model can only be stabilised by environmental heterogeneity expressed as a difference in migration rates between different patches; for a homogenous but subdivided environment the continuous-time model is neutrally stable (Godfray and Pacala 1992). Murdoch *et al.* (1992) offer the general conclusion that the processes affecting stability of metapopulation models are extremely complex, mathematically challenging and little known, primarily because of the several interacting density-dependent factors resulting from spatial asynchrony. In general, however, increasing synchrony reduces stability and vice versa.

Spatial effects: field studies and heterogeneity of risk

Paralleling the development of spatial metapopulation models during the late eighties and early nineties, was the analysis of spatial density-dependence in field populations. A major finding from both field data and theory, which echoed the thinking behind May's model, was a recognition of the crucial significance of heterogeneity of risk between hosts, rather than density-dependent aggregation of parasitoids or predators *per se* (e.g. Chesson and Murdoch 1986). Thus, density-independent or inversely density-dependent aggregation can be as strong a stabilising force as directly density-dependent aggregation (Hassell and Pacala 1990). This led to the '$CV^2 > 1$' rule for stability of spatial parasitoid-host systems, where CV^2 is the coefficient of variation of parasitoid density in the vicinity of each host. Hassell and Pacala (1990) further showed that CV^2 can be partitioned into variation in parasitoid densities independent of and dependent on host density, and that in most field studies the former is the more significant component of the two.

More complex behaviour

In addition to spatial effects and actual mutual interference between parasitoids and predators, continuing ecological work has revealed other mechanisms which may contribute to non-random search and stability, or add further biological dimensions to the basic Nicholsonian model. Some of these have been incorporated into N-B type models and subjected to theoretical analysis, such as multiple-species systems and the effects of prey density on predator rates of increase (e.g. Hassell 1978), density-dependent parasitoid sex-ratios (Hassell *et al.* 1983), mortality arising from host feeding by adult parasitoids (e.g. Kidd and Jervis 1989), and within-host competition between parasitoids (Taylor 1988). However, the theoretical implications of other factors remain to be considered. These include kairomomes which attract parasitoids, dependence of predators and parasitoids on other resources such as nectar, pollen and shelter (e.g. Wratten and van Emden 1995), host-marking as a way of reducing super-parasitism (e.g. Hubbard *et al.* 1987), and 'pseudo-parasitism' (e.g. Munster-Swendsen 1994). The latter involves oviposition-induced host sterility (e.g. Brown and Kainoh 1992), or 'associative mortality' (Goldson *et al.* 1993) in which hosts exposed to parasitoids but not parasitised suffer higher mortality than those not exposed. Such mortality can be significant even under near-natural conditions (Goldson *et al.* 1993). Post-Nicholson work has also involved many specific studies (e.g. Dixon 1959) and models (e.g. Casas *et al.* 1993) aimed at elaborating the behavioural basis for generation-level functional and numerical responses. Of these, Holling's experimental components analysis was probably the most distinctive and thorough, and has already been discussed under 'functional responses' above.

Alternative theory: continuous-time and ratio-dependent models

Compared with the discrete N-B type models, continuous-time predator-prey models have maintained a consistently low profile in the literature. This is changing slightly with a recent revival of interest in logistic-type predator-prey models, resurrected under the new names of 'interferential' (Caughley and Lawton 1981) or 'ratio-dependent' models (e.g. Arditi and Ginzburg 1989; Berryman 1992), and with the development and application of new theory based on delayed differential equations.

May (1974 p.84) gives examples of two typical continuous-time predator-prey models, later described respectively as the laissez-faire and interferential models by Caughley and Lawton (1981):

Laissez-faire:
$$dN/dt = rN(1 - N/K) - kP\left(1 - e^{-cN}\right)$$
$$dP/dt = P\left[-b + a\left(1 - e^{-cN}\right)\right]$$

Interferential:
$$dN/dt = rN\left(1 - \frac{N}{K}\right) - \frac{kPN}{(c+N)}$$
$$dP/dt = gP\left(1 - \frac{hP}{N}\right)$$

where P and N are predator and prey densities, r the intrinsic rate of prey increase, k the maximum prey consumption rate per predator, c a constant describing the shape of the functional response, b the maximum death rate of predators, g the intrinsic rate of predator increase and h the predator-prey ratio at which the predator population is at equilibrium. The functional responses in the host equation are similar but the predator equations are very different. The laissez-faire model assumes that the instantaneous rate of predator increase depends only on prey density. Thus there is a threshold prey density above which the predator increases (without limit if the prey remains above the threshold) and below which it declines. The interferential model, on the other hand, assumes that predators interfere with each other in some way, such that the instantaneous rate of increase is directly density-dependent, relating to densities of both predator and prey. It can be viewed as a logistic equation with predator carrying capacity related to prey density. Thus, there exists a range of predator equilibria corresponding to different prey densities. For this reason the interferential model extends readily to more than one predator species feeding on a single prey (or herbivores feeding on a single vegetation type), whereas the laissez-faire model does not allow coexistence of more than one species consuming the same resource since the threshold resource for one predator's persistence will by definition be different to that for persistence of the other species (Barlow 1985). Although the interferential model is simpler and easier to apply in practice than the laissez-faire and is perhaps more widely used (e.g. Caughley and Lawton 1981; Ludwig *et al.* 1978), its biological basis is somewhat obscure because of the inconsistency between the prey intake function in the prey equation and the prey-dependent predator growth function: intake depends only on prey density whereas increase depends on the predator-prey ratio (Barlow 1985). More logical and realistic would be a model which separated out the limiting effects of prey and predator densities on predator rate of increase, the effect of predator density acting through such factors as territoriality.

Ratio-dependent predator-prey models take the form:

$$dN/dt = N\left(a - \frac{bP}{(w+N)}\right) \tag{8a}$$

$$dN/dt = P\left(c - \frac{fN}{(w+N)}\right) \tag{8b}$$

where *a, b, c, f* and *w* are constants, *a* and *c* being the maximum rates of prey and predator increase (Berryman 1992). Arditi and Ginzburg (1989) justify ratio-dependence or predator interference with prey consumption, on the basis of the different time scales applying to population change and prey consumption: over the interval of a generation there is likely to be prey depletion which will depend on predator density. However, this presumes continuous-time models applied to discrete-generation populations. Strictly they are more relevant to continuously reproducing populations, in which the time scales will be similar although there will be lags between consumption and increase. A further justification for ratio-dependent models (e.g. Berryman 1992) is that they overcome the paradox of enrichment and the so-called 'paradox of biological control', both of which arise from 'classical' predator-prey models. The first paradox is that increasing the prey carrying capacity (e.g. by enrichment of its resources) increases the predator equilibrium density but not that of the prey, while the second is that the models do not allow biological control involving both stability and high levels of host suppression. The second paradox is something of a straw man, since it applies only to continuous-time laissez-faire models. At first sight, ratio-dependent models appear to 'solve' the problems in the sense that they give intuitively more sensible answers because of the different equations, or isoclines, relating predator to prey densities at equilibrium which they generate. However, it is not clear that the inherent biology is any more sensible than that of the interferential model. For example, Eq. 8 implies a relationship between per capita intake (*I*) and per capita rate of increase (*r*) for the predator of the form:

$$r = I\left(p + \frac{q}{N} - \frac{gP}{N}\right)$$

where p, q and g are constants. Thus, when predators are very sparse ($P \to 0$), their per capita rate of increase declines relative to their prey intake as prey density increases, and becomes infinite as the prey disappear. This makes little sense. Furthermore, as a test for ratio-dependence, Arditi and Saiah (1992) use values of m in the predator per capita intake function applied to empirical data sets:

$$\frac{aP^{-m}N}{1+aHP^{-m}N}$$

In the majority of data sets m was close to 1, ostensibly justifying ratio-dependence. However, the test is flawed in that the function applied was essentially arbitrary, and no biological basis for it was offered. A 'classic' functional response with pseudo-interference among the predators (or parasitoids) of the form

$$\frac{aP^{1-m}N}{1+aHN}$$

may have fitted the data equally well. As a general conclusion, and with particular reference to the interferential and ratio-dependent continuous-time models, it would seem that insufficient attention has been devoted to the biological mechanisms underlying the models, the focus being too heavily on their more macroscopic behaviour.

The most recent development in continuous-time models involves the use of delayed differential equations (Gurney *et al.* 1983, Godfray and Hassell 1989, Godfray and Waage 1991). Such models have yielded new insights into the dynamics of parasitoid-host systems, such as the fact that generation cycles of tropical insects could be an outcome of interactions with their parasitoids (Godfray and Hassell 1989), and that phenological asynchrony can stabilise host-parasitoid models (Godfray *et al.* 1994). In addition, Godfray and others have also applied them with some success to specific case studies in biological control (see below).

Application of theory post-Nicholson

Theory and classical biological control: current status

In spite of considerable efforts to develop a theoretical basis for biological control (e.g. Beddington *et al.* 1978; May and Hassell 1988), it is clear that N-B models and even their modern derivatives have been and are being little used in biological control practice (Greathead 1986; Waage 1990). Why should this be? It is certainly not a lack of need. There is a growing recognition worldwide that little progress in practical biological control will occur without a significant increase in understanding and predictive ability (e.g. Greathead 1986; Barlow and Goldson 1993). Such progress is necessary to avoid the well-publicised mistakes of the past and to improve on the modest success rate of the present, which is around 10% for classical biological programmes against insect pests (Waage 1990). The need will be most keenly felt as the explosion of biotechnology vastly increases the availability of choices (Kareiva 1990), and as environmental agencies and the public rightly dictate more stringent screening procedures than in the past. These imperatives mean that we can no longer afford the luxury of ignorance, and they put a premium on the ability to select both a successful and a safe natural enemy. Understanding and prediction are critical but there is also a need to make best use of what is available and to optimise the management of biological control agents, whether this involves single introductions giving permanent results ('classical' biological control), repeated introductions ('augmentative' or 'inundative' control), or enhanced action of existing agents ('natural enemy enhancement').

The reason why theory and models have been so little used may be because they have addressed different questions to those asked by the practitioners (Kareiva 1990), or because any answers obtained are at too strategic a level to have a meaningful impact on tactics (Barlow 1993). Moreover, theory may have led into blind alleys by generating and reinforcing potential myths. One of these is that classical biological control typically implies, and requires, almost complete host suppression with q values (pest density under control/pest density in the absence of control) of less than 5%. Thus, the influential paper by Beddington *et al.* (1978) considered the implications of Nicholsonian-type models and theory for biological control, and showed that substantial and stable host suppression requires either complex density-dependence in the host, such as might result from a sigmoid functional response of polyphagous predators, or aggregation in the parasitoid. Of the two mechanisms, they felt the latter to be the most likely candidate in practice. Yet for biological control to be successful simply requires that the pest is reduced in abundance to below some damage threshold. In cases like the weevil *Sitona discoideus* in New Zealand lucerne, this threshold may be around 60% of K, the pest's natural equilibrium. Consequently, biological control by the introduced parasitoid *Microctonus aethiopoides*, which reduces larval densities to about 40% of K, is considered successful (Barlow and Goldson 1993). At these densities the influence of density-dependence in the host is important, contrary to arguments in the theoretical literature which discount its relevance based on an assumption of low q values. Furthermore, a control agent does not need to be a 'silver bullet' to be useful. A conventionally-defined partially effective agent may be a valuable component of an IPM programme if it helps to erode the compensatory abilities of a pest with a high degree of intra-specific density-dependence (Goldson *et al.* 1994).

A second potential myth is that stability is important. Murdoch *et al.* (1985) take an alternative view to Beddington *et al.* (1978), namely that biological control is more likely to involve local pest extinction and that stability may be largely irrelevant, at least on a local scale. Thus, of the six case studies cited by Beddington *et al.* (1978) as evidence for low, stable equilibria, in only one could the alternative possibility of local pest extinction be ruled out (Murdoch *et al.* 1985). Therefore local pest extinction rather than equilibrium may be a more widespread phenomenon than stability and a better goal in biological control practice. At a lower level, there are difficulties in applying the N-B type models, which underpin so much theory, to real-world problems. In particular, the discrete-generation paradigm is a serious restriction. As a result, Godfray and co-workers have recently departed from the Nicholsonian tradition and begun to apply continuous-time models based on delayed differential equations to real biological control problems. For example, Godfray and Waage (1991) were able to develop useful prospective conclusions from such a model about the likely relative effectiveness of two different parasitoids for control of the mango mealybug, *Rastrococcus invadens*. In effect, there is no reason why these kinds of models embodying finite time lags should portray biological control systems with partially or completely overlapping generations any more effectively than ordinary differential equations with their implicit assumption of exponentially distributed residence times in the various states, equations which are commonly invoked for microparasite-host systems. Nevertheless, the approach does offer an interesting and useful new tool which may prove to be more appropriate for insect populations with age-structure than ordinary differential equations, particularly if the latter require more subclasses within developmental stages to generate realistic distributions of residence times. Moreover, such models allow greater realism than the discrete generation ones for many real-world population systems, but at the same time allow at least some degree of analytical treatment.

Applying the term coined by Perry (1987) to a slightly different context, both discrete-generation Nicholsonian-type models and delayed differential equation ones have been described by Godfray and Waage (1991) as 'models of intermediate complexity', essentially to distinguish them from more detailed and specific simulation models. However, the term should also legitimately include simple simulation models (Barlow and Goldson 1993; Barlow *et al.* 1996), in which the system is deliberately idealised as for an analytical model but which allow more complex functional forms or basic structures as dictated by the biology. In terms of their complexity such models differ little from the delayed differential equation ones. However, they contrast with simulation approaches which deliberately espouse much more of the real-world detail, which is 'black-boxed' in the models of intermediate complexity. The interesting question is the extent to which the semi-analytical delayed differential equation models will be applicable to real-world systems, without losing their capacity for some degree of analytical treatment. For once they lose this, there is essentially no difference between these and simple, 'computer-orientated' simulation approaches, except for the rather greater flexibility of the latter.

If Nicholsonian parasitoid-host models and their derivatives appear to be of limited application as metaphors for complete biological control systems, because of the discrete, synchronous generation paradigm, this does not mean that they have been of no value. Beddington *et al.* (1978) considered analytical elaborations of the Nicholson/Bailey model, extended to include two generations of parasitoids per host generation or vice versa, and a discrete-generation parasitoid attacking a continuously reproducing host. Such modifications potentially expand the applicability of the basic model, though they have not yet been applied to real-world examples. Moreover, as models of attack behaviour the Nicholson-Bailey derivatives typically form vital building blocks for slightly larger 'models of intermediate complexity'. Models embodying a Nicholsonian response to parasitoid density are typically invoked more frequently than the functional response models to prey density so thoroughly studied by Holling and others. In practice, many models of intermediate complexity applied to specific case histories demand the continuous form of the attack equations, but the link to Nicholsonian thinking is retained through their inclusion of parasitoid density and searching behaviour as independent variables. As one example, Godfray *et al.* (1994) used a continuous-time equivalent of May's (1978) negative binomial:

$$dN_a / dt = k ln\left(1 + \frac{aP_t}{k}\right)$$

Theory and classical biological control: future prospects

Potentially, general theory and the right kinds of models offer a number of practical benefits for biological control practice. These include: predicting the outcome and success of a specific introduction; aiding in the selection of the most appropriate agent; predicting the impact of exotic agents on ecosystems and non-target species; increasing understanding of the processes involved; aiding in the identification and interpretation of critical field data; and optimising management of existing or introduced agents. To contribute usefully to these practical aims, general theory may have to address questions of a more modest and bounded nature than the global, strategic issues hitherto targeted, such as the causes of stability in biological control systems. In terms of specific models, although few real biological control systems have been analysed (May and Hassell 1988; Murdoch 1992), the number of such studies and the use of case-specific models is growing. Typically, such models (e.g. Hassell 1980; Hill 1988; Godfray and Waage 1991; Hochberg and Waage 1991; Barlow and Goldson 1993; Barlow *et al.* 1996) have helped in understanding the systems and identifying the reasons for success or failure of biological control, while in some cases they have also addressed relevant management questions. However, only rarely have these models resembled theoretical models of the N-B type, or of any other type. If one were to select any one principal reason for this, it is probably the fact that the structure of the theoretical models, whether in discrete or continuous time, frequently do not match that of real-world systems. Moreover it is often difficult to simplify and mould the real-world systems, involving seasonality, age-structure, temperature-dependence and time-dependent behaviour such as diapause, to fit the format demanded of simple models.

Three examples are given of modest but practical questions for which general theory provides at least a partial answer. We then offer two case studies of specific models, showing the pragmatic approach required to model apparently simple but actually complex biological control systems, in order to address specific questions relating to their management.

Predator pits

The question posed here is: will a reduction in predator numbers disrupt their control of a pest? Although Beddington *et al.* (1978) largely discounted complex host density-dependence as a general mechanism underlying biological control equilibria, there is reason to believe that 'predator pits' resulting from the Type 3 functional response may be responsible for some cases of existing biological control and for the prevention of pest outbreaks. A possible example is regulation of rabbit populations at densities of around 0.4/ha in lowland New Zealand habitats by cats and ferrets. Although there is no direct evidence for a 'predator pit', Pech *et al.* (1992) have demonstrated its likelihood in Australian rabbit populations and it is difficult to envisage any other mechanisms which could be responsible for maintaining such low prey densities in otherwise apparently favourable habitats. Moreover, given the observed predator-prey ratios and relative rates of increase, models (Barlow, in press) suggest that a predator pit is an entirely feasible explanation given the assumption of a Type 3 functional response. This system has the interesting dimension that ferrets are now thought to transmit bovine tuberculosis (Tb), so a question arises as to what would happen if their numbers were reduced. Theory offers a crude answer. If a continuous predator-prey system is regulated entirely by a sigmoid (Type 3) functional response, and if prey population growth (at least at low densities) is described by the single parameter r (the intrinsic rate of increase), then for the prey to escape control by the predator requires that the predator population is reduced by a proportion:

$$1-\frac{2}{c}\sqrt{rN(c-rN)}$$

where N is the initial prey density (i.e. at equilibrium in the predator pit), r the intrinsic rate of prey increase and c the maximum prey consumption rate of all predators initially present. In the example, this means that a 60% reduction in ferrets would allow a rabbit outbreak, which leads to the next question as to whether smaller reductions would have a significant suppressive effect on Tb transmission.

Risks to non-target species

Risk from introduced organisms is becoming increasingly important in a world progressively more aware of past mistakes and their consequences, and to date most proposed introductions have had to satisfy some kind of environmental impact assessment procedure. Typically this involves testing introduced natural enemies against a range of non-target species in the laboratory, often under conditions so unrealistic that the enemy might reasonably be expected to attack its own species. Under more realistic field conditions, Nicholsonian theory has some suggestions to offer. Firstly, much depends on whether the non-target insects share the same habitat as the target pest. If they do, then by and large the news is bad: the pest will serve as a reservoir for the natural enemy independent of the density of the non-target species, so total elimination of the latter will depend entirely on the switching behaviour of the enemy. Consider now the situation where the non-target and target species occupy different habitats. The N-B model predicts that, as for diseases, a certain threshold host density is necessary to sustain the parasitoid or predator (Fig. 2). Given the basic model (Eq. 3b), this threshold is the host density, N, for which the ratio of increase (net reproductive rate) of the parasitoid equals 1 at a parasitoid density tending to zero (i.e. when the parasitoid first invades the habitat). Eq. 3b gives:

$$\frac{P_{t+1}}{P_t}=1=\frac{N_h\left(1-e^{-aP_t}\right)}{P_t}\rightarrow aN_hP_t \ as \ P_t\rightarrow 0$$

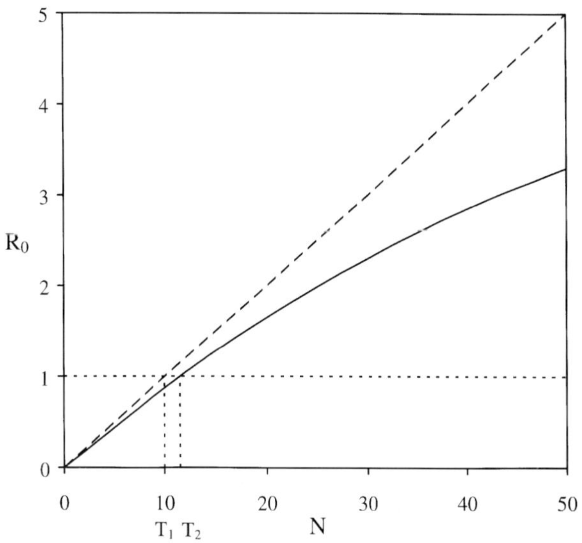

Fig. 2. Relationships between parasitoid net reproductive rate ($R_0 = P_{t+1}/P_t$ as $P_t \rightarrow 0$) and host density (N), showing the host thresholds T_1 and T_2 corresponding to $R_0 = 1$. Dashed line represents parasitism function without handling time, solid line represents parasitism function including handling time.

$$\therefore N_h = \frac{1}{a}$$

or, if handling time is included in the model:

$$N_h = \frac{1}{a(1-H)}$$

where N_h is the threshold host density for parasitoid persistence, H the ratio of handling time to total time or 1/fecundity, a the parasitoid searching efficiency and P the parasitoid density in generations t and $t+1$ respectively. Thus, the non-target species must exceed $1/a$ to be at risk and many non-target species of particular concern will typically exist at low densities. These arguments ignore features of host recognition such as kairomones, and there are clearly practical difficulties in estimating thresholds for parasitoid persistence in a host species it does not currently attack, but theory does indicate the sort of information required and the potential may exist to obtain it, in however crude a form.

Tri-trophic interactions

Tri-trophic interactions are an area of considerable current interest in relation to biological control. In particular, the question arises (J Waage and M Thomas, pers. comm.): can a classical biological agent hitherto regarded as potentially unsuccessful or only partially successful, be combined with a crop cultivar hitherto rejected as being only partially resistant, to give overall satisfactory control of a pest? More generally, what are the likely consequences of combining host plant resistance with an existing natural enemy? We assume for simplicity that plant resistance corresponds to the same proportional reduction in both r and K for the pest and insert this into an N-B model with host density-dependence such as that in Fig. 1 (Model 2 in May *et al.* 1981 with $k \rightarrow \infty$).

The effects are shown in Fig. 3 for different starting points (host-parasitoid interactions in the absence of plant resistance) given by the points A. These represent different combinations of aK and r, giving different initial levels of host suppression (q on the left-hand axis). The dashed lines represent the effects of adding plant resistance, in terms of a 50% reduction in both r and K, with the points B representing the net outcomes of plant resistance combined with parasitism, expressed

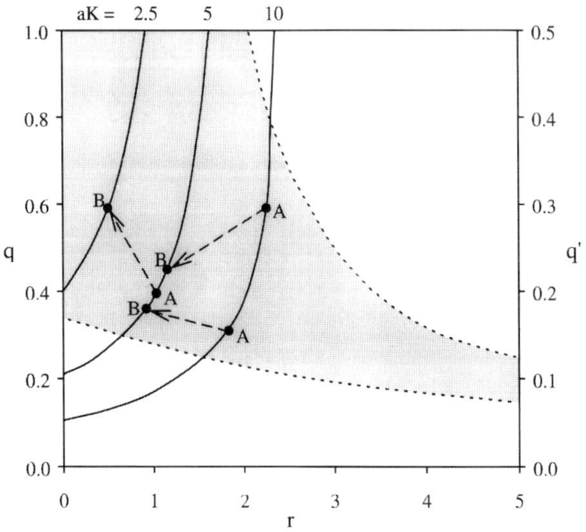

Fig. 3. The effect of combining parasitism, as in Fig. 1a, with plant resistance assumed to reduce both *r* and *K* for the host by 50%. Points A represent situations without plant resistance, and points B connected to them by the dashed lines are the results when resistance is added. The *q* values on the left-hand axis for points B are now relative to the 50% lower *K*; *q* values relative to the original *K* are given by *q'* on the right-hand axis.

in terms of suppression relative to the new, 50% lower K value (q on the left-hand axis) and suppression relative to the original K value (q¢ on the right-hand axis). As the different dashed lines suggest, the combination of the two mechanisms may be synergistic (if point B is lower than point A so that the proportional reduction due to parasitism is increased in the presence of plant resistance), additive in proportional or logarithmic terms (if the points are level and the q values the same), or antagonistic (if point B is higher than point A). In all cases, overall suppression is enhanced, as measured by the *q'* values on the right-hand axis for points B compared with the q values on the left axis for points A, and stability is unaffected (comparing points B with points A relative to the dotted stability boundaries on Fig. 3).

The 'antagonistic' effect (q increases and Point B is higher than Point A) would appear to be as likely as either of the other outcomes, and under these circumstances mortality from parasitism and the proportion of hosts parasitised decreases in the presence of plant resistance. One explanation is the positive feedback inherent in the N-B model: any external factor which reduces host density thereby reduces the density of emerging adult parasitoids, for a given initial proportion of hosts parasitised. The reduced number of searching adult parasitoids translates to a reduced proportion of hosts parasitised in the next generation (Eq. 3b), which leads to reduced output of adult parasitoids, and so on.

Case study 1: Sitona discoideus—Microctonus aethiopoides

The weevil *Sitona discoideus* in New Zealand lucerne crops is univoltine, but the susceptible adults are present virtually year-round. Furthermore, their emergence in summer is prolonged over a period of at least 2 months, and is immediately followed by emigration to nearby aestivating sites, from which the weevils return in autumn. These return migratory flights involve considerable mixing of local populations. The weevil is attacked, and controlled, by the introduced Braconid parasitoid *Microctonus aethiopoides* which has up to 6 generations a year, three of which occur during the summer period of emergence and emigration of the weevils. Other parasitoid generations attack the weevils during times of reproduction, so there is no clear order of events in the life cycles. Moreover, the summer parasitoid generations, which are responsible for the success of biological control, only occur because of the parasitoid's atypical behaviour in the New Zealand environment. This takes the form of a small proportion, about 3%, of each generation completing their development in weevils within the crops. Typically, and elsewhere in the world, the parasitoid's development is arrested at the first instar within emigrating and aestivating weevils (Goldson *et al.* 1990). Clearly such a system does not readily fit a theoretical model framework, and a more pragmatic approach had to be adopted (Barlow and Goldson 1993). Once this was done, the models accounted for the observed decline in *Sitona* over time, part of which occurred before the parasitoid was introduced, they predicted that control would be maintained in the future, and they demonstrated the reasons why biological control succeeded. These were: the atypical development of parasitoids oviposited in newly emerged weevils (< 24 hr old) over summer and the extra parasitoid generations this permits; the high host densities during summer; and the consequent high levels of early autumn parasitism prior to autumn egg-laying by the weevils.

Case study 2: Common wasp – Sphecophaga vesparum vesparum

Another system more complex than it initially appears is the common wasp in New Zealand parasitised by the Ichneumonid *Sphecophaga vesparum vesparum*. The common wasp (*Vespula vulgaris*) is univoltine in terms of sexual generations, but workers are produced continuously throughout the year at an increasing rate. The parasitoid has an unknown number of largely overlapping generations within a year, and its overwintering pupae hatch after a variable period ranging from 2 to 4 years. Again, no existing parasitoid-host model will fit such a system, so methods specific to this system had to be found of simplifying it in a realistic way. Thus, wasp densities were considered in terms of nests rather than individuals, and a simple, one-parameter expression derived to represent the unknown build-up of nest parasitism during the year, on the basis of assumed continuous searching (Barlow *et al.* 1996). Many other details of the interaction were unknown, so these were aggregated where possible into single parameters such as the parasitoid's net reproductive rate (R_0). The model showed that the final level of parasitism and the extent of host suppression depended largely on R_0, which the 6 years of historical data suggested was in the order of 1.3–1.6. Host suppression depended also on nest mortality due to the parasitoid, and several scenarios were considered for this. Under the most optimistic assumptions, future suppression of nest density was unlikely to exceed 10%, largely because of the strong host density-dependence and the delayed emergence of parasitoid overwintering cocoons. Thus, if the average time to cocoon emergence was changed from 2 years to 1 year in the model, the parasitoid changed from an unsuccessful control agent to a highly effective one, suppressing wasp nest densities by around 75%. Changing other components of R_0, such as parasitoid searching efficiency for wasp nests in spring and overwintering survival of parasitoid cocoons, improved its performance but to a lesser extent. In this case the model was effectively a rational way of extrapolating from the parasitoid's past performance over a period of 6 years to its likely future success, and a means of assessing the critical success factors for wasp biological control. Predicting future trends was useful since it allowed an early decision to abandon further parasitoid releases and switch to alternative potential control agents, the ideal one, according to the model, being a microbial agent which sterilises wasp queens without impairing their competitive abilities.

Attempts are also being made to apply theory and models to other classical biological control programmes which are of significant economic importance in New Zealand. These include the use of microbial agents and immunocontraceptive vectors against rabbits and brushtail possums (e.g. Barlow 1994), pathogens against grass grub *Costelytra zealandica* (Barlow and Jackson 1992) and Asian gypsy moth *Lymantria dispar*, and the Braconid parasitoid *Microctonus hyperodae* against Argentine stem weevil *Listronotus bonariensis* (Barlow *et al.* 1994). Only in two of these cases, immunocontraceptive vectors against possums and pathogens against grass grub, could the system be simplified to the extent that standard theoretical models were directly applicable. In the first case these were the continuous-time disease-host models of Anderson and May (1979), and in the second case, because grass grub has a single generation each year, the insect-pathogen system was most easily analysed by an N-B parasitoid-host-type model. In the remaining studies, case-specific 'models of intermediate complexity' had to be devised to handle the awkward but essential detail and resolve the eternal modelling problem of attaining the appropriate mix of simplicity and realism.

Theory and natural enemy enhancement

In discussing the practical application of theory and Nicholsonian ideas, our emphasis so far has been on classical biological control. The field of natural enemy enhancement is in many ways a special case. It is of growing practical significance (e.g. Wratten and van Emden 1995), it has received little, if any, theoretical attention in the past, and it raises some important and to an extent unique questions regarding predator and parasitoid behaviour.

The two distinctive attributes of these systems are the significance of resources other than prey or hosts in determining both the functional and numerical responses of natural enemies, and the importance of their movement and dispersal ability relative to environmental grain. Wratten and van Emden (1995) quote examples. In terms of resources, permanent hedgerows provide shelter for overwintering polyphagous predators between ephemeral crops. Thus the enlargement of cereal fields in Britain and consequent reduction in the boundary/crop area ratio, together with a simultaneous reduction in hedgerow vegetation quality due to increased herbicide use, has reduced spring colonisation of field centres by ground-dwelling predators such as carabid beetles. As a result, aphid predation rates in the centre of the fields are likewise suppressed. At the same time, overall numbers of polyphagous predators may be reduced by the general loss of hedgerow habitat. Another important resource is pollen and nectar from flowers. Female hoverflies (Syrphidae) require pollen to mature the reproductive system, and abundance of hoverflies is enhanced in the vicinity of *Phacelia tanacetifolia* strips planted around cereal fields. More importantly, suppression of aphid numbers (*Sitobion avenae*) has been demonstrated in fields which possess such borders (Wratten and van Emden 1995). This effect is not confined to predators. Wratten and van Emden (1995) list examples in which either pollen or nectar, from weeds, honey plants such as *Phacelia* or honeydew, is believed to have enhanced parasitism in crops, and Foster and Ruesink (1984) showed that access to flowering weeds significantly increased longevity and fecundity of the adult Braconid *Meteorus rubens*.

The adverse impact of increased field size on Carabid abundance in crops, is partly dictated by the beetles' limited capacity for movement and consequent reduced spring immigration rates per unit area of the crop. The development of grassed, raised earth 'beetle banks' within crops (Wratten and van Emden 1995) likewise depends more upon knowledge of the dispersive abilities of predators than on a knowledge of other facets of behaviour more commonly targeted by theory.

The practical strategies for natural enemy enhancement reviewed by Wratten and van Emden (1995) are not based on Nicholsonian theory, but on knowledge of resource needs, habitat requirements and predator dispersal. To accommodate these features, extend their coverage of the spectrum of predator and parasitoid behaviour, and be of value in developing practical strategies for natural enemy enhancement, models will have to consider the effects of 'environmental' resources such as pollen, nectar or shelter on functional and numerical responses, and movement. In particular, special kinds of metapopulation models are required which embody landscape features, including predator-only and predator-prey patches of explicit relative sizes, predator dispersal rates in relation to these patch sizes, and seasonal phenology of pests, enemies and patch characteristics. Corbett and Plant (1993) and Fry (1995) discuss some preliminary approaches of this kind. In essence, the theoretical problem posed by agroecosystem diversification for enhanced biological control simply involves the next logical step of more complex metapopulation models, which can portray, in however idealised fashion, the total population behaviour of natural enemies within and between fields. Such models may also be able to address the question posed by Wratten and van Emden (1995), of whether natural enemy enhancement represents true or global enhancement of their numbers, or simply local redistribution from existing habitats and crops to those in which the natural enemies' resources have been augmented. Clearly the question is most relevant to highly mobile insect such as Syrphids.

Conclusions: Nicholson's Legacy and the Future

Summarising the impact of Nicholson's work, perhaps his most important legacy is his concept of the balance of nature and the associated features of stability and density-dependence, distinguished by a clarity of thought on these issues unmatched by exponents of alternative views at the time. The concept remains equally valid today, modified only by questions of spatial scale and mechanisms. Scarcely less significant were his specific models and descriptions for competition and parasitism, distinguished again by their explicit and precise nature at a time when discursive generalisations were still an ecological norm. Though these models and their derivatives have been little used in practical biological control to date, they remain potentially useful components of more complex and realistic ones, even in the context of the most recent developments in metapopulation theory and self-organised spatial patterns. Two other legacies which persisted were a spatial model, which anticipated both the later negative binomial model and the concept of heterogeneity of risk, and the mathematical analysis of model stability. Above all, in combining experimentation and theory, the latter through highly productive associations with biologically-aware mathematical co-workers, Nicholson set a worthy and still highly relevant example.

Among the many developments in theory post-Nicholson, some highlights can be identified. These include work on: the functional response; host density-dependence; stability and persistence through dispersal in stochastic 'island' patch models; a simple implicit spatial model based on the negative binomial distribution of parasitoid attacks among hosts and the application of this model to a real biological control system; stability resulting from heterogeneity of risk of parasitoid attack, irrespective of the spatial density-dependence of parasitism with respect to hosts, and in particular the criterion for stability that $CV^2 > 1$; stability and self-organisation resulting from deterministic 'stepping-stone' metapopulation models with unstable within-patch dynamics; more complex predator and parasitoid behaviour; and new theory based on continuous-time models and ratio-dependence.

For the future, it seems likely that further advances in theory relevant to classical biological control, particularly to the selection and early evaluation of potential agents, will come less from theory based directly on 'global' models than from theory based more on a steady accumulation of empirical generalities arising from thorough analysis of a greater number of case studies. The role of modelling will be to assist in these analyses and in the development of understanding, initially of specific cases and thence of general phenomena. In the process, it will be necessary to look for paradigms and model frameworks beyond the classical Nicholsonian representations of low, stable subpopulation equilibria. For example, Murdoch *et al.* (1985) suggested that simple modifications to existing models can provide a greater variety of potentially satisfactory biological control solutions. Among these are models for polyphagous predators. Contrary to perceived wisdom, and leaving aside considerations of safety in terms of non-target species, polyphagous predators are unique in allowing stable pest equilibria at zero pest density. Another obvious alternative model is the metapopulation one of Hassell *et al.* (1991), in which spatial interactions have a crucial impact both on the sub-populations and on the metapopulation. More challenging still will be the explicit metapopulation models required to analyse the process of natural enemy enhancement in crop ecosystems, and the consequences for natural enemy populations of increased agroecosystem diversity. In general, more detailed and flexible 'models of intermediate complexity' are likely to be required to achieve an appropriate balance between simplicity and realism in most biological control case studies.

In conclusion, the most important prerequisites for an enhanced future contribution of post-Nicholsonian theory to biological control practice are the targeting of general theory to appropriate practical questions, and the use of appropriate models to analyse and understand more real case-studies.

References

Anderson RM, May RM (1979) Population biology of infectious diseases I. *Nature* **280**:361–367

Andrewartha HG, Birch LC (1954) *The distribution and abundance of animals.* University of Chicago Press, Chicago

Arditi R, Ginzburg LR (1989) Coupling in predator-prey dynamics: ratio dependence. *J Theor Biol* **139**:311–326

Arditi R, Saiah H (1992) Empirical evidence of the role of heterogeneity in ratio-dependent consumption. *Ecology* **73**:1544–1551

Bailey VA, Nicholson AJ, Williams EJ (1962) Interaction between hosts and parasites when some host individuals are more difficult to find than others. *J Theor Biol* **3**:1–18

Barlow ND (1985) The interferential model re-examined. *Oecologia* **66**:307–308

Barlow ND (1993) The role of models in an analytical approach to biological control. In: Prestidge RA (ed), *Proc 6th Australas Conf Grassl Invertebr Ecol.* AgResearch, Hamilton, NZ, pp 318–325

Barlow ND (1994) Predicting the impact of a novel vertebrate biocontrol agent: a model for viral-vectored immunocontraception of New Zealand possums. *J Appl Ecol* **31**:454–462

Barlow ND, Goldson SL (1993) A modelling analysis of the successful biological control of *Sitona discoideus* (Coleoptera: Curculionidae) by *Microctonus aethiopoides* (Hymenoptera: Braconidae). *J Appl Ecol* **30**:165–178

Barlow ND, Jackson TA (1992) A model for the impact of diseases on grass grub (*Costelytra zealandica*) populations in New Zealand. In: Jackson TA, Glare TR (eds) *The use of pathogens in scarab pest management.* Intercept, Andover, UK, pp 127–140

Barlow ND, Goldson SL, McNeill MR (1994) A prospective model for the phenology of *Microctonus hyperodae* (Hymenoptera: Braconidae), a potential biological control agent of Argentine stem weevil in New Zealand. *Biocontrol Sci Technol* **4**:375–386

Barlow ND, Moller H, Beggs JR (1996) A model for the impact of *Sphecophaga vesparum vesparum* as a biological control agent of the common wasp in New Zealand. *J Appl Ecol* **33**:31–44

Beddington JR, Free CA, Lawton JH (1978) Characteristics of successful natural enemies in models of biological control of insect pests. *Nature* **273**:513–519

Berryman AA (1992) The origins and evolution of predator-prey theory. *Ecology* **73**:1530–1535

Brown JJ, Kainoh Y (1992) Host castration by *Ascogaster* spp (Hymenoptera: Braconidae). *Ann Entomol Soc Am* **85**:67–71

Casas J, Gurney WSC, Nisbet R, Roux O (1993) A probabilistic model for the functional response of a parasitoid at the behavioural time-scale. *J Anim Ecol* **62**:194–204

Caughley G, Lawton JH (1981) Plant-herbivore systems. In: May RM (ed) *Theoretical ecology,* second edition. Blackwell Scientific Publications, London, pp 132–166

Chesson P (1981) Models for spatially distributed populations: the effect of within-patch variability. *Theor Popul Biol* **19**:288–325

Chesson PL, Murdoch WW (1986) Aggregation of risk: relationships among host-parasitoid models. *Am Nat* **127**:696–715

Corbett A, Plant RE (1993) Role of movement in the response of natural enemies to agroecosystem diversification: a theoretical evaluation. *Environ Entomol* **22**:519–531

Dixon AFG (1959) An experimental study of the searching behaviour of the predatory coccinellid beetle *Adalia decempunctata* (L.). *J Anim Ecol* **28**:259–281

Foster MA, Ruesink WG (1984) Influence of flowering weeds associated with reduced tillage in corn on a black cutworm (Lepidoptera: Noctuidae) parasitoid, *Meteorus rubens* (Nees von Esenbeck). *Environ Entomol* **13**:664–668

Frazer BD, Gilbert N (1976) Coccinellids and aphids: a quantitative study of the impact of adult ladybirds (Coleoptera: Coccinellidae) preying on field populations of pea aphids (Homoptera: Aphididae). *J Entomol Soc B C* **73**:33–56

Free CA, Beddington JR, Lawton JH (1977) On the inadequacy of simple models of mutual interference for parasitism and predation. *J Anim Ecol* **46**:543–554

Fry G (1995) Landscape ecology of insect movement in arable ecosystems. In: Glen DM, Greaves MP, Anderson HM (eds) *Ecology and integrated farming systems*. John Wiley, Chichester, UK, pp 177–202

Godfray HCJ, Hassell MP (1989) Discrete and continuous insect populations in tropical environments. *J Anim Ecol* 58:153–174

Godfray HCJ, Pacala SW (1992) Aggregation and the population dynamics of parasitoids and predators. *Am Nat* **140**:30–40

Godfray HCJ, Waage JK (1991) Predictive modelling in biological control: the mango mealy bug (*Rastrococcus invadens*) and its parasitoids. *J Appl Ecol* **28**:434–453

Godfray HCJ, Hassell MP, Holt RD (1994) The population dynamic consequences of phenological asynchrony between parasitoids and their hosts. *J Anim Ecol* **63**:1–10

Goldson SL, McNeill MR, Proffitt JR (1993) Unexplained mortality amongst unparasitised *Listronotus bonariensis* in the presence of the parasitoid *Microctonus hyperodae* under caging conditions. In: Prestidge RA (ed), *Proc 6th Australas Conf Grassl Invertebr Ecol*. AgResearch, Hamilton, NZ, pp 355–362

Goldson SL, Phillips CB, Barlow ND (1994) The value of parasitoids in biological control. *NZ J Zool* **21**:91–96

Goldson SL, Proffitt JR, McNeill MR (1990) Seasonal biology and ecology in New Zealand of *Microctonus aethiopoides* (Hymenoptera: Braconidae), a parasitoid of *Sitona* spp. (Coleoptera: Curculionidae), with special emphasis on atypical behaviour. *J Appl Ecol* **23**:703–722

Greathead DJ (1986) Parasitoids in classical biological control. In: Waage JK, Greathead DJ (eds) *Insect parasitoids*. Academic Press, London, pp 290–318

Gurney WSC, Nisbet RM, Lawton JH (1983) The systematic formulation of tractable single-species population models. *J Anim Ecol* **52**:479–495

Gutierrez AP, Baumgaertner JU (1984) Age-specific energetics models: pea aphid *Acyrthosiphon pisum* (Homoptera: Aphididae) as an example. *Can Entomol* **116**:924–932

Hassell MP (1975) Density-dependence in single-species populations. *J Anim Ecol* **44**:283–295

Hassell MP (1978) *The dynamics of arthropod predator-prey systems*. Princeton University Press, Princeton, NJ

Hassell MP (1980) Foraging strategies, population models and biological control: a case study. *J Anim Ecol* **49**:603–628

Hassell MP, May RM (1973) Stability in insect host-parasitoid models. *J Anim Ecol* **42**:693–726

Hassell MP, Pacala SW (1990) Heterogeneity and the dynamics of host-parasitoid interactions. *Philos Trans R Soc Lond B* **330**:203–220

Hassell MP, Varley GC (1969) New inductive model for insect parasites and its bearing on biological control. *Nature* **223**:1133–1137

Hassell MP, Comins HN, May RM (1991) Spatial structure and chaos in insect population dynamics. *Nature* **353**:255–258

Hassell MP, Comins HN, May RM (1994) Species coexistence and self-organising spatial dynamics. *Nature* **370**:290–292

Hassell MP, Waage JK, May RM (1983) Variable parasitoid sex ratios and their effect on host-parasitoid dynamics. *J Anim Ecol* **52**:889–904

Hastings A (1977) Spatial heterogeneity and the stability of predator-prey systems. *Theor Popul Biol* **12**:37–48

Hill MG (1988) Analysis of the biological control of *Mythimna separata* (Lepidoptera: Noctuidae) by *Apanteles ruficrus* (Hymenoptera: Braconidae) in New Zealand. *J Appl Ecol* **25**:197–208

Hestbeck JB (1987) Multiple regulation states in populations of small mammals: a state-transmission model. *Am Nat* **129**:520–532

Hochberg ME, Waage JK (1991) A model for the biological control of *Oryctes rhinoceros* (L.) (Coleoptera: Scarabaeidae) by means of pathogens. *J Appl Ecol* **28**:514–531
Holling CS (1959) Some characteristics of simple types of predation and parasitism. *Can Entomol* **91**:385–398
Holling CS (1965) The functional response of predators to prey density and its role in mimicry and population regulation. *Mem Entomol Soc Can* **45**:3–60
Howard LO, Fiske WF (1911) The importation into the United States of the parasites of the gipsy-moth and the brown-tail moth. *Bull Bur Entomol US Dep Agric* **91**:1–312
Hubbard SF, Marris GC, Reynolds AJ, Rowe GW (1987) Adaptive patterns in the avoidance of superparasitism by solitary parasitic wasps. *J Anim Ecol* **56**:387–401
Ivlev VS (1961) *Experimental ecology of the feeding of fishes*. Yale University Press, New Haven
Kareiva P (1990) Establishing a foothold for theory in biological control practice: using models to guide experimental design and release protocols. In: Baker RR, Dunn PE (eds) *New directions in biological control* Alan R. Liss Inc, New York, pp 65–81
Kidd NAC, Jervis MA (1989) The effects of host-feeding behaviour on the dynamics of parasitoid-host interactions, and the implications for biological control. *Res Popul Ecol* **31**:235–274
Ludwig D, Jones DD, Holling CS (1978) Qualitative analysis of insect outbreak systems: the spruce budworm and forest. *J Anim Ecol* **47**:315–332
May RM (1974) *Stability and complexity in model ecosystems*. Princeton University Press, Princeton, NJ
May RM (1978) Host-parasitoid systems in patchy environments: a phenomenological model. *J Anim Ecol* **47**:833–843
May RM, Hassell MP (1988) Population dynamics and biological control. *Philos Trans R Soc Lond B* **318**:129–169
May RM, Hassell MP, Anderson RM, Tonkyn DW (1981) Density-dependence in host-parasitoid models. *J Anim Ecol* **50**:855–865
Maynard-Smith J (1974) *Models in ecology*. Cambridge University Press, UK
Munster-Swendsen M (1994) Pseudoparasitism: detection and ecological significance in *Epinotia tedella*. *Norwegian J Agric Sci* (Suppl 16) pp 329–335.
Murdoch WW (1992) Ecological theory and biological control. In: Jain SK, Botsford LW (eds) *Applied population biology*. Kluwer Academic Publishers, The Netherlands, pp 197–221
Murdoch WW, Briggs CJ, Nisbet RM, Gurney WSC, Stewart-Oaten A (1992) Aggregation and stability in metapopulation models. *Am Nat* **140**:41–58
Murdoch WW, Chesson J, Chesson PL (1985) Biological control in theory and practice. *Am Nat* **125**:344–366
Murdoch WW, Stewart-Oaten A (1989) Aggregation by parasitoids and predators: effects on equilibrium and stability. *Am Nat* **134**:288–310
Nicholson AJ (1933) The balance of animal populations. *J Anim Ecol* **2**:131–178
Nicholson AJ (1954) An outline of the dynamics of animal populations. *Aust J Zool* **2**:9–65
Nicholson AJ, Bailey VA (1935) The balance of animal populations. Part I. *Proc Zool Soc Lond* **3**:551–598
Pech RP, Sinclair ARE, Newsome AE, Catling PC (1992) Limits to predator regulation of rabbits in Australia: evidence from predator-removal experiments. *Oecologia* **89**:102–112
Perry JN (1987) Host-parasitoid models of intermediate complexity. *Am Nat* **130**:955–957
Rogers DJ (1972) Random search and insect population models. *J Anim Ecol* **41**:369–383
Smith HS (1935) The role of biotic factors in the determination of population densities. *J Econ Entomol* **28**:873–898
Taylor AD (1988) Parasitoid competition and the dynamics of host-parasitoid models. *Am Nat* **132**:417–436
Varley GC (1947) The natural control of population balance in the knapweed gall-fly (*Urophera jaceana*). *J Anim Ecol* **16**:139–187
Waage JK (1990) Ecological theory and the selection of biological control agents. In: Mackauer M, Ehler LE (eds) *Critical issues in biological control*. Intercept, Newcastle, UK, pp 102–136

Watt KEF (1959) A mathematical model for the effect of densities of attacked and attacking species on the number attacked. *Can Entomol* **91**:129–144
Wratten SD, van Emden HF (1995) Habitat management for natural enemies. In: Glen DM, Greaves MP, Anderson HM (eds) *Ecology and integrated farming systems.* John Wiley, Chichester, UK, pp 117–146
Yamada Y (1987) Factors determining the rate of parasitism by a parasitoid with a low fecundity, *Chrysis shanghaiensis* (Hymenoptera: Chrysididae). *J Anim Ecol* **56**:1029–1042

Parasitoid Foraging: The Importance of Variation in Individual Behaviour for Population Dynamics

Louise E.M. Vet

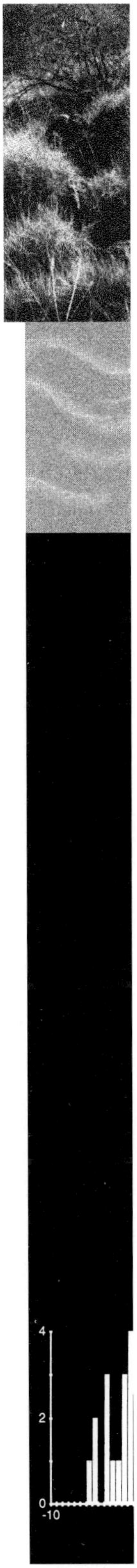

ABSTRACT

The concept of 'searching efficiency' as introduced by Nicholson (1933) greatly influenced both the field of parasitoid searching behaviour and that of population dynamics. The current presentation will highlight recent considerations from empirical behavioural work on parasitoid foraging that have been neglected by theorists working at the population level, but that are fundamental to searching efficiency of parasitoids and so to parasitoid-host interactions. The first consideration is that parasitoids have evolved and function within a multitrophic context. It is becoming increasingly clear that the food plant of the host not only influences the development and behaviour of the herbivore but also that of the parasitoid, the third trophic level. The second consideration is that the behaviour of a parasitoid is a very dynamic characteristic, that is influenced both by the physiological state of the animal and by its experience. This paper focuses on learning as an important source of variability between and within individuals. First, it is shown that learning by parasitoids is much directed towards the food plant of the host, which has ecological and evolutionary consequences for all three trophic levels. Second, evidence is presented that learning essentially changes random search into highly directed search, thereby reducing interpatch travel times. The assumption of 'omniscience' (e.g., used in optimal foraging models) is briefly discussed as well as the role of learning in the parasitoid's assessment of the profitability at the habitat and patch level. A final consideration deals with spatial aspects of parasitoid search. Although predator-prey models show that dispersal and movement between patches has great population dynamic consequences, the exact underlying behavioural mechanisms are rarely known. An example of our present work suggests that parasitoids adapt their searching strategy to specific host distribution patterns, which demonstrates that host population dynamics can affect the optimality of foraging decisions by the individual parasitoid. The paper finishes with a plea for incorporating tritrophic aspects, behavioural variation and the outcome of evolutionary models into population level studies.

Key words: foraging behaviour, tritrophic relationships, learning, spatial, optimisation, evolution

Frontiers of Population Ecology, R.B. Floyd, A.W. Sheppard and P.J. De Barro (eds), CSIRO Publishing, Melbourne, 1996 pp. 245–256

Introduction

In spite of the fact that Nicholson used both a behavioural and a population dynamic approach to parasitoid searching efficiency, both fields evolved quite separately. Behavioural empiricists focussed on the mechanism and function of behavioural traits leading to parasitism, while population dynamicists were concerned with variation in the outcome of this parasitism and its effect on spatial distribution of host mortality and population stability. Nicholson (1933 p. 141) recognised the importance of the searching process as a major factor in population dynamics: "...animals appear to be limited in density, either directly or indirectly, by the difficulty they experience in finding the things they require for existence, or by the ease with which they are found by natural enemies." However, he restricted his interest in behaviour to the population level: "It is important to realise we are not concerned with the searching of individuals, but with that of whole populations" (p. 141). Hence, unlike behavioural empiricists do, Nicholson did not value the fact that individuals differ, and that their searching behaviour can not easily be characterised by a single species-specific constant. Here lies the essence of the separate approaches of population dynamicists and behavioural ecologists. How important is variation in behaviour *between and within individuals* for population level processes? Behavioural studies have taught us that parasitoids search non-randomly in highly heterogeneous environments. They use information from different trophic levels, they learn and alter their decisions accordingly. Their physiological state (hunger level, egg-load) changes during their life time which may affect major foraging decisions. They interact with con- and heterospecifics. Behavioural variation potentially influences population-level processes but the links are rarely explicitly made. The famous Nicholson and Bailey (1935) model, although infamously unstable (Hassell 1978), has inspired population dynamicists to think about and incorporate some behavioural features such as mutual interference between parasitoids and aggregation. However, although these were added to the classic model, the parameters were still based on assumed population averages, freely using a range of possible parameter values. But we know that natural selection acts on variation in individuals. This is the basic assumption of the behavioural ecology approach to animal foraging, that is interested in adaptive individual decision making, using optimality principles. Hence, behavioural parameters used in population dynamic models are subject to change through the process of natural selection, and evolutionary models will predict that some parameter values will be more likely than others. The need for a greater integration of evolutionary ecology and population dynamics is increasingly being expressed (e.g., Koehl 1989; van Baalen and Sabelis 1993; urdoch 1994). Furthermore, there is increasing awareness that models based on real behaviour may be more appropriate for understanding population processes (e.g., Kareiva and Odell 1987; van der Werf *et al.* 1989; Driessen and Hemerik 1991; Bernstein *et al.* 1988, 1991; Kacelnik *et al.* 1992; Sutherland and Dolman 1994). Increasingly powerful computational possibilities will undoubtedly stimulate the incorporation of real behaviour into population models through simulations. Hence there will be an increasing need for realistic biological data on individuals. A stimulating prospect for behavioural empiricists!

Since the last decade there are several exciting developments in the study of parasitoid foraging behaviour. This paper highlights some of these behavioural developments and encourages the incorporation of these individual level considerations into population studies. Although until now practically unnoticed or neglected by population theorists, they are fundamental to searching efficiency of parasitoids and so to parasitoid-host interactions (Fig. 1).

Tritrophic considerations: plant-natural enemy interactions

Traditionally predator-prey or parasitoid-host interactions are studied as a bitrophic system. Since Price and coworkers (Price *et al.* 1980) advocated that natural enemy-herbivore and herbivore-plant interactions can only be fully understood when considered within a tritrophic context, the study of

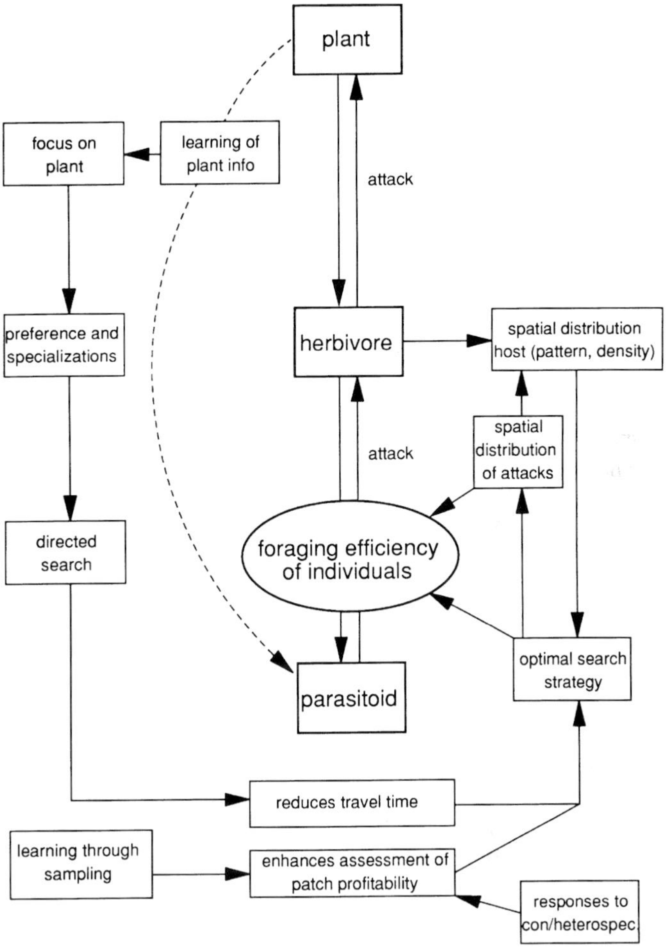

Fig. 1. Parasitoid-host relationships influenced by plant-parasitoid interactions and variation in parasitoid behaviour as discussed in the text. Arrows specify an influence or determination of one organism or process upon another and the direction in which this occurs. See text for additional explanation.

parasitoid foraging behaviour has increasingly expanded to include the plant level and presently plant-natural enemy interactions receive major research effort (see Dicke 1994 and Turlings *et al.* 1995 for reviews). Plants can affect the impact of natural enemies on herbivores both indirectly, (i.e., through effects on herbivore suitability) and directly, by providing natural enemies with food and shelter or by influencing their searching process (reviewed in Boethel and Eikenbary 1986; Dicke 1995). Plants play an important role in providing essential information to foraging natural enemies (see Vet and Dicke 1992 for review). Especially at longer distances, plant stimuli such as volatiles, the production of which is often induced by herbivore damage, guide the parasitoid in the process of host finding. The possibility to use directly host-derived information is limited because of selection against detectability of herbivores (Vet *et al.* 1991). The very fact that plants guide parasitoid host searching has important evolutionary and ecological consequences for all trophic levels. 1. For the plant: the fact that plants can actively 'recruit the enemies of their enemies' (indirect plant-defence; e.g., Dicke and Sabelis 1988a; Turlings *et al.* 1990) links plant selection and foraging efficiency of parasitoids to plant fitness. This will influence the evolution of the plant-defence mechanism. 2. For the herbivore: through plant selection, herbivores can influence their chance of being parasitised. After all, plant species, cultivars, genotypes or even plant parts can vary in their attractivity to parasitoids. A less attractive plant is searched less, which creates a partial refuge for the host, affecting the dynamics of the herbivore population and influencing the evolution of food-plant use by herbivores (Bernays and Graham 1988; Vet and Dicke 1992). As argued later, the fact that parasitoids can learn to focus on profitable plants will enhance these effects. 3. For the parasitoid: plants can affect host suitability and so directly interfere with parasitoid dynamics (e.g., Thorpe and Barbosa 1986). Plant architecture can affect natural enemy foraging efficiency (e.g., Kareiva and Sahakian 1990; Van Lenteren and de Ponti 1990).

The detectability and reliability of plant information affects the foraging success of the parasitoid and so its Darwinian fitness. Consequently we can expect parasitoids to be adapted to the most prevailing and suitable plant-host complexes (Vet *et al.* 1990; Lewis *et al.* 1990). Hence, parasitoid populations can vary in their responses to important foraging cues e.g., plant volatiles. Kester and Barbosa (1991) compared different populations of the parasitoid *Cotesia congregata* that were associated with *Manduca sexta* on tomato and tobacco. The populations not only differed in their survivorship responses to nicotine concentrations in their host's diet, but also in their ovipositional preferences for hosts on tobacco or tomato in the field and in their searching responses to these host plants.

The interaction between the plant and the parasitoid may be even more complex. First of all, plant habit, architecture and plant volatiles may be highly variable in space and time, for example in relation to growth conditions (Visser 1986; Takabayashi *et al.* 1994; Turlings *et al.* 1995). Plant volatiles may quantitatively and qualitatively differ between plant species, populations and individuals. This hampers the evolution of genetically-fixed responses in parasitoids and we expect, and indeed find, plant-related variability in the responses of parasitoids to plants. Responses may also vary due to the parasitoid's physiological state (through its egg load, or whether it needs to forage for food instead of hosts) and, very significantly so, due to the parasitoid's experience. Empirical work on tritrophic aspects that are essential to natural enemy-herbivore relationships is expanding rapidly. This information is increasingly used for population and evolutionary models (e.g., Kareiva and Sahakian 1990; Sabelis *et al.* 1991). Information on how parasitoids perceive host distributions, whether that is from a distance through the plant or after encountering a host aggregation, and respond to them behaviourally is essential for assessing the population dynamic effects of how parasitoids exploit their host resources (e.g., whether parasitoid attack is density dependent).

Learning

Role of the plant

If individual parasitoids have to deal with temporal or spatial changes in foraging cues, flexibility in behavioural response is a necessity, and learning provides an adaptive mechanism (Turlings *et al.* 1993; Vet *et al.* 1995). The mechanisms and function of learning and the different ways in which learning may affect parasitoid foraging behaviour are reviewed in Vet *et al.* (1995). Parasitoids emerge as inexperienced animals with a limited set of responses to major stimuli that they use to locate a host (Vet *et al.* 1990; Lewis *et al.* 1990). Because the environment is variable over and between generations, this initial response profile has limited value but it will determine the chance that a parasitoid is successful in finding a first host. Through associative learning parasitoids can acquire responses to novel stimuli that have little biological meaning initially but appear to be reliable indicators of host presence during the parasitoid's foraging life. Parasitoids have been shown to learn odours, colours, shapes and patterns of the surroundings (plants!) that are associated with the presence of a suitable host. The parasitoid *Microplitis croceipes,* for example, learns to fly to 'targets' with a certain colour, pattern or shape after these are reinforced with a host encounter (Wäckers and Lewis 1994). Furthermore, responses to stimuli can be modified through rewarding and unrewarding experiences, leading to *preference changes and temporary specialisations. Cotesia glomerata*, a parasitoid of *Pieris* caterpillars, learns to fly to the host plant species that is most profitable in terms of host encounter rate (Geervliet *et al.* unpublished). What these and other learning studies essentially show is that parasitoids can learn to focus on key information while foraging and that this key information is often host-plant derived. Learning reinforces the important role of the plant in parasitoid foraging.

Learning and travel time

Through selective increase in responsiveness and induction of preferences, learning can change initially rather random search (guided by few and often unreliable cues), to directed search (guided by an increased set of reliable cues) resulting in increased host-encounter rate. For example while any yeast substrate is attractive to naive *Asobara tabida*, a parasitoid of *Drosophila*, they can learn to distinguish between odour of yeast with and without *Drosophila* host larvae (Vet and van Opzeeland 1984). They can also learn to recognise the aggregation pheromone of the adult flies, the quantity of which is correlated to larval density (Hedlund, Vet and Dicke, unpublished). Another parasitoid of *Drosophila, Leptopilina heterotoma,* can learn to distinguish between different fermenting fruits, between different fruit cultivars and even between yeast with and without an increased level of ethyl-acetate (Vet *et al.* unpublished). Learning can also be used to avoid previously visited host sites where oviposition has taken place, a pronounced deviation from random search (Sheehan *et al.* 1993). Parasitoids stop responding to unprofitable information, their concept of a patch becomes more defined. Hence, in essence, learning reduces interpatch travel time and foraging/travelling costs, an important factor when foraging time is limited. Papaj and Vet (1990) showed a reduction in travel time through learning in field experiments with *Leptopilina heterotoma*. Female wasps, experienced with host-infested food substrates such as mushrooms or fermented apples, were more likely to locate a host-food substrate and found it more quickly than inexperienced wasps. In addition, learning greatly influenced the choice of substrate. Mushroom-experienced females arrived more often at mushroom baits and apple-experienced at apple baits. Learning enhances foraging consistency, leading to a greater foraging efficiency when only one substrate type is profitable.

If learning enhances foraging efficiency, i.e., increases host encounter rate, what consequences does this have for the optimality of the parasitoid's foraging decisions and for population dynamics? In optimal foraging models, the time and costs needed to travel between patches are important parameters for determining the optimal behaviour, and a change in travel time is expected to affect several major foraging decisions such as patch choice, patch residence times, diet choice (host selection) and even the optimal number of eggs laid per host (Stephens and Krebs 1986; Charnov and Stephens 1988; McNamara *et al.* 1993). Optimality models generally predict that a reduction in travel time: 1. decreases the optimal time spent foraging on a patch; 2. reduces the optimal set of patches that should be exploited; 3. increases selectivity in host acceptance and 4. reduces clutch size laid per host.

This reduction in travel time through learning is not only interesting from an optimality, behavioural ecological perspective. It may also have population dynamic consequences when changes in patch and host selection decisions affect spatial heterogeneity in attack rates, a significant factor in the persistence of parasitoid-host interactions. An increase in selectivity through learning can create greater refuges for less suitable host species or stages, which can have important population dynamical consequences. With a model describing predator distribution in relation to prey abundance, Bernstein *et al.* (1991) showed that when travel costs were low, prey depletion was slow. The predators conformed to a predicted so-called 'Ideal Free Distribution', a spatial distribution of a population of predators where each predator maximises its prey-encounter rate. In this case of low travel costs, prey mortality was shown to be density dependent. This in contrast with a situation of higher travel costs, where the population distributed far from the IFD and where host mortality was density independent or negatively density dependent. In summary, through its effect on travel time alone, learning can influence the spatial distribution of host mortality.

Assessing patch profitability

Learning helps parasitoids to assess patch profitabilities. In most optimal foraging models, animals are considered to have complete information about the current and other patches in the habitat and this assumption has been criticised as biologically unrealistic. However, the recently achieved greater insight in the subtle way parasitoids make use of infochemicals (sensu Dicke and Sabelis 1988b) shows that, to some degree, parasitoids may indeed become quite able to estimate 'where to go best'. The examples of *L. heterotoma* given above show how parasitoids can learn to distinguish between different host food substrates with different profitabilities. The parasitoid *C. glomerata* not only learns which cabbage cultivar is most profitable, but can also distinguish in flight between individual cabbage plants infested with different caterpillar densities by using chemical cues (Geervliet *et al.* unpublished). Furthermore, females avoid to fly to plants where their superior competitor *C. rubecula* is already parasitising (Geervliet *et al.* unpublished), which has also been found for *L. heterotoma,* avoiding its competitor *L. clavipes* (Janssen *et al.* 1995). These examples show the power and subtleness of long-distance infochemical use in assessing patch profitability and competition avoidance. It is realistic to assume that, just like any foraging animal, searching parasitoids face a problem of incomplete information (Stephens and Krebs 1986). Hence, also parasitoids need to sample, and experience plays a major role, both in patch selection (where to go) as well as in patch exploitation (how long to stay and which hosts to accept). This is indeed assumed in some more realistic patch models and empirically shown (e.g., Bernstein *et al.* 1988, 1991; Haccou *et al.* 1991; Hemerik *et al.* 1993). Apart from learning about host quantity, parasitoids also use experience to assess the quality of a patch. Experience with different quality hosts (species, age, unparasitised, already parasitised) is important in patch time allocation (e.g., Haccou *et al.* 1991; van Lenteren 1991).

Using a numerical simulation model, Bernstein *et al.* (1988) studied the influence of learning on the distribution of predators among patches. Predators detected precisely the capture rate in their current patch but learned about the average value of the habitat. It was shown that the rate of learning was a crucial factor in determining the distribution of the predators over the patches and the resulting sign of density-dependence (positive or negative) of host mortality. However, in spite of the importance of information processing for parasitoid foraging decisions such as aggregation and dispersal, there is still little empirical data on how parasitoids use and process information relating to the assessment of patch profitability and more behavioural work on learning rules employed is certainly needed.

Spatial aspects

In addition to adaptation of sensory ability for efficient host finding, a parasitoid's foraging strategy is under selection to deal efficiently with specific distributions in space and time of their host population and its food plants. The way natural enemies respond to host distributions and density levels of pest insects is generally considered crucial for their effectiveness as biological control agents (e.g., Kareiva and Odell 1987). Population dynamicists try to understand the role of spatial heterogeneity on population stability. Numerous studies were undertaken to measure resulting spatial patterns of parasitism (reviews: Lessels 1985; Stiling 1987; Walde and Murdoch 1988; Pacala and Hassell 1991). However, the underlying behavioural mechanisms are mostly unknown and can not be generalised easily. The abstract models on the effect of dispersal and patchiness on predator-prey interactions have inspired some researchers to set up experiments with direct behavioural observations and the behavioural data provided essential information for understanding the predator-prey dynamics (e.g., Kareiva 1987; Kareiva and Andersen 1988).

Apart from direct host-defence strategies such as the ability to encapsulate parasitoid eggs, hosts may evolve spatial avoidance or defence strategies. As mentioned above, herbivores may choose to feed from less suitable food plants when these are searched less by their parasitoids. In Japan, *Pieris napi* can avoid parasitisation by laying its eggs on *Arabis* plants that are normally concealed by other plants and so are searched less by the parasitoids. However, there is a cost to this plant selection as these *Arabis* plants are of inferior quality for caterpillar development (Ohsaki and Sato 1994). Another strategy may be to have a highly clumped distribution. If larger aggregations do not attract more natural enemies, aggregation may dilute per capita predation or parasitisation pressure (Stamp and Casey 1993). The creation of some kind of host refuge can lead to the spatial heterogeneity in attack rates, known to stabilise parasitoid-host interactions (Chesson and Murdoch 1986; Murdoch and Stewart-Oaten 1989; Pacala *et al.* 1990).

However, in spite of their importance for population dynamics, behavioural studies on spatial aspects of parasitoid foraging are still limited. In the field of behavioural ecology we see a similar situation. Optimal foraging theory assumes that natural selection acts to maximise life-time fitness, which, in time-limited species, is synonymous with a maximisation of host attack rate. Optimisation of foraging not only demands good host-finding ability (inter-patch processes), but also efficient decision rules for patch leaving (Iwasa *et al.* 1981; McNair 1982; Haccou *et al.* 1991). Although much empirical work has been done on finding ability in parasitoids (Vet and Dicke 1992) there is still little known of how parasitoids behave in time and space after host-infested patches have been found. Theory has rigorously dealt with optimisation of patch residence times (Stephens and Krebs 1986) but support of this theory by empirical data is sparse (Visser *et al.* 1992; Hemerik *et al.* 1993). Data on behavioural variation in response to spatial variation in host distribution is essentially lacking. However, there is some theoretical work available on how individual decisions affect distributions of predators (Bernstein *et al.* 1988).

The following comparative study shows how adaptation to distribution patterns of hosts (clumped versus uniform) may involve changes in movement patterns and optimal patch-leaving rules. It involves two *Cotesia* (Hymenoptera: Braconidae) parasitoid species attacking gregariously or solitarily feeding *Pieris* caterpillars. *Cotesia glomerata* is native to Europe where it attacks mainly young larvae of the large white butterfly (*Pieris brassicae*). This butterfly species lays its eggs in clutches of about 20–50 eggs and the young larvae feed closely together on a leaf. The small white, *Pieris rapae*, lays its eggs singly and so caterpillars feed solitarily. This pierid species is considered a less attacked and less suitable host for *C. glomerata*. It is mainly attacked by its specialist parasitoid *Cotesia rubecula*. When around 1880 *P. rapae* became a pest on cabbage in the USA it was surprisingly *C. glomerata* (and not *C. rubecula*) that was introduced from England into the USA (Le Masurier and Waage 1993). In the absence of *P. brassicae*, which is not present in the USA, *C. glomerata* has persisted on *P. rapae*, ever since. In a comparative study with the American and European *C. glomerata* strain Le Masurier and Waage (1993) showed that the American strain attacked *P. rapae* at a significantly higher rate than did the British strain, but the attack rate was not as high as that of *C. rubecula*.

It was not clear what caused these differences in attack rates. In earlier studies in our laboratory (Wiskerke and Vet 1994) we compared the searching behaviour of Dutch *C. glomerata* and *C. rubecula* when searching in a semi-field set-up of plants infested with either *P. brassicae, P. rapae* or both host species (in their natural distribution patterns). The foraging behaviour of individual females was observed. Among other things, foraging was summarised in ethograms and the time allocation to different behaviours was calculated. There were striking differences between the species

Table 1. Ecological and behavioural comparison of *C. glomerata* (strains) and *C. rubecula*

	Cotesia glomerata	*Cotesia rubecula*
Major host species in Europe	*Pieris brassicae* *P. rapae* is less attacked, less suitable	specialised on *Pieris rapae*
Distribution pattern of major host species	clumped	uniform
Parasitoid behaviour (Wiskerke and Vet, 1994), Dutch strains	area restricted search after oviposition (also on *P. rapae*)	leaves site of attack after oviposition
Host species in USA	*Pieris rapae* only. *P. brassicae* is absent in the USA	(*C. rubecula* only recently introduced into the USA)
Attack rate on *P. rapae*	USA *C. glomerata* > Eur. *C. glomerata*	Eur. *C. rubecula* > USA *C. glomerata*

in the way they foraged, i.e., their movement pattern and patch-leaving decisions. *C. glomerata* was much more efficient in searching for the clustered larvae of *P. brassicae* and exhibited a clear area-restricted search pattern after a first oviposition. When *C. glomerata* was searching in a set-up with the solitarily feeding *P. rapae* only, females, after one oviposition, remained searching on the same leaf in about 60% of the cases. *C. rubecula* was much more efficient in searching for the solitarily feeding *P. rapae* larvae. After oviposition *C. rubecula* moves away from the site of attack and does not show area-restricted search. Le Masurier and Waage (1993) suggested that changes have occurred in the foraging rules of the American *C. glomerata*, to enable it to exploit more efficiently the more sparsely, uniformly distributed *P. rapae* in the USA. We are presently testing this hypothesis by comparing the patch leaving rules of the Dutch and American strains of *C. glomerata*.

A recent study by Driessen *et al.* (1995) with *Venturia canescens*, a larval parasitoid of phycitid flour moths, addresses a similar adaptation to spatial host distribution. In contrast to the classic model and findings of Waage (1979), who found an increase in the tendency of *V. canescens* females to remain in a patch after an oviposition, Driessen *et al.* (1995) found that ovipositions *decreased* the amount of time subsequently spent by the parasitoid on the patch, when tested under low host-density and host kairomone conditions. As in *C. glomerata*, it remains to be investigated whether *V. canescens* is capable of individually changing its patch exploitation rule as a function of host distribution, density and infochemicals. Again, learning could be an adaptive mechanism to deal with such variation.

These examples show how host distribution can affect decisions of individuals, in other words, how population dynamics sets the stage for natural selection to act. We need to know how parasitoids behaviourally deal with spatial variation, what their concept of a patch (or 'elementary foraging unit,' sensu Ayal 1987) is if we aim to understand the population dynamic and evolutionary consequences of spatial variation in parasitoid attack rates.

Future directions

Answering the question of how variation at the level of the individual affects the dynamics of populations is one of the major challenges of population ecology. However, there is still a long way to go and generalisations will be hard to make. To measure the effect of complex individual processes such as plant selection, flexible responses and dispersion on population dynamics may require the use of individual-based models (DeAngelis and Gross 1992; Judson 1994) or, at least, specific models designed for a specific experimental system (Sabelis and van der Meer 1986; Kareiva 1987; Kareiva and Andersen 1988; van der Werf *et al.* 1989). Van Roermund and van Lenteren (unpublished) developed a simulation model to study the population dynamics of greenhouse whitefly and the parasitoid *Encarsia formosa.* They showed that the whitefly population under biological control was strongly affected by variation in the parasitoid's decision to leave a patch (giving up times). When variation in giving up times was excluded from the model, the whitefly population became extinct, whereas when variation was included (as observed by van Roermund *et al.* 1994), populations were much more stable.

One of the lessons we have learned from behavioural studies is that parasitoids are far more 'efficient' and 'sophisticated' foragers than was initially assumed in classical population dynamic models. If this lesson is embodied in individual-based models these models may help us understand why populations can be 'Nicholsonian' instable at the local level. Whether such models will contribute to the quest for understanding overall population stability remains to be seen.

For evolutionary ecologists, the question of how population dynamics influences selection processes and thus shapes the foraging decisions of individuals is probably more interesting than asking the reverse. There are several exciting and powerful developments in the field of behavioural ecology that could be extended to the population level: 1. the use of a game theory approach, (evolutionary stable strategy ESS) to assess the optimal behavioural strategy under certain resource availability conditions and 2. a dynamic state-variable approach to emphasise the role of a changing state of the animal (egg-load, hunger) on the optimality of its foraging decisions (e.g., Mangel and Roitberg 1992). An example of the first approach is given by van Baalen and Sabelis (1993). They combined ecological and evolutionary stability criteria. Using an ESS and IFD approach they questioned whether natural selection favours patch selection strategies (of prey and predators) that promote ecological stability. Using the outcome of evolutionary models as building stones for population models has great potential to make behavioural empiricists and population dynamicists meet. If nothing else, it would please Nicholson.

Acknowledgments

I greatly appreciated the critical reading of a previous draft by Yoram Ayal, Marcel Dicke, Gerard Driessen, Dick Green, Lia Hemerik, Joop van Lenteren, Bill Murdoch, Herman van Roermund and Wopke van der Werf.

References

Ayal Y (1987) The foraging strategy of *Diaretiella rapae* I. The concept of the elementary unit of foraging. *J Anim Ecol* 56:1057–1068

Bernays E, Graham M (1988) On the evolution of host specificity in phytophagous arthropods. *Ecology* 69:886–892

Bernstein C, Kacelnik A, Krebs JR (1988) Individual decisions and the distribution of predators in a patchy environment. *J Anim Ecol* 57:1007–1026

Bernstein C, Kacelnik A, Krebs JR (1991) Individual decisions and the distribution of predators in a patchy environment. II. The influence of travel costs and structure of the environment *J Anim Ecol* 60:205–225

Boethel DJ, Eikenbary RD (eds) (1986) *Interactions of plant resistance and parasitoids and prdators of insects.* Ellis Horwood, Chichester

Charnov EL, Stephens DW (1988) On the evolution of host selection in solitary parasitoids. *Am Nat* 132:707–722

Chesson PL, Murdoch WW (1986) Aggregation of risk: relationships among host-parasitoid models. *Am Nat* 127:696–715

DeAngelis DL, Gross LJ (1992) *Individual-based models and approaches in ecology. Populations communities and ecosystems.* Chapman & Hall, New York

Dicke M (1994) Local and systemic production of volatile herbivore-induced terpenoids: Their role in plant-carnivore mutualism. *J Plant Physiol* **143**:465–472
Dicke M (1995) Direct and indirect effects of plants on performance of beneficial organisms. In: Ruberson JR (ed) *Handbook of Pest Management*, Marcel Dekker Inc, New York (in press)
Dicke M, Sabelis MW (1988a) How plants obtain predatory mites as bodyguards. *Neth J Zool* **38**:148–165
Dicke M, Sabelis MW (1988b) Infochemical terminology: based on cost-benefit analysis rather than origin of compounds? *Funct Ecol* **2**:131–139
Driessen G, Hemerik L (1991) Aggregative responses of parasitoids and parasitism in populations of *Drosophila* breeding in fungi. *Oikos* **61**:96–107
Driessen G, Bernstein C, van Alphen JJM, Kacelnik A (1995) A count down mechanism for host search in the parasitoid *Venturia canescens*. *J Anim Ecol* **64**:117–123
Haccou P, de Vlas SJ, van Alphen JJM, Visser ME (1991) Information processing by foragers: effects of intra-patch experience on the leaving tendency of *Leptopilina heterotoma*. *J Anim Ecol* **60**:93–106
Hassel MP (1978) *The dynamics of arthropod predator-prey systems*. Monographs in Population Biology 13 Princeton University Press, Princeton, New Jersey
Hedlund K, Vet LEM, Dicke M (1996) Generalist and specialist parasitoid strategies of using odours of adult drosophilid flies when searching for larval hosts. *Oikos*, in press
Hemerik L, Driessen G, Haccou P (1993) Effects of intra-patch experiences on patch time, search time and searching efficiency of the parasitoid *Leptopilina clavipes* (Hartig). *J Anim Ecol* **62**:33–44
Iwasa Y, Higashi M, Yamamura N (1981) Prey distribution as a factor determining the choice of optimal foraging strategy. *Am Nat* **117**:710–723
Janssen A, van Alphen JJM, Sabelis MW, Bakker K (1995) Specificity of odour mediated avoidance of competition in *Drosophila* parasitoids. *Behav Ecol Sociobiol* **36**:229–235
Judson OP (1994) The rise of the individual-based model in ecology. *Trends Ecol Evol* **9**:9–14
Kacelnik A, Krebs JR, Bernstein C (1992) The ideal free distribution and predator-prey populations. *Trends Ecol Evol* **7**:50–55
Kareiva P (1987) Habitat fragmentation and the stability of predator-prey interactions. *Nature* **326**:388–390
Kareiva P, Andersen M (1988) Spatial aspects of species interactions: the wedding of models and experiments. *Lect notes Biomath* **63**:35–50
Kareiva P, Odell G (1987) Swarms of predators exhibit 'preytaxis' if individual predators use area-restricted search. *Am Nat* **130**:233–270
Kareiva P, Sahakian R (1990) Tritrophic effects of a simple architectural mutation in pea plants. *Nature* **345**:433–434
Kester KM, Barbosa P (1991) Behavioral and ecological constraints imposed by plants on insect parasitoids: implications for biological control. *Biol Control* **1**:94–106
Koehl MAR (1989) Discussion: from individuals to populations. In: Roughgarden J, May RM, Levin SA (eds) *Perspectives in ecological theory*. Princeton University Press, Princeton, pp 39–53
le Masurier AD, Waage JK (1993) A comparison of attack rates in a native and an introduced population of the parasitoid *Cotesia glomerata*. *Biocon Sci Technol* **3**:467–474
Lessells CM (1985) Parasitoid foraging: Should parasitism be density dependent? *J Anim Ecol* **54**:27–41
Lewis WJ, Vet LEM, Tumlinson JH, van Lenteren JC, Papaj DR (1990) Variations in parasitoid foraging behavior: essential element of a sound biological control theory. *Environ Entomol* **19**:1183–1193
Mangel M, Roitberg BD (1992) Behavioral stabilization of host-parasite population dynamics. *Theor Popul Biol* **42**:308–320
McNair JN (1982) Optimal giving-up times and the marginal value theorem. *Am Nat* **119**:511–529
McNamara JM, Houston AI, Weisser WW (1993) Combining prey choice and patch use – What does rate-maximizing predict? *J Theor Biol* **164**:219–238
Murdoch WW (1994) Population regulation in theory and practice. *Ecology* **75**:271–287
Murdoch WW, Stewart-Oaten A (1989) Aggregation by parasitoids and predators: effects on equilibrium and stability. *Am Nat* **134**:288–310
Nicholson AJ (1933) The balance of animal populations. *J Anim Ecol* **2**:132–178

Nicholson AJ, Bailey VA (1935) The balance of animal populations. Part I. *Proc Zool Soc Lond* **3**:551–598

Ohsaki N, Sato Y (1994) Food plant choice of *Pieris* butterflies as a trade-off between parasitoid avoidance and quality of plants. *Ecology* **75**:59–68

Pacala SW, Hassell MP (1991) The persistence of host-parasitoid associations in patchy environments. II. Evaluation of field data. *Am Nat* **138**:584–605

Pacala SW, Hassell MP, May RM (1990) Host-parasitoid associations in patchy environments. *Nature* **344**:150–153

Papaj DR, Vet LEM (1990) Odor learning and foraging success in the parasitoid, *Leptopilina heterotoma*. *J Chem Ecol* **16**:3137–3150

Price PW, Bouton CE, Gross P, McPheron BA, Thompson JN, Weis AE (1980) Interactions among three trophic levels: influence of plant on interactions between insect herbivores and natural enemies. *Annu Rev Ecol Syst* **11**:41–65

Sabelis MW, Diekmann O, Jansen VAA (1991) Metapopulation persistence despite local extinction: predator-prey patch models of the Lotka-Volterra type. *Biol J Linn Soc* **42**:267–283

Sabelis MW, van der Meer J (1986) Local dynamics of the interaction between predatory mites and two-spotted spider mites. In: Metz JAJ, Diekman O (eds), *Dynamics of physiologically structured populations*. Springer, *Lect Notes Biomath* **68**:322–344

Sheehan W, Wackers FL, Lewis WJ (1993) Discrimination of previously searched, host-free sites by *Microplitis croceipes* (Hymenoptera: Braconidae). *J Insect Behav* **6**:323–331

Stamp NE, Casey TM (1993) *Caterpillars: ecological and evolutionary constraints on foraging*. Chapman and Hall, New York

Stephens DW, Krebs JR (1986) *Foraging theory*. Princeton University Press, Princeton

Stiling PD (1987) The frequency of density dependence in insect host-parasitoid systems. *Ecology* **68**:844–856

Sutherland WJ, Dolman PM (1994) Combining behaviour and population dynamics with applications for predicting consequences of habitat loss. *Proc R Soc Lond B Biol Sci* **255**:133–138

Takabayashi J, Dicke M, Posthumus MA (1994) Volatile herbivore- induced terpenoids in plant-mite interactions: Variation caused by biotic and abiotic factors. *J Chem Ecol* **20**:1329–1354

Thorpe KW, Barbosa R (1986) Effects of consumption of high and low nicotine tobacco by *Manduca sexta* (Lepidoptera: Sphingidae) on survival of gregarious endoparasitoid *Cotesia congregata* (Hymenoptera: Braconidae). *J Chem Ecol* **12**:1329–1337

Turlings TCJ, Loughrin JH, McCall PJ, Rose U, Lewis WJ, Tumlinson JH (1995) How caterpillar-damaged plants protect themselves by attracting parasitic wasps. *Proc Nat Acad Sci* **92**:4169–4174

Turlings TCJ, Tumlinson JH, Lewis WJ (1990) Exploitation of herbivore-induced plant odors by host-seeking parasitic wasps. *Science* **250**:1251–1253

Turlings TCJ, Wäckers FL, Vet LEM, Lewis WJ, Tumlinson JH (1993) Learning of host-finding cues by Hymenopterous parasitoids. In: Papaj DR, Lewis AC (eds) *Insect learning*. Chapman & Hall, New York, pp 51–78

van Baalen M, Sabelis MW (1993) Coevolution of patch selection strategies of predator and prey and the consequences for ecological stability. *Am Nat* **142**:646–670

van der Werf W, Rossing WAH, Rabbinge R, de Jong MD, Mols PJM (1989) Approaches to modelling the spatial dynamics of pests and diseases. In: Cavalloro R, Delucchi V (eds) PARASITIS 88. *Boletin de Sanidad Vegetal, Fuera de Serie* **17**:89–119

van Lenteren JC (1991) Encounters with parasitized hosts: to leave or not to leave a patch. *Neth J Zool* **41**:144–157

van Lenteren JC, de Ponti OMB (1990) Plant-leaf morphology, host-plant resistance and biological control. *Symp Biol Hung* **39**:365–386

van Roermund HJW, Hemerik L, van Lenteren, JC (1994) Influence of intrapatch experiences and temperature on the time allocation of the whitefly parasitoid *Encarsia formosa* (Hymenoptera: Aphelinidae). *J Insect Behav* **7**:483–501

Vet LEM, Dicke M (1992) Ecology of infochemical use by natural enemies in a tritrophic context. *Annu Rev Entomol* **37**:141–172

Vet LEM, Lewis WJ, Cardé RT (1995) Parasitoid foraging and learning. In: Carde RT, Bell WJ (eds) *Chemical Ecology of insects, 2nd edition*. Chapman & Hall, New York, pp 65–101

Vet LEM, Lewis WJ, Papaj DR, van Lenteren JC (1990) A variable-response model for parasitoid foraging behavior. *J Insect Behav* **3**:471–490

Vet LEM, van Opzeeland K (1984) The influence of conditioning on olfactory microhabitat and host location in *Asobara tabida* (Nees) and *A. rufescens* (Foerster) (Braconidae:Alysiinae) larval parasitoids of Drosophilidae. *Oecologia* **63**:171–177

Vet LEM, Wäckers FL, Dicke M (1991) How to hunt for hiding hosts: the reliability-detectability problem in foraging parasitoids. *Neth J Zool* **41**:202–213

Visser JH (1986) Host odor perception by phytophagous insects. *Annu Rev Entomol* **31**:121–144

Visser ME, van Alphen JJM, Hemerik L (1992) Adaptive superparasitism and patch time allocation in solitary parasitoids: an ESS model. *J Anim Ecol* **61**:93–101

Waage JK (1979) Foraging for patchily-distributed hosts by the parasitoid, *Nemeritis canescens*. *J Anim Ecol* **48**:353–371

Wäckers FL, Lewis WJ (1994) Olfactory and visual learning and their combined influence on host site location by the parasitoid *Microplitis croceipes* (Cresson). *Biol Control* **4**:105–112

Walde SJ, Murdoch WW (1988) Spatial density dependence in parasitoids. *Annu Rev Entomol* **33**:441–466

Wiskerke JSC, Vet LEM (1994) Foraging for solitarily and gregariously feeding caterpillars: a comparison of two related parasitoid species. *J Insect Behav* **7**:585–603

Sources of Stability in Host-Parasitoid Dynamics

Andrew D. Taylor

Abstract

The initial challenge for the theory of host-parasitoid dynamics was to discover possible sources of stability to counteract the seemingly intrinsic instability of these systems. The current challenge, however, is to make sense of the wide variety of possible stabilising mechanisms which have been suggested. Realistic modifications of every part of the Nicholson-Bailey model have been proposed and shown to be stabilising. These stabilising mechanisms not only are not mutually exclusive, but often interact with one another: the effect of one factor may depend on the value of some other factor. Furthermore, seemingly similar biological phenomena can have fundamentally different dynamical effects. These points will be illustrated using models of 'aggregation of risk,' of superparasitism and 'within-host competition,' and of non-lethal parasitism.

The task facing theoreticians is to integrate these various parts into a coherent whole; the task for empirical research is to determine which factors are important in particular natural systems. It will be suggested that the key to meeting both these challenges will be to focus on the fundamental Nicholsonian concept of temporally density-dependent population regulation.

Key words: stabilising mechanisms, aggregation of risk, within-host competition, sublethal parasitism

Frontiers of Population Ecology, R.B. Floyd, A.W. Sheppard and P.J. De Barro (eds), CSIRO Publishing, Melbourne, 1996 pp. 257–267

Introduction

The Nicholson-Bailey model of host-parasitoid dynamics (Nicholson and Bailey 1935) has provided the foundation for six decades of productive research, both theoretical and empirical. This is most obviously true in that most later models of host-parasitoid or other arthropod predator-prey interactions – and until very recently nearly all of the most influential models – have been variants of the Nicholson-Bailey model, sharing its basic structure and assumptions. In a more subtle but perhaps more important effect, the model also has set the agenda for the research which followed: because it is unstable (populations undergo rapidly diverging oscillations), the primary goal of theoretical research has been to find factors which might stabilise these interactions; most empirical research has been aimed at testing the resulting theories.

It seems clear, however, that a new research agenda is needed. A wide variety of factors have been identified which theoretically can be stabilising: every part of the Nicholson-Bailey model can be modified in a stabilising way (Table 1), and additional sources of stability have been found in models departing more fundamentally from the Nicholson-Bailey structure (W. W. Murdoch, this volume). While study of the effects of new phenomena will continue to be useful, the initial challenge raised by the instability of the Nicholson-Bailey model has been met.

The agenda now, should be to find ways to effectively deal with the large number and variety of possible sources of stability. The challenge of this arises from the fact that the various factors are not

Table 1. A partial list of sources of stability in Nicholson-Bailey–type parasitoid-host models

Fundamental mechanism (density-dependence(s) affected)	Stabilising processes
Extrinsic host regulation (host dependence on host density)	Resource limitation Other natural enemies
Extrinsic parasitoid regulation (parasitoid dependence on host density) [stability also requires direct host density dependence]	Generalist parasitoid Hyperparasitoids or predators
Parasitism rate dependent on host density, e.g. Type III functional response (host and parasitoid dependence on host density)	Switching Search image formation Constant-number refuge
Parasitoid efficiency inversely dependent on parasitoid density (host and parasitoid dependence on parasitoid density)	Mutual interference Pseudo-interference, aggregation of risk: - host-density-dependent spatial aggregation of parasitoids - host-density-independent spatial aggregation of parasitoids - constant-proportion refuge - variation among host individuals in susceptibility to parasitism, ability to encapsulate parasitoids, phenological exposure to parasitoids, etc.
Intrinsic parasitoid regulation (parasitoid dependence on parasitoid density)	Parasitoid-density-dependent parasitoid sex ratio Density-dependent parasitoid reproduction per parasitised host ('within-host competition')
Impact of parasitism on host reproduction (host dependence on parasitoid density)	Sub-lethal parasitism

mutually exclusive alternatives. Rather, any given real parasitoid-host interaction is likely to involve several of these factors simultaneously. Furthermore, the effects of these factors can combine and interact in complex ways. Studies, typical of much past and current research, which focus on single factors in isolation, cannot adequately address the effects of those factors, whether in a model or in nature. New approaches are needed which are capable of considering, and making sense of, multiple simultaneous sources of stability.

This paper will illustrate these points by discussing three specific factors: 'aggregation of risk' (Chesson and Murdoch 1986), 'within-host competition' (Taylor 1988 and unpub. MS), and 'sublethal parasitism.' Preliminary data will be presented from a parasitoid-host interaction in which all three factors appear to be present, followed by description of the dynamical effects of some combinations of these factors. The paper will conclude with discussion of some of the implications of the reality of multiple co-occurring and interacting sources of stability for theoretical and empirical research.

Aggregation of risk, within-host competition, and sublethal parasitism

Definitions

'Aggregation of risk' (Chesson and Murdoch 1986) is the situation in which the risk of parasitism (the probability of being encountered by and successfully parasitised by a female parasitoid) varies across host individuals. The dynamical effect of this heterogeneity is 'pseudo-interference' (Free *et al.* 1977): the efficiency of the parasitoids (the fraction of the host population parasitised per parasitoid) decreases as parasitoid density increases. This causes host mortality to be less dependent, and parasitoid per capita reproduction to be more strongly inversely dependent, on parasitoid density; these effects usually enhance the stability of the system. Further discussion of the range of mechanisms which can produce aggregation of risk, of how and when it is stabilising, and of the empirical evidence concerning it, is in Taylor (1993b).

'Within-host competition' (Taylor 1988) refers to dynamics in which the number of parasitoid progeny produced from a given host depends on the number of times that host was discovered by parasitoid females. Most models, including the Nicholson-Bailey model, assume this number of progeny is constant. This assumption implies perfectly compensatory within-host competition. In contrast, the number of progeny produced per host will vary with the number of encounters whenever there is at least some superparasitism and larval competition is not perfectly compensating.

'Sublethal parasitism' refers to cases in which a host which is successfully parasitised (i.e. one or more parasitoid progeny are produced) nonetheless itself is able to reproduce. Although quite uncommon among hymenopteran parasitoids, such sublethal effects occur reasonably frequently among dipteran (especially tachinid) parasitoids. In some cases hosts simply survive the emergence of the parasitoid progeny (Timberlake 1916; Clausen 1940; Richards and Walloff 1948; DeVries 1984; English-Loeb *et al.* 1990). Sublethal parasitism appears to more commonly arise, however, from parasitism of adult hosts, which are able to reproduce before eventually being killed by the parasitoid larvae (Clausen 1940; Askew 1971; and the following section).

An example

Aggregation of risk has received a great deal of empirical study, and it appears to occur relatively frequently, though not necessarily at a level sufficient to stabilise models like Eq. 2 (e.g. Hassell and Pacala 1990; Pacala and Hassell 1991; Taylor 1993b and references therein). In contrast there is very little evidence on 'within-host competition' – i.e. on rates of superparasitism or dynamics of larval competition – under natural conditions (see Taylor 1988). Similarly, the data on sublethal parasitism consists of little more than anecdotal observations of survival of or reproduction by parasitised hosts.

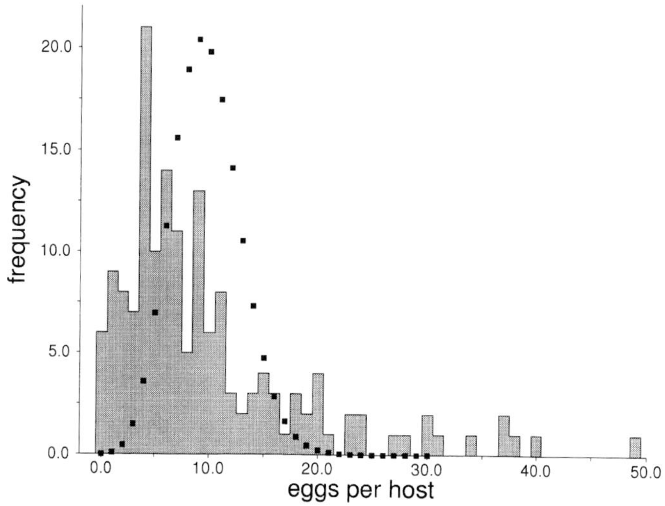

Fig. 1. The distribution of parasitoid (*Trichopoda pennipes*) eggs across hosts (*Nezara viridula*) in a single collection (28 June 1994, Waipio Peninsula, O'ahu, Hawaii, USA). $n = 159$, $\bar{x} = 9.7$, $s^2 = 79.6$. Black squares indicate Poisson distribution with mean = 9.7.

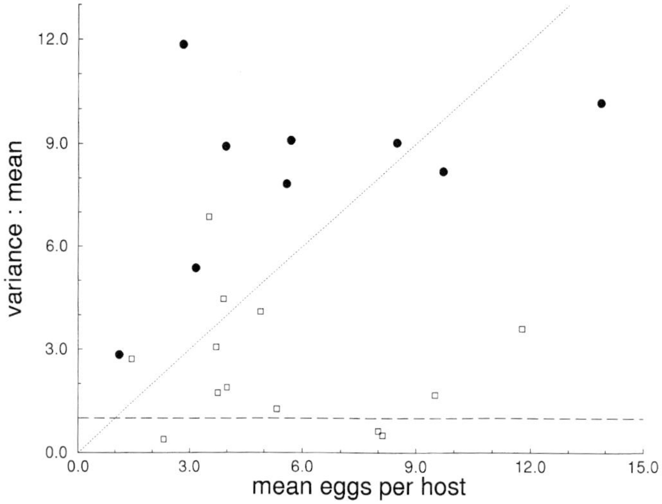

Fig. 2. Variance-to-mean ratios for distributions of parasitoid (*T. pennipes*) eggs across hosts (*N. viridula*), as a function of the mean number of eggs per host. Filled dots represent collections of field-parasitised hosts (as in Fig. 1); hollow squares represent parasitisation in the laboratory during fixed periods (24 or 48 h). Field collections were from the same site as in Fig. 1, from July through November 1994. Points above the horizontal dashed line represent aggregation ($s^2 > 1$); points above the diagonal dotted line satisfy the '$CV^2 > 1$ rule' (Hassell and Pacala 1990; Pacala and Hassell 1991).

The following therefore summarises the evidence that all three phenomena – aggregation of risk, within-host competition, and sublethal parasitism – occur in one particular host-parasitoid interaction, that between *Nezara viridula* (Hemiptera: Pentatomidae; the southern green stink bug) and *Trichopoda pennipes* (Diptera: Tachinidae; the feather-legged fly).

The main evidence concerning aggregation of risk in this system comes from the distribution of egg numbers over hosts. As shown in Figs. 1 and 2, this distribution typically is somewhat aggregated, with variance to mean ratios typically considerably greater than 1. It appears that *T. pennipes* does not discriminate between parasitised and unparasitised hosts (A.D. Taylor unpublished), but if there is some avoidance of superparasitism the true distribution of encounters would be even more aggregated than the egg distributions indicate. Part of the aggregation of eggs in field collections undoubtedly is due to differences in host ages, older hosts having been available longer for

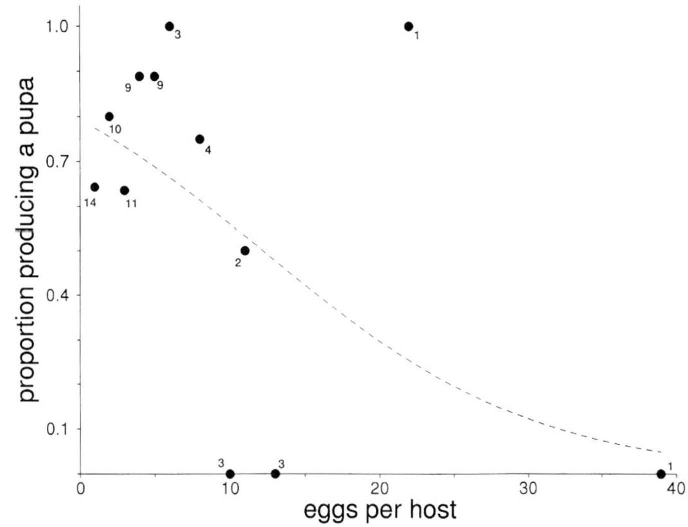

Fig. 3. Survival of parasitoid (*T. pennipes*) larvae, as a function of the number of eggs on the host (*N. viridula*). Parasitised hosts were collected from the natural population used in Fig. 1. The parasitoid eggs on each host were counted and hosts were then reared individually to determine whether a live fly pupa was produced. Data represent several collections from July through November 1994. Numbers by each point indicate the sample size for that egg number. The curved line is a logistic regression best fit, *p*-value for egg effect = 0.0314.

parasitisation. This, however, cannot be the entire explanation since aggregated distributions are produced even by single flies ovipositing on hosts in the laboratory for only 24 hours (Fig. 2). In addition to this aggregation seen within a patch of hosts, we have observed substantial variation in parasitism among host patches at the same time, though we have not yet quantified this component of aggregation of risk.

'Within-host competition' with dynamics different from the 'constant yield' assumption of standard Nicholson-Bailey–type models requires at least some superparasitism, and larval competition which is not perfectly compensatory. Superparasitism clearly occurs in the *T. pennipes–N. viridula* interaction (Fig. 1; also Davis 1964; Shahjahan 1966, 1968; Mitchell and Mau 1971; McLain et al. 1990) and is common in Tachinids (Clausen 1940; Askew 1971). Preliminary data indicate also that larval competition is overcompensating, though not strongly so: the probability of a parasitoid emerging from a given host decreases somewhat as the number of eggs laid on the host increases (Fig. 3; also Shahjahan 1966, 1968). The data in Fig. 3 appear to show a positive effect of egg number on the probability of a pupa being produced, at the lowest egg numbers. Additional data are being obtained which should determine whether this trend is real. A peaked reproduction curve, however, would still be consistent with an interpretation of mild overcompensation, and inconsistent with the usual assumption of constant parasitoid output per host. Combining these effects, the production of parasitoid progeny from a host will be a decreasing function of the number of parasitoid females encountering that host, i.e. within-host competition will be scramble-like.

'Sub-lethal parasitism' occurs in this system because the fly parasitises primarily adult hosts (Jones 1988; Todd 1989). Parasitised hosts remain active and reproductive until the full-grown fly larva emerges from the host some two to three weeks later. Hosts therefore often are able to reproduce before being killed (Table 2; also Shahjahan 1966, 1968; Harris and Todd 1982), and may also have reproduced prior to being parasitised. The effect of parasitism therefore is a shortening of the reproductive period and of total reproductive output, rather than the total elimination of reproduction assumed by standard Nicholson-Bailey–type models.

The emphasis on the preceding three phenomena is not meant to imply that they are the only important factors in the dynamics of this system. The host experiences high levels of egg predation by ants, and egg parasitism by other parasitoid species. Intraspecific competition also is likely among

Table 2. Effect of parasitism by *T. pennipes* on host (*N. viridula*) reproduction

	n	Percent ovipositing	Fecundity of hosts which oviposited Median	Quartiles
Parasitised hosts (fly larva emerged)	33	24.2%	86.0	(28, 90)
Unparasitised hosts (no fly eggs or larvae)	15	66.7%	105.5	(59, 309)

hosts. And the functional response of the parasitoid to variation in host density is almost certainly not linear as assumed in the models below. Reality, therefore, is undoubtedly even more complicated than the three-factor situation described here.

Models

The models considered here all are modifications of the Nicholson-Bailey (1935) formulation:

$$\begin{aligned} H_{t+1} &= FH_t \mathrm{e}^{-aP_t} \\ P_{t+1} &= H_t\left(1-\mathrm{e}^{-aP_t}\right) \end{aligned} \qquad (1)$$

In this and the following models, H_t and P_t represent host and parasitoid densities in generation t, a is the parasitoid's 'area of discovery' (fraction of host population encountered per parasitoid), and F is the host finite rate of increase. This model, and all those below, share the fundamental assumption that host and parasitoid dynamics are determined solely by current densities of the two populations (i.e. they are tightly coupled populations with no 'external' factors, including interactions with other populations, affecting dynamics).

Aggregation of risk will be represented by the well-known negative-binomial model (May 1978), which provides a simple and fairly general representation of the phenomenon. In this model the distribution (over hosts) of encounters with parasitoids follows a clumped negative-binomial distribution, rather than the random Poisson distribution implied by the Nicholson-Bailey model. Assuming no other modifications from Nicholson-Bailey assumptions, the model is

$$\begin{aligned} H_{t+1} &= FH_t\left(1+aP_t/k\right)^{-k} \\ P_{t+1} &= H_t\left[1-\left(1+aP_t/k\right)^{-k}\right] \end{aligned} \qquad (2)$$

The parameter k describes the degree of aggregation: smaller values of k represent greater aggregation of risk and usually produce greater stability (but see Taylor 1993a). As is the case for most stabilising factors, though, this greater stability comes at the price of increased host equilibrium abundance.

Within-host competition will be modelled in its overcompensating ('scramble') form, and in combination with aggregation of risk. Under-compensating, or 'contest', dynamics, have been considered elsewhere. Specifically, the mean number of female parasitoid progeny produced from a host encountered i times is $ci(1-y)^{i-1}$, where y (which must be between 0 and 1) describes the strength of density dependence and c is the number of progeny from a host encountered once. When this submodel is combined with the negative-binomial model for aggregated encounters (as in Eq. 2), the following model is obtained (Taylor 1988):

$$\begin{aligned} H_{t+1} &= FH_t\left(1+aP_t/k\right)^{-k} \\ P_{t+1} &= caP_tH_t\left(1+ayP_t/k\right)^{-(k+1)} \end{aligned} \qquad (3)$$

Stronger competition (larger y) usually gives greater stability, although under extreme conditions excess competition can be destabilising (Taylor 1988) and when k is very small larger y sometimes gives slightly less stability than smaller y (Taylor 1993a). Larger y also gives a higher host equilibrium, but it is noteworthy that the overcompensating parasitoid competition in Eq. 3, when compared with the perfect compensation in the standard negative-binomial model of Eq. 2, can give greater stability and lower host equilibrium simultaneously (Taylor 1988).

The effect of parasitism on a host's reproduction could depend on the intensity of parasitism (e.g. how many parasitoid larvae are developing in it). I will assume, however, that this is not the case, but rather that the reproductive rate of parasitised hosts is the constant G. Incorporating this assumption into the negative-binomial model (Eq. 2) produces the model

$$\begin{aligned} H_{t+1} &= \left\{ F(1+aP_t/k)^{-k} + G\left[1-(1+aP_t/k)^{-k}\right]\right\} \\ P_{t+1} &= H_t\left[1-(1+aP_t/k)^{-k}\right] \end{aligned} \qquad (4)$$

The host equilibrium increases as G increases – as the effect of parasitism on host reproduction decreases – and indeed this model does not have a positive equilibrium if $G > 1$ (i.e. if the host population would grow even if every host were parasitised). For $G < 1$, larger G (less impact of parasitism) is stabilising.

Effects of factors in combination

It was noted above that the three phenomena all generally are stabilising: increased aggregation of risk, stronger within-host competition, and greater sublethality (i.e. reduced impact of parasitism on host dynamics) all enhance stability. The quantitative effect of any of these factors, however, depends on the level of the other factors: there is no general criterion for stability in terms of one factor which is independent of the other factors. Analysis of a model combining all three factors is underway and will be reported elsewhere; here only the two-way interactions of within-host competition with aggregation of risk and of sublethal parasitism with aggregation of risk will be illustrated.

When within-host competition and aggregation of risk are combined (as in Eq. 3), both contribute to stability: stability can be obtained with little aggregation of risk but strong within-host competition, weak competition and substantial aggregation, or intermediate levels of both (Fig. 4). This relationship is not simply additive, however. For either factor to produce stability, the other factor must be present (Taylor 1988): very strong within-host competition can produce stability for any finite k but not with random encounters (a Poisson distribution,), while aggregation of risk cannot produce stability if there is no within-host competition. When both factors are present the effect of either one is greater when the other is weak. A further complication is that stability also is affected by F, the host rate of increase (Fig. 4).

The combination of sublethal parasitism and aggregation of risk (as in Eq. 4) shows a generally similar interaction (Fig. 5). Both factors contribute to stability, so that more of one can compensate for less of the other (with 'more sublethality' meaning less impact of parasitism on hosts, i.e. larger G). The quantitative effect of either factor again is stronger when the other factor is at a lower level. The result is that the contours in Fig. 5 are not straight: intermediate levels of both factors is less effective than having one factor strong and the other weak. One important difference from within-host competition is that sublethal parasitism can produce stability even in the complete absence of aggregation of risk, i.e. with random encounters. Stability again depends also on a third parameter, F the host rate of increase (Fig. 5).

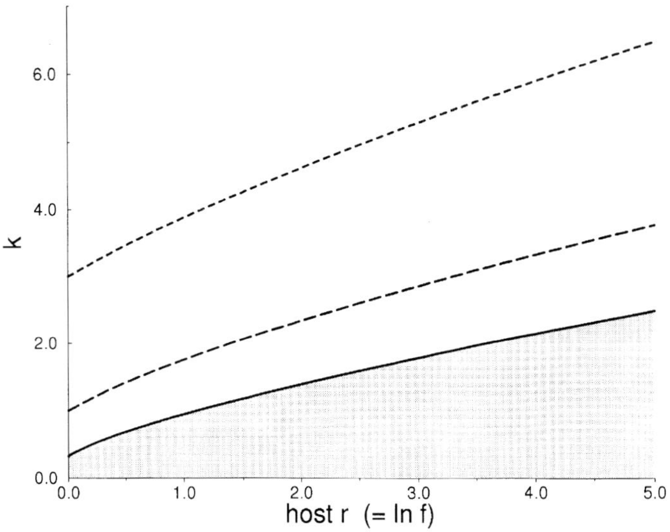

Fig. 4. Stability boundaries for the model in Eq. 3, combining scramble within-host competition and negative-binomial aggregation of risk. Regions below each line have stable equilibria. The lines indicate differing strengths of within-host competition (the parameter *y*): the solid line is weak competition (*y* = 0.25), the long-dashed line is intermediate competition (*y* = 0.5) and the short-dashed line is strong competition (*y* = 0.75). The parameter *r* is the natural log of the host finite rate of increase (*F* in the models in the text), and *k* scales the degree of aggregation (smaller *k* indicating greater aggregation).

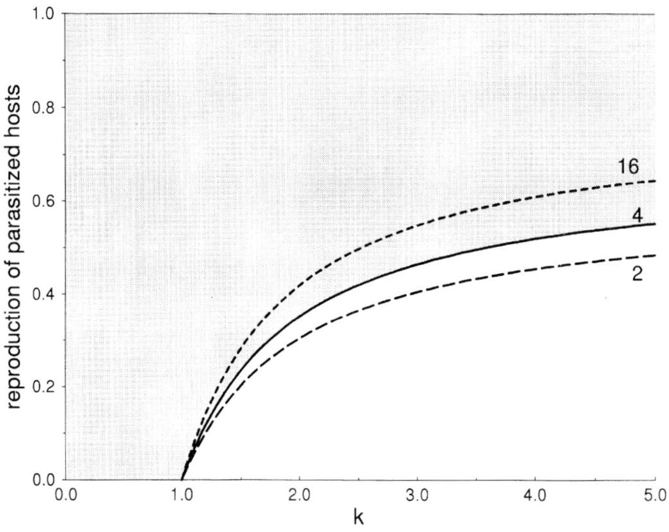

Fig. 5. Stability boundaries for the model in Eq. 4, combining sublethal parasitism and negative-binomial aggregation of risk. Regions above and to the left of each line are stable. The lines are for differing host rates of increase (*F*), as indicated.

Discussion

The complexity illustrated in the preceding sections, with several stabilising factors combining and interacting, is by no means unique to this particular parasitoid-host interaction or these particular dynamical factors. Some of the factors discussed above – especially sublethal parasitism – may indeed be relatively rare. Additional factors not considered above, such as intra- or inter-specific host competition, other enemies, or non-linear functional responses, however, seem likely to affect most parasitoid-host systems (including the *T. pennipes–N. viridula* interaction). While the dynamics of models combining several of such factors have received little attention, it is clear for instance that host density dependence in combination with aggregation of risk (e.g. Hochberg and Lawton 1990), or with aggregation of risk and within-host parasitoid competition, can produce complex results; the same is almost certainly true for other factors as well.

The primary implication of this complicated reality is that it is unlikely that studying a single factor in isolation will ever be fully effective. This was an appropriate approach in the past, when the aim was simply to identify possible sources of stability. Now, though, the aim should be a fuller understanding of the effects and importance of the full set of factors affecting dynamics, and this will require a shift from narrowly focussed to broadly encompassing modes of study.

Theoretical analysis of the dynamical effects of some factor, if conducted in the context of only one set of assumptions concerning other factors, e.g. as an isolated modification of the Nicholson-Bailey model, will not be able to uncover potentially important interactions with other factors, and may lead to conclusions which are not robust in the presence of other factors. Analysis of new phenomena will continue to be useful (witness the results above for sub-lethal parasitism), but there is perhaps a greater need for analysis of the effects of the various already-identified sources of stability in combination with each other. Beyond this, we should strive to develop a more unified, synthetic body of theory which explains, or at least describes and predicts, the dynamical results of particular combinations of factors. A parallel body of theory explaining or predicting which sources of stability will be most important under particular circumstances would be of great value for empirical research, as well as of considerable theoretical interest.

For the same reasons, it will rarely if ever be possible to understand the dynamics of a real parasitoid-host interaction by simply identifying and measuring one source of stability. It will generally be necessary also to identify and measure other dynamical factors which might interact with the first factor. In other words, we should perhaps view the entire body of theory as an aid to understanding empirical systems, by suggesting factors which should be considered, rather than viewing empirical systems as aids to testing a particular theory.

Similarly, it is highly unlikely that any single factor will prove to be the key to successful biological pest control. Rather, when considering a new biological control program or evaluating an existing one, all relevant factors and characteristics should be assessed, with the entire body of available theory (as well as common sense) providing insight as to what factors are likely to be pertinent and what their joint effects are likely to be.

These issues of how to deal with multiple simultaneous influences on dynamics, in my opinion represent the most challenging frontier for parasitoid-host studies today. Since little if any attention has been given to these questions, it is difficult to even guess what approaches will be effective in addressing them. It seems likely, however, that renewed attention to a fundamental concept put forth by Nicholson – temporally density-dependent population regulation – may provide at least a first step towards the sort of comprehensive and integrated studies of parasitoid-host dynamics outlined in the preceding paragraphs.

At the very least, our understanding of how and when particular phenomena affect stability will be strengthened if it is kept in mind that within-generation processes can affect dynamics only by modifying the temporal (between-generation) dependence of parasitoid or host rates of change on the density of their own or an interacting population. It is for this reason, for instance, that aggregation of risk as represented in the negative-binomial model (Eq. 2) can be stabilising only because some within-host parasitoid competition is implicitly assumed, so that the aggregation accentuates density-dependent regulation of the parasitoid population (Taylor 1988, 1993b). The central role of temporal density-dependent regulation also underlies the fundamentally different effects of aggregation of risk dependent only on parasitoid density, and aggregation of risk dependent also on host density (for example, constant-proportion vs. constant-number refuges), and the fact that the latter typically is effective under much more general conditions; it is important to note that these two phenomena, so different dynamically, would be indistinguishable if measured at only one time, without the necessary temporal perspective.

When considering not single factors but the entire range of possible factors, the temporal density dependencies acted on by different factors provide a structure for categorising them (as in Table 1). Such a categorisation might in turn aid in the discovery of general patterns or principles concerning how various factors act in combination. This same categorisation may aid empirical studies in that initial determination of which density dependencies are most important in a given system could indicate which specific phenomena are likely to be present and worth looking for.

It may seem paradoxical to suggest that the way to conquer the current frontier in parasitoid-host studies is to return to the seemingly old-fashioned approach of describing and dissecting the temporal dynamics of populations, rather than to use the sort of short-term experimental tests of particular processes which have been so successful for ecology in recent years. Experimentation can indeed be a powerful way of studying population dynamics and density dependence, but only if either the manipulation or the observed response involves the appropriate temporal dynamics; in many cases an 'old-fashioned' observational approach may be just as useful or may even be all that is possible. The crucial issue, however, is not what methods are used, but what processes and temporal scales are addressed. If the goal is to understand temporal population dynamics it will be necessary to study temporal dynamics explicitly.

References

Askew RR (1971) *Parasitic insects.* American Elsevier Publ Co Inc, New York

Chesson PL, Murdoch WW (1986) Aggregation of risk: relationships among host-parasitoid models. *Am Nat* **127**:696–715

Clausen CP (1940) *Entomophagous insects.* McGraw-Hill Book Co Inc, New York

Davis CJ (1964) The introduction, propagation, liberation, and establishment of parasites to control *Nezara viridula* variety *smaragdula* (Fabricius) in Hawaii (Heteroptera: Pentatomidae). *Proc Hawaii Entomol Soc* **18**:369–37

DeVries PJ (1984) Butterflies and tachinidae: does the parasite always kill its host? *J Nat Hist* **18**:323–326

English-Loeb GM, Karban R, Brody AK (1990) Arctiid larvae survive attack by a tachinid parasitoid and produce viable offspring. *Ecol Entomol* **15**:361–362

Free CA, Beddington JR, Lawton JH (1977) On the inadequacy of simple models of mutual interference for parasitism and predation. *J Anim Ecol* **46**:543–554

Harris VE, Todd JW (1982) Longevity and reproduction in the Southern green stink bug, *Nezara viridula*, as affected by parasitization by *Trichopoda pennipes. Entomol Exp Appl* **31**:409–412

Hassell MP, Pacala SW (1990) Heterogeneity and the dynamics of host-parasitoid interactions. *Philos Trans R Soc Lond B Biol Sci* **330**:203–220

Hochberg ME, Lawton JH (1990) Spatial heterogeneities in parasitism and population – dynamics. *Oikos* **59**:9–14

Jones WA (1988) World review of the parasitoids of the Southern green stink bug, *Nezara viridula* (L.) (Heteroptera: Pentatomidae). *Ann Entomol Soc Am* **81**:262–273

May RM (1978) Host-parasitoid systems in patchy environments: a phenomenological model. *J Anim Ecol* **47**:833–844

McLain DK, Marsh NB, Lopez JR, Drawdy JA (1990) Intravernal changes in the level of parasitization of the Southern green stink bug (Hemiptera: Pentatomidae), by the feather-legged fly (Diptera: Tachinidae): Host sex, mating status, and body size as correlated factors. *J Entomol Sci* **25**:501–509

Mitchell WC, Mau RFL (1971) Response of the female Southern green stink bug and its parasite, *Trichopoda pennipes*, to male stink bug pheromones. *J Econ Entomol* **64**:856–859

Nicholson AJ, Bailey VA (1935) The balance of animal populations. Part 1. *Proc Zool Soc Lond* **3**:551–598

Pacala SW, Hassell MP (1991) The persistence of host-parasitoid associations in patchy environments. II. Evaluation of field data. *Am Nat* **138**:584–605

Richards OW, Walloff N (1948) The hosts of four British Tachinidae (Diptera). *Entomol Mon Mag* **84**:127

Shahjahan M (1966) *Some aspects of the biology of* Trichopoda pennipes *Fabricius (Diptera, Tachinidae), a parasite of* Nezara viridula *(Fabricius) in Hawaii.* MS Thesis, University of Hawaii

Shahjahan M (1968) Superparasitization of the Southern green stinkbug by the tachinid parasite *Trichopoda pennipes pilipes* and its effect on the host and parasite survival. *J Econ Entomol* **61**:1088–1091

Taylor AD (1988) Parasitoid density dependence and the dynamics of host-parasitoid models. *Am Nat* **132**:417–436

Taylor AD (1993a) Aggregation, competition and host-parasitoid dynamics: Stability conditions don't tell it all. *Am Nat* **141**:501–506

Taylor AD (1993b) Heterogeneity in host-parasitoid dynamics: the 'CV2 > 1 rule'. *Trends Ecol & Evol* **8**:400–405

Timberlake PH (1916) Note of an interesting case of two generations of a parasite reared from the same individual host. *Can Entomol* **48**:89–91

Todd JW (1989) Ecology and behavior of *Nezara viridula*. *Annu Rev Entomol* **34**:273–292

Comparative transmission dynamics of two insect pathogens

Robert J. Knell, Michael Begon, David J. Thompson

ABSTRACT

The transmission coefficients of two insect pathogens, *Plodia interpunctella* Granulosis Virus and *Bacillus thuringiensis* were measured using third and fourth instar *Plodia* larvae as hosts. Estimates were higher for the virus than the bacterium, as predicted from laboratory studies of the two host-pathogen systems. In both cases fourth instar larvae gave higher estimates than third instars. This was expected for the bacterium from data on numbers infected in each age class in population experiments but was the opposite of the trend expected for the virus.

Key words: population dynamics, *Bacillus thuringiensis*, granulosis virus, *Plodia interpunctella*

Frontiers of Population Ecology, R.B. Floyd, A.W. Sheppard and P.J. De Barro (eds), CSIRO Publishing, Melbourne, 1996 pp. 269–276

Introduction

The revolution in ecologists' appreciation of the potential importance of pathogens as natural enemies of insects has been mostly driven by theoretical studies (Anderson and May 1981; Holt and Pickering 1985; Hochberg 1989; Bowers and Begon 1990; Begon *et al.* 1992; Bowers *et al.* 1993; Briggs and Godfray 1995) and empirical data on the ecology of host pathogen systems are rare. Some studies have recently been published in which parameters are estimated or short-term, within generation dynamics predicted (Hochberg and Waage 1991; Dwyer 1992; Dwyer and Elkington 1993; Goulson *et al.* 1995).

In this study we discuss how the patterns observed in the long term dynamics and age structures of two simple host pathogen systems might be related to the transmission coefficients of the pathogens, one of the most important parameters used in modelling host-pathogen systems.

The transmission coefficient is a parameter found in almost all epidemiological models. It describes the likelihood of a new infection arising from a contact between an uninfected host and an infectious host, cadaver or free living infectious stage. Although this parameter may not directly effect the equilibrium prevalence of a pathogen (Anderson and May 1981), it is implicated in many other facets of host-pathogen population dynamics. It can for example influence the host threshold density such that the higher the transmission coefficient the lower the threshold density.

We briefly describe the long-term population dynamics of a laboratory host-pathogen system with *Plodia interpunctella* (Hübner) (Lepidoptera: Pyralidae), the Indian meal moth as the host and either the bacterial pathogen *Bacillus thuringiensis* (*Bt*) (*Bacillus thuringiensis* var. Kurstaki strain HD-1 (originally obtained from Dipel®) or *P. interpunctella* granulosis virus (*Pi*GV) or both as the pathogens. This is then related to measured values for the transmission coefficient in a qualitative way.

Population Experiments

A series of laboratory experiments were carried out in which populations of *P. interpunctella* were maintained in the presence of either the *Pi*GV, *Bt*, or both of these pathogens for a year (approximately twelve generations) and monitored once a week. Counts were obtained of dead adult moths in the whole population cage and of larvae infected with the pathogens in a section of one sixth of the total culture medium as it was removed for replacement.

In five out of six experimental populations in which *Bt* was present it became extinct within 25 weeks, and once reintroduced at a higher density underwent a steady decline in prevalence until the experiments were terminated. In all those experimental populations in which *Pi*GV was introduced (four in total) the virus persisted, increasing in prevalence throughout the course of the experiment (Fig. 1). These dynamics have been described before in the case of *Pi*GV (Sait *et al.* 1994b).

During monitoring, the instar of each infected larva found was recorded. As it was impractical to record the total number of healthy larvae present it is not possible to comment on the overall prevalence of the two pathogens. The same number of susceptible larvae were available to both pathogens in the cages where both were present, however, so it is possible to compare the relative prevalances of infected individuals found in each age class, and to draw qualitative conclusions regarding the relative age structures of the two infected populations. The age structures of the populations of infected individuals were different, in that *Pi*GV infected more early than late instar larvae, and *Bt* infected more late instar larvae (Fig. 2).

The two pathogens therefore display quite different dynamics under comparable experimental conditions, both in terms of overall persistence and prevalence and in terms of age structures. It was felt that an estimate of the transmission coefficient for each of the pathogens might provide at least

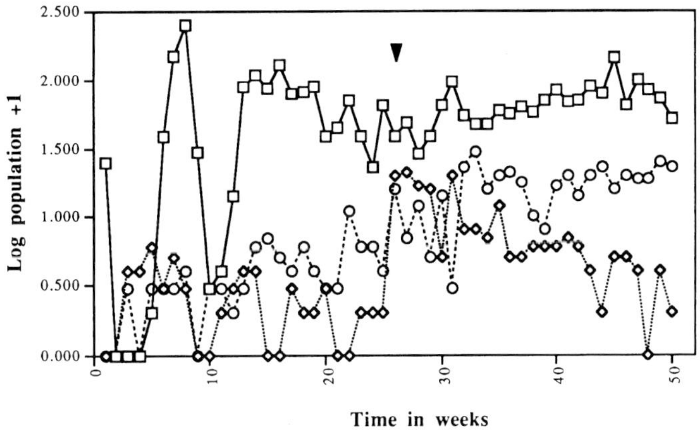

Fig. 1. Population trajectories for a laboratory experiment containing *Plodia interpunctella* (squares, counts of dead adults), larvae infected with *Bacillus thuringiensis* (diamonds) and with *Plodia* Granulosis virus (circles). The arrow indicates where *Bt* was reintroduced having become extinct. Experiments in which only one of the two pathogens were present gave similar results, and the data from a three species experiment is only used here for convenience.

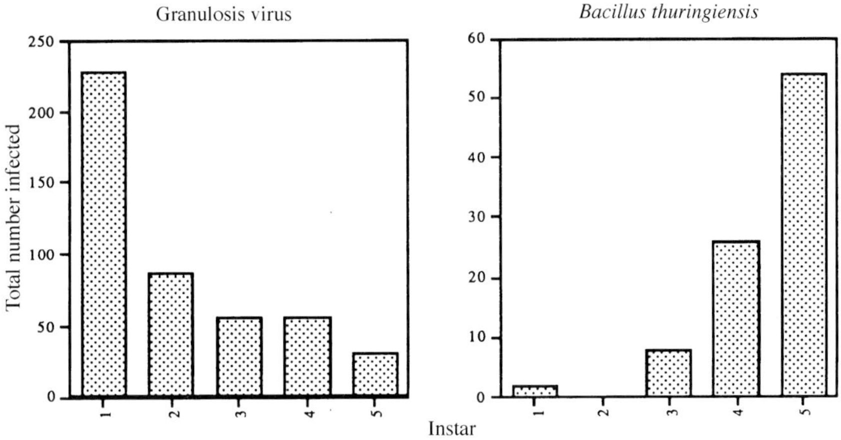

Fig. 2. Total numbers of each larval instar infected by either *Plodia granulosis* virus or *Bacillus thuringiensis* during the course of the population experiment depicted in Fig. 1. Experiments which only contained a single pathogen gave similar results.

a partial explanation for these observed patterns; as *Bt* did not appear to persist stably in the population experiments and *Pi*GV did we might predict that the virus would have a higher transmission coefficient than the bacterium. The higher relative prevalence of *Pi*GV would again lead us to predict a higher transmission coefficient for the virus.

Differences in the transmission coefficient between larval instars might also account for the observed age structures of infected populations: if late instar *P. interpunctella* larvae gave higher values for the transmission coefficient than early instar larvae when exposed to *Bt*, then this would mean that late instar larvae were more likely to contract the infection.

Obtaining an estimate for the transmission coefficient

An experiment was carried out with third and fourth instar *P. interpunctella* larvae, as second and first instar larvae are difficult to handle and fifth instars are highly resistant to the virus (Sait *et al.* 1994a). Ten randomly selected larvae were added to a 50 ml screw top glass jar containing either 25 or 50 mg of culture medium (10:1:1 wheat-bran: brewers' yeast: glycerol) depending on the age of the larvae being used, these amounts being chosen because in both cases they represent the weight of ten larvae, so each larva had its own body weight in food available. These were then maintained for 36 hours to avoid any effects from sorting and transferring the larvae. Numbers remaining in each jar were checked and any animals which had died were replaced with larvae from similar jars randomly included with the experimental containers. Two infected cadavers of the same instar as the susceptible larvae, between 48 and 36 hours after death, were then added to each jar. The host larvae were removed 1,2,3 or 4 hours afterwards, placed in individual cells of a 5 × 5 divided Petri dish with excess food and monitored 7 days later for infection. The experiment was replicated nine times.

An estimate of the transmission coefficient was then obtained using a modification of a method first suggested by Dwyer (1991), and also used by Dwyer and Elkinton (1993) and Goulson *et al.* (1995). The method relies on the assumptions that there is no secondary transmission of the pathogen during the course of the experiment and no mortality due to factors other than the disease in the host population or reproduction. Our experiment was designed so that there was no secondary transmission and no reproduction, the period of time allowed for transmission to take place being far less than the period of time taken for a larva to become infectious once infected or to reproduce. Non-disease mortality during the course of the experiment was negligible (less than 3%). The model also assumes that the transmission coefficient is constant at all host and pathogen densities, and only considers lethal infections.

Under these conditions, we can use a reduced within generation model of transmission which gives us a rate of change for the susceptible host population due to the infection of

$$\frac{ds}{dt} = -\beta' SI, \tag{1}$$

where S=host density, I=density of infectious individuals and β' = the transmission coefficient.

Integrating this between time 0 and t gives us

$$\frac{\log S_t}{\log S_0} = \beta' It. \tag{2}$$

This expression can be used to give an estimate for the transmission coefficient, assuming that the amount of infectious material remains constant. However, during these experiments the amount of infectious material varied as the infectious cadavers were cannibalised. Estimates of β' were therefore obtained by linear regression of the estimate of the transmission coefficient at each time point against the time allowed for the experiment and a value obtained for the Y-intercept, the point at which time was zero and where the density of infectious individuals was known to be two. The data were analysed using analysis of covariance, with instar and pathogen as factors and time as the covariate.

Results

Figure 3 shows the estimates of the transmission coefficient obtained for the two pathogens at the two different ages, and Table 1 gives the results of the analysis. Both pathogen and instar had highly significant effects, with *Pi*GV giving higher values than *Bt*, and third instar larvae having lower values than fourth instars in both cases. That the values obtained for *Pi*GV are higher than those obtained for *Bt* is predictable from the overall dynamics of the two pathogens noted above. In the case of *Bt*, the relative values obtained for the two different instars correspond to those predicted from the age-structure data (see Fig. 2), with the fourth instar larvae giving a higher estimate of the transmission coefficient than the third instar larvae. This pattern is repeated with *Pi*GV although the

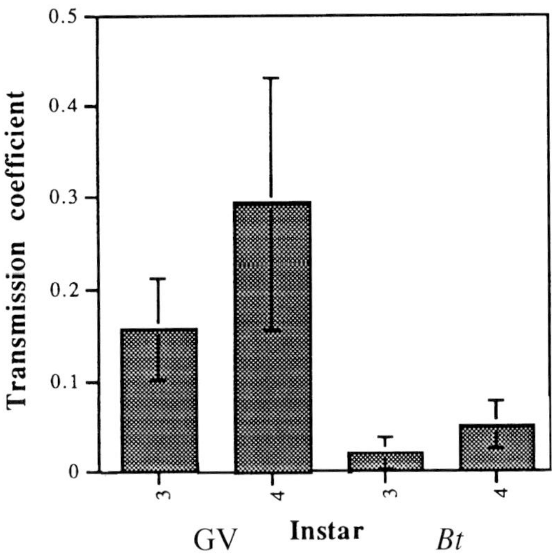

Fig. 3. Transmission coefficients for *Plodia interpunctella* granulosis virus and *Bacillus thuringiensis* infecting third and fourth instar *Plodia* larvae. Error bars represent 95% confidence limits.

Table 1. ANCOVA table for instar and pathogen, with time as covariate. Test for homogeneity of slopes was non-significant ($F = 1.122$, $p = 0.305$).

Source	Analysis of covariance: Sum of squares	*df*	Mean square	*F*	*p*
Instar	0.120	1	0.120	13.425	<0.001
Pathogen	0.618	1	0.618	69.436	<0.001
Instar x pathogen	0.054	1	0.054	6.089	0.015
Time	0.065	1	0.065	7.247	0.008
Error	1.237	139	0.009		

prediction from the age structure data would be that a lower value should be obtained for the older larvae.

The interaction term between pathogen and instar is also significant, indicating that the degree of difference between the results for third and fourth instar larvae is influenced by the pathogen.

Discussion

This is the first time that an attempt has been made to relate parameter estimations for entomopathogens to their long term population dynamics. The differences between our estimates for the transmission coefficients for the two pathogens reflect the observed long-term behaviour of the two pathogens, suggesting that estimates of the transmission coefficient might prove useful in predicting behaviour of real host-pathogen systems.

The conditions under which this experiment were carried out are somewhat different to those found in the long term cultures. To obtain an estimate of the transmission coefficient under more realistic conditions would be very difficult, and as the differences which are being investigated are more qualitative than quantitative, it was felt that the conditions were close enough to allow a comparison.

In the matter of food availability for example, while the amount given during the course of the experiment was fairly restricted, the populations of *Plodia* were themselves food limited, so it would be expected that larvae within the population experiments would themselves experience restricted food availability.

A biological explanation for the difference between the two pathogens may lie in the anti-feeding effect caused by *Bt* toxins (Entwistle *et al.* 1993). In these experiments both *Bt* (Burges and Hurst 1977) and *Pi*GV (personal observation) were largely transmitted by cannibalism. While larvae will nibble *Bt*-infected cadavers, they will not eat much of them. Transmission of the bacterium will depend on a larva eating sufficient toxin to kill it, allowing the *Bt* to reproduce in the cadaver, before the anti-feeding effects begin. Virus-infected cadavers, on the other hand, were rapidly consumed, and it is likely that the high densities of virus particles found in freshly killed victims will be sufficient to infect most individuals that eat even a small amount of the cadaver.

Our population data led to the predictions that transmission coefficients for *Pi*GV should be higher in earlier instar larvae, and that they should be higher in later instar larvae in the case of *Bt*. While this was the case with *Bt*, we found that the results for *Pi*GV were the opposite of what was expected. Why this should be is not known, although an explanation may lie with some process happening in transmission of the virus in population experiments which is not occurring in our transmission experiments, for example release of small quantities of virus into the environment from infected cadavers, by leakage of haemolymph or by physical breakdown of the cadaver. If virus particles were present at low densities this might lead to earlier instar larvae acquiring infection via this route, with later instar larvae being protected by increased resistance, but still acquiring infection through cannibalism of infectious cadavers, which would lead to very high doses of virus particles. The discrepancy between the experimental measurements of the transmission coefficient and the observed relative age structures may also be related to the fact that the infectious cadavers used were of different ages according to the age of susceptible used. This makes it difficult to make a direct comparison between the different ages of susceptible larvae used, and further experimentation is needed to separate effects arising from differences between susceptible larvae of different age classes and differences between infectious cadavers of different age classes.

Age related transmission of entomopathogens has been the subject of some experimental interest recently. While it is usual for theoretical studies to assume, for the sake of mathematical tractability, that likelihood of transmission does not vary with age, this is unlikely to be the case. Age related susceptibility to infection is often measured by obtaining dose-response curves followed by estimation of the LD50 for the pathogen for each larval instar. This invariably shows that resistance to the pathogen increases with age, often considerably (Sait *et al.* 1994a). These results imply that younger larvae would be more likely to contract infection, but there are behavioural changes associated with larval age which can increase the likelihood of transmission in later instars, including feeding rate (Goulson *et al.* 1995), increased larval movement (Dwyer 1991) and increased aggression and likelihood of cannibalism. It is currently unclear whether these changes are likely to 1) compensate for increased resistance with age, thus leading to no change in the probability of infection with instar (Payne *et al.* 1981); 2) be insufficient to compensate such that the probability of transmission decreased with larval instar (Evans 1981) or 3) overcompensate such that probability of transmission increased with larval instar (Dwyer 1991).

There are two published field studies which have investigated the relationship between larval age and transmission using infected animals as the inoculum (Dwyer 1991; Goulson *et al.* 1995), but whereas Dwyer found that fifth instar *Orgyia pseudotsuga* (Douglas-fir tussock moth) were considerably more likely to become infected with a Nuclear Polyhedrosis Virus (NPV) than third instar larvae, Goulson *et al.* found no difference in the likelihood of transmission of *Mamestra brassicae* NPV with age in the field, although laboratory measurements of age-related feeding rate and resistance suggested that transmission rates should increase in older larvae. Using a GV applied as a spray it was found that mid-instar *Pieris rapae* larvae were more likely to become infected (Webb and Shelton 1990). It would seem that generalisation may be difficult.

To summarise, we have obtained the first direct comparison between the transmission coefficient and long term population dynamics of two pathogens under otherwise similar circumstances. The relative values obtained for the transmission coefficient corresponded to those predicted from the population dynamics data, suggesting that this may be a useful parameter for predicting the long-term behaviour of other host-pathogen systems. We have also demonstrated age-dependent differences in transmission coefficient in both pathogens, although the relationship between these values and observed age structures of infected populations is not so clear.

Acknowledgments

Thanks to Tom Heyes and Rob Swain for technical assistance and to Roger Bowers for helpful discussion. RJK was supported by an NERC Research Studentship (No. GT4/92/199/L).

References

Anderson RM and May RM (1981) The population dynamics of microparasites and their invertebrate hosts. *Philos Trans R Soc Lond B Biol Sci* **291**:451–524

Begon M, Bowers RG, Kadianakis N and Hodgkinson DE (1992) Disease and community structure: the importance of host self-regulation in a host-host-pathogen model. *Am Nat* **139**:1131–1150

Bowers RG and Begon M (1990) A host-host-pathogen model with free-living infective stages, applicable to microbial pest control. *J Theor Biol* **148**:305–329

Bowers RG, Begon M and Hodgkinson DE (1993) Host-pathogen population cycles in forest insects? Lessons from simple models reconsidered. *Oikos* **67**:529–538

Briggs CJ and Godfray HCJ (1995) The dynamics of insect-pathogen interactions in seasonal environments. *Am Nat* **145**:855–887.

Burges HD and Hurst JA (1977) Ecology of *Bacillus thuringiensis* in storage moths. *J Invertebr Pathol* **30**:131–139

Dwyer G (1991) The roles of density, stage and patchiness in the transmission of an insect virus. *Ecology* **72**:559–574

Dwyer G (1992) On the spatial spread of insect pathogens: theory and experiment. *Ecology* **73**:479–494

Dwyer G and Elkington JS (1993) Using simple models to predict virus epizootics in gypsy moth populations. *J Anim Ecol* **62**:1–11

Entwistle PF, Cory JS, Bailey MJ and Higgs S (1993). *Bacillus thuringiensis, an environmental biopesticide: Theory and practice.* John Wiley and Sons.

Evans HF (1981) Quantitative assessment of the relationships between dosage and response of the nuclear polyhedrosis virus of *Mamestra brassicae*. *J Invertebr Pathol* **37**:101–109

Goulson D, Hails RS, Williams T, Hirst ML, Vasconcelos SD, Green BM, Carty TM and Cory JS (1995) Transmission dynamics of a virus in a stage- structured insect population. *Ecology* **76**:392–401

Hochberg ME (1989) The potential role of pathogens in pest control. *Nature* **337**:262–265

Hochberg ME and Waage JK (1991) A model for the biological control of *Oryctes rhinoceros* (Coleoptera: Scarabaeidae) by means of pathogens. *J Appl Ecol* **28**:514–531

Holt RD and Pickering J (1985) Infectious disease and species coexistence: a model of Lotka-Volterra form. *Am Nat* **126**:196–211

Payne CC, Tatchell GM and Williams CF (1981) The comparative susceptibilities of *Pieris brassicae* and *P. rapae* to a granulosis virus from *P. brassicae*. *J Invertebr Pathol* **38**:273–280

Sait SM, Begon M and Thompson DJ (1994a) The influence of larval age on the response of *Plodia interpunctella* to a granulosis virus. *J Invertebr Pathol* 63:107–110

Sait SM, Begon M and Thompson DJ (1994b) Long-term population dynamics of the Indian meal moth *Plodia interpunctella* and its granulosis virus. *J Anim Ecol* 63:861–870

Webb SE and Shelton AM (1990) Effect of age structure on the outcome of viral epizootics in field populations of imported cabbageworm (Lepidoptera: Pieridae). *Environ Entomol* 19:111–116

Population regulation in insects used to control thistles: can this predict effectiveness?

Andy W. Sheppard and Tim Woodburn

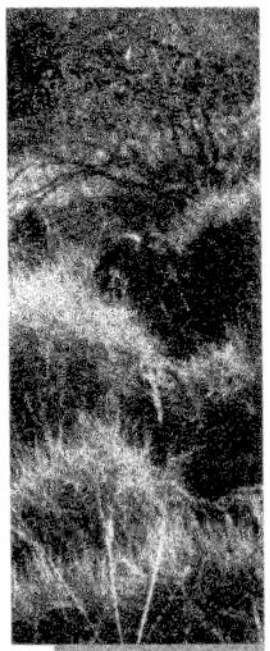

Abstract

It is largely accepted that biological control agents are more effective when freed of their natural enemies. This implies that natural enemies play a key role in the population dynamics of agents in their native range, and therefore release from this mortality will lead to a rapid increase in population growth.

We review and present data on the population dynamics of insects in the flower-heads of thistles and knapweeds. Specifically, we discuss population regulation of two biological control agents released against the thistle *Carduus nutans*, namely a weevil *Rhinocyllus conicus*, and a gall fly *Urophora solstitialis*, in both their native and new regions. The weevil, a successful control agent against this weed in some countries, suffers from a microsporidial parasite in the native range, which reduces fecundity and survivorship. Evidence of spatial density dependent egg parasitism was found in the native range, but this mortality appeared to be replaced by intense larval competition in the new environment and nullified any potential increase in destructiveness. The gall-fly suffers 30 to 90% larval parasitism in the native range, but higher adult mortality due to host plant asynchrony in the new environment has so far blocked any advantage this agent might get in the absence of its natural enemies. While insects in thistle heads appear to be largely regulated by their resources rather than their natural enemies where native, some may have the potential to control their host due to contrasting plant abundance between regions.

Key words: Carduus nutans, biological control, *Urophora, Rhinocyllus*, predator release

Frontiers of Population Ecology, R.B. Floyd, A.W. Sheppard and P.J. De Barro (eds), CSIRO Publishing, Melbourne 1996, pp. 277–290

Introduction

In this paper we discuss a link between Nicholson and Bailey's (1935) theory and biological control dependent on the importance of natural enemy regulation in the population dynamics of herbivorous insects used as control agents of weeds. We review evidence for this in a historically important group in this debate: insects in the flower-heads of thistles and knapweeds (Cardueae, Asteraceae). Studies within this group bring together theoretical and empirical facets of population ecology and biological control. We consider the relevance of their potential to escape such regulation when released as biological control agents. We also reanalyse and present data on the population dynamics of two specialist insect herbivores in flower heads where native and released for the biological control of the weedy thistle *Carduus nutans*. The aim is to investigate whether i) specialist insects in the flower-heads of Cardueae are regulated by natural enemies and ii) whether escape from natural enemies makes these species more effective biological control agents.

Parasitism of biological control agents: Historical perspective

Biological control and Nicholson are inseparably linked by the premise that natural enemies have the potential to regulate their prey populations (cf. Milne 1957). Classical biological control of alien pests operates either by resource-limited agents exploiting their exotically abundant host, or by natural-enemy regulated agents escaping these natural enemies, to bring about density dependent pest control. This escape is largely accepted as a mechanism by which agents can become more effective and is also the most direct implication of Nicholson and Bailey's (1935) theory for biological control. From the start, Nicholson (1933, p. 167) understood the consequence of this escape mechanism for the biological control of insect pests, considering that, while the role of hyperparasites in biological control failure is probably overemphasised: "special natural enemies may prevent primary predators from controlling the densities of their prey".

This basic prediction, and ensuing theoretical debate about the required level of specificity of natural enemies, has not changed much since Nicholson's day (Hassell and Waage 1984; Hassell and May 1986). Hyperparasites are often considered the cause of biological control failure of insect pests (20.3% of cases, Stiling 1993), but only a few early experimental studies (Simmonds 1948; DeBach 1949) and occasional circumstantial evidence (Hassell 1980) suggest that escape from them is actually important. In biological control of weeds the potential negative affect of at least indigenous natural enemies on phytophagous agents is usually considered to be more acute.

> "*...Insect enemies of plants appear generally to be reduced by their own enemies to the threshold level for these, which commonly means that their effect upon plants is not very significant...*" (Nicholson 1954, p 46).

This was later championed by Hairston *et al.* (1960) who argued from "general and widely accepted observations" that herbivores are more likely to be regulated by predators than the predators themselves. Early successes in the biological control of insect herbivore pests also clearly influenced Nicholson and contemporary biocontrol workers (e.g. Thompson 1929; Smith 1935) in their proposals of population regulation. In the biological control of weeds literature, anecdotal evidence for natural enemy interference *following* release is common (41% of cases, Crawley 1986), although we know of only three cases in which there was experimental evidence (Goeden and Louda 1976; Briese 1986). No experimental and few circumstantial studies exist to support a benefit to agents of escape from their *indigenous* natural enemies. Nonetheless we argue that the success of agents in the biological control of weeds may be more dependent on 'escape from natural enemies' as a control mechanism than in biological control of insects (see Biological control of Cardueae).

Gall-formers in Cardueae

Insects in the flower-heads of Cardueae became central to early debate surrounding the Nicholson and Bailey (1935) model. Varley (1947) used the knapweed gall-fly, *Urophora jaceana,* for its first empirical field test. In a life table analysis of the fly at one site over two and a half generations, 56–71% of the larvae were killed by five parasitoids of which the specific endoparasite, *Eurytoma tibialis,* was the most damaging. He concluded that host availability controlled the fecundity of *E. tibialis,* and to a lesser extent *Habrocytus trypetae,* causing them to act as delayed density dependent factors and that:

> "...*Nicholson and Bailey's theory supplies for the first time an analysis of the mutual effect of parasitic and other factors of destruction on the population density of an insect...*" (Varley 1947).

This insect herbivore appeared to provide the first evidence for regulation by natural enemies. This status did not go uncriticised (Kingsland this volume). Milne (1957) forcefully argued that Varley neither tested Nicholson's theory nor solved the problem of what controls the gall-fly. Varley's study did, however, provide a link between theoretical population biology and the practical application of ecological concepts in the practice of biological weed control (Varley and Gradwell 1971; Varley *et al.* 1973). The Nicholson and Bailey model (1935) became the forerunner to a theoretical frame work for biological control (Hassell and Varley 1969; Beddington *et al.* 1975; Murdoch *et al.* 1985; May and Hassell 1987). Insects in the flower heads of Cardueae continue to be used to explore regulation in insect herbivore population dynamics (Romstöck-Völkl 1990; Straw 1991; Redfern *et al.* 1992; Dempster *et al.* 1995a).

Biological control of Cardueae

Many knapweeds and thistles are serious economic weeds outside their native range. Their associated herbivorous insects, including several *Urophora* species, have become potential and actual biological control agents for use against them (Julien 1992). Such biological control agents are usually helped by a greater pest abundance outside their native range. *C. nutans,* for example, is considered less of weed in its native range than in the other four continents it occupies outside its native range (Sheppard *et al.* 1994). In the native range, potential phytophagous agents are more likely to be regulated more often by resource density (i.e. 'bottom-up' processes) than by their natural enemies (i.e. 'top down' processes, see Cappuccino and Harrison this volume), if units of food resource (e.g. host-plants, flower-heads etc.) are at low density. This is because such specialist natural enemies are unlikely to be able to locate their specialist phytophagous hosts more readily than the phytophages can locate their food source. High resource density provided by the weed in the new environment allows resource regulated agents to exploit their abundant resources more effectively (the functional response).

The majority of alien weeds in the Cardueae, e.g. some *Cirsium* and *Centaurea* species are considered to have equal weed status within and outside their native ranges (Holm *et al.* 1979; Sheppard in press). Under such circumstances, where weeds are as abundant locally in their native ranges and where alien, escape from natural enemies becomes more relevant to biological control because it becomes the principle mechanism by which control will be achieved. It therefore becomes important to locate agents that are *regulated* by the natural enemies in the native range. An agent regulated by resources in its native range, in this instance, is perhaps less likely to prove more effective in the new environment if resource density is similar, even if introduction frees the agent population from some degree of natural enemy suppression. The remainder of this paper considers capitulum-feeding agents attacking *C. nutans.*

Insect mortality in *Carduus nutans* capitula

In all Cardueae, stage-specific mortalities of capitulum-feeding insects can be conveniently measured by collecting and dissecting mature capitula (cf. Varley 1947). These mortalities were estimated for the weevil *Rhinocyllus conicus* and the tephritid fly *Urophora solstitialis* in *C. nutans* capitula collected from the native range (southern France), and from Australia where both have been introduced as biological control agents. *Tephritis hyoscyami,* another specialist tephritid (not a gall-former) is also common in *C. nutans* capitula in France and was included in the analysis of mortality, when present, because it shares at least five common parasitoids with *U. solstitialis.*

French *C. nutans* capitula were collected from a 900 m^2 area of one pasture (site 2 described in Sheppard *et al.* 1994) in 1986 (tephritids only) and 1987, and four similarly managed pastures nearby in 1988 (2 sites) and 1989 (2 sites for tephritids only) around La Cavalerie (Aveyron, S. France). All capitula were collected (at least 300 per site and year) from 18–22 flowering plants per site in randomly placed 1 m^2 quadrats. As *C. nutans* capitula fall soon after maturity, *U. solstitialis* galls found to contain live larvae were stored at ground-level in the field until adults emerged. In Australia, capitula collections were made from 6 release sites for *R. conicus* and 5 release sites for *U. solstitialis,* the northern, central and southern highlands of New South Wales from 1991–1995. Fifty capitula were collected at random throughout each of these sites every two weeks during the flowering season.

Key factor analysis was applied to all mortality data obtained to compare results between native and new regions. For *R. conicus,* we also included all similarly detailed published data, i.e. two sites in both Europe (Zwölfer and Harris 1984) and the USA (Smith *et al.* 1984). Our data were also analysed, where possible, to look for evidence of spatial density dependence by regressing proportional mortality of each factor against initial insect density per capitulum, per plant, and per m^2 both within and between sites.

Rhinocyllus conicus, *the nodding thistle receptacle weevil*

This weevil, introduced from France into Canada in 1968, reduced seed production by half and led to the eventual control of the weed in Canada (Harris 1984) and parts of the USA (Kok and Surles 1975). Similar releases have failed to provide any weed control in either Australia (Woodburn and Cullen 1993, in press) or New Zealand (Kelly and McCallum, 1995).

R. conicus is univoltine throughout most of its distribution and has a laboratory measured potential fecundity of over 200 eggs per female. However, under summer field conditions in Australia only 79 ± 14 (s.e.) eggs were laid per female in faecal cases on the developing involucral bracts. The oviposition period is short, mostly covering only the first peak of capitulum production (Woodburn and Cullen 1993). Egg to adult is passed on or in the receptacle of the same capitulum during a 6 week period and the adult overwinters. Overwinter survival of caged adults has been estimated as 52 ± 6% in North America (Kok 1976).

During studies for the release of *R. conicus* in Australia, certain southern French populations were found to have a high incidence of a microsporidial parasite (*Nosema* sp.) passed between generations. This disease was not recorded in the insects introduced elsewhere. No data are available of incidence in relation to weevil density in the field, however such diseases are known to be most damaging in dense host populations (Weiser *et al.* 1976). Caged pairs of diseased adults from a natural population were compared with disease free adults from the same population achieved by selective mating. Disease-free adults laid five times more eggs and survived twice as long after the start of oviposition (Fig. 1). This disease, although not ubiquitous could play a role in the dynamics of this insect (cf. Begon *et al.* this volume).

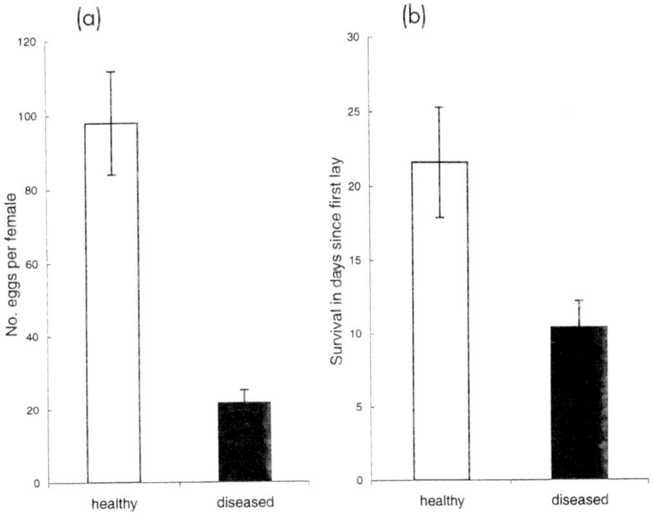

Fig. 1. Effect of infection by the microsporidian *Nosema* sp. on (a) fecundity and (b) survival of adult female *Rhinocyllus conicus* under field conditions (±s.e.)

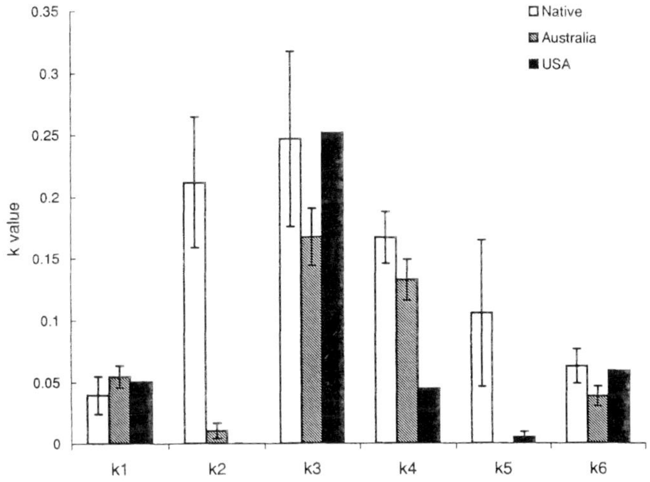

Fig. 2. Summary of key factor analyses (*k*-values ±s.e.) for *Rhinocyllus conicus* egg to adult life-tables from five sites in the native region, six sites in Australia and one site in the USA including all published data: *k*-values; k_1, infertile eggs; k_2, egg parasitism; k_3, 1st-3rd instar death; k_4, 3rd instar to pupal death; k_5, larval parasitism, and k_6, pupal to adult death.

Key factor analysis of all available egg to adult survival data for this weevil are summarised in Fig. 2. This shows a comparison of *k*-values from five sites in the native region (including two sites from Zwölfer and Harris 1984), six sites in Australia and two sites in the USA (Smith *et al.* 1984). Larval death from unknown causes (k_3+k_4) was the key mortality factor (*sensu* Varley *et al.* 1973), including some interspecific competition in the native range (Zwölfer and Harris 1984). Egg parasitism by *Pterandrophysalis levantina* (Hym., Trichogrammidae) (k_2) was the next most important mortality (19–59%) in the native range. Zwölfer and Harris (1984) also recorded some predation by anthocorid bugs which they included here. Larvae are parasitised by a complex of 11 larval parasites and eaten by certain oligophagous moth larvae in the native range, and this mortality (k_5) ranged from 6–43% (cf. Zwölfer and Harris 1984). Outside the native range, egg parasitism has only been recorded in Australia, where attack by a native *Anaphes* sp. (Hym., Mymaridae), ranges from 0–19%, and is increasing. Conversely, larval parasitism has only been recorded in the USA, where a complex of 12 species cause 'unimportant' mortality (Zwölfer and Harris 1984; Wilson and Andres 1986).

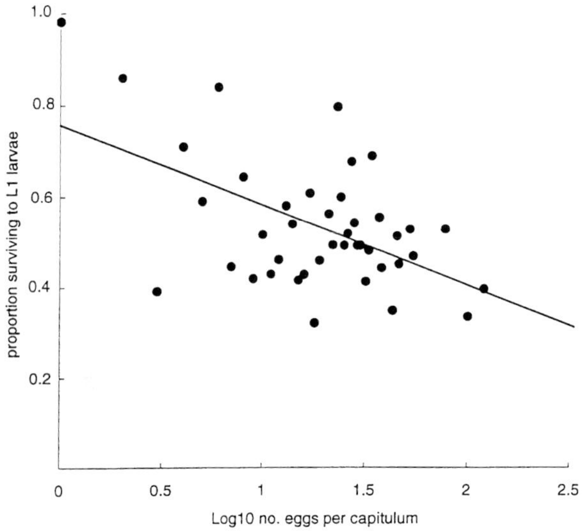

Fig. 3. Proportion of *Rhinocyllus conicus* eggs surviving parasitism by *Pterandrophysalis levantina* (arcsine transformed) relative to the number laid per capitulum at site 1 in southern France in 1987; $y = -0.28x + 1.12$, $r^2 = 0.29$, $p = 0.0002$.

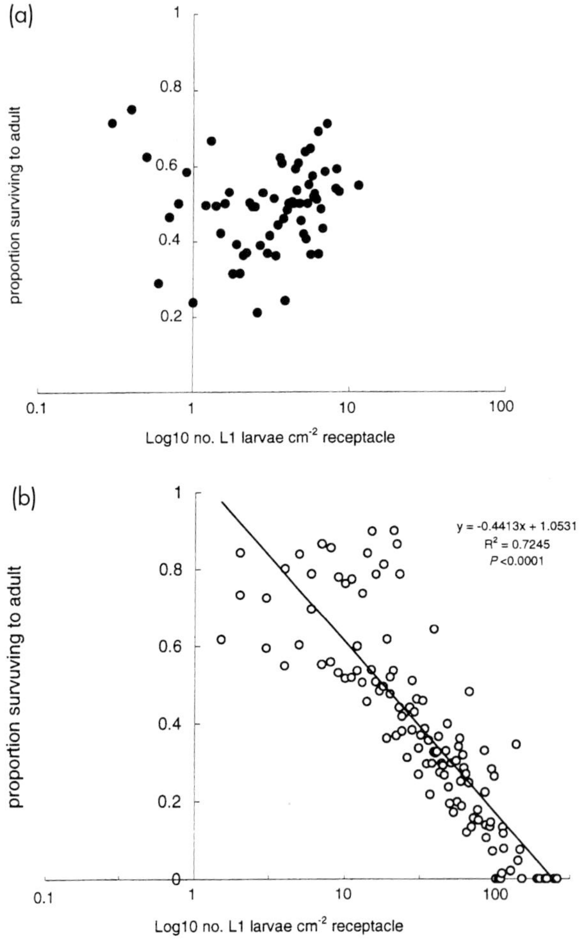

Fig. 4. Proportion of *Rhinocyllus conicus* surviving to adult (arcsine transformed) relative to the density of 1st instar larvae cm^{-2} of thistle receptacle for (a) three sites in the native region (756 capitula) and (b) six sites in Australia (2400 capitula), each point being a mean of all capitula at each larval density.

In the analysis of these mortality factors for spatial density dependence from our sites in the native range, only egg parasitism increased with the number of eggs per capitulum and only at one site (Fig. 3). Overall, egg parasitism was 59% at this site compared to 26% and 34% at the other two sites. Larval death (k_3+k_4, Fig. 4a) and larval parasitism (k_5) were unrelated to first instar larval density per capitulum at our sites, although Zwölfer (1973) found larval parasitism "on average higher where *R. conicus* aggregations were high". Following release in Canada and Australia populations of weevil larvae increased 6 to 18 times their density per capitulum in Europe (Zwölfer and Harris 1984; Woodburn and Cullen 1993). *R. conicus* aggregates its eggs on capitula in the first part of the season and, at these high densities, very often more eggs are laid than the capitulum can support (superparasitism). In the absence of egg parasitism, high larval density per unit area of capitulum results in density dependent larval mortality from first contest then scramble competition with increasing degrees of superparasitism (Fig. 4b). No larvae survive in capitula with more than 100 first instars per cm^2 of capitulum.

Urophora solstitialis, *the nodding thistle gall-fly*

The biology of this gall-fly is typical of its genus (Varley 1947; Möller-Joop 1989), except for a partial second generation in summer. The proportion of larvae that pupate and undergo a second generation decreases dramatically after the longest day of the year (Woodburn 1993). *U. solstitialis* produces multilocular galls containing up to 120 larvae, one per 'cell'. The hymenopterous parasitoid community attacking *Urophora* spp. in Europe is highly predictable (Zwölfer and Arnold-Rinehart 1993). *U. solstitialis* larvae are attacked by *E. tibialis, E robusta* and at least 15 other parasitoids (Vitou *et al.* 1995). Zwölfer (1973) recommended that this agent should follow *R. conicus* as a biological control agent, because it is a stronger direct competitor for capitular resources, and its attack is more aggregated.

Each female causes the formation of 100–200 gall cells estimated from paired flies in field cages and from field observations of fly density. A value of 100 gall cells initiated per female was used to calculate overwinter mortality between years at the one French site and three sites in Australia by comparing the density of gall cells that produced adult flies at the end of summer in the first year to the density of gall cells produced in the subsequent year. Overwinter mortality is a key factor in *Urophora* dynamics (Varley 1947). Estimates of 96% mortality at the French site and >99% at Australian sites combine losses due to predation or environment with losses due to adult dispersal away from the site. High losses were expected in Australia, where established populations are still expanding and where fly and host plant are not yet fully synchronised (see below). Figs. 5 and 6 show the *U. solstitialis* larva to adult mortality *k*-values comparing five sites in both the native and new regions (Fig. 5) and from *U. solstitialis* and *T. hyoscyami* data from one French site from 1986 to 1987 (Fig. 6). This mortality was extremely low in Australia where only 2.2% of larvae failed to develop to adult (Fig. 5).

Endoparasitism by *E. tibialis* (k_2) was the second most important source of mortality and varied from 18–50% across the five French sites. Percent parasitism by this or any other parasitoid did not increase with either the number of *U. solstitialis* larvae per gall or per capitulum, or the number of both hosts per capitulum, at any site. Parasitism was also unrelated to host density at any other spatial scale considered. Parasitism of some *Urophora* species by the less specialised ectoparasites decreases with increasing numbers of larvae per head, but this has not been found for *U. solstitialis* in *C. nutans* (Zwölfer and Arnold-Rinehart 1993).

Life table data from one French site sampled over two years was analysed in more detail to detect evidence of density dependent parasitism by *E. tibialis* between generations (Table 1, Fig. 6). For this analysis it was assumed that *U. solstitialis* was univoltine, which is justified by the late flowering of the thistles in the first year and mid season slashing by the farmer in the second. *T. hyoscyami* is

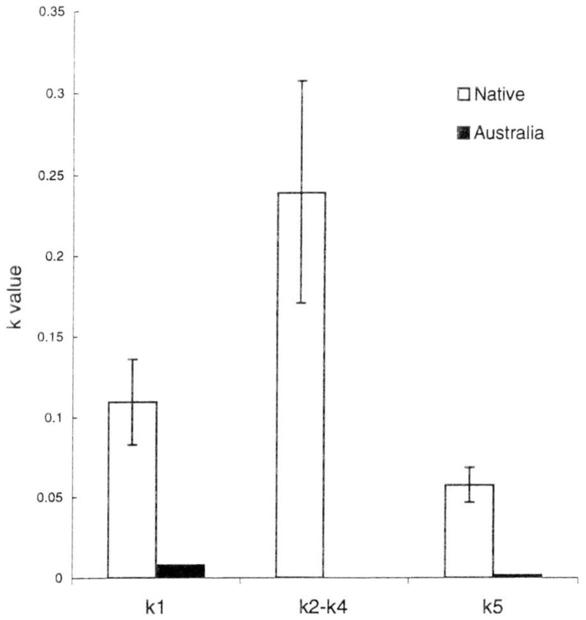

Fig. 5. Summary of key factor analyses (*k*-values ±s.e.) for *Urophora solstitialis* larvae-in-gall to adult life-tables from five sites in both the native region and in Australia, not including overwinter mortality: *k*-values; k_1, larval death (cause unknown); k_2–k_4, larval parasitism; and k_5, pupal to adult death.

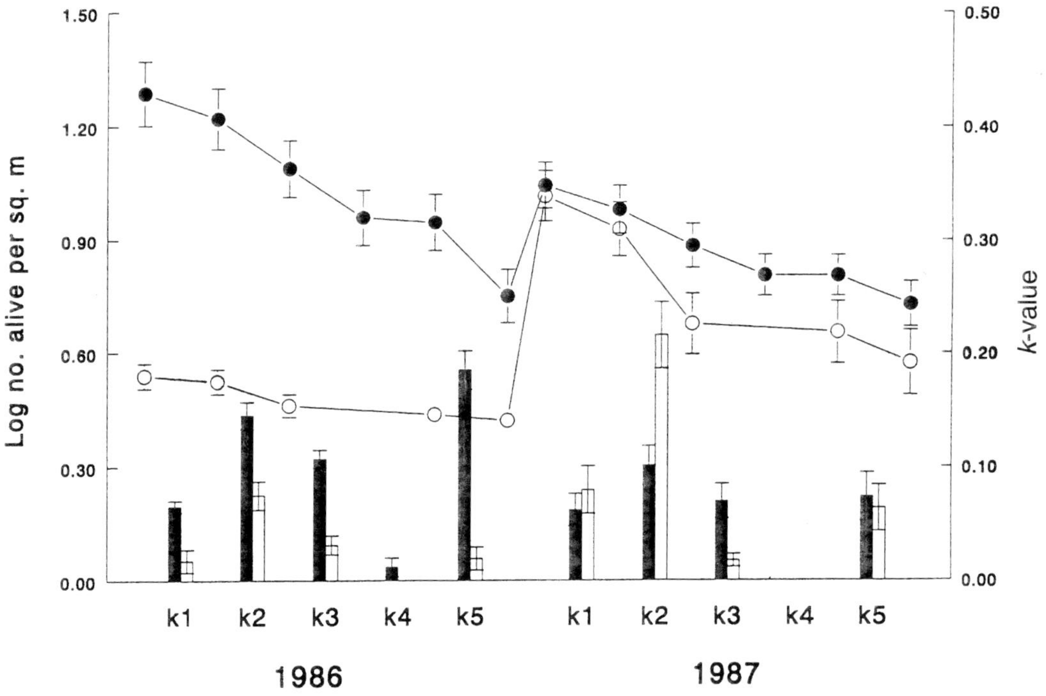

Fig. 6. Densities and *k*-values from key factor analysis (±S.E.) of mid-instar larvae to adult *Urophora solstitialis* (filled bars and symbols) and *Tephritis hyoscyami* (open bars and symbols) a site (site 2 in Sheppard *et al.* 1994) in southern France over two years: *k*-values: k_1, larval death (cause unknown); k_2, endoparasitism by *Eurytoma tibialis*; k_3, parasitism by *Aprostocetus* sp., *Eurytoma robusta* and other hymenopterous parasitoids; k_4, miscellaneous late parasitism; and k_s, pupal to aldult death.

Table 1. Tephritid abundance, attack rate and parasitism levels in *C. nutans* capitula at (a) site 1 in southern France from 1986 to 1987 and (b) at one site, Berridale, NSW Australia

	1986		1987	
(a)	*U. solstitialis*	*T. hyoscyami*	*U. solstitialis*	*T. hyoscyami*
No. larvae per attacked capitulum (±s.e.)	11.8±1.1	4.6±1.1	8.1±1.1	4.6±1.1
Percent of capitula struck by flies	56.0	20.4	8.0	13.7
Percent of fly-struck capitula parasitised by *E. tibialis*	86.5	68.0	68.0	59.1

	U. solstitialis		
(b)	1992/93	1993/94	1994/95
No. larvae per attacked capitulum (±s.e.)	20.5±1.0	17.9±1.0	11.4±1.0
Percent of capitula struck by flies	72.3	59.0	31.0

strictly univoltine, developing to adult (the overwintering stage) without a diapause. Capitulum density increased from 2.6 to 15.0 per m^2 between years, while *U. solstitialis* population density halved and *T. hyoscyami* density trebled (Fig. 6). The corresponding *k*-values of mortality due to *E. tibialis* (k_2) and failed emergence (k_5), between years, decreased for *U. solstitialis* and increased for *T. hyoscyami*. Mortality due to other parasitoids (k_3) changed little between years.

E. tibialis fecundity could not be estimated (cf. Varley 1947). However, the response of the same parasitoid population to two host distributions can be compared over two years to check the assumption that the rate of discovery determines the levels of parasitism observed (Varley 1947). The percent of capitula attacked by these two flies in both years and the number of host larvae per attacked capitulum, compared to the percent of tephritid-containing capitula attacked by *E. tibialis* are given Table 1. These data suggest that searching parasitoids appear to be able to locate *Urophora* galls more easily than capitula attacked by *Tephritis.* Also, as the rate of discovery by *E. tibialis* of capitula attacked by either fly did not differ markedly between years, most of the observed 'density-related' change in mortality between years was due to changing levels of parasitism within individual attacked capitula.

Regulation by natural enemies

In *R. conicus*, slight evidence of some spatial density dependent parasitism of eggs (Fig. 3, r^2=0.29) was observed and associated with high overall rates of parasitism in the native range. No data are available to test for density dependent parasitism between generations. Most native *C. nutans* populations encountered flowered for less than three consecutive years. This limits any temporal analyses of density dependence to so few generations that firm conclusions might not be possible (see below). Zwölfer (1994) proposed that *R. conicus* has a 'redistribution' type population structure to counter this low stability of flower-head resources, whereby several sub-populations from a number of local host plant patches leave the patches to overwinter only to redistribute themselves relative to host plant abundance the following spring. Some support for this comes from a strong positive correlation (r^2=0.83) between oviposition levels per capitulum and capitulum density between but not within thistle subpopulations (Sheppard *et al.* 1994). Such resource-driven density dependent adult dispersal could regulate weevil population dynamics in host-plant populations structured in this way (see Cappuccino and Harrison this volume). This hypothesis is being tested for similar weevils (see Sheppard *et al.* 1991).

Our between generation data set for *U. solstitialis* was short compared to some recent studies, but it was similar in size to that used by Varley (1947). Analyses of density dependence from temporal series have progressed considerably since Varley's time, and even six year time series can be considered not long enough to draw firm conclusions (see Redfern *et al.* 1992). Given the frequency of non-flowering years in these thistle populations, however, analyses over long time series would be atypical even if they were possible. Our data set did allow the comparison of two hosts attacked by a common parasite particularly as Nicholson argued that:

> "*...the effectiveness of a species parasite in controlling the density of a particular host species at a low value is increased if the parasite also attacks other species of hosts...*" (1933, p 156).

Parasitism by *E. tibialis* of each host did vary significantly with the change in host density between generations, however it does not appear to be rate of host discovery that drives the change in parasitism levels. A larger study in space and time is required to determine if the frequency of density dependent parasitism is sufficient to explain population persistence of *U. solstitialis.*

Population regulation has already been studied in eight other species of tephritid that attack the Cardueae, including four *Urophora* spp. Evidence has come out strongly against regulation by natural enemies. Evidence of parasitism being a function of host density between generations was found for two tephritid species (Varley 1947; Redfern *et al.* 1992). For nearly all species, including a reinvestigation of *U. jaceana* at Varley's original sites (Dempster *et al.* 1995a), some evidence was found for regulation through density dependent adult dispersal in relation to available resources (Romstöck-Völkl 1990; Straw 1991; Zwölfer 1994). This dispersal, however, must lead to increased mortality or decreased fecundity to be regulatory (Cappuccino and Harrison this volume). These studies and evidence that the dispersal capacity of a given fly species is dependent on the predictability of its resource (Dempster *et al.* 1995b), tends to suggest a more metapopulation structure (sensu Gilpin and Hanski 1991) to *Urophora* spp. populations than is seen in *R. conicus.* Natural enemies do not have insignificant effects on their host populations. Parasitism caused local sub-populations of the stem-galler, *Urophora cardui,* to go extinct (Zwölfer 1994), but this was parasitism by *E. robusta,* the more generalist ectoparasite (cf. Hassell and May 1986).

Escape from natural enemies and effective biological control of *C. nutans*

R. conicus population density increased, following introduction into other regions without its usual parasites and competitors; causing eggs to become very highly aggregated on early flowering capitula. As a result weevil populations in the new environments are currently regulated at a new equilibrium density by scramble competition between larvae (Fig. 3) which, contrary to previous claims (Zwölfer 1979), is still insufficient to cause measurable increases in seed loss to its host plant compared to the native region (Sheppard *et al.* 1994). Where control has been achieved other factors, such as pasture competition, have contributed to it (Zwölfer and Harris 1984). In Australia and New Zealand, *C. nutans* usually has a longer growing season than elsewhere. This results in the thistle population extending seed production far beyond the activity period of the weevil and overall percent seed losses are significantly lower than in populations studied in the native range (Woodburn and Cullen 1993; Kelly and McCallum 1995). In Australia this led to the introduction of *U. solstitialis* as a second biological control agent (Woodburn 1993).

Two years following release in Australia in the absence of parasitism, *U. solstitialis* pre-winter mortality per capitulum has dramatically decreased (Fig. 5). Greater seed losses per m^2 (43.5±19% (mean±s.e.)), have already been recorded than in the native range (12.2±10%, Sheppard *et al.* 1994). Intraspecific larval competition in this species has not been detected, but this may change if fly populations increase further in the new environment (cf. Redfern and Cameron 1989). If resource density is driving density dependence in this system, this fly, with its second generation, could, like other *Urophora* species before it (Harris 1980), be highly destructive as a biological control agent. One cloud is currently masking this potential in Australia. First generation flies are currently emerging weeks before the first flower buds in Australia causing 70–90% adult mortality prior to oviposition. Such asynchrony has occurred in other *Urophora* species on *Cirsium vulgare* (Redfern and Cameron 1989) and *Centaurea* spp. (P. Harris pers. comm.). In the first instance this varied with locality and in the second flies progressively adapted to the phenology of their hosts.

Discussion

While a role for natural enemies in the population regulation of these two insects (particularly *R. conicus*) can not be ruled out, resource limitation is the more likely regulatory mechanism. If true, this fits the argument that specialist insect herbivores of relatively uncommon plants are also rarely regulated by their natural enemies. *R. conicus* has shown little to no increase in effectiveness (in terms of lost seed production) in the absence of its enemies even where successful control was accomplished. *U. solstitialis* may show greater population increase and cause greater damage in the absence of its enemies. However it will be hard to separate such effects from those caused by higher resource density of *C. nutans* in the new environment, especially as this is most likely a bottom-up regulated agent in its native range.

Comparison with weeds that show similar local abundance and predictability in native and new environments provides some insight. *C. vulgare* fits this category (Holm *et al.* 1979), and where the biological control agent *Urophora stylata* is well synchronised with it, damage levels appear to be higher in the new environment, but full control is not achieved (Redfern and Cameron 1989). Similarly, Cinnabar moth (*Tyria jacobaeae*) the resource limited (Dempster 1971) biological control agent of the ubiquitous weed *Senecio jacobaea*, does not influence weed density in its native range (Crawley and Gillman 1989), but can contribute to biological control in the introduced range if not on its own (McEvoy and Rudd 1993). The best biological control agents for such weeds may prove to be those regulated by their natural enemies in the native range, if they are not too few and far between (Cappuccino and Harrison this volume).

Acknowledgments

We would like to thank Jean-Paul Aeschlimann, Jim Cullen, Mick Neave, Annick Pouchot-Lermans, José Serin, Jean-Louis Sagliocco, Agnes Valin, Janine Vitou and Andrew White for help in data collection and experimental studies. Also David Briese, Margaret Redfern, Tony Wapshere and Helmut Zwölfer for comments on the manuscript. This work was supported by the International Wool Secretariat and the Australian Government.

References

Beddington JR, Free CA, Lawton JH (1975) Modelling biological control: on the characteristics of successful natural enemies. *Nature* 255:58–60

Briese DT (1986) Factors affecting the establishment and survival of *Anaitis efformata* (Lepidoptera: Geometridae) introduced into Australia for the biological control of St John's Wort, *Hypericum perforatum*. II. Field trials. *J Appl Ecol* 23:821–839

Crawley MJ (1986) The population biology of invaders. *Philos Trans R Soc Lond B Biol Sci* 314:711–731

Crawley MJ, Gillman MP (1989) Population dynamics of cinnabar moth and ragwort in grassland. *J Anim Ecol* 58:1035–1050

DeBach P (1949) Populations studies of the long-tailed mealy bug and its natural enemies on citrus trees in Southern California, 1946. *Ecology* 30:14–25

Dempster JP (1971) The population ecology of the Cinnabar moth, *Tyria jacobaeae* L. (Lepidoptera, Arctiidae). *Oecologia* **7**:26–67

Dempster JP, Atkinson DA, Cheesman OD (1995a). The spatial population dynamics of insects exploiting a patchy food resource. I. Population extinctions and regulation. *Oecologia* **104**:340–353

Dempster JP, Atkinson DA French MC (1995b). The spatial population dynamics of insects exploiting a patchy food resource. II. Movements between patches. *Oecologia* **104**:354–362

Gilpin M, Hanski I (eds) (1991) *Metapopulation dynamics: Empirical and theoretical investigations.* Academic Press, London

Goeden RD, Louda SM (1976) Biotic interference with insects imported for weed control. *Annu Rev Entomol* **21**:325–342

Hairston NG, Smith FE, Slobodkin LB (1960) Community structure, population control and competition. *Am Nat* **94**:421–425

Harris P (1980). Effects of *Urophora affinis* Frfld. and *U. quadrifaciatus* (Meig.) (Diptera: Tephritidae) on *Centaurea diffusa* Lam. and *C. maculosa* Lam. (Compositae). *Z Ang Entomol* **90**:190–201

Harris P (1984) *Carduus nutans* L., nodding thistle and *C. acanthoides* L., plumeless thistle (Compositae). In: Kelleher JS, Hulme MA (eds) *Biological control programmes against insects and weeds in Canada 1969–80.* CAB, Slough, pp 115–126

Hassell MP (1980) Foraging strategies, population models and biological control: a case study. *J Anim Ecol* **49**:603–628

Hassell MP, May RM (1986) Generalist and specialist natural enemies in insect predator-prey interactions. *J Anim Ecol* **42**:693–726

Hassell MP, Waage JK (1984) Host-parasitoid population interactions. *Annu Rev Entomol* **29**:89–114

Hassell MP, Varley GC (1969) New inductive population model for insect parasites and its bearing on biological control. *Nature* **223**:1133–1137

Holm L, Pancho JV, Herberger JP, Plucknett DL (1979). *A geographical atlas of world weeds.* Wiley, New York

Julien MHE (1992). *Biological Control of Weeds: a world catalogue of agents and their target weeds, 3rd edition.* CABI, Wallingford

Kelly D, McCallum K (1995). Evaluating the impact of *Rhinocyllus conicus* on *Carduus nutans* in New Zealand. In: Delfosse ES, Scott RR (eds). *Biological control of weeds; Proc VIII Int Symp Biol Contr Weeds.* CSIRO Publishing, Melbourne, pp 205–212

Kok LT (1976) Overwintering mortality of caged thistle head weevils *Rhinocyllus conicus* in Virginia. *Environ Entomol* **5**:1105–1108

Kok LT, Surles WW (1975) Successful biocontrol of musk thistle by an introduced weevil *Rhinocyllus conicus. Environ Entomol* **4**:1025–1027

May RM, Hassell MP (1987) Population dynamics and biological control. *Philos Trans R Soc Lond B Biol Sci* **318**:129–169

McEvoy PB, Rudd NT (1993) Effects of vegetation disturbances on insect biological control of tansy ragwort, *Senecio jacobaea. Ecol Appl* **3**:682–698

Milne A (1957) The natural control of insect populations. *Can Entomol* **89**:193–213

Möller-Joop H (1989) Biosystematisch-ökologische untersuchungen an *Urophora solstitialis* L. (Tephritidiae): Wirtskreis, biotypen und eignung zur biologischen bekämpfung von *Carduus acanthoides* L. (Compositae) in Kanada. Doctoral thesis, University of Bern

Murdoch WW, Chesson J, Chesson PL (1985) Biological control in theory and practice. *Am Nat* **125**:344–336

Nicholson AJ (1933) The balance of animal populations. *J Anim Ecol* **2**:131–171

Nicholson AJ, Bailey VA (1935) The balance of animal populations. *Proc Zool Soc Lond* **3**:551–598

Nicholson AJ (1954) An outline of the dynamics of animal populations. *Aust J Zool* **2**:9–65

Redfern M, Cameron RAD (1989) Density and survival of introduced populations of *Urophora stylata* (Diptera: Tephritidae) in *Cirsium vulgare* (Compositae) in Canada, compared with native populations. In: Delfosse ES (ed) *Proc VII Int Symp Biol Contr Weeds,* 6–11 March 1988, Rome, Italy. Ist Sper Patol Veg (MAF), Rome, pp 203–210

Redfern M, Jones TH, Hassell MP (1992) Heterogeneity and density dependence in a field study of a tephritid-parasitoid interaction. *Ecol Entomol* **17**:255–262

Romstöck-Völkl M (1990) Population dynamics of *Tephritis conura* Loew (Diptera: Tephritidae): determinants of density from three trophic levels. *J Anim Ecol* **59**:251–268

Sheppard AW. The Cardueae: Weeds and the impact of natural enemies for biological control. In: *Compositae, systematics, biology and utilization. Proceedings of the International Compositae Conference.* Kew 24 July to 5th August 1994 (in press)

Sheppard AW, Michalakis Y, Briese DT, Thomann T (1991) Spatial scale in the interaction between seed-head weevils and their host plants (Asteraceae, Cardueae), in a Mediterranean sheep-grazing system. In: Thanos CA (ed) *Proc 6th Int Conf Medit Climate Ecosys, Plant-animal interactions in mediterranean-type ecosystems.* University of Athens, Greece

Sheppard AW, Cullen JM, Aeschlimann J-P (1994) Predispersal seed predation on *Carduus nutans* (Asteraceae) in southern Europe. *Acta Oecol* **15**:529–541

Simmonds FJ (1948) The effective control of parasites of *Schematiza cordiae* Barber in Trinidad. *Bull Entomol Res* **39**:217–220

Smith HS (1935) The role of biotic factors in the determination of population densities. *J Econ Entomol* **28**:873–898

Smith LM, Ravlin FW, Kok LT, Mays WT (1984) Seasonal model of the interaction between *Rhinocyllus conicus* (Coleoptera: Curculionidae) and its weed host, *Carduus thoermeri* (Campanulatae: Asteraceae). *Environ Entomol* **13**:1417–1426

Straw NA (1991) Resource limitation of tephritid flies on lesser burdock, *Arctium minus* (Hill) Bernh. (Compositae). *Oecologia* **86**:492–502

Stiling P (1993) Why do natural enemies fail in classical biological control programs? *Am Entomol* **39**:31–37

Thompson WR (1929) On natural control. *Parasitology* **21**:269–281

Varley GC (1947) The natural control of population balance in the knapweed gall-fly (*Urophora jaceana*). *J Anim Ecol* **16**:139–187

Varley GC, Gradwell GR (1971) The use of models and life tables in assessing the role of natural enemies. In: Huffacker CB (ed) *Biological control.* Plenum Press, NY, pp 93–112

Varley GC, Gradwell GR, Hassell MP (1973) *Insect population ecology: an analytical approach.* Blackwells, Oxford, pp 212

Vitou J, Sheppard AW, Woodburn TL (1995) *Urophora solstitialis* (Dipt.:Tephritidae), a potential biological control agent for *Carduus nutans* (Asteraceae) in Australia. In: Delfosse ES, Scott RR (eds). *Biological control of weeds; Proc. VIII Int. Symp. Biol. Contr. Weeds.* CSIRO Publishing, Melbourne, pp 415–416

Weiser J, Bucher GE, Poinar Jr GO (1976) Host relationships and utility of pathogens. In: Huffaker CB, Messenger PS (eds) *Theory and practice of biological control.* Academic Press, NY, pp 169–188

Wilson RC, Andres LA (1986) Larval and pupal parasites of *Rhinocyllus conicus* (Coleoptera: Curculionidae) in *Carduus nutans* in Northern California. *Pan-Pac Entomol* **62**:329–332

Woodburn TL (1993) Host specificity testing, release and establishment of *Urophora solstitialis* (L.) (Diptera: Tephritidae), a potential biological control agent for *Carduus nutans* L., in Australia. *Biocon Sci Technol* **3**:419–426

Woodburn TL, Cullen JM. Impact of *Rhinocyllus conicus* and *Urophora solstitialis* on seed set in *Carduus nutans* in Australia. In: *Proceedings of the International Compositae Conference*, Kew, 24 July to 5 August (in press)

Woodburn TL, Cullen JM (1993) Effectiveness of *Rhinocyllus conicus* as a biological control agent for nodding thistle, *Carduus nutans,* in Australia. In: *Proc 10th Aust & 14th Asian-Pacific Weed Conf, 6–10 September 1993, Brisbane.* Australia Weeds Society of Queensland, Brisbane, pp 99–103

Zwölfer H (1973) Competition and coexistence in phytophagous insects attacking the heads of *Carduus nutans* L. In: Dunn PH (ed) *Proc. II Int. Symp Biol Contr Weeds Misc Publication No.6.* Commonwealth Institute of Biological Control, Slough, UK, pp 74–77

Zwölfer H (1979) Strategies and counterstrategies in insect population systems competing for space and food in flower heads and plant galls. *Fortschr Zool* **25**:331–353

Zwölfer H (1994) Structure and biomass transfer in food weeds: stability, fluctuations and network control. In: Schulze E-D (ed) *Flux control in biological systems from enzymes to populations and ecosystems*. Academic Press, NY, pp 365–419

Zwölfer H, Arnold-Rinehart J (1993) Parasitoids as a driving force in the evolution of the gall size of *Urophora* on Cardueae hosts. In: Williams MAJ (ed) *Plant galls. Systematics Association Special Volume No. 49*. Clarendon Press, Oxford, pp 245–257

Zwölfer H, Harris P (1984) Biology and host specificity of *Rhinocyllus conicus* (Froel.) (Col.,Curculionidae), a successful agent for biocontrol of the thistle, *Carduus nutans* L. *Z Ang Entomol* 97:36–62

THE DYNAMICS OF DISEASE IN NATURAL PLANT POPULATIONS

Jeremy J. Burdon

ABSTRACT

Plant pathogens are major, although often overlooked, forces affecting the size and genetic structure of plant populations and the composition of whole communities. Studies that focus on the consequences of disease for individual populations generate valuable information regarding the single-plant fitness effects of pathogen epidemics. However, the interactions that occur between individual plant species and their pathogens typically show considerable temporal and spatial variability and only when studies are extended to incorporate several populations simultaneously does the full complexity of such interactions become apparent. Population fragmentation, complemented by restricted movement of host and pathogen may set off a cascade of demographic effects that are impossible to predict from single population studies but which have profound consequences for both the short and long-term dynamics of host-pathogen associations. This is most readily apparent in pathogen populations that are typically more ephemeral than that of their hosts, frequently being reduced to low levels by adverse environmental factors. Such periodic crashes with their inevitable accompaniment of genetic drift or even complete extinction followed by decolonisation, play a vital role in determining the genetic structure of the individual pathogen populations that sequentially utilise a given host patch. In these circumstances, individual host patches may be subject to selective pressures that vary substantially in direction and intensity as distinct pathogen populations follow one another through time. The sum effect of such interactions across many local populations within a metapopulation suggests that the long-term dynamics of host-pathogen associations and their coevolutionary development is a product of local and regional processes in which spatial and temporal variability are of vital importance.

Key words: metapopulation, virulence, resistance, spatial variation, temporal variation

Frontiers of Population Ecology, R.B. Floyd, A.W. Sheppard and P.J. De Barro (eds), CSIRO Publishing, Melbourne 1996, pp. 291–300

Introduction

For decades the invasion of natural plant communities by exotic pathogens has provided a graphic demonstration of the destructive capacity and selective potential of such organisms. Paradoxically though, because of the exotic nature of the interaction, the message embodied in pandemics such as those caused by chestnut blight in North America, Dutch elm disease in Europe or *Phytophthora* die-back in Western Australia has often been ignored. It has only been in more recent years that increasing attention has focussed on the ephemeral endemic diseases of natural systems so that their often insidious role has become more widely appreciated. Such pathogens, by reducing the fecundity and longevity of afflicted individuals (Alexander and Burdon 1984; Parker 1985; Paul and Ayres 1986), may reduce the size and density of host populations (Burdon and Jarosz 1992; Carlsson *et al.* 1990) and generate selective forces responsible for spatial patterns in the distribution of resistance genes (Dinoor 1970; Burdon *et al.* 1983; Hunt and Van Sickle 1984; Jarosz and Burdon 1991). Ultimately pathogen activity may be the driving force behind changes in the rate and direction of successional events (Cook *et al.* 1989; Van der Putten *et al.* 1993) that determine the diversity of whole communities (Holah *et al.* 1993).

To date, evidence for many of these effects is limited and frequently based on the consequences of disease epidemics occurring in single host populations. Without doubt these studies have generated insights into the potential effects of pathogens on their hosts. Equally though, they ignore the possibility of interactions with other populations and all the numeric and genetic consequences such interactions entail. Just how then do the many spatially discrete pathogen demes occurring in an area interact with one another? Are pathogens so sedentary that for all intents and purposes each deme exists in total isolation from those occurring on other populations of the same host – that is, is each interaction potentially unique. Conversely, are pathogens so mobile that individual populations function as if all are drawn from a single, randomly mixed supra-population pool where, within the bounds of random variation, they are epidemiologically and genetically in step and spatial separation has no numeric or genetic consequences? Or does reality lie somewhere between these extremes in temporally fluctuating levels of interaction between individual demes that are spatially separated to lesser and greater extents?

Distinguishing between these possibilities is essential to an understanding of the long-term consequences of diseases in natural plant communities. Here the argument is developed that temporal and spatial phenomena stretching well beyond the bounds of single populations influence the speed and direction of coevolution for very many host plant-fungal pathogen associations. As a result, understanding the long-term dynamics of such coevolving systems requires consideration of the interactive effects of many host and pathogen populations spread irregularly over at least a local geographic area. That is, seeing such coevolving systems as the product of two interacting metapopulations.

Discussion of the epidemiology of fungal pathogens is complicated by their wide range of life histories and modes of interaction with host plants. As a consequence, this paper is largely restricted to a comparison of three well documented systems involving systemic and non-systemic floral and foliar pathogens.

Demographic patterns in single populations

At the level of the individual population, interactions between pathogen and host numbers may be quite limited. The density-dependent interactions detectable in many agricultural associations (Burdon and Chilvers 1982; Burdon 1987) are often far less apparent in natural systems. In the latter situations overlapping generations of both host and pathogen, the presence of alternative inoculum sources, and host densities that are frequently orders of magnitude lower, all limit the long-term maintenance of density-dependent relationships. Like so many interactions involving fungal pathogens, the expression of density-dependence is likely to be strongly influenced by aspects of the life history of both host and pathogen. Certainly examples of positive density-dependence exist as shown by the higher incidence of post-emergence seedling damping-off in high density stands of several neotropical trees than in lower density ones (Augspurger 1984; Augspurger and Kelly 1984). Here though, the limited dispersal ability of the pathogen coupled with a short window of seedling vulnerability, have interacted to 'freeze' the epidemic at a particular stage of development. The majority of interactions between hosts and pathogens are unlikely to be this simple. Rather, density-dependence may be expressed transiently only to be lost as the pathogen spreads to all parts of a population and subsequent auto-infection evens out disease levels. At this stage, time since initial pathogen establishment may become a more significant factor in epidemic development than density. An example of such a pattern is the interaction between *Phacidium infestans* and *Pinus sylvestris* (Burdon *et al.* 1992) where the relationship between disease incidence and host density typically went through three distinct phases during the development of an epidemic. Initially, disease foci were randomly distributed with respect to the host; this was followed by a density-dependent phase when the pathogen was widespread but at low severity. Finally, as the epidemic continued and disease severity rose, disease levels evened out across the population and the relationship with density was lost.

The temporal duration of the density-dependent phase may persist for some years (as in the example above) or may be very fleeting appearing and disappearing again with a matter of a few weeks. The latter situation is particularly likely with airborne pathogens with short generation cycles where initial foci of disease rapidly coalesce and subsequent generations of pathogen spore production will tend to even out differences between individuals.

Demographic patterns in multiple populations

Individual plant species rarely occur as single, large and homogeneous populations. Rather they are typically found as a smattering of small, often ephemeral, groups of individuals of varying number and density dotted across the landscape. Plants occurring within individual patches or demes clearly have greater opportunity for interacting – for competing for resources and for pollen exchange, than do individuals occurring within distinct but adjacent patches. In turn these are more likely to be related than are plants occurring in patches many kilometres apart. These observations may be self-evident for plants but when superimposed on co-occurring pathogens may have profound consequences for the population dynamics and genetics of very many pathogen species.

Temporal variability in pathogen occurrence

The most noticeable common feature of all pathogen epidemics is that they eventually come to an end. Over time pathogen numbers increase but eventually, whether through lack of new susceptible hosts or a change in the suitability of the abiotic environment, numbers reach a peak and then fall, often precipitously, to a low ebb. The time span of these epidemic booms and busts may vary from within individual seasons (frequently the case for discrete foliar diseases) to decades or longer for systemic diseases of naturally long-lived hosts. Fluctuations of this kind have significant theoretical implications for the genetic structure of the pathogen population and the selective pressures they may bring to bear on their hosts.

Whether these potential effects are realised depends on the degree to which individual demes are independent units of an overall metapopulation rather than simply spatially diffuse parts of a common population. This question can be looked at in a number of different ways *viz*: (i) how do disease levels change within populations over time? (ii) is the pattern observed in individual populations in unison or out of step with those in neighbouring demes? (iii) how is the dynamics of

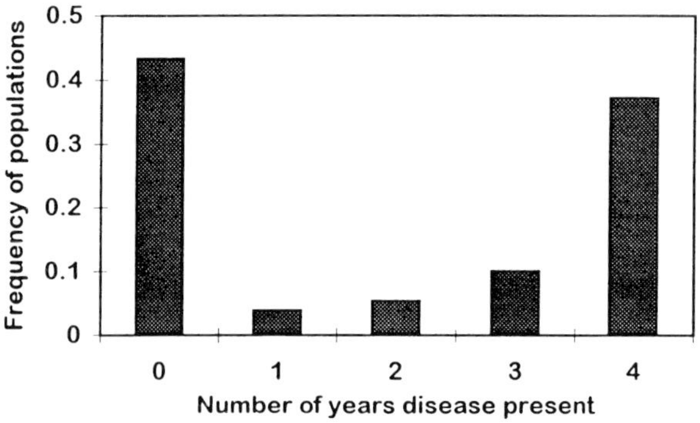

Fig. 1. Frequency of occurrence of the pathogen *Triphragmium ulmariae* over a 4-year period in 129 populations of *Filipendula ulmaria* growing on an archipelago of islands in the northern Baltic (J. Burdon and L. Ericson, unpublished).

disease affected by the size of individual host demes and by the proximity of infected populations? Unequivocal answers to all these questions are currently simply not available. They clearly require intimate knowledge of the dynamics of disease in many populations over a series of years.

Long-term monitoring of the dynamics of pathogens in natural systems is rare. Those studies that do exist indicate considerable temporal and spatial stochasticity – particularly in systems involving discrete lesion forming diseases. Moreover, the spatial patchiness of disease is often fine-grained showing marked differences in incidence and or severity between adjacent host plant populations. Some of this spatial and temporal complexity is seen in two contrasting examples. The first of these involves the incidence of the rust pathogen *Triphragmium ulmariae* in 129 populations of its host *Filipendula ulmaria* growing on an archipelago of islands in the northern Baltic Sea (Burdon *et al.* 1995). Every host population in this archipelago was mapped and their disease status followed over a 4-year period. During that time, some host populations were always infected; some were always free of disease; while in yet others there were complicated patterns in the presence or absence of disease (Fig. 1). Further investigation indicated that the occurrence of disease and, to a lesser extent, its severity within individual populations was influenced by the size of host populations and the local physical environment. In all years the incidence of disease was strongly positively correlated with population size (in all cases $p<0.001$; Burdon *et al.* 1995). However, this relationship was non-linear with a threshold size such that those populations less than approximately 350 plants in size had a low probability of harbouring the pathogen. Conversely the disease was always present in large populations (>800 individuals). Minimum population sizes for the continued maintenance of disease have also been detected in a completely different host-pathogen system involving a systemic anther smut, *Ustilago violacea*, of a perennial host *Viscaria vulgaris*. In that instance the threshold population size was very much smaller (c. 35 individuals; Jennersten *et al.* 1983).

Ustilago violacea is also the pathogen involved in detailed studies of the dynamics of anther smut disease in an extensive set of roadside populations of *Silene alba* in Virginia and island populations of *S. dioica* in Sweden. Like the *Filipendula* study, the incidence of disease within individual roadside populations of *Silene* showed considerable temporal variability. Although in any particular inter-season transition the majority of host populations retained their previous disease status (healthy or diseased), 10–24% of diseased populations became healthy; conversely 6–11% of healthy populations acquired disease (Antonovics *et al.* 1994). These extinction and migration transition rates were considerably higher than those observed among populations of *S. dioica* occurring in the Swedish archipelago. There, comparisons of disease occurrence in 1985 and 1990 showed an annualised rate of migration of *U. violacea* into previously healthy populations of 7% and an extinction rate of 2% (Carlsson and Elmqvist 1992). Differences in the temporal stochasticity of these two closely related systems almost certainly reflects marked differences in the average distance

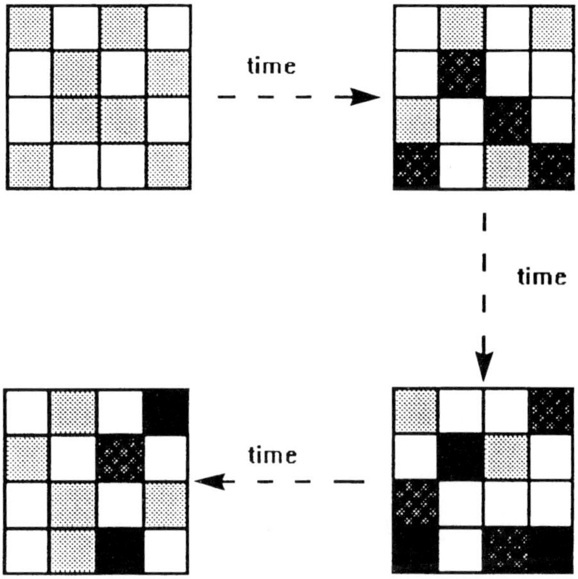

Fig. 2. Hypothetical representation of the spatial and temporal fluctuations that are expected to occur in wild pathogen metapopulations. The presence of the pathogen and the duration of the association between pathogen and host population (shown by varying degrees of shading – increasing intensity indicating an increasing length of time of continuous local association) both vary at least partially independently between different demes.

between neighbouring populations. In the Virginia study, 'populations' were defined as those individuals occurring in 44-yard segments of roadside. As a consequence many populations actually abutted one another. Populations in the Swedish study, on the other hand, were typically separated by hundreds of metres of open water. Furthermore, in the North American study neither host nor pathogen is native and the increased lability of this system may reflect the lower genetic variability of both populations due to migration bottlenecks.

For both *Silene-Ustilago* systems large host populations again had a greater probability of infection than small ones. However, unlike the results reported by Jennersten and his colleagues for the *Viscaria-Ustilago* system or for the *Filipendula-Triphragmium* system above, there was no evidence for a minimum population size for the maintenance of the disease. Instead, although smaller populations were less likely to be diseased than large ones, when disease was present the proportion of individuals affected was consistently higher in small populations.

Spatial variability in pathogen occurrence

The other particularly striking feature of these and many other natural host-pathogen systems is the spatial variability that occurs in both the incidence and severity of disease (see also Dinoor and Eshed 1990; Carlsson *et al.* 1990). At any given time adjacent host populations can show marked differences in disease severity. This may simply reflect continuing static differences in habitat quality (as often seen in changes in the physical microclimate between shaded and exposed parts of individual populations [e.g. Jarosz and Burdon 1988]) or may be more dynamic in nature. In both the *Silene-Ustilago* and the *Filipendula-Triphragmium* interactions, adjacent populations separated by only tens (*Silene*; Antonovics *et al.* 1994) or hundreds of metres (*Filipendula*; Burdon *et al.* 1995) showed differences in disease occurrence. In some cases these differences persisted over the length of the study; in others they were more ephemeral persisting for just one or two seasons.

At a larger scale there was evidence for some structuring within the entire *Filipendula* system such that in two of the four years of the study there was a significant relationship between the presence or absence of disease in a population, its size and the degree of isolation from the nearest diseased population (Burdon *et al.* 1995). Large populations close to diseased neighbours had a significantly greater chance of being infected than did isolated large populations or all small populations. However, this relationship was not apparent in other years nor was it by any means universal in the years when it was apparent.

The *Silene* and *Filipendula* studies give a clear picture of the consequences of host population sub-division for the dynamics of associated pathogen populations. Essentially sub-division of the host population leads to marked instability in the dynamics of the association – we observe instances of local pathogen extinction, followed by variable periods of pathogen absence, before migration and pathogen re-establishment eventually occurs (Fig. 2).

The two examples discussed in detail above represent distinctly different combinations of host and pathogen life histories and, not surprisingly, show differences in the frequency of extinction and recolonisation events. Indeed, in a general sense this is to be expected as the occurrence of such events will be greatly influenced by a myriad of life history features of both host and pathogen. Of particular and obvious importance are the efficiency of pathogen transmission and the mechanism whereby the pathogen survives through harsh 'off-season' conditions. Here insect-mediated spore transmission and high off-season survival in the *Silene* system (within infected individuals) contrasts with the vagaries of wind dispersal and survival in the *Filipendula* system where resting spores (teliospores) remain on dead plant remains that are often stripped away by fierce winter storms.

Genetic changes in multiple pathogen populations

The numerical interactions involving the *Ustilago* and *Triphragmium* pathogen populations on their *Silene* and *Filipendula* hosts respectively show many of the characteristics expected of a metapopulation structure. A lack of temporal synchrony between the dynamics of different pathogen demes within the metapopulation as a whole; the existence of a limited degree of spatial structuring in disease incidence; and extensive population turnover "the hallmark of genuine metapopulation dynamics" (Hanski and Gilpin 1991) have all been shown to occur on a temporal and spatial scale that was previously unknown. The question that now arises is what effects do these fluctuations in pathogen numbers actually have on the underlying genetic structure and diversity of pathogen populations?

At the level of the individual population, massive reductions in population size at the end of each season obviously enhance the chance of substantial drift and hence significant genetic differentiation between individual pathogen demes (Wright 1969). Summed over all local demes where drift, migration and selection will all be operating at varying and different levels of intensity, this has enormous implications for the overall structure of the pathogen metapopulation. Furthermore, these changes have the effect of introducing abrupt shifts in the direction of pathogen-induced selection within individual demes of the host metapopulation – as do instances of migration following local extinction events. Theoretically at least, the direction and intensity of these shifts may vary between adjacent demes thus leading to further distinction between both host and pathogen demes in the interacting metapopulations.

Knowledge of the dynamics of the genetic structure of wild pathogen populations is even more limited than that concerning their demography. However, while such information has not been gathered for either the *Triphragmium* or the *Ustilago* systems considered above, it is available for the interaction occurring between another rust pathogen *(Melampsora lini)* and its host *(Linum marginale)*. On the other hand, this study was restricted to only nine well-dispersed populations covering a range of approximately 120 kilometres in southern New South Wales, Australia and hence can provide relatively little information concerning the degree to which the dynamics of pathogen demes within a metapopulation are related. However, at the individual population level this pathogen shows considerable within and among season variation in population size. In years

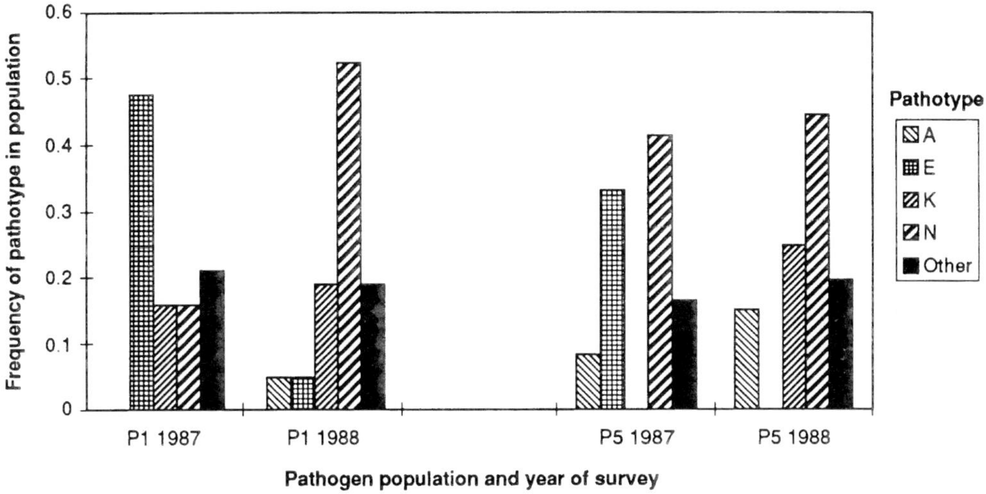

Fig. 3. Temporal variation in the pathotypic structure of populations of the rust pathogen *Melampsora lini* growing in two separate populations (P1 and P5) of the host plant *Linum marginale*. The four common pathotypes occurring at these two sites are shown separately; all other pathotypes are combined in a single others category.

favourable for pathogen growth and development, numbers may rise rapidly as disease epidemics sweep the associated host population. At the end of each season plants typically die back to ground level and the pathogen is forced to survive the subsequent winter on a few protected shoots.

Temporal variation

Year-to-year fluctuations in *M. lini* are very large with epidemic years being followed by others in which the pathogen is present at very low levels or even totally missing (Jarosz and Burdon 1992; J. Burdon unpublished). As a consequence, temporal variation in pathotypic structure was apparent in many pathogen demes (Fig. 3). Within any given host population the frequency of particular pathotypes showed marked variation from one year to the next.

For example, over the winter of 1987 the frequency of pathotype E in the pathogen population at Patch 1 fell from 47 to 5% while the frequency of a different pathotype (N) rose from 16 to 52% (Fig. 3). At the same time, in Patch 5 (approximately 65 kilometres away and hence too far for close association), pathotype E disappeared in 1988 while pathotype K, missing in 1987, occurred at a frequency of 25% in 1988. The actual causes of these changes are unknown but, as the host population was relatively unchanged, the simplest explanation is one of drift resulting from end-of-season crashes in population size. Obviously, they may be relatively rapidly reversed during the subsequent season (particularly if the host population exerts differential selection). However, in many instances such changes persisted for more than one season.

A more drastic effect resulting from the crash of population numbers is the local extinction of individual pathotypes or even entire populations. Extinction and migration are important forces for change, and temporal patterns occurring in one population of *M. lini* at Kiandra provide further illustration of the dynamic nature of the genetic structure of many pathogen populations. Over the period 1987–91 the structure and diversity of this pathogen population underwent considerable change. Initially, the pathogen population was composed of three pathotypes with frequencies of greater than 5 % (A, 71%; N, 8%; H, 5%) and a range of other more minor pathotypes (Fig. 4).

In 1988 the three common pathotypes were still present but a new pathotype complex (AL/AR) with a novel virulence combination was detected for the first time. Isozyme analysis confirmed the

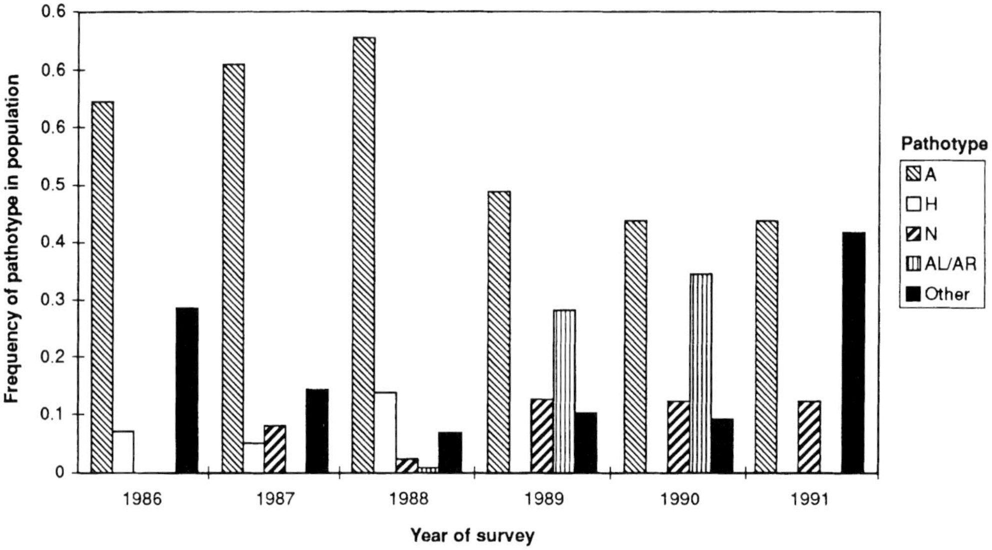

Fig. 4. Temporal changes in the structure of the pathogen population occurring at Kiandra, NSW showing examples of extinction and migration events in the *Melampsora lini* – *Linum marginale* system (Burdon and Jarosz 1992; J. Burdon, unpublished). Four common pathotypes are shown separately; all others are combined in a single category.

unique nature of this pathotype with respect to the rest of the pathogen population – that is, it was a confirmed migration event. Over the subsequent two years the frequency of this pathotype rapidly increased so that by 1990 it rivalled the long-term common pathotype at the site (A) in frequency. Pathotype AL/AR disappeared during the following population crash. In the meantime pathotype H had also disappeared during an earlier end-of-season population crash in 1988. Neither of these pathotypes has since been detected at this site despite continued monitoring.

Spatial variation

The spatial distribution of the populations monitored in the *Melampsora* study makes it impossible to consider the relationships between the genetic structure of adjacent populations in detail (most populations are separated by many kilometres in which other unmonitored populations existed). However, two adjacent populations (Kiandra and Patch 1) which are only 300 metres apart showed consistent differences over many years while simultaneously showing distinct within-site temporal variation. Over the years 1987–91, the pathogen population at Kiandra was heavily dominated by pathotypes A (43–76%) and AL/AR (0–34%) with lesser contributions from pathotypes N (2–12%), H (0–14%) and a range of minor pathotypes (Fig. 4). By way of contrast, at Patch 1 pathotype A was absent in two of the five years and only exceeded 10% in 1990. The population at this site was dominated by pathotypes N, E and K whose frequencies ranged between 0–52%, 0–47% and 16–80% respectively. Of the latter two pathotypes, K was never detected at Kiandra while the frequency of E was always less than 5%.

Conclusions

The view developed here sees natural pathogen populations not as single all-embracing populations, but as a series of demes of an overall metapopulation. Each deme is isolated to some degree from its neighbours such that the presence or absence of disease, the stage of the demographic cycle at which the deme is, and the magnitude of the subsequent population crash, all may differ. The consequences of these features (drift), coupled with differing degrees of inter-demic connectedness (migration) interact with demes of the host metapopulation (selection) to shape the genetic structure of pathogen and host populations and thus of the co-evolutionary interaction itself.

The pathogen populations quoted as examples, are the only ones for which considerable empirical evidence is currently available and are quite different to each other. One is an ephemeral local lesion disease with up to circa 10–12 generations per year and poor off-season survival, whereas the other is a perennial systemic disease with circa 1 generation per year and relatively high off-season survival. However, an even greater range of pathogen life histories exist. These range from those with short generation times and high off-season survival through to others that occur very infrequently, spreading as a front through an area killing nearly all host individuals before disappearing. Some of these species may initially appear very unpromising as candidates for interactions operating at a metapopulation level. However, many of these concerns disappear when viewed on a time scale that is calibrated according to the generation times of both host and pathogen. For short-lived associations in which many generations occur per year, the time scale of extinction and recolonisation events will be substantially shorter than for interactions where generation times are measured in decades or possibly even centuries.

References

Alexander HM, Burdon JJ (1984) The effect of disease induced *by Albugo candida* (white rust) and *Peronospora parasitica* (downy mildew) on the survival and reproduction of *Capsella bursa-pastoris* (shepherd's purse). *Oecologia* **64**:314–318

Antonovics J, Thrall PH, Jarosz AM, Stratton D (1994) Ecological genetics of metapopulations: the *Silene-Ustilago* plant-pathogen system. In: Real LA (ed) *Ecological genetics.* Princeton University Press, Princeton, New Jersey, pp 146–170

Augspurger CK (1984) Seedling survival of tropical tree species: interactions of dispersal distance, light-gaps, and pathogens. *Ecology* **65**:1705–1712

Augspurger CK, Kelly CK (1983) Pathogen mortality of tropical tree seedlings: experimental studies of the effects of dispersal distance, seedling density, and light conditions. *Oecologia* **61**:211–217

Burdon JJ (1987) *Diseases and plant population biology.* Cambridge University Press, Cambridge, UK

Burdon JJ, Chilvers GA (1982) Host density as a factor in plant disease ecology. *Annu Rev Phytopathol* **20**:143–166

Burdon JJ, Oates JD, Marshall DR (1983) Interactions between *Avena* and *Puccinia* species. I. The wild hosts: *Avena barbata* Pott ex Link, *A. fatua* L. and *A. ludoviciana* Durieu. *J Appl Ecol* **20**:571–584

Burdon JJ, Wennstrom A, Ericson L, Muller WJ, Morton R (1992) Density-dependent mortality in *Pinus sylvestris* caused by the snow blight pathogen *Phacidium infestans. Oecologia* **90**:74–79

Burdon JJ, Ericson L, Muller WJ (1995) Temporal and spatial relationships in a metapopulation of the rust pathogen *Triphragmium ulmariae* and its host, *Filipendula ulmaria. J Ecol* **83**:979–989

Burdon JJ, Jarosz AM (1992) Temporal variation in the racial structure of flax rust *(Melampsora lini)* populations growing on natural stands of wild flax *(Linum marginale)*: local versus metapopulation dynamics. *Plant Pathol* (Oxf) **41**:165–179

Carlsson U, Elmqvist T, Wennstrom A, Ericson L (1990) Infection by pathogens and population age of host plants. *J Ecol* **78**:1094–1105

Carlsson U, Elmqvist T (1992) Epidemiology of anther-smut disease *(Microbotryum violaceum)* and numeric regulation of populations of *Silene alba. Oecologia* **90**:509–5 17

Cook SA, Copsey AD, Dickman AW (1989) Responses of *Abies* to fire and *Phellinus.* In: Brock J, Linhart YB (eds) *The evolutionary ecology of plants.* Westview, Boulder, Colorado, pp 363–392

Dinoor A (1970) Sources of oat crown rust resistance in hexaploid and tetraploid wild oats in Israel. *Can J Bot* **48**:153–161

Dinoor A, Eshed N (1990) Plant diseases in natural populations of wild barley *(Hordeum spontaneum).* In: Burdon JJ, Leather SR (eds) *Pests, pathogens and plant communities.* Blackwell Scientific Publications, Oxford, pp 169–186

Hanski I, Gilpin M (1991) Metapopulation dynamics – Brief history and conceptual domain. *Biol J Linn Soc* **42**:3–16

Holah JC, Wilson MV, Hansen EM (1993) Effects of a native forest pathogen, *Phellinus weirii,* on Douglas-fir forest composition in western Oregon. *Can J For Res* **23**:2473–2480

Hunt RS, Van Sickle GA (1984) Variation in susceptibility to sweet fern rust among *Pinus contorta* and *P. banksiana. Can J For Res* **14**:672–675

Jarosz AM, Burdon JJ (1988) The effect of small-scale environmental changes on disease incidence and severity in a natural plant-pathogen interaction. *Oecologia* **75**:278–281

Jarosz AM, Burdon JJ (1991) Host-pathogen interactions in natural populations of *Linum marginale* and *Melampsora lini:* II. Local and regional variation in patterns of resistance and racial structure. *Evolution* **45**:1618–1627

Jarosz AM, Burdon JJ (1992) Host-pathogen interactions in natural populations of *Linum marginale* and *Melampsora lini:* III. Influence of pathogen epidemics on host survival and flower production. *Oecologia* **89**:53–61

Jennersten O, Nilsson SG, Wastljung U (1983) Local plant populations as ecological islands: the infection of *Viscaria vulgaris by* the fungus *Ustilago violacea. Oikos* **41**:391–395

Parker MA (1985) Local population differentiation for compatibility in an annual legume and its host-specific pathogen. *Evolution* **39**:713–723

Paul ND, Ayres PG (1986) The impact of a pathogen *(Puccinia lagenophorae* Cooke) on populations of groundsel *(Senecio vulgaris* L.) overwintering in the field. I. Mortality, vegetative growth and the development of size hierarchies. *J Ecol* **74**:1069–1084

Van der Putten WH, Van Dijk C, Peters BAM (1993) Plant-specific soil-borne diseases contribute to succession in foredune vegetation. *Nature* **362**:53–55

Wright S (1969) *Evolution and the genetics of populations, Vol. 2.* University of Chicago Press, Chicago

THE ROLE OF MUTUALISMS IN PLANT POPULATION DYNAMICS

Andrew R. Watkinson, Kevin K. Newsham and Alastair H. Fitter

ABSTRACT

The fungicide benomyl has been used to investigate the role that arbuscular mycorrhizal (AM) fungi play in the population dynamics of the winter annual grass, *Vulpia ciliata*. Inevitably using benomyl to reduce AM colonisation also has consequences for other fungi, such as pathogens, that infect the roots. Reduction in AM colonisation at different sites and in different years has variable effects on the finite rate of population increase of the plant. Initial estimates indicated that the benefit of AM fungi to the fitness of the plant might be in excess of 30%. It has subsequently been shown that the reduction in AM colonisation has no consequences for the mineral nutrition of the plant, and that the chief benefit of mycorrhizal fungi to the plant might be through providing protection against pathogens, which have been shown to reduce fecundity in the field by up to 50%. Recent transplant experiments, with colonised and non-colonised plants, have confirmed that AM fungi do indeed protect plants from pathogenic attack. It has also been shown that AM fungi may have an impact on the abundance of plants and play a major role in determining the composition of plant communities. Quantifying the role of mycorrhizal plants in the population dynamics of plants, nevertheless, remains a daunting problem as the fungal flora, just as the plant population, shows considerable variation both in time and space.

Key words: arbuscular mycorrhiza, benomyl, fungi, fungicides, *Vulpia*

Frontiers of Population Ecology, R.B. Floyd, A.W. Sheppard and P.J. De Barro (eds), CSIRO Publishing, Melbourne, 1996 pp. 301–309

Introduction

The flux of individuals within populations is dependent upon six controlling processes: the intrinsic rate of increase of the plant, intraspecific competition for resources, interspecific competition, natural enemies, mutualism and refuge effects such as the immigration of seeds from other populations. In considering the role of population interactions as determinants of population size, most attention has focused on competition (Grace and Tilman 1990) and natural enemies, such as herbivores (Crawley 1983, 1988) and pathogens (Burdon 1987; Augspurger 1988), while the role of mutualists has been largely ignored (Law 1988). There is not even a sound theoretical framework for the treatment of mutualistic interactions. One of the reasons for this is that the Lotka-Volterra models that have formed the basis for theoretical studies of competition and predator-prey interactions lead to both populations undergoing unbounded exponential growth. To counter this, most models of mutualism involve the density of the host plant species being limited by factors other than the mutualistic partner (e.g. mycorrhiza, pollinators), whereas the equilibrium density of the latter is directly proportional to the density of the former (Law 1988). In contrast, Crawley (1986) includes mutualism in his model as a negative term, arguing that the full capacity for population growth by the plant, λ, can only be exhibited when the individual has its full complement of obligate mutualists. When these mutualists are in limited supply the actual rate of increase is reduced.

In addition to the difficulties of making mathematical models that capture the essence of mutualistic interactions, there are also major problems in demonstrating empirically that mutualistic interactions play a part in the population dynamics of plants. It has been suggested that a great many plant invasions probably fail for want of specialised pollinators or mycorrhiza (Crawley 1986), and the difficulties sometimes experienced in growing host populations without effective mycorrhizal endophytes suggest that mycorrhizal fungi have important effects on host dynamics (Law 1988), especially in the case of ectomycorrhizal associations (Allen 1991). There is, however, very little information in the literature concerning the effects of mutualists on the dynamics of their host plant populations. For mycorrhizal fungi, this lack of data is in part due to the technical problems involved in studying the dynamics of an interaction that occurs below ground and the difficulty of experimentally manipulating the system. The effects of arbuscular mycorrhiza (AM) on the growth of host plants in pot culture experiments using sterilised soil have been intensively studied, showing that phosphorus uptake per unit root length is typically enhanced by colonisation. This enhancement leads to a greater growth rate in phosphorus deficient soils. A striking example of the effect of mycorrhizal fungi in phosphorus deficient soils is provided by the greenhouse study of Hartnett *et al.* (1993) in which the prairie grass *Andropogon gerardii* was grown at a range of densities to examine the effects of AM fungi and phosphorus availability on intraspecific competition. In treatments in which plants were grown in phosphorus deficient soil the suppression of mycorrhizal activity reduced biomass by almost three orders of magnitude. In contrast, the less mycorrhizal dependent *Elymus canadensis* showed little response to the suppression of mycorrhizal activity.

Under field conditions, however, the evidence for increased growth is less convincing (Fitter 1990). Two basic approaches have been used to examine the effect of mycorrhizal fungi in the field on plant growth. The first, involving the attempted inoculation of plants, has produced contradictory results (McGonigle 1988). In some cases the increase in the growth of the host plants has been considerable, whereas in others the gains have been small or negligible (Law 1988 and references therein). Colonisation of plants in control plots by naturally occurring inoculum is one explanation of these results. An alternative approach to the use of inoculation is to remove mycorrhizal fungi from a proportion of the population by fumigation, sterilisation or fungicides. Again the results have been contradictory (e.g. Carey *et al.* 1992), but it can be argued that fungicides in particular offer the only currently available method of directly comparing plants growing with and without mycorrhiza in the field (West *et al.* 1993a).

The impact of fungicides on soil fungi

Insecticides and herbicides have been used extensively to gain an understanding of the role that herbivores (e.g. Morrow and La Marche 1978) and competitors (e.g. Silander and Antonovics 1982) play in the population dynamics of plants. Alternative methods involving the exclusion of herbivores and the physical removal of potential plant competitors also exist for examining the dynamics of competition and herbivory. Some of these techniques can be used to remove specific species, but in the case of soil fungi the only option available to date is the application of non-specific fungicides. A range of fungicides has been used to manipulate soil fungi, primarily with the aim of manipulating mycorrhizal fungi (West *et al.* 1993a). Different fungicides vary in their effectiveness in suppressing mycorrhizal fungi, but typically benomyl has been found to be amongst the most effective (Fig. 1A). The reduction in mycorrhizal colonisation may typically be in the range of 50–100%. While there would appear to be no direct physiological effect on the plant (Paul *et al.* 1989; West *et al.* 1993a), benomyl and other fungicides do inevitably have a significant impact on other soil fungi (Fig. 1B). The application of benomyl to the soil will, therefore, result in comparisons being made of plants growing in environments differing in terms not only of mycorrhizal fungi, but also of other fungi and possibly also bacteria and animals that are affected by the fungicide. Understanding the role of mycorrhiza in natural field communities, therefore, requires an understanding of the role of microbial associates.

The impact of fungicide application on natural plant populations

Fungicides have now been used on several occasions to control mycorrhizal fungi in the field with the specific aim of manipulating demographic (e.g. Hartnett *et al.* 1994) and community responses (e.g. Newsham *et al.* 1995c). We have recently been investigating the role that AM fungi play in the population dynamics of the winter annual grass *Vulpia ciliata* ssp. *ambigua* growing on sandy soils in the east of England by applying the fungicide benomyl to reduce AM colonisation. In the first field experiment, the application of benomyl produced very different results at different sites (Carey *et al.* 1992). The application of the fungicide at one site reduced the fecundity of plants by 24%, whereas at another, fecundity was increased by 344%. At the site where benomyl reduced fecundity, this appeared to be linked to a significant reduction in AM colonisation in the plant roots, while at the site where there was an increase in fecundity, there was only a negligible amount of AM colonisation, even in the roots of control plants, and consequently the application of benomyl had

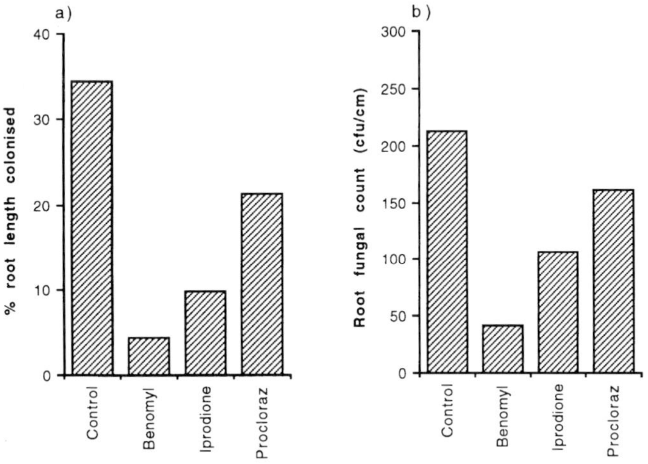

Fig. 1. The effect of three fungicides applied to field populations of *Vulpia ciliata* on a) mycorrhizal colonisation in roots and b) total fungal counts (all species summed) in rhizosphere soil. From West *et al.* (1993a).

no effect on AM colonisation. It was suggested, therefore, that the increase in fecundity with the application of benomyl at this site may have resulted from a reduction in infection, presumably by pathogenic fungi.

On the basis of the decrease in fecundity with the reduced abundance of AM fungi and pathogens at one site and the increase in fecundity with the reduction of pathogens alone at the other, Carey *et al.* (1992) estimated that the benefit of AM fungi to the fitness of the plant might be in excess of 31% as the result of protection from pathogens. Undoubtedly such an analysis was too simplistic. Subsequent studies have shown that we can discount any major benefit to *V. ciliata* from increased P uptake (West *et al.* 1993b). That is not to deny the benefit to other plant species from increased P uptake, but the chief benefit of AM fungi to *V. ciliata* appeared to be rather through protection against pathogens (West *et al.* 1993a).

In the study of *V. ciliata*, only fecundity was found to be affected by benomyl application and mycorrhizal activity. This may reflect the fact that mycorrhizal colonisation is very low for much of the vegetative stage of the life cycle during the autumn and winter (Carey *et al.* 1992). In a range of tall grass prairie species, Hartnett *et al.* (1994) found that benomyl application not only affected flowering but also seedling emergence amongst some of the C_3 grasses; there were no effects of benomyl application on seedling emergence of the C_4 grasses. Whilst their data can be interpreted in terms of mycorrhiza, the role of other fungi cannot be discounted.

The interaction between mycorrhizal and pathogenic fungi

In an attempt to elucidate further the interaction between mycorrhizal and pathogenic fungi in determining the fitness of plants, we have applied benomyl to natural populations of *V. ciliata* growing at three sites in eastern England (Newsham *et al.* 1994). Benomyl resulted in significant reductions in AM colonisation at all three sites, but had no effect on P inflow, plant biomass or fecundity. This result might be taken to suggest that the fungi were neutral in their impact on the plant, except that the fungicide also reduced the level of root infection of several pathogenic fungi, notably *Fusarium oxysporum* and *Embellisia chlamydospora*. When the fecundity of the *V. ciliata* plants was plotted as a function of the degree of infection of these fungi (and a *Phoma* sp. at one of the sites) a very striking effect was revealed: fecundity was a simple negative correlate of colonisation (Fig. 2); in other words there was *prima facie* evidence that various root-inhabiting fungi were having large effects on fecundity (and hence in this species on fitness), despite the fact that the plants were apparently quite asymptomatic. Moreover, the reductions in fecundity were large enough (30–48%) to have a major impact on the population dynamics of *V. ciliata*, as it has been shown that the abundance of plants in populations can be directly related to fecundity (Carey *et al.* 1995).

The poor relationship between plant fecundity and benomyl application in this experiment contrasted markedly with the clear effects of benomyl on root pathogenic and AM fungi, and with the clear impact of the root pathogenic fungi on plant fecundity. The most likely explanation for this apparent paradox was that the two groups of fungi were in some way interactive, and that when both groups were reduced in abundance, the resultant effects on the plants were neutral (Newsham *et al.* 1994). The analysis also indicated that the positive impact of the AM fungi was only expressed in the presence of the pathogens. This discovery led us to design contest experiments (Newsham *et al.* 1995a) in which plants were inoculated in the laboratory with a factorial combination of the *Fusarium oxysporum* and an AM fungus, a *Glomus* sp., before being transplanted into a natural population of *V. ciliata*. The results were striking (Fig. 3). Both 62 and 90 days after transplantation into the field, inoculation with *Glomus* sp. did not increase plant P concentrations or growth, but protected the plants from deleterious effects of *F. oxysporum* infection. These results were further confirmed by the significant monotonic relationship between the intensity of AM colonisation in

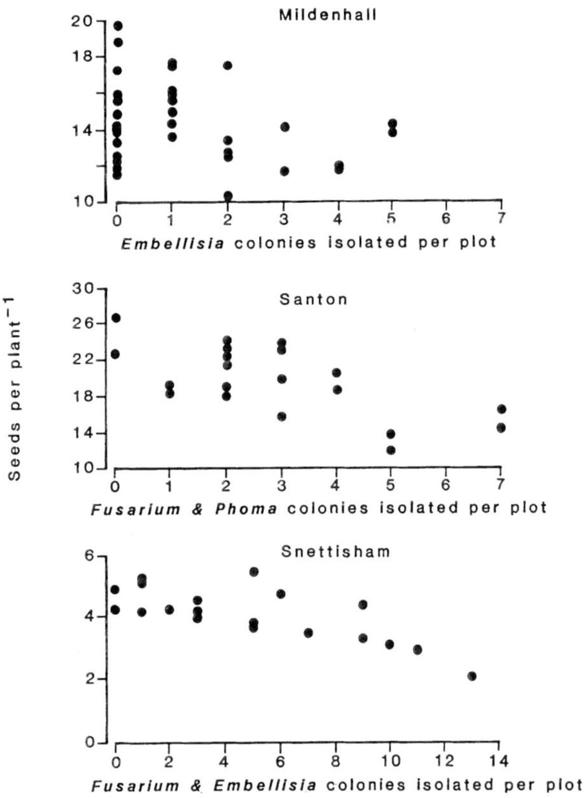

Fig. 2. The effect of *Fusarium oxysporum*, *Embellisia chlamydospora* and a species of *Phoma*, alone and in combination, on the fecundity of *Vulpia ciliata* at three natural populations (Mildenhall, Santon, Snettisham) in East Anglia. From Newsham *et al.* (1994).

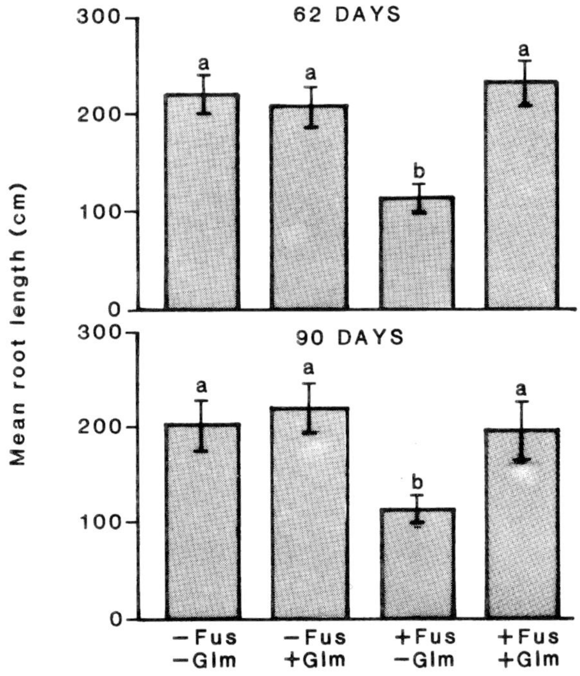

Fig. 3. The effects of factorial combinations of *Fusarium oxysporum* (Fus) and *Glomus* sp. (Glm) on the mean root length (± sem) of *Vulpia ciliata* plants sampled from the field 62 and 90 days after transplantation. From Newsham *et al.* (1995a).

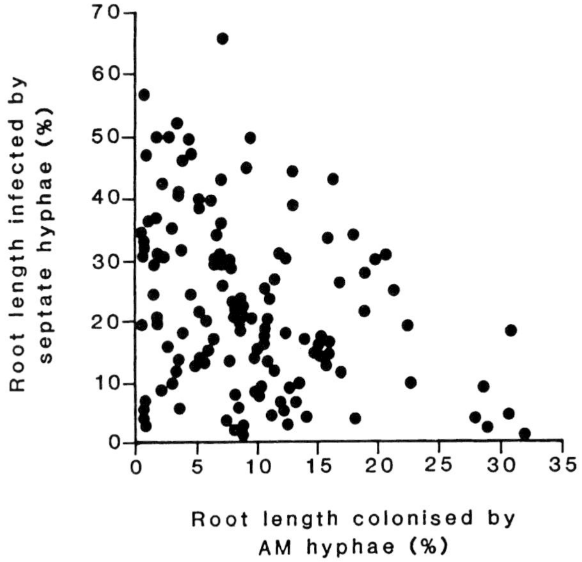

Fig. 4. Percentage of root length of *V. ciliata* infected by septate hyphae as a function of percentage root length colonised by arbuscular mycorrhizal (AM) hyphae after 62 and 90 days growth in the field. From Newsham *et al.* (1995a).

the roots and the infection of roots by septate hyphae; plants inoculated with *Glomus* were infected less frequently by septate hyphae and roots heavily colonised by AM fungi were always weakly infected with pathogens (Fig. 4). Moreover, when plants inoculated with *Glomus* only were compared with controls (no inoculation), it was found that washed root pieces plated into agar media developed fewer colonies of both *Fusarium oxysporum* and *Embellisia chlamydospora.*

Although we are not certain as to how AM fungi protect *V. ciliata* against fungal root infections, the large portions of roots which are typically uncolonised by either AM or pathogenic fungi suggest that the protective effect is systemic, i.e. from the AM fungus *via* the plant, such as by the induction of chitinase activity (Dehne and Schönbeck 1978) or the increased lignification of cell walls (Dehne and Schönbeck 1979b). The phenomenon of AM fungi protecting plants from fungal root pathogens is known from studies made on horticultural and agricultural species (Dehne and Schönbeck 1979a; Hwang *et al.* 1992), but what makes these data novel is that they are the first definitive evidence that the phenomenon occurs in the field, and that AM fungi can protect plants from pathogens in a situation where their effects on plant nutrition can be discounted.

The impact of mycorrhizal fungi on populations and communities

The results from our studies are one of the few clear demonstrations in a temperate ecosystem that AM fungi can benefit plants growing under natural conditions and the first to show that the benefit in the field can be unrelated to P nutrition. They also demonstrate that the benefit of AM fungi to *V. ciliata* in protection from pathogens may be in excess of the 31% reported by Carey *et al.* (1992). In annual species such as *V. ciliata* where abundance is directly related to fecundity (Carey *et al.* 1995), AM fungi may therefore play an important role in determining the dynamics of populations. However, it should be noted that we are only just beginning to understand the nature of the mutualistic interaction between the AM fungi and the plant, and that with fungicides we have only a very blunt tool for exploring their interaction in the field. Quantifying the role of mycorrhizal fungi in the dynamics of *V. ciliata* remains a daunting problem, especially as the fungal flora, just as the plant population, shows considerable variation both in time and space (Newsham *et al.* 1995b).

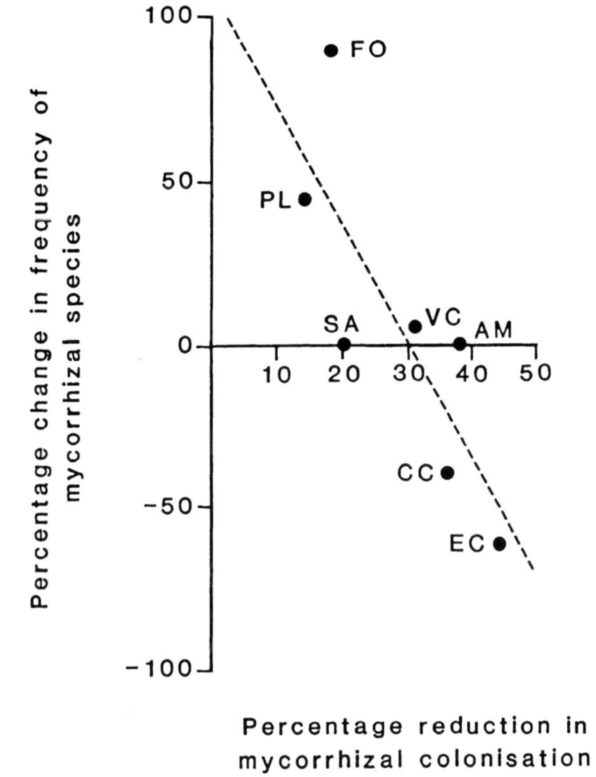

Fig. 5. The relationship between the change in frequency in the vegetation of mycorrhizal species in a long-term fungicide experiment, and the reduction in colonisation of roots by AM fungi brought about by benomyl. Abbreviations used: EC, *Erodium cicutarium*; CC, *Crepis capillaris*; AM, *Achillea millefolium*; VC, *Vulpia ciliata*; FO, *Festuca ovina*; PL, *Plantago lanceolata*; SA, *Sedum acre*. From Newsham *et al.* (1995c).

That mycorrhizal fungi have an impact on the abundance of plants can be seen from an analysis of the effects of benomyl on the communities in which *V. ciliata* occurs (Newsham *et al.* 1995c). Perhaps not surprisingly the long-term application of the fungicide resulted in the elimination of the lichens (predominantly *Cladonia rangifomis*) in the community. This resulted in a large increase in one of the moss species, *Ceratodon purpureus*, implying that the lichen was competitively dominant to the moss. There were also changes in the abundance of several higher plant species that could be related directly to mycorrhizal colonisation. Of the angiosperm species, two showed significant declines (*Erodium cicutarium* and *Crepis capillaris*) and two significant increases (*Rumex acetosella* and *Arenaria serpyllifolia*). Strikingly the first two are mycorrhizal and the latter two non-mycorrhizal, and further analysis revealed that the percentage change in frequency of the species that did form mycorrhiza at the site was a simple function of the percentage reduction in mycorrhizal colonisation brought about by fungicide treatment (Fig. 5). Other studies (Koide *et al.* 1988; Gange *et al.* 1990, 1993) have similarly concluded from field based fungicide experiments that AM fungi may modify plant community structure. In particular, Gange *et al.* (1990) found that the fungicide iprodione depressed AM fungal colonisation and resulted in reduced cover abundance of four mycorrhizal species. A number of studies have shown that AM fungi can influence the outcome of competition between species (Fitter 1977; Hartnett *et al.* 1993). Interestingly the results from our study reveal that two groups of symbiotic fungi are important regulators of plant community structure: the lichen *C. rangiformis* was a keystone organism in the community and appeared to be able to suppress the moss *C. purpureus*, and AM fungi appeared to play a significant role in determining the interspecific competitive abilities of higher plant species. It can, therefore, be expected that other symbioses, in addition to those formed by the AM fungi and lichens, may be important determinants of structure and function in a range of communities. The role of such mutualism in determining plant population and community dynamics remains a largely unexplored area of research.

References

Allen MF (1991) *The Ecology of Mycorrhizae*. Cambridge University Press, Cambridge

Augspurger CK (1988) Impact of pathogens on natural plant populations. In: Davy AJ, Hutchings MJ, Watkinson AR (eds) *Plant Population Ecology*. Blackwell Scientific Publications, Oxford

Burdon JJ (1987) *Diseases and Plant Population Biology*. Cambridge University Press, Cambridge

Carey PD, Fitter AH, Watkinson AR (1992) A field study using the fungicide benomyl to investigate the effect of mycorrhizal fungi on plant fitness. *Oecologia* **90**:350–355

Carey PD, Watkinson AR, Gerard FFO (1995) The determinants of the distribution and abundance of the winter annual grass *Vulpia ciliata* subsp. *ambigua*. *J Ecol* **83**:177–187

Crawley MJ (1983) *Herbivory. The Dynamics of Animal-Plant Interactions*. Blackwell Scientific Publications, Oxford

Crawley MJ (ed) (1986) *Plant Ecology*. Blackwell Scientific Publications, Oxford

Crawley M J (1988) Herbivores and plant population dynamics. In: Davy AJ, Hutchings MJ, Watkinson AR (eds) *Plant Population Ecology*. Blackwell Scientific Publications, Oxford, pp 367–392

Dehne HW, Schönbeck F (1978) Untersuchungen zum einfluss der endotrophen mycorrhiza auf planzenkrankheiten. III. Chitinase activität und orthinzyklus. *Z Pflanzenkr Pflanzenschutz* **85**:666–678

Dehne HW, Schönbeck F (1979a) Untersuchungen zum einfluss der endotrophen mycorrhiza auf planzenkrankheiten. I. Ausbreitung von *Fusarium oxysporum* f. sp. *lycopersici* in tomaten. *Phytopathol Z* **95**:105–110

Dehne HW, Schönbeck F (1979b) Untersuchungen zum einfluss der endotrophen mycorrhiza auf planzenkrankheiten. II. Phenolstoffwechsel und lignifizierung. *Phytopathol Z* **95**:210–216

Fitter AH (1977) Influence of mycorrhizal infection on competition for phosphorus and potassium by two grasses. *New Phytol* **79**:119–125

Fitter AH (1990) The role and ecological signficance of vesicular-arbuscular mycorrhizas in temperate ecosystems. *Agric Ecosyst Environ* **29**:137–151

Gange AC, Brown VK, Farmer LM (1990) A test of mycorrhizal benefit in an early successional plant community. *New Phytol* **115**:85–91

Gange AC, Brown VK, Sinclair GS (1993) Vesicular-arbuscular mycorrhizal fungi: a determinant of plant community structure in early succession. *Funct Ecol* **7**:616–622

Grace JB, Tilman D (eds) (1990) *Perspectives on Plant Competition*. Academic Press, San Diego

Hartnett DC, Hetrick BAD, Wilson GTW, Gibson DJ (1993) Mycorrhizal influence on intra- and interspecific neighbourhood interactions among co-occurring prairie grasses. *J Ecol* **81**:787–795

Hartnett DC, Samenus RJ, Fischer LE, Hetrick BAD (1994) Plant demographic responses to mycorrhizal symbiosis in tallgrass prairie. *Oecologia* **99**:21–26

Hwang SF, Chang KF, Chakravarty P (1992) Effects of vesicular-arbuscular mycorrhizal fungi on the development of *Verticillium* and *Fusarium* wilts of alfalfa. *Plant Dis* **76**:239–243

Koide RT, Huenneke LF, Hamburg SP, Mooney HA (1988) Effects of applications of fungicide, phosphorus and nitrogen on the structure and productivity of an annual serpentine plant community. *Funct Ecol* **2**:335–344

Law R (1988) Some ecological properties of intimate mutualisms involving plants. In: Davy AJ, Hutchings MJ, Watkinson AR (eds) *Plant Population Ecology*. Blackwell Scientific Publications, Oxford, pp 315–342

McGonigle TP (1988) A numerical analysis of published field trials with vesicular arbuscular mycorrhizal fungi. *Funct Ecol* **2**:473–478

Morrow PA, La Marche VC (1978) Tree ring evidence for chronic insect suppression of productivity in subalpine Eucalyptus. *Science* **201**:1224–1226

Newsham KK, Fitter AH, Watkinson AR (1994) Root pathogenic and arbuscular mycorrhizal fungi determine fecundity of asymptomatic plants in the field. *J Ecol* **82**:805–814

Newsham KK, Fitter AH, Watkinson AR (1995a) Arbuscular mycorrhiza protect an annual grass from root pathogenic fungi in the field. *J Ecol* **83**:991–1000

Newsham KK, Watkinson AR, Fitter AH (1995b) Rhizosphere and root-infecting fungi and the design of ecological field experiemnts. *Oecologia* **102**:230–237

Newsham KK, Watkinson AR, West HM, Fitter AH (1995c) Symbiotic fungi determine plant community structure: changes in a lichen-rich plant community induced by fungicide application. *Funct Ecol* **9**:442–447

Paul DN, Ayres PG, Wyness LE (1989) On the use of fungicides for experimentation in natural vegetation. *Funct Ecol* **3**:759–769

Silander JA, Antonovics J (1982) Analysis of interspecific interactions in a coastal plant community – a perturbation approach. *Nature* **298**:557–560

West HM, Fitter AH, Watkinson AR (1993a) The influence of three biocides on the fungal associates of the roots of *Vulpia ciliata* ssp. *ambigua* under natural conditions. *J Ecol* **81**:345–350

West HM, Fitter AH, Watkinson AR (1993b). Response of *Vulpia ciliata* ssp. *ambigua* to removal of mycorrhizal infection and to phosphate application under field conditions. *J Ecol* **81**:351–358

ECOLOGICAL CONSTRAINTS TO THE DEPLOYMENT OF ARTHROPOD RESISTANT CROP PLANTS: A CAUTIONARY TALE

Joanne C. Daly and Paul W. Wellings

ABSTRACT

About 300 insect-resistant crop varieties are grown in different parts of the world and the majority of these are used in major cereal production systems. Until now development of these cultivars has been the product of selection programs that use germplasm derived from geographical regions where the insect is endemic or the plant originated. Recent advances in plant molecular biology have the potential to change this approach as the technology offers the opportunity to insert novel sources of insect resistance in the host plant through the introduction of foreign primary gene products. This technology will influence both the rate at which new cultivars can be developed and, eventually, the range of novel resistance characteristics that can be incorporated into plants. These changes could have a profound impact on pest management practices. However, the impact of synchronous use of resistant cultivars, based on single genes, across regions will require careful management if this approach is to be sustainable. In addition, transgenic plants are only one component of integrated management practices and their influence on preferred agronomic characters and beneficial organisms needs careful evaluation.

In this paper we examine a range of ecological issues that need to be addressed if the potential of resistant host plants is to be fully exploited. These include

a. how to develop deployment strategies that maintain susceptibility in the pest population
b. the potential consequences of ontogenetic, seasonal and environmentally-mediated expression of resistance traits
c. the impact of the introduced foreign genes on plant fitness and the way 'costs of resistance' and pest dynamics interact, and
d. determining the potential consequences for multi-trophic interactions.

These questions are explored through examination of the literature on insect-plant interactions and models of the dynamics of insect populations.

Key words: insecticide resistance management, host plant resistance, *Helicoverpa*, *Bacillus thuringiensis*, toxins, antixenosis, tolerance, antibiosis

Frontiers of Population Ecology, R.B. Floyd, A.W. Sheppard and P.J. De Barro (eds), CSIRO Publishing, Melbourne, 1996 pp. 311–323

Introduction

Contemporary strategies for pest management vary among the world's major crops. Over half the insecticides used worldwide are in the cotton and horticultural industries (Boulter 1993). These are for the control of major pests such as Lepidoptera (Boulter 1993) in the genera, *Helicoverpa, Heliothis, Plutella* and *Spodoptera.* In contrast, insecticide usage in cereals, such as rice and wheat, accounts for less than 20% of world consumption. In these crops, pest control has been more reliant on traditional breeding programs to produce cultivars with enhanced natural resistance to Hemipteran pests, such as aphids and planthoppers. Plant breeding for insect resistance has been less important in horticultural crops, due in part to the demand for blemish free produce (Smith 1989).

Concerns over environmental contamination by insecticide residues, and the development of resistance to insecticides in major pests, has led to renewed interest in the use of host plant resistance (HPR) for insect control. Traditionally, breeding for insect resistant crop cultivars formed the backbone of crop protection (Zadoks 1991). Desirable genes within different cultivars of a plant species, or within the genome of the wild progenitor, are increased in frequency by selection programs. More recently, the technology of genetic engineering has extended the utility of HPR for pest control, as it permits the introduction of foreign genes between unrelated species, not necessarily plants. It can also give direct control over the regulation of the expression of those genes.

These factors, coupled with the initial difficulties of engineering cereal crops, is leading, in the first instance, to the development of transgenic plants in crops normally dependent on chemical pest control. The first commercial releases of insect resistant crops will be of cotton for *Heliothis* spp. and *Helicoverpa* spp. control, and of potatoes for the control of Colorado potato beetle, and of maize for the control of European corn borer, all within the next two years. In these crops, the inserted toxin gene is expressed at levels sufficient to kill the target insect. This use of transgenic plants as substitutes for insecticides (Gould 1988), has had a major influence on their development and deployment. Table 1 contrasts the use of transgenic versus natural host plant resistance (HPR) for insect control. In this paper we explore a number of these, in particular, the use in transgenic plants as alternatives to chemical insecticides – 'magic bullets' reliant on one or two resistance mechanisms that are expressed at high levels. Such an approach requires the development of resistance management strategies, akin to those developed for insecticides. Additionally, transgenic plants may have a yet unknown yield penalty and high expression may be difficult to maintain under field conditions. This situation contrasts with natural resistance which is often complex and incomplete in expression. We argue that there are considerable benefits in deploying engineered genes that confer only partial resistance to insects, as part of an integrated pest management approach (van Emden 1990, 1991; de Ponti and Mollema 1991; Wellings and Ward 1994).

Mechanisms of resistance

Mechanisms of host plant resistance are generally grouped into three categories (Smith 1989 for review):

a. *Antixenosis* or *non-preference* which are mechanisms that influence insect behaviour, such as feeding or oviposition.

b. *Antibiosis* refers to the mechanisms that influence the insect's biology – they reduce the capacity of the insect to use the plant as a host. Often these mechanisms are antimetabolic and involve toxins.

c. *Tolerance* refers to the plant's capacity to withstand or recover from insect damage.

Each of these three categories may contain both physical or chemical mechanisms. For example, a physical character such as leaf morphology can affect the insect's attraction to the plant or its ability

Table 1. A comparison of natural and engineered host plant resistance (HPR)

Category	Natural HPR	Engineered HPR
Mechanisms	Antibiosis, antixenosis, tolerance	Antibiosis
Basis of resistance	Diverse chemical and physical	Chemical – antimetabolic
Target pests	Hemiptera > Diptera > Lepidoptera	Lepidoptera > Hemiptera > Coleoptera
Impact on mortality	Mortality rates variable	High mortality
Cost of Resistance	Possibility of yield penalties	Possibility of yield penalties
Expression	Variable	Constitutive
Tri-trophic interactions	Complex interactions with parasitoids and predators	Unknown but possibly simple
Resistance management	May be required	Required
Social Issues	Simple	Complex

to utilise the plant as food. Hairy leaves on cotton seem to reduce oviposition by the whitefly, *Bemisia tabaci* (Wilson and George 1986). Adult densities of this insect are highest on cotton plants with intermediate numbers of trichomes (Butler *et al.* 1991). Further, numbers of whitefly are also reduced at high trichome densities. In tomatoes, hairy leaves trap the female whitefly, a factor that may reduce the transmission of plant viruses (Kisha 1981). Domatia on leaves can provide protection for predatory mites (Walter and O'Dowd 1992). These small pits or chambers are found on many commercial trees, shrubs and vines (O'Dowd and Willson 1989). Leaf toughness is a key factor in the suitability of the host food for the monophagous Richmond birdwing butterfly, *Troides richmondius* (D.P.A. Sands *pers. comm.*) and for redlegged earth mites, *Halotydeus destructor* (Jiang and Ridsdill-Smith 1996).

Plants may evolve or be selected for a range of these mechanisms against a single insect. For example, plant breeding programs with Australian cultivars of cotton, *Gossypium hirsutum,* have developed partial resistance to the insect, *Helicoverpa armigera,* in a variety of ways. Oviposition is reduced on plants with okra-shaped leaves (GP Fitt *pers. comm.*). Other plants with enhanced HPR are glabrous, have frego (open) bracts or are nectariless (Fitt 1994) or the leaves produce an array of secondary compounds that are toxic to larvae (Fitt *et al.* 1994a). Cotton plants are also extremely tolerant of insect feeding and will shed, then replace, damaged buds or fruits. This tolerance is used as part of pest management – economic thresholds early in the season are three times that of mid-to late season growth (Hearn and Fitt 1992).

In contrast, engineered resistance is very simple. To date, insect-resistant plants nearing commercial release, have had inserted into their genome, a modified form of the Cry1A(c) δ-endotoxin gene from the bacterium, *Bacillus thuringiensis* (*Bt*) that codes for a crystal protein with insecticidal activity (Perlak *et al.* 1990; Fitt *et al.* 1994b). There are a wide variety of crystal proteins, each which has an activity generally restricted to within an insect order (Hofte and Whitely 1989). While these endotoxins may have antifeedant effects at low concentration (antixenosis), the level of expression in transgenic plants is intended to kill the insect (antibiotic). Other potential insecticidal genes for genetic engineering, such as enzyme inhibitors, lectins, and lectin-like proteins (Gatehouse *et al.* 1992) and viruses (K. Gordon and T. Hanzlik *pers. comm.*) are being assessed. This bottleneck of suitable genes appears to result from the expectation that engineered resistance must produce high mortality of the insect (review in Gatehouse *et al.* 1992). If this requirement was relaxed the number of suitable genes could well be much larger than currently is available.

Natural resistance involves a wide range of chemicals, including not only proteins, but also a wide range of plant secondary compounds such as phenols, tannins, alkaloids, terpenoids (Bernays and Chapman 1994). Many of these compounds have been selectively bred out of agriculturally important plants because of their impact on yield (van Emden 1991) and because of their toxicity to grazing mammals. They are not suitable candidates for genetic engineering because they either are the product of complex biochemical pathways or are toxic to the plants when expressed at high levels.

The reliance on a single mechanism for pest control has a major impact. It requires that the plants have high expression of the toxin with possible effects on yield. It also increases the probability that resistance will evolve to the transgenic plants. We explore both these factors below. Further, antimetabolic mechanisms may not be the appropriate defence mechanism for commercial crops (Kennedy *et al.* 1987) when even minor feeding by an insect renders these agricultural products unsuitable for market, such as feeding on potatoes by whitefringed weevils, *Graphognathus leucoloma,* or the African black beetle, *Heteronychus arator* (Matthiessen and Learmonth 1995); or feeding by the citrus pest, adult fruit piercing moths, *Othreis fullonia,* (DPA Sands *pers. comm*). Similarly, Delanney *et al.* (1989) observed that transgenic tomato plants expressing *Bt* toxins had less feeding damage from lepidopteran pests compared with control plants. However, the larvae fed primarily by boring into the fruit and feeding internally, so while feeding damage was less, the fruit was not commercially acceptable.

As a consequence of the small number of genes available, and the limitations on the benefits of antimetabolic resistance mechanisms, few transgenic plants are engineered presently for insect resistance. Between 1986–89, 7% of applications for field trials were with plants engineered for insect resistance, compared with 39% for herbicide resistance (OECD 1993). Data from European countries in the year 1991–93 indicate little change in these figures (Tachmintzis 1994).

Plant fitness and the cost of resistance

The development of transgenic plants with resistance to herbivores has been driven by the need to substitute pesticide control measures with alternative technologies that may be more sustainable and have fewer non-target effects. In developing these technologies, we have assumed that the agronomic characteristics of these varieties will be conserved and be closely related to the parental types, but with the benefit of the inserted resistant traits. In order to be economically viable, the yield of such plants should be at least, on average, the economic yield of their non-transgenic relatives (i.e., the yield that gives the highest economic return on expenditure; Zadoks 1980) less the average expenditure on pesticides. However, this can be a complex problem as yield depends on the production environment, production level and arthropod/environment interactions (Wellings *et al.* 1989).

Where these is a cost of resistance, Wellings and Ward (1994) suggest that the major factors influencing whether partial or completely resistant crops have the higher expected yield include:

a. the form of the trade-off between yield and degree of resistance;
b. the size of the pest population;
c. the degree of density-dependence in regulating mechanisms and the time at which resistance has its effect.

There are very limited data on the relative yield of insect-resistant transgenic plants compared with their non-transgenic parents. However, there are concerns that they will have reduced performance. What is not clear is the form of the relationship between yield and relative resistance. Even in the absence of a direct 'cost of resistance' in transgenic plants, the proposed strategies such as setting aside refugia of susceptible plants or marketing seed mixes in order to reduce the selection pressure on the herbivores (Mallet and Porter 1992; Caprio 1994; Gould 1994) will automatically generate a similar effect in the cropping system: Expected average economic yields will be reduced as a result of the losses experienced by the susceptible lines.

Where there are costs of resistance, partial resistance can be more profitable than complete resistance (van Emden 1991). This outcome is, however, highly dependent on average pest density and the mechanism of resistance. Wellings and Ward (1994) show that if resistance is based on non-preference, then complete resistance should be favoured, irrespective of the underlying regulatory mechanisms in the herbivore population. In contrast, partial resistance may be favoured when resistance is based on antibiosis.

Expression of resistance

Natural HPR is not always a constitutive character but can vary in response to many factors to do with the plant, the insect or the environment (Smith 1989; Waterman and Mole 1989). Some of the variables are as follows:

a. Plant – density, height, tissue age and type, prior damage, infection, cultivar;
b. Insect – age, sex, activity period, infestation level, biotypes, density;
c. Environment – water, temperature, light, soil fertility, agrochemicals, relative humidity, season (after Smith 1989).

For example, many cultivars of barley are suitable hosts for the cereal aphid, *Metopolophium dirhodum*, at all developmental growth phases. This is in contrast to wheat, which expresses the natural insecticide, DIMBOA, in young seedlings. Aphids reared on young wheat seedlings produce small adults that are less fecund than aphids reared on older wheat or on barley at all growth stages (Niemeyer *et al.* 1993). Cultivars of cowpea have been bred for resistance to the bruchid pest, *Callosobruchus maculatus*. In susceptible plants, 100% of seedlings are infested are 100 days, compared with only about 80% of resistant seedlings after 180 days (Singh *et al.* 1985). This difference in infestation rate is enough to seriously reduce the pest problem.

In addition, natural HPR mechanisms may be overwhelmed by insect behaviour. Bark beetles survive and breed on their host plants, conifers, only if they kill their host. Yet many of the compounds found naturally in conifers are toxic to the beetles at naturally occurring levels (Raffa 1986). The beetles overcome the plant defenses by mass attack. Host monoterpenes are metabolised by the beetles to produce an aggregation pheromone (Conn *et al.* 1984). The rapidly formed feeding aggregations disrupt and drain the resin ducts of the tree rendering it susceptible to the beetles (Raffa 1986). Raffa and Berryman (1983) were able to demonstrate experimentally that when the number of females that arrived at a tree was restricted, death of the tree was prevented and reproduction in the beetles was reduced.

Hughes and Hughes (1988) observed a breakdown of HPR by spotted clover aphid, *Therioaphis trifolii*, in resistant cultivars of lucerne. They observed aphid infestations on autumn regrowth of lucerne over five years. In one of the years the aphids built up to damaging levels on the resistant cultivar. They attributed this to the prolonged and massive immigration of aphids into the crops coupled with lower temperatures in that year.

Just as natural resistance is not always constitutive, Gould (1988) and others discuss the advantages of variable expression of toxin genes in transgenic plants. In particular, it has been proposed as one way to reduce the evolution of insecticide resistance in target pests (Gould 1988; Roush 1996). Such plants, by expressing the engineered toxins in the vulnerable parts of the plants, or only after feeding damage, would mimic the action of natural host plant resistance. However, these alternative strategies appear to not be feasible in the first generation of transgenic plants because of the difficulties of getting 'all-or-none' expression.

As a result, the first generation of commercial varieties of cotton and potato will use constitutive promoters that are intended to give high expression of toxin at all stages of plant growth. Daly (1994) discussed the possibility for attenuation of expression in such transgenic plants. She proposed that two factors are likely to be most critical: the effect of water or temperature stress or plant age, on the constitutive expression of proteins (Sachs and Ho 1986; Vierling 1991; Jones *et al.* 1989; Basra 1993) and the possibility of sequestration of protein toxins by plant tannins (Navon *et al.* 1993). Each of these factors is capable of influencing protein expression and thus, has the potential to reduce the concentration of an engineered toxin. Consequently, field performance of transgenic crops will be difficult to predict throughout their distribution. To date, there have been very limited opportunities to investigate these issues in field trials of transgenic plants. One exception is that of the assessment of the field performance of Australian cultivars of transgenic cotton, described below.

Seasonal variation in the expression of the *Bt* toxins in transgenic cotton

The two pest species *Helicoverpa armigera* and *H. punctigera* are serious pest of broadacre and horticultural crops in Australia (Zalucki *et al.* 1986). Traditionally, they are controlled in high value crops by frequent applications of chemical insecticides. In cotton, *Gossypium hirsutum*, the pests can be present at damaging densities throughout the plant's growth so that season-long protection from insects is required. To overcome the dependence on chemical insecticides, Australian cotton cultivars have been genetically engineered to express one of a variety of δ-endotoxin genes from *B. thuringiensis kurstaki* (Fitt *et al.* 1994b). The two *Helicoverpa* species are the primary targets of the *Bt*-cottons, although other lepidopteran species are also susceptible to the transgenic plants (Fitt *et al.* 1994b). The promoter has constitutive expression (Perlak *et al.* 1990; Llewellyn *et al.* 1992), thus, it was expected that the plants would express high levels of *Bt* toxin to give season long control of *Helicoverpa* larvae (Llewellyn *et al.* 1992; Fitt *et al.* 1994b).

Fitt *et al.* (1994b) reported on the first field trials of transgenic cottons expressing a truncated construct of the Cry1A(b) δ-endotoxin. They observed little or no survival of 1st instar larvae on young *Bt*-cotton plants in November and December, and survivors did not grow. However, from mid-summer onwards, both the survival of 1st instar larvae, and their subsequent growth, increased steadily so that by the end of the cropping season, *H. armigera* pupae were found in the soil (GP Fitt *pers. comm.*). Fitt and his colleagues (*pers. comm.*) have found the same decline in efficacy of transgenic cotton expressing the Cry1A(c) gene. Daly and Olsen (unpublished data) have estimated that the decline in the efficacy of *Bt*-cotton plants from seedlings to boll-set is of the order of 3-fold. The cause in the seasonal decline in the efficacy of these plants is not known (Fitt *et al.* 1994b) but it is possible that older plants may be producing less toxin or the toxin may be sequested by the tannins (Navon *et al.* 1993; Daly 1994) that increase as the plants mature (Wilson 1994).

Such seasonal decline in efficacy of the plants has important consequences for resistance management and pest control. Fitt *et al.* (1994b) proposed that strategies for the deployment of the transgenic plants emphasise alternative pest control measures late in the season, when the plants no longer control *H. armigera.*

Tri-trophic interactions and resistance

The interactions between plant resistance, insect herbivores and the biological control agents remain poorly understood. Host plant resistance has a major role in two classes of tri-trophic interaction:

a. effects on the herbivore population caused by direct interaction between the resistant plant and the natural enemies;

b. effects on natural enemies of the resistant plant mediated via the herbivore (Bergman and Tingey 1979; van Emden and Wratten 1990).

Direct effect of the host plant include the influence of plant architecture (such as leaf surface waxes and trichomes) (Bergman and Tingey 1979) and volatile plant allelochemicals (e.g. Nordlund 1987; Vinson *et al.* 1987). Indirect effects include historical and present exposure of parasitoids to host cues (Kester and Barbosa 1994; also see Vet, this volume) and changes in the fitness of natural enemies as a result of feeding on/attacking herbivores reared on resistant plants (Wellings 1988).

Biological control and host plant resistance are generally thought to be compatible component technologies in integrated pest management. In an operational setting, however, this will depend on the relative effects of plant resistance in the populations of herbivores and natural enemies (Gutierrez 1986; Wellings 1988). The final outcomes of these interactions depends on a large number of environmental and biological variables (e.g., Orr and Boethel 1986; Wellings 1986; Krischik *et al.* 1988). We lack a comprehensive understanding of possible effects of tri-trophic interactions on the population dynamics of herbivores and need case studies on these problems built around variation in the resistance status of the host plants.

Management of resistance

Arthropods such as aphids and planthoppers develop resistance to natural host plant resistance in commercial crop cultivars (Smith 1989), particularly when resistance is a simple genic character and cultivars are grown in monoculture (de Ponti and Mollema 1991). Strategies for the deployment of natural HPR emphasise the use of partial resistance (Parlevliet 1991), pyramiding of genes (the deployment of more than one resistance gene in a cultivar), and the utilisation of both antixenotic and antibiotic forms of resistance (de Ponti and Mollema 1991). Certainly, in cases in which natural HPR has involved a single major gene, resistance in the target insect has developed rapidly (Smith 1989).

Arthropod pests certainly develop resistance to insecticides (Georghiou 1986) and foliar applications of *Bt* (Tabashnik 1994a). Considerable concerns have now being expressed about the possibility that resistance will evolve in arthropods from exposure to transgenic plants (Gould 1988; Roush 1994, 1996). These concerns centre on the use of a single toxin, the potential that the crops will be grown as large areas of monoculture, the possibility that the toxin (the selecting agent) is present continuously, and more recently, concerns of the fluctuating expression in apparently 'high expression' plants. The last two factors together increase the chance that the concentration of toxin present in the plant will fall in the range that can discriminate between susceptible and resistance genotypes, that is, at concentrations at which selection can occur (Roush and McKenzie 1987).

The management of resistance in arthropods to transgenic plants is being developed from the theory of insecticide resistance management (see Tabashnik 1994a). This is not surprising given that the initial use of transgenic plants was as an alternative to pesticides. Table 2 lists possible elements in strategies to manage resistance (see also McGaughey and Whalon 1992). The different elements work by reducing selection pressure or by enhancing gene flow: (a)–(c) reduce, either directly or indirectly, the discrimination between susceptible and resistant individuals; (d)–(g) all increase the fitness of susceptible individuals, either by reducing mortality of susceptible individuals below 100% (d), or by creating 'toxin-free' refuges in the crop (e) in space or time (g), or outside the crop (h).

While Georghiou and Taylor (1977) anticipated the value of refugia for resistance management to insecticides, the focus with insecticides has been to moderate selection pressure through altering insecticide use (Roush and Daly 1990). It is only more recently with transgenic plants that the emphasis has switched to altering gene flow by the provision of artificially created refugia for susceptible individuals (Gould 1986; Roush 1989). This may reflect early optimism that engineered resistance would have uniform expression.

Table 2. Strategies for managing resistance to transgenic plants

Strategy
a. high expression of toxin
b. pyramiding of multiple toxins
c. use of natural enemies in integrated pest management
d. low expression of toxin at sub-lethal levels
e. seed mixtures of non-transgenic and transgenic plants
f. rotations of plants containing unrelated toxins
g. temporal, tissue specific, or damage induced promoters
h. refuge habitats outside the crop

The tactics in Table 2 are not mutually exclusive approaches to resistance management of transgenic plants. Strategies are a mixture of the above (Roush and Daly 1990). However, some key choices need to be made regarding the level of expression, the number of toxin genes to be integrated and whether to use pure stands or mixes of transgenic and non-transgenic plants (Roush 1989, 1996; Denholm and Rowland 1992).

Low versus high expression, or partial versus full resistance

Theory suggests that resistance can be delayed either by exposing pest populations to very high or to low concentrations of a toxin. In either case, the population is not exposed to concentrations that discriminate between susceptible and resistant insects – a necessary condition for evolution of resistance to occur (Roush and McKenzie 1987). Gould *et al.* (1991) and van Emden (1991) argue that resistance management becomes less of a problem when a toxin is used at sub-lethal levels. In the case of engineered resistance, it is used as an adjunct to other mechanisms of HPR or as part of an integrated pest management system.

As discussed above, however, the high expression strategy is preferred when transgenic plants are used as a substitute for insecticides. Advocates for this approach combine it with the provision of refuges for susceptible individuals (Denholm and Rowland 1992; Roush 1994, 1996). Resistance in the arthropod pest is more likely to occur with high, rather than low, expression and a number of conditions must be met to ensure the delay of the onset of resistance.

Firstly, Roush (1994, 1996) has argued that the dose (or concentration) of toxin in the tissue to be digested by the herbivore should be sufficient to kill > 95% of heterozygous individuals. Values less than this will see rapid onset of resistance. The capacity to achieve this level of expression in the field will depend on the pest and the resistance genes as insect species can vary in their sensitivity to toxins (Hofte and Whitely 1989; Gould 1994).

Secondly, expression about 10-fold above the LC_{99} for susceptible insects will not delay resistance in cases where genes confer higher levels of resistance in heterozygotes. Early reports in the literature have assumed that arthropod resistance to the *Bt* toxins will be recessive traits based on the notion that the most likely mechanism for resistance would be a modification to the binding site (Van Rie 1991). These mechanisms are usually recessive (Roush and McKenzie 1987). However, a variety of mechanisms is possible and field deployment of transgenic plants will select for those that confer higher levels of protection in arthropods (Heckel 1994).

Thirdly, plants may not be able to achieve constitutively high levels of expression, such as in the case of Australian cultivars of cotton (see above).

Seed mixtures versus refugia

There has been considerable debate over the choice of seed mixtures as a strategy for pest control (Mallet and Porter 1992; Tabashnik 1994b; Caprio 1994; Roush 1994, 1996). Seeds from non-transgenic and transgenic plants are mixed in known proportions so that susceptible insects can survive within the crop. The alternative is to use refuges of non-transgenic crops grown elsewhere (see Gould 1994 for review). Seed mixtures have advantages for practical reasons. In particular, the seed company can mix the seed before sale to the grower so that compliance with the strategy is ensured. Emergence of susceptible and possible resistant insects is more likely to be in synchrony than if susceptible individuals are produced in refuges. This increases the chance of random mating between all genotypes (Tabashnik 1994b). The disadvantages of seed mixtures is that a proportion of plants must be susceptible to effectively delay resistance, a situation that may result in unacceptable yield losses. Further, the ratio of susceptible to resistant seeds will vary according to the target arthropod pest, and will depend on factors such as their life history, reproductive rates (Caprio 1994), dispersal of feeding stages (Roush 1996), overwintering sites and adult dispersal (Whalon and Wierenga 1994).

Refuges may be provided naturally in areas in which a number of crop hosts are grown at the same time of year, and the crops are a mixture of non-transgenic and transgenic plants. Given that the first genes to be deployed are those from *Bt*, it also assumes that *Bt* will not be used extensively as a foliar spray on non-transgenic plants (Forrester 1994; Roush 1994, 1996). Alternatively, growers could plant refuge crops (Fitt *et al.* 1994a) or trap crops (Whalon and Wierenga 1994). Voluntary compliance of growers to produce such refuges may be difficult to achieve, although it is under active consideration for the deployment of Australian cotton cultivars (GP Fitt *pers. comm.*).

Pyramiding

This strategy recognises an underlying flaw of transgenic plants – the exposure of insects to a single toxin. Resistance is delayed when an insect is exposed to two or more toxins simultaneously (Gould 1986, 1994; Van Rie 1991; McGaughey and Whalon 1992; Roush 1994, 1996). The word 'pyramiding' has been borrowed from traditional plant breeding (Kennedy *et al.* 1987) and refers to the addition into a plant genome of more than one HPR gene. It is equivalent to the insecticidal resistance management strategy of mixtures, in which two pesticides are sprayed or applied simultaneously (Daly 1994). The problem with this approach is the unavailability of different genes, limited effectively at present to different crystal protein genes from *Bt*. Deployment of other genes is being trialed (Hilder *et al.* 1990; Gatehouse *et al.* 1992; Hallahan *et al.* 1992) but none appear to be nearing commercial release. Controversy surrounds the use of different *Bt* genes as part of a pyramiding strategy because of the possibility of cross-resistance occurring among the different *Bt* toxins (Van Rie 1991; McGaughey 1994).

Conclusions

The ecological literature on insect-host plant interactions shows that natural host plant resistance is a complex suite of mechanisms, and plants generally do not rely on a single mechanism to defend themselves against insects. In contrast, engineered resistance is being introduced to mimic pest control using insecticides in a 'high kill' strategy. There are already some cautionary messages for us regarding transgenic plants, in particular, the potential for plant resistance to break-down under field conditions, as is observed with cotton crops. While such a strategy is acceptable for the first generation of transgenic plants, this approach is unlikely to be sustainable both because of resistance in the pests, and because of likely yield impacts. The future for genetic engineering in plants may be more secure if we can draw on our understanding of insect-plant interactions and mimic some of the complexity of natural host plant resistance.

Acknowledgments

Thanks are due to all our colleagues who provided lively discussion in the development of these ideas. In particular, it is a pleasure to thank Rick Roush, Gary Fitt, Seamus Ward. Paul De Barro provided valuable comments on earlier drafts.

References

Basra AS (1993) *Stress-induced gene expression in plants*. Harwood Academic Publishers, pp 287

Bergman JM, Tingey WM (1979) Aspects of interaction between plant genotypes and biological control. *Entomol Soc Am Bull* 25:275–279

Bernays EA, Chapman RF (1994) *Host-selection by phytophagous insects*. Chapman Hall, New York, pp 312

Boulter D (1993) Insect control by copying nature using genetically engineered crops. *Phytochemistry* 34:1453–1466

Butler GD Jr, Wilson FD, Fishler G (1991) Cotton leaf trichomes and populations of *Empoasca lybica* and *Bemisia tabaci*. *Crop Prot* 10:461–464

Caprio MA (1994) *Bacillus thuringiensis* gene deployment and resistance management in single- and multi-tactic environments. *Biocontrol Sci Technol* 4:487–498

Conn JE, Borden JH, Hunt DWA, Holman J, Whitney HS, Spanier OJ, Pierce HD Jr, Oehlschlager AC (1984) Pheromone production by axenically reared *Dendroctonus ponderosae* and *Ips paraconfusus* (Coleoptera: Scolytidae). *J Chem Ecol* 10:281–290

Daly JC (1994) Ecology and resistance management for *Bacillus thuringiensis* transgenic plants. *Biocontrol Sci Technol* 4:563–571

Delanney X, LaVallee BJ, Proksch RK, Fuchs RL, Sims SR, Greenplate JT, Marrone PG, Dodson RB, Augustine JJ, Layton JG, Fischhoff DA (1989) Field performance of transgenic tomato plants expressing the *Bacillus thuringiensis* var. *kurstaki* insect control protein. *Biocontrol Sci Technol* 7:1265–1269

Denholm I, Rowland MW (1992) Tactics for managing pesticide resistance in arthropods: Theory and practice. *Ann Rev Entomol* 37:91–112

de Ponti OMB, Mollema C (1991) Emerging breeding strategies for insect resistance. In: Stalker HT, Murphy JP (eds) Plant Breeding in the 1990s. *Proceedings of the Symposium on Plant Breeding in the 1990s*. CAB International, Wallingford, UK, pp 323–353

Fitt GP (1994) Cotton pest management: Part 3. An Australian perspective. *Ann Rev Entomol* 39:543–562

Fitt GP, Mares CL, Thomson NJ (1994a) Evaluation of resistance to insects in Australian cotton varieties. *Proceedings 7th Australian Cotton Growers Research Conference*, Surfers Paradise, pp 135–144

Fitt GP, Mares CL, Llewellyn DJ (1994b) Field evaluation and potential ecological impact of transgenic cottons (*Gossypium hirsutum*) in Australia. *Biocontrol Sci Technol* 4:535–548

Forrester NW (1994) Resistance management options for conventional *Bacillus thuringiensis* and transgenic plants in Australian summer field crops. *Biocontrol Sci Technol* 4:549–554

Gatehouse AMR, Boulter D, Hilder VA (1992) Potential of plant-derived genes in the genetic manipulation of crops for insect resistance. In: Gatehouse AMR, Hilder VA, Boulter D (eds) *Plant genetic manipulation for crop protection*. CAB International Wallingford, UK, pp 155–181

Georghiou GP (1986) The magnitude of the resistance problem. *In: Pesticide resistance: Strategies and tactics for management*. National Academy Press, Washington DC, pp 14–43

Georghiou GP, Taylor CE (1977) Genetic and biological influences in the evolution of insecticide resistance. *J Econ Entomol* 70:320–323

Gould F (1986) Simulation models for predicting durability of insect-resistant germ plasm: a deterministic diploid, two-locus model. *Environ Entomol* 15:1–10

Gould F (1988) Genetic engineering, integrated pest management and the evolution of pests. *Biocontrol Sci Technol*, Suppl, 3:15–18

Gould F (1994) Potential and problems with high-dose strategies for pesticidal engineered crops. *Biocontrol Sci Technol* 4:451–463

Gould F, Kennedy GG, Johnson MT (1991) Effects of natural enemies on the rate of herbivore adaptation to resistant host plants. *Entomol Exp Appl* **58**:1–14

Gutierrez AP (1986) Analysis of the interactions of host plant resistance, phytophagous and entomophagous species. In: Boethel DJ, Eikenbury RD (eds) *Interaction of plant resistance and parasitoids and predators of insects*, Ellis-Horwood, UK, pp 198–215

Hallahan DL, Pickett JA, Wadhams LJ, Wallsgrove RM, Woodcock CM (1992) Potential of secondary metabolites in genetic engineering of crops for resistance. In: Gatehouse AMR, Hilder VA, Boulter D (eds) *Plant genetic manipulation for crop protection.* CAB International, Wallingford, England, pp 215–248

Hearn AB, Fitt GP (1992) Cotton cropping systems. In: Pearson CJ (ed) *Field crop ecosystems of the world.* Elsevier, Amsterdam Vol 5:85–142

Heckel DG (1994) The complex genetic basis of resistance to *Bacillus thuringiensis* toxin in insects. *Biocontrol Sci Technol* **4**:405–417

Hilder VA, Gatehouse AMR, Boulter D (1990) Genetic engineering of crops for insect resistance using genes of plant origin. In: Lycett GW, Grierson D (ed) *Genetic engineering of crop plants.* Butterworths, London, pp 51–66

Hofte H, Whitely HR (1989) Insecticidal crystal proteins of *Bacillus thuringiensis. Microbiol Rev* **53**:242–255

Hughes RD, Hughes MA (1988) Temporary loss of antibiosis in plants of a lucerne cultivar selected for resistance to spotted alfalfa aphid. *Entomol Exp Appl* **49**:75–82

Jiang Y, Ridsdill-Smith TJ (1996) Examination of the involvement of mechanical strength in antixenotic resistance of subterranean clover cotyledons to the redlegged earth mite (*Halotydeus destructor*) (Acarina: Penthaleidae). *Bull Entomol Res* **86**:263–270

Jones HG, Flowers TJ, Jones MB (eds) (1989) *Plants under stress: Biochemistry, physiology and ecology and their application to plant improvement.* Cambridge University Press, Cambridge, pp 257

Kennedy GG, Gould F, de Ponti OMB, Stinner RE (1987) Ecological, agricultural, genetic and commercial considerations in the deployment of insect-resistant germplasm. *Environ Entomol* **16**:327–338

Kester KM, Barbosa P (1994) Behavioural responses to host foodplants of two populations of the insect parasitoid *Cotesia congregata* (Say). *Oecologia* **99**:151–157

Kisha JSA (1981) Observations on the trapping of the whitefly *Bemisia tabaci* by glandular hairs on tomato leaves. *Ann Appl Biol* **97**:123–127

Krischik VA, Barbosa P, Riechelderfer CF (1988) Three trophic level interactions: allelochemicals, *Manduca sexta* (L), and *Bacillus thuringiensis* var *kurstaki* Berliner. *Environ Entomol* **17**:476–482

Llewellyn DJ, Brown M, Cousin Y, Hartweck L, Last D, Mathews A, Murray F, Thistleton J (1992) The science behind transgenic cotton plants. In: *Proceedings of 6th Australian Cotton Growers Research Conference*, Surfers Paradise, pp 365–379

Mallet J, Porter P (1992) Preventing insect adaptation to insect-resistant crops: are seed mixtures or refugia the best strategy? *Proc R Soc Lond B Biol Sci* **250**:165–169

Matthiessen JN, Learmonth SE (1995) Impact of soil insects African black beetle, *Heteronychus arator* (Coleoptera: Scarabaeidae) and whitefringed weevil, *Graphognathus leucoloma* (Coleoptera: Curculionidae), on potatoes and effects of insecticide treatments in south-western Australia. *Bull Entomol Res* **85**:101–111

McGaughey WH (1994) Implications of cross-resistance among *Bacillus thuringiensis* toxins in resistance management. *Biocontrol Sci Technol* **4**:427–435

McGaughey WH, Whalon ME (1992) Managing insect resistance to *Bacillus thuringiensis* toxins. *Science* **258**:1451–1455

Navon A, Hare JD, Federici BA (1993) Interactions among *Heliothis virescens* larvae, cotton condensed tannin and the Cry1A(c) δ-endotoxin of *Bacillus thuringiensis. J Chem Ecol* **19**:2485–2499

Niemeyer HM, Givovich G, Copaja SV (1993) Hydroxamic acids: chemical defences in wheat against aphids. In: Corey S, Dall D, Milne W (eds) *Pest control and sustainable agriculture.* CSIRO Publications, Melbourne, pp 39–43

Nordlund DP (1987) Plant produced allelochemicals and their involvement in the host selection behaviour of parasitoids. In: Corey S, Dall D, Milne W (eds) Insects-plants. *Proceedings of the 6th International Symposium on Insect-Plant Relationships.* Dr W Junk, Dordrecht, pp 103–107

O'Dowd DJ, Willson MF (1989) Leaf domatia and mites on Australasian plants: ecological and evolutionary implications. *Biol J Linn Soc* **37**:191–236

OECD (1993) *Field releases of transgenic plants, 1986–1992 – an analysis.* OECD Publications, Paris, France, pp 39

Orr DB, Boethel DJ (1986) Influence of plant antibiosis through four trophic levels. *Oecologia* **70**:242–249

Parlevliet JE (1991) Selecting components of partial resistance. In: Stalker HT, Murphy JP (eds) *Proceedings of a Symposium on Plant Breeding in the 1990s.* CAB International, Wallingford, UK, pp 281–239

Perlak FJ, Deaton RW, Armstrong TA, Fuchs RL, Sims SR, Greenplate JT, Bischhoff DA (1990) Insect resistant cotton plants. *Biocontrol Sci Technol* **8**:939–943

Raffa KF (1986) Devising pest management tactics based on plant defense mechanisms, theoretical and practical considerations. In: Brattsen LB, Ahmad S (eds) *Molecular aspects of insect-plant associations.* Plenum Press, New York, pp 301–327

Raffa KF, Berryman AA (1983) The role of host plant resistance in the colonization behavior and ecology of bark beetles (Coleoptera: Scolytidae). *Ecol Monogr* **53**:27–49

Roush RT (1989) Designing resistance management programs: how can you choose? *Pest Sci* **26**:423–441

Roush RT (1994) Managing pests and their resistance to *Bacillus thuringiensis:* Can transgenic crops be better than sprays? *Biocontrol Sci Technol* **4**:501–516

Roush RT (1996) Can we slow adaptation by pests to insect-resistant transgenic crops?. In: Persley G (ed) *Biotechnology for integrated pest management.* CAB International, Wallingford, UK (in press)

Roush RT, Daly JC (1990) The role of population genetics in resistance research and management. In: Roush RT, Tabashnik B (eds) *Pesticide resistance in arthropods.* Chapman Hall, New York, pp 97–153

Roush RT, McKenzie JA (1987) Ecological genetics of insecticide and acaricide resistance. *Ann Rev Entomol* **32**:361–380

Sachs MM, Ho T-HD (1986) Alteration of gene expression during environmental stress in plants. *Ann Rev Plant Physiol* **37**:363–376

Singh BB, Singh SR, Adjadi O (1985) Bruchid resistance in cowpea. *Crop Sci* **25**:736–739

Smith CM (1989) Plant resistance to insects: A fundamental approach. John Wiley and Sons, New York, pp 286

Tabashnik BE (1994a) Evolution of resistance to *Bacillus thuringiensis. Ann Rev Entomol* **39**:47–79

Tabashnik BE (1994b) Delaying insect adaptation to transgenic plants: seed mixtures and refugia reconsidered. *Proc R Soc Lond B Biol Sci* **255**:7–12

Tachmintzis J (1994) EEC regulatory provisions for the release of transgenic plants. *Biocontrol Sci Technol* **4**:591–592

van Emden HF (1990) The interaction of host plant resistance to insects with other control measures. In: *Proceedings of Brighton Crop Protection Conference, Pests and Diseases.* The British Crop Protection Council **3**:939–948

van Emden HF (1991) The role of host plant resistance in insect pest mis-management. *Bull Entomol Res* **81**:123–126

van Emden HF, Wratten SD (1990) Tri-trophic interactions involving plants in the biological control of aphids. In: Peters DC, Webster JA, Chlouber CS (eds) *Aphid-plant interactions: Populations to molecules,* USDA-ARS & Okalahoma State University, pp 29–43

Van Rie J (1991) Insect control with transgenic plants: resistance proof? *Trends Biotechnol* **9**:177–179

Vierling E (1991) The roles of heat shock proteins in plants. *Ann Rev Plant Physiol Plant Mol Biol* **42**:579–620

Vinson SB, Elzen GW, Williams HJ (1987) The influence of volatile plant allelochemicals on the third trophic level (parasitoids) and their herbivorous hosts. In: Labeyrie V, Fabres GKL, Lachaise D (eds) Insect-plants. *Proceedings of the 6th International Symposium on Insect-plant Relationships.* Dr W Junk, Dordrecht, pp 109–114

Walter DE, O'Dowd DJ (1992) Leaf morphology and predators: effects on leaf domatia on the abundance of predatory mites (Acari: Phytoseiidae). *Environ Entomol* **21**:478–484

Waterman PG, Mole S (1989) Extrinsic factors influencing production of secondary metabolites in plants. In: Bernays EA (ed) *Insect-plant interactions.* Vol 1. CRC Press, Florida, pp 107–134

Wellings PW (1986) Assessing the effectiveness of *Aphidius ervi* as a biological control agent of the blue-green lucerne aphid *Acyrthosiphus kondoi*: a parasite inclusion experiment. In: Hodek I (ed) *Ecology of Aphidophaga.* Academia, Prague, Dr W Junk, Dordrecht, pp 385–390

Wellings PW (1988) Sex ratio variations in aphid parasitoids: Some influences on host-parasitoid systems. In: Niemczyk E, Dixon AFG (eds) *Ecology and effectiveness of Aphidophaga.* SPB Academic Publishing, The Hague, pp 243–248

Wellings PW, Ward SA (1994) Host-plant resistance to herbivores. In: Leather SR, Walters KFA, Mills NJ, Watt AD (eds) *Individuals, populations and patterns in ecology.* Intercept, Andover, pp 199–211

Wellings PW, Ward SA, Dixon AFG, Rabbinge R (1989) Crop loss assessment. In: Minks AK, Harrewijn, P (eds) *Aphids, their biology, natural enemies and control.* Elsevier, Amsterdam, Vol C, pp 49–64

Whalon ME, Wierenga JM (1994) *Bacillus thuringiensis* resistant Colorado potato beetle and transgenic plants: some operational and ecological implications for deployment. *Biocontrol Sci Technol* **4**:555–561

Wilson LJ (1994) Plant quality effects on life history parameters of two-spotted spider mites (Acari: Tetranychidae) in cotton. *J Econ Entomol* **87**:1665–1673

Wilson FD, George BW (1986) Smoothleaf and hirsute cottons response to insect pests and yield in Arizona. *J Econ Entomol* **79**:229–232

Zadoks JC (1980) Yields, losses and costs of crop protection – three views, with special reference to wheat growing in The Netherlands. In: Teng PS, Krupa SV (eds) *Crop loss assessment.* Miscellaneous Publications, University of Minnesota, Agricultural Experiment Station, No 7 pp 17–22

Zadoks JC (1991) A hundred and more years of plant protection in The Netherlands. *Neth J Plant Prot* **97**:3–24

Zalucki MP, Daglish G, Firempong S, Twine P (1986) The biology and ecology of *Heliothis armigera* (Hübner) and *H. punctigera* Wallengren (Lepidoptera: Noctuidae) in Australia: What do we know? *Aust J Zool* **34**:779–814

INFLUENCE OF VARIABLE QUALITIES OF *EUCALYPTUS* AND OTHER HOST-TREES ON SAP-FEEDING INSECTS

Roger A. Farrow and Robert B. Floyd

ABSTRACT

Significant differences in susceptibility to attack by sap-feeding, cardiaspine psyllids occur between populations and individuals of *Eucalyptus* species in south-east Australia. These differences are related to the effects on psyllid performance of host-tree variations in phenology, morphology and biochemistry at the species, population and individual level. The effects of variations in host-tree qualities on the performance of sap-feeding pests of eucalypts in Australia and of forest and woodland tree species in the northern hemisphere are reviewed with special reference to the concepts of environmental stress and host plant vigour. The strategies and adaptations that sap-feeders have evolved to overcome the resistance of plants to herbivory and to maximise uptake of nutrients are discussed for the principle forest and woodland pest species. Although the action of sap-feeding can induce responses in the host-tree that may defend the tree against further feeding, other responses, such as recovery from insect attack by refoliation, may make trees more suitable for sap-feeders. Some sedentary sap-feeders have adapted, through natural selection, to the characteristics of individual host-trees and exist as clonal colonies on individual trees. This is an inevitable response to the heterogeneity of host-tree populations. Although some plant structures and metabolites reduce the performance of insect herbivores and appear to act as defenses to insect feeding, most of these characteristics appear to have evolved as adaptive responses to the plant's physical environment, rather than as specific defences to insect herbivores, and rarely prevent counter-adaptation by insect herbivores. Sap-feeding populations appear to be limited more by the phenological, morphological and biochemical variability of host-tree populations than by the effects of any individual plant compounds or structures.

Key words: nutrients, metabolites, stress, resistance, variability

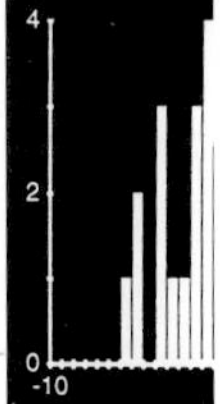

Frontiers of Population Ecology, R.B. Floyd, A.W. Sheppard and P.J. De Barro (eds), CSIRO Publishing, Melbourne, 1996 pp. 325–338

Introduction

The majority of sap-feeding insect herbivores belong to the order Hemiptera and have specialised piercing and sucking mouthparts. They extract sap from vascular conducting cells of the phloem and xylem, from intact or macerated cells in the parenchyma, including mesophyll and cortex, and/or from intra-cellular spaces (Raven 1983; Miles and Taylor 1994) (Table 1). Sap-feeding insects, like all herbivores, are exposed to a range of plant structures and chemical products, depending on feeding site, that may vary between species, populations and individuals of the plant host (Denno and McClure 1983). Many plant products are supposed to have evolved as defences against insect feeding (Rhoades and Gates 1976) through a process of co-evolution (Erhlich and Raven 1964). However, such products do not generally protect plants from pest outbreaks or from the ravages of introduced herbivores and many plant metabolites have little effect on insect feeding and fitness (Bernays 1981, 1982). It has been suggested that the so-called plant secondary compounds or allelochemicals play a primary metabolic role (Siegler 1977), rather than being mere by-products of metabolism (Whittaker and Feeny 1971). Furthermore, many plant characteristics appear to be primarily adapted to the physical environment although they may act as insect deterrents (Bernays and Graham 1988). Herbivores can overcome putative plant defences because of the potential for rapid selection and adaptation during the fast generation cycling of insect herbivores compared with their host-trees (Bernays and Graham 1988). The low abundance of most insect herbivores also suggests there may be a distinct lack of selective pressure on host-plants of the kind needed to develop the co-evolutionary arms race between insects and their host-plants (Feeny 1976; Lawton and Strong 1981, Strong *et al.* 1984).

We found that the level of necrosis caused by the psyllid *Cardiaspina retator* to the leaves of five, widely separated provenances of *Eucalyptus camaldulensis,* planted together in a fully replicated trial, varied significantly between provenances as shown in Fig.1 (Floyd *et al.* 1994). These variations are due to the effects of phenotypic variation in the host on psyllid population dynamics. The most susceptible provenance is the local one (Whitehead's Creek) to which, possibly, the local psyllid population is the most closely adapted. The variation within provenances was also considerable, suggesting that inter-tree phenotypic variation was also influencing psyllid populations dynamics (R. Floyd and R. Farrow, unpublished). In a natural population of *E. blakelyi* we found similar variations in susceptibility to attack between individual trees to a second species of psyllid, *C. albitextura.* In this review we relate our observations of the effects of host-tree variability in eucalypts on cardiaspine psyllids and other sap-feeders in Australia, with the variability found in conifers and deciduous trees in the forests of the northern hemisphere, and its effect on a range of pests in the sap-feeding guild (Table 2).

Plant phenology

The availability and quality of food supply for insect herbivores are affected by variations in host-plant phenology as insect herbivore life cycles need to coincide with periods of shoot and leaf flushing to maximise fitness (Faeth 1987). Variations in the timing of flushes between host-tree species, populations and individuals will affect herbivore performance, particularly when growth and foliage flushes are rapid so that favourable food sources are present for only brief periods. When defoliation of flushes occurs as a result of high herbivory, replacement growth can sustain additional generations of herbivores and cause pest outbreaks (e.g., tip-feeding bugs on eucalypts). In herbivores that feed on mature foliage such as cardiaspine psyllids, replacement flushes can have a negative effect on populations because of the absence of suitable mature foliage for oviposition and outbreaks may collapse (Clark 1962). The timing of abscission can affect the completion of life-cycles particularly in sedentary species feeding on deciduous trees (Faeth 1987).

Table 1. Feeding sites of representative sap-feeding groups (after Raven 1983, Speight & Wainhouse 1989)

Tissue	Strategy	Insect Group
Xylem	Vascular	Cercopids Cicadas
Phloem	Vascular	Aphids Cardiaspine psyllids* Eriococcids Eurymelids
Parenchyma	Maceration	Adelgids Coccids Cryptococcids Diaspidids Margarodids
	Laceration	Mirids* Pentatomids
	Cavity	Coreids*

* indicates necrosis inducing insects

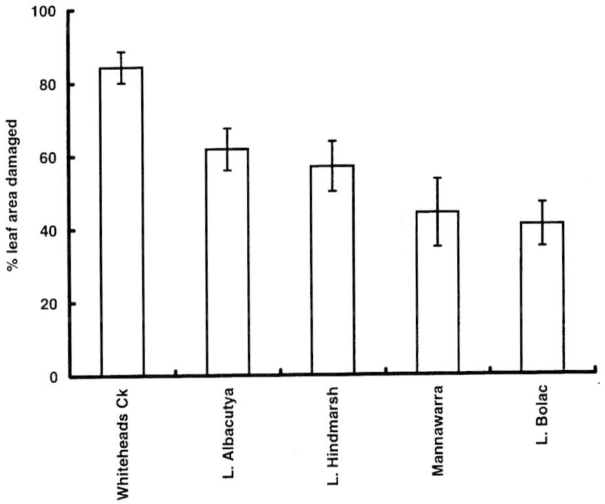

Fig. 1. Maximum leaf damage (± s.e.) caused by *Cardiaspina retator* on 5 provenances of *Eucalyptus camaldulensis*, between February 1992 and August 1993. These were selected from 29 provenances of *E. camaldulensis* planted in a fully replicated trial in 1981 at Whitehead's Creek near Seymour, Victoria. Leaf damage = % leaf necrosis + premature leaf fall. (Adapted from Floyd *et al.* 1994).

The physical and chemical qualities of leaves and shoots also vary over time in connection with seasonal plant phenology, as discussed in later sections, although it is often difficult to separate the effects of biochemical from physical changes on herbivores as plant structures age. The preference for plant structures of a particular age is exemplified by the cardiaspine psyllids which prefer to oviposit and feed on fully expanded leaves of intermediate hardness rather than on young expanding leaves preferred by many chewing insects (R. Floyd and R. Farrow, unpublished).

Plant morphology

The leaves and stems of plants vary in terms of hardness, surface waxes, hairs, spines and degree of lignification which contributes to toughness. These attributes can often be related to aspects of the

Table 2. List of principle outbreak species of sap-feeders on forest and woodland trees

Species	Family	Common name	Host	Geographic range	Attachment site	Feeding site	Reference
Adelges picea	Adelgidae	Balsam woolly aphid	*Abies fraseri* + other spp (*A. alba*)	Introduced NA Native EUR	Branches	Parenchyma	Hain 1988
Amorbus spp	Coreidae	Tip-feeding bug Stinkbug	*Eucalyptus* spp	Native AUS	Shoot tips	Parenchyma	Miles & Taylor 1994
Cardiaspina albitextura *C. retator*	Psyllidae	White lace lerp Red gum basket lerp	*Eucalyptus blakelyi* *E. camaldulensis*	Native AUS Native AUS	Leaf	Phloem	Clark 1962, 1964
Cinara pini *C. pinea*	Aphididae	Pine aphid	*Pinus sylvestris*	Native EUR	Needle	Phloem	Kidd 1985
Cryptococcus fagi	Cryptococcidae	Beech scale	*Fagus sylvatica* *Fagus* spp	Native EUR Introduced NA	Trunk	Parenchyma	Wainhouse & Gate 1988
Drepanosiphum platanoides	Aphididae	Sycamore aphid	*Acer pseudoplatanus*	Native EUR	Leaf underside	Phloem	Dixon 1970
Elatobium abietinum	Aphididae	Green spruce aphid	*Picea abies* *P. sitchensis*	Native NA Introduced EUR	Twigs	Phloem	Major 1990
Eriococcus confusus *E. coriaceus*	Eriococcidae	Eucalypt scale	*Eucalyptus* spp	Native AUS Introduced NZ	Branches	Phloem	Patel 1971
Fiorina externa *Nuculaspis tsugae*	Diaspididae	Hemlock armoured scale	*Tsuga canadiensis* (*T. sieboldii*)	Introduced NA Native Japan	Needle	Phloem	McClure 1983
Matsucoccus feytaudi	Margarodidae	Cochineal scale	*Pinus maritima*	Native EUR	Trunk	Parenchyma	Schvester 1988
Pulvinaria regalis	Coccidae	Horse chestnut scale	*Aesculus* spp + other deciduous trees	Native EUR Introduced UK	Branches	Parenchyma	Speight & Wainhouse 1989

Where NA, North America; EUR, Europe; AUS, Australia; NZ, New Zealand; UK, United Kingdom plant's environment, such as the aridity or nutrient status (e.g. waxy, sclerophyllous leaves of eucalypts are primarily a xeric adaptation). However, the presence of trichomes and waxes on leaves and shoots can hinder sap-feeders from settling, attaching to the cuticle, attaching eggs or from penetrating their stylets (Levin 1973). This type of defence has been termed *neutral resistance* (Edwards 1989). For example, the holm oak is less susceptible to attack by the aphid, *Tuberculoides annulatus,* because of the presence of a dense mat of trichomes on the leaf (Kennedy 1986). The presence of stone cells appears to inhibit stylet penetration by the scale (*Cryptococcus fagi*) into its host, the beech *(Fagus sylvatica*), in European forests (Wainhouse and Howell 1983).

In our study of resistance to cardiaspine psyllids in two species of woodland eucalypt in southeast Australia, we measured leaf hardness, wax thickness and wax structure in the leaves of relatively resistant and susceptible trees in relation to oviposition and nymphal survival (R. Floyd and R. Farrow, unpublished). The two leaf attributes that were most consistently related to psyllid survival and to numbers of eggs laid were the nature of the cuticle wax architecture and the size of the stomatal aperture. Leaves that were covered with cuticular waxes that formed large plates rather than a dense fibrous matrix and had larger stomatal apertures were less susceptible to psyllid feeding and oviposition. It is unlikely that these traits are a direct defence to psyllid feeding but form part of the neutral resistance syndrome of the plant. The populations of river red gum *(E. camaldulensis*) that were more resistant to the psyllid, *C. retator*, by virtue of their wax plates, originated from inland Australia where they are not naturally exposed to this psyllid.

There is increasing evidence that inter-tree variations in morphological attributes can affect the herbivore abundance. In the northern hemisphere individual beeches resistant to the scale (*C. fagi*) have more stone cells than susceptible trees (Wainhouse and Howell 1983). Individual trees with resistance to the psyllid (*C. albitextura*) are relatively rare in naturally occurring populations of Blakely's red gum (*E. blakelyi*) but again the level of resistance was correlated with the degree of waxiness (R. Floyd and R. Farrow, unpublished).

Plant nutrients

Nutrient levels and composition

Living plants do not represent an optimal food resource for insects, principally because plant matter is very low in protein compared with insect tissue (Southwood 1973; Mattson 1980) and has different proportions of amino acids which vary between parenchyma, phloem and xylem (Table 3). The amounts of these compounds also vary seasonally according to seasonal growth patterns and mobilisation and translocation of reserves (Speight and Wainhouse 1989) and spatially according to the nutrient status of the soil (Judd *et al.* 1991). Fluctuations in soil moisture levels through drought and waterlogging can adversely affect the tree's ability to maintain its water balance and cause the conversion of polysaccharides and proteins to soluble carbohydrates, amino acids and amides in the phloem (White 1969, 1984; Brodbeck and Strong 1987).

Sap-feeders have to cope with low nitrogen diets especially on host-trees growing in low nutrient sites. In forest and farmland eucalypts, leaf nitrogen varies between 0.5 and 2.5% (Fox and Macauley 1977; R. Floyd and R. Farrow, unpublished) similar to that of conifer needles (Mattson 1980) but less than that of deciduous trees such as oak, 2–5%, (Feeny 1970). A threshold value of 1% nitrogen is required for the survival of leaf chewers on eucalypts (Fox and Macauley 1977; Ohmart *et al.*. 1987) and sap-feeders such as psyllids appear to have low survival where leaf nitrogen falls below 1.4% (R. Floyd and R. Farrow, unpublished). Sap-feeders increase nitrogen uptake by a number of mechanisms including: high rates of sap-uptake through the development of strong pharyngeal pumps (Horsfield 1978) (e.g., cercopids); injection of hydrolytic and proteolytic enzymes to cause the cell walls in the leaf parenchyma to break down and the contents to autolyse and provide access

Table 3. Relative amounts of nutrients and metabolites in intra-cellular leaf tissue and in sap (after Raven 1983)

		Tissue		Sap	
		Leaf	Parenchyma	Phloem	Xylem
Nitrogen	Form	protein	amino-acids	amides	amides
	Concentration	1-5 %	low (> 1%)	low (> 1%)	very low(> .01%)
	Variability	high	high	low	low
Carbon	Form	cellulose	sugars	sugars	sugars
	Concentration	moderate	low	high	very low
	Variability	moderate	low	low	low
Metabolites	Form	tannins etc	phenols etc	phenols etc	phenols etc
	Concentration	high	low	low	very low
	Variability	high	low	low	low

to the chloroplast proteins (Huffaker 1982) (e.g., cardiaspine psyllids); colony formation which causes the local sap flow to increase, with nutrients drawn in from adjacent branches (Way and Cammell 1970) (e.g. stem-feeding eriococcids); and induction of nurse cells around the stylets to assist solute transfer (Rohfritsch 1988) (e.g. stem-feeding adelgids).

The low nutrient levels found in most plant tissues can also operate as a form of innate or neutral resistance to insect feeding (Edwards 1989), slowing development and increasing exposure to the effects of natural enemies (Moran and Hamilton 1980). In our studies of *C. albitextura* on *E. blakelyi*, we found that those trees with the lowest levels of psyllid attack had the lowest leaf nitrogen (R. Floyd and R. Farrow, unpublished). The psyllid nymphs developing on this foliage also grew slower and showed higher mortality than on foliage with more nitrogen although there were no differences recorded in terms of parasitisation.

The presence and proportion of amino acids could be potentially more important for insect nutrition than total nitrogen (House 1969). Kidd *et al.* (1990) have shown that pine aphids (*Cinara* spp) are more dependent for growth and survival on suites of specific amino acids present in the sap, rather than total nitrogen, although in the sycamore aphid (*Drepanosiphum platanoides*) changes in total amino acid titre had more impact on reproduction than individual acids (Dixon *et al.* 1993). This difference could be due to some species having endosymbionts to supply essential amino acids (Dadd 1973; Mattson 1980). Where there is dependence on a suite of host-tree amino acids, sap-feeders will tend to become host adapted and to speciate among host-tree species, depending on host-tree chemistry. This may explain why cardiaspine psyllids and tree-feeding aphids are often restricted to single or closely related host-tree species (Taylor 1962). In contrast, stem feeders such as the eriococcids and eurymelids are less host specific, possibly because of the greater uniformity of their food source.

Seasonal tree phenology may affect the flow and composition of phloem and xylem sap and composition of parenchyma solutes (Speight and Wainhouse 1989). Most sap-feeder populations effectively track seasonal trends in nitrogen or amino acid concentrations in the leaves or phloem, depending on feeding site, suggesting that populations are mainly resource limited and are not in a classical Nicholsonian dynamic equilibrium, even at low densities (Kidd *et al.* 1990). For example, two peaks of reproduction are observed in spring and autumn in sycamore aphids (*Drepanosiphum plantanoides*) corresponding to the two peaks of nitrogen in the phloem Dixon (1970). On conifers, the pine aphid *C. pini* moves from young distal internodes in late spring to the older internodes in late summer in response to changes in nitrogen (Kidd 1985). In eucalypts where leaf abscission occurs in mid-summer between the spring and autumn flushes, mobilisation of nutrients could be more extended but its effects on sap-feeders have not been determined.

Stress

Host-plant stress is induced by moisture fluctuations (White 1969) and by drought (Mattson and Haack 1987). It may cause changes in nutrient and metabolite levels (Rhoades 1983; White 1969) and resource allocation (Speight and Wainhouse 1989) resulting in insect outbreaks. There are a number of examples where increased insect damage has occurred on stress-affected trees. For example water-stressed firs or those species of fir which are more susceptible to water stress, suffer greater damage from the introduced balsam woolly aphid (Hain 1988).

White (1969, 1984) showed that outbreaks of *C. albitextura* on *E. camaldulensis* could be correlated with extreme fluctuations in seasonal rainfall resulting in waterlogging followed by drought, although such variation is often normal in the environment of eucalypts. He proposed that this adversely affected the eucalypt's water balance and caused an increase in soluble metabolites to maintain turgor and minimise water loss from transpiration (Speight and Wainhouse 1989). The increased availability of soluble amino acids, he suggested, improved the survival of the first instar psyllids and led to population increases and outbreaks. Although the release of soluble nitrogen has been observed in other studies (Brodbeck and Strong 1987), it has not been confirmed in eucalypts (Landsberg 1990a), nor has it been shown that the amino acids released, mostly proline, would benefit herbivore survival (Cockfield 1988) and supporting evidence is still required (Larsson 1989). An alternative cause for psyllid outbreaks was favoured by Clark (1962, 1964) which related to the production of better quality foliage by rainfall and reflects the *plant vigour/climate release hypothesis* (Price 1991). However, biochemical and ecological evidence is still needed to confirm which of these two hypotheses is correct.

Inter-tree variation

Variations in foliar nutrient levels between trees can be due to environmental variations in soil fertility (Landsberg *et al.* 1990; Judd *et al.* 1991) or possibly to the localised effects of climatic stress (Landsberg 1990a). In our studies of the nutritional status of resistant and susceptible Blakely's red gum (*E. blakelyi*) to white lace lerp (*C. albitextura*) we found that resistant trees occurred at a frequency of about 1% and were scattered throughout a 500 ha open woodland (Floyd *et al.* 1994). In addition to the morphological differences previously discussed these trees were very low in nitrogen with an average composition of 1.3% compared to 1.7% in the susceptible trees. It is unlikely that these differences were directly due to environment variation. However, it has not been possible to separate cause and effect because the foliage of heavily attacked trees is more juvenile, as such trees are frequently replacing necrotic foliage, and juvenile foliage contains more nitrogen (Floyd *et al.* 1994; Landsberg 1990b). Even when leaves were matched for age this difference persisted suggesting a whole tree difference in nitrogen level (Fig. 2), which may have resulted from repeated defoliation cycles. An alternative hypothesis is that the trees with low lerp populations have inherently low nitrogen levels due to genetic differences between trees.

Effects of fertiliser application

Early studies suggested that fertiliser applied to forest trees increased populations of sap-feeding insects and decreased those of chewing insects (Stark 1965) but subsequent studies reveal a more complex situation depending on what form the nitrogen was added and on the overall response of the species to fertilisation (Carrow and Betts 1973). Soil nitrogen levels have substantially increased over the lifetime of the eucalypts remaining in farmland, as a result of nitrate release from legumes introduced in pasture improvement programs and fertiliser application (Landsberg 1990a; Farrow and Floyd 1995). Eucalypts planted in agricultural land can have up to twice as much leaf nitrogen

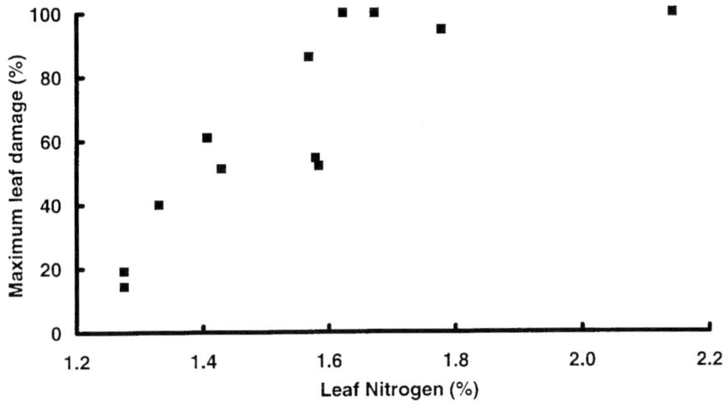

Fig. 2. Maximum leaf damage caused by *Cardiaspina albitextura* on 12 individual *Eucalyptus blakelyi* trees in relation to leaf nitrogen. The trees were part of a natural stand on farmland. Leaf damage = % leaf necrosis + premature leaf fall.

as those planted on ex-forest sites (Judd *et al.* 1991). Although trees growing in agricultural land appear to be suffering from heavier levels of insect attack than in closed woodland (Landsberg *et al.*.1990) there is little evidence that plantations on farmland are more attacked by sap-feeders than those on forest sites or that the increased nitrogen levels are causative. Eucalypts are widely fertilised in plantations and in some cases irrigated with sewage effluent which in our studies resulted in an increase of only 0.2% of leaf nitrogen (R. Floyd and R. Farrow, unpublished). No consistent differences have been found in the level of feeding by sucking insects or other insect herbivores between trees irrigated with water and those with effluent (R. Floyd and R. Farrow, unpublished).

Plant metabolites

Chemistry and function

Plants contain a range of metabolites such as alkaloids, anthocyanins, glycosides, phenols, tannins, and terpenes, among others (Bernays and Chapman 1994). Although most are stored intracellularly or in glands and specialised canals, some are present in low concentrations in phloem and xylem sap (Raven 1983). The production of metabolites has been related to carbon/nitrogen resource allocation (Tuomi *et al.* 1988). Where growth is nitrogen limited on nutrient-poor sites, it is proposed that the surplus carbon produced by photosynthesis cannot be used for growth and is stored in long chain carbon-based compounds, such as terpenoids and phenolics. Eucalypts are adapted to low fertility sites and this is associated with an abundance of phenols and tannins which may make up to 40% and 27% of leaf dry weight, respectively, in the species measured by Macauley and Fox (1980) and of terpenes which may make up to 20% of fresh leaf weight (Morrow and Fox 1980). However, there is little evidence that variations in site fertility affect the levels of terpenoids and phenolics in eucalypt foliage. The performance of some insect herbivores of eucalypts is relatively unaffected by high levels of phenolics, tannins and terpenoids in their diet (Fox and Macauley 1977) although in others feeding behaviour is influenced by the composition of terpenoids (Edwards *et al.* 1993). In many monophagous species which have become well adapted to their host-plant, the metabolites often act as host attractants and phagostimulants (van Emden 1978; Bernays and Graham 1988). In general, sap-feeders can escape the effects of many of these compounds as they avoid feeding on bound proteins and insoluble compounds and on terpenoids located in oil glands and resin canals. This may make their interactions with soluble nutrients more apparent than in chewing insects.

Phenological effects

Early studies suggested that plant metabolites such as tannins increase with leaf age (Feeny 1970) but more recent work, including studies on eucalypts (Macauley and Fox 1980; Landsberg 1990b) show that tannins and phenols are highest in newly expanded foliage and decline with leaf age. Young needles of conifers are better endowed with resin canals containing terpenoids and contain more phenolics than older needles and there is a trade-off between amino acid availability and level of plant metabolites. As a consequence, sap-feeding aphids find current and one-year old growth less suitable as a food source, despite the advantage conferred by the high amino acid levels in young needles (Kidd *et al.* 1990). Similar hypotheses could be applied to the lerp-forming psyllids in the genera *Cardiaspina, Glycaspis, Creeis* etc, which all feed on fully expanded leaves of intermediate age, however, other free-living psyllids in the genus *Ctenarytaina* all feed in the shoot tips, protected among the unfolding leaves, where they would be presumably exposed to high levels of plant metabolites. Tip-feeding bugs in the genus *Amorbus* also feed on the first centimetre or so of shoot by the method of 'osmotic pump' feeding, which would expose them to relatively high concentrations of inter- and intra-cellular products (Miles and Taylor 1994). The high production of metabolites in the shoot tips also reflects the high rates of metabolism in growing tissue and is not necessarily a response to protect such tissues from herbivory. Herbivores may also avoid the shoots tips because of increased exposure to natural enemies and it is noteworthy that tip-feeding bugs are well protected from predation by the emissions from their repugnatorial glands. The preferences of herbivores for different aged sites on host-trees has also been demonstrated by insect transfer experiments, for example, the preference of a gall-forming aphid for mature cottonwood trees (*Populus angustifolia*) rather than juveniles (Kearsley and Whitham 1989) and the effect that this transfer had on aphid fitness.

Stress

Production of plant metabolites are often expected to be reduced during periods of stress (Speight and Wainhouse 1989). However, several studies have shown that concentration of metabolites such as terpenes and alkaloids may actually increase in leaves as a result of stress (Mattson and Haack 1987), due possibly, to the re-allocation of resources from growth to storage (Gershenzon 1984). The composition of chemical groups may also change as a result of stress. Stress-induced changes in terpene composition influenced the performance of the green spruce aphid, *Elatobium abietum*, (Major 1990), but there have been no studies of these effects in eucalypts.

Inter-tree variation

The carbon-based compounds usually occur in arrays of closely related compounds whose synthesis is largely under genetic control. In populations of some tree species these arrays of chemicals form polymorphic series e.g. the terpenoids in conifers (Smith 1964), in which each tree could be regarded as an individual with respect to terpenoid composition (Hanover 1967). Up to 1000 metabolites occur in an infinite number of combinations in individual trees and are the major source of chemical diversity and variability (Hanover 1975). In eucalypts, which have arrays of similar terpenoids, the differences between individuals are generally smaller (Boland *et al.* 1991) although some species are dimorphic or polymorphic for the concentration of particular terpenes such as cineole (Boland *et al.* 1991; Edwards *et al.* 1993) as well as polyphenols (Hillis and Isoi 1965), anthocyanins (Wilcox 1982) and probably other metabolites. Individual and inter-provenance differences in the terpenoids of two species of eucalypt had no impact on sap-feeding cardiaspine psyllids (R. Floyd and R. Farrow, unpublished). This was not surprising since the terpenoids are primarily concentrated in discrete glands in the leaf and are isolated from the psyllid feeding sites. Quantitative and qualitative variations in plant metabolites can also affect insect behaviour because of their role as phagostimulants, as discussed earlier. In a study of the performance of two diaspidid species on two species of hemlock *Tsuga canadensis* and *T. sieboldi*, McClure and Hare (1984) showed that fecundity was positively correlated with acyclic terpenes in one species of scale but the opposite was true for a second species. Ge (1981) found that resistance of individual pines to attack by the margarodid (*Matsucoccus matsumurae*) was due to the absence of three resin acids, again suggesting a phagostimulatory role.

Induced responses

The very act of feeding by sap-feeders elicits a range of responses from the plant host. Penetration of stylets into plant tissue by sap-feeders may induce biochemical and physical wound responses by the host-plant. In the beech scale (*C. fagi*) which feeds on parenchyma tissue, the formation of gall-like cell complexes around the stylets enhances the nutritive value of the tissues (Lonsdale 1983). On the other hand, the heavy bark fissuring and production of procyanidans at the stylet sites can prevent probing and settlement of following generations of *C. fagi* (Lunderstädt and Eisenwinder 1989). Adelgids such as *A. picea* cause wound responses in host-trees. In resistant trees these responses are restricted to the periderm and inhibit adelgid development, but in susceptible trees a hypersensitive response occurs which spreads from the periderm into the conducting tissue and interferes with the conduction process (Hain 1988). Feeding by the pine aphid (*Cinara pinea*) results in a lignification of the cortex at the stylet sites and a reduction in stem quality in the year following attack and these sites are subsequently avoided by the aphids (Kidd *et al.* 1990). Higher concentrations of terpenes may also be found at feeding sites but these may be involved in repair rather than increased defence (Kidd *et al.* 1990). The response to feeding by pine aphid, *Shizolachnus pineti*, in Scots pine includes increases in phenolics and total nitrogen and a change in composition of amino acids which is decidedly unfavourable to this aphid (Kidd *et al.* 1990). Aphids in the genus *Elatobium* induce needle chlorosis and the mobilisation of amino acids from protein breakdown which, in contrast to the preceding species improves nutrition and performance (Fisher 1987). Senescing needles have a high nutritive value and are sought by these aphids (Kidd *et al.* 1990). Feeding by cardiaspine psyllids causes necrosis and premature senescence of host leaves which favours herbivore development. Induced responses which deter insect feeding are true plant defences but as we have shown, some responses are overcome by the herbivore and used to its advantage; the necrotisation response to cardiaspine psyllids presumably evolved in this way.

Discussion

Plants exhibit an array of chemical and morphological characteristics which affect feeding behaviour and performance of sap-feeders as well as induced responses to the act of sap-feeding. Sap-feeders have evolved a range of adaptations to counteract the effects of these characteristics in order to maximise their survival and reproduction. Such adaptation can occur quite rapidly. In a study of the introduced hemlock armoured scales (*Fiorina externa* and *Nuculaspis tsugae*) McClure (1983) concluded that it took about 12 years for these species to adapt to the specific attributes of the indigenous host species. The western yellow pine (*P. ponderosa*) shows considerable inter-tree variations in its terpenoids (Smith 1964) and is colonised by the black pineleaf scale (*Nuculaspis californica*). Founding colonies progressively adapt to their individual host-tree and their performance is eventually unaffected by the host-tree terpenoids. This adaptation may extend over the life of the tree which corresponds to about 200 insect generations. Most individuals cannot then survive on adjacent trees with different genotypes and terpene compositions. A similar adaptation to individual host-trees is found in beech scale (*C. fagi*) (Wainhouse and Howell 1983).

Adaptation to the chemistry of individual host-trees by sedentary herbivores is not a surprising development in view of the strong degree of heterozygosity and genetic variation between individuals trees (Edmunds and Alstad 1978). It is possible that the large infestations of the witch's broom causing armoured scale (*Maskellia globosa*, Diaspididae), which build up in individual eucalypt trees and may eventually result in tree death, are a consequence of this type of adaptation. As Edmunds and Alsted (1978) have pointed out, the trend for sedentary sap-feeder demes to increase their fitness by adapting to the chemical and morphological profile of their host and to form clones, makes such clones ill adapted to spread through a polymorphic host population without further adaptation and selection. More recent studies have critically examined local adaptation and found that it operates at a range of scales and intensities depending on the extent of dispersal and of gene flow (Unroh and Luck 1987; Cobb and Whitham 1993; Hanks and Denno 1994).

Such extreme adaptation cannot occur in most sap-feeders because of their mobility and there is probably a trade-off in most insect herbivores between adaptation at the individual and at the population level with respect to host plant variability. Our studies on cardiaspine psyllids suggest that a number of host-tree characteristics may determine the susceptibility of a tree to psyllid damage and that these are represented in a spatial mosaic of tree phenotypes of varying morphological and biochemical characteristics, superimposed on which are the effects of environmental variables. In more general terms, the examples presented in this review suggest that trees are genetically and phenotypically individually variable over a range of spatial scales and for a range of characteristics which can include shoot and leaf flushing time, morphological features of the plant surface, chemical structure and abundance of metabolites and nutrients and intensity and mode of operation of induced responses. It is this variability that prevents populations of sap-feeding species from maximising their reproduction and survival and overcoming host-plant resistance at the population level. There is little evidence that the factors in the host-tree that influence resistance are primarily related to insect defence but rather to other environmental factors and to the basic physiological processes of production, mobilisation and storage of nutrients and metabolites.

References

Bernays EA (1981) Plant tannins and insect herbivores: an appraisal. *Ecol Entomol* 6:353–360

Bernays EA (1982) *The insect on the plant – a closer look.* Proc 5th int Symp Insect-Plant Relationships, Wageningen. Pudoc, Wageningen, pp 3–17

Bernays EA, Chapman RF (1994) *Host-plant selection by phytophagous insects.* Chapman and Hall, London

Bernays E, Graham M (1988) On the evolution of host specificity in phytophagous arthropods. *Ecology* 69:886–892

Boland DJ, Brophy JJ, House APN (1991) *Eucalyptus leaf oils – use, chemistry, distillation and marketing.* Inkata, Sydney

Brodbeck B, Strong D (1987) Amino acid nutrition of herbivorous insects and stress in host-plants. In: Barbosa P, Schultz JC (eds) *Insect outbreaks.* Academic Press, San Diego, pp 347–364

Carrow RI, Betts RE (1973) Effects of different foliar applied nitrogen fertilisers on balsam woolly aphid. *Can J For Res* 3:122–139

Clark LR (1962) The general biology of *Cardiaspina albitextura* (Psyllidae) and its abundance in relation to weather and parasitism. *Aust J Zool* 10:537–86

Clark LR (1964) The population dynamics of *Cardiaspina albitextura* (Psyllidae). *Aust J Zool* 12:362–380

Cobb NS Whitham TG (1993) Herbivore deme formation on individual tree: a test case. *Oecologia* 94:496–502

Cockfield SD (1988) Relative availability of nitrogen in host-plants of invertebrate herbivores: three possible nutritional and physiological definitions. *Oecologia* 77:91–94

Dadd RH (1973) Insect nutrition: current developments and metabolic implications. *Annu Rev Entomol* 18:381–420

Denno RF, McClure MS (1983) Variability: a key to understanding plant-herbivore interactions. In: Denno RF, McClure (eds) *Variable plants and herbivores in natural and managed ecosystems.* Academic Press, New York, pp 1–12

Dixon AFG (1970) Quality and availability of food for a sycamore aphid population. In: Watson A (ed) *Animal populations in relation to their food resources.* Blackwell, Oxford, pp 271–287

Dixon AFG, Wellings PW, Carter C, Nichols JFA (1993) The role of food quality and competition in shaping the seasonal cycle in the reproductive activity of the sycamore aphid. *Oecologia* **95**:89–92

Edmunds GF, Alstad DN (1978) Coevolution in insect herbivores and conifers. *Science* **199**:941–945

Edwards PB, Wanjura WJ, Brown WV (1993) Selective herbivory by Christmas beetles in response to intraspecific variation in *Eucalyptus* terpenoids. *Oecologia* **95**:551–557

Edwards PJ (1989) Insect herbivory and plant defence theory. In: Grubb PJ, Whittaker, JB (eds) *Toward a more exact ecology.* Blackwell scientific Pubs, Oxford, pp 275–298

Ehrlich PR, Raven PH (1964) Butterflies and plants: a study in coevolution. *Evolution* **18**:586–608

Faeth SH (1987) Community structure and folivorous insect outbreaks: the role of vertical and horizontal interactions. In: Barbosa P, Schultz JC (eds) *Insect outbreaks.* Academic Press, San Diego, pp 135–171

Farrow RA, Floyd RB (1995) Effects of changing landuse on eucalypt dieback in Australia in relation to insect phytophagy and tree re-establishment. In: Harrington R, Stork NE (eds) *Insects and a changing environment, Proc R Entomol Soc Symp No 17.* Rothamsted, 7–12 Sept 1983. Academic Press, London, pp 456–460

Feeny PP (1970) Seasonal changes in oak leaf tannins and nutrients as a cause of spring feeding by winter moth caterpillars. *Ecology* **51**:565–581

Feeny PP (1976) Plant apparency and chemical defence. *Recent Adv Phytochem* **10**:1–40

Fisher M (1987) The effect of previously infested spruce needles on the growth of the green spruce aphid *Elatobium abietum* and the effect of the aphid on the amino acid balance of the host-plant. *Ann Appl Biol* **111**:33–41

Floyd RB, Farrow RA, Neumann FG (1994). Inter-and intra-provenance variation in resistance of red gum foliage to insect feeding. *Aust For* **57**:45–48

Fox LR, Macauley BJ (1977) Insect grazing on *Eucalyptus* in response to variations in leaf tannins and nitrogen. *Oecologia* **29**:145–162

Ge ZH (1981) A study on the resistance of several pine species to the bast scale (*Matsucoccus matsumurae* Kuwana). *Rep Chin Acad For* **4**:400–405

Gershenzon J (1984) Changes in the level of plant secondary metabolites under water and nutrient stress. In: Timmerman BN, Steelink C, Loewus FA (eds) *Phytochemical adaptation to stress.* Plenum, New York, pp 273–320

Hain FP (1988) The balsam woolly adelgid in North America. In: Berryman A (ed) *Dynamics of forest insect populations.* Plenum, New York, pp 88–109

Hanks RM, Denno RF (1994) Local adaptation in the armoured scale insect *Pseudaulacaspis pentagona* (Homoptera: Diaspididae). *Ecology* **75**:2301–2310

Hanover JW (1967) Genetics of monoterpenes 1. Gene control of monoterpene levels in *Pinus monticola* Dougl. *Heredity* **21**:73–84

Hanover JW (1975) Physiology of tree resistance to insects. *Annu Rev Entomol* **20**:75–95

Hillis WE, Isoi K (1965) Variation in the chemical composition of *Eucalyptus sideroxylon. Phytochemistry* **4**:541–550

Horsfield D (1978) Evidence for xylem feeding by *Philaenus spumarious* (L.) (Homoptera: Cercopidae). *Entomol Exp Appl* **24**:95–99

House HL (1969) Effects of different proportions of nutrients on insects. *Entomol Exp Appl* **12**:651–669

Huffaker RC (1982) Biochemistry and physiology of leaf proteins. In: Boulter D, Partier B (eds) *Nucleic acids and proteins in plants, structure, biochemistry and physiology of proteins. Encyclopaedia of plant physiology.* Springer-Verlag, New York, pp 370–394

Judd TS, Atiwell PM, Adams MA (1991) Foliar diagnosis of plantations of *Eucalyptus globulus* and *E. nitens* in southeastern Australia. In: Ryan PA (ed) *Productivity in perspective,* Third Australian Forest Soils and Nutrition Conference, 7–11 October 1991 Melbourne. Forestry Commission NSW Sydney, pp 162–163

Kearsley MC, Whitham TG (1989) Development changes in resistance to herbivory: implications for individuals and populations. *Ecology* **70**:422–434

Kennedy CEJ (1986) Attachment may be a basis for specialisation in oak aphids. *Ecol Entomol* **11**:292–300

Kidd NAC (1985) The role of host-plant in the population dynamics of the large pine aphid, *Cinara pinea*. *Oikos* **44**:114–122

Kidd NAC, Smith SDJ, Lewis GB, Carter CI (1990) Interactions between host-plant chemistry and the population dynamics of conifer aphids. In: Watt AD, Leather, SR, Hunter, MD, Kidd NAC (eds) *Population dynamics of forest insects*. Intercept, Andover, pp 183–193

Landsberg J (1990a) Dieback of rural eucalypts: the effects of stress on the nutritional quality of foliage. *Aust J Ecol* **15**:97–107

Landsberg J (1990b) Dieback of rural eucalypts: Response of foliar dietary quality and herbivory to defoliation. *Aust J Ecol* **15**:89–96

Landsberg J, Morse J, Khanna P (1990) Tree dieback and insect dynamics in remnants of native woodlands on farms. *Proc Ecol Soc Aust* **16**:149–165

Larsson S (1989) Stressful time for the plant stress-insect performance hypothesis. *Oikos* **56**:277–283

Larsson S, Ohmart CP (1988) Leaf age and larval performance of the leaf beetle *Paropsis atomaria*. *Ecol Entomol* **13**:19–24

Lawton JH, Strong DR (1981) Community patterns and competition in folivorous insects. *Am Nat* **118**:317–333

Levin DA (1973) The role of trichomes in plant defence. *Q Rev Biol* **48**:3–15

Lonsdale D (1983) Wood and bark anatomy of young beech in relation to *Cryptococcus* attack. In: Houston D, Wainhouse D (eds) *Proc. IUFRO beech bark disease working party conference*. USDA Forest Service Gen Tech Rep WO-37, pp 43–49

Lunderstädt von J, Eisenwinder U (1989). Physiological basis of the population dynamics of beech scale, *Cryptococcus fagi* Lind (Coccidae, Coccina) in beech (*Fagus sylvatica*) stands. *J Appl Entomol* **107**:248–260

Macauley BJ, Fox LR (1980) Variation in total phenols and condensed tannins in *Eucalyptus*: leaf phenology and insect grazing. *Aust J Ecol* **5**:31–35

Major EJ (1990) Water stress in Sitka spruce and its effect on the green spruce aphid. In: Watt AD, Leather, SR, Hunter, MD, Kidd NAC (eds) *Population dynamics of forest insects*. Intercept, Andover, pp 169–182

Mattson WJ (1980) Herbivory in relation to plant nitrogen content. *Annu Rev Ecol Syst* **11**:515–522

Mattson WJ, Haack RA (1987) The role of drought in provoking outbreaks of phytophagous insects. In: Barbosa P, Schultz JC (eds) *Insect outbreaks*. Academic Press, San Diego, pp 365–407

McClure MS (1983) Reproduction and adaptation of exotic hemlock scales (Homoptera: Diaspididae) on their new and native hosts. *Environ Entomol* **12**:1811–1815

McClure MS, Hare JD (1984) Foliar terpenoids in *Tsuga* species and the fecundity of scale insects. *Oecologia* **63**:185–193

Miles PW, Taylor, GS (1994) 'Osmotic pump' feeding by coreids. *Entomol Exp Appl* **73**:163–173

Moran VC, Hamilton WD (1980) Low nutritive quality as defence against herbivores. *J Theor Biol* **80**:289–306

Morrow PA, Fox LR (1980) Effects of variation in *Eucalyptus* essential oil yield on insect growth and grazing damage. *Oecologia* **45**:209–219

Ohmart CP, Thomas JR, Stewart LG (1987) Nitrogen, leaf toughness and the population dynamics of *Paropsis atomaria* (Coleoptera: Chrysomelidae): a hypothesis. *J Aust Entomol Soc* **26**:303–307

Patel JD (1971) Morphology of the gumtree scale *Eriococcus coriaceus* Maskell (Homoptera: Eriococcidae) with notes on its life history and habits near Adelaide, South Australia. *J Aust Entomol Soc* **10**:43–56

Price PW (1991) The plant vigour hypothesis and herbivore attack. *Oikos* **62**:244–251

Raven JA (1983) Phytophages of xylem and phloem: a comparison of animal and plant sap-feeders. *Adv Ecol Res* **13**:135–234

Rhoades DF (1983) Herbivore population dynamics and plant chemistry. In Denno RF, McClure MS (eds) *Variable plants and herbivores in natural and managed systems*. Academic Press, New York, pp 155–220

Rhoades DF, Gates R (1976) Towards a general theory of plant anti-herbivore chemistry. *Recent Adv Phytochem* **10**:168–213

Rohfritsch O (1988) A resistance response of *Picea excelsa* to the aphid, *Adelges abietis* (Homoptera: Aphidoidea). In: Mattson WJ, Levieux J, Bernard-Dagan C (eds) *Mechanisms of woody plant defenses against insects – search for pattern.* Springer-Verlag, New York, pp 253–266

Schvester D (1988) Variations in susceptibility of *Pinus maritima* to *Matsucoccus feytaudi* (Homoptera: Margarodidae). In: Mattson WJ, Levieux J, Bernard-Dagan C (eds) *Mechanisms of woody plant defenses against insects – search for pattern.* Springer-Verlag, New York, pp 267–275

Siegler DS (1977) Primary roles for secondary compounds. *Biochem Syst Ecol* **5**:195–199

Smith RH (1964) Variation of the monoterpenes of *Pinus ponderosa* Laws. *Science* **143**:1337–1338

Southwood TRE (1973) The insect/plant relationship – an evolutionary perspective. In: van Emden HF (ed) *Insect/plant relationships.* Blackwell, Oxford, pp 3–30

Speight MR, Wainhouse D (1989) *Ecology and management of forest insects.* Clarendon Press, Oxford

Stark RW (1965) Recent trends in forest entomology. *Annu Rev Entomol* **10**:303–324

Strong DR, Lawton JH, Southwood TRE (1984) *Insects on plants: community patterns and mechanisms.* Blackwell, Oxford

Taylor K (1962) The Australian genera *Cardiaspina* Crawford and *Hyalinaspis* Taylor (Homoptera: Psyllidae). *Aust J Zool* **10**:307–348

Tuomi J, Niemelä P, Chapin FS, Bryant JP, Sirén S (1988) Defensive responses of trees in relation to their carbon/nutrient balance. In: Mattson WJ, Levieux J, Bernard-Dagan C (eds) *Mechanisms of woody plant defenses against insects – search for pattern.* Springer-Verlag, New York, pp 57–72

Unroh TR, Luck RF (1987) Deme formation in scale insects : a test with pinyon needle scale and a review of other evidence. *Ecol Entomol* **12**:439–449

van Emden HF (1978) Insects and secondary plant substances – an alternative viewpoint with special reference to aphids. In: Harbourne JB (ed) *Biochemical aspects of plant and animal coevolution.* Academic Press, London, pp 309–323

Wainhouse D, Gate IM (1988) The beech scale. In: Berryman A (ed) *Dynamics of forest insect populations.* Plenum, New York, pp 67–85

Wainhouse D, Howell RS (1983) Intraspecific variation in beech scale populations and in susceptibility of their host *Fagus sylvatica. Ecol Entomol* **8**:351–359

Way MJ, Cammell M (1970) Aggregation behaviour in relation to food utilisation by aphids. *Br Ecol Soc Symp* **10**:229–247

White TCR (1969) An index to measure weather-induced stress of trees associated with outbreaks of psyllids in Australia. *Ecology* **50**:905–909

White TCR (1984) The abundance of invertebrate herbivores in relation to the availability of nitrogen in stressed food plants. *Oecologia* **63**:90–105

Whittaker RH, Feeny PP (1971) Allelochemics: chemical interactions between species. *Science* **171**:757–770

Wilcox MD (1982) Anthocyanin polymorphism in seedlings of *Eucalyptus fastigata* Deane and Maid. *Aust J Bot* **30**:645–657

Indirect effects of grazing gastropods on recruitment of Sydney rock oysters

Marti J. Anderson

ABSTRACT

Populations do not exist in isolation but usually interact with populations of other species in an assemblage. Indirect effects can occur if the presence or activity of a third species changes the effect that one species has on another. In temperate marine habitats, experimental work has shown the importance of the direct effects of gastropod grazers on algal assemblages and concomitant indirect effects of grazing on recruitment of other invertebrate animals, including oysters, barnacles and mussels. Oyster farmers in New South Wales, Australia depend on natural recruitment of Sydney rock oysters, *Saccostrea commercialis*, but the ecology of populations of this species on commercial oyster leases, including potential indirect effects of other species, has not been studied. In this study, grazing marine gastropod snails (*Austrocochlea porcata* and *Bembicium auratum*) were experimentally excluded from concrete panels attached to commercial oyster leases to determine their effect on recruitment of Sydney rock oysters. The presence of grazers significantly increased recruitment of oysters in experiments initiated in the early spring (October 1993). The growth of algae inhibited oyster recruitment by pre-empting available space; therefore grazers indirectly increased oyster settlement by removing this algae. This effect did not occur in experiments initiated in summer (January 1994), because algae did not grow sufficiently to pre-empt much of the space before oysters recruited in February/March 1994. These results showed that grazers can have important indirect effects on populations of oysters and that their influence depends on the timing of initiation of experiments with respect to the settlement of oysters.

Key words: seasonal effects, interaction chain

Frontiers of Population Ecology, R.B. Floyd, A.W. Sheppard and P.J. De Barro (eds), CSIRO Publishing, Melbourne, 1996 pp. 339–347

Introduction

Prediction of the dynamics of a population depends upon the knowledge of potential interactions with other species which can influence that population. Indirect effects can occur if the presence of a third species changes the effect that one species has on another. Indirect effects can take many forms and can be defined in several ways (see the concepts of 'centrum' and 'web' described in Andrewartha and Birch 1984; see also Miller and Kerfoot 1987; Strauss 1991; Wootton 1993; Billick and Case 1994). In this paper, indirect effects or interactions are used in the sense of Billick and Case (1994), synonymous with 'interaction chains' in Wootton (1993). By this definition, indirect effects can be predicted from knowledge of the direct interactions of species pairs (i.e. A affects B, B affects C, therefore A indirectly affects C). As such, they have been considered to be more predictable than 'higher order interactions' (Billick and Case 1994) or 'interaction modifications' (Wootton 1993), where observations of direct interactions in species pairs are insufficient to predict the dynamics of three or more species together. Experimental removals or manipulations of densities of certain species allow for hypotheses to be tested concerning potential direct and indirect interactions among species in an assemblage.

In temperate marine habitats, experimental work has demonstrated the importance of the direct effects on algal assemblages of gastropod grazers in intertidal areas (Hawkins and Hartnoll 1983; Lubchenco 1983; Moreno and Jaramillo 1983; Underwood *et al.* 1983; Underwood and Jernakoff 1984; Jernakoff 1985b) and gastropod and echinoid grazers in subtidal areas (Paine and Vadas 1969; Sammarco 1982; Himmelman *et al.* 1983; Fletcher 1987; Andrew 1988; Chapman and Johnson 1990; Underwood and Kennelly 1990). Effects of grazers on other invertebrate species are, however, not always readily observed.

Since most marine invertebrates are broadcast spawners and therefore have open populations, initial recruitment may often dictate population dynamics at a particular place and time (Underwood and Denley 1984). Correspondingly, direct and indirect effects of intertidal gastropods on invertebrate populations are likely to be most important during periods of recruitment by the target species. For example, gastropods can negatively affect recruitment of barnacles directly by 'bulldozing' or ingesting recently settled barnacles (Dayton 1971; Denley and Underwood 1979; Petraitis 1983, 1987; Farrell 1988; Miller and Carefoot 1989; Turner and Todd 1991). Alternatively, grazers can enhance recruitment of barnacles indirectly by removing algae which would otherwise pre-empt available space or smother small barnacles (Sousa 1979; Creese 1982; Petraitis 1983; Underwood *et al.* 1983; Jernakoff 1985a; Dungan 1986, 1987; van Tamelen 1987). In contrast to the enhancement of recruitment of barnacles by grazers, studies of indirect effects of grazers on bivalves have shown that grazers indirectly inhibit recruitment of oysters and mussels by removing algae which are required for their recruitment (Dayton 1971; Sousa 1984; Menge *et al.* 1986; Petraitis 1990).

All of the above studies were done on relatively exposed coastlines. There has been almost no examination of potential indirect effects of gastropod grazers in protected or estuarine habitats (but see Bertness 1984). Effects of grazers in estuarine intertidal zones might be expected to differ from their effects in rocky intertidal zones, due to differences in the assemblages and physical variables in the two habitats. For example, experiments in New South Wales have demonstrated that the behaviour (in terms of movement) of the snail, *Bembicium auratum*, located in mangrove habitats differs from the behaviour of populations on rocky shores (T. Crowe, unpublished).

Underwood and Anderson (1994) noted the presence of the grazing snails *Bembicium auratum*, *Austrocochlea porcata* and limpet *Patelloida mimula* on commercial oyster leases in estuaries in New South Wales, Australia. Oyster farmers depend on natural recruitment of Sydney rock oysters, *Saccostrea commercialis*, but the ecology of populations of this species on oyster leases has not been studied. Given that grazers are abundant on oyster leases and have been shown to indirectly influence recruitment of bivalves in other habitats, it is likely that recruitment of *S. commercialis* may be controlled by indirect effects.

This study was designed to test the null hypothesis that the presence of these grazing gastropods has no effect on recruitment of Sydney rock oysters. Previous work on these assemblages has established that a) oysters settle primarily from February to April and b) most algal species settle and grow from October to March (Underwood and Anderson 1994). Thus, due to the potential importance of differences in the timing of recruitment of oysters and algae. The null hypothesis of no difference in the magnitude or direction of the effect of gastropod grazers on recruitment of oysters in experiments initiated in different seasons was also tested.

Materials and methods

The experiments were done on a commercial oyster lease at Quibray Bay (34°01'S, 151°11'E), part of Botany Bay, south of Sydney, Australia. The experiment made use of the existing structure of the commercial oyster lease to support the panels. This structure consisted of two parallel beams (5 × 2.5 cm in thickness) of tar-covered hardwood, each about 10 m long and situated approximately 1 m apart. The beams were supported by several timber posts (8 × 5 cm in thickness) embedded in the sand such that the tidal height of the beams was ca. 0.5 m above Indian Spring Low Water.

Concrete panels (10 cm × 10 cm × 1 cm) were used for the experiment because concrete has been shown to be a good substratum for sampling recruitment of oysters (Anderson and Underwood 1994). Four panels were attached 26.6 cm apart on to 180-cm long, tar-covered hardwood sticks (2.5 × 2.5 cm in thickness) using self-tapping stainless steel screws. Sticks were then placed perpendicularly across the two parallel beams with surfaces of panels facing downwards between the beams. The sticks were attached to the beams with plastic cable ties approximately 30–40 cm apart.

Effects of the grazing snails, *Austrocochlea porcata* and *Bembicium auratum*, were tested by three treatments: 1) caged sticks, excluding grazers; 2) open sticks, subject to normal grazing; and 3) cage control, where sticks were caged but grazers were put inside cages at their natural frequency on these structures (4 snails per stick) to control for effects of the cages not caused by the removal of snails. The limpet *Patelloida mimula* was not included in cage controls due to the fact that its natural frequency was less than one limpet per stick. Cages consisted of 6 mm plastic mesh cylinders (915 mm long × 300 mm diameter) around each stick. The cages were scrubbed free of accumulated detritus and fouling every 2 weeks to prevent large differences in water flow or light inside cages.

Experiments were done over a six-month period in each of two seasons: October 1993 to April 1994 and January to July 1994. There were three sticks in each treatment from October to April and five sticks in each treatment from January to July. There was a random arrangement of sticks of all treatments on the oyster lease structure.

After six months, each panel was examined under a dissecting microscope and the number of oysters and percent cover of algae were recorded. Where an oyster had not survived, there was a noticeable scar left by the bottom valve of the shell on panels and these were also counted. A grid of 25 intersecting points made of transparent nylon wire mesh was placed over each panel and percent cover of algae was estimated as the total occurrences of algae underneath each point multiplied by four.

Results

In each season, where grazers were excluded, panels developed an assemblage of algae composed primarily of ephemeral species: *Oscillatoria* sp., *Rhizoclonium* sp., *Ulvaria oxysperma*, *Enteromorpha prolifera* and *Caloglossa leprieurii*. Detritus also accumulated amongst the algal strands on these panels.

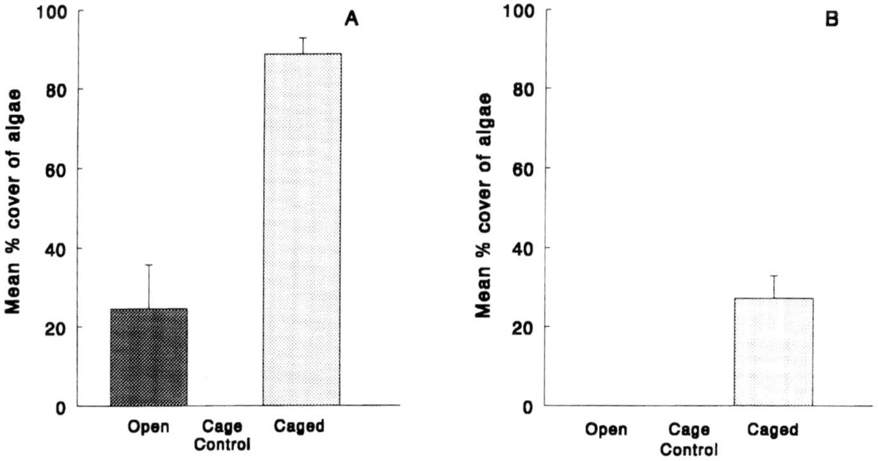

Fig. 1. Mean (+s.e.) percent cover of algae on experimental panels in each of three treatments from A) October 1993 to April 1994 (n = 12 panels in each treatment); and B) January to July 1994 (n = 20 panels in each treatment). Open panels and cage controls had grazing snails.

Table 1. Analyses of variance and results of Student-Newman-Keuls tests for the number of oysters on concrete panels (n = 4) on sticks (nested, random factor) in grazing treatments for experiments done in each of two seasons

A. October 1993 to April 1994 (Data were transformed to $x' = \log_e(x + 1)$, Cochran's test not significant, $p > 0.05$)

Source	*df*	Mean square	*F*	*p*
Grazing	2	5.10	20.64	.0020
Sticks(Grazing)	6	0.25	2.77	.0315
Residual	27	0.09		
Total	35			

SNK test results ($p < 0.05$) → Cage Control = Open > Caged

B. January to July 1994 (No transformation, Cochran's test not significant, $p > 0.05$)

Source	*df*	Mean square	*F*	*p*
Grazing	2	1321.02	11.48	.0016
Sticks(Grazing)	12	115.12	2.42	.0160
Residual	45	47.53		
Total	59			

SNK test results ($p < 0.05$) → Cage Control > Caged > Open

Grazers significantly decreased the percentage cover of algae from October to April (Fig. 1A, Analysis of variance, $F_{2,6} = 10.13$, $p < 0.05$, Student-Newman-Keuls (SNK) tests) and from January to July (Fig. 1B, ANOVA, $F_{2,12} = 13.76$, $p < 0.001$, SNK tests). There was greater cover by algae over the period October to April than from January to July (Fig. 1).

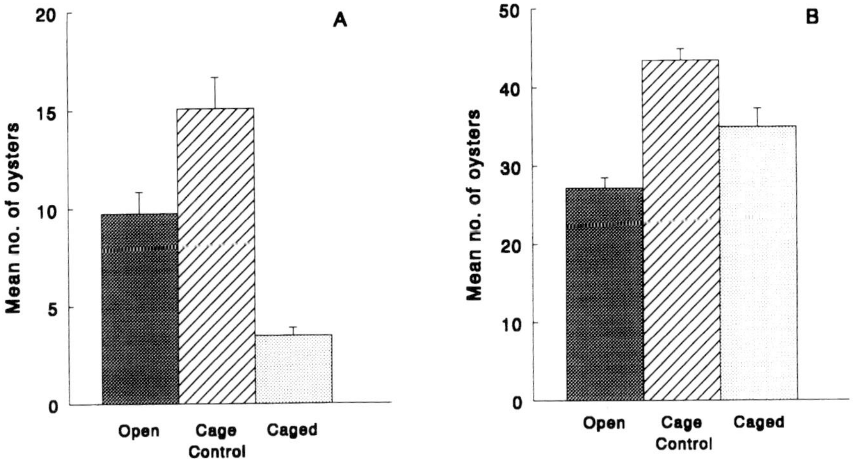

Fig. 2. Mean (+s.e.) number of oysters on experimental panels in each of three treatments from A) October 1993 to April 1994 (n = 12 panels in each treatment); and B) January to July 1994 (n = 20 panels in each treatment). Open panels and cage controls had grazing snails.

The presence of grazers significantly increased the number of oysters on panels from October to April (Table 1A, Fig. 2A). In contrast, from January to July there were significantly greater numbers of oysters on panels in cages and cage controls than on open panels (Table 1B, Fig. 2B).

The nested factor of 'Sticks' was significant in analyses of numbers of oysters in both seasons (Table 1A, B), indicating significant small scale spatial variability from stick to stick within treatments.

There were never any oyster scars on panels in cages or cage controls. Open panels had a mean (+s.e.) number of oyster scars of 1.08 (+0.23) from October to April (n = 12) and 2.20 (+0.40) from January to July (n = 20).

Discussion

Grazers enhanced recruitment of oysters from October to April by clearing away algae and detritus which would have otherwise dominated the space over this period. In contrast, when the same experiment was done beginning in January, grazers apparently had no effect, because algae did not grow substantially enough to pre-empt much space before oysters recruited. Thus, the timing of experiments with respect to seasonal differences in recruitment of algae and oysters determined whether indirect effects of grazers occurred.

Apart from differences in seasonal recruitment of oysters and algae, there are other possible reasons for the difference in results in the two seasons (Underwood and Jernakoff 1984). There was greater overall recruitment of oysters from January to July than from October to April. Although there was similar average percent cover of algae on open panels from October to April as on caged panels from January to July (Fig. 1), the latter had over 3 times as many oysters on average (Fig. 2). It may be that when oysters are settling in greater numbers, they simply swamp available surfaces, regardless of whether algae are present or not. This result indicates that there is not a simple relationship between the percent cover of algae and the number of oysters that recruit and suggests that variations in recruitment can interact with potential indirect effects.

Greater numbers of oysters in caged panels and on cage control panels from January to July may have been caused by cages changing physical variables (such as light, temperature, humidity or water flow) in a way which enhanced survival of oysters. Artefacts due to cages have been well documented elsewhere (e.g. see Hawkins and Hartnoll 1983; Underwood 1983; Schmidt and Warner 1984; Kennelly 1991) and may have varied seasonally. In addition, there could have been seasonal differences in behaviour and activity of grazers (e.g. see Cubit 1984). Also, the absence of oyster scars on panels in any caged treatments indicated that predation by fish on juvenile oysters on open panels may have confounded results. This predation may have varied seasonally, but its effects were of little magnitude compared to effects of grazers. It is more reasonable to conclude that there were important changes in indirect effects of grazers with differences in recruitment of affected species.

In addition to recruitment and seasonal effects, other factors have been shown to interact with indirect effects. There can be important changes in the direction or magnitude of indirect effects with variations in densities of herbivores (Underwood *et al.* 1983; Fletcher 1987) and in different habitats (Lubchenco 1983; Underwood 1983; Dethier and Duggins 1984; Wootton 1992). This study demonstrated that any predictions concerning populations of Sydney rock oysters, including those for management of this commercial fishery, must consider temporally variable indirect effects of grazers on recruitment.

The potential positive indirect effect of grazers on recruitment of oysters found here differed from results reported in exposed rocky intertidal areas, where grazers have a negative indirect effect on recruitment of mussels and oysters (Dayton 1971; Menge *et al.* 1986; Petraitis 1990). This is apparently because in wave-swept habitats bivalves require certain filamentous or foliose algae to recruit (Dayton 1971; Menge *et al.* 1986; Petraitis 1990).

In protected estuaries of New South Wales there is little hard substratum and the absence of grazers not only causes initial increases in algal growth, but also results in the accumulation of mud and detritus. Although the pre-emption of space by algae appears to be the most likely mechanism of an indirect effect of grazers on oysters, grazers may also clear away detritus and sedimentation which could otherwise accumulate and negatively affect recruitment or survival of oysters. Bertness (1984) found that removal of the snail *Littorina littorea* from sheltered shores in New England resulted in accumulation of sediment and of sediment-dwellers, at the expense of barnacles, encrusting algae and hard substratum.

Apart from the surfaces of oyster leases, the only other conspicuous hard surfaces at low tide are mangrove trees and clumps of oysters. It has been shown that *Bembicium auratum* and *Patelloida mimula* are found in significantly greater numbers on oyster shells than elsewhere in these habitats (Underwood and Barrett 1990; T. Crowe pers. comm.). Perhaps by clearing space for settlement of oysters, grazers indirectly increase a microhabitat (oysters) that positively affects their own survival. Further study of these assemblages would be required to determine this.

This study does not provide conclusive evidence of an indirect effect. Further study (including manipulation of algae in the presence and absence of grazers) is required to ensure that effects were, in fact, indirect (i.e. due to negative effects of grazers on algae and negative effects of algae on oysters) and not rather due to some direct positive effect of grazers on oysters. For example, the mucous trail left by grazing gastropods on a surface may directly enhance recruitment of oysters.

The results of this study and others suggest that the ultimate effect of a species on a population via an 'interaction chain' (*sensu* Wootton 1993) may not be simply predictable by knowledge of the interactions between species-pairs at some point in space or time, but may be influenced by specific aspects of the life histories of the interacting species. It is not argued here that the absence of a predictable indirect effect in certain circumstances necessarily indicates the presence of higher order interactions (*sensu* Billick and Case 1994), for no specific mathematical models were tested here. It is emphasised, rather, that indirect effects can be spatially and/or temporally variable, which weakens their predictability. The timing of interactions with respect to life-history events important to the population, such as recruitment in this example, may be of more importance in predicting effects of interactions. While this has been shown here to be especially true for broadcast-spawning marine invertebrates, it is expected that it may also be generally true for other organisms with open populations.

Acknowledgments

This study forms part of my Ph.D. dissertation at the University of Sydney and was supported by an Australian Postgraduate Award. I would like to thank R.L. Vadas, M. Schreider and T. Crowe for help in the field and stimulating discussion. I thank T. Minchinton for his critique of the manuscript, my supervisor A.J. Underwood for his considerable contributions throughout the study and B.E. Wingett for his continued support.

References

Anderson MJ, Underwood AJ (1994) Effects of substratum on the recruitment and development of an intertidal estuarine fouling assemblage. *J Exp Mar Biol Ecol* **184**:217–236

Andrew NL (1988) Ecological aspects of the common sea urchin, *Evichinus chloroticus*, in northern New Zealand: a review. *NZ J Mar Freshwater Res* **22**:415–426

Andrewartha HG, Birch LC (1984) *The ecological web: more on the distribution and abundance of animals.* University of Chicago Press, Chicago, pp 506.

Bertness MD (1984) Habitat and community modification by an introduced herbivorous snail. *Ecology* **65**:370–381

Billick I, Case TJ (1994) Higher order interactions in ecological communities: what are they and how can they be detected? *Ecology* **75**:1529–1543

Chapman ARO, Johnson CR (1990) Disturbance and organisation of macroalgal assemblages in the Northwest Atlantic. *Hydrobiologia* **192**:77–121

Creese RG (1982) Distribution and abundance of the Acmaeid limpet, *Patelloida latistrigata*, and its interaction with barnacles. *Oecologia* **52**:85–96

Cubit JD (1984) Herbivory and the seasonal abundance of algae on a high intertidal rocky shore. *Ecology* **65**:1904–1917

Dayton PK (1971) Competition, disturbance, and community organization: the provision and subsequent utilization of space in a rocky intertidal community. *Ecol Monogr* **41**:351–389

Dethier MN, Duggins DO (1984) Indirect commensalism between marine herbivores and the importance of competitive hierarchies. *Am Nat* **124**:205–219

Denley EJ, Underwood AJ (1979) Experiments on factors influencing settlement, survival, and growth of two species of barnacles in New South Wales. *J Exp Mar Biol Ecol* **36**:269–293

Dungan ML (1986) Three-way interactions: barnacles, limpets, and algae in a Sonoran desert rocky intertidal zone. *Am Nat* **127**:292–316

Dungan ML (1987) Indirect mutualism: complementary effects of grazing and predation in a rocky intertidal community. In: Kerfoot WC, Sih A (eds) *Predation: direct and indirect impacts on aquatic communities.* University Press of New England, Hanover New Hampshire, pp 188–200

Farrell TM (1988) Community stability: effects of limpet removal and reintroduction in a rocky intertidal community. *Oecologia* **75**:190–197

Fletcher WJ (1987) Interactions among subtidal Australian sea urchins, gastropods, and algae: effects of experimental removals. *Ecol Monogr* **57**:89–109

Hawkins SJ, Hartnoll RG (1983) Grazing of intertidal algae by marine invertebrates. *Oceanogr Mar Biol Annu Rev* **21**:195–282

Himmelman JH, Cardinal A, Bourget E (1983) Community development following removal of urchins, *Strongylocentrotus droebachiensis*, from the rocky subtidal zone of the St. Lawrence Estuary, Eastern Canada. *Oecologia* **59**:27–39

Jernakoff P (1985a) The effect of overgrowth by algae on the survival of the intertidal barnacle *Tesseropora rosea* Krauss. *J Exp Mar Biol Ecol* **94**:89–97

Jernakoff P (1985b) An experimental evaluation of the influence of barnacles, crevices and seasonal patterns of grazing on algal diversity and cover in an intertidal barnacle zone. *J Exp Mar Biol Ecol* **88**:287–302

Kennelly SJ (1991) Caging experiments to examine the effects of fishes on understorey species in a sublittoral kelp community. *J Exp Mar Biol Ecol* **147**:207–230

Lubchenco J (1983) *Littorina* and *Fucus*: effects of herbivores, substratum heterogeneity, and plant escapes during succession. *Ecology* **64**:1116–1123

Menge BA, Lubchenco J, Ashkenas LR (1986) Experimental separation of effects of consumers on sessile prey in the low zone of a rocky shore in the Bay of Panama: direct and indirect consequences of food web complexity. *J Exp Mar Biol Ecol* **100**:225–269

Miller KM, Carefoot TH (1989) The role of spatial and size refuges in the interaction between juvenile barnacles and grazing limpets. *J Exp Mar Biol Ecol* **134**:157–174

Miller TE, Kerfoot WC (1987) Redefining indirect effects. In: Kerfoot WC, Sih A (eds) *Predation: direct and indirect impacts on aquatic communities.* University Press of New England, Hanover New Hampshire, pp 33–37

Moreno CA, Jaramillo E (1983) The role of grazers in the zonation of intertidal macroalgae of the Chilean coast. *Oikos* **41**:73–76

Paine RT, Vadas RL (1969) The effects of grazing by sea urchins, *Strongylocentrotus* spp., on benthic algal populations. *Limnol Oceanogr* **14**:710–719

Petraitis PS (1990) Direct and indirect effects of predation, herbivory and surface rugosity on mussel recruitment. *Oecologia* **83**:405–413

Petraitis PS (1987) Factors organizing rocky intertidal communities of New England: herbivory and predation in sheltered bays. *J Exp Mar Biol Ecol* **109**:117–136

Petraitis PS (1983) Grazing patterns of the periwinkle and their effect on sessile intertidal organisms. *Ecology* **64**:522–533

Sammarco PW (1982) Effects of grazing by *Diadema antillarum* Philippi (Echinodermata: Echinoidea) on algal diversity and community structure. *J Exp Mar Biol Ecol* **65**:83–105

Schmidt GH, Warner GF (1984) Effects of caging on the development of a sessile epifaunal community. *Mar Ecol Prog Ser* **15**:251–263

Sousa WP (1979) Experimental investigations of disturbance and ecological succession in a rocky intertidal algal community. *Ecol Monogr* **49**:227–254

Sousa WP (1984) Intertidal mosaics: patch size, propagule availability, and spatially variable patterns of succession. *Ecology* **65**:1918–1935

Strauss SY (1991) Indirect effects in community ecology: their definition, study and importance. *Trends Ecol & Evol* **6**:206–210

Turner SJ, Todd CD (1991) The effects of *Gibbula cineraria* (L.), *Nucella lapillus* (L.) and *Asterias rubens* L. on developing epifaunal assemblages. *J Exp Mar Biol Ecol* **154**:191–213

Underwood AJ (1983) Spatial and temporal problems in the design of experiments with marine grazers. In: Baker JT, Carter RM, Sammarco PW, Stark KP (eds) *Proceedings of the Great Barrier Reef Conference, Townsville, 29 August – 2 September.* James Cook University and Australian Institute of Marine Science

Underwood AJ, Anderson MJ (1994) Seasonal and temporal aspects of recruitment and succession in an intertidal estuarine fouling assemblage. *J Mar Biol Assoc UK* **74**:563–584

Underwood AJ, Barrett G (1990) Experiments on the influence of oysters on the distribution, abundance and sizes of the gastropod *Bembicium auratum* in a mangrove swamp in New South Wales, Australia. *J Exp Mar Biol Ecol* **137**:25–45

Underwood AJ, Denley EJ (1984) Paradigms, explanations, and generalizations in models for the structure of intertidal communities on rocky shores. In: Strong DR, Simberloff DS, Abele LG, Thistle AB (eds) *Ecological communities: conceptual issues and the evidence.* Princeton University Press, Princeton New Jersey, pp 151–180

Underwood AJ, Denley EJ, Moran MJ (1983) Experimental analyses of the structure and dynamics of mid-shore rocky intertidal communities in New South Wales. *Oecologia* **56**:202–219

Underwood AJ, Jernakoff P (1984) The effects of tidal height, wave-exposure, seasonality and rock-pools on grazing and the distribution of intertidal macroalgae in New South Wales. *J Exp Mar Biol Ecol* **75**:71–96

Underwood AJ, Kennelly SJ (1990) Ecology of marine algae on rocky shores and subtidal reefs in temperate Australia. *Hydrobiologia* **192**:3–20

van Tamelen PG (1987) Early successional mechanisms in the rocky intertidal: the role of direct and indirect interactions. *J Exp Mar Biol Ecol* **112**:39–48

Wootton JT (1992) Indirect effects, prey susceptibility, and habitat selection: impacts of birds on limpets and algae. *Ecology* **73**:981–991

Wootton JT (1993) Indirect effects and habitat use in an intertidal community: interaction chains and interaction modifications. *Am Nat* **141**:71–89

SECTION

3

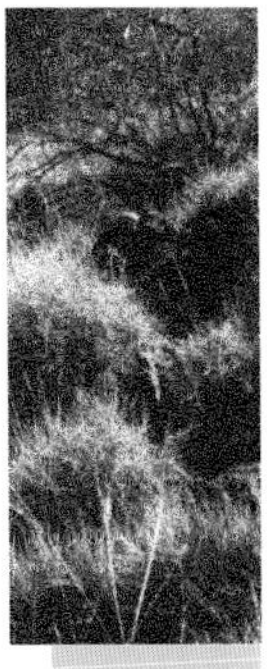

SPATIAL PROCESSES IN POPULATION DYNAMICS

The consequences of variation or heterogeneity in the spatial dynamics of populations on their long-term stability was recognised soon after the concept of population regulation was formalised earlier this century. Both Nicholson and William Thompson realised, from early attempts at biological control resulting in successful control of some pests to stable low levels, that apparent stability was due to significant variation in the dynamics between habitat patches. Nicholson (1954) used the term "fragmented oscillations" to describe how large scale stability appeared out of locally unstable predator-prey dynamics. He also used the terms "spotty distribution" (Nicholson 1957) of differing subpopulations with out-of-phase dynamics and "density-induced migration" to explain how the dramatic decline of, for example, prickly-pear caused by *Cactoblastis cactorum* reached a new lower equilibrium density after the cactus population crashed. Nicholson described this stability resulting from spatial effects as a natural consequence of population fragmentation following the influence of a major mortality factor. Stability from spatial processes was also the essence of Thompson's theory albeit without a need for density dependence (Milne 1957). These theories encouraged studies of spatial processes which led to the concept of metapopulation dynamics and a questioning of the need for balance or equilibria in local dynamics to achieve apparent stability, particularly at large geographic scales (Strong 1989).

In the lead paper, Peter Chesson neatly describes the theoretical mechanism driving stability in metapopulation dynamics as a process where the average of non-equilibrium density dependant dynamics (from limit-cycles to chaos) in several subpopulations can generate increased stability for the larger population or regional scales, provided that local dynamics between sub-populations are asynchronous. Increasing stability with decreasing resolution can not be generated theoretically in this way without density dependence in the system as this 'scale transition' effect operates through density dependent dynamics at finer scales. To clarify the mechanism involved, Chesson provides a global definition of density dependent population dynamics, at the finest level, as the effect of experience of the individual on its descendants, thus providing an ideal against which operational definitions of density dependence in empirical systems can be judged. State transition dynamics can generate regional persistence in all forms of density dependent interaction. As such, it can theoretically support, for example, Nicholson's (1957) belief that with a "mosaic of conditions" it is a genetic difference (i.e. different species), and not "Gause's hypothesis" of difference in competitive ability or ecology, that is the primary prerequisite for the coexistence of two species.

Tony Underwood follows by providing a series of examples from the intertidal world where contrasting population processes are observed at different localities and different scales. Estimation of this variation is critical. He warns of the serious consequences of sampling too few localities and sampling at the wrong spatial scale. The same organism may also experience different processes in separate habitats or biogeographic regions making the appropriate scale hard to discern even within a given species. The grain in the dispersion pattern of organisms will be determined by the population processes involved and the scale at which these are density dependent will determine the pattern of variance between scales. Detailed understanding of ecological processes will help determine this in the same way that some understanding of how variances change in time and space will help determine the scale at which density dependence acts. This is an extension of the grain in Nicholson's mosaic of habitat favourability (Nicholson 1957) and how this changes with density through time, and with habitat through space.

Mathematics has provided many tools which help to understand not only the dispersion pattern of organisms, but also how to manage populations of rare organisms that have had their distribution reduced to a few patches. Hugh Possingham explores the use of one such tool, Markovian decision theory, in the conservation of endangered species, where the metapopulation structure is so small that the risk of species extinction by chance is dangerously high. Species distributed in this way require more patches to provide a basis of longer term persistence. The conservation management dilemma, of whether to assist dispersal by moving individuals to new suitable habitat patches or create suitable patches within the dispersal arena of the population to expand the size of the metapopulation, is particularly pertinent here. Resources for such activities are always scarce so these mathematical techniques can aid decision making once the state of the metapopulation is known. These models should also suggest the minimum ecological knowledge base necessary to make such decisions with reasonable confidence. Unfortunately, only an extreme minority of cases will have comprehensive ecological understanding available on which to base conservation management (see examples in Morton this volume).

In order to manage animal populations it is important to account for behaviour patterns of individuals as they move within their physical and biological environment. Two contributions discuss specific models of pest species that incorporate variation between individuals stemming from either nature or nurture. In the first, Alistair Drake and Gavin Gatehouse present a conceptual model of a migrating insect where two of the four components of the model; the phenotypic and genotypic basis of the initiation and maintenance of migration, stress the role of variation between individuals for providing long-term stability in the behavioural response. Because of inherent difficulties, long-distance migration is a relatively new area of research for exploring population regulation. Evidence that migratory behaviour is density-related is scarce. Migration ecology has long provided an area of population ecology where it is perhaps hardest to invoke examples that support the need for density dependent mechanisms to ensure long-term persistence. Support for the overall need of some degree of density dependence now largely comes from a disbelief of the alternative and an acceptance that density related processes are more likely to set diffuse variation limits rather than create equilibria (Strong 1989). Will it ever be possible to detect such mechanisms that will be rare in space and time? Belief in a mechanism that is currently untestable will be a continuing problem for migration ecologists.

The model of Murray Efford is an encouraging example of how to incorporate 1990s technology in a spatial model of individual dispersal. Geographic Information System (GIS) data from satellite imagery of habitat favourability are combined with an individual-based model to predict the effect of controlling the brushtail possum (an alien pest in New Zealand) over a defined area on the density dependant dispersal back into that area. The potential applications of such models are great, provided that the behavioural ecology of the subject in the recognisable habitat types is well understood. It will be similarly applicable to conservation problems involving the creation of wildlife corridors or the spread of potential biological control agents and provides a clear application of population theory to practical ecological problems based on animal movement.

In the final part of this section, Paul Walker attempts to look beyond this millennium to the types of GIS systems and analytical tools that are becoming available to population ecologists, but are still under-utilised. There is always a risk that the application of such technology becomes decoupled from ecological theory as the complex technical expertise needed to use it discourages mainstream ecologists. Only time will reveal to what degree Walker's call for population ecologists to take advantage of such systems will be heeded.

References

Mine A (1957) Theories of natural control of insect populations. *Cold Spring Harbor Symp Quant Biol* **22**:253–271

Nicholson AJ (1954) An outline of the dynamics of animal populations. *Aust J Zool* **2**:965
Nicholson AJ (1957) The self-adjustment of populations to change. *Cold Spring Harbor Symp Quant Biol* **22**:153–172
Strong DR (1989) Density-vague population change. *Trends Ecol & Evol* **1**:39–42

MATTERS OF SCALE IN THE DYNAMICS OF POPULATIONS AND COMMUNITIES

Peter Chesson

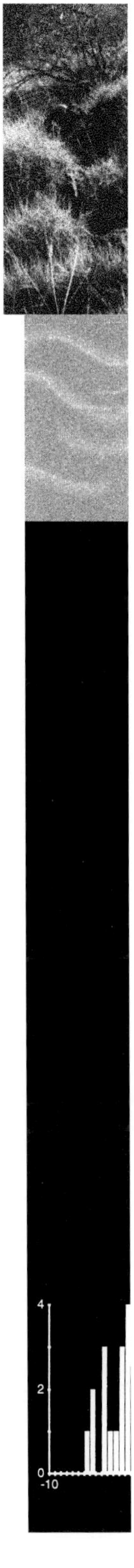

ABSTRACT

It is now well established theoretically that the stability of a population or community on a large spatial scale may depend on heterogeneities, fluctuations or instabilities on smaller scales. Asynchrony of local population fluctuations is often viewed as an alternative to density dependence in achieving long–term regional population stability. However, it is shown here that fluctuations achieve this effect only through their interaction with density–dependent processes. This observation relies on the fact that large spatial scales reflect the average of small scale variation. When local population dynamics are nonlinear (a reflection of density dependence), the variance on the small scale affects the average at the larger scale. This nonlinear averaging that links small and larger scales may have stabilising or diversity promoting effects in many systems. How nonlinear averaging leads to these effects can be understood by a simple graphical technique.

Key words: Density depenence, non–linear averaging, spatial scale, scale transition, model selection

Frontiers of Population Ecology, R.B. Floyd, A.W. Sheppard and P.J. De Barro (eds), CSIRO Publishing, Melbourne, 1996 pp. 353–368

Introduction

There is growing recognition of the need to consider spatial scale in ecological studies. On small spatial scales, populations are invariably *open*, i.e. there is appreciable immigration from outside the population. At larger spatial scales, the contribution from immigration declines in relation to dynamic processes within a population, until at some scale a population is effectively *closed*, i.e. eliminating immigration altogether would have a negligible effect on population dynamics. It may not be possible to understand an open population, or a community consisting of open populations, without an understanding of immigration and emigration. For example, Pulliam (1988) has suggested that many local populations may actually be *sinks*, i.e. insupportable without an excess of immigration over emigration. Thus, the persistence of such a population depends on factors beyond the boundary of the population itself.

Most ecological concepts and hypotheses, however, have been developed for closed populations. Applying them to local populations that are in fact open could be quite misleading. For example, it can be shown theoretically that a constant rate of immigration from outside a local population can be highly stabilising (Hughes 1990; Stone 1993), as it is equivalent mathematically to the 'constant number refuge' discussed in predator-prey models (Hassell 1978; Murdoch *et al.* 1995). Attempts to explain the stability of an open population based on the many stability concepts for closed populations may be pointless because the stability of such a population may depend entirely on stable immigration. Similarly, explaining species diversity of an open community suffers from the problem that a diverse local community can be sustained by immigration whether or not local conditions favour maintenance of high diversity (Sale 1977).

In both of the above examples, the question shifts from the characteristics of local population dynamics to the characteristics of immigration. We must ask respectively, why is immigration stable, and why is the immigrant pool diverse? These questions then shift to a larger spatial scale from the which the immigrants are derived. Only when the system is effectively closed can we be sure that explanations for the system's properties are to be found within the system. Moving to a large scale of an effectively closed system, however, does not mean lower scales can or should be profitably ignored. Lower scales derive importance from heterogeneity on those scales (Chesson 1991). Much ecological theory, however, has been developed assuming that populations are not only closed but also internally spatially homogeneous.

If a system is spatially homogeneous in terms of the physical environment and its fluctuations over time, and in terms of the biological species and their fluctuations over time, then an open subset of such a system could be studied as if it were closed as observations anywhere would be indicative of the workings of the system as a whole. Thus, the various concepts and hypotheses developed from models of closed systems would be perfectly applicable. In many field studies, the implicit assumption is not this extreme spatial homogeneity but unimportance of the spatial variation that is present. The variation is treated as noise – a nuisance that requires larger samples for statistical significance (Chesson 1986). But a growing body of theoretical and empirical evidence sees danger in putting spatial variation in the category of noise.

As an initial guide to whether the observed spatial variation is noise, one might first ask the question, Are the important ecological processes qualitatively similar in most localities? The answer, "no", to this question means that one ignores spatial variation to one's peril. If the answer is "yes" however, one asks, Are the important ecological processes quantitatively similar in space? Quantitative dissimilarity in the presence of qualitative similarity can often be important but need not be. One approach to answering this question is given below in the section on *The scale transition.*

Variation in space often has a temporal component, and so, to the questions above, one can add the question, Are temporal fluctuations spatially synchronous or asychronous? It is generally unwise to ignore asynchronous fluctuations, but synchronous fluctuations that vary spatially in strength can be of major significance also (Chesson 1990).

The reasons for these answers have to do with the fact that on a large scale the appearance of processes is affected in major ways by variation on smaller scales. How this occurs depends on biological processes that cause various kinds of density dependence.

Density-dependence and the spreading of risk

Asynchronous local population fluctuations and density dependence are often seen as alternative means of achieving population stability. Indeed, some authors suggest that population asynchrony dominates over density dependence in nature (Den Boer 1981; Andrewartha and Birch 1984). Theoretical evidence, however, implies that fluctuations derive their importance through their interaction with density-dependent processes (Chesson 1991). To see that this is so, it is first necessary to develop a clear notion of the phenomenon of density dependence.

Clear definitions of density dependence are hard to come by in the literature. However, most commonly, density dependence means that the per capita growth rate is a decreasing function of population density (Murdoch and Walde 1989). For example, the logistic model is a simple expression of this idea. In the logistic model, the per capita rate of population growth has the maximum value r, at effectively zero population density, and declines linearly to 0 as population density approaches the carrying capacity K, as follows:

$$\frac{1}{N}\cdot\frac{dN}{dt} = r\left(1-\frac{N}{K}\right). \tag{1}$$

The logistic model, however, represents a very special kind of density dependence, because it assumes that the per capita growth rate responds instantly to changes in population density – the per capita growth rate at time t is expressed directly as a function of population density at time t. Such a representation must invariably be an approximation only closely applicable when density dependence is caused by direct effects of a population on itself, for example by cannibalism or interference competition. When intraspecific competition is exploitative, Eq. 1 is best replaced by an equation of the form

$$\frac{1}{N}\cdot\frac{dN}{dt} = f(R), \tag{2}$$

where R is resource availability, and f is some increasing function relating resource availability to per capita population growth. A separate equation describes the dynamics of resources, for example

$$\frac{dR}{dt} = g(R) - h(R)N, \tag{3}$$

where $g(R)$ is the rate of resource renewal and $h(R)$ is rate of consumption of resource per individual in the population.

Eqs. 2 and 3 describe a feed back loop of a species on itself through the resource. Density dependence is not evident from Eq. 2 alone, but only from the pair of Eqs. 2 and 3 where the negative impact of the population on the resource is clear. However, the density-dependent feedback of the population on itself through the resource does not lead to instantaneous changes, but is often regarded as working with a time lag. The effect of an increase in the population is to decrease the resource growth rate, lowering resource densities over time with a consequent gradual lowering of the population per capita growth rate Eq. 2.

From this discussion, it is evident that the rate of response of per capita population growth to changes in population density depends on the speed of resource dynamics. MacArthur (1970) pointed out that when resource dynamics are fast relative to population dynamics, the resource can be assumed to be at an equilibrium dependent on N. This equilibrium is the solution of the equation $g(R^*) - h(R^*)N = 0$ for R^*. This value R^* is a function of N describing the level of resource availability determined by a particular population density. It represents the balance between consumption and renewal of the resource for that particular population density. Substituting R^* for R in Eq. 2 leads to an equation for population per capita growth rate of the form

$$\frac{1}{N} \cdot \frac{dN}{dt} = q(N) \tag{4}$$

where $q(N)$ is some function of N expressing density dependence once again as an instantaneous change in per capita population growth with a change in the population density. Depending on the functions f, g and h, the function $q(N)$ can be linear in N, so that Eq. 4 is the logistic equation once more (MacArthur 1970). More importantly, but unsurprisingly is the result that fast adjustment of resource availability to population density leads to a fast adjustment of per capita population growth to population density.

Although Eqs. 1 and 4 are the standard way of expressing density dependence mathematically, they are best viewed as approximations or limiting cases of more complex equations such as the pair of Eqs. 2 and 3. Equation 2 alone, however, is not informative about density dependence. Indeed, population growth written in the form of Eq. 2 is sometimes thought of as density-independent simply because N does not appear explicitly on the right hand side of the equation. The results here show this perception to have limited utility.

In discrete time, rather than focus on the per capita rate of change, it is more natural to express dynamics in terms of the ratio of population sizes at successive points in time, $N(t+1)/N(t)$. The analogue of Eq. 4 in that case is

$$N(t+1)/N(t) = Q(N(t)), \tag{5}$$

where $Q(N(t))$ is some decreasing function of $N(\mathrm{t})$. For example, in the Ricker model (May and Oster 1976),

$$Q(N(t)) = Re^{-\alpha N(t)} \tag{6}$$

where R is a net reproductive rate without competition and α is an intraspecific competition coefficient. This model leads to the discrete time iteration

$$N(t+1) = RN(t)e^{-\alpha N(t)}. \tag{7}$$

Such discrete-time density dependence can represent exploitative competition with less stringent conditions than its continuous time counterpart Eq. 4. Resource dynamics must be on a shorter timescale than population dynamics, but Eq. 4 can hold simply if resources are renewed independently of population density in every unit of time, with no carryover of resources from one time to the next. Indeed, there are a number of important classes of populations where these conditions are applicable, for example, insects whose larvae develop in fruit, carrion or other ephemeral food sources (Shorrocks and Rosewell 1987; Ives 1991).

Density dependence through exploitative competition for resources is an indirect interaction between individuals in the same population. Resources are the medium of the this indirect interaction. A similar indirect interaction occurs through predators, pathogens and parasites. Holt (1977) has termed this indirect interaction 'apparent competition' in the situation where predators lead to indirect interactions between species. However, the idea that predators lead to density dependence within their prey populations, championed by Nicholson (1933), has a long history (see Kingsland this volume). Such density dependence can occur in a number of ways. Predators may modify their behaviour in response to prey density, and they may also increase in abundance or individual size and thus increase prey per capita mortality (Murdoch and Bence 1987). Some of these changes may be fast, so that an equation like Eq. 4 can summarise the impact on the prey, but those relying on changes in predator abundance will lead to a pair of equations analogous to Eqs. 2 and 3. In such cases, density dependence may not be apparent from the formula for the per capita growth rate of the prey population alone. Just as in the case of indirect interactions through resources, the density dependence only becomes apparent when the equation for the mediator of the indirect interaction is considered.

All of the above are examples of density dependence, but can density dependence be given a clear definition? Fundamentally, density dependence concerns how an individual is affected by the density of other individuals. Per capita growth rates are about average effects on individuals, and are not fundamental, but derivative. A formal definition is as follows:

> "*...The dynamics of a population are density dependent if the lines of descent from individual members of the population are affected by population density...*"

By focusing on lines of descent, this definition allows for arbitrary time lags. For example, if the number of great grandchildren of an individual is affected by the current density, then dynamics are density dependent. Normally, the term density dependence refers to negative effects of current density on future population numbers. The opposite case, which occurs when there is an Allee effect or a type II functional response (Hassell 1978), is commonly called inverse density dependence. The definition here includes both the usual and the inverse sort of density dependence.

The difficulty of defining density dependence in terms of per capita growth rates is apparent in stochastic models that incorporate demographic stochasticity (May 1973) or within-individual variability (Chesson 1978, 1990). In such models, the per capita rates of change can never be independent of density. When a population is not density-dependent according to the definition above, it is nevertheless true that the variance over time of the per capita growth rate is inversely proportional to population density (Nisbet and Gurney 1982). However, the idea that the per capita growth rate should depend on density provides a useful operational definition of density dependence, especially in large populations where fluctuations due to demographic stochasticity are small. In this regard, it is important to note that the definition above is an ideal definition for use theoretically. Indeed, it is exactly the general and precise definition of the concept that is needed for theoretical analysis, discussed below, of the possible sorts of population dynamics in highly structured but density-independent systems. However, the definition is difficult to apply to empirical data.

For empirical studies, operational definitions of density dependence must be constructed for particular systems, and indeed such operational definitions are implicit in many empirical tests of density dependence (e.g. Turchin 1990; Dennis and Taper 1994; see also Cappuccino and Harrison this volume). The role of the definition given above is to provide the ideal which is to be approximated operationally. The concept of density dependence is so general that we should expect its expression to vary greatly from system to system, and therefore particular circumstances need more restricted definitions of the idea, so that the idea becomes useful in practice for the system or data at hand. The ideal definition above then provides a point of reference for judging the adequacy of an operational definition given the assumptions applying for particular circumstances.

Many ecologists have felt that most populations in nature must have density-dependent dynamics (see Cappuccino and Harrison this volume). In spite of this, density dependence is often difficult to detect empirically (Murdoch and Walde 1989; Turchin 1990; Dennis and Taper 1994; Murdoch 1994; see also Fox and Ridsdill-Smith this volume). And if density dependence is weak, or occurs only some of the time, one may legitimately question the emphasis on density dependence in modern ecology (Andrewartha and Birch 1984; Strong 1986). To address this debate, the consequences of strict density independence of population dynamics are examined.

The simplest model of population dynamics without density dependence is the geometric growth equation,

$$N(t+1) = RN(t) \tag{8}$$

which can be obtained from Eq. 7 by setting $\alpha = 0$. The key feature is that population size at time $t + 1$ is simply proportional to population size at time t. It is well-known that Eq. 8 leads to exponential population growth, with unlimited increase if $R > 1$, decline to extinction if $R < 1$, and constancy if $R = 1$. The first two situations cannot describe persisting populations in nature if the idea of density independence is to be retained. Therefore, density-independent population persistence must be described by the last situation with $R = 1$. In this case, however, it is commonly argued that stochastic fluctuations will lead to a random walk, with extinction being the ultimate result. The idea of spreading of risk (den Boer 1968, 1981; Andrewartha and Birch 1984) is a response to this argument. According this idea, it is very important to recognise that populations are structured in space into local populations (Andrewartha and Birch 1984) or interaction groups (den Boer 1981). The principal conclusion is that these local populations fluctuate asynchronously over time and lead to regional population fluctuations of a smaller order. Coupled with dispersal between local populations, such asynchrony and reduced regional variance is expected to lead to prolonged regional persistence.

To test this idea mathematically, we must generalise the density-independent model in Eqn 8 to a structured population (see Appendix). No matter how complex we make population structure, there is an inescapable conclusion: Eq. 8 remains a good summary of regional population dynamics in the long run. The actual value of R applying in complex settings is a complicated space-time average of the parameters of the system, which may fluctuate in various ways in space and time. However, the inescapable conclusion is that highly structured density-independent populations still grow exponentially in the long run. The only hope for long-term persistence of such a population is for R to equal 1 or very nearly. It does not seem likely that this could apply to many populations in nature. If it did apply, and there was any phenotypic variation leading to differences in average fitness, then natural selection would quickly ensure that R became greater than 1. Thus, stable regional populations with density-independent dynamics seem quite implausible, in agreement with previous arguments (Murdoch and Walde 1989; Murdoch 1994) and the stronger claim that density dependence is a logical necessity for persisting populations (Godfray and Hassell 1992).

These conclusions, however, do not mean that the concept of spreading of risk is unimportant. It is most important when population dynamics are density-dependent. Indeed, there is an important interaction between density dependence and variation in space that helps reconcile their respective roles. This interaction is the topic of the next section.

The scale transition

I use the term *scale transition* for the changes that take place in population dynamics when the view shifts from one scale in space or time to another (Chesson 1991). Understanding this scale transition is crucial to understanding the effects of variation on population dynamics. The development here has general lessons that apply in some form to most populations or communities. However, the analysis uses just a single-species, discrete-time, density-dependent model. To make the discussion concrete, it is worthwhile to have some examples in mind. Ideal examples are insects such as drosophilids (Shorrocks and Rosewell 1987) or carrion flies (Ives 1991) that colonise ephemeral patches of food in which they develop for just one generation before dispersal. Annual plants without a seed bank may also satisfy difference equations of these sorts (Watkinson 1980; Pacala and Silander 1985), although generally with milder density dependence, and with more local dispersal. The basic principles are the same. In both cases, organisms interact with each other locally in space but form part of a much larger population over which dispersal takes place. Although the specific development fits best in these cases, the lessons are general for populations in which density dependence occurs on the scale of a local population, which is open and therefore influenced significantly by dispersal.

In the above situation, What happens when we pass from the scale of a local population, where interactions occur within a generation, to the scale of a regional, effectively closed population? Total regional population size is simply the sum of all the local population sizes. Population density as number per unit area, is an average of the local population densities. This average is a simple arithmetic average if local populations all occupy the same area (e.g. by defining local populations in terms of equal sized areas), but otherwise it is a weighted average of local population densities, with weights equal to the areas occupied by the local populations. Thus, when we pass from the scale of the local population to the scale of the regional population we are taking an arithmetic average of the population densities.

Arithmetic averages may seem straightforward, but they can be rather surprising, especially when local population dynamics are density-dependent. Consider population dynamics in discrete time with discrete periods of dispersal, such that $N_j(t)$ is population density in patch j before dispersal, $N_j(t+h)$ is this local population density immediately after dispersal but before population growth (t to $t+h$ is the dispersal period), and $N_j(t+1)$ is local population density after local population growth and before dispersal ($N_j(t)$ one unit in time later). As remarked above, regional population density is the same as average population density (appropriately weighted), and can be denoted $\bar{N}(t)$. Although dispersal commonly involves mortality, for simplicity, we shall assume that dispersal merely leads to a rearrangement of the population without mortality. Thus, regional population density is the same after dispersal, i.e. $\bar{N}(t+h) = \bar{N}(t)$. This assumption has no effect on the conclusions below if dispersal is density-independent. Let us now assume that local population growth can be described by an equation of the form

$$N_j(t+1) = F(N_j(t+h)) \tag{9}$$

where F is some function of the local population density. If there were no variation in space, regional population dynamics would be described by this equation also. In other words

$$\bar{N}(t+1) = F(\bar{N}(t)) \tag{10}$$

For example, taking F from the Ricker model Eq. 7, Eq. 10 becomes

$$\bar{N}(t+1) = R\bar{N}(t)e^{-\alpha\bar{N}(t)} \tag{11}$$

With positive values of α, there is a point equilibrium which is stable provided $\ln R$ is $\leqslant 2$. For values of $\ln R > 2$, the equilibrium is unstable and a regional population satisfying Eq. 11 fluctuates over time (May and Oster 1976). These fluctuations can be in the form of regular cycles but for sufficiently large values of R, the population exhibits large chaotic fluctuations (Fig. 1a).

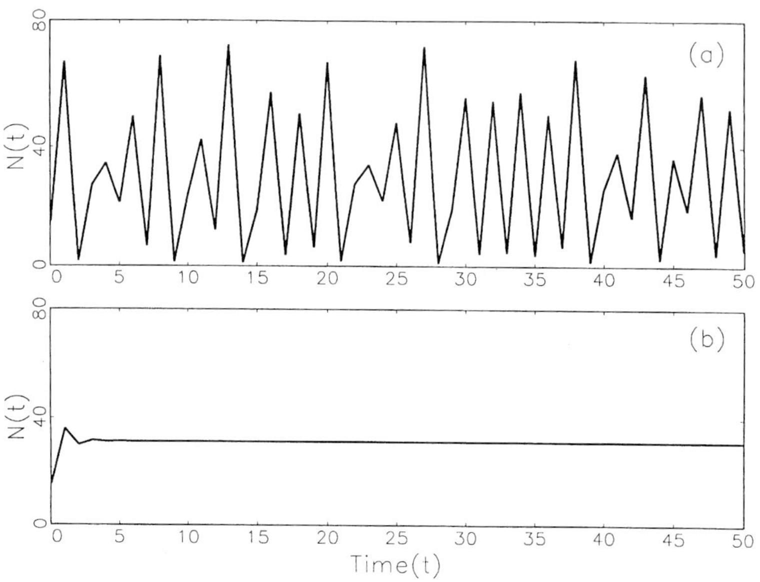

Fig. 1. (a) Simulation of the Ricker model, Eq. 11, with $\ln R = 3$ and $\alpha = 0.1$. (b) is a simulation of regional dynamics given by Eq. 13 with $k = 0.5$, and R and α as in (a).

Spatial variation can prevent this chaotic behaviour. That in itself is not surprising because if fluctuations in different patches are out of phase, the overall fluctuations are bound to be less. However, something more subtle and important is going on. With spatial variation, regional population dynamics are not described by Eq. 10 but by the equation

$$\bar{N}(t+1) = \frac{\sum_{j=1}^{k} F\big(N_j(t+h)\big)}{k} = \overline{F\big(N(t+h)\big)}. \tag{12}$$

The important point here is that the average $\overline{F(N(t+h))}$ differs from $F(\bar{N}(t))$ whenever there is variation in space. The reason for this difference is not a difference between $\bar{N}(t)$ and $\bar{N}(t+h)$ because we have assumed they are the same. The difference arises instead from the nonlinearity of the function F – the fact that F specifies a curved relationship between $N_j(t+1)$ and $N_j(t+h)$. Nonlinearity in turn arises from density dependence, as the density-independent relationship 8 is not curved, but a straight line. The important interaction between nonlinearity and spatial variation is explained in Fig. 2.

The thick line in Fig. 2, gives the Ricker curve. The points A, B and C represent different spatial locations with different values of $N_j(t+h)$ and corresponding values of $F(N_j(t+h))$. Suppose for the moment that half the local populations are represented by point A and the other half by point B, then the average value of $F(N_j(t+h))$ $(= \bar{N}(t+1))$ corresponds to the y coordinate of the point D, the midpoint of the line joining A and B. The x coordinate of the point D is $\bar{N}(t)(= \bar{N}(t+h))$. These facts can be deduced from the geometry of the figure noting that $\bar{N}(t+1)$ $(= \overline{F(N(t+h))})$ lies half way between the two values of $F(N_j(t+h))$, and $\bar{N}(t)$ lies halfway between the two values of $N_j(t+h)$.

This diagram shows how spatial variation alters regional population dynamics. Note first of all that $F(\bar{N}(t))$ corresponds to the y coordinate of the point E, directly above $\bar{N}(t)$, showing that here, $F(\bar{N}(t+h)) > \overline{F(N(t+h))}$. Although this may initially be counter intuitive, the diagram makes it

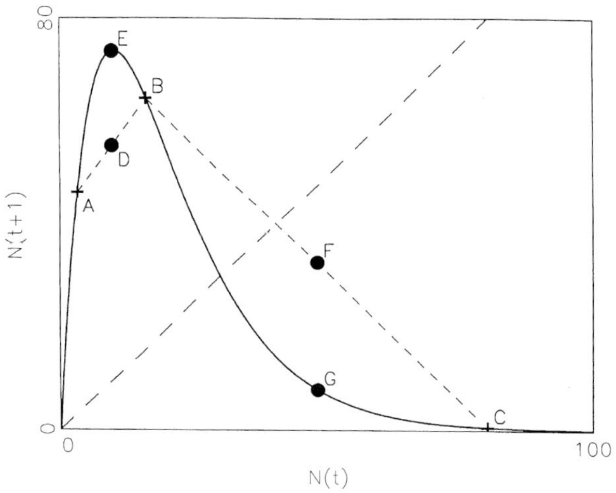

Fig. 2. Averaging the Ricker curve in space as explained in the text. (Parameters as for Fig. 1)

clear that in this case there is no other possibility as $F(\overline{N}(t+h))$ is greater than both values of $F(N_j(t+h))$ and so certainly must be larger than their average.

The inequality $F(\overline{N}(t+h)) > \overline{F(N(t+h))}$ holds in all places where the curve is bending over (is concave), while the reverse inequality $F(\overline{N}(t+h)) < \overline{F(N(t+h))}$ applies whenever the curve is bending upwards (is convex). For example, if half the patches correspond to the point B and the other half to the point C, the new regional population density is the y coordinate of the point F midway between B and C. This point F has x and y coordinates $\overline{N}(t+h))$ and $\overline{F(N(t+h))}$ and lies above the point G with x and y coordinates $\overline{N}(t+h))$ and $F(\overline{N}(t+h))$. The important thing to note is that this spatial variation has led to a moderation of the density-dependent response on the larger spatial scale. The point D being less than the peak value E shows moderation by spatial variation of the maximum achievable density the next time period. Similarly, the point F being above G shows that the extent of the crash at high population densities is reduced also.

Applying the procedure illustrated in Fig. 2 to a continuous range of two-point spatial distributions allows the construction of a relationship between $\overline{N}(t+1)$ and $\overline{N}(t)$. Thus, in Fig. 3 each chord on the Ricker curve joins two different values of local population density; and the midpoints of these chords (large dots) then give the relation between $\overline{N}(t+1)$ and $\overline{N}(t)$ arising with spatial variation. As can be seen, the density-dependent response is much more moderate on the larger scale in the presence of spatial variation.

The simple two-point spatial distributions used in these figures illustrate well the phenomenon of nonlinear averaging, but they are not very realistic. Local density variation in nature, however, is often well-approximated by a negative binomial distribution (May 1978). For this spatial distribution, the exact value of $\overline{N}(t+1)$ is

$$\overline{N}(t+1) = \frac{R'\overline{N}(t)}{\left(1+\alpha'\overline{N}(t)/k\right)^{k+1}}, \tag{13}$$

where $R' = Re^{-\alpha}$ and $\alpha' = 1 - e^{-\alpha}$ (de Jong 1979; Hassell and May 1985). These curves are plotted along with the Richer curve in Fig. 4. Clearly shown is the strong flattening of the density-dependent relationship for small k corresponding to high spatial heterogeneity. The steep peak of the Ricker equation leads to a large build up of population density followed by a crash, and thus the highly oscillatory population dynamics shown in Fig. 1a. With spatial heterogeneity, the peak is

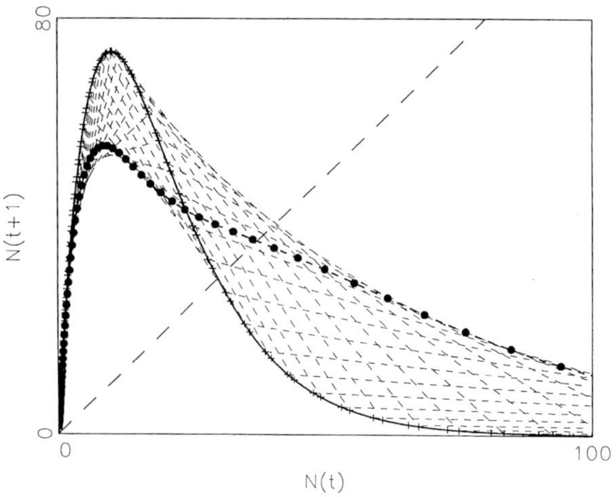

Fig. 3. Continuous averaging of the Ricker curve in space as explained in the text. (Parameters as for Fig. 1)

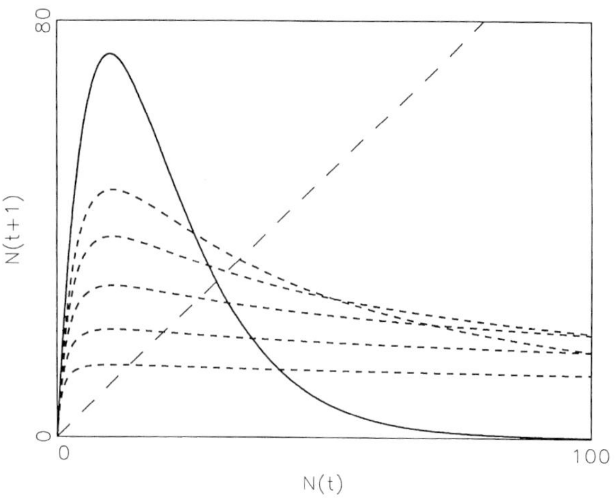

Fig. 4. The Ricker curve and Eq. 13 plotted for k = 0.1 to 1 in an ascending order. (Parameters as for Fig. 1)

reduced at the regional level (at the local level nothing has changed) and the decline after the peak is quite mild. Indeed, with high levels of heterogeneity, the response is nearly completely flat, which means that a regional equilibrium is achieved very rapidly and is very stable. This stability is a product of a high level of spatial variation (Hassell and May 1985).

The above discussion of the dynamics of Eq. 13 assumes that *k*, the clumping parameter of the negative binomial distribution, remains constant over time. This is a reasonable proposition in some situations, for example, if local populations are ephemeral habitat patches that are colonised each time period based on their physical attributes without any density-dependent interactions between the organisms colonising a patch. In such cases, the negative binomial arises from heterogeneous Poisson sampling (a 'mixed' Poisson distribution; Johnson and Kotz 1969; Southwood 1978), and implies a constant value of *k*. Such heterogeneous Poisson sampling is potentially applicable to insects that colonise dead organisms or fruits (Ives 1991; Shorrocks and Rosewell 1987). With a constant value of *k*, Eq. 13 provides a sufficient description of the dynamics of regional density.

Other cases of colonisation of ephemeral habitats may involve some density dependence in the colonisation process. This is to be expected any time the number of individuals already present at a site affects subsequent arrivals. In such instances, k should vary with density but not necessarily otherwise with time. Equation 13 is then modified by having a function of density substituted for k. There is no reason, however, to restrict application of these ideas to the case of ephemeral habitats (Hastings and Higgins 1994; Kareiva and Wennergren 1995), but in more general situations, patchiness will be affected by local population dynamic processes in addition to migration. Although the negative binomial distribution may remain a reasonable approximation, the value of k may vary over time in a manner that is not tied to population density. Even so, the graphical analysis above suggests that the stabilising tendency of variation should still be applicable provided variation in k is not too large. An example is instructive. From Fig. 3 we can see that provided k never varies outside the range 0.1 to 1, no matter how k changes with time in this range, regional densities cannot fluctuate more than four fold, which contrasts with the more than eighty fold fluctuation of the corresponding Ricker model in a spatially homogeneous system. Such stability from spatial variation is amply illustrated by the Ricker model of Hastings and Higgins (1994) and Kareiva and Wennergren (1995), who consider continuous migration in one dimension. Although they make the point that their model may take a very long time to settle into its final pattern of fluctuation, the dampening effect of spatial variation is immediately evident.

The stabilising effect of spatial variation seen here does not depend on fluctuations being out of phase in the different localities, although such out of phase fluctuations appear to be a feature in Hastings and Higgins (1994) and Kareiva and Wennergren (1995). Indeed, the spatial pattern of densities could be fixed. Such a fixed pattern of variation might arise, for example, because some localities are more difficult for migrating organisms to find than others. Intuitively, spatial variation leads to stability by the fact that patches with high numbers are balanced by patches with low numbers such that the severe decline experienced in the high density localities is buffered by more moderate decline or even increases in other localities. There is no requirement for fluctuations to be out of phase in different localities for this to occur. There is also no requirement here for particular dispersal distances relative to the scale of local interactions so long as spatial heterogeneity is maintained within the regional population, as the results of Hastings and Higgins (1994) attest.

This paper has shown how this stabilising effect involves nonlinear averaging: the dynamics of regional density are given by $\overline{F(N(t+h))}$ which differs from $F(\overline{N}(t+h))$. This inequality is the key to the scale transition, or change in the dynamical behaviour when passing from local dynamics to regional dynamics because consideration of local population dynamics alone, without regard to spatial variation, predicts that regional dynamics should be given by $F(\overline{N}(t+h))$. Quantitative and qualitative differences between local and regional dynamics arise quite generally from nonlinear averaging in both single-species and multispecies settings (Chesson 1981; Durrett and Levin 1994), with both spatial and temporal variation (Chesson 1990). Moreover, such nonlinear averaging can occur on small scales as well as large scales (Chesson 1981; Durrett and Levin 1994), leading to large-scale dynamics that can be viewed as the outcome of an accumulation of scale transitions involving nonlinear averaging (Chesson and Huntly 1993). There is no necessity that nonlinear averaging should lead to stability as it has in the illustration given above. Instability from nonlinear average is possible also (Chesson 1990).

Discussion

Although variation in space may lead to lower temporal fluctuation at the regional scale than the local scale, in the density-independent case, population dynamics at the regional scale are not qualitatively different from dynamics at the local scale. In essence geometric growth occurs on both scales, discounting dispersal on the local scale. With density dependence, local and regional dynamics can be qualitatively different. For example, we have seen how chaos at the level of a local population can change to very strong stability at the regional population scale as a consequence of the combined effects of variation and locally nonlinear dynamics. The reason for this qualitative change is the nonlinear nature of averaging occurring with the transition between scales.

In studies of metapopulation models (Gilpin and Hanski 1991), there is often such a strong emphasis on density-independent processes of colonisation and extinction (Caswell 1978) that there is a danger of assuming that they are examples of regional stabilisation with density-independent dynamics. However, the usual assumption in such models that local communities can be described adequately by the list of species present is an implicit assumption of strong density dependence locally in space. Although many kinds of density dependence will suffice, this simplest such assumption is that upon colonisation, a local population rises quickly to a carrying capacity, about which it fluctuates until extinction occurs. Hence strong density dependence is in fact assumed, and stabilisation of regional dynamics in such models can be viewed also as an outcome of nonlinear averaging (Chesson 1981).

Stabilisation of population fluctuations, however, is just one possible effect of nonlinear averaging in the scale transition. In models of competing species, nonlinear averaging can enable coexistence in spatially variable environments (Chesson 1985; Comins and Noble 1985; Shmida and Ellner 1985; Pacala 1987; Durrett and Levin 1994) and in temporally variable environments (Abrams 1984; Ellner 1984; Chesson 1986, 1990; Loreau 1992). Nonlinear averaging can also lead to persistence of predator-prey, and host-parasitiod associations (Comins and Hassell 1987; Hassell *et al.* 1991). The variation on which these effects depend can arise from fixed patterns of spatial (Chesson 1985) or temporal environmental variation (Loreau 1992; Chesson and Huntly 1993) or randomly varying patterns (Comins and Noble 1985), or it may be generated largely by intrinsic population fluctuations (Armstrong and McGehee 1980; Hastings and Higgins 1994).

In some cases, nonlinear averaging can create a stable equilibrium at the regional scale (Hassell and May 1985; Chesson 1985; Gilpin and Hanski 1991; Murdoch 1994) or on a large temporal scale (Chesson 1984; Chesson and Huntly 1993), where no equilibrium exists on smaller scales. This is most apparent in competition models where in the absence of spatial and temporal variation, only one species could persist in the long run and therefore stable equilibria have just a single species at positive density, with other species at zero density. Spatial and temporal variation may have the effect of creating stable equilibria with many species at positive densities. Murdoch (1994) has posed the question of whether long-term persistence must always involve an equilibrium of some form. The answer to this question is "yes" in all essentials. Indeed, it is essentially tautological that long-term persistence on particular spatial and temporal scales involves an equilibrium on the relevant spatial and temporal scales.

This fact is commonly used in the standard invasibility analysis for species coexistence (Turelli 1978; Levins 1979; Chesson 1994), which involves an argument of the following sort. Ruling out the possibility of indefinite increases in density, and assuming that persistence means that the population does not converge to zero or show ever stronger population crashes, the long-term average change in log population size must be approximately zero. This near zero average change can be regarded as implying that the system is in equilibrium or approximately so. In this sense, equilibrium is inevitable on a sufficiently long timescale, and is almost a rephrasing of persistence as the idea that on a long timescale population density does not change much relative to the time involved. Thus, it makes no sense to expect long-term population persistence without an equilibrium on a long timescale. Such an equilibrium may not involve precise densities (a point equilibrium), however, but a probability distribution of densities, being the stochastic analogue of an equilibrium point (Turelli 1981; Ellner 1989).

Density-independent systems are no exception to the equilibrium requirement for long-term persistence. As remarked above, they must have R close to one (on a log scale, $\ln R$ close to 0). However, such an equilibrium cannot be stable in the sense that perturbations of density will lead to a return to some range or distribution of densities. Indeed, systems with density-independent dynamics are neutral to perturbations of density. To suggest otherwise is to contradict the assumed density independence. In density-dependent systems, however, the long-term equilibrium can be stable, which is observed as a long-term trend to increase or decrease when a population is perturbed from its normal range or distribution of fluctuations (Ellner 1984; Chesson and Huntly 1993; Chesson 1994). But, it is important to keep in mind that a stable equilibrium on a long timescale may be an emergent property of heterogeneity and nonlinear dynamics on smaller scales. Small scale heterogeneity may stabilise the equilibrium on a larger spatial scale, even though an equilibrium need not be present on the smaller scale for it to emerge on the larger scale. It can be created on the larger scale by nonlinear averaging in the transition between scales.

Appendix

The population projection matrix approach (Caswell 1989), commonly used for stage or age structured populations can also be generalised to model density-independent population growth for a population with any spatial structure or stage structure, or combinations of both stage and spatial structure: elements of the matrix can represent density-independent migration rates between localities in addition to transition rates between stages. The elements of this matrix can also vary stochastically in time, and thus represent environmental variability of great complexity. Population dynamics are then given by the equation

$$\mathbf{N}(t+1) = \mathbf{M}(t)\mathbf{N}(t) \tag{14}$$

where the components of $\mathbf{N}$ correspond to the numbers of individuals in each stage, age and place combination, and the elements of $\mathbf{M}(t)$ express the rates of change between these possible states of an individual organism. This iteration was studied mathematically many years ago (Furstenberg and Kesten 1960), and generalised more recently (Heyde 1985) with the conclusion that the elements of $\mathbf{N}(t)$ exhibit a common exponential rate of growth with stochastic fluctuations about this rate. More specifically, there is a constant R such that total population size $N(t)$ satisfies the following equation:

$$N(t) = N(0)R^{t(1 + Z(t))} \tag{15}$$

where $Z(t)$ is a random variable that converges to 0 as t becomes large. The individual elements of $\mathbf{N}(t)$ all satisfy similar equations.

This analysis does not include demographic stochasticity, but branching process theory (Jagers 1975) shows that the addition of demographic stochasticity does not alter the essence of these conclusions.

References

Abrams P (1984) Variability in resource consumption rates and the coexistence of competing species. *Theoret Pop Biol* 25:106–124

Andrewartha HG, Birch LC (1984) *The Ecological Web: More on the Distribution and Abundance of Animals.* Chicago University Press, Chicago. 566 pp

Armstrong RA, McGehee R (1980) Competitive exclusion. *Am Nat* 115:151–170

Caswell H (1978) Predator-mediated coexistence: a nonequilibrium model. *Am Nat* 112:127–154

Caswell H (1989) *Matrix Population Models: construction, analysis, interpretation.* Sinauer Associates, Sunderland, Massachussetts. 328 pp

Chesson PL (1978) Predator-prey theory and variability. *Annu Rev Ecol Syst* **9**:323–347

Chesson PL (1981) Models for spatially distributed populations: the effect of within-patch variability. *Theoret Pop Biol* **19**:288–325

Chesson PL (1984) The storage effect in stochastic population models. In: Levin SA, Hallam TG (eds) *Mathematical Ecology: Trieste Proceedings, Lecture Notes in Biomathematics* No. 54 pp 76–89

Chesson PL (1985) Coexistence of competitors in spatially and temporally varying environments: a look at the combined effects of different sorts of variability. *Theoret Pop Biol* **28**: 263–287

Chesson PL (1986) Environmental variation and the coexistence species. In: Diamond J, Case T (eds) *Community Ecology*, Harper and Row, pp 240–256

Chesson P (1994) Multispecies competition in variable environments. *Theoret Pop Biol* **45**:227–276

Chesson P (1990) Geometry, heterogeneity and competition in variable environments. *Phil Trans Roy Soc Lond B* **330**:165–173

Chesson P (1991) Stochastic population models. In: Kolasa J, Pickett STA (eds) *Ecological Heterogeneity, Ecological Studies: analysis and synthesis*, Volume 86. Springer-Verlag, New York, pp123–143

Chesson P, Huntly N (1993) Temporal Hierarchies of Variation and the Maintenance of Diversity. *Plant Sp Biol* **8**:195–206

Comins HN, Hassell MP (1987) The dynamics of predation and competition in patchy environments. *Theoret Pop Biol* **31**:393–421

Comins HN, Noble IR (1985) Dispersal, variability and transient niches: species coexistence in a uniformly variable environment. *Am Nat* **126**: 706–723

de Jong G (1979) The influence of the distribution of juveniles over patches of food on the dynamics of a population. *Netherl J Zoo* **29**:33–51

den Boer PJ (1968) Spreading of risk and stabilisation of animal numbers. *Acta Biotheoretica* **18**:165–194

den Boer PJ (1981) On the survival of populations in a heterogeneous and variable environment. *Oecologia* **50**:39–53

Dennis B, Taper ML (1994) Density dependence in time series observations of natural populations: estimation and testing. *Ecol Monogr* **64**:205–224

Durrett R, Levin S (1991) The importance of being discrete (and spatial). *Theoret Pop Biol* **46**:363–394

Ellner S (1984) Asymptotic behavior of some stochastic difference equation population models. *J Math Biol* **19**:169–200

Ellner S (1989) Convergence to stationary distributions in two-species stochastic competition models. *J Math Biol* **27**:451–462

Furstenberg H, Kesten H (1960) Products of random matrices. *Annal Math Stats* **31**:457–469

Gilpin M, Hanski I (eds) (1991) *Metapopulation Dynamics: Empirical and Theoretical Investigations.* Academic Press, London. 336 pp

Godfray HCJ, Hassell MP (1992) Long time series reveal density dependence. *Nature* **359**:673–674

Hassell MP (1978) *The Dynamics of Arthropod Predator-Prey Systems.* Princeton University Press, Princeton, USA 237pp

Hassell MP, May RM, Pacala SW, Chesson PL (1991) The persistence of host-parasitoid associations in patchy environments. I. A general criterion. *Am Nat* **138**:568–583

Hassell MP, May RM (1985) From individual behaviour to population dynamics. In: Sibly RM, Smith RH (eds) *Behavioural Ecology: Ecological Consequences of Adaptive Behaviour.* Brit Ecol Soc Symp, Blackwells, Oxford, pp 3–33

Hastings A, Higgins K (1994) Persistence of transients in spatially structured ecological models. *Science* **263**:1133–1136

Heyde CC (1985) An asymptotic representation for products of random matrices. *Stochastic Processes Applic.* **20**:307–314

Holt RD (1977) Predation, apparent competition, and the structure of prey communities. *Theoret Pop Biol* **12**:197–229

Hughes TP (1990) Recruitment limitation, mortality, and population regulation in open systems: a case study. *Ecology* **71**:12–20

Ives AR (1991) Aggregation and coexistence in a carrion fly community. *Ecol Monogr* **61**:75– 96

Jagers P (1975) *Branching Processes with Biological Applications.* John Wiley, London, 268 pp

Johnson NL, Kotz S (1969) *Distributions in statistics: discrete distributions.* Houghton Mifflin, Boston, 328 pp

Kareiva P, Wennergren U (1995) Connecting landscape patterns to ecosystem and population processes. *Nature* **373**:299–302

Levins R (1979) Coexistence in a variable environment. *Am Nat* **114**:765–783

Loreau M (1992) Time scale of resource dynamics, and coexistence through time partitioning. *Theoret Pop Biol* **41**:401–412

MacArthur RH (1970) Species packing and competitive equilibria for many species. *Theoret Pop Biol* **1**:1–11

May RM (1973) Stability and Complexity in Model Ecosystems. Princeton University Press, Princeton, 265pp

May RM (1978) Host-parasitoid systems in patchy environments: a phenomenological model. *J Anim Ecol* **47**:833–44

May RM, Oster GF (1976) Bifurcations and dynamic complexity in simple ecological models. *Am Nat* **110**:573–599

Murdoch WW (1994) Population regulation in theory and practice. *Ecology* **75**:271–287

Murdoch WW, Bence J (1987) General predators and unstable prey populations. In: Kerfoot W, Sih A (eds) *Predation: Direct and Indirect Effects on Aquatic Communities.* University Press of New England, Hanover, USA, pp 17–30

Murdoch WW, Luck RF, Swarbrick SL, Walde S, Dickie SY, Reeve JD (1995) Regulation of an insect population under biological control. *Ecology* **76**:206–217

Murdoch WW, Walde SJ (1989) Analysis of insect population dynamics. In: Whitaker J, Grubb PJ (eds) *Toward a more exact ecology.* Brit Ecol Soc Symp Blackwells, Oxford. pp 113–140

Nisbet RM, Gurney WSC (1982) *Modelling Fluctuating Populations.* J. Wiley, Chichester, UK, 379 pp

Pacala SW (1987) Neighborhood models of plant population dynamics. 3. Models with spatial heterogeneity in the physical environment. *Theoret Pop Biol* **31**:359–392

Pacala SW, Silander JA Jr (1985) Neighborhood models of plant population dynamics. I. Single-species models of annuals. *Am Nat* **125**:385–411

Pulliam HR (1988) Sources, sinks and population regulation. *Am Nat* **132**:652–661

Reeve JD (1988) Environmental variability, migration, and persistence in host-parasitoid systems. *Am Nat* **132**:810–836

Sale PF (1977) Maintenance of high diversity in coral reef fish communities. *Am Nat* **111**:337–359

Shmida A, Ellner SP (1985) Coexistence of plant species with similar niches. *Vegetatio* **58**:29–55

Shorrocks B, Rosewell J (1987) Spatial patchiness and community structure: coexistence of drosophilids on ephemeral resources. In: Gee JHR, Giller PS (eds) *Organization of communities: past and present.* Blackwells, Oxford, pp 29–51

Southwood TRE (1978) *Ecological methods with particular reference to the study of insect populations.* Chapman and Hall, London, 524 pp

Stone L (1993) Period-doubling reversals and chaos in simple ecological models. *Nature* **365**: 617–620

Strong DR (1986) Density-vague population change. *Trends Ecol Evol* **2**:39–42

Turchin P (1990) Rarity of density dependence or population regulation with lags? *Nature* **344**:660–663

Turelli M (1978) Does environmental variability limit niche overlap? *Proc Natl Acad Sci USA* **75**:5085–5089

Turelli M (1981) Niche overlap and invasion of competitors in random environ-ments I. Models without demographic stochasticity. *Theoret Pop Biol* **20**:1–56
Watkinson AR (1980) Density-dependence in single-species populations of plants. *J Theor Biol* **83**:345–357

SPATIAL PATTERNS OF VARIANCE IN DENSITY OF INTERTIDAL POPULATIONS

A.J. Underwood

ABSTRACT

Abundances of many populations fluctuate from time to time and often these temporal fluctuations are very different from one patch of habitat to another. The statistical interaction between temporal and spatial variability is the focus of attention for detecting or estimating the magnitude of environmental perturbations. It deserves better understanding and the development of predictive capacity for environmental and ecological studies. Such understanding requires research programmes to investigate and explain processes influencing spatial and temporal variances; a preliminary account of some results of this research is illustrated here.

Examples of patterns in space in the abundance of marine intertidal invertebrates are presented. These include the variability at small spatial scales of densities of the snail, *Littorina unifasciata*, the spatial patterns of distribution of the barnacle, *Chamaesipho tasmanica*, on sheltered and wave-exposed shores and the patterns of spatial variance of the barnacle, *Tesseropora rosea*, at different tidal heights on the shore.

The examples demonstrate the complexity of spatial variation from one scale to another and between different habitats or pats of habitats occupied by the same species. This is discussed with respect to design and interpretation of ecological studies, including the need to ensure that all estimates of relationships between variances and means of densities of populations are properly replicated in space and time. Because spatial variance is not constant from place to place and time to time, it requires careful interpretations at different scales. Spatial (or temporal) variance is a complex function of behaviour and previous history of the organisms, confounded with physical aspects of habitat and distribution of resources. Manipulative experiments are needed to unravel this complexity.

Key words: Intertidal animals, patchiness, spatial scale, variance

Frontiers of Population Ecology, R.B. Floyd, A.W. Sheppard and P.J. De Barro (eds), CSIRO Publishing, Melbourne, 1996 pp. 369–389

Introduction

One of the major impediments to more rapid progress in ecology is the large variability from place to place in the intensity, magnitude and frequency of disturbances and processes of recruitment that affect patterns of abundance. There are differences from place to place and, more importantly, differences in the time-courses of these processes among places to place. These result in considerable interaction among sites and times of sampling in the abundance of individual species, interactions among species and structures of local assemblages (e.g. Chesson 1978; Underwood and Denley 1984; Underwood and Fairweather 1989; Underwood *et al.* 1983).

This matters in terms of the science of ecology, making general models difficult to apply (e.g. Foster 1990), causing difficulties in the development of general models for the structure of populations (Roughgarden and Iwasa 1986) and creating considerable complexity in the analytical framework in which populations are studied (Andrew and Mapstone 1987). Much more importantly, it makes predictions about populations very difficult (Peters 1991) and causes immense difficulty in the detection and estimation of magnitude of human disturbances (i.e. environmental impacts)(see particularly Underwood and Peterson 1988; Underwood 1989, 1990, 1991, 1992, 1993, 1994a).

There have been several approaches to development of theory about interactions of space and time. First, Pielou (1969) recommended modelling and analytical procedures which allowed examination of spatial patterns or temporal dynamics as separate entities. These were methods for dealing with what Chesson (1985) called 'pure spatial' and 'pure temporal' environmental variation. This approach ignores interactive processes. Second, Taylor (1961) and Taylor and Taylor (1977), in the context of density-dependent regulation of large-scale populations, examined variances in abundance as a function of mean abundance. Temporal patterns were, however, lost because data from different years were plotted without regard to their order. Thus, spatial variance was shown to fluctuate as a function of mean abundance through time, but the temporal pattern (or lack of it) was unimportant. Third, Steele (1985) compared variances in environmental variables in terrestrial and marine systems. He concluded that the distribution of variances in processes operating at different rates caused major differences in patterns in the two environments. His approach provided insights into the role of different types of temporal variance in affecting populations. There was, however, no prospect of learning anything about spatial differences in abundance within habitats.

Chesson (1985) has done more by developing models for coexistence of putatively competing species, based on environmental variability. He considered spatial and temporal variance separately and in conjunction with one restrictive class of interaction (so-called pure spatiotemporal environmental variation). Results of these models demonstrated inadequacies of existing methods of documenting stochastic processes regulating populations (Chesson 1985; Warner and Chesson 1985). Thus, Grossman (1982) considered that temporal patterns of relative abundance of species would allow estimation of the importance of stochastic events. Chesson (1985) demonstrated that asynchronous local fluctuations (i.e. interactions in space and time) would be masked unless spatial scales were carefully chosen.

Chesson's (1985) analysis, however, only considered spatial-temporal interactions that retained constant spatial averages and spatially consistent temporal averages. Observations on rocky shores suggest that there is also 'impure' spatiotemporal variability – patches in space fluctuate independently in time, but spatial averages vary in time (and temporal averages are not spatially consistent). There is increasing evidence that the spatial scale of patterns and the temporal scales of the processes that cause them vary in complex ways from one spatial or temporal scale to another (see, for example, Morrisey *et al.* 1992a, b). As a result, particularly in sampling to detect environmental disturbances, there need to be data at several hierarchical temporal (Underwood 1991) and spatial (Underwood 1992) scales.

Before much progress can be made in interpreting or predicting variability in trajectories of populations, there needs to be careful investigation of the nature of each of spatial and temporal variance in its own right. Otherwise, their interactions will not be understood. Many published accounts of spatial variation assume that it is constant from time to time, independent of the scale of sampling and invariant from one to another part of an organism's habitat.

To demonstrate the validity of such assumptions requires replicated measures of variation from time to time and place to place at several scales. Such studies are described here as a preliminary account of a larger research programme on the nature of spatial and temporal variation in populations of marine animals.

Measures of variability among places or times of sampling typically include true population variability and a measure of sampling error. Calculation of components of variation from hierarchical analyses of variance allows separation of sampling error from estimates of true variability (Gaston and McArdle 1994). The sampling programme for the studies described here was designed to allow spatial variability to be identified at scales varying from quadrats within a site, among sites within a shore, among heights on a shore and among shores separated by different distances. Components of variation from these data will allow characterisation of the four different types of variance-mean relationships identified by Gaston and McArdle (1993, 1994) at a number of different spatial and temporal scales for a number of co-existing species in the same habitat, thus allowing intra- and interspecific comparisons of these relationships.

There is evidence that closely-related taxa show more similar temporal variances than distantly-related taxa, but these comparisons are probably meaningless because they have been made, for example, by comparing insects and birds, which probably fluctuate at very different temporal scales (Gaston and McArdle 1994). The data in the present study include comparisons of the same species in different habitats and, because related species (e.g. different species of gastropods or barnacles) are being examined, it will eventually be possible to do direct tests of this prediction by testing the relationships between spatial and temporal variances and relatedness of taxa for a range of intertidal species.

There are, however, some considerable impediments in the use of variance components where these are calculated from a single set of data, so that only one estimate of each component or source of variation is available for any site. There are problems with determining the degrees of freedom for each variance (Burdick and Graybill 1992; Gaylor and Hopper 1969; Satterthwaite 1946; Searle *et al.* 1992; Tukey 1956) and, because estimates of variances are sometimes negative, with biases in their calculation (Scheffe 1959). These are also more mundane illogicalities in the general use of mean square estimates of variance components where experiments or sampling are done with different sizes of samples, different structures and designs and in different ways. Underwood and Petraitis (1993) reviewed many of these problems with the associated logical difficulties caused by interactions among component and the sheer illogicality of calculating independent estimates of variances, but then confounding them by expressing them as percentages.

To avoid many of these problems, independent measures of spatial variability can be obtained by sampling populations several independent times. If independently sampled, each time of sampling can therefore provide an independent, replicate estimate of spatial variance for any chosen scale and will therefore allow more coherent (and usually statistically simpler) comparisons from one site or habitat or species to another. That the times and methods of sampling used here are statistically independent will be demonstrated elsewhere.

By having replicated samples, each furnishing an independent estimate of variance at some spatial scale, another problem can be solved. In any hierarchical sampling design done at one time, there are more degrees of freedom (i.e. greater precision) for estimated variances at smaller than at larger spatial scales (see Table 1). Consequently, comparisons among spatial scales are difficult to interpret.

If, in contrast, each time of sampling provides a replicate, the average variance at each scale of the hierarchy can be estimated with the same number of degrees of freedom (a function solely of the number of times of sampling). Although each replicate estimate of variation at small spatial scales is still more precise than those at larger spatial scales, the average variance can be estimated in a similar and directly comparable manner at each scale of sampling.

The same argument holds for estimates of temporal variation. They can be replicated by sampling numerous populations in a given habitat. Each site provides an independent estimate of spatial differences. Long-term sampling allows replicated estimation of interactive variances between spatial differences and temporal trajectories.

In this paper, some preliminary results are described to indicate how replicated estimation of variances in space (or time) can provide much valuable information about processes in populations. The results also warn against simple assessment of spatial structure at only one time or site. The examples are of snails and barnacles on rocky intertidal shores, because these animals are fairly easy to sample and much is known of their general ecology and the processes influencing densities (reviewed by Underwood 1985, 1994b; Underwood and Chapman 1995).

The present paper is simply introductory to the approach used and the data being obtained. The examples described here illustrate spatial patterns in density of a snail at various scales and that these vary from site to site; the spatial patterns in density of a barnacle and how these indicate different processes operating on sheltered and exposed coast-lines and the complex relationships between spatial variance and mean density of a different species of barnacle at different heights on an intertidal gradient.

Methods and study-sites

Sites along the coast of New South Wales were sampled at different times for various purposes (reviewed in Underwood 1994b). Generally, sheltered coast-lines in bays and estuaries and the usually steeper exposed shores on the open coast can easily be stratified into an upper level dominated by abundant small snails, *Littorina unifasciata.* At mid-shore levels, barnacles or gastropods dominate, often depending on wave-action (Underwood 1981a; Fairweather and Underwood 1991). At the lower levels, algae, tube-worms or large solitary ascidians dominate the habitat (Dakin *et al.* 1948).

The data sets described here were collected from 1972 to 1994 in different studies. The details of sites and the abundances of the organisms cannot possibly be provided here and examples are chosen that are not site-dependent. In all cases, data come from counts of randomly-thrown quadrats (50 × 50 cm) in areas of shore about 3 × 5m or 4 × 4m. Areas were sampled on subsequent occasions, but quadrats were not fixed in place. Consequently data from one time to another are generally independent and can be treated as such in analyses. *L. unifasciata* were sampled in a different manner in studies at Cape Banks; the details are provides with the Results. Individual animals were counted in five sub-quadrats within each quadrat. These are 4.5 × 4.5 cm and approximately 20 cm^2.

Results presented here largely come from analyses of variance. The models of analyses (fixed or random factors, nested components, expected values of mean squares) were established using the Cornfield and Tukey (1956) procedures as summarized in detail in Underwood (1981b). All data were used without transformations—the analyses were not to test hypotheses (which require assumptions of homoscedasticity), but to determine the magnitudes of components of mean squares. These calculations followed the procedures in Winer *et al.* (1991).

Generally, where a component of variation was estimated to be negative, it was used as a negative estimate, assuming that the real variance was small or zero and was being underestimated in sampling. As an alternative, such components can be set to zero, for convenience, even though this procedure biases estimates (Scheffe 1959, Winer *et al.* 1991). It was preferred here to retain unbiased estimates, particularly where comparisons were to be made across several replicate estimates of each component of variation.

In the case of some analyses of the barnacle *Chamaesipho tasmanica*, components of variation were consistently large and negative, indicating that they were not simply sampled underestimates of a small or zero variance. Under these circumstances, two alternative procedures exist to estimate the other components in an analysis. The first is to eliminate the relevant component from the model and to re-analyse the data, thus forcing the variance components for the remaining terms to be positive. The second procedure is to use residual maximal likelihood estimators instead of analysis of variance, least-squares estimates. The latter force all components of variation in any model to be calculated as positive values (Robinson 1987). For balanced designs, as used everywhere here, the two procedures produce identical results, but the former are much simpler to compute. In the case of data for the barnacle *C. tasmanica* from sheltered shores, large-scale variation was always estimated to be negative, implying zero variation from one location to another. The largest scale of variability was eliminated from the analytical models to ensure the most reliable estimators of variance for the all other scales (see Results).

For many analyses of relationships between variances and means, the variances and means are routinely transformed to logarithms, as recommended in many texts (see the discussion in McArdle and Gaston 1992). Where this was done here, the variances were nevertheless calculated from analyses of variance on untransformed data. Converting estimates of variances to logarithms is impossible if any of the estimates are zero or negative. As a result, it is not a practicable procedure for comparisons of variances estimated from hierarchical analyses of variance where the higher components (the larger scales) are estimated by subtraction from smaller scales (see Table 1). The only cases where transformation to logarithms is appropriate are analyses or comparisons of variances at a single (and therefore the smallest or lowest) scale of sampling in any hierarchy. This transformation was used here only for such comparisons, involving the barnacle *Tesseropora rosea.*

Results

Density of the snail Littorina unifasciata

The small grazing snail *L. unifasciata* has been studied in 30 sites at the Cape Banks Scientific Marine Research Area, Botany Bay (NSW) as part of experimental studies of its ecology (e.g. Chapman 1986; Underwood and Chapman 1985, 1989, 1992). At numerous times over about 12 years, these populations were sampled, to provide information about spatial variations in density at very small (subquadrats, 20 cm^2 about 20–30 cm apart) and larger scales (quadrats, pairs as in Fig. 1). Sites were mostly 5–100s of metres apart and spanned the gradients of height (i.e. tidal rise and fall) and wave-action (horizontally along the shore).

In general, densities of snails fluctuated from time to time and differed from place to place. Data from 7 of the 30 sites studied at Cape Banks show that there are great differences from height to height, more so than occurs among sites along the shore (Fig. 2). Some sites persistently have large densities and some stay relatively small. Although densities are generally small, bear in mind that a density of 4 per quadrat of 20 cm^2 is approximately generally 2000 snails per m^2, so even a small density represents many individuals in a site.

Finally, the data show convincingly that fluctuations through time are not concordant among sites, so there is considerable interaction in analyses and their interpretations.

For any given site, sampled independently through time, the estimates of variance components can be extracted from an analysis of the form illustrated in Table 1. Note that degrees of freedom must obviously increase with decreasing spatial scale because of the choice of a hierarchical sampling scheme. In examples illustrated here, the spatial variances at each chosen scale were estimable for any site for each of the times of sampling (as in Table 1). Thus, there were 30 estimates for each site from each of 30 times.

Across the 30 sites at Cape Banks, temporal variations in mean abundance were positively correlated with abundance (Fig. 3), which is illustrated by the trends in Figure 2. Thus, where snails are numerous, they have greater temporal change in density than where they are sparse.

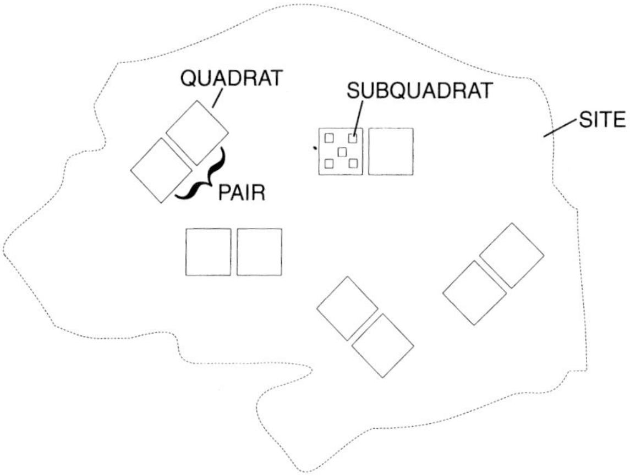

Fig. 1. Sampling design for the snail *Littorina unifasciata*. A site (approximately 4 × 4 or 5 × 3 m) is sampled by five pairs of 50 × 50 cm quadrats. The pairs are scattered at random and approximately 190 cm apart. The two quadrats in each pair are centred 50 cm apart. In each are five replicate subquadrats (4.5 × 4.5 cm), approximately 18 cm apart on average). For sampling the barnacles, there were only single quadrats with five subquadrats.

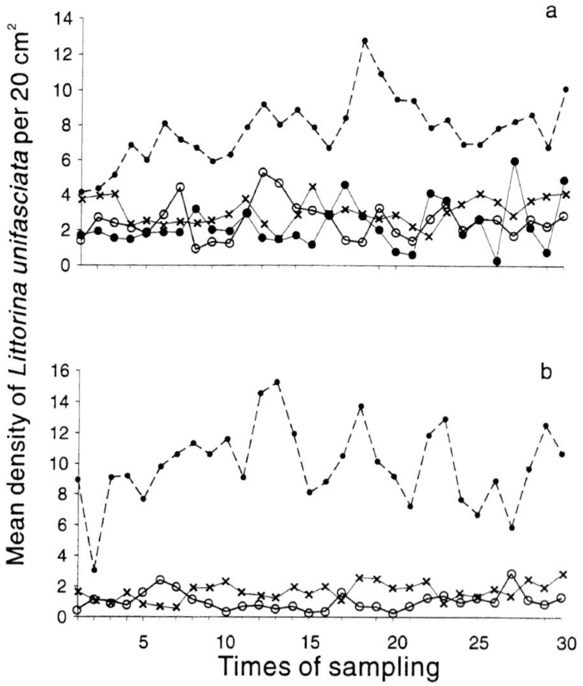

Fig. 2. Mean density (n = 50 subquadrats) of the snail *Littorina unifasciata* at seven sites in the Cape Banks Scientific Marine Research Area at 30 independent times of sampling over 8–10 years (times not shown at correct intervals). (*a*) four heights on the shore, sites Flat A 1–4; (*b*) three heights on the shore, sites Wreck B 1–3 (details of the sites do not matter here).

Table 1. Calculation of components of variation from analyses of variance. All factors are random. In the examples illustrated, there are 30 times of sampling each site; in each site, there are 5 Pairs, each of 2 Quadrats and animals are counted in each of 5 Subquadrats in each Quadrat (see also Figure 1).

(a) Variation among times of sampling: analysis for each site

Source of variation		Degrees of freedom	Mean square estimates
Among Times	= T	29	$\sigma_R^2 + 50\sigma_T^2$
Among all subquadrats = Residual	= R	1470	σ_R^2

Estimated variance among times = $\hat{\sigma}_T^2 = \dfrac{\text{Mean square T} - \text{Mean square R}}{50}$

(b) Spatial variation at different spatial scales: analysis at each site for each time

Source of variation		Degrees of freedom	Mean square estimates
Among Pairs	= P	4	$\sigma_S^2 + 5\sigma_Q^2 + 10\sigma_P^2$
Between Quadrats (Pair)	= Q	5	$\sigma_S^2 + 5\sigma_Q^2$
Among Subquadrats = Residual	= S	40	σ_S^2

Estimated variance among Pairs = $\hat{\sigma}_P^2 = \dfrac{\text{Mean square P} - \text{Mean square Q}}{10}$

Estimated variance among Quadrats = $\hat{\sigma}_Q^2 = \dfrac{\text{Mean square Q} - \text{Mean square S}}{5}$

Estimated variance among Subquadrats = $\hat{\sigma}_S^2$ = Mean square S

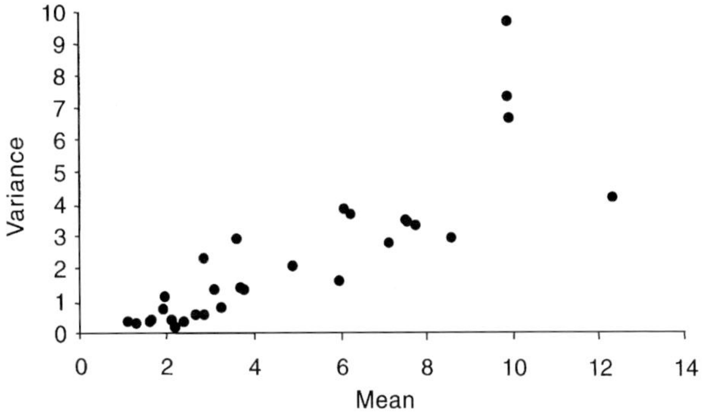

Fig. 3. Positive correlation between temporal variance ($\hat{\sigma}_T^2$ in Table 1) among mean densities in 30 independent times of sampling and mean density averaged over all times. Each point represents results from a single site; there are 30 sites.

Spatial variability in density of *L. unifasciata*

The choice of spatial scale of sampling is often arbitrary or, at best, based on some understanding of the behaviour or spatial arrangement of use of resources of animals. As a result, spatial structure or variability is often measured at arbitrary scales.

To examine the effect of spatial scale on variances, the components of variance for three scales were examined. These were as described in Figure 1. The first major finding was that the greatest variance was at the smallest spatial scale (subquadrats about 18–25 cm apart). This is shown in Table 2 for several sites.

Table 2. Spatial variation at different spatial scales for three sites at Cape Banks (Point 1, see Fig. 4; Scrapings, see Fig. 5; Wreck B 2, see Fig. 6). Data are from analyses as in Table 1 for 30 independent times of sampling in each site.

	Point 1	Scrapings 1	Wreck B 2
Variation among Pairs ($\hat{\sigma}_P^2$ in Table 1). Mean (S.E.)	5.7 (1.2)	2.5 (0.5)	3.8 (0.9)
Variation among Quadrats ($\hat{\sigma}_Q^2$ in Table 1). Mean (S.E.)	1.6 (0.7)	1.0 (0.5)	2.5 (1.1)
Variation among Subquadrats ($\hat{\sigma}_S^2$ in Table 1). Mean (S.E.)	24.1 (2.4)	15.8 (2.0)	57.3 (5.2)

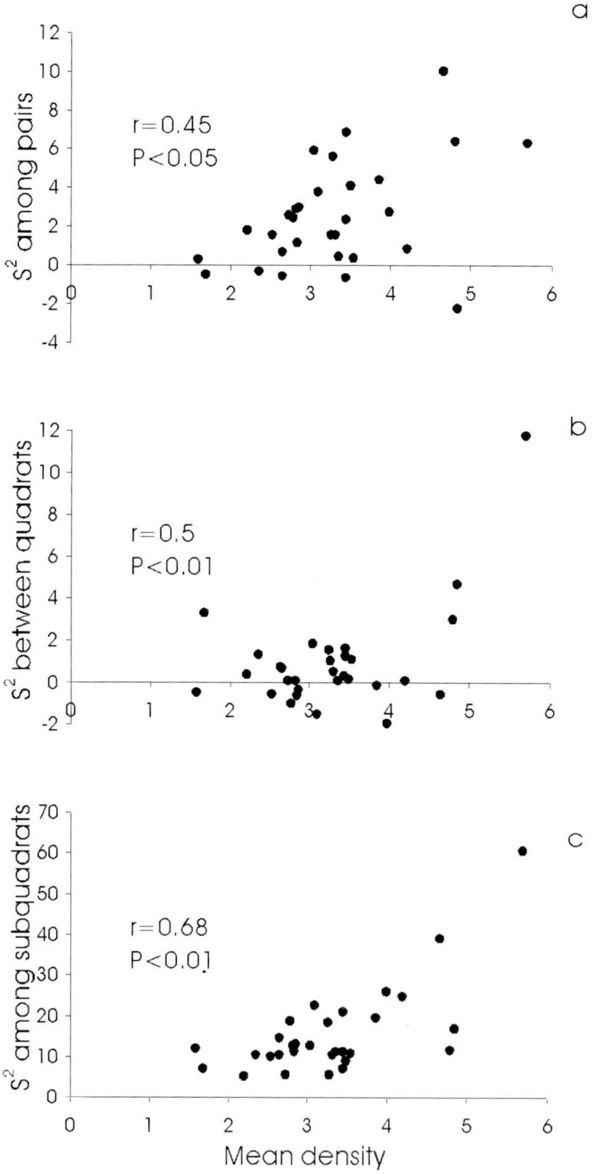

Fig. 4. Relationships between spatial components of variance and mean density of *L. unifasciata* for a single site ('Scrapings' site 1) at Cape Banks. Data are from 30 independent times of sampling. a) variance among pairs, b) variance between quadrats, c) variance among subquadrats; $\hat{\sigma}_P^2$, $\hat{\sigma}_Q^2$ and $\hat{\sigma}_S^2$, respectively, in Table 1.

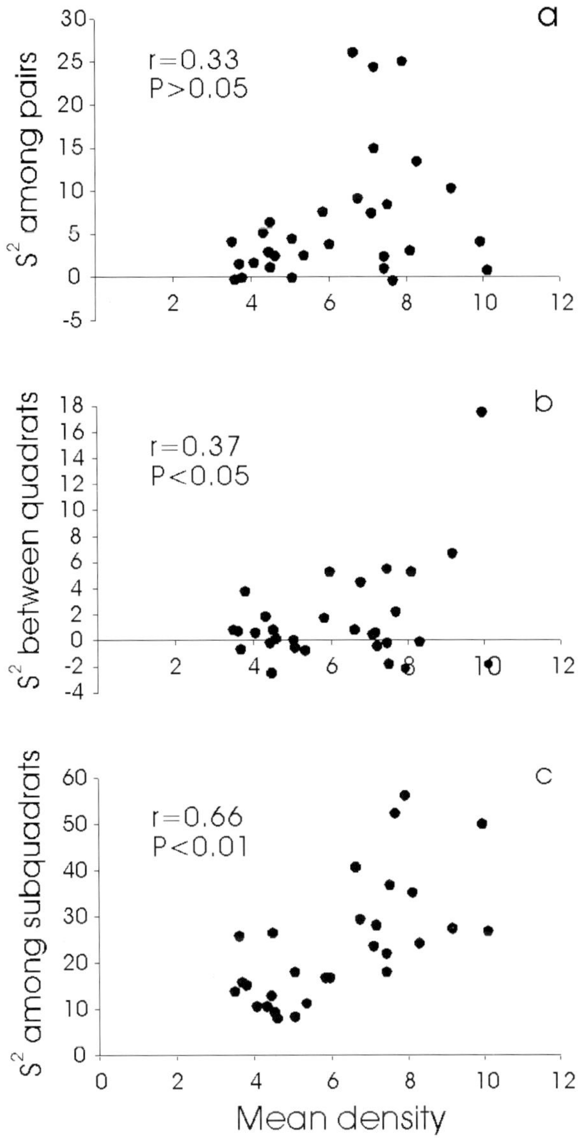

Fig. 5. Relationship between spatial components of variance and mean density of *L. unifasciata* for the 'Point' Site 1 at Cape Banks. All other details as in Fig. 4.

A second important feature is that typically there was a linear increase in variance with increasing mean density (Fig. 4) for variances at the smallest spatial scale. This trend was sometimes also present for the variance at other scales (quadrats centred 50 cm apart and pairs of quadrats about 190 cm apart, on average). Thus, at some sites, variances at all three scales were correlated with means (Fig. 4) . At other sites, there was no correlation at larger scales, even though there was at smaller ones (Fig. 5). In yet other sites, there was no significant correlation between variance at any scale and mean (Fig. 6).

This is an important result because it demonstrates not only that choice of scale will lead to different outcomes of any analysis of spatial variance, but that results from a single site cannot typify the pattern for a species with a patchy distribution.

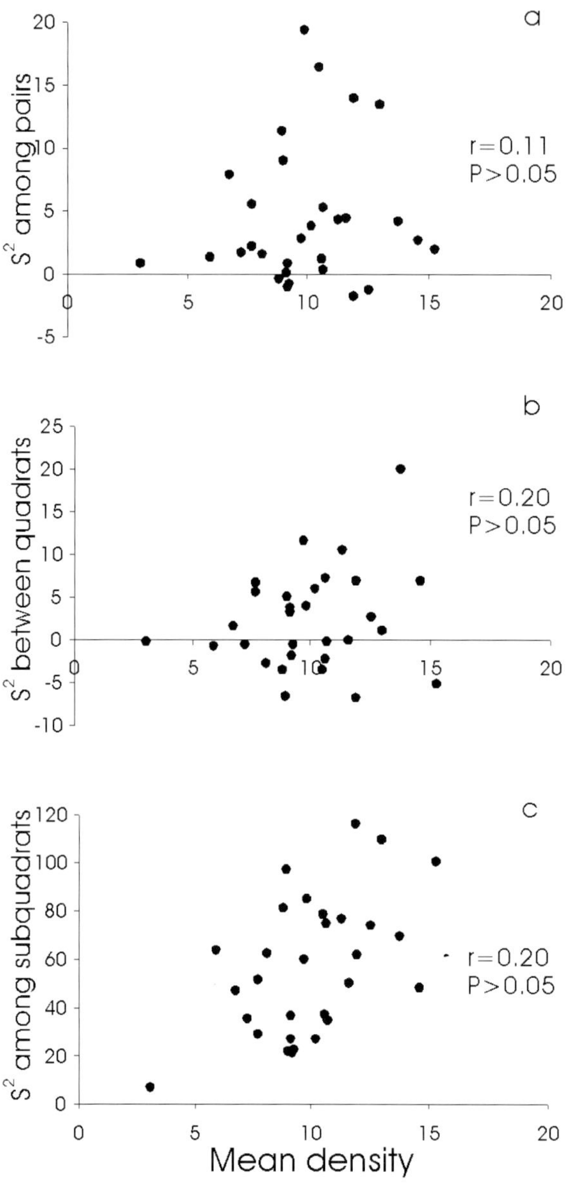

Fig. 6. Relationship between spatial components of variance and mean density of *L. unifasciata* for the 'Wreck B' Site 2 at Cape Banks. All other details as in Fig. 4.

The results from the three sites shown suggested that the overall trend was for a positive trend of variance where the mean density was over a small range (as in Fig. 4), but no significant correlation where mean density was larger (as in Fig. 6). This general pattern leads to the hypothesis that the correlation of variance with mean should itself decrease with increasing mean density. This hypothesis was tested using the variances at the smallest spatial scale (subquadrats as in Fig. 4c). The correlation of variance with mean for the thirty times of sampling was plotted against the mean density over all times for each site (Fig. 7). The outcome was a very definite negative correlation. Thus, the trend of increasing variance with increasing mean at any site decreases with increasing mean density among sites.

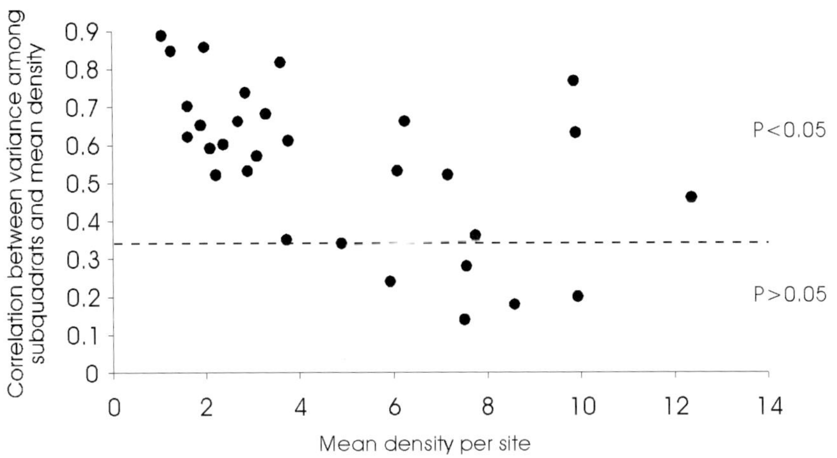

Fig. 7. Correlation between variance among subquadrats ($\hat{\sigma}^2_S$ in Table 1) and mean density for each of 30 sites, plotted against overall mean density in each site. Note that each point is the correlation coefficient from 30 times of sampling (shown as *r* for one site in Fig. 4c). Also note that some of the individual correlations for sites with relatively large mean densities were not significant as *P* = 0.05 (the critical value with 28 degrees of freedom is shown as a dashed line).

Patterns of variation in density of the barnacle *Chamaesipho tasmanica*

Chamaesipho tasmanica is a small barnacle, rarely larger than a few mm across, that occurs in vast numbers on much of the mid-portion of sheltered shores and in large densities at higher levels on exposed coasts (Underwood 1981a; Underwood and Denley 1984). Because the same species can be found in similar abundances on the two types of shore, it was predicted that it would show similar trends of spatial variation and relationships to mean abundances in the two habitats.

Barnacles were sampled on five open coastal locations along the coast of New South Wales at three to four-monthly intervals for several years from early 1980. The details of locations will be described with a more complete treatment of the data elsewhere. At each location, two sites were sampled, some tens to hundreds of metres apart. *Chamaesipho* were mostly abundant in one level of the shore, so the data from that are considered here.

In addition and for different reasons, the same species was sampled on sheltered shores around Ulladulla and around Jervis Bay from 1988 to the present. The shores chosen were those with the largest consistent densities of *Chamaesipho*, because these were more obviously comparable to the sheltered shores.

A comparable set of data was available from four locations around Jervis Bay for the period from 1988 to 1991. This poses a problem for the comparison of sheltered and open coast locations because of the several years of difference in the period sampled. To attempt to unconfound the comparison, a second set of data from sheltered shores was examined – three locations around Ulladulla and its environs from 1990 to 1993. This differed by about the same period as the difference between the open coast (to 1983) and sheltered shores (from 1988).

The populations on sheltered coasts were very variable from time to time (Fig. 8), reflecting periods of recruitment and mortality, probably largely due to predatory whelks (Fairweather *et al.* 1984). In contrast, the open coast populations were generally declining in abundance during the period surveyed (Fig. 9). The mean densities on the shores were, however, not particularly different from those found on sheltered shores, so the spatial variances were considered to be comparable.

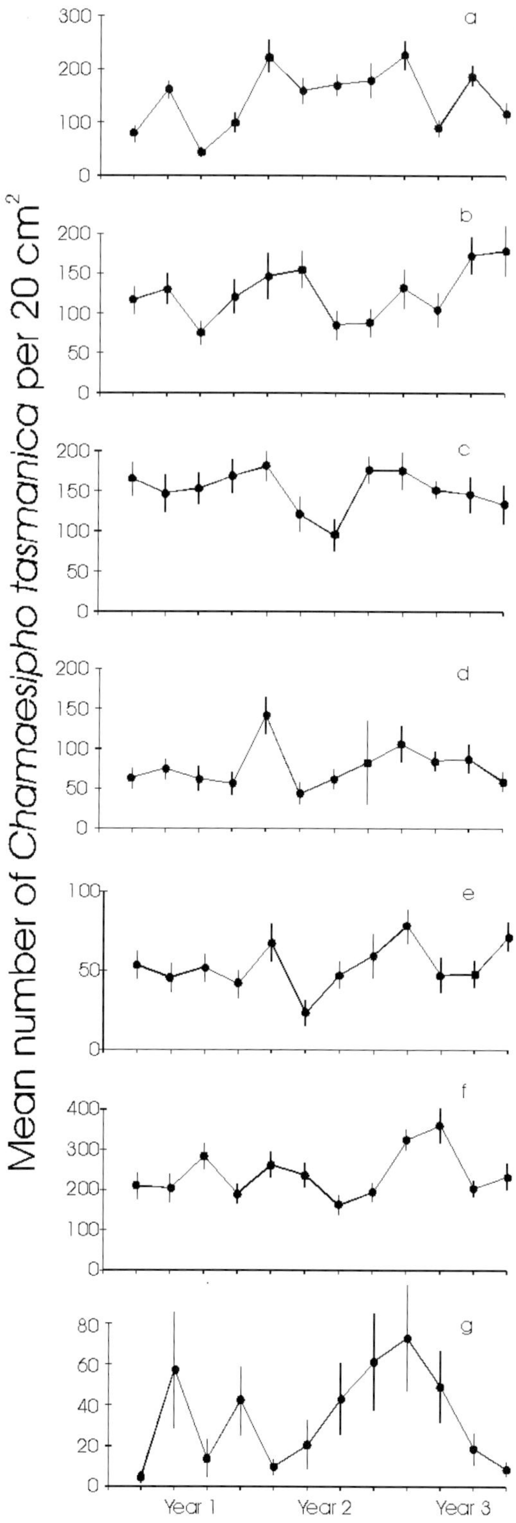

Fig. 8. Mean density of the barnacle *Chamaesipho tasmanica* (*n* = 25 subquadrats) at seven sheltered sites around Jervis Bay from 1988 to 1991. a, b Green Point, Sites 3.1 and 3.2; c, d Honeymoon Bay, Sites 2.2 and 3.1; e, Long Beach South, Site 3.1; f, g Plantation Point, Sites 3.1 and 3.2.

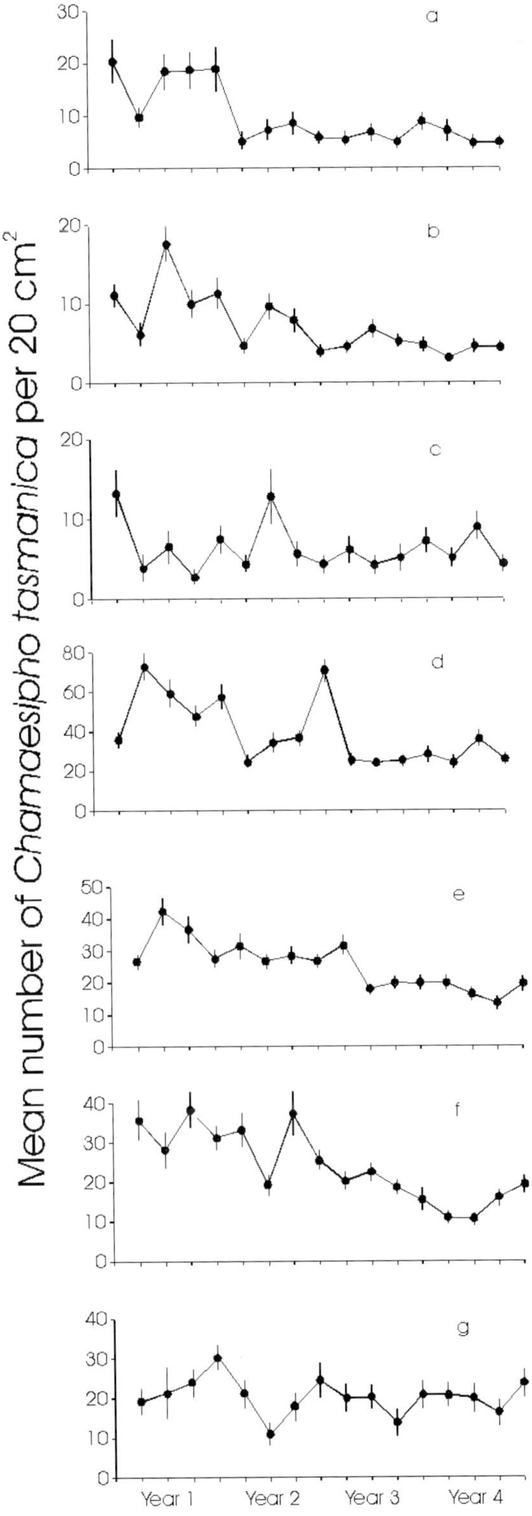

Fig. 9. Mean density of the barnacle *Chamaesipho tasmanica* (n = 25 subquadrats) at seven exposed shores on the coast of NSW from 1980 to 1983. a, b, c Blueys Head, Sites 2.2, 3.1 and 3.2; d, e Tura Head, Sites 2.1 and 2.2; f, g Flat Rock Island, Sites 2.1 and 2.2.

Table 3. Spatial variation at different scales for the barnacle *Chamaesipho tasmanica* in sheltered or exposed coastal locations. There were two sites at each location; five quadrats were sampled in each site and five subquadrats were sampled in each quadrat. In Jervis Bay, the locations were Honeymoon Bay, Green Point, Plantation Point and Long Beach South. Around Ulladulla, locations were Mollymook, Ulladulla Harbour and Mollymook Golf Course. The exposed locations were at Bluey's Head, Cape Banks, Flat Rock Island and Tura Head, but the details are not provided here. Each set of data is averaged from five independent times of sampling during a period of about four years. For sheltered shores, variance components for Locations were negative and have been set to zero (see Methods).

	Mean (S.E.) variation among:			
	Locations	Sites	Quadrats	Subquadrats
Sheltered Shores				
Jervis Bay 1988–1991 (4 locations)	0	6424 (1476)	3548 (752)	6373 (417)
Ulladulla region 1990–1993 (3 locations)	0	5537 (1325)	2277 (648)	3,046 (433)
Exposed shores 1980–1983 (4 locations)	220 (59)	51 (32)	72 (17)	146 (24)

For the comparisons of spatial variances made here, data from each of the three sets (open coast and two sheltered coastal sets) were chosen so that they spanned the same time-period – about four years. Also, the same number of times of sampling were analysed. So, for the four open coast locations, analyses were of a total of five times of sampling from 1980 to 1983. For four locations in the Jervis Bay data, five times of sampling from 1988 to 1991 were used. For three locations round Ulladulla, five times of sampling from 1990 to 1993 were chosen.

The most striking feature of the analyses was the complete difference in the scales of spatial variance in sheltered and open coastal locations. Both sets of sheltered shores showed very large variance at the smallest scale (subquadrats within quadrats), as in Table 3. There was also considerable variance among quadrats and sites, but no difference among locations; estimates of variance among locations were always negative. In the formal analyses of variance of differences among shores, none of the *F*-ratio tests for differences among locations was ever significant (Chapman and Underwood, unpublished). The components of variation for the other scales in Table 3 were therefore calculated from analyses where differences among locations were eliminated from the model (see Methods). The general pattern is of similar densities from location to location and vast small-scale variability in numbers of barnacles. This was consistent over the two periods and the various times of sampling and over the many shores examined. The data here are quite typical.

In contrast, the variability on wave-exposed coastal shores was quite different. Except for variation among locations, variances were much smaller at all scales than on the sheltered shores (Table 3). There was large-scale variation from location to location and this was often a significant source of variation in formal tests of the data, in complete contrast to the situation on sheltered shores.

The consistency of results from the two periods sampled on sheltered shores strongly supports the notion that there is a substantial difference between sheltered and exposed coasts, although this conclusion would obviously have been much strengthened if there had been another comparable set of data from exposed coasts during a period some years after the first set. The other influence on the results is that several of the sheltered coasts did have larger mean densities during the periods analysed than were found on the exposed coasts. If only those sheltered shores with a similar range of densities (e.g. locations a, e, g in Fig. 8 and locations d, e, f, g in Fig. 9) were compared, the outcome and conclusions would be identical. The differences found are a feature of the different habitats and not a feature of different densities.

Spatial variation at different heights on the shore: the barnacle *Tesseropora rosea*

The final example illustrated is the surf-barnacle *Tesseropora rosea*, a conspicuous and abundant occupant of mid-shore levels on wave-exposed shores (Denley and Underwood 1979; Underwood *et al.* 1983). The barnacle reaches a diameter of up to 4 cm and sometimes settles profusely, blanketing patches of the shore. Unlike *Chamaesipho*, however, *Tesseropora* tend not to settle and survive in the immediate vicinity of others, so there is little crowding or touching.

The barnacles were sampled on the exposed coasts, as described in the previous section. As found previously with *Littorina unifasciata*, there was always a striking positive correlation between variance among subquadrats and the mean number of barnacles at a site (shown after transformation to natural logarithms in Fig. 10). Densities were somewhat smaller in the upper half of the barnacle's distribution (note range in the lower two examples in Fig. 10).

Apart from this typical pattern, there was a noticeable (and statistically significant at $P<0.05$) difference between the slope of the regression of log variance on log mean between the two heights. There was a greater increase in variance for a given increase in mean in the lower half of the distribution (mean slope of the two examples shown = 1.43 compared with 1.17 in the upper half of the distribution).

This means that, at a given small density, the spatial variability among subquadrats is similar in the lower and upper parts of the barnacle's distribution. At larger densities, the variability increases, but more so in the lower than upper half of the range of distribution of barnacles (as illustrated in Fig. 11). Thus, there is a subtle, but nevertheless real difference in the spatial pattern of distribution of the barnacle at two heights along the tidal gradient. None of the extensive previous work on the ecology of the species can currently help to explain what are the processes or mechanisms causing this pattern. The most likely processes are to do with settlement from the planktonic stage of life-history and influences of predatory whelks, both of which on these exposed coasts are more prevalent lower on the shore. How these (or other processes) cause the observed pattern is not at all clear.

Discussion

The examples are presented as a preliminary account of the sorts of patterns of spatial variation of density shown by intertidal animals. Three general issues emerge. First, patterns are complex and not well revealed by inadequate sampling. The examples have all shown that. The data for *L. unifasciata* convincingly demonstrated the lack of consistency from one site to another in the trends of variance against mean for the three spatial scales examined in each site. Often, however, there is only one set of data from one time of sampling and general inferences cannot be clearly supportable.

Furthermore, the examples of *L. unifasciata* and *T. rosea* provide overwhelming evidence of the great spatial variation at small scales. This information would be lost if data were collected in larger units (e.g. per quadrat here). So, care must be taken to ensure that relevant spatial scales are used to estimate density and its variance. Because the most relevant spatial scale cannot usually be known prior to the collection of data, a set of scales should be used. This is best (most efficiently) achieved by a nested or hierarchical design, enabling simultaneous independent estimation of variances at several spatial scales.

In contrast, sometimes the relevant spatial (or temporal) scales are defined by the hypotheses being tested (see ecological examples in Underwood 1990, 1991). In these cases, it is crucial to be sure that

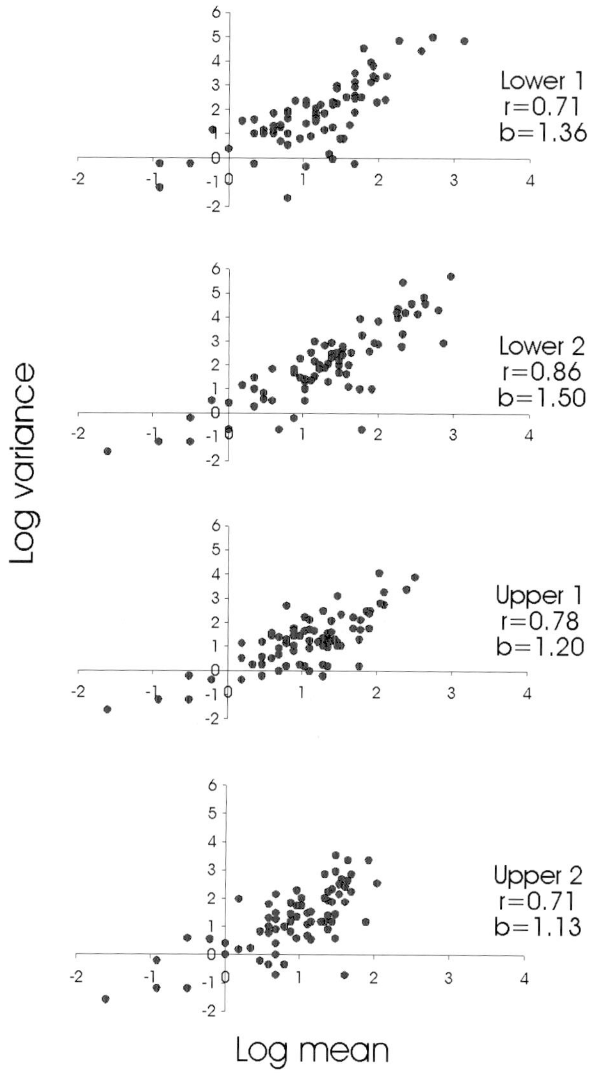

Fig. 10. Correlations between log variance among subquadrats and log mean density of the barnacle *Tesseropora rosea*. Each point represents the data for a single quadrat at a single time of sampling (n = 5 subquadrats). There are five such quadrats at each of 16 times of sampling, giving 80 data points for each site. Data are for the lower and upper halves of *Tesseropora's* vertical range at Flat Rock Island, sites 1 and 2. For each set of data, the correlation coefficient (r) and slope of the regression (b) are shown.

the scales at which variation is being measured are the correct or relevant ones. The results presented here demonstrate how misleading it would be to base conclusions on spatial variances measured at another scale. Also on this point, estimates of spatial variance at some chosen scale of sampling are often used in cost-benefit analyses or in consideration of sampling designs to achieve some pre-determined power of tests (e.g. Cohen 1977, Underwood 1981). The results here demonstrate the care that must be taken to get pilot or preliminary estimates at the appropriate scale. Extrapolation from other scales is unwise or misleading.

The second issue is that the spatial variances of density and the relationships between variance and mean density are different in different habitats (e.g. *C. tasmanica* in exposed and sheltered locations) or different parts of an environmental gradient (e.g. *T. rosea* at different heights on a shore). This is important in interpretation of ecological studies because the processes creating different patterns of spatial variance at a given density are not yet known. Development of explanations for these patterns

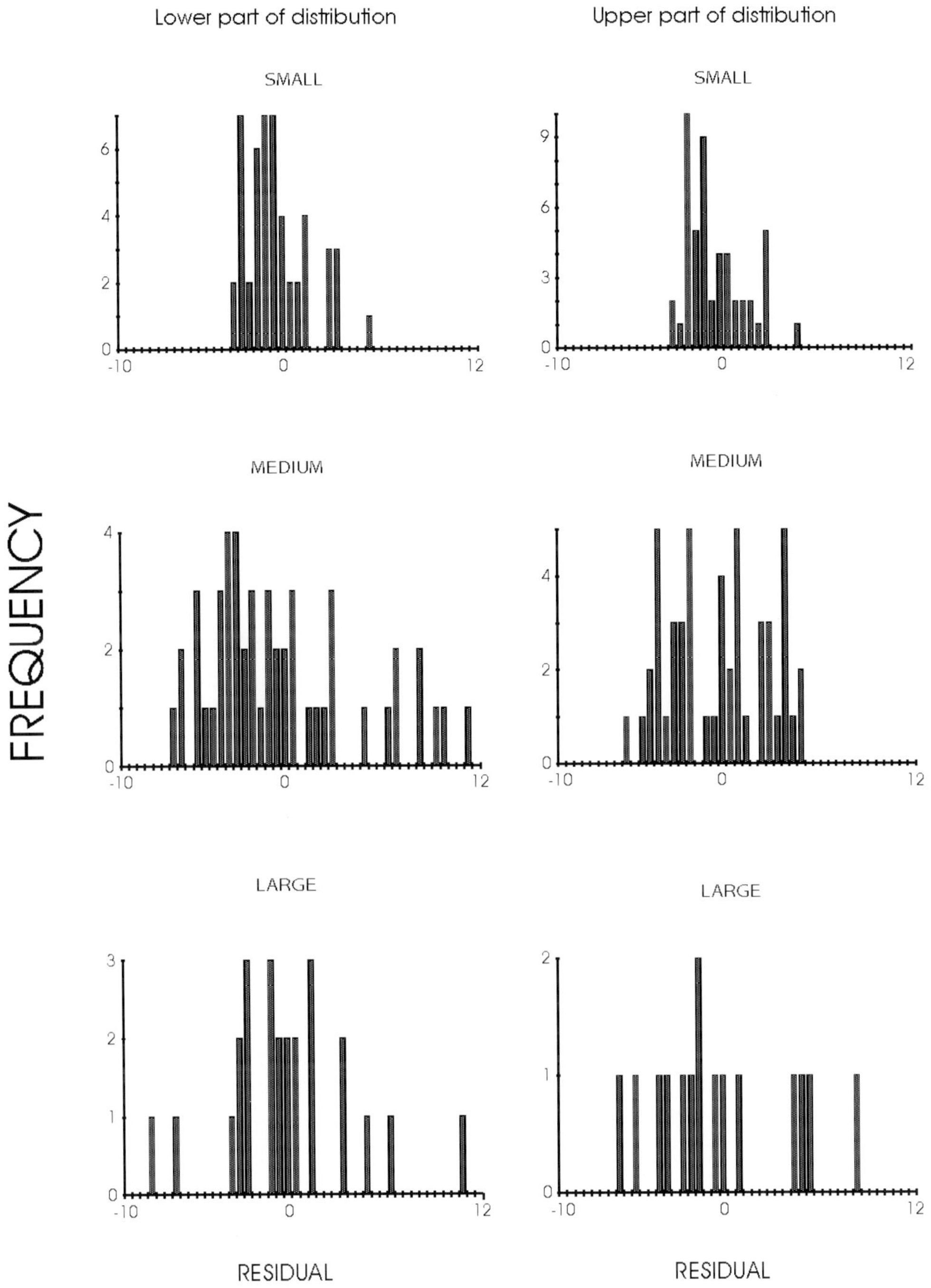

Fig. 11. Illustration of variance among subquadrats in the density of *Tesseropora rosea* at different heights on the shore. Frequency distributions of residuals (i.e. density in each subquadrat minus mean density in the quadrat) for lower and upper halves of the vertical range of the barnacle. Data are plotted for numerous quadrats with a small (1–3 barnacles per subquadrat), medium (4–6 per subquadrat) or large (8–12 barnacles per subquadrat) mean density. As predicted from Figure 10, variance increases with increasing mean density, but the increases are greater in the lower regions occupied by *Tesseropora*.

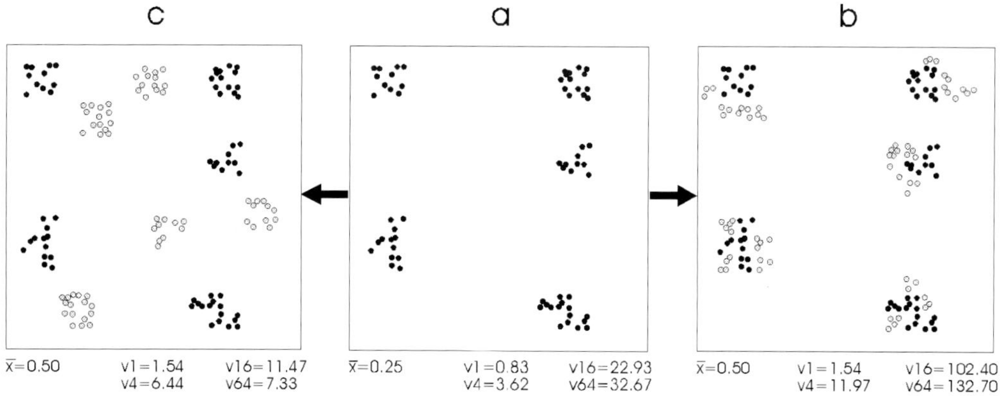

Fig. 12. Illustration of difference between intensity and grain of spatial variance. In (a), mean density of aggregated animals is 0.25 per unit area. The variances among quadrats of 1, 4, 16 and 64 units are shown (as V1, V4, V16, V64). In (b), the density is doubled, but individuals clump in the same aggregations increasing intensity of pattern; variances at all sizes of quadrats are increased. In (c), the density is doubled, but the individuals form clumps in new parts of the habitat, increasing the grain of pattern. Variances are greater for small quadrats (V1 and V4), but decreased for larger quadrats (V16, V64).

is a current research need. Note that Gaston and McArdle (1993) have drawn attention to the distinction between spatial variation estimated for abundance, as opposed to density of a population. Here, the focus is on densities and the results and patterns are a consequence of different processes affecting variation in numbers in different patches or parts of a species' range. So, it is important to be able to replicate patches or sites to prevent confounding of interpretations. Thus, a difference in patterns of variance between two sites or habitats may be due to a real difference in the ecological processes operating or simply a geographical difference due to stochastic variability from one site to another. It is no less important to acquire replicate estimates of variance than in any other type of ecological investigation (Hurlbert 1984).

Results here demonstrate that there are small-scale influences on distribution that lead to small-scale variations in density. We need to move on from being satisfied that there is a relationship between variance and mean density to an interest in why there might be. The organisms considered here have a repertoire of behavioural responses to cues in their environments. These exert considerable influence on the settlement of the animals from the plankton, when they first enter the intertidal habitats (Denley and Underwood, 1979). These behaviours can only operate over the small scale of perception of individual larvae.

The snails typically move over relatively short distances, but are known to respond to small-scale features of their habitats (Underwood and Chapman, 1992) and to aggregate in response to cues from the habitat and each other (Chapman and Underwood unpublished). As a result, a change or difference in density in any area can potentially result in quite different responses and patterns of variance. This is illustrated for a theoretical aggregated population in Fig. 12. If extra animals move into the area, they may respond to existing aggregations or to the same cues of habitat that caused the original aggregations to form (Fig. 12b). This would lead to increased spatial variance at any scale of sampling-unit. Thus, the intensity of pattern is increased (e.g Pielou 1969) and there is a positive relationship between variance and mean at all sampled scales.

In contrast, the extra animals may respond to other sites in the area – perhaps because they avoid the existing aggregations or because the sites where aggregations form are simply randomly scattered. Thus, new aggregations form (Fig. 12c) and the grain of the pattern is altered. Under these circumstances, at the smaller scales of sampling units, there will be increased variance with increased density. At the larger scales, variances will, in contrast, be smaller:

The responses of animals to small-scale features of habitat and to each other are likely themselves to be density-dependent in complex, non-linear ways. This is why a change in density may increase or decrease variances at some scales (and do the opposite at other scales). Without appropriate understanding of the various processes and their interactions, it is unlikely that predictions from simple spatial and temporal models (e.g. Gaston and McArdle 1994) will be realistic.

This is a particularly important point in the context of environmental impacts. To make coherent predictions about potential changes to populations in response to disturbances requires understanding of interactive variances. An impact is definable (and only measurable) in terms of change in the variance component that is associated with differences between abundances of populations in disturbed and control sites before and after the disturbance (Green 1979; Underwood 1992, 1993, 1994a). To understand these variances requires examination of previous environmental disturbances as experiments (Underwood 1989) and development of appropriate models and theory.

As a first step, much more needs to be understood about spatial variation, how it changes from time to time under natural conditions and then how and why time-courses of abundance of populations vary from one place to another. This will not be achieved without rigorous and intense study.

To sort out the reasons for any given pattern of variance or relationship between variances and means requires increased experimentation. Manipulations of densities to test specific hypotheses about effects on spatial variances are needed. These need careful controls and appropriate design (Chapman, 1986; Chapman and Underwood, 1992; Underwood and Chapman 1989). Development of models from which such hypotheses are derived requires careful interpretation of observations on spatial and temporal variance. The examples presented here indicate how these observations can be obtained, the complexities they contain and the challenge for ecologists to deal with the processes of change and difference and to study variances. The issues require a break from the mean-mindedness of the past!

Acknowledgments

This study has been supported (since 1973) by the Australian Research Council and (since 1984) by the Institute of Marine Ecology of the University of Sydney. I am grateful to many colleagues for discussion but, above all, to Dr M.G. Chapman who is a partner in this project and whose ideas, enthusiasm and skills underpin the analyses and concepts here. I thank Ms V. Mathews for help in preparation of the paper and Drs D. Fletcher and M.J. Keough for helpful revisions to the manuscript.

References

Andrew NL, Mapstone BD (1987) Sampling and the description of spatial pattern in marine ecology. *Ann Rev Oceanog Mar Biol* **25**:39–90

Burdick RK, Graybill FA (1992) Confidence intervals on variance components. Marcel Dekker, New York

Chapman MG (1986) Assessment of some controls in experimental transplants of intertidal gastropods. *J Exp Mar Biol Ecol* **103**:181–201

Chapman MG, Underwood AJ (1992) Experimental designs for analyses of movements by molluscs. In: Grahame J, Mill PJ, Reid DG (eds) *Proceedings of the third international symposium on littorinid biology.* The Malacological Society of London, London, pp 169–180

Chesson PL (1978) Predator-prey thoery and variability. *Ann Rev Ecol Syst* **9**:323–347

Chesson PL (1985) Coexistence of competitors in spatially and temporally varying environments: a look at the combined effects of different sorts of variability. *Theor Pop Biol* **28**:263–287
Cohen J (1977) *Statistical power analysis for the behavioural sciences*. Academic Press, New York
Cornfield J, Tukey JW (1956) Average values of mean squares in factorials. *Ann Math Stat* **27**:907–949
Dakin WJ, Bennett I, Pope E (1948) A study of certain aspects of the ecology of the intertidal zone of the New South Wales coast. *Aust J Sci Res B* **1**:176–230
Denley EJ, Underwood AJ (1979) Experiments on factors influencing settlement, survival and growth of two species of barnacles in New South Wales. *J Exp Mar Biol Ecol* **36**:269–293
Fairweather PG, Underwood AJ (1991) Experimental removals of a rocky intertidal predator: variations within two habitats in the effects on prey. *J Exp Mar Biol Ecol* **154**:29–75
Fairweather PG, Underwood AJ, Moran MJ (1984) Preliminary investigations of predation by the whelk *Morula marginalba*. *Mar Ecol Prog Ser* **17**:143–156
Foster MS (1990) Organization of macroalgal assemblages in the Northeast Pacific: the assumption of homogeneity and the illusion of generality. *Hydrobiologia* **192**:21–34
Gaston KJ, McArdle BH (1993) Measurement of variation in the size of populations in space and time: some points of clarification. *Oikos* **68**:357–360
Gaston KJ, McArdle BH (1994) The temporal variability of animal abundances: measures, methods and patterns. *Phil Trans Roy Soc B* **345**:335–358
Gaylor DW, Hopper FN (1969) Estimating the degrees of freedom for linear combinations of mean squares by Satterthwaite's formula. *Technometrics* **11**:691–706
Green RH (1979) *Sampling design and statistical methods for environmental biologists*. Wiley, Chichester
Grossman GD (1982) Dynamics and organisation of a rocky intertidal fish assemblage: the persistence of taxocene structure. *Am Nat* **119**:611–637
Hurlbert SJ (1984) Pseudoreplication and the design of ecological field experiments. *Ecol Monogr* **54**:187–211
McArdle BH, Gaston KJ (1992) Comparing population variabilities. *Oikos* **64**:610–612
Morrisey DJ, Howitt L, Underwood AJ, Stark JS (1992) Spatial variation in soft-sediment benthos. *Mar Ecol Progr Ser* **81**:197–204
Morrisey DJ, Underwood AJ, Howitt L, Stark JS (1992) Temporal variation in soft-sediment benthos. *J Exp Mar Biol Ecol* **164**:233–245
Peters RH (1991) *A critique for ecology*. Cambridge University Press, Cambridge
Pielou EC (1969) *An introduction to mathematical ecology*. Wiley Interscience, New York
Robinson DL (1987) Estimation and use of variance components. *The Statistician* **36**:3–14
Roughgarden J, Iwasa Y (1986) Dynamics of a meta-population with space-limited subpopulations. *Theor Pop Biol* **29**:235–261
Satterthwaite FE (1946) An approximate distribution of estimates of variance components. *Biom Bull* **2**:110–114
Scheffe H (1959) *The analysis of variance*. Wiley, New York
Searle SR, Casella G, McCulloch CE (1992) *Variance components*. Wiley, New York
Steele JH (1985) A comparision of terrestrial and marine ecological systems. *Science* **313**:355–358
Taylor LR (1961) Aggregation, variance and the mean. *Nature* **189**:732–735
Taylor LR, Taylor RAJ (1977) Aggregation, migration and population mechanics. *Nature* **265**:415–421
Tukey JW (1956) Variance of variance components. 1. Balanced designs. *Ann Math Stat* **27**:722–736
Underwood AJ (1981a) Structure of a rocky intertidal community in New South Wales: patterns of vertical distribution and seasonal changes. *J Exp Mar Biol Ecol* **51**:57–85
Underwood AJ (1981b) Techniques of analysis of variance in experimental marine biology and ecology. *Ann Rev Oceanogr Mar Biol* **19**:513–605
Underwood AJ (1985) Physical factors and biological interactions: the necessity and nature of ecological experiments. In: Moore PG, Seed R (eds) *The ecology of rocky coasts*. Hodder and Stoughton, London, pp 371–390
Underwood AJ (1989) The analysis of stress in natural populations. *Biol J Linn Soc* **37**:51–78

Underwood AJ (1990) Experiments in ecology and management: their logics, functions and interpretations. *Aust J Ecol* **15**:365–389

Underwood AJ (1991) Beyond BACI: experimental designs for detecting human environmental impacts on temporal variations in natural populations. *Aust J Mar Freshwat Res* **42**:569–587

Underwood AJ (1992) Beyond BACI: the detection of environmental impact on populations in the real, but variable, world. *J Exp Mar Biol Ecol* **161**:145–178

Underwood AJ (1993) The mechanics of spatially replicated sampling programmes to detect environmental impacts in a variable world. *Aust J Ecol* **18**:99–116

Underwood AJ (1994a) On beyond BACI: sampling designs that might reliably detect environmental disturbances. *Ecol Appl* **4**:3–15

Underwood AJ (1994b) Rocky intertidal shores. In: Hammond LS, Synnot R (eds) *Marine biology*. Longman-Cheshire, Melbourne, pp 273–296

Underwood AJ, Chapman MG (1985) Multifactorial analysis of directions of movement of animals. *J Exp Mar Biol Ecol* **91**:17–43

Underwood AJ, Chapman MG (1989) Experimental analyses of the influences of topography of the substratum on movements and density of an intertidal snail, *Littorina unifasciata*. *J Exp Mar Biol Ecol* **134**:175–196

Underwood AJ, Chapman MG (1992) Experiments on topographic influences on density and dispersion of *Littorina unifasciata* in New South Wales. In: Grahame J, Mill PJ, Reid DG (eds) *Proceedings of the third international symposium on littorinid biology*. The Malacological Society of London, London, pp 181–195

Underwood AJ, Chapman MG (1995) Rocky shores. In: Underwood AJ, Chapman MG (eds) *Coastal marine ecology of temperate Australia*. New South Wales University Press, Sydney, pp 55–82

Underwood AJ, Denley EJ (1984) Paradigms, explanations and generalizations in models for the structure of intertidal communities on rocky shores. In: Strong DR, Simberloff D, Abele LG, Thistle AB (eds) *Ecological communities: conceptual issues and the evidence*. Princeton University Press, New Jersey, pp 151–180

Underwood AJ, Denley EJ, Moran MJ (1983) Experimental analyses of the structure and dynamics of mid-shore rocky intertidal communities in New South Wales. *Oecologia (Berlin)* **56**:202–219

Underwood AJ, Fairweather PG (1989) Supply-side ecology and benthic marine assemblages. *Trends Ecol & Evol* **4**:16–20

Underwood AJ, Peterson CH (1988) Towards an ecological framework for investigating pollution. *Mar Ecol Progr Ser* **46**:227–234

Underwood AJ, Petraitis PS (1993) Structure of intertidal assemblages in different locations: how can local processes be compared? In: Ricklefs RE, Schluter D (eds) *Species diversity in ecological communities: historical and geographical perspectives*. University of Chicago Press, Chicago, pp 38–51

Warner RR, Chesson PL (1985) Coexistence mediated by recruitment fluctuations – a field guide to the storage effect. *Am Nat* **125**:769–787

Winer BJ, Brown DR, Michels KM (1991) *Statistical principles in experimental design, Third edition*. McGraw-Hill, New York

Decision theory and biodiversity management: How to manage a metapopulation

Hugh P. Possingham

ABSTRACT

An objective of nature conservation is to manage threatened species so that their chance of extinction is minimised. This paper explores the application of Markov decision theory as a tool to choosing between management options for a threatened species. Specifically, the problem of minimising the extinction probability of a metapopulation is considered. A presence–absence stochastic metapopulation model is used to model the dynamics of the population. The two management options considered are to make a new patch of habitat, or to reintroduce the species to a suitable but empty patch. The best strategy employed will depend on the current state of the metapopulation. This management problem is an illustration of the application of decision theory to making optimal state-dependent decisions for conservation in an uncertain environment. If applied conservation biology is to be successful it is essential that more conservation theory within an explicit decision-making framework is developed.

Key words: Markov decision theory, metapopulation dynamics, conservation biology, biodiversity management

Frontiers of Population Ecology, R.B. Floyd, A.W. Sheppard and P.J. De Barro (eds), CSIRO Publishing, Melbourne, 1996, pp. 391–398

Introduction

Many ecological theories and concepts, e.g. island biogeography theory, metapopulation theory, theories of coexistence, and the notion of a minimum viable population size, have been developed to help us make decisions in applied nature conservation. Generally these theories are not particularly useful because they are not enclosed within a management framework. Most existing theories only indicate which management strategies are useful, without enabling those strategies to be ranked. For example island biogeography tells us that big patches of suitable habitat are better than small patches, and corridors linking habitat patches are useful. However, island biogeography theory does not indicate how to trade off reserve size with connectedness, that is, should a revegetation program concentrate on making corridors between patches or on increasing the size of existing patches.

Where time and money are limited, it is vital to choose the best management options within the constraints of those resources. To help managers achieve the best result with the resources available, existing theories of population ecology need to be merged with decision theory tools. This union of decision theory and ecological theory can lead to a theory of applied conservation management. This paper gives one example of how that union can occur.

Over the past decade there has been increasing interest in modelling the dynamics of spatially structured populations. One class of ideas and models used to understand the dynamics of populations that are not spatially homogeneous is metapopulation theory (Hanski 1991; Verboom *et al.* 1991; Mangel and Tier 1993; Adler and Nuernberger 1994; Hanski 1994). Metapopulations are made up of a number of local populations. The simplest kind of metapopulation model follows the dynamics of the number of patches that are occupied, ignoring local population dynamics (Levins 1969). In this case the two processes of local extinction and colonisation drive the dynamics of the metapopulation (Fig. 1). This is often referred to as a presence–absence metapopulation model and is the kind of metapopulation model used in this paper (Richter-Dyn and Goel 1972).

This paper explores the application of Markov decision theory to decision making in applied nature conservation (Intriligator 1971), in particular the optimal management of a metapopulation. Using decision theory tools in nature conservation is not new, but it is rare (Maguire 1986; Ralls and

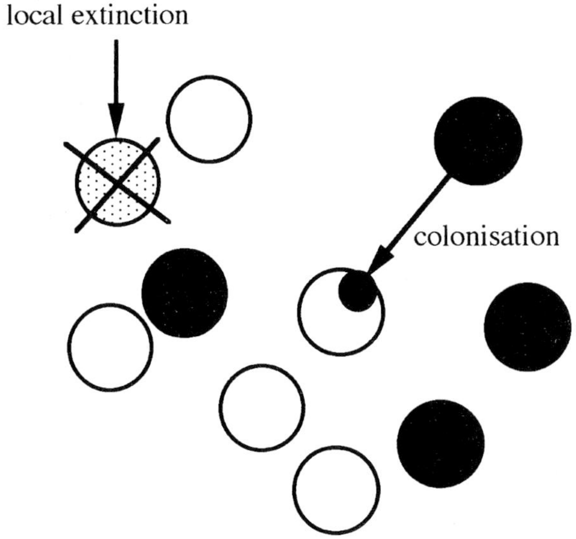

Fig. 1. A schematic diagram of a metapopulation in which the species is either present in (●), or absent from (**O**), a patch. The dynamics of the metapopulation is driven by the extinction and colonisation processes.

Starfield 1995). Moore (1990) outlined a process where Markov decision theory could be used to manage vegetation succession, representing the only application of this tool to nature conservation.

In this example the objective is to minimise the likelihood of extinction of the metapopulation, the standard objective in applied threatened-species management (Burgman *et al.* 1993). Two real management options for minimising the likelihood of metapopulation extinction are to make more patches of suitable habitat or reintroduce the species to suitable empty patches. I show how Markov decision theory can be used to choose between these two management options and enable optimal management decisions to be made in an uncertain environment. Hopefully, this paper will be a step towards accepting the utility of these methods.

Model

Solving a management problem using Markov decision theory (Intriligator 1971) has two parts. First the stochastic dynamics of the system are modelled, in this case the dynamics of a metapopulation. This is the simplest stochastic metapopulation model possible – a presence-absence model that is not spatially explicit. Second is the decision theory, where the stochastic dynamic programming equations need to be formulated and rewards defined for the outcome of the management.

The metapopulation model

It will be assumed that the interest is in a metapopulation where all the patches are identical and it is equally likely that colonisation of a patch occurs from any occupied patch. This type of non-spatial model has been used widely to explore aspects of metapopulation dynamics (Levins 1969; Richter-Dyn and Goel 1972) and although it is somewhat unrealistic the lack of a formal spatial structure may be relatively unimportant. It will also be assumed that patches are either occupied or unoccupied. A presence–absence model like this ignores the possibility that different patches may contain different numbers of individuals – however where local population dynamics operate rapidly on the time scale of the extinction/colonisation processes that determine the dynamics of metapopulations this is a reasonable simplification (Hanski and Thomas 1994).

A discrete time and discrete state space stochastic model is used for this sort of metapopulation. In any time step (henceforth referred to as years) our state variables will be the number of occupied patches, i, and the number of suitable patches, m, which may be occupied or unoccupied. The total number of patches that could be occupied is not constant, because we allow for the possibility of creating suitable new habitat. This variation in the number of suitable patches is not normally incorporated into metapopulation models. Each year any patch that is unoccupied may be colonised, and any patch that is occupied may experience a local extinction and become unoccupied. Without loss of generality, it is also assumed that the annual sequence of events is colonisation followed by extinction. The following equations describe the dynamics of this metapopulation in the absence of management.

Let β be the per patch colonisation rate – namely the probability that an empty patch is colonised by one of the occupied patches. The probability that an empty patch is colonised in one year is then

$$c = 1-(1-\beta)^{i} . \qquad (1)$$

Let e be the probability that an occupied patch becomes unoccupied in one year, that is the local population extinction rate. It is also assumed that the probability of local extinction is independent of the state of the other patches, such that there is no rescue effect (Hanski and Gyllenberg 1993).

For a system with a constant number of patches, m, the state of the system at a particular time is simply the probability there are i full patches at time t, denoted as $p_i(t)$. To determine the probability of moving from one state to another, let A be the transition matrix that defines the probability of going from one state to another in each year so that

$$\mathrm{p}(t+1)=\mathrm{p}(t)\mathrm{A} \;, \tag{2}$$

where $\mathbf{p}(t)$ is the vector describing the probability of being in any state, with components $p_i(t)$ as defined above. The elements of A; $a_{ij}(m)$ are the probabilities of moving from i occupied patches to j occupied patches in one year, assuming there are m suitable patches in the system. The equation for the transition probabilities in terms of the colonisation and extinction parameters is

$$\begin{aligned} a_{ij}(m) &= \sum_{k=0}^{m-i} \binom{m-i}{k} c^k (1-c)^{m-i-k} \binom{i+k}{i+k-j} e^{i+k-j}(1-e)^j \\ \textit{for} \quad & i+k-j \geq 0 \,. \end{aligned} \tag{3}$$

The stochastic metapopulation dynamics are fully described by Eqns 2 and 3. The probability of being in any state at some future time, given an initial state, can be calculated by iterating Eqn 2.

Dynamic programming equations

The objective of Markov decision theory is to determine the state-dependent optimal decision for controlling a stochastic process. To use this method the 'pay-off' values for achieving a certain state of the system need to be defined. Also, the management options need to be described mathematically, and the dynamic programming equation needs to be defined. In this example the pay-off will be 1 if the population persists (is not extinct) and 0 otherwise. The optimal management strategy is found by back-stepping through time, choosing the optimal decision for each year assuming that subsequent decisions are made optimally.

At any year a value is assigned to being in a particular state. Let $J_t(i,m)$ be the value of being in state (i,m) at time t (i of the m patches occupied). To find the value of being in that state, values first need to be given to the state of the system at some terminal time T at which success is measured. If the metapopulation survives to year T the value is 1, and 0 otherwise, so

$$\begin{aligned} &J_T(0,m)=0 \\ &J_T(i,m)=1 \qquad \textit{for} \;\; 0<i\leq m \,. \end{aligned} \tag{4}$$

To calculate the optimal strategy for the penultimate time, T–1, the value of being in state (i,m) is expressed in terms of the value of being in state (i,m) at the terminal time T, weighted by the chance of moving to each of those future states. This enables the best strategy for the penultimate time to be chosen and hence the value of each state at the penultimate time. The process is iterated backwards through time using the general equation relating present value to future value

$$J_T(i,m)=r\sum_{j=0}^{m} a_{i+1j}(m)J_{t+1}(j,m)+(1-r)\sum_{j=0}^{m+1} a_{ij}(m+1)J_{t+1}(j,m) \;, \tag{5}$$

where r = 1 if a reintroduction is done and r = 0 if a new patch is made instead. It is assumed that reintroductions and patch construction are always successful and they have equal cost because only one or the other can be done in a single year. The first summation term in Eqn 5 is the expected value of being in state (i,m) in year t if we chose to reintroduce the species to an empty patch, while the second summation term is the expected value of being in state (i,m) in year t if we chose to make a new patch. Equation 5 requires two logical restrictions. The number of patches occupied after a colonisation cannot be more than the number of suitable patches m, and the total number of suitable patches after a patch is created, m + 1, cannot be more than n, a ceiling set on the number of suitable patches that can exist.

From Eqn 5 the value of r that maximises $J_t(i,m)$ for every year and state (i,m) can be found. Because we are interested in a long-term optimal strategy, that is a strategy that minimises the long-term extinction probability, the process was backstepped to a terminal time T = 100. All the results presented here are 'equilibrium' optimal strategies, that is they are the strategies that maximise long-term persistence of the metapopulation. The results are found numerically because the Monte Carlo simulation is not required.

Results

First a particular example is examined in detail, before considering the consequences of worse environmental conditions on the optimal decision. Unless otherwise stated there is a maximum of 12 patches (n = 12), the final time step being T = 100, the per patch extinction probability is e = 0.2, and the patch colonisation rate parameter is β = 0.05.

Figure 2 displays the optimal strategy choice for the baseline parameter choice. All the possible combinations of total patch number, and number of patches occupied, are shown with an r indicating that the optimal strategy is to do a reintroduction, otherwise the optimal strategy is to make a new patch. By thinking about the problem some of the results can be quickly explained. When m = 12 no more patches can be made so the optimal strategy will always be to reintroduce the species. When i = m (the leading diagonal of the results diagram), all the patches are occupied, reintroduction is not possible, and the optimal strategy is to make a new patch. When i = m = 12 neither strategy is possible. In between these extremes either strategy is possible. In general the reintroduction strategy is optimal when the number of occupied patches is small or the fraction of occupied patches is low.

Besides considering the optimal strategy at a particular time it is useful to follow a likely trajectory of strategies. Consider a specific circumstance, say with m = 5 and i = 4. Then the optimal strategy is to begin by making a new patch, forcing m = 6. The number of patches can only increase so, in the absence of extinction the state of the system will inevitably move to the right (more patches). Whether the number of occupied patches increases or decreases depends on chance. Consequently a management scenario for a particular population will generally involve both strategies in the short

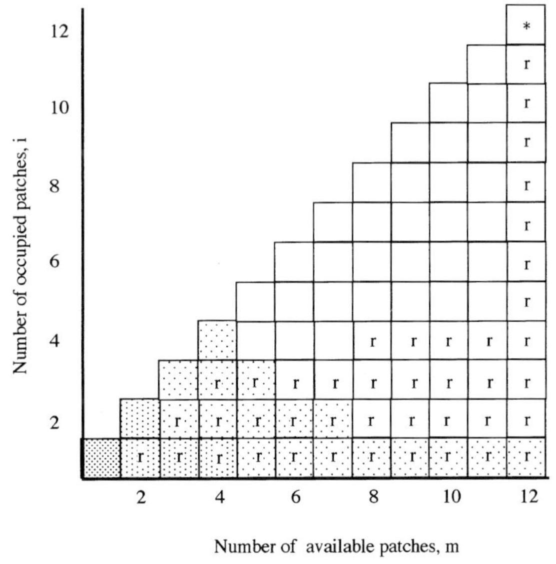

Fig. 2 The optimal strategy for the baseline example: n = 12, e = 0.2, β = 0.05 and T = 100. States in which the optimal strategy is to reintroduce the species to an empty patch are indicated with an r, otherwise the best strategy is to make a new patch (except for the state (12,12) where neither strategy is possible). The shading indicates the probability of extinction within the next 100 years if the population starts in that state and assuming we continue to make optimal decisions. Darker shading is an indication of a higher likelihood of extinction.

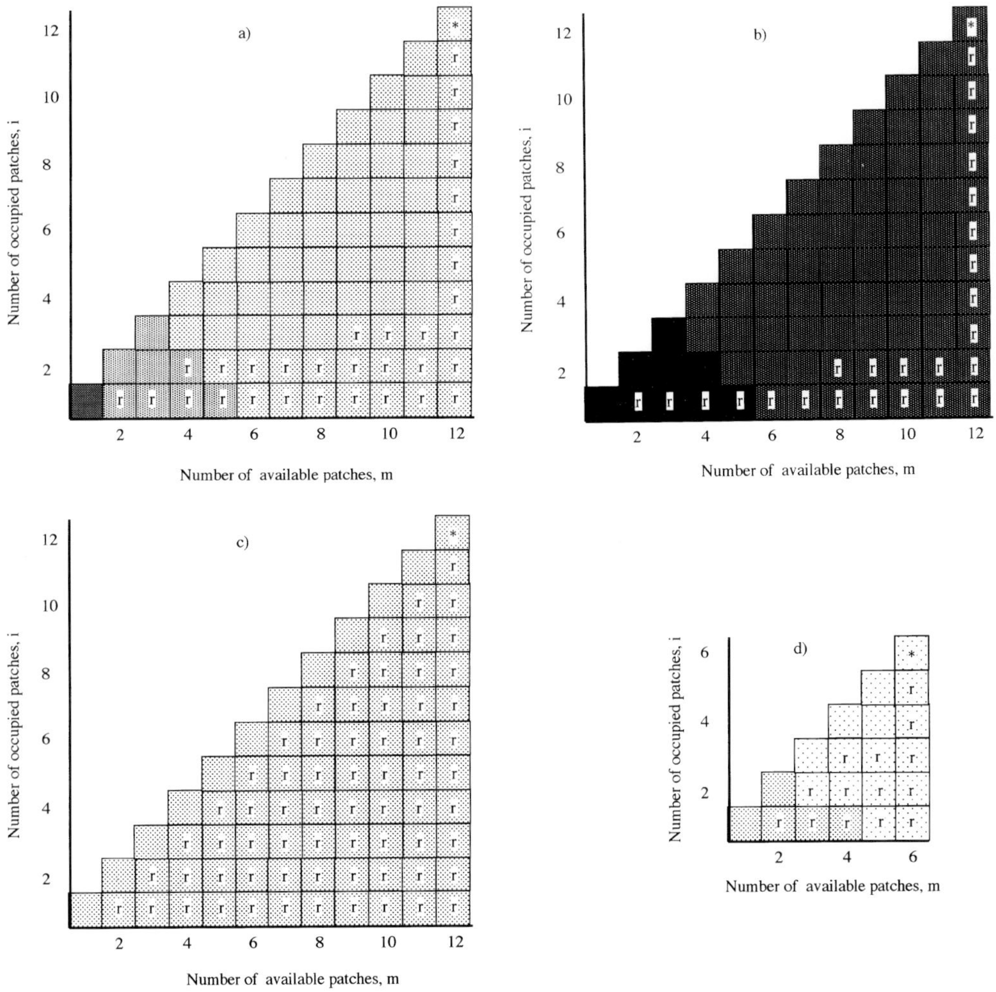

Fig. 3. Optimal strategies for a range of environments that are worse than the baseline case: a) $e = 0.3$, the per patch extinction rate is higher, b) $e = 0.4$, a further increase in extinction rate, c) $\beta = 0$, no natural colonisations and d) $n = 6$, a maximum of only six patches. As with Fig. 2, darker shading is an indication of a higher likelihood of extinction.

term, with the patch construction option becoming less frequent as time passes until the only available strategy is to do repeated reintroductions.

The long-term extinction probability increases as we move down (fewer occupied patches) and to the left (fewer suitable patches). This is indicated by the darker shading in the lower left corner of Fig. 2.

Figure 3 shows the response of the optimal strategy to worse environmental conditions. If the extinction rate increases, reintroductions become a less favoured option. At the highest extinction rate (Fig. 3b), it is only profitable to do a reintroduction when one or two of the patches are occupied, otherwise it is best to make a new patch if possible. Under these circumstances the extinction rate is high and the best chance for the population to persist for a substantial period is if there are many patches. Note that making more patches is a long-term strategy.

Another way of making the environment worse for the metapopulation is to allow no natural colonisations by setting $\beta = 0$. In the absence of management this would mean that the metapopulation is doomed. However, because we can artificially colonise patches, the metapopulation has a reasonable chance of persisting over 100 years. In this case a worse environment means reintroduction is generally the best option, the reverse effect to increasing the extinction rate. Finally the environment can be made worse in the model by halving the maximum number of patches that can be made (reducing n, see Fig. 3d). Although extinction rates increase slightly, the optimal strategy remains virtually unchanged. When the maximum number of patches is doubled to $n = 24$ the optimal strategy also remains unchanged.

Discussion

These brief examples show that it is difficult to make generalisations about what is the best management strategy. No simple rules were derived by exploring numerous examples. Every situation needs to be modelled to find the best management policy.

The example presented here can be extended in several ways. Reintroductions are rarely assured, and it is possible to associate a risk of failure with either the reintroduction or patch construction strategy. It is also possible to relax the assumption that only one action can be done in a single year. We could allow for several reintroductions and/or patch constructions. This example explored the trade-off between only two management options. There are other options that can be modelled, for example, a temporary reduction in the extinction rate (reflecting say predator control) or an increase in the colonisation rate by corridor construction. We can even include an explicit monetary budget, where the project receives income and that money can be used for different strategies including saving.

There are many possible extensions to the existing model. Day and Possingham (1996) explore a metapopulation model with explicit spatial structure. Optimal strategies for this sort of model could be determined for specific cases using their model coupled to the Markov decision theory process outlined here. Greater realism could be added to our example in many ways. For example, it is possible to make the reintroduction process uncertain, allow more than one reintroduction, put a time delay on our ability to create a new patch, and allow suitable patches to become unsuitable with a certain probability.

Markov decision theory enables explicit decisions to be made about management of a metapopulation that depend on the current state of the metapopulation. An enormous number of different problems can be explored in the Markov decision theory framework. The most profitable paths for future research will involve specific examples (e.g. Morton this volume pp. xx) and searches for generalisations and rules of thumb that provide robust solutions in most situations.

Acknowledgments

I am indebted to John Lawton for providing a stimulating environment for my study leave at Imperial College where these ideas were developed. Mark Burgman, Jemery Day and Nick Nicholls provided useful comments on the manuscript.

References

Adler FR, Nuernberger B (1994) Persistence in patchy irregular landscapes. *Theoret Pop Biol* **45**:41–75

Burgman MA, Ferson S, Akçakaya HR (1993) *Risk assessment in conservation biology*. Chapman and Hall, London

Day J, Possingham HP (1996) A stochastic metapopulation model with variability in patch size and position. *Theoret Pop Biol* **48**:333–360

Hanski I (1991) Single-species metapopulation dynamics: concepts models and observations. *Biol J Linn Soc* **42**:17–38
Hanski I (1994) A practical model of metapopulation dynamics. *J Anim Ecol* **63**:151–163
Hanski I, Gyllenberg M (1993) Two general metapopulation models and the core-satellite hypothesis. *Am Nat* **142**:17–41
Hanski I, Thomas CD (1994) Metapopulation dynamics and conservation: a spatially explicit model applied to butterflies. *Biol Conserv* **68**:167–180
Intriligator MD (1971) *Mathematical optimisation and economic theory.* Prentice-Hall, Englewood Cliffs, USA
Levins R (1969) Some demographic and genetic consequences of environmental heterogeneity for biological control. *Bull Entomol Soc Am* **15**:237–240
Maguire LA (1986) Using decision analysis to manage endangered species populations. *J Environ Manage* **22**:345–360
Mangel M, Tier C (1993) Dynamics of metapopulations with demographic stochasticity and environmental catastrophes. *Theoret Pop Biol* **44**:1–31
Moore AD (1990) The semi-Markov process: a useful tool in the analysis of vegetation dynamics for management. *J Environ Manage* **30**:111–130
Ralls K, Starfield AM (1995) Choosing a management strategy: two structured decision-making methods for evaluating the predictions of stochastic simulation models. *Conserv Biol* **9**:175–181
Richter-Dyn N, Goel NS (1972) On the extinction of a colonising species. *Theoret Pop Biol* **3**:406–433
Verboom J, Lankester K, Metz JAJ (1991) Linking local and regional dynamics in stochastic metapopulation models. *Biol J Linn Soc* **42**:39–55

Population Trajectories Through Space and Time: A Holistic Approach to Insect Migration

V.A. Drake and A.G. Gatehouse

ABSTRACT

The capacity of mobile organisms to migrate adds a new dimension to Nicholson's concept of population regulation through density dependence. Migrants can vacate a habitat as conditions there become limiting, and colonise another that may provide, at least temporarily, a more favourable environment for population growth. Such processes occur frequently among the insects, where they have been the subject of concerted research because many of the migratory species are serious pests. A series of insights gained over the last few decades now begins to allow a holistic description of a 'migration system'.

The migration system of one extensively studied insect population, the oriental armyworm in east Asia, is described in terms of the physical and biotic 'arena' the population inhabits, the reticulate 'trajectory' it follows, the behavioural and physiological migration 'syndrome' that steers it along this trajectory, and the 'genetic complex' that underlies this syndrome. The system is seen to function through a series of processes, including the action of contemporary natural selection, that link and maintain the system components. This example provides insights into the persistence and regulation of highly mobile organisms in environments in which relatively rapid spatio-temporal variations occur.

Key words: Population ecology, spatio-temporal heterogeneity, insect migration, Lepidoptera, Noctuidae, *Mythimna separata*

Frontiers of Population Ecology, R.B. Floyd, A.W. Sheppard and P.J. De Barro (eds), CSIRO Publishing, Melbourne, 1996, pp. 399–408

Introduction

In an earlier paper (Drake *et al.* 1995), we presented a conceptual model of insect migration that endeavours to recognise all the physical and biological components of a migration system and the interlinkages between them. The model is concerned primarily with population processes, and especially the maintenance of the population through natural selection on a contemporary time scale; it necessarily incorporates both a spatial and a genetic dimension. This holistic approach builds upon Southwood's (1962, 1977) association of migration with temporary habitats by recognising the significance of specific landscapes, weather patterns, natural-enemy complexes, etc. in maintaining a population's particular suite of adaptations.

In this contribution, we apply the Drake *et al.* (1995) conceptual model to a specific migration system, and discuss how it can be used to advance our understanding of the role of migration in insect population processes.

A migration system: the oriental armyworm *Mythimna separata* in east Asia

The extensive temperate-zone regions of the northern-hemisphere continents have highly seasonal climates in which conditions for insect life are alternately favourable and unfavourable. In inland North America and central and east Asia, winters at middle and high latitudes are severe and lethal to many insects, while in spring and summer the generally warm weather and rapidly growing vegetation provides favourable conditions for insect growth, development, and reproduction. In these environments, some insects have evolved migration systems that allow them to exploit these seasonally abundant resources by colonising them from lower latitudes each spring, with a return to lower latitudes in autumn (Johnson 1995; Chen *et al.* 1995; Kisimoto and Sogawa 1995). In this section we examine one such migration system, that of the oriental armyworm *Mythimna separata*, in east Asia; this population was selected primarily because its migrations and adaptations for migration have been studied extensively.

In the conceptual model (Drake *et al.* 1995), a migration system has four linked components: the *migration arena*, the *population trajectory*, the *migration syndrome*, and the *genetic complex* that underlies the migration syndrome (Fig. 1a); some of the known features of these for the population of *M. separata* that migrates through China are illustrated in Figs. 1b–d. The *migration arena* is the best documented component; it comprises the region within which the population migrates and incorporates all physical and biotic features that influence the fitness of population members. For *M. separata*, important elements of the arena include the disposition of land and sea, the location and phenology of the main hosts (wild and pasture grasses, wheat, rice, and maize), the seasonal patterns of photoperiod and temperature and their variation with latitude, the wind systems on which long-range migrations occur, and the distribution and seasonal incidence of natural enemies (Fig. 1b). The *population trajectory* of *M. separata* has been established through a sustained research programme involving population surveys, studies of survival and development times, and mark-and-capture and radar investigations of adult movement (summarised in Chen *et al.* 1995). The primary feature (Fig. 1b,c) is an annual cycle in which the population relocates northwards in spring and summer in three stages and then apparently returns to the overwintering areas in the south of China and Indo-China in two or three stages during autumn (e.g. Rainey 1989, pp. 239–240). Subpopulations that reach the extremities of the species' breeding range in the far northeast of China, Korea, and Japan (Chen *et al.* 1989; Lee and Uhm 1995; Hirai 1995), but are unable (at least in some years) to return south before winter arrives, are represented by trajectory spurs. The *migration syndrome* incorporates not only the essential apparatus for sustained flight (wings, fuel reserves, etc.) but also the mechanisms that determine when migratory flight is initiated and how long it is maintained (Gatehouse and Zhang 1995). Migration in *M. separata* appears to be exclusively pre-reproductive and two key components of migratory potential (Gatehouse and Zhang 1995) can be recognised: the pre-reproductive period (PRP, i.e. the number of nights on which the adult moth is physiologically capable of migration) and the flight capacity (FC, i.e. the time spent flying each night). These have been investigated for Chinese *M. separata* in the laboratory (Han and Gatehouse 1991a, 1993), and it is evident that the migratory potential of these moths is highly variable, being very substantial in some individuals (Fig. 1d). The *genetic complex* underlying the migration syndrome comprises the genes that code for the syndrome traits, their modes of inheritance, interactions (dominance, epistasis), correlations (including pleiotropy), and the mechanisms (e.g. genotype × environment interactions) affecting their phenotypic expression. PRP and FC in *M. separata*, in common with most migration-syndrome traits in insects (Dingle 1986, 1991; Gatehouse 1989), are quantitative traits, inherited polygenically. Loci on both autosomes and the X-chromosome have been shown to influence PRP in oriental armyworm moths (Han and Gatehouse 1991a). Nothing is yet known of the other elements of the genetic complex in this species.

The Drake *et al.* (1995) conceptual model recognises that a migration system functions through a series of processes, and that it is through these processes that the four components of the system are linked. For an established migration system, such as that of *M. separata* in east Asia (where there is historical evidence suggesting the species has been an agricultural pest for thousands of years, Zou 1956), the outcome of these processes is that the states of the components (e.g. resource quality, population size and location, distribution of migratory potential) fluctuate, but that the fluctuations remain within ranges that allow the population to survive indefinitely. The model posits that this self-maintenance of the system depends ultimately on natural selection acting on a contemporary timescale, and that the characteristics of the system components and processes, and the way that they change in space and time, have arisen through this evolutionary process.

For the *M. separata* east Asian migration system, many processes can be recognised. These include the interactions of climate and host plants (modulated, in this case, by the agricultural activities of man) that provide a seasonally changing disposition of larval food resources (Fig. 1b), the normal life-history processes of growth, development, and reproduction, and the mechanisms of inheritance of migration syndrome traits. Of more specific interest are the direct effects of weather and climate on the population trajectory: e.g. the modulation of development rates by changing temperatures, and winter mortality in subpopulations that fail to return from high latitudes in autumn. The destination regions of migratory flights are determined largely by the changing winds associated with passing large-scale weather systems: some migrants will be carried to regions where host plants are just becoming favourable for colonisation, others to regions where there is no possiblity of survival (e.g. out to sea, Asahina and Turuoka 1969), and some to regions where habitats are temporarily favourable but where autumn winds offer no prospect of escape from winter (Chen *et al.* 1989). Photoperiod and temperature modulate the incidence and duration of migratory flights through their interaction with the migration syndrome, e.g. by providing cues and thresholds for the initiation of flight and by influencing migratory potential. The temperature threshold for flight in Chinese *M. separata* is 8°C, and moths fly well at temperatures in the range 11–32°C (Zhang and Li 1985, Fig. 1d). Processes involving the genetic complex underlying the migration syndrome include modulation of the phenotypic expression of component traits through genotype × environment interactions, the adjustment by contemporary natural selection of allele frequencies at migration-syndrome loci, and genetic partitioning according to migratory potential as a result of migration itself. In Chinese *M. separata*, PRP is increased by declining photoperiod and temperature during development (Han and Gatehouse 1991b, Fig. 1e), and the PRPs of moths from samples taken in northern China in summer were longer than those of samples from intermediate latitudes (Han and Gatehouse 1991a, Fig. 1d).

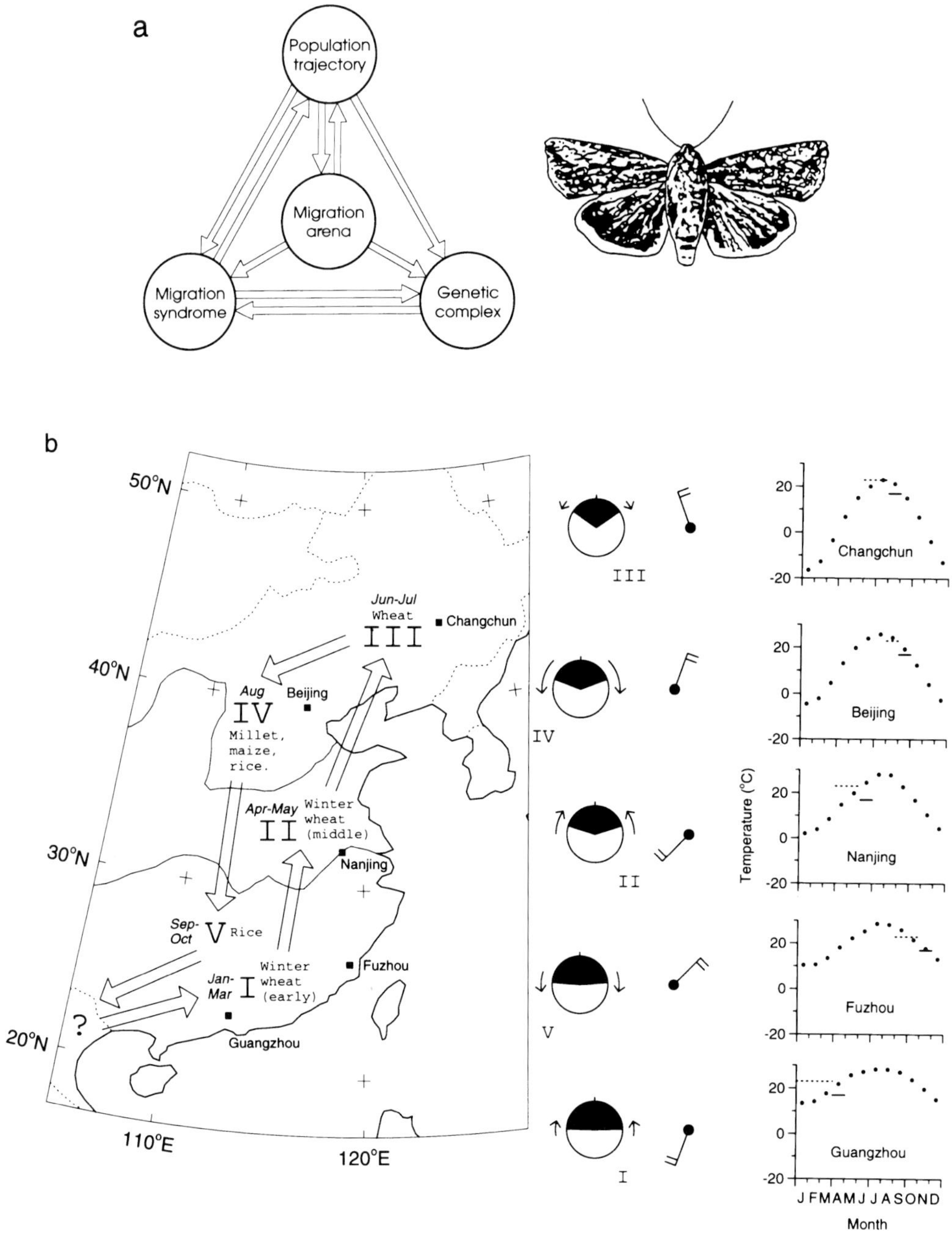

Fig. 1. Some known aspects of the migration system of the east Asian population of *M. separata*. (a) a conceptual model of a migration system (reproduced from Drake *et al.* 1995, with permission). (b) Arena, with population trajectory superimposed and principal crop hosts indicated. Photoperiods at time of development, 1500 m average winds at time of adult stage, and annual cycle of average temperatures are shown at right for each of the five generations. Temperatures favourable for development (23°C, dotted line) and flight (17°C, solid line) are indicated for the larval and adult periods of each generation (adapted from Chen *et al.* 1995, Domrös and Peng 1988, Zhao 1986).

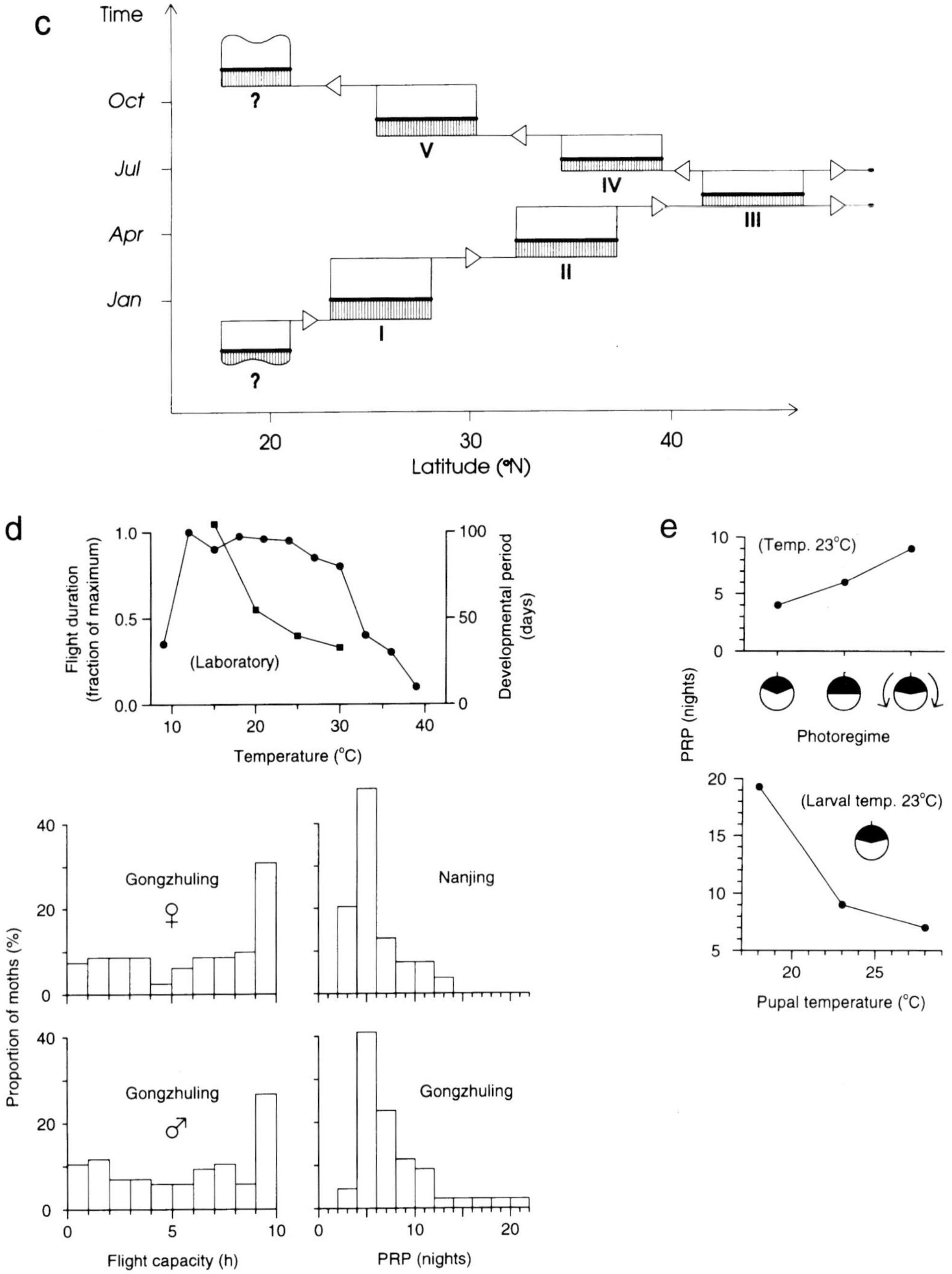

Fig. 1. (cont'd) (c) Schematic form of population trajectory (adapted from Drake *et al.* 1995). (d) Migration syndrome: variation of developmental period (■, adapted from Hirai and Santa 1983) and flight duration (●, adapted from Zhang and Li 1985) with temperature, and PRPs and FCs for moths collected in spring at Nanjing and Gongzhuling (near Changchun) (adapted from Han 1988). (e) Genetic complex: variation of PRP with temperature and photoperiod for different sibling groups (adapted from Han and Gatehouse 1991b).

This holistic approach allows us to begin to perceive how the migration system functions. Migration allows *M. separata* to exploit temporarily abundant resources during the growing season in central and northern China and provides the means of escape from lethal winter temperatures by a return in autumn to lower latitudes where climatic conditions allow continued breeding. The high reproductive rates resulting from breeding by successive generations on host plants that are at growth stages favourable for armyworm larval development, in a succession of habitats where competition for food resources and the impact of specialist natural enemies are both likely to be low, lead to high potential fitness for individuals capable of migrating long distances northwards in spring. A warm southwesterly airflow prevails over southern and central China in April and over central and northeastern China in June, providing a transport system for successive colonisation of the second and third generation areas respectively (Zhao 1986, Fig. 1b); average air temperatures in this flow are above the 8°C flight threshold at altitudes up to ca. 1–2 km. Wind speeds of ca. 10 m/s enable the new habitats to be reached in only a few nights of flight, and moths are typically 2–5 days old on arrival (Chen *et al.* 1989, 1995). Because the southwesterlies also bring the conditions required for crop growth, they transport the migrants into the appropriate destination region. As autumn approaches, conditions at higher latitudes become unfavourable as crops mature and temperatures start to fall. The northerly airflows extend gradually southwards throughout autumn (Zhao 1986), and provide southward transport into, successively, the fourth and fifth generation areas and possibly eventually into Indo-China, the Philippines, and the Indonesian archipelago. Only migrants reaching the southern overwintering areas, together with individuals of limited migratory potential from subpopulations that have remained there throughout the year, will contribute to the genetic complex of the following season's generations. Subpopulations producing moths that fail to reach regions where overwintering is possible, because suitable winds are not available to transport them there or because environmental conditions do not induce migratory behaviour in time, can be regarded as 'victims of the Pied Piper' (McNeil 1987). Such losses are evidently sustainable, and do not prevent the complex adaptation manifested in the migration syndrome from maintaining the population as a whole from year to year.

The effect of genetic partitioning according to migratory potential during the northwards migrations of spring and summer is that the further north a subpopulation is established in each generation, the greater will tend to be the migratory potential of the moths it produces (i.e. the higher the frequencies of alleles coding for traits that favour long-distance migration). A high proportion of moths from subpopulations at the highest latitudes will therefore have migratory potentials that 'pre-adapt' them for the long return journey in autumn, assuming suitable winds are available to transport them. The decreasing photoperiods and falling temperatures of autumn induce extended PRPs, further enhancing migratory potential and assisting these later generations to make their escape southwards before winter sets in (Han and Gatehouse 1991b).

The spatial distribution of favourable habitat patches available to migrating *M. separata* moths is to some extent unpredictable, depending for example on the phenologies of the host plants (and thus in turn on the sequence of preceding weather) in the particular year. Rather more unpredictable are the variations in windspeed and direction, within and between nights on which migratory flights are made, that influence the destinations that migrants with particular FC and PRP values will reach. Thus, the pattern of distribution of habitat patches, together with the stochastic effects of the wind on the extent and direction of displacements, must result in selection for extensive variation in migratory potential at each generation (Gatehouse 1994). The laboratory studies of Han and Gatehouse (1991a, 1993) do indeed show that there is extensive variation in both FC and PRP in Chinese *M. separata.* The polygenic inheritance of these traits will ensure that high levels of variation will be found among the offspring of the successful migrants, mainly as a result of recombination at each reproduction but also by mechanisms such as multiple mating (which is common in migratory noctuids, Drummond 1984). There is also reason to believe that FC and PRP are inherited independently in these windborne species (Gatehouse 1994), so that the range of variation in migratory potential (the composite of these two traits) among offspring will extend beyond that of the parent migrants reaching a particular favourable habitat patch. Variation in migratory potential is, therefore, on the one hand a consequence of the unpredictability of the locations of favourable habitat patches and of the winds that transport the migrants, and on the other a key attribute that enables subsequent generations to survive and exploit this same unpredictability. Similarly, unpredictability in the effects of weather and climate on both the rates of development of the migrants and the rate of deterioration of their habitats generates variation in other migration-syndrome traits (such as thresholds of response to the environmental cues inducing migration), and thus ensures, through the process of polygenic inheritance, that subsequent generations are able to survive the unpredictable environmental conditions they will encounter. These arguments apply to individual migrants, the fitness of which depends on the presence of variation in the migration-syndrome traits among their own offspring.

The model of the east Asian *M. separata* migration system outlined here is undoubtedly an oversimplification. The population trajectory is almost certainly more complex than that indicated by the available data which relate almost exclusively to high-density populations. Little is known about non-outbreak populations and how their migrations relate to the main strand of the trajectory, so it is not possible to assess their significance. Even the important southward migration in autumn is poorly documented and understood. Furthermore, no consideration has been given to the specific topography of the region and the probabilities of migrants being carried into unfavourable regions, but presumably these also play a role in the process of maintaining the population's genetic complex through natural selection.

Discussion

The Drake *et al.* (1995) model has its origins in the observation of Johnson (1960) that 'migration [is] an evolved adaptation rather than…a current reaction to adversity' and the hypothesis proposed shortly afterwards by Southwood (1962) that 'the prime evolutionary advantage of migratory movement lies in its enabling a species to keep pace with the changes in the locations of its habitats'. It incorporates both Kennedy's (1961) concept of migration as an active process achieved through specific behavioural and physiological adaptations, and Taylor and Taylor's (1977) view, presented so graphically with their 'fern stele' model, of a migratory population as a reticulum in space and time. It also treats explicitly the genetic processes that occur during migration (e.g. Comins 1977) and that underlie it (Dingle 1984). Like Taylor and Taylor's (1983) 'paradigm', it recognises and seeks to integrate these behavioural, physiological, genetic, and ecological components of migration within a Darwinian framework.

The basic model (Drake *et al.* 1995) provides a generic framework on which specific models, such as that outlined above for *M. separata* in east Asia, can be built. Such models are *analytical* (Southwood 1975), i.e. they are concerned with the causal mechanisms underlying population changes and represent component processes in some detail. They can be used to test hypothesised forms of these mechanisms by comparing model predictions of their components, as well as of the total population size, with field data. They allow an examination of some of the most fundamental questions of migration biology, such as how the population persists, how migration contributes to fitness, and why the migration syndrome takes the particular form it does in a particular arena. A series of such models, e.g. for populations subjected to similar evolutionary and ecological constraints (i.e. of the same or closely related species) but exploiting differing types of arena, might provide considerable insight into the role played by, and the significance of, each component process; it should also further reveal how the arena acts as a 'templet for ecological strategies' (Southwood 1977).

This agenda will require specific models to be formulated in a much more sophisticated way than the simple descriptive outline given here for *M. separata*. A computer-based model appears appropriate, and should incorporate a realistic representation of the arena's topography and the distribution of host plants. Climate, winds, host phenology, and photoperiod would be modelled for each region and season, with some measure of variance included for all but the last of these. Each temporary subpopulation would be modelled in terms of the distributions among its members of migration syndrome and relevant demographic traits (temperature thresholds for development, reproduction, and flight; responses to environmental cues of habitat favourability; PRPs and FCs; fecundity; schedules of reproduction; etc.) and possibly also other traits of particular interest or that might serve as markers (e.g. resistance to insecticides). On running the model, the interactions of these traits with a random or historical weather sequence would produce a simulated population trajectory in which each subpopulation has its own development rate, phenology, migration outcome, and mortality; branching and coalescence of trajectory strands would occur in the 'fern stele' manner (see Taylor and Taylor 1977). Distributions of the key traits would in turn be derived from a simplified model of the genetic complex in which each quantitative trait would be represented by a set of (presumed) contributing loci. Allele frequencies would alter as the model population developed and migrated, with the interaction of each subpopulation's migration syndrome with the arena causing differential mortality, genetic partitioning, and the merging of temporarily separated subpopulations. The output of the model, in the form of subpopulation locations and phenologies, and the distributions of individual traits, would provide a set of observables that could be compared with field data and used to determine which model components are realistic and which need revision.

As might be expected for a model concerned with populations in rapidly-changing environments, this framework (Drake *et al.* 1995) reflects the viewpoint of Andrewartha and Birch (1954), that population processes are dominated by environmental factors, rather than that of Nicholson, who emphasised regulation through density dependence (Kingsland this volume). In migrant populations, density-dependent dispersal offers a mechanism for deferring mortality and avoiding competition (Taylor and Taylor 1983), and it has been argued that another noctuid species, the African armyworm *Spodoptera exempta*, is specifically adapted to maintain low population densities by frequent dispersal during windborne migration (Gatehouse 1987). There is, however, no evidence of a significant influence of larval density on migratory potential in *M. separata* (Hill and Hirai 1986; Han 1988). Nevertheless, density-dependent mortality probably does occur in some circumstances (Strong 1984), perhaps during localised outbreaks or dispersal from them. The proposed simulation model provides one means of exploring the role of density-dependent processes in population regulation, with model results perhaps being used to guide the design of field studies.

There has been a remarkable surge of interest in spatial processes in ecology and population dynamics over the last decade or two (e.g. Hanski and Gilpin 1991; Hansson *et al.* 1995; Kareiva 1990; Renshaw 1991). Migration, a key spatial process, has been especially well studied in insects over a long period (e.g Johnson 1969; Danthanarayana 1986; Rainey 1989; Drake and Gatehouse 1995), but this work features only marginally in the broader literature, which appears to be focussed on less rapidly changing environments and population distributions. Integration of the findings of entomologists on migration into a more general consideration of population processes in a spatially extensive and heterogeneous environment seems desirable, and it is hoped that the approach outlined here may help to advance that process.

References

Andrewartha HG, Birch LC (1954) *The distribution and abundance of animals.* University of Chicago Press, Chicago

Asahina S, Turuoka Y (1969) Records of the insects visited a weathership located at the Ocean weather station 'Tango' on the Pacific. III. *Kontyu* **37**:290–304 [In Japanese, English summary]

Chen RL, Bao XZ, Drake VA, Farrow RA, Wang SY, Sun YJ, Zhai BP (1989) Radar observations of the spring migration into northeastern China of the oriental armyworm moth, *Mythimna separata*, and other insects. *Ecol Entomol* **14**:149–162

Chen RL, Sun YJ, Wang SY, Zhai BP, Bao XZ (1995) Migration of the Oriental Armyworm *Mythimna separata* in East Asia in relation to weather and climate. I. Northeastern China. In: Drake VA, Gatehouse AG (eds) *Insect migration: tracking resources through space and time.* CUP, Cambridge UK, pp 93–104

Comins HN (1977) The development of insecticide resistance in the presence of migration. *J Theor Biol* **64**:177–197

Danthanarayana W (ed) (1986) *Insect flight: dispersal and migration.* Springer-Verlag, Berlin

Dingle H (1984) Behavior, genes, and life histories: complex adaptations in uncertain environments. In: Price PW, Slobodchikoff CN, Gaud WS (eds) *A new ecology: novel approaches in interactive systems.* Wiley, New York, pp 169–194

Dingle H (1986) Evolution and genetics of insect migration. In: Danthanarayana W (ed) *Insect flight: dispersal and migration.* Springer-Verlag, Berlin, pp 83–93

Dingle H (1991) Evolutionary genetics of animal migration. *Amer Zool* **31**:253–64

Domrös M, Peng GB (1988) *The climate of China.* Springer-Verlag, Berlin

Drake VA, Gatehouse AG (eds) (1995) *Insect migration: tracking resources through space and time.* CUP, Cambridge UK

Drake VA, Gatehouse AG, Farrow RA (1995) *Insect migration: a holistic conceptual model.* In: Drake VA, Gatehouse AG (eds) *Insect migration: tracking resources through space and time.* CUP, Cambridge UK, pp 427–457

Drummond BA III (1984) Multiple mating and sperm competition in the Lepidoptera. In: Smith RL (ed) *Sperm competition and the evolution of animal mating systems.* Academic Press, Orlando, pp 291–370

Gatehouse AG (1987) Migration and low population density in armyworm (Lepidoptera: Noctuidae) life histories. *Insect Sci Appl* **8**:573–580

Gatehouse AG (1989) Genes, environment, and insect flight. In: Goldsworthy GJ, Wheeler CH (eds) *Insect flight.* CRC Press, Boca Raton Florida, pp 115–138

Gatehouse AG (1994) Insect migration: variability and success in a capricious environment. *Res Pop Ecol* **36**:165–171

Gatehouse AG, Zhang XX (1995) Migratory potential in insects: variation in an uncertain environment. In: Drake VA, Gatehouse AG (eds) *Insect migration: tracking resources through space and time.* CUP, Cambridge UK, pp 193–242

Han EN (1988) *Laboratory studies on the regulation of migration in the oriental armyworm,* Mythimna separata *(Walker) (Lepidoptera: Noctuidae).* PhD Thesis, University of Wales

Han EN, Gatehouse AG (1991a) Genetics of precalling period in the oriental armyworm, *Mythimna separata* (Walker) (Lepidoptera: Noctuidae), and implications for migration. *Evolution* **45**:1502–1510

Han EN, Gatehouse AG (1991b) Effect of temperature and photoperiod on the calling behaviour of a migratory insect, the oriental armyworm *Mythimna separata*. *Physiol Entomol* **16**:419–427

Han EN, Gatehouse AG (1993) Flight capacity: genetic determination and physiological constraints in a migratory moth *Mythimna separata*. *Physiol Entomol* **18**:183–188

Hanski I, Gilpin M (1991) Metapopulation dynamics: brief history and conceptual domain. *Biol J Linn Soc* **42**:3–16

Hansson L, Fahrig L, Merriam G (eds) (1995) *Mosaic landscapes and ecological processes.* Chapman and Hall, London

Hill MG, Hirai K (1986) Adult responses to larval rearing density in *Mythimna separata* and *Mythimna pallens* (Lepidoptera: Noctuidae). *Appl Entomol Zool* **21**:191–202

Hirai K (1995) Migration in the oriental armyworm *Mythimna separata* in East Asia in relation to weather and climate. III. Japan. In: Drake VA, Gatehouse AG (eds) *Insect migration: tracking resources through space and time.* CUP, Cambridge UK, pp 117–129

Hirai K, Santa H (1983) Comparative physio-ecological studies on the armyworms, *Pseudaletia separata* Walker and *Leucania loreyi* Duponchel (Lepidoptera: Noctuidae). *Bull Chugoku Natl Agric. Exp. Stn Ser E* (Environ Div) **21**:55–101

Johnson CG (1960) A basis for a general system of insect migration and dispersal by flight. *Nature* **186**:348–350

Johnson CG (1969) *Migration and dispersal of insects by flight.* Methuen, London

Johnson SJ (1995) Insect migration in North America: synoptic-scale transport in a highly seasonal environment. In: Drake VA, Gatehouse AG (eds) *Insect migration: tracking resources through space and time.* CUP, Cambridge UK, pp 31–66

Kareiva P (1990) Population dynamics in spatially complex environments: theory and data. *Phil Trans R Soc Lond B* **330**:175–190

Kennedy JS (1961) A turning point in the study of insect migration. *Nature* **189**:785–791

Kisimoto R, Sogawa K (1995) Migration of the Brown Planthopper *Nilaparvata lugens* and the White-backed Planthopper *Sogatella furcifera* in East Asia: the role of weather and climate. In: Drake VA, Gatehouse AG (eds) *Insect migration: tracking resources through space and time.* CUP, Cambridge UK, pp 67–91

Lee JH, Uhm KB (1995) Migration in the oriental armyworm *Mythimna separata* in East Asia in relation to weather and climate. II. Korea. In: Drake VA, Gatehouse AG (eds) *Insect migration: tracking resources through space and time.* CUP, Cambridge UK, pp 105–116

McNeil JN (1987) The true armyworm, *Pseudaletia unipuncta*: a victim of the 'Pied Piper' or a seasonal migrant. *Insect Sci Applic* **8**:591–597

Rainey RC (1989) *Migration and meteorology. Flight behaviour and the atmospheric environment of locusts and other migrant pests,* OUP, Oxford UK

Renshaw E (1991) *Modelling biological populations in space and time.* CUP, Cambridge UK

Southwood TRE (1962) Migration of terrestrial arthropods in relation to habitat. *Biol Rev* **37**:171–214

Southwood TRE (1975) *The dynamics of insect populations.* In: Pimentel D (ed) *Insects, science and society.* Academic Press, New York, pp 151–199

Southwood TRE (1977) Habitat, the templet for ecological strategies? *J Anim Ecol* **46**:337–365

Strong DR (1984) Density-vague ecology and liberal population regulation in insects. In: Price PW, Slobodchikoff CN, Gaud WS (eds) *A new ecology: novel approaches in interactive systems.* Wiley, New York, pp 313–327

Taylor LR, Taylor RAJ (1977) Aggregation, migration and population mechanics. *Nature* **265**:415–421

Taylor LR, Taylor RAJ (1983) Insect migration as a paradigm for survival by movement. In: Swingland IR, Greenwood PJ (eds) *The ecology of animal movement.* Clarendon Press, Oxford UK, pp 181–214

Zhang ZT, Li GB (1985) A study on the biological characteristics of the flight of the oriental armyworm [*Mythimna separata* (Walker)] moth. *Acta Phyt Sin* **12**:91–100 [In Chinese, English summary.]

Zhao SJ (1986) Relation between long-distance migration of oriental armyworms and seasonal variation of general circulation over East Asia. *Advances in Atmospheric Sciences* (Beijing) **3**:215–226

Zou SW (1956) [Review of the armyworm damage and control in the historical record in China.] *Kunchong Zhishi* [Entomological Knowledge] **2**:241–246 [In Chinese]

Simulating spatially distributed populations

Murray G. Efford

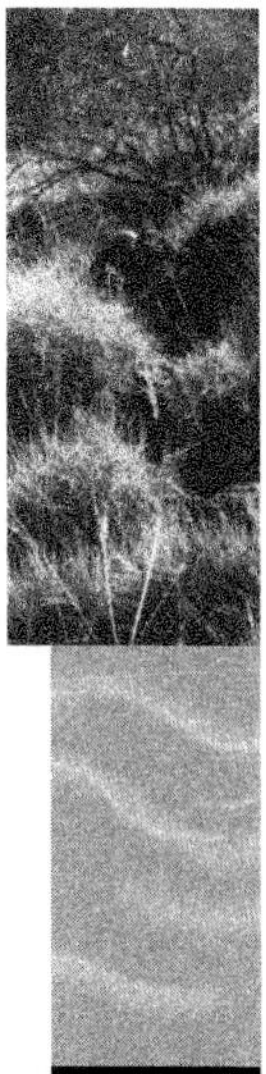

ABSTRACT

This approach to spatial population modelling extends the classical view of density dependence and incorporates individual dispersal behaviour. The model has been developed to understand the dynamics of the Australian brushtail possum (*Trichosurus vulpecula*) introduced to New Zealand, where it now seriously affects native plants and animals, and transmits bovine Tb to cattle and farmed deer. Since eradication is currently impossible, the aim is to reduce local populations sufficiently to protect selected conservation values, to reduce disease transmission to livestock, and to prevent disease spread. A realistic model must integrate *in situ* demography, continuous habitat variation and dispersal, while also allowing for possibly complex pest management scenarios. These requirements were met by coupling an individual-based demographic model to a raster GIS. Birth, death and dispersal were stochastic and density dependent. Calculations of local density dependence used a continuous local population surface $N(x,y)$ based on home range overlap. Carrying capacity, $K(x,y)$, and intrinsic rate of increase, $\lambda_0(x,y)$, were site dependent in the model (i.e. allowed to vary continuously in space, limited only by the pixel size of the map). Animals 'dispersed' according to probabilistic behavioural rules. The probability that a disperser would settle increased with the number of vacancies at a site, measured by the difference between $K(x,y)$ and $N(x,y)$. The model can be used to develop rational pest management strategies. It is applied in an example to a 45 x 55 km habitat map derived from Landsat TM imagery.

Key words: Trichosurus vulpecula, dispersal, spatially explicit population models, simulation, pest management

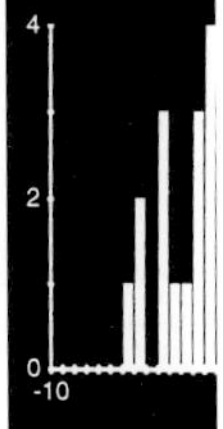

Frontiers of Population Ecology, R.B. Floyd, A.W. Sheppard and P.J. De Barro (eds), CSIRO Publishing, Melbourne, 1996, pp. 409–418

Introduction

The prediction of population responses to management, whether for conservation, harvesting or pest control, is likely to be substantially improved by the explicit inclusion of space. As noted by Van Horne (1991): 'by making models spatially explicit, even in a rather simplistic fashion, we remove one of the largest sources of error in existing modelling procedures'. In addition, many questions in vertebrate pest management include location: Where should control be done? Is it more important to stem the flow of immigrants reinfesting an area or to repeat control *in situ*? Conservation biologists now frequently use the 'metapopulation' metaphor in assessments of population viability (e.g. Gilpin and Hanski 1991; Lahaye *et al.* 1994), and large-scale spatial models have been developed for insect pests such as spruce budworm (Clark and Holling 1979; see also Fleming this volume). However, spatial modelling has so far made relatively little impact on vertebrate pest management (e.g. Caughley and Sinclair 1994).

I suggest that the lag in application of spatial modelling to vertebrate pests reflects the fundamental difficulty of allowing for individual-scale behaviour when modelling abundant and widely distributed species. Given the complexity of 'real landscapes' and our ignorance of how to map this onto simpler abstract descriptions, we are obliged at present to work with representations of the actual habitat geometry. Technologies such as satellite remote sensing and geographic information systems (GIS) now make possible the capture and processing of large-scale spatial data (Haslett 1990; Johnston 1993; Liebhold *et al.* 1993; Roberts *et al.* 1993), while individual behaviour is increasingly seen as the unifying focus of ecological research (Huston *et al.* 1987). The promised linkage of individual-based simulation models with real landscapes in a GIS (DeAngelis and Gross 1992; Johnston 1993) has been limited by computing constraints and uncertainty about the appropriate level of detail and the means of linking spatial scales (Wiens 1989; Hastings 1990; Van Horne 1991; Wiens *et al.* 1993).

This paper outlines a general scheme for spatial population simulation in discrete time that integrates individual and landscape scales. The model is used to predict populations of a single pest species, the brushtail possum (*Trichosurus vulpecula*). Possums are now common in farmland and native forests throughout New Zealand, where they threaten native flora and fauna and form a major wildlife reservoir for the stock disease bovine tuberculosis (Cowan 1990). The ultimate aim of modelling is to find the combination of localised control measures that will most cost-effectively reduce possum numbers in sensitive areas.

General framework

The model represents a set of individuals, each located at a point in 2-D and subject to certain probabilities of birth, death and dispersal. The number of individuals in any small area is assumed to be regulated by density-dependent feedback on *in situ* demography and/or movement, rather than by interaction with an explicit resource (cf Caughley and Sinclair 1994; McCarty this volume). Time is treated in discrete (annual) steps, which is, arguably, appropriate for a strongly seasonal animal such as the possum (Cowan 1990).

In situ demography should reflect location in two different senses. The first is 'spatial dependence' as used in geostatistics and elsewhere to refer to the tenet of 'locality': each organism is primarily affected by others in its neighbourhood (DeAngelis and Gross 1992). Density dependence is redefined below in terms of a continuous local population surface based on home-range overlap. In the second sense account needs to be taken of how habitat, conceived as an external driving variable, may vary with location (Southwood 1977). The term 'site dependence' is suggested for the spatial variation of demographic parameters.

Three aspects of the model structure will now be examined in more detail: spatial density-dependence, the parameterisation of habitat variation, and rule-based individual dispersal.

Spatial density-dependence and local population size

For a homogeneously mixing population, a demographic rate s is density-dependent when:

$$s = f(N)\, s_0, \tag{1}$$

where s_0 is the 'intrinsic' or 'density-independent' rate and $f(N)$ is a decreasing function of total population number N. In a stochastic model we understand s to be a probability between 0 and 1. The function $f(N)$ is a surrogate for the combined intraspecific competitive effects acting on an individual (cf Caughley and Sinclair 1994). Total population N may be seen as a sum of the pairwise interactions, all of equal (unit) magnitude, between a notional additional animal and the existing members of the population. This formulation is awkward for the non-spatial case, but it helps with the spatial case when we wish to substitute for N a quantity $N(x,y)$ representing the local population size. In a spatially distributed population, the intensity of interaction $I(a,b)$ between two animals located at $\mathbf{x_a} = (x_a,y_a)$ and $\mathbf{x_b} = (x_b,y_b)$ is presumed to decline with the radial distance $r = |\mathbf{x_a}-\mathbf{x_b}|$ between them:

$$I(a,b) = g(r) \tag{2}$$

where $g(r)$ represents the interaction neighbourhood. Each animal is assumed to use space differentially according to the probability density of its home range distribution (van Winkle 1975). To a reasonable approximation, $g(r)$ is the quantitative overlap between the ranges of two animals or, equivalently, their probability of encounter (Waser and Wiley 1979; Addicott *et al.* 1987; Czarin and Bartha 1992). Note that $g(r)$ differs subtly from the radial distribution of activity within each home range. Supposing the activity of each animal has a circular bivariate Gaussian distribution with variance σ^2 the overlap is a Gaussian function of r with variance $2\sigma^2$ (Rappoldt and Hogeweg 1980; van den Bosch *et al.* 1990).

We can now define the local population size $N(x,y)$ at $x_0 = (x,y)$ in such a way that it can be used directly in Eq. 1:

$$N(x,y) = \sum_i g\left(|x_0 - x_i|\right) \tag{3}$$

In words, the local population size at a point is the sum of the distance-weighted interactions with all other members of the population that would be experienced by a new individual located precisely at that point. Local population size is defined for all x,y regardless of whether the site is occupied. The $N(x,y)$ surface is a spatial property of the population analogous to a field in physics that substitutes 'force at a point' for 'action at a distance' between particles (Hockney and Eastwood 1981). A major computational advantage of $N(x,y)$ is that the response of an individual at x,y may be predicted by comparing the local population with the local carrying capacity, without each time conducting a spatial search for potential competitors.

It is convenient to scale the measure of overlap $g(r)$ so that:

$$N = \int_{\text{All space}} N(x,y) \tag{4}$$

This is achieved by representing the contribution of each individual as a bivariate probability density distribution. In practice, one estimates $N(x,y)$ numerically by summing individual contributions. Estimation resembles spatial smoothing or bivariate density estimation with a Gaussian kernel (Silverman 1986).

It is assumed that home-range size is not site dependent.

Site-dependent carrying capacity and intrinsic rate of increase

Assume that, in the absence of movement, *in situ* population processes yield a finite rate of increase $\lambda(x,y)$ for individuals located at x,y. By analogy with the simple logistic model we can define the local carrying capacity $K(x,y)$ such that:

$$\begin{aligned} &\lambda(x,y) = 1 \quad \text{when } N(x,y) = K(x,y) \\ &\lambda(x,y) > 1 \quad \text{when } N(x,y) < K(x,y) \\ &\lambda(x,y) < 1 \quad \text{when } N(x,y) > K(x,y). \end{aligned} \tag{5}$$

Extending the analogy, we can also define a spatial version of the intrinsic rate of increase $\lambda_0\,(x,y)$ as the limit of $\lambda(x,y)$ as $N(x,y) \to 0$.

Together, $K(x,y)$ and $\lambda_0(x,y)$ describe spatial variation in habitat. They are assumed for the present model to be constant in time, although that restriction is not conceptually necessary. This parameterisation is preferable to indices of 'habitat suitability' or 'habitat quality' because it makes explicit the significance of the variation for population dynamics and avoids the ambiguity of these terms (Van Horne 1983).

Dispersal

Dispersal movements link local dynamics across space (McCarty, this volume). Possums, like many other vertebrates, are usually sedentary and disperse only over a short period when young, if at all. During dispersal an individual is assumed to be 'vagrant', with no attachment to a home site (Lidicker 1975). Taylor and Taylor (1977) repudiated the common treatment of dispersal movements as random in direction and distance, proposing instead that dispersal is a programmed search for a site to breed. Random dispersal remains the norm in analytical population models for reasons of tractability (Durrett and Levin 1994), but in simulations the individual state variable (i state) paradigm allows vastly greater realism (e.g. Folse *et al.* 1989; DeAngelis and Gross 1992). In an i-state configuration model each individual is a discrete entity with its own state description (e.g. x,y coordinates, age, time since dispersal began).

The dispersal algorithm in the present model comprises for each animal a 'decision' to abandon the home site (emigration), an indefinite series of discrete movements ('jumps'), and an ultimate decision to settle (settlement). Both emigration and settlement are likely to be guided by rules based on the animal's 'assessment' of its fitness in the current habitat given the local population density (e.g. Rosenzweig 1991). The translation of habitat selection theory into individual dispersal rules is unclear, and I have relied on an arbitrary functional form for each. Emigration was considered to be driven by crowding in the same manner as other *in situ* processes (Eq. 1). Settlement was also considered to involve a comparison of current density with carrying capacity. [The assumption that dispersion is determined directly by the dispersion of resources is justified for female mammals, but it is less plausible for males, which may choose to settle where they have access to females; see Davies 1991.] However, it would be illogical to use a function of $N(x,y)/K(x,y)$ as this would not allow the disperser to discriminate vacant high-quality habitat from vacant low-quality habitat. The stopping rule was based instead on the number of 'vacancies' ($K(x,y) - N(x,y)$):

$$\begin{aligned} P(STOP) &= 1 - \frac{1}{1 + \frac{K-N}{c}} \quad \text{when } N < K \\ &= 0, \ \textit{otherwise} \ . \end{aligned} \tag{6}$$

The parameter c ($c > 0$) is the number of vacancies at which 50% of dispersers settle. Increasing values of c correspond to more selective settlement behaviour.

The actual route taken by a disperser could be modelled in great detail as a search path. A reasonable compromise is to consider the mean direction of the next jump and its angular variance as site-dependent. The model could also include mortality as a function of the hazards dispersers encounter.

Implementation

The implementation of the proposed spatial model for brushtail possums in New Zealand must take account of the relevant scales. Possum home ranges are usually only 0.01–0.10 km^2, but dispersing possums have been known to travel more than 20 km (Cowan 1990). This suggests that regional possum control plans should cover large areas, say upwards of 2000 km^2, corresponding to equilibrium populations >400 000 female possums (assuming an average density of 4 ha^{-1} and an even sex ratio). Precise data against which the model may be calibrated are likely to come from small-scale experiments (0.1–1.0 km^2), so it was desirable to program the model in such a way that it could be applied at a range of scales with a minimum of modification.

For efficient storage and processing, the population was represented mostly as a 2-D map of individual locations. The alternative representation, a list of individuals with their *x,y* coordinates and other attributes, was used only for dispersing animals. Population and habitat maps were stored in spatially discrete form as raster GIS files. Pixel size could be varied between simulations independently of other parameters, but was no larger than 100 × 100 m (0.01 km^2). The particular GIS used (EPPL7 Release 2.1, Anon. 1992) stores 1-byte integer map values (0–255) in a compressed file format. This was adequate for storing the integer number of individuals per pixel, typically <5, but values of continuous site-dependent parameters had to be scaled and rounded before storage. The virtue of the medium was that it placed no constraint on map size: simulations with maps of 250 000 pixels were quite feasible.

The overall structure of the model as implemented is shown in Fig. 1. The model was programmed in Turbo Pascal. Only females were considered. Yearlings and adults were mapped separately to enable age-specific modelling of dispersal, mortality and reproduction. A pooled local population surface (defined above) was derived from the individual locations by summing the yearling and adult maps and applying a Gaussian smoother with variance equal to twice the estimated scale of home range movement. Edge effects were circumvented by assuming all processes are reflected at the boundaries, which is equivalent to assuming that the map lies in an infinite tessellation of identical, reflected maps.

Births, deaths and emigration occurred *in situ* with independent binomial probabilities. Emigrating individuals joined a list of dispersers that was processed separately. To allow for independent spatial variation in each age-specific birth and death rate would have required excessive detail. Rather, global reference values of the four age-specific rates were scaled logarithmically by a common factor until their combined effect, λ, matched the mapped value of λ_0. A similar procedure was used to find the component rates that gave $\lambda = 1$ when $N = K$. Rate of increase λ was estimated by the dominant eigenvalue of the corresponding projection matrix (Caswell 1989).

Density dependence followed Eq. 1 with:

$$f(N)=\frac{1}{1+a\left(\frac{N}{K}\right)^{\theta}} \qquad (7)$$

where N and K were spatially indexed and $K > 0$ (cf. Gilpin and Ayala 1973; Pielou 1977; Bellows 1981). The constant θ varies the shape of the density dependence; a value of 2–3 is commonly used for possums (e.g. Barlow and Clout 1983; Barlow 1991). The parameter a is specific to the process

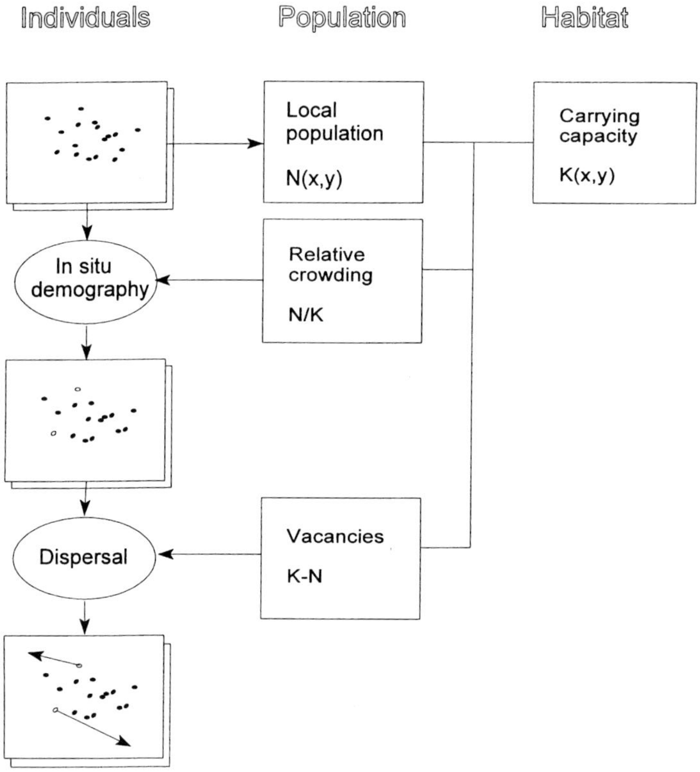

Fig. 1. Spatial possum model during one time step. Rectangles represent 2-D maps stored as raster GIS files. Adults and yearlings are mapped separately. The local population surface is derived from the summed point location maps by applying a smoother scaled to the common home range radius. Maps of site-dependent intrinsic growth rate λ_0 and constraints on dispersal are omitted for graphical clarity (see text).

being modelled ($a = s_0/s_K - 1$, where s_0 is the intrinsic rate and s_K is the rate when $N = K$). It was convenient to construct an intermediate map of *(N/K)* ('relative crowding') that was common to all *in situ* density-dependent processes. In any landscape there are areas of 'non-habitat' (e.g. lakes and rivers) in which it is inconceivable that any animal would settle. These were identified separately by an 'offsite' code rather than setting $K(x,y) = 0$, which would be incompatible with the equation for density dependence.

Jump direction was treated as a von Mises pseudo-random variate (Ripley 1987) with site-dependent parameters. Jump length was drawn from a pseudo-random gamma distribution (Press *et al.* 1989).

The model was used to simulate the effects of localised possum control by exposing all individuals to a mapped (site-dependent) binomial probability of death before any other calculations.

Example

A brief example is given of the application of the model. The vegetation of a 45 × 55 km window of the central North Island, New Zealand, was mapped from a Landsat Thematic Mapper image classified into broad vegetation categories (bare ground, water, pasture, pine forest, *Leptospermum* scrub, and native forest) by reference to a conventional 1:250 000 digital forest class map. Thirty-nine percent of the vegetated landscape was native forest, mostly Department of Conservation land in the Hauhungaroa Range, but also including numerous remnants surrounded by pasture.

Possum carrying capacity was mapped by applying mean densities for broad vegetation categories (e.g. $K(x,y)$ = 3 females ha^{-1} in native forest, B. Warburton and G. Nugent pers. comm.). The carrying capacity of forest and scrub within 1 km of pasture was increased to mimic the edge effect shown by Coleman *et al.* (1980). Each 100 × 100 m pixel was initially populated with a Poisson number of possums, with mean equal to the nominal carrying capacity, and the model was run for 20 years to equilibrate. The finite intrinsic rate of increase was assumed to be uniform across the landscape. Mean dispersal direction for sites in pasture was towards the nearest forest, and within forest was up the local gradient of increasing carrying capacity; the von Mises variance parameter k was set to 20 for pasture (50% of jumps within ±9° of the mean) and to 2 for forest (50% of jumps within ±30° of the mean). Control was modelled by removing 90% of possums living in forest or scrub within 1 km of pasture, a total area of 666 km^2. Other parameter values were: home range σ = 100 m; λ_0= 1.35; θ = 2; dispersal probability when N=K; yearlings = 0.20; adults = 0.05; mean jump length 1 km (exponential). N is approximately 500 000 females at equilibrium. The simulation of population changes over 1 year took 4.9 ± 0.1 minutes on a 66 MHz 486 PC, including the dispersal of about 42 000 individuals.

Recovery of total population size within the 1 km controlled zone was simulated under various values of c, the selectivity of settlement (Fig. 2). Simulated net dispersal distances immediately after control (mean ± SD) ranged from 1.2 ± 1.2 km (c = 1) to 3.8 ± 3.0 km (c = 8). Population recovery in the controlled zone was more rapid when dispersers were more selective about the sites in which they settled, because the pool of potential colonists was not then limited to those living close to the vacant habitat. The ultimate equilibrium density predicted by the spatial model in the controlled

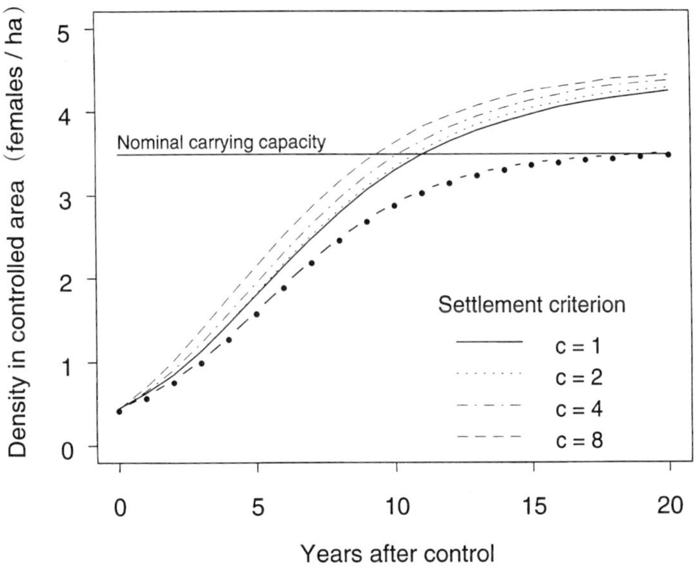

Fig. 2. Simulated recovery of a controlled possum population in 666 km^2 of forest and scrub within 1 km of pasture, central North Island, New Zealand, under four hypothetical scenarios of dispersal behaviour. The horizontal line indicates the nominal carrying capacity $K(x,y)$ within the controlled area. Heavy dots show the pattern of recovery when all dispersal is suppressed.

area when $c = 2$ was 27% above that expected from *in situ* dynamics alone (i.e. $K(x,y)$), and outside the controlled area was 11% below the non-spatial expectation. These effects did not cancel; the overall density increased by 9%.

Discussion

The development of software for spatial population modelling must go hand in hand with conceptual developments in population ecology. Approaches driven by the technology of GIS (Johnston 1993) or artificial intelligence (Folse *et al.* 1989) run the risk of losing contact with the mainstream of population ecology. A good example is provided by the Nicholsonian concept of density dependence, which has long been a cornerstone of population theory. In some senses the concept is redundant in a truly individual-oriented spatial model, for which the 'population' is a distant and arbitrary construct. However, as shown here, density dependence may be recast in a spatial form that is consistent with models of individual behaviour. The benefits are various: we retain some of the parameterisation and comfortable concepts of non-spatial models, and are better able to focus on reasons for the discrepancies between spatial and non-spatial results.

Source/sink effects (Pulliam and Danielson 1991) are likely to be a common cause of mismatch with non-spatial models, although we are only slowly gathering a working understanding (Watkinson and Sutherland 1995). In the example provided here, the area selected for control was also 'preferred' habitat that received a net inflow of dispersers because of the rules governing settlement. Much field research is needed to elucidate these rules and their population-level consequences. An interesting first step would be to map the difference between the *in situ* carrying capacity $K(x,y)$ and the modelled equilibrium population $N^*(x,y)$ for different habitat geometries and dispersal rules.

Acknowledgments

Thanks to Meryll White and Mark Smale for assistance in preparing the Hauhungaroa vegetation map, and to Bastow Wilson for discussion. Bruce Warburton, Bill Lee, Phil Cowan and Joanna Orwin made helpful comments on a draft. This work was funded by the New Zealand Animal Health Board and by the Foundation for Research, Science and Technology from the Public Good Science Fund contract C09301.

References

Addicott JF, Aho JM, Antolin MF, Padilla DK, Richardson JS, Soluk DA (1987) Ecological neighbourhoods: scaling environmental patterns. *Oikos* 49:340–346

Anon. (1992) *EPPL7 Environmental planning and programming language Version 7 Release 2.1 User's Guide.* State of Minnesota Land Management Information Center, St Paul MN

Barlow ND (1991) A spatially aggregated disease/host model for bovine Tb in New Zealand possum populations. *J Appl Ecol* 28:777–793

Barlow ND, Clout MN (1983) A comparison of 3-parameter, single-species population models in relation to the management of brushtail possums in New Zealand. *Oecologia* 60:250–258

Bellows TS Jr (1981) The descriptive properties of some models for density dependence. *J Anim Ecol* 50:139–156

Caswell, H (1989) *Matrix population models.* Sinauer, Sunderland Mass.

Caughley G, Sinclair ARE (1994) *Wildlife ecology and management.* Blackwells, Boston

Clark WC, Holling CS (1979) Process models, equilibrium structures, and population dynamics: on the formulation and testing of realistic theory in ecology. *Fortschr Zool* 25:29–52

Coleman JD, Gillman A, Green WQ (1980) Forest patterns and possum densities within podocarp/mixed hardwood forests on Mt Bryan O'Lynn, Westland. *NZ J Ecol* 3:69–84

Cowan PE (1990) Brushtail possum. In: King CM (ed) *Handbook of New Zealand Mammals.* Oxford, Auckland, pp 68–98

Czarin T, Bartha S (1992) Spatiotemporal models of plant populations and communities. *Trends Ecol Evol* **7**:38–42

Davies NB (1991) Mating systems. In: Krebs JR, Davies NB (eds) *Behavioural ecology*, 3rd ed. Blackwells, Oxford, pp 263–294

DeAngelis DL, Gross LJ (eds) (1992) *Individual-based models and approaches in ecology.* Chapman & Hall, New York

Durrett R, Levin S (1994) The importance of being discrete (and spatial). *Theor Pop Biol* **46**:363–394

Folse LJ, Packard JM, Grant, WE (1989) AI modelling of animal movements in a heterogeneous habitat. *Ecol Model* **46**:57–72

Gilpin ME, Ayala FJ (1973) Global models of growth and competition. *Proc Natl Acad Sci USA* **70**: 3590–3593.

Gilpin M, Hanski I (eds) (1991) *Metapopulation dynamics: empirical and theoretical investigations.* Academic Press, London

Haslett JR (1990) Geographic information systems: a new approach to habitat definition and the study of distributions. *Trends Ecol Evol* **5**:214–218

Hastings A (1990) Spatial heterogeneity and ecological models. *Ecology* **71**:426–428

Hockney RW, Eastwood JW (1981) *Computer simulation using particles.* McGraw-Hill, New York

Huston M, DeAngelis D, Post W (1988) New computer models unify ecological theory. *BioScience* **38**:682–691

Johnston, CA (1993) Introduction to quantitative methods and modeling in community, population and landscape ecology. In: Goodchild MF, Parks BO, Steyaert LT (eds) *Environmental modelling with GIS.* Oxford, New York, pp 276–283

Lahaye WS, Gutiérrez RJ, Akçakaya HR (1994) Spotted owl metapopulation dynamics in Southern California. *J Anim Ecol* **63**:775–785

Liebhold AM, Rossi RE, Kemp WP (1993) Geostatistics and geographic information systems in applied insect ecology. *Ann Rev Entomol* **38**:303–327

Lidicker WZ (1975) The role of dispersal in the demography of small mammals. In: Golley FB, Petrusewicz K, Ryskowski L (eds) *Small mammals: their production and population dynamics.* CUP, Cambridge, UK.

Pielou EC (1977) *Mathematical ecology.* J. Wiley, New York

Press WH, Flannery BP, Teulosky SA, Vetterling WT (1989) *Numerical recipes in Pascal.* CUP, Cambridge, UK.

Pulliam HR, Danielson BJ (1991) Sources, sinks and habitat selection: a landscape perspective on population dynamics. *Am Nat* **137**:S50–S66

Rappoldt C, Hogeweg P (1980) Niche packing and number of species: *Am Nat* **116**:480–492

Ripley B (1987) *Stochastic simulation.* J. Wiley, New York

Roberts EA, Ravlin FW, Fleischer SJ (1993) Spatial data representation for integrated pest management programs. *Am Entomol* **39**:92–107

Rosenzweig ML (1991) Habitat selection and population interactions: the search for mechanism. *Am Nat* **137**:S5–S28

Silverman BW (1986) *Density estimation for statistics and data analysis.* Chapman & Hall, London

Southwood TRE (1977) Habitat, the templet for ecological strategies? *J Anim Ecol* **46**:337–365

Taylor LR, Taylor RAJ (1977) Aggregation, migration and population mechanics. *Nature* **265**:415–421

Van den Bosch F, Metz JAJ, Diekmann O (1990) The velocity of spatial population expansion. *J Math Biol* **28**:529–565

Van Horne B (1983) Density as a misleading indicator of habitat quality. *J Wildl Manage* **47**:893–901

Van Horne B (1991) Spatial configuration of avian habitats. *Acta XX Congr Int Ornithol*, pp 2313–2319

Van Winkle W (1975) Comparison of several probabilistic home range models. *J Wildl Manage* **39**:118–123

Waser PM, Wiley RH (1979) Mechanisms and evolution of spacing in animals. In: Marler P, Vandenbergh JG (eds) *Social behavior and communication (Handbook of behavioral neurobiology, vol 3)*. Plenum, New York pp 159–223

Watkinson AR, Sutherland WJ (1995) Sources, sinks and pseudo-sinks. *J Anim Ecol* **64**:126–130

Wiens JA (1989) Spatial scaling in ecology. *Funct Ecol* **3**:385–397

Wiens JA, Stenseth NC, van Horne B, Ims RA (1993) Ecological mechanisms in landscape ecology. *Oikos* **66**:369–380

SPATIAL MODELLING AND POPULATION ECOLOGY

Paul A. Walker

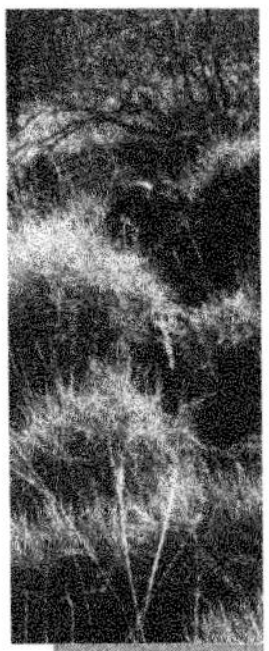

ABSTRACT

Geographers concerned with spatial patterns and processes have developed techniques for analysing spatial data and for searching for spatial patterns. Geographic information systems (GIS) are among the tools routinely used; this tool is increasingly being supplemented with a range of other modelling techniques.

Researchers in population ecology are largely unfamiliar with the range of tools for analysing their data. Giving greater emphasis to temporal and spatial considerations may refine or even broaden the questions that they are addressing. For example, how would a spatial modeller address questions concerned with the temporal dynamics of populations, where there is a requirement to identify spatial relationships between habitat and species composition, with questions concerned with environmental corridors or taxanomic structure? What lessons would a spatial modeller communicate about the spatial implications of scale and measurement techniques or about the implications of sampling method and survey bias on delineating species distributions and abundance.

Spatial modelling involves finding the right tool to answer the right question. Because of the lack of appreciation of spatial considerations or about spatial modelling, many data sets lie under-utilised. This paper seeks to identify some of the analytical modelling options available to population ecologists.

Key words: GIS, modelling, ecology

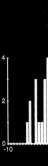

Frontiers of Population Ecology, R.B. Floyd, A.W. Sheppard and P.J. De Barro (eds), CSIRO Publishing, Melbourne, 1996, pp. 419–429

Introduction

Geographers are concerned with spatial patterns and processes, and have developed techniques for analysing spatial data and for searching for spatial patterns. Geographic information systems (GIS) are one of the tools routinely used. Increasingly, the mapping and modelling techniques of GIS are being supplemented with a range of more analytically oriented techniques.

Research in population ecology rarely employs techniques for analysing spatially referenced data. Giving greater emphasis to temporal and spatial considerations may refine or even broaden the questions ecologists are addressing. This paper identifies tools that could be of interest to ecologists working with spatially referenced data and discusses the strengths and weaknesses of GIS and modelling strategies.

Geographic Information Systems

GIS can assist in the collection, storage and analysis of spatially referenced data. Many ecological and resource management questions have benefited from the use of a GIS, provided that data sets and the tools for exploring these questions are appropriate (Efford this volume).

For some people and agencies, the GIS is an elaborate and expensive way of producing pretty maps or a solution to all their data base, mapping and modelling problems. These agencies are often unable to see beyond the glossy mapping procedures and the ever-increasing demands for more hard disks and better output. Consequently they require an ever-increasing GIS budget.

The real power of GIS lies in analysis, and extends beyond this limited perception of GIS as a mapping and data-storage tool. GIS could become a spatial exploration tool which acts as a pivot for a range of other modelling tools concerned with both spatial and temporal processes and patterns.

GIS capabilities

The key capabilities that users expect in a GIS include visualisation, organisation, analysis, interrogation, integration of disparate spatial data, and pattern detection based on spatial and temporal data.

GIS vary enormously in their implementation of these capabilities (see Berry 1987; Goodchild 1989; Walker and Moore 1988): some focus on map storage and map display, whereas others focus on providing analytical capabilities. Almost all address different types of features (point, line, area, surface, network) and some extend beyond 2-dimensional space into 3 dimensions. Structures used to represent data also vary. Some systems use a vector or topological model, in which features are defined by edges and relationships. In this model, regions are defined by a set of edges, each of which has a mathematical relationship (e.g. adjacency, node connection) to other edges and areas. Alternatively, the raster model represents spatial data using a grid or matrix. Each grid cell corresponds to an area on the Earth's surface. As grid cells become smaller, the data volume increases and the smaller the feature than can be represented in the grid. More recently, systems using an object-oriented structure (e.g. a river) are being developed.

Most GIS include a wide range of functions for co-ordinate transformations, area calculation, scale, stretch, Boolean retrieval based on attributes, classification, overlay and intersection, and many forms of neighbourhood analysis, including spreading, contours, theissen polygons, and trend surfaces. In addition, most GIS include 'standard' procedures for producing statistical summaries, calculation of proximity measures (to roads, streams, places, etc.), overlay and clipping of one layer to another, report generation, contouring, deriving mathematical indices using map attributes, etc.

Simple analytical operations may require only the application of a re-classification scheme to one or more map layers (e.g. vegetation cover) to generate a new map layer (e.g. cleared land). More complex operations overlay many map layers to form composite maps containing new attributes. In some systems, many lines of mathematical formula can generate new geographic entities and attributes. Questions concerned with the distribution and adjacency of areas, points and lines can be modelled by some systems.

While the full range of techniques included in commercial GIS are not mentioned in this paper, two are worth discussing briefly.

Multi-criteria modelling

Multi-criteria modelling (also known as cartographic modelling, map algebra, weighted map overlay or map indexing) (Tomlin *et al.* 1983; Burrough 1986), has become a trade-mark of many environmental studies. Essentially, this involves applying a set of GIS functions to assess the suitability or attractiveness of locations for a nominated purpose (e.g. development or conservation purposes). The technique is often used in habitat-suitability modelling, where the user wants to identify where the most suitable habitat patches for a species are located and identify which of these are most vulnerable in some sense.

A GIS might contain mapped information on vegetation types, temperatures, rainfall, slope, elevation and land-use patterns. The user could reclassify each of the map layers to a common scale, i.e. to an arbitrary suitability score, where a higher score implies that the location is more suitable for the activity (prime habitat). Several re-classified map layers (Fig. 1) can contribute to an overall index. A weight may be applied to each map layer to reflect its contribution to the model.

The procedure is often presented as 'map algebra' in which arithmetic operations are performed on map layers as though they were attributes. For instance, a command 'ADD Proximity-to-roads to Proximity-to-rivers to Proximity-to-urban-areas FOR constraint-accessibility' would calculate a new thematic map layer, the elements of which are calculated by the sum of the elements in the three component maps.

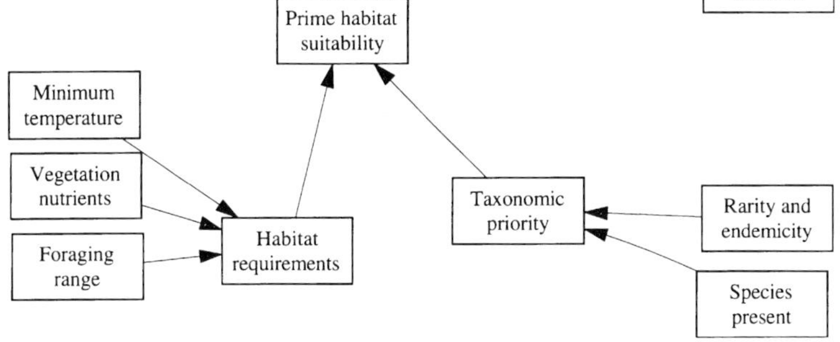

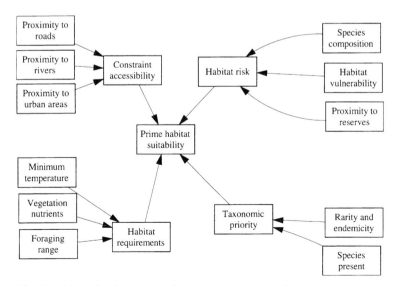

Fig. 1. Relationships between map layers in a multi-criteria model.

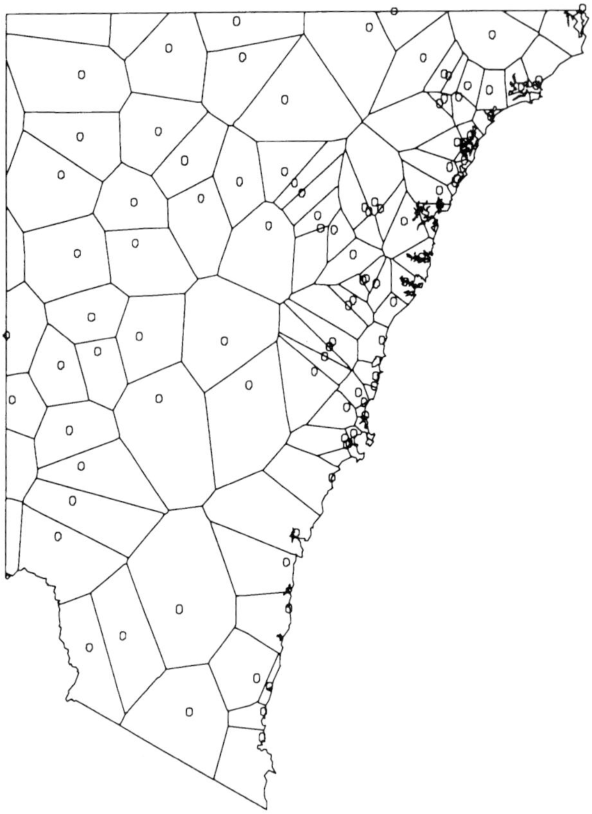

Fig. 2. Voronoi diagram used to delineate regions around base camps.

Such simple modelling strategies raise concern (Walker 1992; Burrough 1986). Its strength is its simplicity and hence flexibility; it is, after all, very easy to both learn and implement. Despite reservations about these procedures, multi-criteria modelling does provide many users with an opportunity to explore spatial data and to undertake simulation-like exercises within a GIS framework.

Proximity analyses

GIS often contain a range of tools for proximity analyses, including buffers, zones of influence, topological analyses, neighbourhood functions, and proximate analyses.

A set of buffer zones can be delineated around a point or line feature, where the size of each buffer is based on distance from the feature. A typical question is to identify all the species sightings within 10 km of a town or river. Simple or graded buffers (e.g. buffers at 1 km, 2.5 km and 10 km) can be constructed. Zones of influence can be delineated in many ways, but one of the most common is through the use of a Voronoi diagram (Fig. 2) (also known as Thiessen polygons). Given a set of points, a Voronoi diagram would show a set of polygons such that all areas within each polygon are closer to one point than to any other point. A typical question for this type of analyses is to identify the areas that would be most easily accessed for field survey from a set of base camps (although ignoring transport routes).

A GIS can also be used to determine which features (point, line, area) are within or adjacent to other nominated features. This can be used, for instance, to determine all areas of suitable habitat which are adjacent to existing reserve areas. Neighbourhood functions are also available in GIS, where the objective is to calculate an attribute for each location not just as a function of its current location but as a function of the attributes of a neighbourhood or region. This can be used to identify locations where the location attribute is not the same as the neighbourhood attribute. This technique can be used, for instance, to search for local anomalies in spatial patterns and areas of high neighbourhood or regional variability.

Modelling with spatial data

In a recent publication, Morrison *et al.* (1992) attempted to categorise the types of models used in examining wildlife–habitat relationships. They itemised a number of types of models for forecasting or hindcasting, which can be analytic, univariate, multivariate, deterministic, stochastic, descriptive, conceptual, diagrammatic, mathematical, statistical, probabilistic, spatial, temporal, global, regional, specific, generic, not to mention dynamic and locally adaptive. This probably explains why many people are overwhelmed when faced with the range of modelling options.

We are often unable to predict responses in even relatively simple systems, let alone the complex world of ecological relationships. Just like the process of scientific discovery (Gould 1993), modelling is a process of discovery, of false starts, and excursions into trial and error. Modellers are equally prone to exaggerate the strengths and dismiss the limitations of their models.

How a model is to be used and why it was formulated in the first place provides a key to whether or not the process of modelling was useful. Users of models need to understand the assumptions and realise that models are not exact solutions, but are the result of hypotheses concerning a generalised view of the relationships being studied. This is particularly important when model results are extrapolated in geographic scale, from region to region, from time period to time period and from one scale to another, as happens all the time with spatially referenced models.

While models are only as good as the data they contain, many models with years of scientifically based data collection behind them have been consigned to the bookshelf. Too many modellers have become consumed by 'realism', continually attempting to improve the 'accuracy' by adding more and more complexity. Modelling is about *explanation with simplicity.* It is about producing a formulation, a generalisation of some process, but in a way which the user (or client) can relate to.

The problem often comes when models are applied outside the domain within which they were developed. This same problem occurs with spatial data. Most GIS applications use data that has been collected by other agencies for some other purpose. At the same time data sets collected by ecologists, lie under-utilised because of the lack of appreciation of spatial considerations or about spatial modelling.

Modelling occurs in a number of inter-connected domains (after Aspinall and Lees 1994), (Fig. 3). GIS operate almost exclusively in geographic space, although they are increasingly being asked to focus on temporal aspects. Time series analysis and systems dynamics fit clearly in the temporal domain, while spatial dynamics crosses temporal and geographic domains. Other analytical tools, such as pattern detection packages and statistical tools, frequently operate using data from environmental space, but then link to geographic space using GIS. Some tools operate in taxonomic, environmental and geographic space.

Some of the key issues that ecologists are addressing include dynamics of population relationships (see Efford this volume), species composition, species distribution and abundance, conservation corridors, taxonomic prioritisation, sampling procedures and pattern identification.

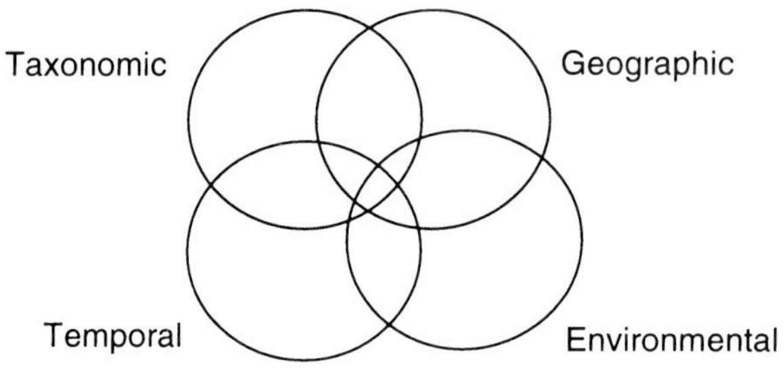

Fig. 3. Modelling at the intersection of multiple domains.

These issues imply questions concerned with temporal changes in population relationships, community composition and structure, response functions and their dynamics, short and long-term population trends and viability, spatial fragmentation and regional-scale processes, evaluation methods and habitat suitability for management strategies, and distribution (e.g. potential niche, abundance).

Tools for model interpolation are relevant to the task of estimating the distribution or abundance of a wildlife species. Simulation tools may be relevant to looking at population dynamics and learning tools could be used to explain patterns observed in species occurrence and abundance. (see Efford this volume) Tools for assessing similarity and diversity of objects could be used for prioritizing locations and/or species.

The potential use of five modelling techniques with ecological data are discussed below.

Some relevant tools

Model interpolation – extrapolation

While many observers record the presence of a species, few remember to record absences. Observational data only tells us where the species have been observed. When using these data in modelling exercises, it is important to ensure that the data is well distributed spatially (i.e. in the geographic domain) as well as providing coverage of all species (i.e. in the taxonomic domain).

Tools, such as HABITAT (Walker and Cocks 1991) provide a linkage from environmental space to geographic space. It can be used to identify the relationships between environmental attributes and the presence of a species. For instance, the relationship between temperature and rainfall and data base records where a species has been observed or not observed can be evaluated. In Fig. 4, locations were compared in terms of temperature and rainfall, to see how similar they were to locations where the species had been recorded.

Some of these locations were evaluated as being 'outside' the environmental polytope (domain) covering the presence sightings; whereas other locations where the species had not been recorded (no-record) were seen as quite similar, in environmental conditions, to recorded sightings. Such analysis can be useful in exploring bias in the observational data and in identifying parts of both geographic and environmental space that have not been sampled.

Model simulation

Simulation tools, including systems dynamics and cellular-automata, are increasingly being used by modellers (Efford this volume) as a framework encompassing spatial interactions, how these interactions affect a region, and how they might change over time.

A simulation model could involve creating an influence diagram, for instance, of rabbit–fox interactions (Fig. 5). While the model in Fig. 5 is rather complicated, it does show how such a system can be used to identify positive and negative feedbacks in a modelling framework. The user

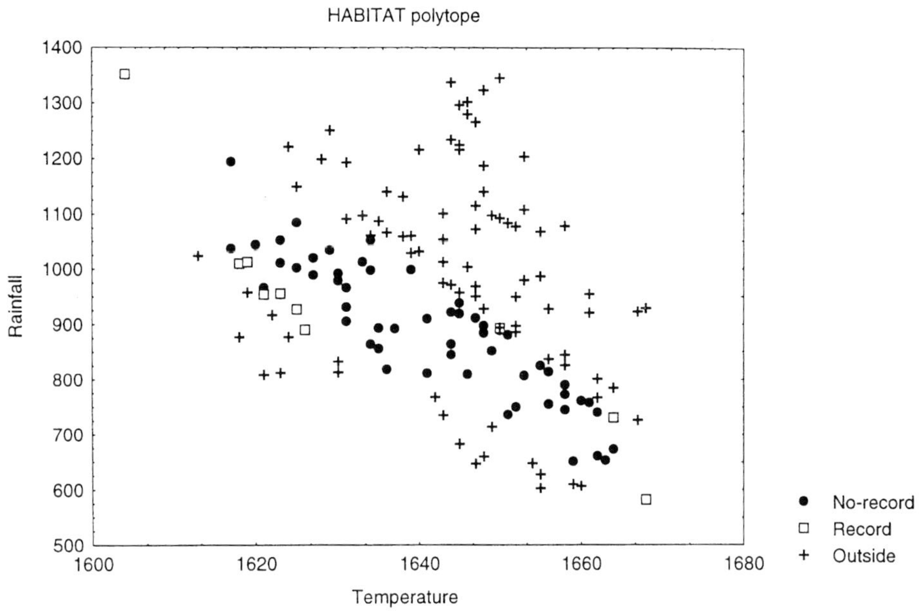

Fig. 4. An environmental envelope based on recorded sightings.

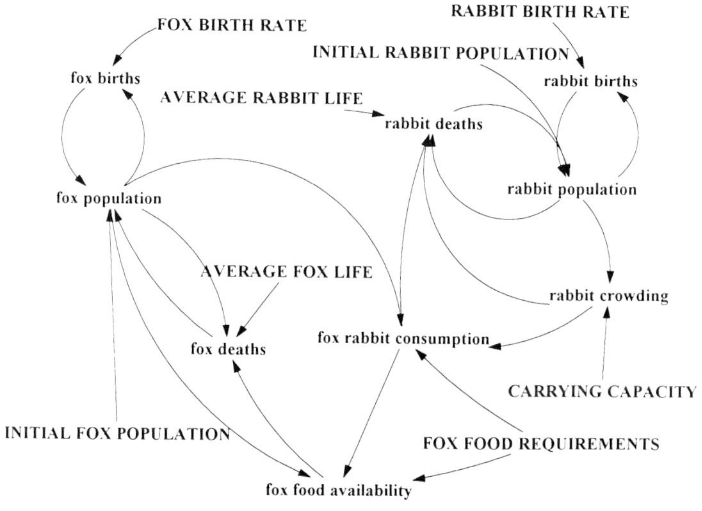

Fig. 5. Dynamics of species interactions (Ventana 1994).

can then simulate the model through time, intervene in the model when desired (for instance to impose a management strategy), identify relationships and feedbacks in the model and identify how relationships between components of the model and outcomes change though time. Components of the model can be assigned spatial characteristics and the models can be linked to a GIS.

The user seeks to identify how the parts of the model work together, and how some action in one part might benefits and/or create opportunities in another. These tools are best suited to tasks where the dynamics of the inter-relationships are complex, involving many feedbacks, and in which the goal is to understand the behaviour of the complex system through time.

Inductive learning

Current GIS lack a capability for developing new hypotheses about spatially varying relationships between features. The process of developing spatial relationships (e.g. between environmental attributes and the distribution of species; Walker 1990) can be treated as an inductive modelling task. Inductive learning is a process in which we learn by example.

One increasingly popular method of developing these relationships is through the computer induction of classification trees. More sophisticated learning systems include expert systems and neural nets. Classification trees can be used to help explain why some areas are seen as different from others. What, for instance, distinguishes the species-rich locations from those that are not? Are these patterns related to topography, climate, land-use, soil type, lithology?

Classification trees (see Breiman *et al.* 1984) produce a systematic prediction of class membership (i.e. presence) of an object (i.e. locations) by examining a learning sample consisting of a set of measurement vectors or attributes (i.e. climate) for objects of known classes (i.e. presence) and expressing the relationship as a set of decision rules. Figure 6 shows output from a classification tree package. In this example, when the land use had a value of 2 (see arrow), then 68% of 1375 observations had the highest species richness index.

The goal with classification trees is 'explanation with simplicity'. Because of the algorithms they use, classification trees tend to be best suited to data sets where the explanatory variables are not smooth or continuous; that is where a variable (e.g. rainfall) exhibits natural break points (e.g. vegetation type changes due to elevation, distance from coast, mountain ranges etc.).

Assessing similarity

Because it is costly to collect data, it makes sense to design efficient surveys for data collection and also to invest time to extract the maximum value from these data. This means giving some consideration to how the spatial data is to be analysed before data survey commences. Unfortunately,

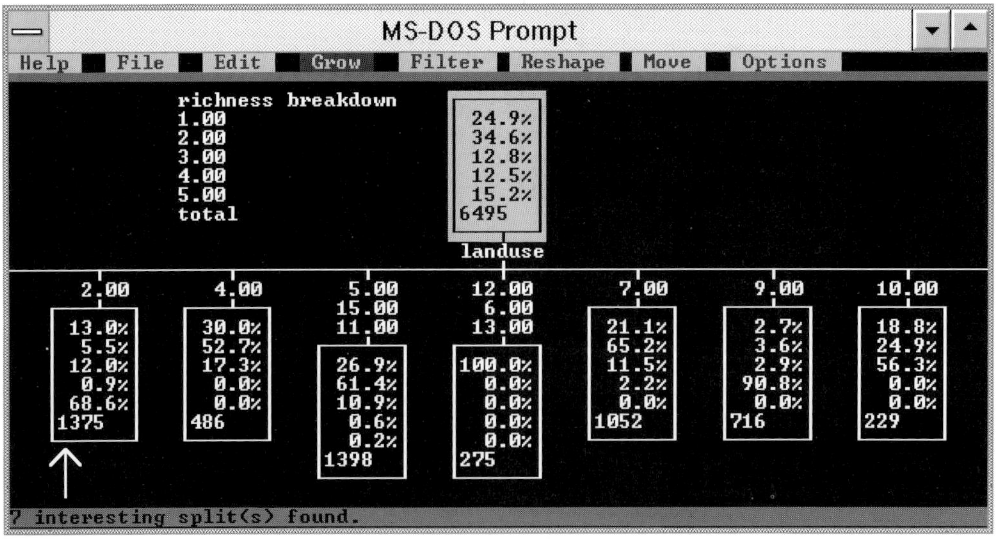

Fig. 6. Typical classification tree output showing relationship between species richness and land use class.

there is often no option but to use survey data with imprecise spatial referencing and which has been collected in an opportunistic (and often unknown) manner.

Species data, in particular, pose special problems. Frequently, species presence (and not absence) is recorded. Often, a map showing species-distribution data in relation to accessibility (roads, rivers) can be more illuminating about how the observer travelled (foot, car, boat) than about the species distribution. Given these problems, it is often easier to delineate species habitat using environmental factors than to systematically survey regions for occurrences of species.

Locations which resemble a species-rich location in terms of environmental attributes may also resemble the same location in terms of species present. The greater the similarity between these locations, the more likely we are to infer the characteristics of these locations from survey data.

Tools such as PATN (Belbin 1993) are designed to discover the structure or pattern in data and identify how and why some objects are similar to others. Figure 7 shows a typical use of similarity measures. PATN was used to identify geographic regions having a similar species composition. The input data consisted of the presence (or absence) of more than 860 individual species. The output is a set of geographically referenced regions sharing a similar composite of species.

Assessing diversity

Increasingly, we are observing debates about 'diversity', what it means and how we might measure it. Diversity, in its most general sense means 'the variability among living organisms .. including ... the ecological complexes of which they are part' (footnote 1 Convention on Biological Diversity – as published *Biology Internatl.* 25, 1992 p. 22). Because diversity is the variety of living organisms, it will by definition be difficult to summarise in any one measure.

Fig. 7. Regionalisation of species composition (provided by L. Belbin). Topographic map sheets having similiar composition of species. Data input consisted of the species present for each 1:100 000 topographic map sheet.

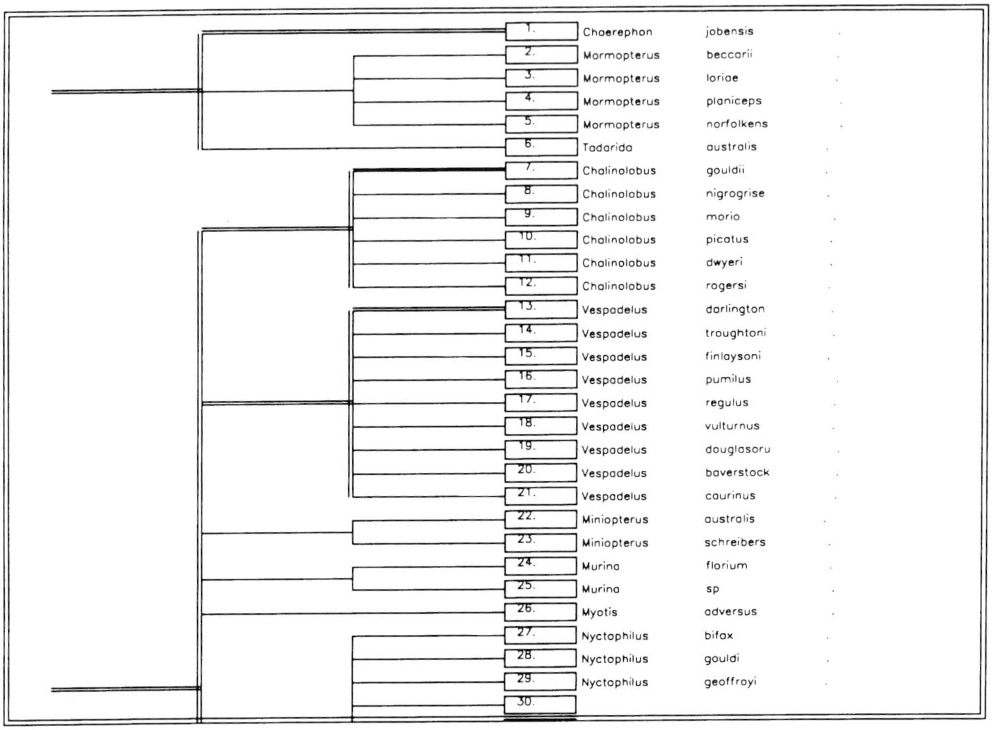

Fig. 8. Species associated with selected locations.

Recently, a number of software packages have been developed to address issues of diversity; packages that merge taxonomic, environmental and geographic domains (e.g. Faith and Walker 1994). A given set of sites will have greater taxonomic diversity if the sites in some sense span as much of the taxonomic space as possible. Similarly, a given set of sites will have greater environmental diversity to the extent that they span as much of the environmental space as possible. Exactly, how one measures the amount spanned is one of the key questions.

DIVERSITY-PD measures the span of taxonomic space using a phylogenetic diversity measure, such that a subset of species that spans a greater part of a tree connecting all taxa in the subset is more diverse than one that spans a lesser portion of the tree.

Figure 8 shows part of the taxonomic representation of a subset of species observed in a set of three geographic locations.

DIVERSITY-ED asserts that the environmental space is well represented by a subset of sites if the distance from any point in the space to its nearest member site is small. Environmental diversity will be greatest when the environmental redundancy of any subset of sites is small.

Conclusion: need for multi-disciplinary linkages

Much of the biological data encountered is inadequate in some respects for spatial or temporal modelling. Often the geographic referencing is incomplete. Inadequate or imprecise geographic referencing can be avoided. Similarly, ensuring that geographically referenced data are placed in the context of the sampling method is needed; for instance, in terms of method (e.g. aerial survey or ground survey, electronic sensor), and time (e.g. duration of sampling, day, night, summer, winter). The most obvious omission is that the observational data generally only records presence of species.

Despite the considerable reduction in the price of computer technology, the construction of GIS remains relatively expensive. The real costs are often not with the technology but with manpower training, data collection and interchange. This paper has briefly noted a number of tools routinely used to complement the modelling capabilities of GIS.

GIS development, however, is being criticised by administrators responsible for policy development as providing insufficient return on investment. At the state and national level, resource and environmental policy advisers seek quick responses to strategic questions. Systems that integrate economic, ecological and social considerations across space and through time are needed. Models must cross and intersect the temporal, spatial, environmental and taxonomic domains. The challenges in building geographic information systems and modelling systems that transcend several disciplines are immense.

References

Aspinall RJ, Lees BG (1994) Sampling and analysis of spatial environmental data. In: Waugh TC Healey RG (eds) *Proceedings of Spatial Data Handling 1994: Advances in GIS research*. University of Edinburgh pp 1086–1098

Belbin L (1993) Environmental representativeness: regional partitioning and reserve selection. *Biol Conserv* **66**:223–230

Belbin L (1987) The use of non-hierarchical clustering methods in the classification of large sets of data. *Aust Comp J* **19**:32–41

Berry JK. (1987) Fundamental operations in computer-assisted map analysis. *Int J GIS* **1**:119–136

Burrough PA (1986) *Principles of Geographical Information Systems for Land Resources Assessment.* Clarendon Press, Oxford

Breiman L, Friedman JH, Olshen RA, Stone CJ (1984) *Classification and Regression Trees*, Wadsworth Statistics / Probability Series, Belmont, Ca.: Wadsworth

Faith DP, Walker PA (1994) *DIVERSITY: Reference and users guide.* CSIRO Division of Wildlife and Ecology, Canberra

Goodchild MF (1989) A spatial analytical perspective on geographical information systems. *Int J GIS* **1**:327–334

Gould SG (1993) *Eight little piggies.* Penguin Books, UK

Morrison ML, Marcot BG, Mannan RW (1992) *Wildlife–Habitat relationships: concepts and applications.* University of Wisconsin Press

Tomlin CD, Berwick SH, Tomlin SM (1983) Cartographic analysis of deer habitat utilization. In: Teicholz E, Berry JK (eds) *Computer graphics and environmental planning.* Prentice-Hall, Englewood Cliffs, NJ, pp 141–150

Ventana (1994) *Vensim: Vensim Simulation Environment. Users Guide.*Ventana Systems Inc, Belmont MA USA

Walker PA (1990) Modelling wildlife distributions using a geographic information system: kangaroos in relation to climate. *J Biogeog* **17**:279–289

Walker PA (1992) Spatial exploration systems for natural resource/environmental problems. *Proceedings of AURISA 91.* Australasian Urban and Regional Information Systems Association pp 153–169

Walker PA, Cocks KD (1991) HABITAT: a procedure for modelling a disjoint environmental envelope for a plant or animal species. *Global Ecology and Biogeography Letters* **1**:108–118

Walker PA, Moore DM (1988) SIMPLE: an inductive modelling and mapping tool for spatially-oriented data. *Int J GIS* **2**:347–363

SECTION 4

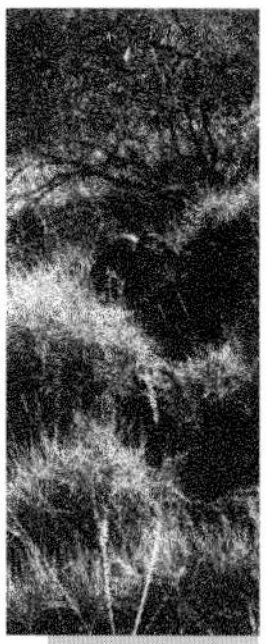

POPULATION EVOLUTION AND MOLECULAR ECOLOGY

INTRODUCTION

One of the major advances in the study of population ecology in recent years has been the development and application of molecular techniques to the investigation of a wide range of theoretical and practical issues including parentage and relatedness, current and historic population structure, and analysis of population size. Craig Moritz and Shane Lavery open this section by drawing attention to some of the temporal, spatial and practical sampling limitations of traditional approaches to population ecology and highlighting the ability of 'molecular ecology', the blend of the tools of molecular biology and the theory of population genetics, to effectively compliment traditional studies. The relative merits of various molecular techniques are described in terms of their ability to provide information on allele frequency, molecular differences between alleles and the degree of resolution of molecular differences. Moritz and Lavery provide a series of examples of the useful application of molecular ecology to diverse population ecological questions on taxa ranging from crabs to cetaceans and *Drosophila* to dunnocks.

One of the key advantages of a molecular approach to population ecology is the ability to investigate long-term population processes and provide a long-term perspective on current trends. This strength of molecular approaches clearly illustrates the need for both current (ecological) and historical (molecular) processes to be studied together so that both sources of data become far more informative. The remaining papers in this section provide examples of the synergy between the two approaches.

Andrew Cockburn's study of the social evolution of the Corvida, a large group of oscine passerines, provides a well argued example of the combination of behavioural observation, molecular analysis of relatedness and phylogenetic analysis to examine the origins of co-operative breeding. Social systems and in particular, cooperative breeding have traditionally been viewed as a labile response to availability of breeding resources, mates and breeding vacancies, however recent studies have suggested that they may also be influenced by phylogenetic history. In the Corvida, cooperative breeding may be an ancestral trait according to phylogenetic relationships clarified by molecular analysis. Cockburn concludes that the evolution of complex social systems in birds may depend on mutualism and phylogenetically constrained aspects of life history, rather than being simply a density-dependent response to habitat availability.

SECTION

4

A review of population evolution and molecular ecology would hardly be complete without a contribution on *Drosophila*. Bill Ballard and Nora Galway review the use of mitochondrial DNA for studying the evolutionary biology of *Drosophila* and demonstrate how these data can link the disciplines of systematics, population genetics and population ecology. Much of the paper focuses on the dangers of using inappropriate or inadequate sampling strategies and analysis techniques resulting in misleading phylogenetic inferences which could affect the study of systematics and conservation biology.

Allozyme analysis was used by Paul Sunnucks and Graham Stone in their study of the genetic structure of the invasion of a gallwasp in Western Europe. This study very neatly illustrates how genetic data can be very informative in the study of invading populations and in this case, clearly demonstrated that the invasion occurred in a generally linear directional 'stepping-stone' pattern. The genetic subdivision and small founder population size generated diverse new genetic combinations in the invaded range.

The final example of the application of molecular ecology to population studies is the investigation of host dependant stratification of grain aphid populations by Paul De Barro, Thomas Sherratt and Norman Maclean. This study demonstrates that variation in fitness of herbivorous insects feeding on various hosts may have a strong genetic component. DNA fingerprinting and RAPD-PCR techniques were used to characterise populations of grain aphids feeding on wheat and wild grasses and showed population stratification which was further supported by reciprocal host transfer studies.

The papers in this section have reviewed the potential of molecular techniques in the study of population ecology and given a number of specific examples. As Moritz and Lavery pointed out, the potential application to population ecology is enormous and in many cases, molecular ecology is the only way to gain a deeper understanding of population processes. A number of papers in the next part of this book on managing populations, give further examples of the practical use of molecular techniques in population ecology.

MOLECULAR ECOLOGY: CONTRIBUTIONS FROM MOLECULAR GENETICS TO POPULATION ECOLOGY

Craig Moritz and Shane Lavery

ABSTRACT

Using traditional ecological methods, it is usually difficult to identify the boundaries of functionally independent populations and to measure the amount of migration between these. On a finer scale, estimates of relatedness from observations are time-consuming and prone to error. 'Molecular Ecology' employs the tools of molecular biology to address these and other questions in population ecology. DNA fingerprinting provides precise identification of parentage, particularly where hypotheses derived from observations can be tested. It remains difficult, however, to estimate pedigrees or relationships other than first-order for unknown populations. The analysis of population structure from the geographic distribution of alleles has a long history; with DNA sequences we can now infer allelic relationships as well as distribution. The uses of this information are still being explored through development of theory and application to natural populations. Early indications are that inclusion of molecular differences between alleles may increase power to detect population subdivision, but may also reflect historical more than contemporary population processes. Measures of genetic diversity and allele phylogenies can be used to estimate long-term effective population size and, possibly, to investigate historical demography. The combination of traditional ecological and genetic methods and the newer molecular approaches may therefore allow comparisons between long-term vs. current trends in migration and population size, providing a new, historical perspective for ecologists.

Key words: Population structure, gene flow, historical ecology, population genetics

Frontiers of Population Ecology, R.B. Floyd, A.W. Sheppard and P.J. De Barro (eds), CSIRO Publishing, Melbourne 1996, pp. 433–450

Introduction

Traditional approaches to population ecology use intensive field work and monitoring of marked individuals to estimate population size, construct life tables and estimate rates of immigration and emigration. The strength of this approach is that, potentially, it can provide detailed information on current demography and on proximate influences on reproduction, mortality and movement. This can then be used to investigate mechanisms of population regulation, model population viability (e.g., Boyce 1992) or predict the outcomes of different management strategies (e.g., Lindenmayer *et al.* 1993; McCallum *et al.* 1995). However, there are also significant limitations:

1. Ecological studies are usually snapshots in time and, with some exceptions, usually do not encompass rare effects such as drought or fire that have significant long-lasting effects on community organisation, particularly in the Australian environment;
2. Ecological studies are often very limited in space, of necessity being focussed on a small part of the range of a species or community and potentially vulnerable to problems of scale. For large animals tracking by radio or satellite enables movements or mortality to be monitored over a greater area, but, typically, only a small number of individuals can be monitored;
3. As a corollary of (2), ecological studies are often concerned with arbitrarily delimited 'populations' for which it is very difficult to distinguish immigration and emigration from birth and death; and
4. Ecological studies based on marked individuals may be restricted to specific age, sex or size classes to which permanent tags or marks can be applied. For example, ecological studies of marine turtles, species that are long-lived and migratory, have tended to focus on breeding females because only these are accessible and readily tagged (e.g., Limpus *et al.* 1992).

The theme of this paper is that Molecular Ecology, a blend of the tools of molecular biology and the theory of population genetics (Hoelzel and Dover 1991; Burke 1994) is an effective complement to traditional field studies and can obviate some of the above limitations. There is a synergy between the two approaches; information from field studies is important in the design of genetic studies and interpretation of the results; whereas genetic studies can provide a picture of the broad-scale and long-term population processes (Avise 1994). This longer term perspective provided by molecular data may be particularly significant in providing insights into palaeo-ecology and ultimate influences on population and community structure.

Molecular Ecology is a young and rapidly evolving discipline. Applications range from determining parentage and relationships to investigations of historical biogeography and long-term population processes. The detection of cryptic species is another application relevant to ecologists (e.g., Baverstock *et al.* 1994; Hopper 1994). In the last decade, several new techniques have been introduced that have revolutionised the study of natural populations (Hillis *et al.* 1996) and led to the development of new population genetics theory (Hudson 1990). In the following we consider some of the strengths and limitations of Molecular Ecology; but first, the tools.

The tools of molecular ecology

Traditional population genetics has been based on analysis of codominant allele frequencies such as revealed by allozyme electrophoresis (Richardson *et al.* 1986; Murphy *et al.* 1996). While this remains a cost-effective tool for some problems (e.g., species boundaries, mating systems), the analysis of variation in DNA is superior for many applications and is enabling investigation of new questions (Table 1). Key technical innovations have been the development of multilocus minisatellite probes (Jeffreys *et al.* 1985; Burke 1989) and the use of the Polymerase Chain Reaction (PCR, Saiki *et al.* 1988) to assay variation in DNA via gene amplification. Because only a minute

Table 1. A general guide to the tools of Molecular Ecology. Symbols: na , not applicable; * to ***, increasing effectiveness.

Method	Types of information			Applications					
	Genotype frequency	Allele frequency	Allele genealogy	Parentage	Relatedness	Hybridisation	Population structure Current	Population structure Historic	Population size (historical)
Allozyme electrophoresis	Yes	Yes	No	*	**	***	**	*	*
Multilocus[1] minisatellites	No	No	No	***	**	na	*	*	**
Single locus[2] microsatellites	Yes	Yes	No	***	***	*	***	*	**
RAPDs[2,3]	No	Yes	No	*	*	**	**	?	?
RFLPs[2]	Yes	Yes	Yes	*	*	***	***	**	**
Sequences[2]	No	Yes?[4]	Yes	*	na	*[4]	**[4]	***	***

[1] Genotype and allele frequencies at individual loci cannot be calculated, but average heterozygosity can be estimated from coefficients of band sharing given several assumptions (Lynch 1988).

[2] These methods can make use of PCR and thus be applied to a broader range of tissues/organisms

[3] RAPDs are typically dominant and thus need greater sample sizes (Lynch and Milligan 1994); they also need to be used and interpreted with considerable caution (Hadrys *et al.* 1992; Dowling *et al.* 1995).

[4] RFLP analysis or some other indirect methods are typically more efficient, but can be combined by sequencing with great effect (e.g., Lessa and Appelbaum 1993; Slade *et al.* 1993).

amount of tissue is needed, PCR can be used to analyse variation in virtually any size or form of organism and in extant or extinct populations (e.g., Thomas *et al.* 1990; Wayne and Jenks 1991; Taylor *et al.* 1994). Different types of variation appropriate to different problems can be assayed by PCR. Individual hypervariable loci (microsatellites; Weber and May 1989) can be used to determine parentage, estimate relatedness, or examine fine-scale population structure (Queller *et al.* 1993; Bruford and Wayne 1993). Sequence variation can be assessed directly (e.g., Baker *et al.* 1993), or indirectly (through methods such as RFLPs, RAPDs or gradient gels: Hadrys *et al.* 1992; Lessa and Appelbaum 1993; Dowling *et al.* 1996) for a variety of purposes, including analysis of long-term population processes (Table 1).

An important distinction between the methods is whether or not they provide information on the genealogy of alleles as well as their distribution, as the former can provide unique insights (Avise 1989; Hudson 1990, and see below). Allozymes, RAPDs and microsatellites provide information on allele frequency only, whereas DNA sequencing and RFLP analysis also reveal the molecular differences among alleles. Another consideration is the level of resolution; minisatellites and microsatellite loci typically have higher heterozygosity and more alleles than allozyme loci and DNA sequencing provides finer resolution of the molecular differences than RFLPs. Whether the high resolution is an advantage or hindrance depends on the nature of the question and the results of pilot studies (Baverstock and Moritz 1996).

Applications of molecular ecology

Parentage and relatedness

Animal ecologists have usually inferred parentage from detailed observation of matings or, more usually, territorial or mating behaviours. Interpretations based on such observations, and the increasingly elaborate sexual selection theory developed to accommodate them, assume that parentage, and thus fitness, has been accurately determined. It is now clear from molecular comparisons, first with allozymes (e.g., Brooker *et al.* 1990) and subsequently with DNA fingerprints (e.g., Burke *et al.* 1989), that parentage has been incorrectly inferred in many studies of species with complex mating systems (reviewed by Burke 1989; Avise 1994). Indeed, in a study of the Pukeko, Lambert *et al.* (1994) found that success in defending territories did not predict the number of copulations which, in turn, did not predict parentage.

Despite the apparent limitations of the field data, parentage testing is most powerful where hypotheses derived from long-term population studies and behavioural observations are tested. Typically, the alleles present in a series of offspring are compared to those in a small number of potential parents suggested by ecological studies (e.g., Burke *et al.* 1989). The presence of alleles in the offspring that are absent from the putative parents excludes one or both of the latter (barring mutation) and may indicate extra-pair fertilisation or brood parasitism. If the allelic profiles match, then the null hypothesis that these are the parents is accepted, but the likelihood that such a match occurs by chance must also be considered. The power of the test increases with number of alleles and unlinked loci examined (Burke *et al.* 1991; Chakraborty and Kidd 1991). Thus, hypervariable minisatellite and microsatellite loci are extremely powerful in this regard (e.g., Gibbs *et al.* 1990; Amos *et al.* 1993).

The ideal is to be able to construct a pedigree for a population or estimate relatedness among randomly sampled individuals. In practice, this has proved extremely difficult because of the large number of combinations possible (e.g., Morin *et al.* 1994a) and theoretical and technical constraints on measures of band-sharing; (Lynch 1988; Chakraborty and Kidd 1991; Brookfield and Parkin 1993). Using multi- and single-locus DNA fingerprints, it has sometimes proved possible to distinguish between relatives and unrelated individuals where the hypervariable loci have been

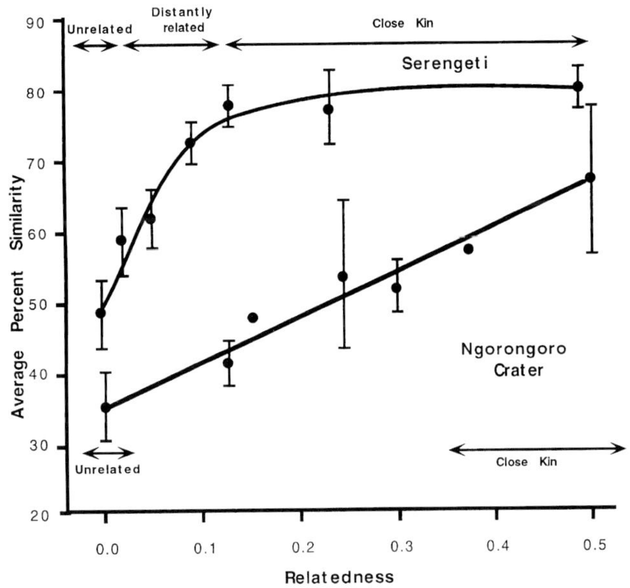

Fig. 1. Assessment of relatedness of lions from observed levels of band sharing in multilocus DNA fingerprints in two populations, the large Serengeti population and the recently bottlenecked Ngorongoro Crater population (redrawn from Gilbert *et al.* 1991). Having calibrated each system using lions from pedigreed social groups, observed levels of band sharing can be used to classify unknown individuals as closely related or unrelated.

calibrated for the species in question. For example, Gilbert *et al.* (1991) were able to distinguish three classes of relatedness among Serengeti lions (Fig. 1); unrelated, distantly related, and closely related (1st cousins, siblings). Overall genetic similarity was higher in the Ngorongoro Crater population established from a few related individuals, but even here it was possible to distinguish first-order relatives from unrelated individuals (Fig. 1). This level of discrimination depends on a wide range in levels of band-sharing and this may not always be the case, especially where multilocus probes derived from other species are used or populations are small and isolated (e.g., Degnan 1993).

The development of single-locus microsatellite analysis is extremely promising in this regard. Large numbers of highly variable and precisely scorable loci can be isolated and these can often be applied across related species (e.g. cetaceans, Schlötterer *et al.* 1991; birds, Hanotte *et al.* 1994; marine turtles, FitzSimmons *et al.* 1995; macropod marsupials, C Moritz unpublished). Microsatellite loci typically have large numbers of alleles and the average relatedness of individuals within social groups can be estimated from the proportion of alleles shared between individuals or related measures (Queller and Goodnight 1989; Amos *et al.* 1993). For example, Morin *et al.* (1994b) examined the proportion of alleles shared among 8 microsatellite loci in wild chimpanzees and found that males within the community were more closely related (≈half-sibs) than were females. Based on analysis of 6 microsatellite loci in pilot whales, Amos *et al.* (1993) concluded that pods, social groups of 50–200 individuals, typically consist of matrilines with females mating males from outside of the pod.

Contemporary and historical population structure

The assessment of population structure via genetic methods can be qualitative, inferring the type of population structure, or quantitative, e.g., testing for subdivision or estimating the amount of gene flow. If the aim is to determine the geographic scale or spatial boundaries of random mating populations, measures of genetic difference or correlation can be plotted against geographic separation of samples (Kimura and Weiss 1964; Richardson *et al.* 1986; Slatkin 1993). Three outcomes are possible (Fig. 2): (i) no significant divergence, irrespective of distance, suggesting random mating or panmixia; (ii) a step function indicating an island model with limited migration, and (iii) a positive linear correlation between genetic divergence and geographic separation indicative

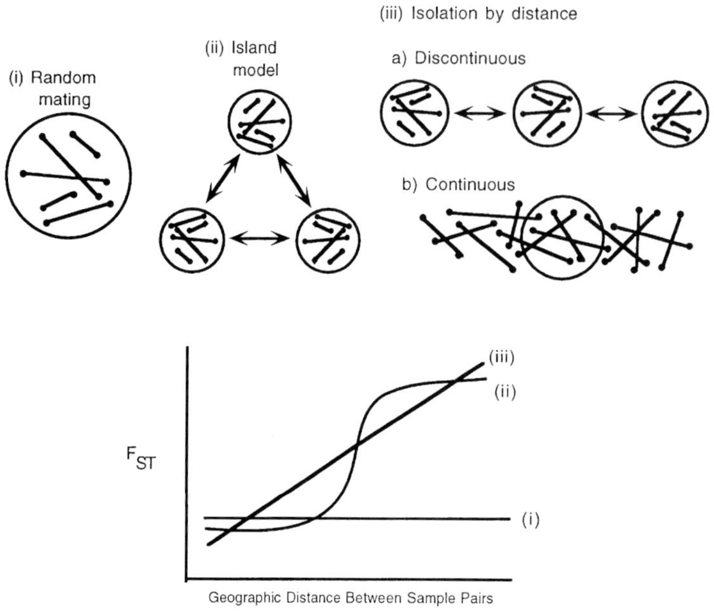

Fig. 2. Schematic of different types of population structure and expected relationship between genetic divergence (in this case, standardised variance in allele frequency – F_{st}) and geographic distance between samples. The bars indicate the net distance moved between birth and reproduction and the circles indicate the area of a random mating population. For the continuous isolation-by-distance model (iiib) this corresponds to the neighbourhood area.

of isolation by distance among discontinuous or continuous populations. The latter can arise where the distance moved by individuals between birth and reproduction is small relative to the geographic area occupied and has major implications for studies of population ecology.

An ecologist studying a non-migratory, discontinuously distributed species usually has a good intuition, perhaps supported by analysis of lifetime movements, about whether an occupied patch is likely to comprise a functionally independent population, in which case the question might be how much exchange occurs between the patches making up the total population (see below). However, continuously distributed (or migratory) species pose a problem because the spatial boundary of populations cannot be estimated *a priori*, making it difficult to establish the appropriate geographic scale for field studies or management. Genetic analyses can help by estimating the area of a 'genetic neighbourhood' (Fig. 2), i.e., the geographic area over which the parents of a centrally located individual can be drawn at random (Wright 1969). The larger population can then be considered as a set of overlapping neighbourhoods and studied on this scale.

The type of population structure perceived within a species can vary in time and space. In the coconut crab, a terrestrial species with a short (2–4 week) pelagic larval phase, analyses of isozymes (Lavery *et al.* 1995) and mtDNA (Lavery *et al.* 1996a) revealed a step function in genetic divergence between Pacific Ocean localities and Christmas Island in the Indian Ocean (i.e., island model), isolation by distance among distant localities within the Pacific Ocean, and random mating among localities within archipelagos, e.g., PNG to Vanuatu (Fig. 3). Based on the genetic divergence observed between replicate samples from a single locality (Richardson *et al.* 1986), Lavery *et al.* (1995) estimated the neighbourhood size within the Pacific to be of the order of 2000 km, a value

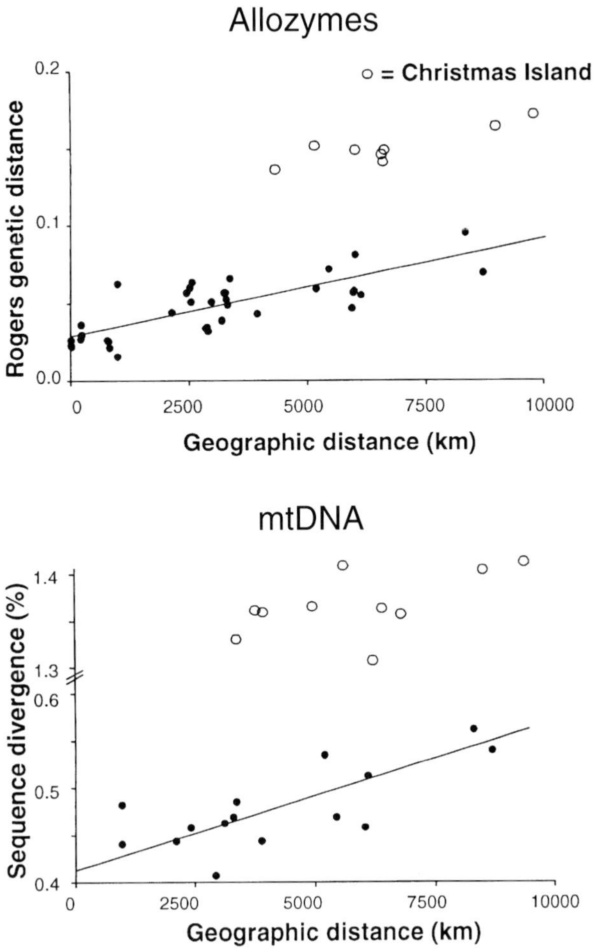

Fig. 3. Observed relationship between genetic differentiation and geographic distance for populations of coconut crabs sampled from Pacific islands (●) and between Pacific islands and Christmas Island in the Indian Ocean (**O**). The former are consistent with isolation-by-distance, whereas the latter are more consistent with an island model. See Lavery *et al.* (1995; 1996a) for further details.

consistent with potential dispersal estimated from current flows and the length of the larval period. The extreme genetic divergence of the Christmas Island population, which is much greater than predicted by the isolation-by-distance pattern within the Pacific (Fig. 3) is presumably the result of historical isolation.

More quantitative applications include testing for population subdivision and estimating rates of gene flow; in both cases difficulties arise with distinguishing between the effects of current and historical gene flow and in estimating rates of immigration that are meaningful in large populations (e.g., 1%, Slatkin 1987). For neutral genes, allele frequencies within populations are expected to diverge by genetic drift at a rate determined by both their effective population size (N_e) and rate of migration (m). Analytical theory and simulations suggest that for an island model (Fig. 2) at equilibrium, the same array of alleles will be maintained across populations where the product $N_e m > 1$ (Wright 1969); the critical value of $N_e m$ is somewhat higher where migration (i.e., gene flow) occurs predominantly among neighbouring populations in stepping stone models (Crow and Aoki 1984). However, there may be significant heterogeneity of allele frequency even with high migration rates. For example, Allendorf and Phelps (1981) observed significant heterogeneity in allele frequency among 20 populations in 60% of simulations with sample sizes of 50, populations of N = 100 exchanging 50 migrants per generation (i.e., Nm = 50)! Comparing just two populations, significant heterogeneity of allele frequencies was observed in < 20% of simulations for $N_e m > 10$.

With the above limitations in mind, genetic methods can be used to test for population subdivision. Traditionally, such comparisons have been based on divergence in allele frequencies tested using heterogeneity tests or estimates of F_{st}, the standardised variance in allele frequency (Wright 1951), depending in part on whether the aim is to make inferences about the specific populations or the species as a whole (Weir 1995). More recently, measures analogous to F_{st} have been developed which are based on DNA sequences and include information on the sequence divergence among alleles as well differences in frequency (e.g., Takahata and Palumbi 1985; Lynch and Crease 1990; Excoffier *et al.* 1992). Another approach using information from sequences was introduced by Maddison and Slatkin (1991); this tests whether the geographic distribution of alleles in relation to their phylogeny (i.e., the phylogeography) departs from randomness.

The relative power of these different approaches for detecting population subdivision remains to be determined. Hudson *et al.* (1992a) found that tests for heterogeneity of allele frequency were more powerful than those incorporating information on sequence divergence where there was no recombination, mutation rate (i.e., sequence diversity) was low, and sample sizes were large. Excoffier *et al.* (1992) observed a small increase in inferred subdivision of human populations assayed for mtDNA variation when sequence divergences were included in the estimates of F_{st}, whereas Lavery *et al.* (1996a) found the opposite for coconut crab mtDNA. This is an area ripe for further theoretical and empirical study.

The different perspectives on population subdivision provided by allele phylogeography and frequency are illustrated by patterns of mtDNA variation in the red kangaroo (S Clegg unpublished). Phylogenetic analysis of sequences from 29 individuals sampled from across the range did not reveal any geographic structure (Fig. 4); the Maddison and Slatkin (1991) test indicated panmixia across states and the F_{st} (ϕ_{ct}, as defined by Excoffier *et al.* 1992) value was close to zero. However, analysis of frequencies of five alleles (i.e., mtDNA haplotypes) detectable using restriction enzymes (and representing different clades, Fig. 4) for larger samples revealed highly significant differences, even between some localities separated by less than 80 km. The two results are not strictly comparable because of the differences in sample size. Nonetheless, the conclusions drawn are obviously very different. One possible reason for the difference is that the two approaches are reflecting gene flow on different time scales; the lack of phylogeographic structure (Fig. 4) reflecting long-term gene flow across the range, in contrast to local heterogeneity in allele frequencies arising through transient population subdivision. For an ecologist interested in kangaroo population dynamics, both perspectives are relevant.

Measures of genetic divergence among populations, e.g., F_{st}, are often used to estimate the average number of migrants ($N_e m$) among populations under the assumptions that (i) mutation rate is low, and (ii) the populations are at equilibrium between migration and drift (Slatkin 1987). This is possible because of the expectation that, for a large number of populations:

$$F_{st} \approx 1/(4N_e m + 1) \qquad (1)$$

(Wright 1951) and because F_{st} is expected to approach equilibrium relatively rapidly (e.g., Fig. 5, see also Varvio *et al.* 1986), although this may still be a long time in an ecological context. Several studies have pointed out the dangers in assuming that observed F_{st} reflect current rate of migration (e.g., Varvio *et al.* 1986; Boileau *et al.* 1992: Sunnucks and Stone this volume). For example, Larson *et al.* (1984) found much lower F_{st} values among populations of salamanders in glaciated than non-glaciated areas of North America and suggested that the former are still influenced by past gene flow accompanying the post-glacial range expansions. Slatkin (1993) suggests that it might be possible to

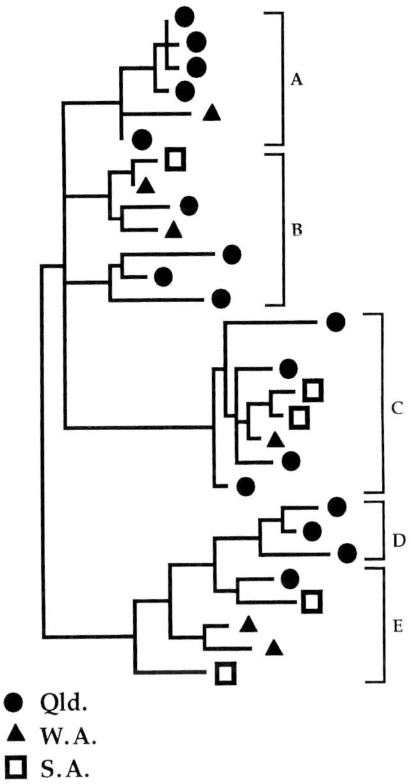

Fig. 4. Phylogeny of mtDNAs from Red Kangaroos sampled from across their geographic range (i.e., through Queensland, Qld; South Australia, S.A.; and Western Australia, W.A.). The phylogeny is based on parsimony analysis of control region sequences from 29 individuals (S Clegg unpublished). The groups A–E indicate different phylogenetic groups of alleles diagnosable using restriction enzymes and screened for variation in allele frequencies (see text).

distinguish species having equilibrium patterns of isolation-by-distance from those with recent range expansion by plotting $log(N_e m)$ against *log*(km).

The amount of migration can also be estimated from allele phylogeography by mapping locations onto a phylogeny of sequences obtained from all individuals, estimating the minimum number of migration events (s) that could give rise to this pattern, and from this estimating $N_e m$ (Slatkin and Maddison 1989). For example, for the kangaroo data (Fig. 4), $s = 9$ and $N_e m > 40$. With respect to DNA sequence data, Hudson *et al.* (1992b) suggest that sequence analogues of F_{st} provide more accurate estimates for low values of $N_e m$, whereas the phylogenetic method performs better at high rates of migration. The rate of approach of these sequence-based methods to equilibrium remains to be determined and, intuitively, they should be regarded as estimates of migration averaged over evolutionary time (i.e., multiples of N_e generations; see also Barton and Wilson 1995).

The $N_e m$ parameter can be interpreted as the number of immigrants that successfully reproduce, independent of population size and can be estimated in a variety of ways (Slatkin and Barton 1989). However, for an ecologist it may be the proportion of individuals that are immigrants (m) that is important, rather than the absolute number and the former can only be estimated from allele frequency data if effective population size is known. It should also be noted that a small rate of migration between large populations (e.g., $N_e = 1000$, $m = 0.01$, $N_e m = 10$) and a large rate of migration among small populations (e.g., $N_e = 50$, $m = 0.2$, $N_e m = 10$) have the same genetic signature in an island model.

The genealogical information in DNA sequences is especially powerful for detecting long-term population subdivision and, thus, providing information on historical biogeography (i.e.,

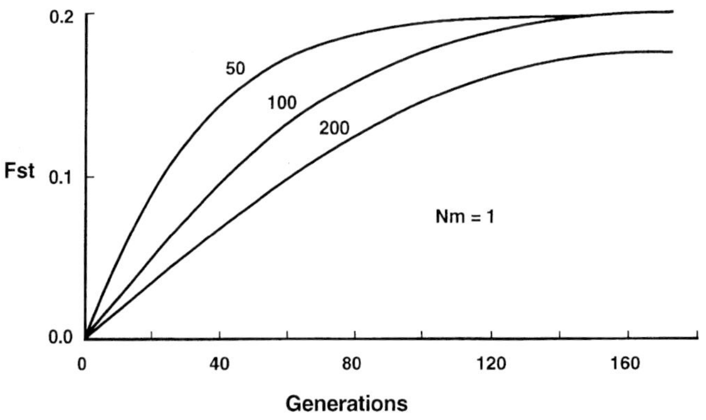

Fig. 5. Rate of approach of F_{st} to equilibrium for an island model with subpopulation sizes of N_e = 50, N_e = 100 and N_e = 200 and a migration rate of *Nm* = 1. Note that smaller subpopulations (exchanging correspondingly higher proportions of migrants '*m*') approach equilibrium faster, but it still takes approximately N_e generations for equilibrium to be approached. Redrawn from Allendorf and Phelps (1981).

'phylogeography', Avise *et al.* 1987). Through simulations, Neigel and Avise (1986) found that it usually takes approximately $4N_e$ generations from the time that two populations are isolated for the alleles within each of the populations to become reciprocally monophyletic, i.e., to establish strong phylogeographic structure. This sort of structure was observed for mtDNA from the ghost bat *Macroderma gigas*, where each regional population was found to have not only fixed allelic differences, but monophyletic alleles (Worthington-Wilmer *et al.* 1994). This suggests that each population has been isolated, at least as far as females are concerned, for a long period (probably > $4N_e$ generations) and that this isolation preceded the range contraction observed over the past 80 or so years.

Where the long-term disjunctions among populations arise because of extrinsic barriers to gene flow, it is expected that other species with similar ecologies should show congruent patterns of phylogeography (Bermingham and Avise 1986; Avise 1992). A dramatic example of this was provided by analyses of mtDNA phylogeography in vertebrates living in the montane rainforests of northeast Queensland (Joseph *et al.* 1995). Here, four rainforest specialist species (three birds and one skink) showed major phylogenetic disjunction between populations north and south of an area currently connected by rainforest but predicted to have been unsuitable for rainforest during the glacial maximum 18000 years ago (Nix 1991), although the extent of genetic differences suggested much earlier vicariance events for most species. The levels of sequence divergence between the southern and northern populations were broadly related to the presumed susceptibility to rainforest contraction based on current ecology. Species less restricted to rainforest did not show strong genetic disjunctions and, oddly, neither did a high altitude rainforest specialist which may have been through a population bottleneck. These data provide strong support for the role of climate-induced rainforest contraction in promoting diversity in rainforests. They also support Avise's (1992) suggestion that comparative phylogeography can give insights into long-term fluctuations in community membership and organisation.

Analysis of population size

A fertile but somewhat controversial application of molecular data is to estimate population size. The parameter under consideration in genetic studies, effective population size (N_e), can be defined in various ways but is usually considered as the size of an ideal population that experiences genetic change at the same rate as the population under consideration. N_e is usually smaller than census population size (*N*) due to factors such as unequal sex ratio and differential breeding success (see Lande and Barrowclough 1987). N_e can be estimated from demographic data, but it is notoriously difficult to do so in natural populations (but see Grant and Grant 1992). Ecological estimates of N_e are often much larger than those derived from genetic data (Frankham 1995; Easteal and Floyd 1986), perhaps because the former typically apply to short time intervals and because it is extremely difficult to estimate variance in reproductive success. In the context of conservation, Nunney and Elam (1994) suggested that ecological estimates of N_e are more relevant because the effects of specific management actions or environmental variables can be gauged. By contrast, Frankham (1995) argues that the lower genetic estimates are preferable because they take all relevant factors into account and more accurately reflect the rate of inbreeding or loss of genetic diversity.

Despite the inherent difficulties, estimates of N_e can provide useful insights. In species where individuals cannot be counted readily (e.g., marine species) or those that are highly numerous, the physical census of N can be a difficult or impossible task, and an estimate of N_e may be a useful surrogate. Where N is already known or estimated, the discrepancy between it and N_e provides valuable information about the level of inbreeding in a population, or may indicate whether a bottleneck (severe reduction in population size) has reduced genetic variation in a population. Both matters may have considerable conservation implications. In species with non-overlapping generations, the estimate of N_e actually provides the effective number of breeders (N_b), which is a very important (and often inestimable) parameter in applications such as fisheries modelling.

The various genetic estimators of N_e fall into two broad categories: those that estimate a long-term effective population number, and those that estimate the current size (Waples 1991). The long-term estimates apply over evolutionary time, i.e., over many ($>N_e$) generations and are based on the levels of genetic variation in a population. Estimates can be made from allozyme (or microsatellite) data using heterozygosity (Kimura and Crow 1964) or the number of alleles present (Watterson 1975), or the equivalents for DNA data; nucleotide diversity (Nei and Tajima 1981; Avise *et al.* 1988) and the number of segregating (polymorphic) sites (Watterson 1975). More recently, estimators have been developed that are based on the phylogenetic relationships of the alleles in a population, specifically, the coalescence times of those alleles. These have been developed using either maximum-likelihood methods (Felsenstein 1992) or general linear models (Fu 1994a, b). These new methods are better in that they make fuller use of the information and, accordingly, have lower variance.

There are a number of assumptions and limitations in the use of these estimators of long-term N_e. Assumptions include (i) that the population is at equilibrium, (ii) that there is no population subdivision (see Marjoram and Donnelly 1994), and (iii) that selection is not acting on the genetic markers. Perhaps the greatest limitation is that these methods require an accurate estimate of mutation rate (μ) as it is the product, $N_e\mu$, that is estimated from the data. In all but the most intensively studied species this can introduce substantial error. Waples (1991) suggested that estimates of long-term N_e based on heterozygosity are most accurate for species with large population size.

Estimators of current N_e provide a value that may be an average over the last few generations. Estimates can be made using either the temporal changes in allele frequencies due to random genetic drift (Easteal 1985; Waples 1989), or the level of gametic disequilibrium (Waples and Smouse 1990; Bartley *et al.* 1992). Assumptions include (i) that the samples are random, and (ii) that there is no immigration or selection. These estimates are independent of mutation rate and can be made precise by intensive sampling of individuals and loci. In contrast to the long-term N_e estimators, the current N_e estimators are best suited to small populations.

The main application of the theory has been to compare population sizes in one way or another. One use is to make qualitative inferences about past changes in size by comparing the estimated long-term N_e with current population size (e.g., Dallas *et al.* 1995). Another is to infer past changes in population size by comparing two measures of genetic variation that will be affected differently by changes in N_e. For example, the levels of nuclear and mitochondrial DNA variation may respond quite differently to a bottleneck (e.g., Moritz 1991; Takahata 1993). Also, the number of alleles (or segregating sites in DNA) responds very differently from heterozygosity (or nucleotide diversity) to changes in population size (Maruyama and Fuerst 1985; Tajima 1989). A third use is to compare estimates of N_e from two closely related species or populations and relate this back to difference in population demography or history.

In practice, a number of studies have interpreted comparatively low genetic diversity as evidence for past bottlenecks, using a variety of genetic systems. The comparatively low levels of mtDNA variation in humans has been used to argue that our species passed through a bottleneck some 100 000 to 150 000 years ago (Di Rienzo and Wilson 1991), although exactly when and where has proved contentious. The analysis of mtDNA variation in a variety of other species including elephant seals (Hoelzel *et al.* 1993), eels, catfish and blackbirds (Avise *et al.* 1988) has revealed levels of diversity much lower than that expected from populations of their current size, suggesting long-term effective sizes orders of magnitude lower than respective current population sizes. Several studies of nuclear gene systems have revealed reduced variation consistent with known or suspected population reductions (e.g., minisatellites, Gilbert *et al.* 1990; Degnan 1993; major histocompatibility genes, Yuhki and O'Brien 1990). In some instances, species may have such low levels of detectable genetic variation that only the most sensitive methods can be used to make inferences about N_e. One such case is the northern hairy-nosed (NHN) wombat, where microsatellite analysis was necessary to detect sufficient variation (Taylor *et al.* 1994). In an interesting twist, the power of PCR was used to its full by using museum specimens to compare the last remaining NHN wombat population with an extinct population of the species. The results showed that both these populations appeared to have considerably lower effective sizes than that of a closely-related southern hairy-nosed wombat population, and thus had probably experienced prolonged bottlenecks.

Temporal changes in N_e can also be studied using information derived from allele phylogenies. Unlike a stable population, an exponentially growing population leaves behind a star-like phylogeny of alleles and a Poisson frequency distribution of pairwise genetic differences (Slatkin and Hudson 1991). This occurs largely because there is little stochastic elimination of lineages in a rapidly-expanding population. The distribution of pairwise differences has since been explored in more depth under various models of population growth and decline (Rogers and Harpending 1992). These analyses suggest that it may be possible to estimate the magnitude of change in population size, the rate of growth, and when population growth commenced. Also, it has been suggested that a range of different population demographic changes can be detected by examining graphs of the number of allelic lineages through time (Nee *et al.* 1995). However, the accuracy and reliability of these methods remains to be determined and a number of substantial limitations have been identified (Marjoram and Donnelly 1994; and see below).

One limitation in making inferences about changes in N_e from allele phylogenies is that there is a considerable time lag before demographic changes become apparent in the genetic signature of a population. A stark example of this concerns the coconut crab (Fig. 6). Here, the phylogenetic data has the signature of an exponentially increasing population, whereas, in fact, the population has crashed dramatically over at least the last few hundred years (Lavery *et al.* 1996b). The genetic pattern appears to be a distant relic of a rapid post-glacial population expansion in the Pacific.

A second limitation that we are just coming to appreciate concerns the effects of episodic selection. Several recent studies have suggested that natural selection is affecting patterns of diversity of mtDNA (e.g., Ballard and Kreitman 1994; Nachman *et al.* 1994; Rand *et al.* 1994) as well as nuclear genes (Gillespie 1991). Because mtDNA does not recombine, any advantageous mutation will cause the entire genome to sweep through the population, reducing allelic and nucleotide diversity (a 'selective sweep'; Maruyama and Birky 1991). This illustrates that the presence of selection, either positive or balancing, will bias our estimates of the timing of past events and our interpretations of coalescence

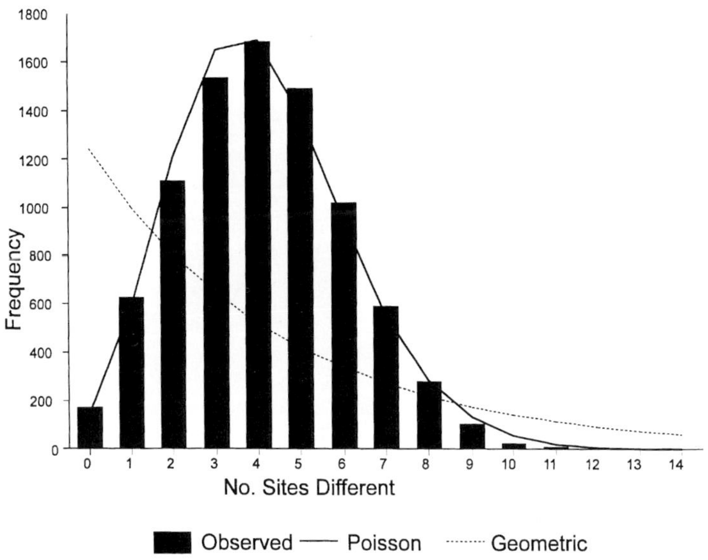

Fig. 6. Distribution of pairwise restriction site differences between mtDNAs from Pacific island coconut crabs (Lavery *et al.* 1996b). The observed distribution is a close fit to a Poisson distribution with the same mean suggesting a history of exponential population growth. This is in stark contrast to the recent population declines in the species.

times (Fig. 7; Rand *et al.* 1994). These points highlight two factors: (i) the need, wherever possible, to address the issue of selection rather than ignoring it, and (ii) the need to avoid reliance on just one genetic marker. Regrettably, for all its advantages, mtDNA is a single locus! To obtain reliable estimates of $N_e\mu$, it will therefore be necessary to obtain robust allele phylogenies from multiple nuclear genes; data of this sort are now beginning to accumulate for a small number of species.

Historical ecology

In many respects (other than parentage testing), molecular data are more powerful for investigating long-term population processes than making inferences about contemporary processes. This applies in particular to methods that make use of allele genealogy or sequence divergence statistics. This has been highlighted in this review by examples of studies examining changes in population size, population structure, and community biogeography. In some ways, this may be disappointing to those ecologists who perceived molecular analysis as promising a panacea for many intractable questions about current population processes. However, the long-term bias of molecular studies highlights two features. The first is the absolute interdependency of molecular and ecological approaches. In many ecological questions, data from one source becomes far more meaningful in conjunction with results from the other (e.g., De Barro *et al.* this volume). The second feature is the great opportunity that exists to contrast the historical (molecular) and current (ecological) processes governing the way individuals and species interact within communities (see also Moritz 1995). As ecological studies tend to provide information on only a snapshot of time (the present), it can be difficult to determine if the observed processes are characteristic of the species or are aberrant (and perhaps due to human influence). That is, molecular studies may provide a long-term perspective, in the light of which current trends can be interpreted. Differences between the two perspectives may indicate the role of disturbance or fluctuations in population history and community organisation. In some sense, molecular studies may also be predictive in nature. If currently healthy populations

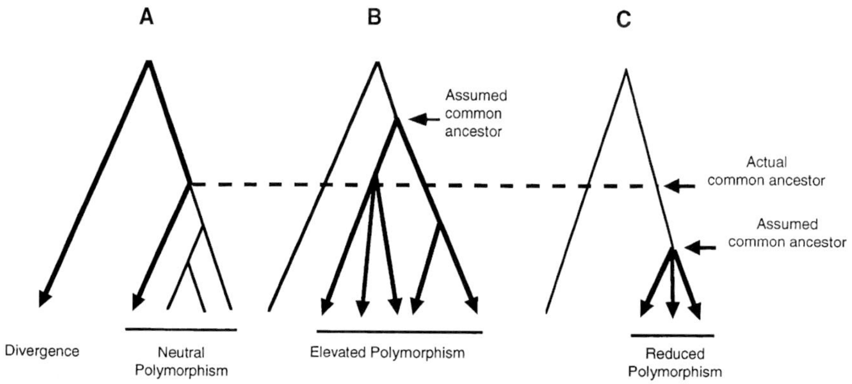

Fig. 7. Expected effects of selection on estimated time since common ancestry of sequences and hence estimates of $N_e\mu$. (A) represents a phylogeny of neutral alleles, (B) the genealogy under some form of diversifying selection, and (C) the effects of a recent selective sweep. Redrawn from Rand *et al.* (1994).

are shown genetically to have survived periods of severe bottleneck or inbreeding, this suggests that they may be tolerant to such events in the future. Alternatively, populations or species that are shown by molecular studies to be relicts of historically larger entities may be especially vulnerable to future environmental changes. The field of Molecular Ecology therefore broadens the scope of population ecology and may contribute to the merging of ecology and palaeo-ecology (Davis 1994).

References

Allendorf FW , Phelps SR (1981) Use of allele frequencies to describe population structure. *Can J Fish Aquat Sci* **38**:1507–1514

Amos B, Schlötterer C, Tautz D (1993) Social structure of Pilot Whales revealed by analytical DNA profiling. *Science* **260**:670–672

Avise JC (1989) Gene trees and organismal histories: a phylogenetic approach to population biology. *Evolution* **43**:1192–1208

Avise JC (1992) Molecular population structure and the biogeographic history of a regional fauna: a case history with lessons for conservation biology. *Oikos* **63**:62–76

Avise JC (1994) *Molecular markers, natural history and evolution.* Chapman & Hall, New York

Avise JC, Arnold J, Ball RM, Bermingham E, Lamb T, Neigel JE, Reeb CA, Saunders NC (1987) Intraspecific phylogeography: the mitochondrial DNA bridge between population genetics and systematics. *Ann Rev Ecol Syst* **18**:489–522

Avise JC, Ball RM, Arnold J (1988) Current versus historical population sizes in vertebrate species with high gene flow: a comparison based on mitochondrial DNA lineages and inbreeding theory for neutral mutations. *Mol Biol Evol* **5**:331–344

Baker CS, Parry A, Bannister JL, Weinrich MT, Abernathy RB, Calambokidis J, Lien J, Lambertsen RH, Urban J, Vasquez RO, Clapham PJ, Alling A, O'Brien SJ, Palumbi SR (1993) Abundant mitochondrial DNA variation and world-wide population structure in humpback whales. *Proc Natl Acad Sci (USA)* **90**:8239–8243

Ballard JW, Kreitman M (1994) Unravelling selection in the mitochondrial genome of *Drosophila*. *Genetics* **138**:757–772

Bartley D, Bagley M, Gall G, Bentley B (1992) Use of linkage disequilibrium data to estimate effective size of hatchery and natural fish populations. *Conserv Biol* **6**:365–375

Barton N, Wilson I (1995) Genealogies and geography. *Philos Trans R Soc Lond Ser B* **349**:49–59

Baverstock PR, Moritz C (1996) Project design. In: Hillis DM, Moritz C, Mable B (eds) *Molecular systematics, 2nd edn.* Sinauer Associates Inc., Sunderland, pp 17–28

Baverstock PR, Joseph L, Degnan S (1994) Units of management in biological conservation. In: Moritz C, Kikkawa J (eds) *Conservation biology in Australia and Oceania*. Surrey Beatty and Sons, Chipping Norton, pp 287–293

Bermingham E, Avise JC (1986) Molecular zoology of freshwater fishes in southeastern United States. *Genetics* **113**:939–965

Boileau MG, Herbert PND, Schwartz SS (1992) Non-equilibrium gene frequency divergence: persistent founder events in natural populations. *J Evol Biol* **5**:25–39

Boyce MS (1992) Population viability analysis. *Annu Rev Ecol Syst* **23**:481–506

Brooker MG, Rowley I, Adams M, Baverstock PR (1990) Promiscuity: an inbreeding avoidance mechanism in a socially monogamous species. *Behav Ecol Sociobiol* **26**:191–199

Brookfield JFY, Parkin DT (1993) Use of single-locus DNA probes in the establishment of relatedness in wild populations. *Heredity* **70**:660–663

Bruford MW, Wayne RK (1993) Microsatellites and their application to population genetic studies. *Curr Opin Genet Dev* **3**:939–943

Burke T (1989) DNA fingerprinting and other methods for the study of mating success. *Trends Ecol & Evol* **4**:139–144

Burke T (1994) Spots before the eyes: molecular ecology. *Trends Ecol & Evol* **9**:355–356

Burke T, Davies MW, Bruford M, Hatchwell BJ (1989) Parental care and mating behaviour of polyandrous dunnocks *Prunella modularis* related to paternity by DNA fingerprinting. *Nature* **338**:249–251

Burke T, Hanotte O, Bruford MW (1991) Multilocus and single locus minisatellite analysis in population biological studies. In: Burke T, Dolf G, Jeffreys AJ, Wolff R (eds) *DNA fingerprinting: approaches and application*. Birkhäuser Verlag, Basel, Switzerland, pp 155–168

Chakraborty R, Kidd KK (1991) The utility of DNA typing in forensic work. *Science* **254**:1735–1739

Crow JF, Aoki K (1984) Group selection for a polygenic trait: estimating the degree of population subdivision. *Proc Natl Acad Sci (USA)* **81**:6073–6077

Dallas JF, Dod B, Boursot P, Prager EM, Bonhomme F (1995) Population subdivision and gene flow in Danish house mice. *Mol Ecol* **4**:311–320

Davis MB (1994) Ecology and paleoecology begin to merge. *Trends Ecol & Evol* **9**:357–358

Degnan S (1993) Genetic variability and population differentiation inferred from DNA fingerprinting in silvereyes (Aves: Zosteropidae). *Evolution* **47**:1105–1117

Di Rienzo A, Wilson AC (1991) Branching pattern in the evolutionary tree for human mitochondrial DNA. *Proc Natl Acad Sci (USA)* **88**:1597–1601

Dowling TE, Moritz C, Palmer JD, Riesenberg LH (1996) Nucleic acids II: comparisons of restrictions sites and fragments. In: Hillis DM, Moritz C, Mable B (eds) *Molecular systematics,* 2nd edn. Sinauer Associates Inc., Sunderland, pp 249–320

Easteal S (1985) The ecological genetics of introduced populations of the giant toad *Bufo marinus*. II. Effective population size. *Genetics* **110**:107–122

Easteal S, Floyd RB (1986) The ecological genetics of introduced populations of the giant toad, *Bufo marinus* (Amphibia: Anura): dispersal and neighbourhood size. *Biol J Linn Soc* **27**:17–45

Excoffier L, Smouse PE, Quattro JM (1992) Analysis of molecular variance inferred from metric distances among DNA haplotypes: application to human mitochondrial DNA restriction data. *Genetics* **131**:479–491

Felsenstein J (1992) Estimating effective population size from samples of sequences: inefficiency of pairwise and segregating sites as compared to phylogenetic estimates. *Genet Res* **59**:139–147

FitzSimmons NN, Moritz C, Moore SS (1995) Conservation and dynamics of microsatellite loci over 300 million years of marine turtle evolution. *Mol Biol Evol* **12**:432–440

Frankham R (1995) Effective population size/adult population size ratios in wildlife – a review. *Genet Res* **66**:95-107

Fu YX (1994a) Estimating effective population size or mutation rate using the frequency of mutations of various classes in a sample of DNA sequences. *Genetics* **135**:1377–1386

Fu YX (1994b) A phylogenetic estimator of effective population size or mutation rate. *Genetics* **136**:685–692

Gibbs HL, Weatherhead PJ, Boag PT, White BN, Tabak LM , Hoysak DJ (1990) Realized reproductive success polygynous red-winged blackbirds revealed by DNA markers. *Science* **250**:1394–1397

Gilbert DA, Lehrman N, O'Brien SJ, Wayne RK (1990) Genetic fingerprinting reflects population differentiation in the California channel island fox. *Nature* **344**:764–767

Gilbert DA, Packer C, Pusey AE, Stephens JC, O'Brien SJ (1991) Analytical DNA fingerprinting in lions: parentage, genetic diversity and kinship. *J Heredity* **82**:378–386

Gillespie JH (1991) *The causes of molecular evolution.* Oxford University Press, New York

Grant PR, Grant R (1992) Demography and genetically effective sizes of two populations of Darwin's Finches. *Ecology* **73**:766–784

Hadrys H, Balick M, Schierwater B (1992) Applications of random amplified polymorphic DNA(RAPD) in molecular ecology. *Mol Ecol* **1**:55–63

Hanotte O, Zanon C, Pugh A, Greig C, Dixon A, Burke T (1994) Isolation and characterization of microsatellite loci in a passerine bird: the reed bunting. *Mol Ecol* **3**:529–531

Hillis DM, Moritz C, Mable B (eds) (1996) *Molecular systematics, 2nd edn.* Sinauer Associates Inc., Sunderland

Hoelzel AR, Dover AR (1991) *Molecular genetic ecology.* Oxford University Press, London

Hoelzel AR, Halley J, O'Brien SJ, Campagna C, Arnbom T, Le Boeuf B, Ralls K and Dover GA (1993) Elephant seal genetic variation and the use of simulation models to investigate historical population bottlenecks. *J Heredity* **84**:443–449

Hopper SD (1994) Plant taxonomy and genetic resources: foundations for conservation. In: Moritz C, Kikkawa J (eds) *Conservation Biology in Australia and Oceania.* Surrey Beatty and Sons, Sydney, pp 269–285

Hudson RR (1990) Gene genealogies and the coalescent process. *Oxf Surv Evol Biol* **7**:1–44

Hudson RR, Boos DD, Kaplan NL (1992a) A statistical test for detecting geographic subdivision. *Mol Biol Evol* **9**:138–151

Hudson RR, Slatkin M , Maddison WP (1992b) Estimation of levels of gene flow from DNA sequence data. *Genetics* **132**:583–589

Jeffreys AJ, Wilson V, Thein SL (1985) Individual-specific 'fingerprints' of human DNA. *Nature* **316**:76–79

Joseph L, Moritz C, Hugall A (1995) Molecular support for vicariance as a source of diversity in rainforest. *Proc R Soc Lond Ser B* **260**:177–182

Kimura M, Crow JF (1964) The number of alleles that can be maintained in a finite population. *Genetics* **49**:725–738

Kimura M, Weiss GH (1964) The stepping stone model of population structure and the decrease of genetic correlation with distance. *Genetics* **49**:561–576

Lambert DM, Millar CD, Jack K, Anderson S, Craig JL (1994) Single and multilocus DNA fingerprinting of communally breeding pukeko – do copulations or dominance ensure reproductive success? *Proc Natl Acad Sci (USA)* **91**:9641–9645

Lande R, Barrowclough GF (1987) Effective population size, genetic variation and their use in population management. In: Soule M (ed) *Viable populations for conservation.* Cambridge University Press, New York, pp 87–124

Larson A, Wake DB, Yanev KP (1984) Measuring gene flow among populations having high levels of genetic fragmentation. *Genetics* **106**:293–308

Lavery S, Moritz C, Fielder DR (1995) Changing patterns of population structure and gene flow at different spatial scales in the coconut crab (*Birgus latro*). *Heredity* **74**:531–541

Lavery S, Moritz C, Fielder DR (1996a) Indo-Pacific population structure and evolutionary history of the coconut crab (*Birgus latro*). *Mol Ecol* (in press)

Lavery S, Moritz C, Fielder DR (1996b) Genetic patterns suggest exponential population growth in a declining species. *Mol Biol Evol* (in press)

Lessa EP, Appelbaum G (1993) Screening techniques for detecting allelic variation in DNA sequences. *Mol Ecol* **2**:119–129

Limpus CJ, Miller JD, Parmenter CJ, Reimer D, McLachlan N and Webb R (1992) Migration of green (*Chelonia mydas*) and loggerhead (*Caretta caretta*) turtles to and from eastern Australian rookeries. *Aust Wildl Res* **19**:347–358

Lindenmayer DB, Lacy RC, Thomas VC, Clark TW (1993) Predictions of impacts of changes in population size and environmental variability on Leadbeater's possum, *Gymnobelideus leadbeateri* Mccoy (Marsupialia: Petauridae) using population viability analysis: an application of the program VORTEX. *Wildl Res* **20**:67–86

Lynch M (1988) Estimation of relatedness by DNA fingerprinting. *Mol Biol Evol* **5**:584–599

Lynch M, Crease TJ (1990) The analysis of population survey data on DNA sequence variation. *Mol Biol Evol* **7**:377–394

Lynch M, Milligan BG (1994) Analysis of population genetic structure with RAPD markers. *Mol Ecol* **3**:91–100

Maddison WP, Slatkin M (1991) Null models for the number of steps in a character on a phylogenetic tree. *Evolution* **45**:1184–1197

Marjoram P, Donnelly P (1994) Pairwise comparisons of mitochondrial DNA sequences in subdivided populations and implications for early human evolution. *Genetics* **136**:673–683

Maruyama T, Fuerst PA (1985) Population bottlenecks and non-equilibrium models in population genetics. II. Number of alleles in a small population that was formed by a recent bottleneck. *Genetics* **111**:145–163

Maruyama T, Birky CW (1991) Effects of periodic selection on gene diversity in organelle genomes and other systems without recombination. *Genetics* **127**:449–451

McCallum HI, Timmers P, Hoyle S (1995) Modelling the impact of predation on reintroductions. *Wildl Res* **22**:163–171

Moritz C (1991) The origin and evolution of parthenogenesis in *Heteronotia binoei* (Gekkonidae): evidence for recent and localized origins of widespread clones. *Genetics* **129**:211–219

Moritz C (1995) Uses of molecular phylogenies for conservation. *Philos Trans R Soc Ser B* **349**:113–118

Morin PA, Wallis J, Moore JJ, Woodruff DS (1994a) Paternity exclusion in a community of wild chimpanzees using hypervariable simple sequence repeats. *Mol Ecol* **3**:469–478

Morin PA, Moore JJ, Chakraborty R, Li J, Goodall J, Woodruff DS (1994b) Kin selection, social structure, gene flow and the evolution of chimpanzees. *Science* **265**:193–1201

Murphy RW, Sites JW Jr, Buth DG, Haufler CH (1996) Proteins: isozyme electrophoresis. In: Hillis DM, Moritz C, Mable B (eds) *Molecular systematics, 2nd edn*. Sinauer Associates Inc., Sunderland, pp 51–120

Nachman MW, Boyer SN, Aquadro CF (1994) Nonneutral evolution at the mitochondrial NADH dehydrogenase subunit 3 gene in mice. *Proc Natl Acad Sci (USA)* **91**:6364–6368

Nee S, Holmes EC, Rambaut A, Harvey PH (1995) Inferring population history from molecular phylogenies. *Philos Trans R Soc Ser B* **349**:25–31

Nei M, Tajima F (1981) DNA polymorphism detectable by restriction endonucleases. *Genetics* **97**:145–163

Neigel J, Avise JC (1986) Phylogenetic relationships of mitochondrial DNA under various demographic models of speciation. In: Nevo E, Karlin S (eds) *Evolutionary processes and theory*. Academic Press, New York, pp 515–534

Nix HA (1991) Biogeography: patterns and process. In: Nix HA , Switzer M (eds) *Rainforest animals. Atlas of vertebrates endemic to Australia's wet tropics*. Australian National Parks and Wildlife Service, Canberra, pp 11–39

Nunney L, Elam DR (1994) Estimating the effective population size of conserved populations. *Conserv Biol* **8**:175–184

Queller DC, Goodnight KF (1989) Estimating relatedness using genetic markers. *Evolution* **43**:258–275

Queller DC, Strassman JE, Hughes CR (1993) Microsatellites and kinship. *Trends Ecol & Evol* **8**:285–288

Rand DM, Dorfsman M, Kann LM (1994) Neutral and non-neutral evolution of *Drosophila* mitochondrial DNA. *Genetics* **138**:741–756

Richardson BJ, Baverstock PR, Adams M (1986) *Allozyme electrophoresis: a handbook for animal systematics and population studies*. Academic Press, Sydney

Rogers AR, Harpending H (1992) Population growth makes waves in the distribution of pairwise genetic differences. *Mol Biol Evol* **9**:552–569

Saiki RK, Gelfand DH, Stoffel S, Scharf SJ, Higuchi R, Horn GT, Mullis KB, Erlich HA (1988) Primer directed enzymatic amplification of DNA with a thermostable DNA polymerase. *Science* **239**:487–491

Schlötterer C, Amos B, Tautz D (1991) Conservation of polymorphic simple sequence loci in cetacean species. *Nature* **354**:63–65

Slade RW, Moritz C, Heideman A, Hale PT (1993) Rapid assessment of single-copy nuclear DNA variation in diverse species. *Mol Ecol* **2**:359–373

Slatkin M (1987) Gene flow and the geographic structure of animal populations. *Science* **236**:787–792

Slatkin M (1993) Isolation by distance in equilibrium and non-equilibrium populations. *Evolution* **47**:264–279

Slatkin M, Barton NH (1989) A comparison of three indirect methods for estimating average levels of gene flow. *Evolution* **43**:1349–1368

Slatkin M, Maddison WP (1989) A cladistic measure of gene flow inferred from phylogenies of alleles. *Genetics* **123**:603–613

Slatkin M, Hudson RR (1991) Pairwise comparisons of mitochondrial DNA sequences in stable and exponentially growing populations. *Genetics* **129**:555–562

Tajima F (1989) The effects of change in population size on DNA polymorphism. *Genetics* **123**:597–601

Takahata N (1993) Allelic genealogy and human evolution. *Mol Biol Evol* **10**:2–23

Takahata N, Palumbi SR (1985) Extranuclear differentiation and gene flow in the finite island model. *Genetics* **109**:441–457

Taylor AC, Sherwin WB, Wayne RK (1994) Genetic variation of microsatellite loci in a bottlenecked species: the northern hairy-nosed wombat *Lasiorhinus krefftii*. *Mol Ecol* **3**:277–290

Thomas WK, Paabo S, Villablanca FX, Wilson AC (1990) Spatial and temporal continuity of kangaroo rat populations shown by sequencing mitochondrial DNA from museum specimens. *J Mol Evol* **31**:101–112

Varvio S-L, Chakraborty R, Nei M (1986) Genetic variation in subdivided populations and conservation genetics. *Heredity* **57**:189–198

Waples RS (1989) A generalized approach for estimating effective population size from temporal changes in allele frequency. *Genetics* **121**:379–391

Waples RS (1991) Genetic methods for estimating the effective size of cetacean populations. In: Hoelzel AR (eds) *Genetic ecology of whales and dolphins*. International Whaling Commission, Cambridge, pp 279–300

Waples RS, Smouse PE (1990) Gametic disequilibrium analysis as a means of identifying mixtures of salmon populations. *Am Fish Soc Symp* **7**:439–458

Watterson GA (1975) On the number of segregating sites in genetics models without recombination. *Theor Popul Biol* **7**:256–276

Wayne RK, Jenks SM (1991) Mitochondrial DNA analysis implying extensive hybridisation of the endangered red wolf *Canis rufus*. *Nature* **351**:565–568

Weber JL , May PE (1989) Abundant class of human DNA polymorphisms which can be types using the polymerase chain reaction. *Am J Hum Genet* **44**:388–396

Weir BS (1996) Genetic population structure. In: Hillis DM, Moritz C, Mable B (eds) *Molecular systematics, 2nd edn*. Sinauer Associates Inc., Sunderland. pp 385–406

Worthington-Wilmer J, Moritz C, Hall L, Toop J (1994) Extreme population structuring in the threatened Ghost Bat, *Macroderma gigas*: evidence from mitochondrial DNA. *Proc R Soc Lond* **257**:193–198

Wright S (1951) The genetical structure of populations. *Ann Eugenics* **15**:323–354

Wright S (1969) *Evolution and the Genetics of Populations. Vol 2. The Theory of Allele Frequencies*. Chicago University Press, Chicago

Yuhki N, O'Brien SJ (1990) DNA variation of the mammalian major histocompatibility complex reflects genomic diversity and population history. *Proc Natl Acad Sci (USA)* **87**:836–840

Why do so many Australian birds cooperate: social evolution in the Corvida?

Andrew Cockburn

ABSTRACT

The most widely accepted models which attempt to account for both the diversity of avian mating systems in general, and the occurrence of cooperative breeding in particular, attribute importance to variation in habitat quality. Social systems are viewed as labile responses to the availability of breeding resources for females, availability of mates to males, and availability of breeding vacancies to young birds. Recent demonstrations that both mating systems and cooperative breeding are distributed patchily between taxa challenge this view, and suggest that the lability of reproductive behaviour is strongly influenced by phylogenetic history. Phylogenetic bias towards cooperative breeding is most pronounced among the Corvida, a large group of oscine passerines that originated and first radiated in Australia, and then spread throughout the world. Cooperative breeding and complex mating systems are massively over-represented among this group, and this over-representation occurs elsewhere in the world as well as in Australia. The major ecological pattern in the dispersion of cooperative breeding in this group is the tendency of non-cooperative species to be found in dense visually occluded habitats, such as rainforest canopy, dense riparian thickets, and coastal scrub. This distribution is not predicted by habitat saturation models of cooperative breeding. Further, both the dispersion of other traits that are usually presumed to be rare (such as duetting by monogamous pairs) and direct cladistic analysis suggest that living in unassisted pairs is often a derived state. I argue that the extreme lifespan of the Corvida makes retention of a small group of offspring in the natal territory highly advantageous to parents, but leads to conflicts of interest which generate diverse evolutionary outcomes. As in the social insects, the evolution of complex social systems in birds may depend on mutualism and phylogenetically constrained aspects of life history, rather than being simply a density-dependent response to habitat availability.

Key words: Corvida, cooperative breeding, evolution, phylogeny

Frontiers of Population Ecology, R.B. Floyd, A.W. Sheppard and P.J. De Barro (eds), CSIRO Publishing, Melbourne 1996, pp. 451–472

Introduction

There is general support for the notion that social organisation in birds and mammals varies in response to proximate habitat constraints. For example, compare the synthetic diagrams offered by Davies (1991) and Emlen (1991) to explain variation in avian mating systems and the development of cooperative breeding respectively (Fig. 1). Both authors point to the variation in the dispersion of resources as the principal determinant underlying variation in social organisation. The idea of lability receives support from many sources. At a broad comparative scale, closely related species often exhibit vastly different social systems (e.g. anthropoid primates; Short 1979). At a more local scale, it is often possible to change social organisation by manipulating food resources. In a classic experiment, Davies (1985) was able to change the modal mating system in a small population of dunnocks merely by adding food.

This confidence in the importance of proximate determinants of social organisation sits uncomfortably with the evidence that some social systems appear to be distributed non-randomly with respect to phylogeny. In a simple example, among homoeothermic vertebrates, most mammals are polygynous, and most birds are not (Davies 1991). One particular form of polygyny (harem-defence polygyny) is quite common in mammals, but has only recently been confirmed in birds (Webster 1994). The evolution of cooperative breeding among birds has only recently been considered in this light. In his major review, Brown (1987, p. 34) concluded that 'Communal breeding is clearly not a trait whose phylogeny can be usefully analysed along phylogenetic lines' and 'In the end ... curiosities in the taxonomic distribution tell us little'. Challenges to this view are increasingly frequent (Russell 1989; Peterson and Burt 1992; Edwards and Naeem 1993, 1994; Ligon 1993; McLennan and Brooks 1993; Ketterson and Nolan 1994), and have been prompted by two important developments. The first is increased attention to the need to incorporate phylogenetic history into properly designed comparative studies (Harvey and Pagel 1991; Harvey and Purvis 1991). The second is the radical change in our view of avian phylogeny suggested by the massive work of Sibley and Ahlquist using DNA hybridisation (Sibley and Ahlquist 1985, 1990; Sibley and Monroe 1990, 1993).

The taxa most radically affected by Sibley and Ahlquist's revision are the Australian passerines. Traditionally viewed as a polyphyletic assemblage formed by repeated colonisation of predominantly Eurasian groups (e.g. Mayr 1944), the morphological similarity which led to this view has been shown to result from convergence, and many of the Australian passerines have been shown to belong to a single clade (the Corvida) which in turn has proved to be the cradle of radiations which have taken place in Africa, Eurasia and the Americas (Sibley and Ahlquist 1985). Although Australia has always been recognised as a hotspot for cooperative breeding (Rowley 1976; Dow 1980; Trivers

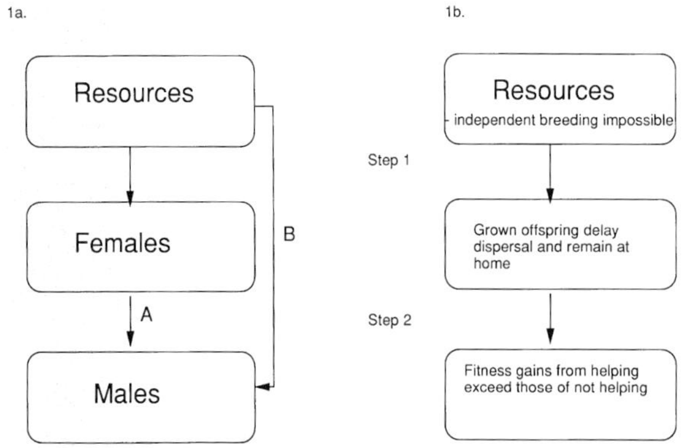

Fig. 1. The conventional pathways hypothesised to lead to the evolution of a) vertebrate mating systems (after Davies 1991) and b) cooperative breeding (after Emlen 1991). Diagram 1a views resources as influencing the dispersion of females. Although resources also determine the distribution of males (path B), males should adjust their dispersion to increase access to females (path A). Diagram 1b views cooperation as resulting from constraints on resources which prevent independent reproduction. Prolonged philopatry provides the conditions for the evolution of helping behaviour.

1985; Brown 1987), Russell (1989) was the first to highlight the importance of this reinterpretation for cooperative breeding. Using the data of Brown (1987), she observed that 22% of the Corvida in Australia are cooperative, while none of the Laurasian sister group (the Passerida) which have invaded Australia secondarily is cooperative. This stands in stark contrast to the estimate of Brown (1987), that cooperation occurs in 2.5% of birds world-wide.

In this contribution I expand on this observation, and examine the significance of the phylogenetic bias for the traditional density-dependent hypothesis of cooperative breeding, and for attempts to explain the prevalence of cooperative breeding among Australian birds. I do this by an exploration of social evolution in the Corvida, a group of more than 1000 species now breeding on all continents but Antarctica.

Density-dependent models of cooperative breeding

Cooperative breeding initially challenged behavioural ecologists because of the apparently maladaptive behaviour of birds which forgo independent reproduction and instead help other birds rear young. Most authors agree that this behaviour is best dissociated into two questions; why do young birds defer dispersal and any attempt to breed independently, and second, why do those birds which stay behind help? Because the reasons for helping behaviour have been well catalogued elsewhere (Emlen *et al.* 1991), this paper comments on the first step, the evolution of prolonged philopatry. All models emphasise ecological constraints on independent breeding, either because all available habitat is saturated (habitat constraint models; Emlen 1982; Brown 1987), or because habitat is so variable, that it pays to wait for a good site rather than attempt to breed on a low value vacant site (benefits-of-philopatry models; Stacey and Ligon 1991). These models have received strong support from elegant single-species studies, which either use long-term demographic data to infer the limits to colonisation (e.g. Woolfenden and Fitzpatrick 1984), or manipulate the resource presumed to limit colonisation (e.g. Pruett-Jones and Lewis 1990). In the most elegant study to date, Komdeur (1992) has been able to use the entire population of the endangered Seychelles warbler (*Bebronis sechellensis*) to provide compelling evidence for both habitat constraints and benefits-of-philopatry arguments. He showed both by monitoring the recovery of the original population of this species, and by observing translocated individuals on a new island, that cooperative behaviour was density-dependent, only emerging once all available habitat was filled with unassisted pairs. However, there was habitat variability on the original island. When he looked at the effect of the vacancies created by removing individuals for translocation, it became clear that young helpers in habitat with superior resources were much less likely to disperse than those in habitat where food supply was poor.

Despite this strong support from single-species studies, the predictive ability of ecological constraints models is less clear. Ideally, we should be able to deduce which species will exhibit cooperative behaviour *a priori*, but many species showing all the hallmarks of ecological constraints do not exhibit cooperative breeding (e.g. Smith 1994). While there have been attempts to generalise models to encompass a range of alternative ways of dealing with constraints (e.g. formation of non-breeding flocks; Koenig *et al.* 1992; Zack and Stutchbury 1992), these models still lack any predictive power (Heinsohn *et al.* 1990; Mumme 1992a).

One possibility for strengthening these models would be to adopt a comparative approach by identifying circumstances where cooperation develops repeatedly, rather than attempting to generalise from the idiosyncrasies of individual species. Generalities have been offered. For example, Trivers (1985, p 184) comments that 'The species are almost always tropical, subtropical or Australian'. The former two correlates are of dubious importance, as the overwhelming majority of birds are tropical or subtropical in distribution (in direct contrast to the distribution of well-studied species). However, the prevalence of cooperative breeding in Australia has long been recognised (e.g. Rowley 1976; Dow 1980; Brown 1987) and seems to demand some special explanation. Early hypotheses focused on the variability and unpredictability of climate (Rowley 1965; Harrison 1969), though a more comprehensive review of climatic variables by Dow (1980) concluded that 'no overwhelmingly clearcut result emerges from this systematic analysis of possible effects of environmental factors'. Ford *et al.* (1988) have recently elaborated upon a hypothesis of Rowley (1968), that the lack of seasonality is a critical stimulus to cooperative breeding. Rowley (1968) emphasised the lowered cost in retaining young on the territory because winter was not associated with a dramatic decline in food resources. By contrast, Ford *et al.* (1988) highlighted the difficulty in 'acquiring the additional food necessary to breed, especially if inexperienced', in environments where there is no huge flush of food resources during the breeding season. Thus analysis has tended from the viewpoint that Australian environments are unpredictable, to the view that there is no special pattern, to the view that they are predictably unseasonal. This trend maps the rise in importance of ecological constraints models.

The comparative biology of cooperative breeding

The research approach which has dominated the study of cooperative breeding is the use of long-term studies of the demography of single species (Stacey and Koenig 1990), occasionally augmented by experiments on features of interest (Hannon *et al.* 1985; Mumme 1992b; Komdeur 1992). This paper adopts the use of phylogenetically-based comparative methods to ask questions about the geographic and taxonomic distribution of cooperative breeding.

By any standards, such comparative studies should be more easily accomplished using birds as study organisms. First, they are unambiguously monophyletic, and their supraspecific classification is dissected to an unusual extent, allowing greater depth in phylogenetic comparison. Indeed, in the case of the phylogeny proposed by Sibley, Ahlquist and Monroe (Sibley and Monroe 1990), the phylogeny is based on common methods of DNA hybridisation and clearly articulated assumptions. Second, a wealth of data gathered by enthusiastic amateur ornithologists allows a greater proportion of taxa to be compared than is usually the case for such a large clade.

Nonetheless, some of the previous comparative studies of cooperative breeding are plagued by several difficulties. These difficulties are obvious and clearly stated by both Dow (1980) for Australian birds and Brown (1987) for all birds. However, they have rarely been formally taken into account, and are therefore worth elaborating in some detail.

The first set of problems arise because of the uncertain meaning which can be attributed to the absence of a definite record of helping behaviour. Comparative analyses have sometimes proceeded on the assumption that any species not listed by Brown (1987) or Dow (1980) does not show cooperative behaviour (with occasional amendments based on recently-published studies of new species in the mainstream ornithological or behavioural literature), and hence exhibits typically exclusive pair breeding (e.g. see debate in Edwards and Naeem 1993, 1994; McLennan and Brooks 1993). Even among those who are sensitive to this difficulty, Brown (1987) presents proportions of species exhibiting cooperation in each taxon, and Ford *et al.* (1988) enumerate cooperative species according to broad habitat categories and foraging mode. The reasons that this could be misleading are obvious. While negative records are probably reliable for species in the well-studied post-glacial landscapes of the northern hemisphere, they do not mean very much for a species living in inaccessible habitats, for which the nest may have never been seen, let alone any population being subject to a detailed colour-banding study. Second, the scoring of negatives is based on the assumption that cooperation is sufficiently rare that it is best to presume that pair-dwelling is the conservative state. Within the Corvida, this is hardly a parsimonious assumption for many groups.

For example, in the genus *Malurus*, Schodde's (1982) brilliant monograph provides evidence that all but two species (whose nesting behaviour has never been recorded) of *Malurus* and its sister genus *Sipodotus* were cooperative breeders. In this case, it is conservative to assume that all species are cooperative, or at best that only the species which have been studied (100% cooperative) should be scored in any compilation. In addition, it is important to recognise that cooperative breeding is not the only 'bizarre' manifestation of social behaviour in the Corvida, and pair-dwelling may be absent as a consequence. In many members of some taxa (lyrebirds, scrub-birds, bowerbirds and birds-of-paradise), there is uniparental care by the female, so the absence of help is almost trivial. It may be appropriate to factor these species out of any analysis when comparing the Corvida with its sister group the Passerida, where uniparental care is exceptionally rare. It is thus worrying when Ford *et al.* (1988) conclude that within Australia cooperative behaviour is more common in *Eucalyptus* woodland than in rainforest and desert, when detailed population studies in *Eucalyptus* woodland vastly outnumber those in other habitats.

A second related problem is the assumption that evolution inevitably proceeds from the presumed common state (pair-dwelling; Lack 1968) to cooperation. Not only is this cladistically unsound (potential outgroups such as reptiles and ratites are only rarely monogamous; see also Wesolowski 1994), but in the case of cooperative breeding, there is already convincing evidence that evolution proceeds in both directions. For example, in the *Aphelocoma* jays, a genus which has been subject of intense behavioural and phylogenetic studies, Peterson and Burt (1992) show that cooperative behaviour has been lost in a single clade (*Aphelocoma insularis* and *A. californica*), so the evolution of pair-dwelling requires special explanation (Fig. 2a). In direct contrast, the recent phylogeny suggested by Christidis and Schodde (1993) for the meliphagine honeyeaters suggests that cooperative breeding has evolved several times, but always as a terminal state (Fig. 2b). Where cooperation is an ancestral trait common to the ancestor of a clade, it is necessary to be wary of assuming that it evolved in response to contemporary conditions. For example, it seems unlikely that *Eucalyptus* woodlands and forests were common when many of the major cooperative clades evolved and diversified in Australia (Heinsohn *et al.* 1990; Cockburn 1991), yet some hypotheses attribute importance to the food supply found in these habitats (Ford *et al.* 1988). Any phylogenetic inertia will also tend to lead to the retention of cooperation in inappropriate habitats, rather than the converse bias towards the retention of pair-living.

Last, there has been a tendency to assume that cooperative behaviour is a unitary phenomenon (see the critique by McLennan and Brooks 1993), perhaps because there is a smugness that its proximate basis is well understood. Yet even cursory examination of the literature suggests that it takes radically different forms.

In an attempt to circumvent these difficulties, the literature on the Corvida was examined, adopting the taxonomy and phylogeny proposed by Sibley and Monroe (1990, 1993). While there remains strong resistance to the appropriateness of this treatment (Mayr and Bock 1994), there is growing evidence that it is broadly correct with respect to treating the Corvida as a monophyletic group (Baverstock *et al.* 1992; Christidis and Schodde 1991; Sibley 1994), and of great value in comparative analysis (Mooers and Cotgreave 1994). For sources on behaviour, the major compilations on the avifauna for each region or taxon were used (Table 1), augmented by my own non-exhaustive perusal of the primary literature. This treatment is necessarily very uneven, as data on African and Asian species are particularly poorly described. Each species was classified according to the following schema:

- Taxa where parental care is exclusively by the female, and where cooperation and potential pair living are rare (this is most obvious in the lekking birds-of-paradise, where the transition between lekking and monogamy is well treated by Beehler and Pruett-Jones 1983).

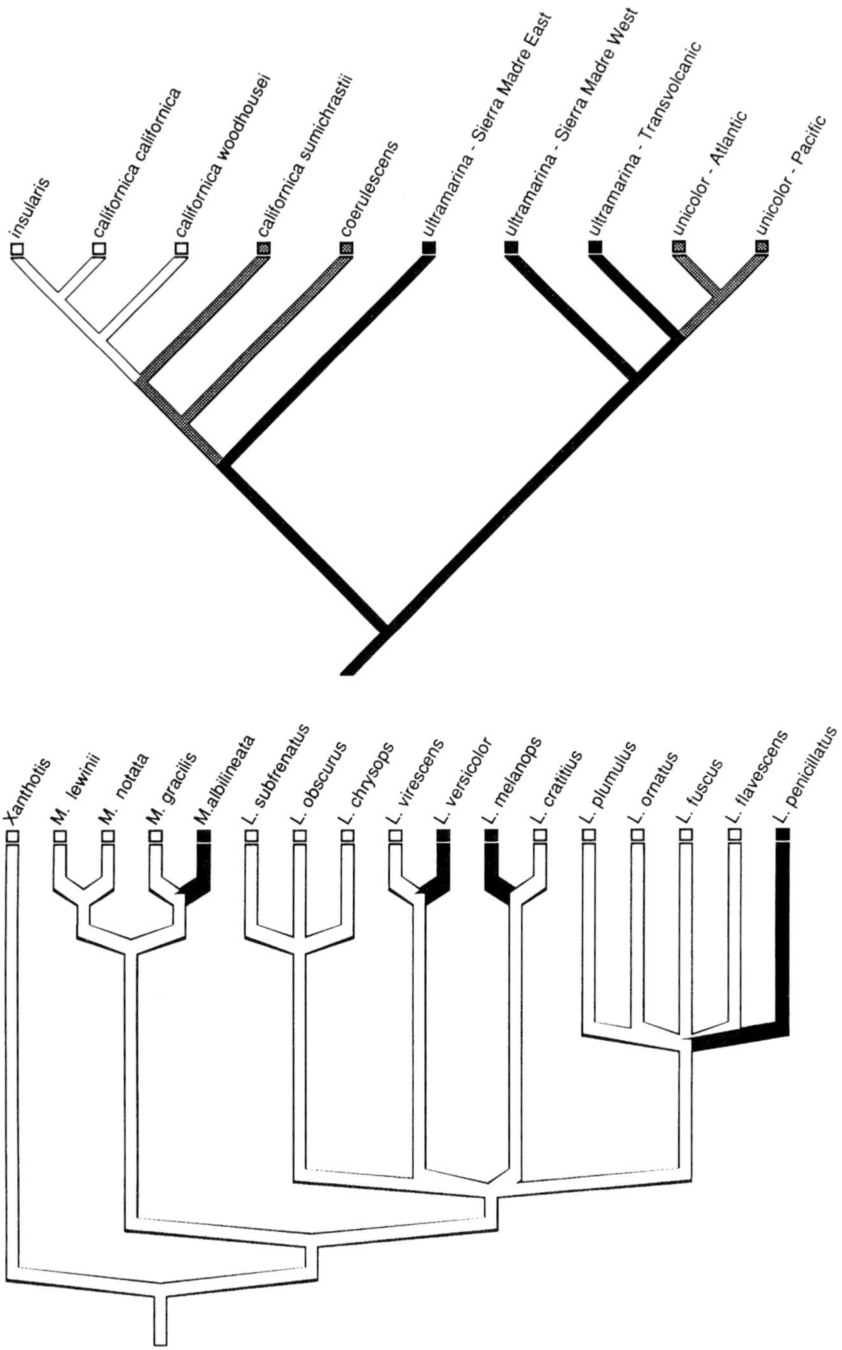

Fig. 2. The distribution of cooperative breeding in two groups within the Corvida where the phylogeny is well resolved. (a) Distribution in the *Aphelocoma* jays. Revised from Peterson and Burt (1992) to reflect the nomenclature used by Sibley and Monroe (1990) (filled bars – cooperative behaviour and plural breeding; shaded bars – cooperative behaviour and singular breeding; unfilled bars – pair dwelling). (b) Distribution in the meliphagine subgroup of honeyeaters. Cladogram follows Christidis and Schodde (1993) though species with poorly resolved social behaviour have been deleted from the cladogram (filled bars – cooperative breeding; open bars – pair dwelling).

- Taxa where breeding behaviour has been sufficiently well documented to conclude that cooperation is absent or extremely rare. In order to include a bird in this category, the required evidence was that some nests had been observed by naturalists, or that the birds generally were solitary or lived in pairs. It is likely that a small number of species in this category will prove to be cooperative. For example, although accounts of the New Zealand avifauna are unanimous in describing *Notiomystis* as monogamous, recent unpublished research shows that this species has a complex mating system including polygyny, polyandry and polygynandry (Castro, Minot, Fordham and Birkhead, *pers. comms.*).
- Taxa where cooperative behaviour has been recorded. Cooperative behaviour was defined as any case where more than two birds contribute to one of the following: nest-building, incubation, or feeding of nestlings or fledglings. Two records which are difficult to interpret were dismissed. First, following Dow (1980), *Epthianura tricolor* was excluded where the record suggests that an immature was seen with food near a nest but did not deliver the food to the nestlings. Second, Coates (1990) reports an observation by the late H.L. Bell of two pairs of *Rhipidura hyperythra* feeding a single fledgling. Because Bell had almost unique experience with a wide range of cooperative species, it seems likely that he would have explicitly described this as cooperative breeding if that was his interpretation of the behaviour. This category unfortunately contains a very wide range of intensity and form of cooperative behaviour. For example, Harris and Arnott's (1988) observation of helping behaviour in the typically pair-dwelling *Dryoscopus cubla* is very different from the extreme and obligate sociality observed in *Corcorax melanorhamphos* (Rowley 1978; Heinsohn 1992). However, the primary literature is so fragmentary that it is not yet possible to assign enough species along a graded series of intensity, frustrating the global comparisons which are one of the chief aims of this analysis.
- Species where it is impossible to tell, or where it seems premature to conclude that a higher taxon does not exhibit cooperative behaviour.

Table 1. Major literature summaries used in compilation of breeding data

Region	Taxon	Sources
Global	Tribe Corvini	Madge and Burn (1994)
Australia	Family Maluridae	Schodde (1982)
	Meliphagidae	Longmore (1991)
	Epthianura	Serventy (1982); Major (1992)
	Pardalotinae	Woinarski (1985b)
	Dasyornithinae	Serventy (1982)
	Acanthizinae	Serventy (1982)
	Corcoracinae	Rowley (1978); Heinsohn (1992)
	Oriolini (*Coracina, Lalage*)	Strahan (1994)
	All other taxa	Blakers *et al.* (1984); Boles (1988)
Papua New Guinea	All taxa	Coates (1990)
Fiji, Samoa and Tonga	All taxa	Watling (1982)
New Zealand	All taxa	Robertson (1985)
India and neighbouring areas	All taxa	Ali and Ripley (1971, 1972)
Africa	Laniidae; Malaconotinae	Harris and Arnott (1988)
	Coracina, Campephaga	Keith *et al.* (1992)
	All other taxa	Mackworth-Praed and Grant (1973); Du Plessis *et al.* (1995); Ginn *et al.* (1989)
Europe	All taxa	Cramp and Perrins (1993, 1994)
Americas	Vireonidae	Skutch (1960)

The data were compiled into summaries of the number of species in each category, the number of genera in each category, organised by the lowest suprageneric taxon recognised by Sibley and Ahlquist (1990). For example, they dissect the Family Corvidae into many subfamilies and tribes, but leave the Family Meliphagidae undissected. Genera were occasionally classified as unknown if all the well-studied species were not cooperative, but there were observations of group-living well into the breeding season providing a *prima facie* case that cooperation will be detected in other species with further study.

I traced character evolution within taxa using visual representations generated by the MacClade package (Maddison and Maddison 1992), and conventional methods (outgroup analysis and parsimony) to identify the direction of evolution where possible.

With respect to specific hypotheses, the use of available computer packages has been avoided (e.g. Purvis 1991), because of the fragmentary and uneven nature of the data. Instead, my aim was to identify the number of contrasts which were informative with respect to each test. For example, to identify whether cooperative breeding was different between rainforest and other habitats, firstly each suprageneric taxon was identified that was invariant with respect to behaviour, and analysed whether it could be exclusively associated with a particular habitat. All taxa (including genera) which were variable with respect to behaviour were enumerated and scored as informative if they exhibited some correlation associated with the habitat axis under consideration. These informative contrasts are described in detail rather than by summary statistics.

Distribution of cooperative breeding in the Corvida

The first conclusion is that the extent of cooperative behaviour in the Corvida has been spectacularly underestimated, and that if anything, the important phylogenetic correlation identified by Russell (1989) was understated. In this analysis I was able to add 41 species of Corvida for which at least some level of cooperative behaviour can be reliably inferred to the compilation by Brown (1987) (Table 2). Cooperative behaviour occurs in 24 of the 36 (67%) higher taxa (Tribe or above) used by Sibley and Monroe (1990) (Table 3). In four of the 13 non-cooperative taxa, the predominant mating system involves complete male emancipation from paternal care (Menuridae, Atrichornidae, Ptilorhynchidae and Paradisaeini), often coupled with display site defence promiscuity (*sensu* Davies 1991). The behaviour exhibited by these birds is itself a radical departure from the quintessential monogamous bird portrayed by Lack (1968), and has no counterpart elsewhere in the oscine passerines, though it is well represented in the suboscine passerines.

At lower taxonomic levels, cooperation is found in at least 61 genera (of 140 for which reliable data are available; 44%) and 128 species (of 439; 29%) in the groups without male emancipation (Table 3). This is clearly not an exhaustive list. I suspect further study will add many additional cooperative species and some additional cooperative genera in most geographic regions. For example, group-living which apparently persists into the breeding season has been reported in some species of *Cyanolyca* (Corvini, Americas; Madge and Burn 1994); *Clytomyias* (Maluridae, New Guinea; Frith and Frith 1992); *Stipiturus* (Maluridae, Australia; Schodde 1982); *Batis* and *Malaconotus* (Malaconotinae, Africa; Harris and Arnott 1988) and *Falculea* (Vangini, Madagascar; Langrand 1990).

Indeed, the evidence that Corvida exhibit dramatically different behaviour outside Australia than they do within Australia is, at best, weak. The proportion of genera in clades which originated outside Australia that contain at least one cooperative species is similar to the pattern in Australia/Papua New Guinea (Table 4). Although the proportion of cooperative species found in Australia is higher, at least some of this over-representation may be because Australian ornithologists are attuned to the behaviour.

Table 2. Species from the Corvida recognised to be at least occasionally cooperative in this analysis and which were not in the compilation provided by Brown (1987), together with the source on which that recognition is based. Note that some of these changes reflect adherence to the Sibley and Monroe taxonomy.

Family	Species	Source
Maluridae	*Amytornis woodwardi*	Noske (1992)
	A. housei	Freeman (1970)
	Sipodotus wallacii	Schodde (1982)
	Malurus alboscapulatus	Schodde (1982)
	M. melanocephalus	Schodde (1982)
	M. cyanocephalus	Schodde (1982)
	M. coronatus	Rowley and Russell (1993)
	M. amabilis	Schodde (1982)
Meliphagidae	*Entomyza cyanotis*	Boles *et al.* (1981)
	Lichenostomus versicolor	Longmore (1991)
	Meliphaga albilineata	Longmore (1991)
	Melithreptus affinis	Longmore (1991)
	M. laetior	Longmore (1991)
	M. gularis	Longmore (1991)
	M. validorostris	Longmore (1991)
	Notiomystis cincta	I. Castro and E. Minot (pers. comm.)
Pardalotidae	*Acanthiza murina*	Coates (1990)
	Aphelocephala leucopsis	Sandbrink and Robinson (1994)
	A. nigricincta	Blakers *et al.* (1984)
	Chthonicola sagittatus	Dow (1980)
Petroicidae	*Melanodryas vittata*	Boles (1988)
Pomatostomidae	*Pomatostomus ruficeps*	Coates (1990)
Corvidae	*Artamus maximus*	Coates (1990)
	Cracticus cassicus	Coates (1990)
	Aphelocoma californica	Peterson and Burt (1992)
	Calocitta colliei	Madge and Burn (1994)
	Cyanocorax violaceus	Madge and Burn (1994)
	C. cristatellus	Madge and Burn (1994)
	C. affinis	Madge and Burn (1994)
	Urocissa caerulea	Severinghaus (1987)
	Corvus corone	Richner (1990)
	Specotheres viridis	Woodall (1980)
	Grallina cyanoleuca	Aston (1988)
	Platysteira peltata	Harris and Arnott (1988)
	Dryoscopus cubla	Harris and Arnott (1988)
	Schetba rufa	Yamagishi *et al.* (1992, 1995)
	Mohoua albicilla	Gill and McLean (1992)
	M. ochrocephala	Robertson (1985)
	Daphaenositta miranda	Coates (1990)
	Terpsiphone viridis	Grimes (1976)
	Prionops alberti	Mackworth-Praed and Grant (1973)

Despite the obvious weaknesses in the data, it is possible to ask whether any of the habitat or life history correlations identified by other authors are supported by formal phylogenetic analysis. There is no support from the Corvida for the suggestion by Ford *et al.* (1988) that cooperative breeding in Australia is less likely in arid zone species. I first classified species according to the distributions reported by Blakers *et al.* (1984). Of four genera which have strongly arid-zone centred distributions, three are monospecific (*Acanthagenys, Pyrrholaemus, Ashbyia*) and non-cooperative, as

Table 3. Distribution of cooperative breeding within the Corvida. For each of the higher taxa recognised by Sibley & Ahlquist (1990), I classified genera and species according to whether there was adequate data in the literature sources available to me, and then whether the species exhibited some cooperative behaviour. The likely area of origin was assigned by determining the area in which most genera were concentrated from the sources in Table 1. The likely area of origin of the cosmopolitan Corvini is unclear

Superfamily	Family	Subfamily	Tribe	Genera			Species			Geographic
				Recognised	Good data	Cooperative	Recognised	Good data	Cooperative	centre
Menuroidea	Climacteridae			2	2	1	7	5	4	AustraloPapuan
	Menuridae	Menurinae		1	1	0	2	2	0	AustraloPapuan
		Atrichothorninae		1	1	0	2	2	0	AustraloPapuan
	Ptilonorhynchidae			7	7	0	20	20	0	AustraloPapuan
Meliphagoidea	Maluridae	Malurinae	Malurini	3	2	2	15	12	12	AustraloPapuan
			Stipiturini	1	0	0	3	1	0	AustraloPapuan
		Amytornithinae		1	1	1	8	3	2	AustraloPapuan
	Meliphagidae			42	26	11	181	75	21	AustraloPapuan
	Pardalotidae	Pardalotinae		1	1	1	4	4	1	AustraloPapuan
		Dasyornithinae		1	1	0	3	2	0	AustraloPapuan
		Acanthizinae	Sericornithini	10	8	2	26	15	3	AustraloPapuan
			Acanthizini	4	4	4	35	19	11	AustraloPapuan
Corvoidea	Petroicidae			14	8	2	46	21	5	AustraloPapuan
	Irenidae			2	1	0	10	3	0	Asian
	Orthonychidae			1	1	1	2	2	1	AustraloPapuan
	Pomatostomidae			1	1	1	5	5	5	AustraloPapuan
	Laniidae			3	3	3	30	11	4	African
	Vireonidae			4	2	0	51	16	0	American
	Corvidae	Cinclosomatinae		6	2	1	15	6	1	AustraloPapuan
		Corcoracinae		2	2	2	2	2	2	AustraloPapuan
		Pachycephalinae	Neosittini	1	1	1	2	2	2	AustraloPapuan
			Mohouini	1	1	1	3	3	2	New Zealand
			Falcunculini	3	2	1	3	2	1	AustraloPapuan
			Pachycephalini	9	4	0	51	17	0	AustraloPapuan
		Corvinae	Corvini	25	22	10	117	78	25	Asian/American
			Paradisaeini	17	16	0	45	43	0	AustraloPapuan
			Artamini	6	4	3	24	14	9	AustraloPapuan
			Oriolini	8	6	2	111	24	2	Asian

	Dicrurinae	Rhipidurini	1	1	0	42	12	0	Asian
		Dicrurini	2	1	1	24	9	1	African
		Monarchini	18	13	4	98	34	4	African
	Aegithininae		1	1	0	4	2	0	Asian
	Malaconotinae	Malaconotini	8	6	1	48	20	1	African
		Vangini	19	9	4	58	17	8	African
incertae cedis	Callaeatidae		3	2	0	3	2	0	New Zealand
	Picathartidae		2	1	1	4	1	1	African
Totals			231	164	61	1104	506	128	

Table 4. Dispersion of cooperative breeding by continent (Asia and America are combined because of the uncertain geographic placement of the Corvini). Each clade was assigned an epicentre of radiation (Table 3) and all members of that clade summed to give the proportion of well understood species and genera which can be considered cooperative. The lekking species are included.

	Genera			Species		
	Known	Cooperate	Proportion	Known	Cooperate	Proportion
Australo-Papuan	95	34	0.36	274	80	0.29
Non-Australian	69	27	0.39	232	48	0.21
New Zealand	3	1	0.33	5	2	0.40
Asia and Americas	33	12	0.36	135	27	0.20
Africa	33	14	0.42	92	19	0.21

appear to be their closest relatives from more mesic habitats. One genus (*Aphelocephala*) contains more than one species, and it is likely that most of those species are cooperative (Table 2). Among genera where appropriate contrasts should be possible because there are both mesic and arid species, several retain non-cooperative habits throughout their range (*Chlamydera, Psophodes, Corvus, Certhionyx, Epthianura*), some show a pattern which cannot be aligned along the proposed gradient (*Pardalotus, Cinclosoma, Lichenostomus, Phylidonyris*), some it is premature to comment on whether the desert species are cooperative or not (*Conopophila, Amytornis, Stipiturus*), and the remainder show cooperative habits at least as well developed in arid-centred species as in more mesic species (*Artamus, Cracticus, Malurus, Manorina, Pomatostomus*). I have developed similar arguments elsewhere to show that there are no obvious correlations between foraging mode and the prevalence of cooperation (Cockburn 1991), as was originally suggested by Ford *et al.* (1988).

My analysis defies conventional wisdom on cooperative breeding (e.g. Trivers 1985; Brown 1987), but strongly supports the observations of Bell (1985) and Ford *et al.* (1988), that cooperation is relatively uncommon in rainforest. Among the groups (Tribe and above) which lack cooperation, all but two (Malaconotini and Dasyornithinae) are predominantly frugivores or insectivores which have radiated and live in wet, heavily vegetated, tropical habitats (Table 5; habitat assignments from sources in Table 1). Of the two exceptions, the Dasyornithinae live in dense, wet coastal scrub, and many members of the Malaconotini are also inhabitants of dense thickets, riparian vegetation and rainforest. By contrast, there are no exclusively cooperative clades which have radiated in or are largely confined to wet tropical habitats. Informative intrageneric comparisons occur in *Acanthiza, Conopophila, Meliphaga* and *Cracticus.* Informative between-genera comparisons are possible in the Climacteridae, Vangini in both Africa and Madagascar, New World jays and Asian magpies. Every one of these contrasts supports the view that cooperation is a phenomenon of open habitats, while pair-living occurs in species in dense, visually-occluded habitats. Although I am sympathetic to Brown's (1987) and Ford *et al.*'s

Table 5. Major taxa (Tribe and above) in which cooperative behaviour has not been recorded

Behavioural or habitat groupings	Major taxa
Display site defence promiscuity predominates	Menurinae
	Ptilonorhynchidae
	Paradisaeini
Other male emancipation system predominates	Atrichothorninae
Insectivores and frugivores of wet forests	Irenidae
	Vireonidae
	Pachycephalini
	Rhipidurini
	Aegithininae
	Callaeatidae[1]
Dense thickets and scrub	Dasyornithinae
	Malaconotini[2]

[1] Data on social behaviour in this group are unreliable, as all species are extinct or close to extinction
[2] Except for an isolated record in *Dryoscopus cubla*

(1988) suggestions that cooperation may not have been reported in rainforest because of the extreme difficulty in observing birds under those conditions, the contrasts I have used in establishing the correlations include only taxa where both rainforest and open habitat species have been quite well studied.

There are two published explanations for this pattern. Ford *et al.* (1988) attributed particular importance to the absence of seasonality in *Eucalyptus* woodlands relative to rainforest habitats in Australia. I have pointed out the difficulty of reconciling this explanation with the biogeographic history of the spread of *Eucalyptus* (Heinsohn *et al.* 1990; Cockburn 1991). The explanation is further weakened by the observation that it occurs on continents where *Eucalyptus* is not a component of the native vegetation, and in the absence of life history differences (e.g. increased clutch size) which might indicate greater seasonality in rainforests (Poiani and Jermin 1994). Indeed, Gaston (1978) has argued that cooperation is rare in rainforests because they are aseasonal. By contrast, Bell (1985) argued that group cohesion was difficult to maintain in these visually occluded habitats. Support for this view comes from the occurrence of another behavioural trait which is rare among birds but well-developed in the Corvida: duetting, or the precise alternation or synchronisation of song elements between paired males and females. The distribution of duetting in the Corvida suggests a broadly complimentary distribution to that of cooperative breeding (Table 6), and raises the possibility that the occurrence of conventional monogamy in visually-occluded habitats may itself be a highly derived state involving special patterns of communication between the partners involved (Thorpe 1961; Kunkel 1974; Farabaugh 1982). This in turn heightens the possibility that monogamy is often a derived and specialised state, and that, as suggested above, monogamy may often evolve from group-living.

Examination of the cladistic distribution of mating systems in the Corvida supports this view (Fig. 3). Monogamous clades form the twigs of the cladogram and scarcely ever the major branches, or the taxa that diverge early among particular clades. The early branches are typically taxa which display either a high level of cooperative breeding, or alternatively, often show complete male emancipation. Consider the following specific examples. Within the Meliphagoidea, all taxa except the terminal Dasyornithinae show some level of cooperative behaviour. The first branch is to the Maluridae, the most cooperative group. Within the true shrikes (Laniidae), there are three genera,

Table 6. Members of the Corvida where pairs show either simultaneous or antiphonal duet singing. Unless otherwise stated, data are derived from compilations in Thorpe (1961), Diamond (1972), Kunkel (1974), and Farabaugh (1982)

Major taxa	Species	Source
Meliphagidae	*Melithreptus laetior*	
	Myzomela pectoralis	
	Melidectes belfordi	
	M. ochromelas	
	Foulehaio carunculata	
	Gymnomyza viridis	
	Philemon argenticeps	
	P. corniculatus	
	P. novaeguineae	
	Acanthochaera chrysoptera	
	Acanthogenys rufogularis	
	Conopophila rufogularis	
	Phylidonyris albifrons	
	Lichmera indistincta	
	Lichenostomus fasciogularis	
	L. fuscus	
	L. melanops	
	L. flavescens	
	L. unicolor	
Dasyornithinae	*Dasyornis longirostris*	Serventy (1982)
Orthonychidae	*Orthonyx temmincki*	Cockburn (pers. obs.)
Pomatostomidae	*Pomatostomus temporalis*	Boles (1988)
Vireonidae	*Vireo latimeri*	Skutch (1960)
	Hylophilus aurantiifrons	
	H. decurtatus	
	H. ochraceiceps	
Cinclosomatinae	*Psophodes olivaceus*	
	P. nigrogularis	Boles (1988)
	P. cristatus	Boles (1988)
Pachycephalini	*Pachycephala rufiventris*	
	P. pectoralis	Brown and Brown (1994)
	Pitohui kirhocephalus	
	Pitohui incertus	Diamond and Raga (1978)
Artamini	*Cracticus sp.*	
	Gymnorhina tibicen	
Oriolini	*Coracina montana*	
	Campochaera sloetii	
	Hemipus picatus	
	Oriolus cholorocephalus	
	O. brachyrhynchus	
Monarchini	*Grallina cyanoleuca*	
Aegithininae	*Aegithina tiphia*	T.D. Price (pers. comm.).
Malaconotini	*Lanioturdus torquatus*	
	Nilaus afer	
	Laniarius barbarus	
	L. aethiopicus	
	L. erythrogaster	
	L. ferrugineus	
	L. turatii	
	L. bicolor	

continued over page

Table 6. continued

Major taxa	Species	Source
	L. ruficeps	
	L. funebris	
	L. atroflavus	
	L. mufumbiri	
	L. atrococcineus	
	L. fulleborni	
	L. luehderi	
	L. brauni	
	L. amboimbensis	
	L. poensis	
	L. leucorhynchus	
	Dryoscopus cubla	
	Tchagra tchagra	
	T. senegala	
	T. australis	
	T. minuta	
	Telophorus zeylonus	
	T. nigrifrons	
	T. sulfureopectus	
Vangini	*Platysteira cyanea*	
	Schetba rufa	Langrand (1990)
	Vanga curvirostris	Langrand (1990)
	Xenopirostris damii	Langrand (1990)
	X. polleni	Langrand (1990)
Callaeatidae	*Callaeas cinerea*	Robertson (1985)

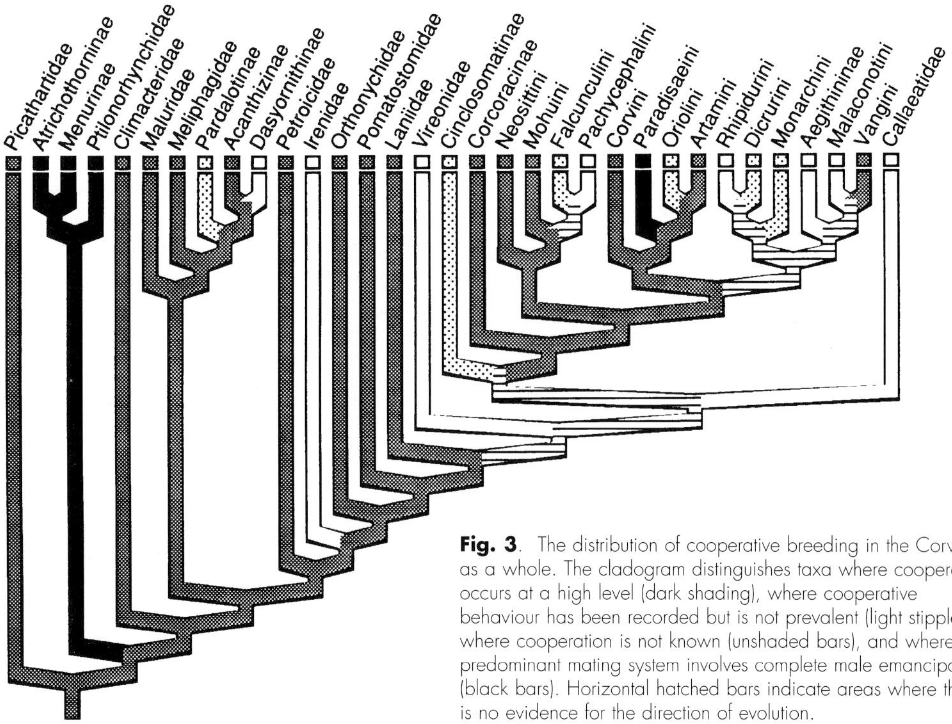

Fig. 3. The distribution of cooperative breeding in the Corvida as a whole. The cladogram distinguishes taxa where cooperation occurs at a high level (dark shading), where cooperative behaviour has been recorded but is not prevalent (light stipple), where cooperation is not known (unshaded bars), and where the predominant mating system involves complete male emancipation (black bars). Horizontal hatched bars indicate areas where there is no evidence for the direction of evolution.

two of which are confined to Africa, the presumed source of radiation for the Family. Both endemic genera are highly cooperative, and cooperation occurs in the third genus (*Lanius*) within Africa. However, cooperation has been lost in *Lanius*, where many of the well-studied species are migratory, or occur outside Africa. Within the Corvidae, the Pachycephalinae contains the largest exclusively pair-dwelling Australian taxon (Pachycephalini). However, both the early branches on this clade (the Australo-Papuan Neosittini and the New Zealand Mohuini) are highly cooperative, suggesting that the state in Pachycephalini may be derived. The only exception to this almost ubiquitous pattern is the clade containing the dircrurines, aegithinines, and malaconotines, where the pattern in the ancestral taxa is obscure. Among all birds, it is within this group that duetting is best developed.

The observation that many of the deepest branches on the tree are cooperative initially seems to sit uncomfortably with the view that rainforest is an environment less favourable to the development of cooperative breeding than open habitats, as for much of its history, Australia has been covered with rainforest. Indeed, some scenarios for life history evolution of these birds explicitly attempt to explain their habits as retention of behaviour that developed in rainforest (Yom-Tov 1987). However, this interpretation depends critically on the time at which the passerines diversified. The presence of a large radiations of passerines on Laurasia (Passerida), Australia (Corvida) and South America (Tyrannides) as well as groups scattered on New Zealand (Acanthisittides) and India/Madagascar (Eurylaimides) is consistent with a vicariant model, with each of the Gondwanan radiations taking place *in situ* after the fracturing of the continents towards the end of the Cretaceous. However this view is not supported by the fossil record, which suggests that while passerines may have occurred in Australia in the early Eocene (Boles 1995), passerines did not radiate extensively until the late Oligocene and Miocene (Feduccia 1995), when some modern forms also became recognisable in Australia (Boles 1993). By this stage diversity of Australian vegetation forms was increasing, first by the establishment of gradients of aridity away from the coast, but then more dramatically by a marked decline in rainforest, which was replaced initially by other forest types, and then by savannas (Martin 1994; McPhail *et al.* 1994; Kershaw *et al.* 1994), with concomitant effects on the fauna (Archer *et al.* 1994). This raises the intriguing possibility not just that most passerines are of Australian origin (see also Christidis and Schodde 1991), but also that the primary diversification of the Corvida took place in habitats other than rainforest, including environments conducive to the development of cooperative breeding.

Implications for the study of cooperative breeding

Finally, it is appropriate to ask whether this study identifies any recurrent patterns that may have been missed by the single-species approach to cooperative breeding. There are two observations that will repay further study. First, it is clear that observations of the behaviour of many of these species (including the most intensively cooperative) cannot be reconciled with the habitat saturation model, or the peculiar habitat or social constraints faced by the species which were used to derive that model (Florida scrub jay and acorn woodpecker). Many Australian species are habitually cooperative throughout their range, occurring in a wide variety of densities and habitats. The same seems likely to be true elsewhere, with studies of both African (Grimes 1980) and American Corvida (Williams *et al.* 1994) showing that group-living appears to be habitual in some species, even when vacant habitat for colonisation is available. Thus although the habitat saturation model has been supported by elegant experimental work on some species (see above), these studies may represent a special case among cooperative breeders as a whole. Further, there is no way of reconciling the absence of cooperation in rainforest taxa and other dense (and often patchily distributed) vegetation with the predictions of the habitat saturation model.

Second, very many species show extended natal philopatry without developing helping behaviour, from equatorial conditions (e.g. most *Coracina, Campephaga* and *Campochaera*; Bell 1982; Keith *et al.*

1992) to above the Arctic circle (e.g. *Perisoreus*; Ekman *et al.* 1994). Thus the advantages of prolonged parental investment and juvenile philopatry may be general in many or most of the non-cooperative Corvida as well. Indeed, it is reported increasingly often that the failure of offspring to stay and help is not a consequence of the offspring deciding to disperse (the critical question at the heart of the ecological constraints model), but instead of their being forced to do so by persistent parental aggression. For example, Noske (1991) reports that the salient difference between treecreepers in the cooperative genus *Climacteris* and the non-cooperative sister genus *Cormobates* was parental attacks on the young in *Cormobates*. Mulder (1995) has shown that the difference between *Malurus splendens*, where females are frequently philopatric and help (Russell and Rowley 1993a) or breed plurally (Rowley *et al.* 1989), and the sister species, *Malurus cyaneus*, where all females disperse, was intense maternal aggression against juvenile females at the start of the breeding season in the latter species. Collectively, these observations suggest that philopatry and helping will sometimes be determined by parental intervention as well as by offspring decisions. Offspring may often prefer to remain philopatric but be unable to do so. A richer theory of cooperative breeding will no longer assume that young are free to do as they choose, but will instead focus explicitly on the costs and benefits received by both adults and young.

While the benefits and costs to young have been well explored (Brown 1987; Emlen *et al.* 1991), what might be the costs and benefits to adults, and where might there be mutual interests and conflicts? Further, why should this balance be different in the Corvida relative to other taxa? Several authors have argued that there are consistent life history and demographic differences between the old endemic Australian passerines (Corvidae), new invaders to Australia, and other passerine birds in general. Collectively, these studies identify a syndrome of high survival and long lifespans (Fry 1980; Woinarski 1985a, 1989; Rowley and Russell 1991; but see Yom-Tov *et al.* 1992; Brawn *et al.* 1995), low clutch sizes (Yom-Tov 1987), long breeding seasons (Wyndham 1986) and prolonged reproductive investment (Woinarski 1985a, 1989), and indicate the possibility that these demographic differences may be associated with the distribution of cooperative breeding (Russell 1989). However, an explicit link between these habits and cooperation has not been suggested.

Current analyses do not support the view that cooperative Corvida in Australia have different demography to the non-cooperative taxa (Poiani and Jermin 1994), but this does not mean that life history parameters might not influence the propensity to develop cooperative behaviour. For example, while the study of the evolution of cooperation in eusocial insects was initially dominated by measurements of patterns of kinship and conflict within these societies, recent theoretical attention has also turned to the importance of demographic factors in the evolution of sociality (e.g. Gadagkar 1990, 1991), and mutualistic benefits to colony members (Seger 1991; Ito 1993). Gadagkar (1991) pays particular attention to the advantages of assured fitness returns for subordinate individuals. Similar considerations may lie at the heart of the prevalence of cooperative breeding in Australia.

In any environment where there is no obvious episodic period of mortality (such as migration) and breeders enjoy long lifespans, there are two difficulties. The first, faced by juveniles, is the well recognised difficulty that the young have in finding vacancies, increasing the adaptive advantage of staying within their own territory either in the hope of inheriting the vacancy (Wiley and Rabenold 1984), or of prospecting for high quality vacancies nearby (Zack 1990). This is one of the ideas that lies at the heart of the ecological constraints model, and particularly the benefits-of-philopatry version of this hypothesis.

The second difficulty is experienced by adults. Reproductive success for adult territorial birds will be derived from three sources; dispersal of progeny reared on the bird's territory to breeding vacancies elsewhere, egg-dumping and extra-pair mating in neighbouring territories, and having young inherit a breeding vacancy on the territory on the death of one of the breeding pair. Successful dispersal of young into adjacent territories will be highly stochastic and risk-prone. Explicit models of life history in environments of this sort consider the tradeoff between maximising reproductive effort, and minimising the risk of extinction of the genotype, assuming that there is a mortality cost associated with increasing fecundity. If the risk of total failure is minimised, Stearns and Crandall (1981) predict very modest rates of reproduction and great longevity, which are the characteristics attributed to the Corvida. These life history predictions could interact with the presence of helping in a variety of ways. First, as fecundity drops, the direct costs in terms of lowered food availability will decline, reducing one cost of helping behaviour. Second, the retention of helpers will reduce the vacancies available to dispersers, because many of those vacancies will be inherited directly. Counter to this, the retention of helpers will increase the possibility of successfully dispersing a young into a neighbouring territory if a vacancy does become available (Zack 1990). Analytical models investigating these effects are likely to depend on very precise quantification of dispersal probabilities, which are currently unavailable for any species (Stearns 1992). Current evidence suggests that extra-pair paternity and egg-dumping are highly variable among cooperative breeders, with species having both the highest (Mulder *et al.* 1994) and among the lowest known values (Westneat and Webster 1994). By contrast, ensuring territorial inheritance is comparatively risk averse, and although data are rare, at least one study suggests that it may be the most common path to avoiding total reproductive failure (Russell and Rowley 1993b). Partitioning of the sources of variation in lifetime success in birds may show this to be the central advantage to adults in tolerating helpers, and also a powerful stimulus to philopatry on the part of young. This model explicitly links all the features of the Corvida life history.

However, like other models, it suggests an advantage that should be general among many habitats, and does not obviously predict the rarity of cooperation in rainforests. The explanation for this correlation may come from the potential costs of helpers, of which reproductive conflict (notably sperm competition) will be among the most potent. The study of such conflict is in its infancy, but a burgeoning literature using molecular techniques to resolve genealogy suggests diverse and at times counterintuitive outcomes (e.g. Rabenold *et al.* 1990; Davies *et al.* 1992; Piper and Slater 1993; Haig *et al.* 1994; Gibbs *et al.* 1994; Jamieson *et al.* 1994; Lambert *et al.* 1994; Mulder *et al.* 1994; Dunn *et al.* 1995). In visually occluded habitats, control of group members may be more difficult for dominant birds (see Davies 1985), and the costs of juvenile retention may be intolerable.

Acknowledgements

I am grateful to the numerous authors who have paved the way for a comparative review of this sort. In particular, I do not believe that I would have had the courage to attempt this work without Brian Coates' (1990) magnificent summary of the literature on the passerines of Papua New Guinea. Isabel Castro, Ed Minot and Trevor Price shared their recent unpublished data on species that I would otherwise have been unable to discuss in this compilation. Comments on the manuscripts and ideas expressed herein were generously offered by David Curl, Ian Rowley, and the Behavioural Ecology Discussion Group at the ANU. My research on cooperative breeding has been supported by the Australian Research Council.

References

Ali S, Ripley SD (1971) *Handbook of the birds of India and Pakistan. Vol. 6. Cuckoo-shrikes to baboxes.* Oxford University Press, Bombay

Ali S, Ripley SD (1972) *Handbook of the birds of India and Pakistan. Vol. 7. Laughing thrushes to the mangrove whistler.* Oxford University Press, Bombay

Archer M, Hand SJ, Godthelp H (1994) Patterns in the history of Australia's mammals and inferences about palaeohabitats. In: Hill RS (eds) *History of the Australian vegetation: Cretaceous to Recent.* Cambridge University Press, Cambridge, pp 80–103

Aston HI (1988) Communal breeding by the Australian magpie-lark. *Emu* **88**:112–114

Baverstock PR, Schodde R, Christidis L, Krieg M, Birrell J (1992) Evolutionary relationships of the Australasian mud-nesters (Grallinidae, Corcoracidae): immunological evidence. *Aust J Zool* **40**:173–190

Beehler B, Pruett-Jones SG (1983) Display dispersion and diet of birds of paradise: a comparison of nine species. *Behav Ecol Sociobiol* **13**:229–38

Bell HL (1982) Are the cuckoo-shrikes *Coracina boyeri, C. melaena* and *Campochaera sloetii* cooperative breeders. *PNG Bird Soc Newsletter* **191–192**:8–11

Bell HL (1985) The social organization and foraging behaviour of three syntopic thornbills *Acanthiza* spp. In: Keast A, Recher HF, Ford H, Saunders D (eds) *Birds of Eucalyptus forests and woodlands: ecology, conservation and management.* RAOU and Surrey Beatty & Sons, Sydney, pp 151–163

Blakers M, Davies SJJF, Reilly PN (1984) *The atlas of Australian birds.* Melbourne University Press, Melbourne

Boles WE (1988) *The robins and flycatchers of Australia.* Angus & Robertson, Sydney

Boles WE (1993) A logrunner *Orthonyx* (Passeriformes, Orthonychidae) from the Miocene of Riversleigh, North-Western Queensland. *Emu* **93**:44–49

Boles WE (1995) The world's oldest songbird. *Nature* **374**:21–22

Boles WE, Longmore W, Lindsey TR (1981) Auxiliary at the nest of the blue-faced honeyeater. *Corella* **5**:36

Brawn JD, Karr JR, Nichols JD (1995) Demography of birds in a neotropical forest: effects of allometry, taxonomy, and ecology. *Ecology* **76**:41–51

Brown JL (1987) *Helping and communal breeding in birds: ecology and evolution.* Princeton University Press, Princeton.

Brown RJ, Brown MN (1994) Matched song and duetting by a breeding pair of golden whistlers *Pachycephala pectoralis. Emu* **94**:58–59

Christidis L, Schodde R (1991) Relationships of Australo-Papuan songbirds – protein evidence. *Ibis* **133**:277–285

Christidis L, Schodde R (1993) Relationships and radiations in the meliphagine honeyeaters, *Meliphaga, Lichenostomus* and *Xanthotis* (Aves, Meliphagidae): protein evidence and its integration with morphology and ecogeography. *Aust J Zool* **41**:293–316

Coates BJ (1990) *The birds of Papua New Guinea. Volume II. Passerines.* Dove Publications, Alderley, Queensland

Cockburn A (1991) *An introduction to evolutionary ecology.* Blackwell Scientific Publications, Oxford

Cramp S, Perrins CM (1993) *Handbook of the birds of Europe, the Middle East and North Africa. Volume VII. Flycatchers to shrikes.* Oxford University Press, Oxford

Cramp S, Perrins CM (1994) *Handbook of the birds of Europe, the Middle East and North Africa. Volume VIII. Crows to finches.* Oxford University Press, Oxford

Davies NB (1985) Cooperation and conflict among dunnocks, *Prunella modularis,* in a variable mating system. *Anim Behav* **33**:628–48

Davies NB (1991) Mating systems. In: Krebs JR, Davies NB (eds) *Behavioural ecology: an evolutionary approach.* 3rd ed. Blackwell Scientific Publications, Oxford, pp 263–294

Davies NB, Hatchwell BJ, Robson T, Burke T (1992) Paternity and parental effort in dunnocks *Prunella modularis*: how good are chick-feeding rules? *Anim Behav* **43**:729–745

Diamond JM (1972) Further examples of dual singing by southwest Pacific birds. *Auk* **89**:180–183

Diamond JM, Raga MN (1978) The mottle-breasted pitohui *Pitohui incertus. Emu* **78**:49–53

Dow DD (1980) Communally breeding Australian birds with an analysis of distributional and environmental factors. *Emu* **80**:121–140

Du Plessis MA, Siegfried WR, Armstrong AJ (1995) Ecological and life-history correlates of cooperative breeding in South African birds. *Oecologia* **102**:180–188

Dunn PO, Cockburn A, Mulder RA (1995) Fairy-wren helpers often care for young to which they are unrelated. *Proc R Soc Lond B Biol Sci* **259**:339–343

Edwards SV, Naeem S (1993) The phylogenetic component of cooperative breeding in perching birds. *Am Nat* **141**:754–789

Edwards SV, Naeem S (1994) Homology and comparative methods in the study of avian cooperative breeding. *Am Nat* **143**:723–733

Ekman J, Sklepkovych B, Tegelstrom H (1994) Offspring retention in the Siberian jay (*Perisoreus infaustus*): the prolonged brood care hypothesis. *Behav Ecol* **5**:245–253

Emlen ST (1982) The evolution of helping. I. An ecological constraints model. *Am Nat* **119**:29–39

Emlen ST (1991) Evolution of cooperative breeding in birds and mammals. In: Krebs JR, Davies NB (eds) *Behavioural ecology: an evolutionary approach 3rd ed.* Blackwell Scientific Publications, Oxford, pp 301–337

Emlen ST, Reeve HK, Sherman PW, Wrege PH, Ratnieks FLW, Shellman-Reeve J (1991) Adaptive versus nonadaptive explanations of behaviour: the case of alloparental helping. *Am Nat* **138**:259–270

Farabaugh SM (1982) The ecological and social significance of duetting. In: Kroodsma DE, Miller EH (eds) *Acoustic communication in birds.* Academic Press, New York, pp 85–124

Feduccia A (1995) Explosive evolution in Tertiary birds and mammals. *Science* **267**:637–638

Ford HA, Bell H, Nias R, Noske R (1988) The relationship between ecology and the incidence of cooperative breeding in Australian birds. *Behav Ecol Sociobiol* **22**:239–49

Freeman DJ (1970) The rediscovery of the black grass wren *Amytornis housei* with additional notes on the species. *Emu* **70**:193–195

Frith C, Frith D (1992) Annotated list of birds in Western Tari Gap, southern highlands, Papua New Guinea, with some nidification notes. *Aust Bird Watcher* **14**:262–276

Fry CH (1980) Survival and longevity among tropical landbirds. In: Johnson DN (ed) *Proceedings of the Pan-African Ornithological Congress* 1976, pp 333–343

Gadagkar R (1990) Evolution of eusociality: the advantage of assured fitness returns. *Philos Trans R Soc Lond B Biol Sci* **329**:17–25

Gadagkar R (1991) Demographic predisposition to the evolution of eusociality: a hierarchy of models. *Proc Natl Acad Sci USA* **88**:10993–10997

Gaston AJ (1978) The evolution of group territorial behaviour and cooperative breeding. *Am Nat* **112**:1091–1100

Gibbs HL, Goldizen AW, Bullough C, Goldizen AR (1994) Parentage analysis of multi-male social groups of Tasmanian native hens (*Tribonyx mortierii*): genetic evidence for monogamy and polyandry. *Behav Ecol Sociobiol* **35**:363–371

Gill BJ, McLean IG (1992) Population dynamics of the New Zealand whitehead (Pachycephalidae) – a communal breeder. *Condor* **94**:628–635

Ginn PJ, McIlleron WG, Milstein PlS (1989) *The complete book of South African birds.* Struik Winchester, Cape Town

Grimes LG (1976) The occurrence of cooperative breeding behaviour in African birds. *Ostrich* **47**:1–15

Grimes LG (1980) Observations of group behaviour and breeding biology of the yellow-billed shrike *Corvinella corvina. Ibis* **122**:166–192

Haig SM, Walters JR, Plissner JH (1994) Genetic evidence for monogamy in the cooperatively breeding red-cockaded woodpecker. *Behav Ecol Sociobiol* **34**:295–303

Hannon SJ, Mumme RL, Koenig WD, Pitelka F (1985) Replacement of breeders and within-group conflict in the cooperatively breeding acorn woodpecker. *Behav Ecol Sociobiol* **17**:303–12

Harris T, Arnott G (1988) *The shrikes of southern Africa.* Struik Winchester, Cape Town

Harrison CJO (1969) Helpers at the nest in Australian birds. *Emu* **69**:30–40

Harvey PH, Pagel MD (1991) *The comparative method in evolutionary biology.* Oxford University Press, Oxford

Harvey PH, Purvis A (1991) Comparative methods for explaining adaptations. *Nature* **351**:619–624

Heinsohn RG (1992) Cooperative enhancement of reproductive success in white-winged choughs. *Evol Ecol* **6**:97–114

Heinsohn RG, Cockburn A, Mulder RA (1990) Avian cooperative breeding: old hypotheses and new directions. *Trends Ecol & Evol* **5**:403–407

Ito Y (1993) *Behaviour and social evolution in wasps: the communal aggregation hypothesis.* Oxford University Press, Oxford

Jamieson IG, Quinn JS, Rose PA, White BN (1994) Shared paternity is a result of an egalitarian mating system in a communally breeding bird, the pukeko. *Proc R Soc Lond B Biol Sci* **257**:271–277

Keith S, Urban EK, Fry CH (1992) *The birds of Africa. Volume IV.* Academic Press, London

Kershaw AP, Martin HA, McEwen Mason JRC (1994) The Neogene: a period of transition. In: Hill RS (eds) *History of the Australian vegetation: Cretaceous to Recent.* Cambridge University Press, Cambridge, pp 299–327

Ketterson ED, Nolan V (1994) Male parental behavior in birds. *Annu Rev Ecol Syst* **25**:601–628

Koenig WD, Pitelka FA, Carmen WJ, Mumme RL, Stanback MT (1992) The evolution of delayed dispersal in cooperative breeders. *Q Rev Biol* **67**:111–150

Komdeur J (1992) Importance of habitat saturation and territory quality for evolution of cooperative breeding in the Seychelles warbler. *Nature* **358**:493–495

Kunkel P (1974) Mating systems of tropical birds: the effects of weakness or absence of external reproduction-timing factors, with special reference to prolonged pair bonds. *Z Tierpsychol* **34**:265–307

Lack D (1968) *Ecological adaptations for breeding in birds.* Chapman & Hall, London

Lambert DM, Millar CD, Jack K, Anderson S, Craig JL (1994) Single- and multilocus DNA fingerprinting of communally breeding pukeko: do copulations or dominance ensure reproductive success? *Proc Natl Acad Sci USA* **91**: 9641–9645

Langrand O (1990) *Guide to the birds of Madagascar.* Yale University Press, New Haven

Ligon JD (1993) The role of phylogenetic history in the evolution of contemporary avian mating and parental care systems. In: Power DM (ed) *Current ornithology.* Plenum Press, New York, pp 1–45

Longmore W (1991) *Honeyeaters and their allies.* Angus & Robertson, Sydney

Mackworth-Praed CW, Grant CHB (1973) *Birds of west central and central Africa. African handbook of birds. Series III, Volume II.* Longman, London

Maddison WP, Maddison DR (1992) MacClade: *Analysis of phylogeny and character evolution. Version 3.0.* Sinauer, Sunderland, Massachusetts

Madge S, Burn H (1994) *Crows and jays.* Christopher Helm, London

Major RE (1992) Mate guarding in a population of white-fronted chats, *Ephthianura albifrons* Jardine & Selby (Passeriformes, Ephthianuridae) – a response to group living and a male-skewed sex ratio. *Aust J Zool* **40**:401–409

Martin HA (1994) Australian Tertiary phytogeography: evidence from palynology. In: Hill RS (eds) *History of the Australian vegetation: Cretaceous to Recent.* Cambridge University Press, Cambridge, pp 104–142

Mayr E (1944) Timor and the colonization of Australia by birds. *Emu* **44**:113–130

Mayr E, Bock WJ (1994) Provisional classifications vs standard avian sequences: heuristics and communication in ornithology. *Ibis* **136**:12–18

McLennan DA, Brooks DR (1993) The phylogenetic component of cooperative breeding in perching birds: a commentary. *Am Nat* **141**:790–795

McPhail MK, Alley NF, Truswell EM, Sluiter IRK (1994) Early Tertiary vegetation: evidence from spores and pollen. In: Hill RS (eds) *History of the Australian vegetation: Cretaceous to Recent.* Cambridge University Press, Cambridge, pp 189–261

Mooers AO, Cotgreave P (1994) Sibley and Ahlquist's tapestry dusted off. *Trends Ecol & Evol* **9**:458–459

Mulder RA (1995) Natal and breeding dispersal in a cooperative, extra-group-mating bird. *J Avian Biol* **26**:234–240

Mulder RA, Dunn PO, Cockburn A, Lazenby-Cohen KA, Howell MJ (1994) Helpers liberate female fairy-wrens from constraints on extra-pair mate choice. *Proc R Soc Lond B Biol Sci* **255**:223–229

Mumme RL (1992a) Delayed dispersal and cooperative breeding in the Seychelles warbler. *Trends Ecol & Evol* **7**:330–331

Mumme RL (1992b) Do helpers increase reproductive success – an experimental analysis in the Florida scrub jay. *Behav Ecol Sociobiol* **31**:319–328

Noske RA (1991) A demographic comparison of cooperatively breeding and non-cooperative treecreepers (Climacteridae). *Emu* **91**:73–86

Noske RA (1992) The status and ecology of the white-throated grasswren *Amytornis woodwardi*. *Emu* **92**:39–51

Peterson AT, Burt DB (1992) Phylogenetic history of social evolution and habitat use in the *Aphelocoma* jays. *Anim Behav* **44**:859–866

Piper WH, Slater G (1993) Polyandry and incest avoidance in the cooperative stripe-backed wren of Venezuela. *Behaviour* **124**:227–247

Poiani A, Jermin LS (1994) A comparative analysis of some life history traits between cooperatively and non-cooperatively breeding Australian passerines. *Evol Ecol* **8**:1–18

Pruett-Jones SG, Lewis MJ (1990) Sex ratio and habitat limitation promote delayed dispersal in superb fairy-wrens. *Nature* **348**:541–542

Purvis A (1991) *Comparative analysis by independent contrasts*. V. 1.2. Oxford University, Oxford

Rabenold PP, Rabenold KN, Piper WH, Haydock J, Zack SN (1990) Shared paternity revealed by genetic analysis in cooperatively breeding tropical wrens. *Nature* **348**:538–540

Richner H (1990) Helpers-at-the-nest in carrion crows *Corvus corone corone*. *Ibis* **132**:105–108

Robertson CJR (1985) *Complete book of New Zealand birds*. Reader's Digest, Sydney

Rowley I (1965) The life history of the superb blue wren (*Malurus cyaneus*). *Emu* **64**:251–297

Rowley I (1968) Communal species of Australian birds. *Bonn Zool Beitr* **19**:362–368

Rowley I (1976) Co-operative breeding in Australian birds. In: Frith HJ, Calaby JH (eds) *Proceedings of the 16th International Ornitholgical Congress Australian Academy of Science*, Canberra, pp 657–666

Rowley I (1978) Communal activities among white-winged choughs *Corcorax melanorhamphos*. *Ibis* **120**:178–97

Rowley I, Russell E (1991) Demography of passerines in the temperate southern hemisphere. In: Perrins CM, Lebreton J-D, Hirons GJM (eds) *Bird population studies*. Oxford University Press, Oxford, pp 22–44

Rowley I, Russell E (1993) The purple-crowned fairy-wren *Malurus coronatus*. 2. Breeding biology, social organisation, demography and management. *Emu* **93**:235–250

Rowley I, Russell E, Brown R, Brown M (1988) The ecology and breeding biology of the red-winged fairy-wren *Malurus elegans*. *Emu* **88**:161–176

Rowley I, Russell E, Payne RB, Payne LL (1989) Plural breeding in the splendid fairy-wren, *Malurus splendens* (Aves: Maluridae), a cooperative breeder. *Ethology* **83**:229–247

Russell EM (1989) Cooperative breeding – a Gondwanan perspective. *Emu* **89**:61–62

Russell EM, Rowley I (1993a) Philopatry or dispersal: competition for territory vacancies in the splendid fairy-wren, *Malurus splendens*. *Anim Behav* **45**:519–539

Russell EM, Rowley I (1993b) Demography of the cooperatively breeding splendid fairy wren, *Malurus splendens* (Maluridae). *Aust J Zool* **41**:475–505

Sandbrink J, Robinson D (1994) An observation of communal breeding by southern whitefaces. *Corella* **18**:88

Schodde R (1982) *The fairy-wrens: a monograph of the Maluridae*. Lansdowne, Melbourne

Seger J (1991) Cooperation and conflict in social insects. In: Krebs JR, Davies NB (eds) *Behavioural ecology: an evolutionary approach 3rd ed*. Blackwell Scientific Publications, Oxford, pp 338–373

Serventy VN (1982) *The wrens and warblers of Australia*. Angus & Robertson, Sydney

Severinghaus LL (1987) Flocking and cooperative breeding of Formosan blue magpie. *Bull Inst Zool Acad Sin (Taipei)* **26**:27–37

Short RV (1979) Sexual selection and its component parts, somatic and genital selection as illustrated by man and the great apes. *Adv Study Behav* **9**:131–58

Sibley CG (1994) On the phylogeny and classification of living birds. *J Avian Biol* **25**:87–92

Sibley CG, Ahlquist JE (1985) The phylogeny and classification of the Australo-Papuan passerines. *Emu* **85**:1–14

Sibley CG, Ahlquist JE (1990) *Phylogeny and classification of birds: a study in molecular evolution*. Yale University Press, New Haven

Sibley CG, Monroe BL (1990) *Distribution and taxonomy of birds of the world*. Yale University Press, New Haven

Sibley CG, Monroe BL (1993) *A supplement to distribution and taxonomy of birds of the world.* Yale University Press, New Haven

Skutch AF (1960) *Life histories of Central American birds. II. Families Vireonidae, Sylviidae, Turdidae, Troglodytidae, Paridae, Corvidae, Hirundibidae and Tyrannidae.* Cooper Ornithological Society, Berkeley, California

Smith SM (1994) Social influences on the dynamics of a northeastern black-capped chickadee population. *Ecology* **75**:2043–2051

Stacey PB, Koenig WD (1990) *Cooperative breeding in birds: long-term studies of ecology and behavior.* Cambridge University Press, Cambridge

Stacey PB, Ligon JD (1991) The benefits-of-philopatry hypothesis for the evolution of cooperative breeding: variation in territory quality and group size effects. *Am Nat* **137**:831–846

Stearns SC (1992) *The evolution of life histories.* Oxford University Press, Oxford

Stearns SC, Crandall RE (1981) Bet-hedging and persistence as adaptations of colonizers. In: Scudder GGE, Reveal JL (eds) *Evolution today.* Hunt Institute, Philadelphia, pp 371–383

Strahan R (1994) *Cuckoos, nightbirds and kingfishers of Australia.* Angus & Robertson, Sydney

Thorpe WH (1961) *Bird song: the biology of vocal communication and expression in birds.* Cambridge University Press, Cambridge

Trivers R (1985) *Social evolution.* Benjamin/Cummings, Menlo Park, California

Watling D (1982) *Birds of Fiji, Tonga and Samoa.* Millwood Press, Wellington, New Zealand

Webster MS (1994) Female-defence polygyny in a neotropical bird, the *Montezuma oropendola. Anim Behav* **48**:779–794

Wesolowski T (1994) On the origin of parental care and the early evolution of male and female parental roles in birds. *Am Nat* **143**:39–58

Westneat DF, Webster MS (1994) Molecular analyses of kinship in birds: interesting questions and useful techniques. In: DeSalle R, Wagner GP, Shierwater B, Streit B (eds) *Molecular approaches to ecology and evolution.* Birkhauser Verlag, Basel, pp 91–126

Wiley RH, Rabenold KN (1984) The evolution of cooperative breeding by delayed reciprocity and queuing for favorable social positions. *Evolution* **38**:609–21

Williams DA, Lawton MF, Lawton RO (1994) Population growth, range expansion, and competition in the cooperatively breeding brown jay, *Cyanocorax morio. Anim Behav* **48**:309–322

Woinarski J (1985a) Breeding biology and life history of small insectivorous birds in Australian forests: response to a stable environment. *Proc Ecol Soc Aust* **14**:159–168

Woinarski J (1985b) Foliage gleaners of the treetops, the pardalotes. In: Keast A, Recher HF, Ford H, Saunders D (eds) *Birds of Eucalyptus forests and woodlands: ecology, conservation and management.* RAOU and Surrey Beatty & Sons, Sydney, pp 165–175

Woinarski J (1989) Some life history comparisons of small leaf gleaning bird species of south-eastern Australia. *Corella* **13**:73–80

Woodall PF (1980) Communal breeding in the figbird. *Sunbird* **11**:73–75

Woolfenden GE, Fitzpatrick JW (1984) *The Florida scrub jay: demography of a cooperative-breeding bird.* Princeton University Press, Princeton

Wyndham E (1986) Length of birds' breeding seasons. *Am Nat* **128**:155–64

Yamagishi S, Urano E, Eguchi K (1992) The social structure of the rufous vanga (*Schetba rufa*) in Ampijoroa, Madagascar. In: Yamagishi S (ed) *Social structure of Madagascar higher vertebrates.* Osaka City University, Osaka, pp 46–52

Yamagishi S, Urano E, Eguchi K (1995) Group composition and contributions to breeding by rufous vangas *Schetba rufa* in Madagascar. *Ibis* **137**:157–161

Yom-Tov Y (1987) The reproductive rates of Australian passerines. *Aust Wildl Res* **14**:319–330

Yom-Tov Y, McCleery R, Purchase D (1992) The survival rate of Australian passerines. *Ibis* **134**:374–379

Zack S (1990) Coupling delayed breeding with short-distance dispersal in cooperatively breeding birds. *Ethology* **86**: 265–286

Zack S, Stutchbury BJ (1992) Delayed breeding in avian social systems: the role of territory quality and 'floater' tactics. *Behaviour* **123**:194–219

ASSESSING EVOLUTIONARY HYPOTHESES GENERATED FROM MITOCHONDRIAL DNA: INFERENCES FROM *DROSOPHILA*

J.W.O. Ballard and N.J. Galway

ABSTRACT

Analysis of metazoan mtDNA permits and perhaps even demands an extension of phylogenetic thinking to the microevolutionary level and this approach can be directed towards understanding ecological questions. As such, data from mtDNA can provide a link between systematists, population geneticists and ecologists. In this paper we investigate some of the forces that shape mtDNA evolution. The influence of sampling strategies and analysis techniques is discussed as both can have profound effects on phylogenetic inference and the resultant formulation of conservation policies and systematic revisions. We also investigate the effect of sampling small regions of mtDNA and suggest that while longer sequences have the potential for greater phylogenetic resolution rigorous analysis can minimise the generation of incorrect genealogies.

Key words: Mitochondrial DNA, *Drosophila*, population genetics, systematics

Frontiers of Population Ecology, R.B. Floyd, A.W. Sheppard and P.J. De Barro (eds), CSIRO Publishing, Melbourne 1996, pp. 473–483

Introduction

There has been a dramatic increase in the application of genetic markers to problems in population genetics and systematics in the last decade and it is likely that this trend will flow into ecology during the next few years. A number of methods of genetic analysis are available, each with concomitant benefits and costs. Direct comparison of sequences offers extremely high resolution; this can be converted to estimates of sequence divergence and the data can be directly investigated using a multitude of statistical tests. An alternative assay for sequence variation involves comparison of the number and size of fragments produced by digestion of the DNA with restriction endonucleases. Restriction fragment length polymorphism (RFLP) analysis is cheaper than direct sequence comparisons, but offers less information on the evolution of the sequence itself. DNA fingerprinting is a powerful approach for inferring individual relatedness but like randomly amplified polymorphic DNA (RAPD's), should be restricted to inferences about close relatives. For many studies of mating systems, population structure, and heterozygosity estimates, isozyme electrophoresis remains the best technique available if sufficient variation can be found. In this paper we specifically investigate some of the pros and cons of employing mitochondrial DNA (mtDNA) as an evolutionary marker.

Mitochondria contain multiple copies of their own genome (mtDNA's). These copies are inherited independently of the nuclear genome mostly through cytoplasms of female gametes in higher plants and animals. Mitochondria detected in progeny are usually exclusively maternally derived although in some species sperm mitochondria enter an egg when fertilised (Dawid and Blacker 1972 but see Kondo *et al.* 1990). Consequently, if there is no 'leakage' of paternal mtDNA, a breakdown of the reproductive barriers between species will result in the offspring of an interspecific cross having the mtDNA of the maternal ancestor. Thus, investigation of mtDNA specifically tests maternal inheritance. If there are 10 successive backcrosses, less than 0.1% of the nuclear genome is derived from the female progenitor, while 100% of the mtDNA should remain unchanged. Painter *et al.* (1993) used their knowledge of the mode of mitochondrial inheritance to investigate the purity of the Melbourne Zoo mandrills (*Mandrillus sphinx*). Facial coloring of females suggested that the Zoo's founding female was a drill-mandrill hybrid. The founding female whose mother was suspected to be a drill (*Mandrillus leucophaeus*) is the only individual to have contributed mtDNA to the population. Painter *et al.* (1993) amplified and sequenced a 307-base pair region of the mtDNA cytochrome *b* locus from Melbourne Zoo animals and from known specimens of both mandrill and drill. The results obtained confirmed that all current members of the 'mandrill' population possess drill mtDNA, supporting the belief that the original female founder was a hybrid. This study shows how the theory behind the inheritance of mtDNA can be employed to unravel specific genetic hypotheses, in this case, the purity of the Melbourne Zoo mandrills. It also suggests caution be taken in transplanting and mixing of genetic 'stocks' within (or in this case between) species until the evolutionary relationships of the individuals are assessed.

An under-appreciated feature of mitochondrial population genetics is that there is only one genealogical history for the entire molecule because it does not normally recombine. This means that each haplotype in a population will have a particular history and all the genes (or alleles) on it should share that history. Thus, the phylogeny derived from one locus should be consistent with the 'true' genealogy – unless there is a bias in codon usage or a non-neutral pattern of amino acid replacements (Stewart and Wilson 1987; Shields *et al.* 1988; Ballard and Kreitman 1994; Rand *et al.* 1994).

The literature is rife with studies employing mtDNA as an evolutionary marker for biogeographical reconstructions, ethology, conservation issues and systematic questions. An example that links these three disciplines is that of Waldman *et al.* (1992). These workers examined the mitochondrial genetic structure of the toad *Bufo americanus* at five breeding ponds in Massachusetts using 21 different restriction enzymes and found that only 2 of 86 copulating pairs could possibly be siblings.

This result led the authors to suggest that siblings recognise and avoid their kin. Their finding demonstrates the versatility of recently developed molecular techniques in addressing questions concerning the heritability of behavioural traits in natural populations.

This paper presents the utility of mtDNA in understanding an organism, its history and environment and addresses some methodological issues that can effect the outcome of the analyses. We consider the evolutionary implications of a possible case of intraspecific introgression in the *Drosophila melanogaster* subgroup and use this example to illustrate the need for rigorous testing of phylogenetic hypotheses. We conclude by suggesting that the genealogy of multiple individuals from divergent populations should be assessed prior to invoking any conservation policies or systematic revisions and show how knowledge of the dynamics of the molecule may aid resolution of the controversy surrounding a common female ancestor of modern humans, the human mitochondrial 'eve'.

The Drosophila melanogaster *subgroup*

In *Drosophila* there are three distinct types of mtDNA's in *D. simulans* (*si*I, -II and -III) and two in *D. mauritiana* (*ma*I and -II) (Solignac and Monnerot 1986). The *si*III haplotype is endemic and confined to Madagascar where the three *D. simulans* haplotypes are sympatric (Solignac and Monnerot 1986). The *si*I haplotype is known to occur only on the Seychelles Islands, New Caledonia, Polynesia, Hawaii and Madagascar (Lachaise *et al.* 1988). The *si*II haplotype has a worldwide distribution but has not been collected on any Indian or Pacific Islands (Baba-Aïssa *et al.* 1988). *D. mauritiana* is confined to Mauritius where it is endemic (Tsacas and David 1974).

Wolbachia is a maternally inherited rickettsia-like microorganism that infects both *D. simulans* and *D. mauritiana*. *Wolbachia* has been shown to reduce gene flow among the three *D. simulans* haplotypes but not the two *D. mauritiana* haplotypes. Infected *D. simulans* can mate with any male and produce progeny, but uninfected females produce few offspring when they mate with infected males (Hoffmann *et al.* 1986; Turelli and Hoffmann 1991). Here the mtDNA is carried passively as the microorganism sweeps through the population – a parasite induced selective sweep. This results in a reduction in haplotypic diversity (Turelli *et al.* 1992). The potential for external agents, such as microorganisms, to directly influence the evolutionary dynamics of mtDNA have not been fully appreciated to date.

Methods

In this study we employ parsimony analyses with the cladistic permutation probability (PTP) test (Archie 1989; Faith and Cranston 1991), the topology-dependent PTP (T-PTP) test (Faith 1991), the bootstrap test (Efron 1982; Felsenstein 1985) and successive weighting (Farris 1969).

PTP tests investigate whether there is any hierarchical structure in the complete data set. T-PTP tests provide tests for specific hypotheses of monophyly or nonmonophyly by randomising in-group character states (e.g. Ballard *et al.* 1992) or by randomising all characters if no outgroup can be assumed *a priori* (D. Faith, *pers. comm.*).

Bootstrapping provides an imprecise measure of repeatability but is generally a highly conservative estimate of the probability of correctly inferring the corresponding clades (Hillis and Bull 1993). In the current paper a bootstrap proportion of greater than or equal to 80% is considered supportive (Hillis and Bull 1993).

The successive weighting method of Farris (1969) is used because cladistically related characters should be correlated with each other in a non-linear hierarchically correlated fashion. If there is only one true phylogeny, cladistically unrelated neutral characters should only be hierarchically related by chance. This expectation is appropriate for mtDNA as it does not normally recombine. Thus, Farris's

(1969) successive approximations approach to character weighting is appropriate for mtDNA where hypervariable 'blinking' sites may be expected to be highly homoplasious.

Results

In this paper we reanalyse a 2527 base pair region from seven haplotypes within the *D. melanogaster* subgroup: *D. melanogaster* , *D. sechellia*, *D. simulans si*I, -II and -III and *D. mauritiana ma*I and -II (Satta and Takahata 1990). Like Satta and Takahata (1990) the *D. yakuba* sequence (Clarey and Wolstenholm 1985) is employed as the outgroup (Lachaise *et al.* 1988). The 2527 base pair region (2531 including alignment gaps) includes three tRNA genes, most of the NADH dehydrogenase subunit 2 and cytochrome oxidase I. Satta and Takahata (1990) consider that their study revealed i) long persistence times of distinct haplotypes within a species, ii) variable substitution rates and, iii) the low saturation level of transitional differences. In this paper we consider points one and two by investigating the evolutionary implications of sampling strategies and considering the potential for intraspecific introgression. We have not retained the nomenclature of Satta and Takahata (1990), but employ the nomenclature of Baba-Aïssa and Solignac (1984) and Solignac *et al.* (1986) because it has priority.

We investigated the mean number of differences between *D. yakuba* and each of the seven other taxa by sliding a 20 base window along the 2531 base data set (Fig. 1). Parsimony analysis with *a priori* T-PTP testing and bootstrapping support monophyly of the *D. simulans* clade (*D. mauritiana, D. simulans* and *D. sechellia*) (Fig. 2. I). However, the *D. mauritiana* and *D. simulans* haplotypes are not monophyletic. Rather *D. mauritiana ma*II is basal, *D. simulans si*I and *D. sechellia* have sister mitochondrial genotypes and *D. simulans si*II, -III and *D. mauritiana ma*I cluster together (Fig. 2. I).

One explanation for polyphyly of the *D. mauritiana* haplotypes is the retention of ancestral polymorphisms (Avise *et al.* 1984). However, inspection of the data shows *D. simulans si*III and *D. mauritiana ma*I differ by a single substitution. Moreover, they have no restriction site length polymorphism differences (Solignac and Monnerot 1986). *D. mauritiana* (males) and *D. simulans* (females) can be easily crossed in the laboratory resulting in sterile male and fertile female F1 hybrids (Robertson 1983). Thus, it is possible that a female of *D. simulans si*III was blown or migrated from Madagascar to Mauritius where it mated with a male of *D. mauritiana ma*II resulting in the *D. mauritiana ma*I haplotype. Sequencing autosomal genes from known haplotypes has the potential to clarify this hypothesis. Surprisingly, Satta and Takahata (1990) do not consider this alternative.

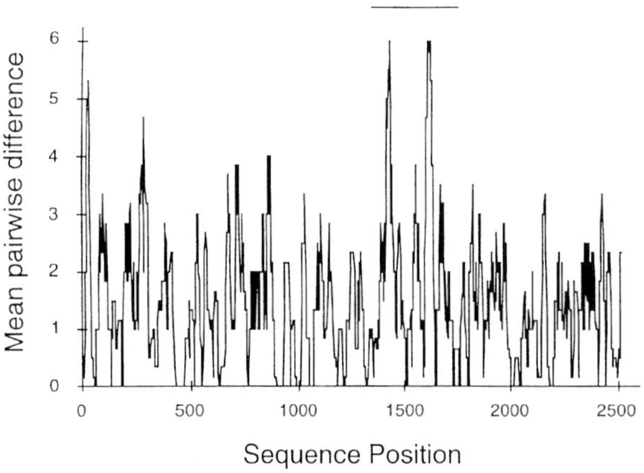

Fig. 1. Twenty base sliding-window of the mean pairwise difference between *D. yakuba* (as outgroup) and the other taxa in a 2531 base region of the mitochondrial genome. The region includes three tRNA genes and most of the NADH dehydrogenase subunit 2 (positions 1–825) and cytochrome oxidase (1034–2531) genes. The bar above the sliding-window shows the region chosen to illustrate the effect of sampling relatively small divergent regions of DNA.

If the *D. mauritiana ma*I mtDNA sequence is the result of interspecific introgression it should either be recognised as *D. simulans* mtDNA or it should be removed from the analysis. In either case parsimony analysis with *a priori* T-PTP testing and bootstrapping both support monophyly of *D. simulans si*II and -III (Fig. 2. I). Further analysis supports monophyly of *D. sechellia* and *D. simulans si*I (Fig. 2. I). These data suggest that *D. mauritiana* is the outgroup to *D. simulans* and *D. sechellia, D. simulans si*I has retained an ancient mitochondrial haplotype and *D. simulans si*II and -III have sister mitochondrial genotypes. This proposal is consistent with the hypotheses that, i) the three *D simulans* haplotypes originated on Madagascar (Monchamp-Moreau *et al.* 1991) where they are currently sympatric (Lachaise *et al.* 1988), ii) *D. simulans si*I migrated east colonising the Indian-Pacific Islands including the Seychelles Islands where *D. sechellia* is endemic (Lachaise *et al.* 1988), and iii) *D. simulans si*II migrated west and now has a worldwide distribution (Baba-Aïssa *et al.* 1988). *D. mauritiana* is endemic to Mauritius (Tsacas and David 1974). The direct implication from this interpretation is that calculations of the genetic diversity of *D. mauritiana* should not include *D. mauritiana ma*I.

Sampling and possible intraspecific introgression of mtDNA

Sub-sampling taxa

A variety of problems may arise from sub-sampling taxa. These may include i) changes in evolutionary hypotheses, ii) changes in the inferred support for monophyly of an assemblage and, iii) changes in evidence for introgression. The first two of these are general phenomena while the latter may be an unappreciated additional problem. In the first example we show that pruning a phylogenetic tree may not change the tree topology but it may change the inferred evolutionary relationships, in this case, because intraspecific introgression is unlikely to be considered. In the second example changes to the degree of support for monophyly are documented. While the degree of support required for significant monophyly is arbitrary (as is any significance level) inferences gained from phylogenetic reconstructions can influence decisions relating to a variety of conservation and systematics issues.

As a first example we shall consider the influence of taxon sampling and intraspecific introgression of mtDNA. If *D. simulans si*I and -III and *D. mauritiana ma*II are removed from the analysis (ie. not sampled) the tree topology does not change. However, the phylogenetic inferences obtained from this topology differ from that obtained when all the taxa are included in the analysis. This may sound counter intuitive but it should be stressed that a phylogenetic tree is an hypothesis about systematic relationships. In this particular example *D. simulans si*II and *D. mauritiana ma*I are monophyletic haplotypes (Fig. 2. II). We suggest that it is unlikely that introgression would be considered a possibility in this example because *D. simulans si*III was not sampled. Thus, the sequence similarity between *D. simulans si*III and *D. mauritiana maI* would not have been observed. As a result the systematic hypothesis generated from this example would be that *D. simulans si*II and *D. mauritiana ma*I are sister taxa and not *D. simulans* and *D. sechellia* . Thus, taxon sampling in concert with intraspecific introgression may not change the tree topology but can alter the phylogenetic inferences generated from that tree.

As a second example consider the removal of *D. mauritiana ma*I and subsequent independent tests for the monophyly of each *D. simulans* haplotype with *D. sechellia.* In this case parsimony analysis with T-PTP testing and bootstrapping support monophyly of *D. simulans si*I and *D. sechellia* but not of *D. simulans si*II or -III and *D. sechellia.* (Fig. 2. III).

Sampling small regions of mtDNA

Direct sequencing of mtDNA is clearly preferable to restriction fragment length polymorphism profiles because it helps to circumvent many of the difficulties of interpreting the particular molecular changes responsible for the observed restriction profiles. However, it is clearly not feasible

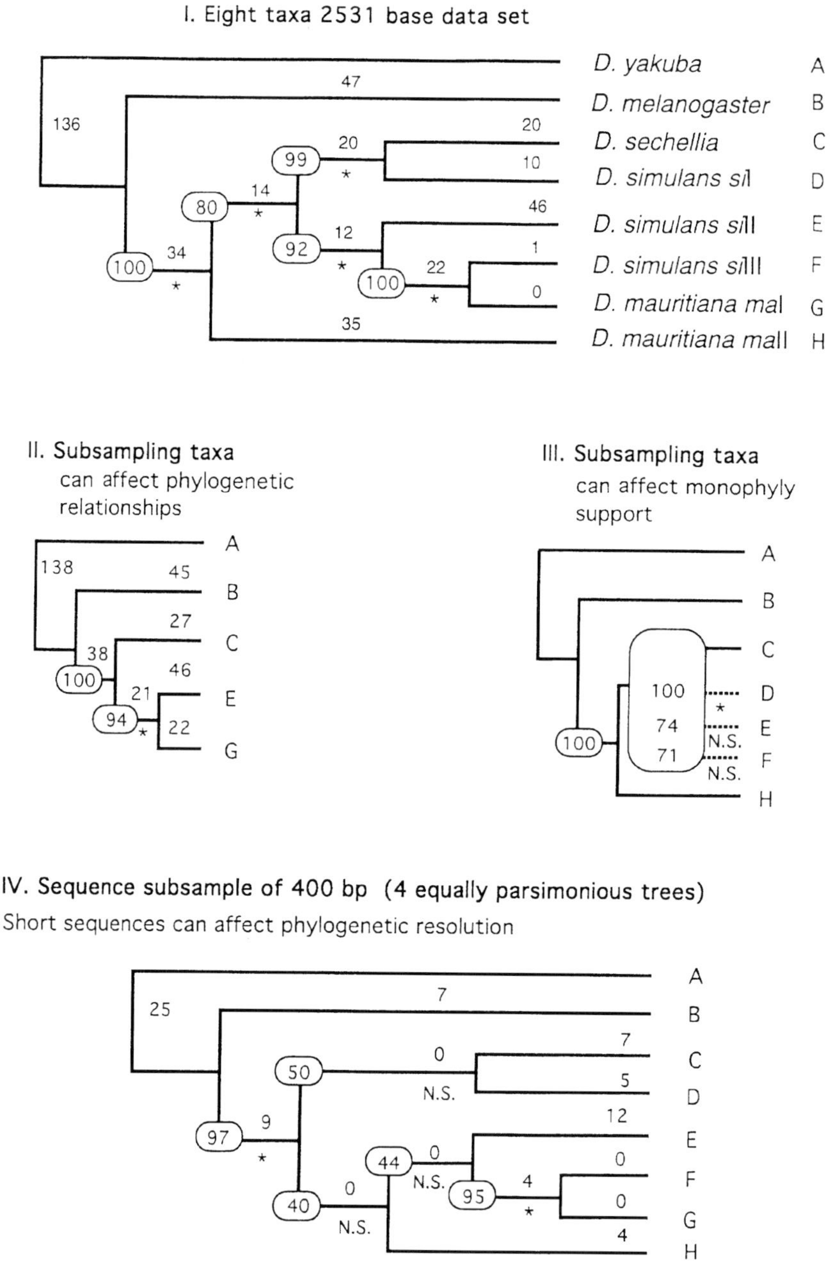

Fig 2. I: Phylogenetic analysis using PAUP (Swofford 1993) of the eight taxa 2531 base data set from Satta and Takahata (1990) generates a single most parsimonious tree of 397 steps (Consistency Index (CI) = 0.85). PTP testing shows significant structure to the 103 informative sites (PTP = 0.001 from 999 randomized trees). Branch lengths are recorded above each line, bootstrap proportions from 1000 pseudosamples are shown in circles and T-PTP results are shown beneath each line (a star indicates significance). Parsimony analysis with *a priori* T-PTP testing and bootstrapping supports monophyly of *D. mauritiana, D. simulans* and *D. sechellia* (T-PTP = 0.001, 19 steps – 100%), *D. sechellia, D. maritiana maI, D. simulans siI*, -II and -III (T-PTP = 0.001, 3 steps – 80%), *D. simulans*

Caption continued on next page ➤

to sequence the complete mitochondrial genome for every individual in a study. Thus, one specific question that arises is what is the effect of sampling small regions of mtDNA?

The phylogenetic signal of a region of mtDNA is dependent upon the question that is being considered. Regions under strong functional constraint will exhibit lower levels of polymorphism and divergence than regions with low functional constraint. Intuitively it may be expected that hypervariable sites may be homoplasious, however, if the mtDNA is evolving as a neutral-linked marker there should only be one 'true' phylogenetic history. Thus, in theory small divergent regions may generate a number of equally parsimonious trees but one of these should be the 'true' mitochondrial genealogy (this may not be the species tree).

To investigate this problem we returned to our *Drosophila* example and selected a 400 base region that had a high pairwise difference between *D. yakuba* and the seven other taxa (Fig. 1). For this data set, parsimony analysis generates four equally parsimonious trees one of which has the same topology as the eight taxa 2531 base data set (Fig. 2. IV). Reduction of non-significant nodes defined either by T-PTP testing or bootstrapping generates a tree that is simply less resolved than the hypothesised true tree.

Application of successive weighting (Farris 1969) to the 400 base pair data set using the rescaled consistency index with a base weight of 1000 using a program for phylogenetic analysis using parsimony (PAUP, Swofford 1993) generates two trees of length 6500 steps after two iterations. One of these trees has the same topology as the 2531 base data set. The other places *D. sechellia* and *D. simulans si*I outside *D. mauritiana ma*I and -II and *D. simulans si*II and -III. In this case small regions of mtDNA have not given an incorrect genealogy. Theory suggests that neutral hypervariable characters should not give a consistent hierarchical signal, thus the successive approximations approach to character weighting is appropriate for mtDNA (Farris 1969). However, it may not be appropriate in regions where recombination is high and genealogies can become uncoupled.

In direct contrast to these data, Satta and Takahata (1990) suggest that positions 316–1516 and 2003–2564 and positions 38–315 and 1517–2002 give distinct UPGMA (unweighted pair group method using arithmetic averages) tree topologies. We have analysed these positions and do not find any differences in the topologies of these regions generated by parsimony. Thus the differences Satta and Takahata (1990) observed were artifacts of using UPGMA. UPGMA assumes a molecular clock hypothesis and is known to be an inconsistent technique for phylogenetic reconstructions (Huelsenbeck and Hillis 1993).

Caption continued from previous page

*si*I, *si*II and *D. mauritiana ma*I (T-PTP = 0.001, 7 steps – 92%) and *D. sechellia* and *D. simulans si*I (T PTP = 0.001, 11 steps – 99%). Removal of *D. mauritiana ma*I because the haplotype may be the result of introgression does not change the phylogenetic relationships of the remaining taxa or alter the significance of any T-PTP test. II: Sub-sampling taxa can influence the interpretations of phylogenetic inference. For example, parsimony analysis supports monophyly of *D. simulans* and *D. mauritiana* if *D. simulans si*I and *D. mauritiana ma*II are excluded (*a priori* T-PTP = 0.002, 7 steps – 94%). III: Sub-sampling taxa can influence monophyly support. For example, the alternate inclusion of one *D. simulans* haplotype effects the inferences of *D. simulans* and *D. sechellia* monophyly if *D. mauritiana ma*I is excluded (*D. simulans si*I and *D.sechellia* ,*a priori* T-PTP = 0.001, 13 steps – 100%; *D. simulans si*II and *D. sechellia*, *a priori* T-PTP = 0.053, 3 steps – 74%; *D. simulans si*III and *D. sechellia a priori* T-PTP = 0.068, 3 steps – 71%). IV: Short sequences can affect phylogenetic resolution. Including 400 of the 2531 bases (positions 1450–1859) generates four equally parsimonious trees of length 68 steps (CI= 0. 91). This is a bootstrap consensus showing all compatible groups. PTP testing indicates there is significant structure in the 21 informative sites (PTP = 0.001). The 400 base data set supports monophyly of *D. mauritiana*, *D. simulans* and *D. sechellia* (*a priori* T PTP = 0.001, 4 steps – 97%) but there is little resolution within the *D. simulans* clade. The tree generated from the 2531 base and the four trees generated from the 400 base data sets are not significantly different (comparing the two topologies T-PTP = 0.12, 0 steps using the 2531 base data set).

Discussion

This simple case study has demonstrated that phylogenetic inferences can be directly influenced by sampling strategies. Thus, the genealogical relationships of multiple individuals from divergent populations should be assessed prior to invoking any conservation policies or systematic revisions. But is the clade, the species, the haplotype or the population the fundamental unit of conservation? Avise and colleagues (for example Avise 1992; Ball and Avise 1992) argue that many species should not be viewed as monotypic entities but rather a series of geographically differing populations with a hierarchical and sometimes deep genetic and historic structure. As an example, Bowen *et al.* (1993) employed 17 restriction enzymes to assess population genetic structure and evolutionary relationships among 113 samples from four nesting beaches of loggerhead turtles, *Caretta caretta*, in the northwestern Atlantic Ocean and from one nesting beach in the Mediterranean Sea. Significant differences in haplotype frequency between nesting populations in Florida and in Georgia/ South Carolina, and between both of these and the Mediterranean nesting colony prompted these workers to suggest that nesting populations should be managed as demographically independent units.

The value of this approach can be seen by returning to our case study. In this study the three *D. simulans* mitochondrial haplotypes do not form a monophyletic assemblage relative to *D. sechellia* and *D. mauritiana ma*I. Moreover, Ballard and Kreitman (1995) have shown an uncoupling in the evolutionary rates among the three *D. simulans* mitochondrial haplotypes that cannot plausibly be explained by selection. These data prompted an investigation of the mating behaviour of three *D. simulans* haplotypes. Preliminary evidence from pairwise comparisons shows assortative pre-mating behaviour among the *D. simulans* haplotypes suggesting that they are at least partially genetically distinct (Ballard unpubl. data). This is not to say the haplotypic divergence causes the assortative mating. It is more likely that mtDNA is a simple molecular marker. These data, along with those of Painter *et al.* (1993) on the Melbourne Zoo mandrill and Bowen *et al.* (1993) on the loggerhead turtle, suggest caution be taken in transplanting and mixing of genetic 'stocks' until the evolutionary relationships of the haplotypes, populations or individuals are assessed.

Data presented here supports the intuitive notion that longer mtDNA sequences have the potential for stronger phylogenetic signal (Saitou and Nei 1986) but rigorous analysis should minimise the generation of incorrect genealogies. However, the most appropriate locus for a specific question is still equivocal and is in need of attention.

The mtDNA control region, the major non-coding region containing the origin of the heavy strand of replication and the displacement loop (D loop) in mammals or the A+T rich region in insects is often considered to be the most variable part of the animal mitochondrial genome (Brown 1985). For this reason it has been used for investigating closely related mitochondrial lineages within species. For example, Edwards (1993) studied the mitochondrial gene genealogy and gene flow among island and mainland populations of the grey-crowned babblers *Pomatostomus temporalis* in northern Australia using D-loop mtDNA sequence data. Phylogenetic analyses of the sequence variation observed among the 44 individuals suggest that the number of times lineages in one population trace back to ancestors of a different population was too high to be compatible with a model supporting population divergence solely by genetic drift, thus, ongoing gene flow is probably occurring. However, the high rate of D-loop variation has caused problems in other circumstances. For example, Vigilant *et al.* (1991) presented a highly controversial data set and concluded that their data supported a common human mtDNA ancestor some 200,000 years ago because i) the most basal splits of a maximum parsimony tree are among purely African lineages and ii) this branch has the greatest mutational depth of all branches. Unfortunately, there are serious flaws with both of these points. Firstly, both Templeton (1991) and Hedges *et al.* (1991) found parsimony trees shorter than that reported by Vigilant that do not have Africans as a basal group. Secondly, Templeton

(1993) has shown that the diversity of the African and Asian populations is not significantly different. One of the major problems with this data set is that there are a number of hypervariable sites in the data set and as would be expected in parsimony analysis, when the number of sequences (136 humans) is larger than the number of characters (117 informative sites) there is a large (and in this case unknown) number of parsimony trees. We suggest that the controversy concerning human origins (reviewed by Templeton 1993) could be resolved if sequences from linked loci could be employed to impose *a priori* structure on the D-loop data. This is not as problematic as one might first suppose because Marzuki *et al.* (1991) have already begun to compile a database of normal variants of human mtDNA. In that study Marzuki *et al.* (1991) sequenced the entire mtDNA of 13 Caucasians (Ed Byrne *pers. comm.*) and report 128 nucleotide variants: 21 of which were in the ND5 region alone. Thus, the monophyly of Africans, for example, could be investigated by sequencing the ND5 region. Significant structure obtained by bootstrapping or T-PTP testing of the ND5 data could then be imposed on the more variable D-loop sequence data.

MtDNA has provided the most extensive and readily accessible DNA data that permits a strong conceptual bridge between the nominally separate disciplines of systematics, population genetics and ecology. In this paper some of the forces that shape mtDNA evolution have been investigated. Properties unique to mtDNA have enormous potential for providing insight at a molecular level to answer ecological and genetic problems. Mitochondrial haplotypes are useful for addressing the female component of the population structure but they leave parallel questions about male dispersal unresolved. In the future, studies should integrate mtDNA and autosomal data to provide a more rounded picture of the evolutionary dynamics of a species.

Acknowledgements

Thanks to Kirrie Ballard and the Coopers and Cladistics discussion group for their constructive comments. Financial support to J.W.O.B. was supplied by a C.J. Martin Postdoctoral Fellowship from the Australian National Health and Medical Research Council. N.J.G. is supported by Cooperative Research Center for Plant Science, ANU and CSIRO.

References

Archie JW (1989) A randomization test for phylogenetic information in systematic data. *Syst Zool* 38:219–252

Avise JC (1992) Molecular population structure and the biogeographic history of a regional fauna: A case history with lessons for conservation biology. *Oikos* 63:62–76

Avise JC, Neigel JE, Arnold, J (1984) Demographic influences on mitochondrial lineage survivorship in animal populations. *J Mol Evol* 20:99–105

Baba-Aïssa F, Solignac M (1984) La plupart des populations de *Drosophila simulans* ont problement pour ancêtre une femelle unique dans un passé récent. *CR Acad Sci (Paris)* 299:289–292

Baba-Aïssa F, Solignac M, Dennebouy N, David JR (1988) Mitochondrial DNA variability in *Drosophila simulans*: Quasi absence of polymorphism within each of the three cytoplasmic races. *Heredity* 61:419–426

Ball RM, Avise JC (1992) Mitochondrial DNA phylogeographic differentiation among avian populations and the evolutionary significance of subspecies. *Auk* 109:626–636

Ballard JWO, Kreitman M (1994) Unraveling selection in the mitochondrial genome of *Drosophila*. *Genetics* 138:757–772

Ballard JWO, Kreitman M (1995) Is mitochondrial DNA a strictly neutral marker. *Trends Ecol & Evol* 10:485–488

Ballard JWO, Olsen GJ, Faith DP, Odgers WA, Rowell DM, Atkinson PW (1992) Evidence from 12S ribosomal RNA sequences that onychophorans are modified arthropods. *Science* 258:1345–1348

Bowen B, Avise JC, Richardson JI, Meylan AB, Margaritoulis D, Hopkins-Murphy SR (1993) Population structure of loggerhead turtles (*Caretta caretta*) in the Northwestern and Mediterranean Sea. *Conserv Biol* 7:834–844

Brown WM (1985) The mitochondrial genome of animals. In: MacIntyre RJ (ed) *Molecular evolutionary genetics*. Plenum Press, New York, pp 95–130
Clarey DO, Wolstenholm DR (1985) The mitochondrial molecule of *Drosophila yakuba*: Nucleotide sequence, gene organization and genetic code. *J Mol Evol* **22**:252–271
Dawid IB, Blacker AW (1972) Maternal and cytoplasmic inheritance of mitochondrial DNA in *Xenopis*. *Dev Biol* **29**:152–161
Edwards SV (1993) Mitochondrial gene genealogy and gene flow among island and mainland populations of a sedentary songbird the grey-crowned babbler (*Pomatostomus temporalis*). *Evolution* **47**:1118–1137
Efron B (1982) The jackknife the bootstrap and other resampling plans. *Conf Board Math Sci Soc Ind Appl Math* **38**:1–92
Faith DP (1991) Cladistic permutation tests for monophyly and nonmonophyly. *Syst Zool* **40**:366–375
Faith DP, Cranston PS (1991) Could a data set this short have arisen by chance alone?: On permutation tests for cladistic structure. *Cladistics* **7**:1–28
Farris JS (1969) A successive approximations approach to character weighting. *Syst Zool* **18**:374–385
Felsenstein J (1985) Confidence limits on phylogenies: An approach using the bootstrap. *Evolution* **39**:783–791
Hedges SB, Kumar S, Tamura K, Stoneking M (1991) Human origins and analysis of mitochondrial DNA sequences: reply. *Science* **255**:737
Hillis DM, Bull JJ (1993) An empirical test of bootstrapping as a method for assessing confidence in phylogenetic analysis. *Syst Biol* **42**:182–192
Hoffmann AA, Turelli M, Simmons GM (1986) Unidirectional incompatibility between populations of *Drosophila simulans* . *Evolution* **40**:692–701
Huelsenbeck JP and Hillis DM (1993) Success of phylogenetic methods in the four taxon case. *Syst Biol* **42**:427–264
Kondo R, Satta Y, Matsuura ET, Ishiwa H, Takahata N, Chigusa SI (1990) Incomplete maternal transmission of mitochondrial DNA in *Drosophila*. *Genetics* **126**:657–663
Lachaise D, Cariou M-L, David JR, Lemeunier F, Tsacas L, Ashburner M (1988) Historical biogeography of the *Drosophila melanogaster* subgroup. *Evol Biol* **22**:159–225
Marzuki S, Noer AS, Lertrit P, Thyagarajan D, Kapsa K, Utthanapol P, Byrne E (1991) Normal variants of human mitochondrial DNA and translation products: the building of a reference data base. *Hum Genet* **88**:139–145
Monchamp-Moreau C, Ferveur J-F, Jacques M (1991) Geographic distribution and inheritance of three cytoplasmic incompatibility types in *Drosophila simulans*. *Genetics* **129**:399–407
Painter JN, Crozier RH, Westerman M (1993) Molecular identification of a *Mandrillus* hybrid using mitochondrial DNA. *Zool Biol* **12**:359–365
Rand D, Dorfsman M, Kann LM (1994) Neutral and non-neutral evolution of *Drosophila* mitochondrial DNA. *Genetics* **138**:741–756
Robertson HM (1983) Mating behavior and the evolution of *Drosophila mauritiana*. *Evolution* **37**:1283–1293
Saitou N, Nei M (1986) The number of nucleotides required to determine the branching order of three species with special reference to the human-chimpanzee-gorilla divergence. *J Mol Evol* **24**:189–204
Satta Y, Takahata N (1990) Evolution of *Drosophila* mitochondrial DNA and the history of the *melanogaster* subgroup. *Proc Natl Acad Sci USA* **87**:9558–9562
Shields DC, Sharp PM, Higgins DG, Wright F (1988) 'Silent' sites in *Drosophila* genes are not neutral: Evidence of selection among synonymous codons. *Mol Biol Evol* **5**:704–716
Solignac M , Monnerot M (1986) Race formation and introgression within *Drosophila simulans D mauritiana* and *D sechellia* inferred from mitochondrial DNA analysis. *Evolution* **40**:531–539
Solignac M, Monnerot M, Mounolou J-C (1986) Mitochondrial DNA evolution in the melanogaster species subgroup of *Drosophila*. *J Mol Evol* **23**:31–41
Stewart C -B , Wilson AC (1987) Sequence convergence and functional adaptation of stomach lysozymes from foregut fermenters. *Cold Spring Harbor Symp Quant Biol* **52**:891–899

Swofford DL (1993) *PAUP: Phylogenetic analysis using parsimony. Version 3.1.* Computer program distributed by the Illinios Natural History Survey, Champaign, Illinios
Templeton AR (1991) Human origins and analysis of mitochondrial DNA sequences. *Science* 255:737
Templeton AR (1993) The 'eve' hypothesis: A genetic critique and reanalysis. *Am Anthropol* 95:51–72
Tsacas L, David J (1974) *Drosophila mauritiana* n sp du groupe *melanogaster* de l'île Maurice (Dipt Drosophilidae). *Bull Soc Entomol Fr* 79:42–46
Turelli M, Hoffmann AA (1991) Rapid spread of an inherited incompatibility factor in California *Drosophila. Nature* 353:440 442
Turelli M, Hoffmann AA, McKechnie SW, (1992) Dynamics of cytoplasmic incompatibility and mtDNA variation in natural *Drosophila simulans* populations. *Genetics* 132:713–723
Waldman B, Rice JE, Honeycutt RL (1992) Kin recognition and incest avoidance in toads. *Am Zool* 32:18–30
Vigilant L, Stoneking M, Harpending H, Hawkes K, Wilson AC (1991) African populations and the evolution of human mitochondrial DNA. *Science* 253:1503–1507

Genetic structure of invading insects and the case of the knopper gallwasp

Paul Sunnucks and Graham N. Stone

ABSTRACT

Genetic subdivision and small population size can have considerable evolutionary consequences, as illustrated by studies of the cynipid gallwasp *Andricus quercuscalicis* which has recently colonised large areas of Western Europe. Population processes associated with colonisation were investigated by allozyme analysis of 1400 wasps in 57 populations. Patterns of genetic variation showed that very severe serial genetic subsampling occurred in a westerly direction which generated diverse new genetic combinations in the invaded range. Colonisation appeared to occur with very small numbers of founders reaching suitable habitat patches, and invasion progressed in a generally linear directional 'stepping-stone' pattern with topography dictating the routes of colonisation. The genetic structure of colonising or subdivided insect populations is reviewed.

Key words: Colonisation, phylogeography, population genetics, genetic variation, population subdivision, *Andricus quercuscalicis*, gallwasp

Frontiers of Population Ecology, R.B. Floyd, A.W. Sheppard and P.J. De Barro (eds), CSIRO Publishing, Melbourne 1996, pp. 485–495

Introduction

Genetic variation in small or subdivided populations is of central importance in evolutionary biology; in evolution of new characters and ultimately on speciation (Mayr 1982; Berlocher 1984; Endler 1992). Populations may lose genetic variation by decline through human interference, occupation of marginal or transitory habitat or specialised niches, and formation of new groups during colonisation. Relevant studies on insects represent diverse interests including experimental evolutionary biology (Bryant and Meffert 1992), metapopulation dynamics (Seitz and Komma 1984; Eber *et al.* 1992; Eber and Brandl 1994; McCauley 1991; Whitlock 1992), conservation genetics (Frankham 1995a,b; Brakefield and Saccheri 1994), and many applied situations including pest invasions (Gasperi *et al.* 1991; Puterka *et al.* 1993) and changes occurring during colonisations (Parker *et al.* 1977; Bryant *et al.* 1981; Berlocher 1984; Ross and Trager 1990).

We describe here a study of the genetic effects of a recent natural range expansion by the cynipid gallwasp, *Andricus quercuscalicis* (Schönrogge *et al.* 1995; Stone and Sunnucks 1993; Sunnucks *et al.* 1994), and set the results in the context of other studies of subdivided insect populations.

Invasion of Western Europe by *Andricus quercuscalicis*

The life cycle and biogeography of Andricus quercuscalicis

The ecology and biology of *A. quercuscalicis* is described briefly below, based on a number of recent reviews (Hails 1994; Sunnucks *et al.* 1994; Schönrogge *et al.* 1995). *A. quercuscalicis* has two generations per year with an asexual generation on pedunculate oak, *Quercus robur*, and an alternating sexual generation on Turkey oak, *Q. cerris*. Pedunculate oak is now common throughout western Europe, as a result of post-glacial range expansion, and occurs naturally with Turkey oak in southeastern Europe and Turkey. However, in the last 300–400 years, Turkey oak has been planted outside its native range and is now common in southern and western Germany and southern Britain, although it is relatively rare in northern France, Belgium, the Netherlands, northern Britain and Ireland. In its introduced range, distances between Turkey oak patches are much larger and the number of oaks in each patch are far lower than in its native range.

The native range of *A. quercuscalicis* is limited to areas where both Turkey oak and pedunculate oak occur together in central and southeastern Europe and Turkey. Following human dispersal of Turkey oak, *A. quercuscalicis* has invaded areas up to 1600 km from its native range. It reached eastern Germany by 1631, progressed through western Europe, and was first recorded in Britain in the 1950s. It has since spread rapidly throughout England and Wales, and reached southern and eastern Ireland in the late 1980s.

Genetic studies of the invasion of A. quercuscalicis

Genetic variation in agamic female *A. quercuscalicis* reared from galls was detected at seven allozyme loci, and scored in 823 wasps from 39 locations through the species' range (Stone and Sunnucks 1993). The data set has since been expanded to around 1400 wasps from a total of 58 locations (Fig. 1) (P. Sunnucks and G. Stone unpublished). The population genetic analysis of these samples has been described in Stone and Sunnucks (1993).

Linkage disequilibrium (LD) measures how co-occurrences of alleles at two loci diverge from random. From allelic correlations we estimated genetically effective population size, N_e (Bartley *et al.* 1992). High LD caused by founder effects is expected to decline in just a few generations in sexual populations through recombination and segregation. Consequently, it has been predicted that LD should be highest in recently formed populations in the invaded parts of a species' range (Berlocher 1984).

The invasion process resulted in loss of genetic variation in the newly-colonised range of *A. quercuscalicis* with heterozygosity and allelic diversity declining linearly with distance west from

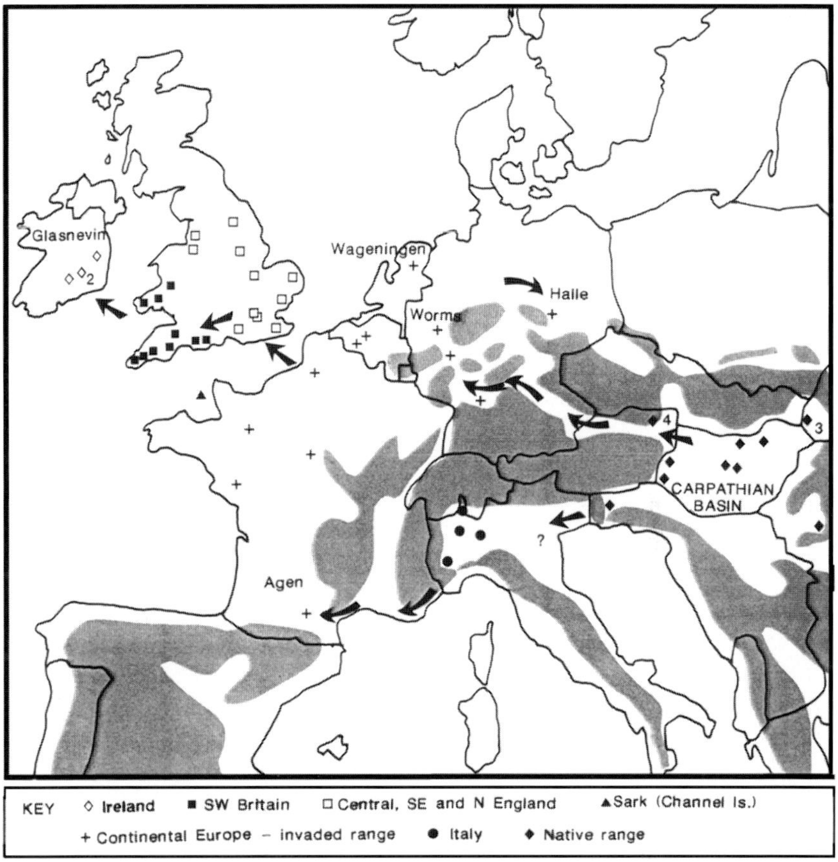

Fig. 1. Map of 58 sampling sites of *A. quercuscalicis* from various regions of Europe. Numbers next to symbols denote multiple sites. Shaded areas represent elevation above 400 m. Arrows denote routes of invasion indicated by genetic data, and '?' denotes a possible invasion route. Sites mentioned specifically in the text are named.

Hungary (Stone and Sunnucks 1993). For ease of presentation, the populations in the extended data set have been grouped into regions (Fig. 1) as suggested by earlier analyses. Genetic variation in each region declined in the order, native range > Italy > other continental Europe > Central/SE/N England > SW Britain > Ireland. Non-parametric tests showed that in most cases, genetic variation was significantly lower in populations that were further from the native range (Fig. 2).

Colonisation was associated with an increase in genetic isolation of populations in the invaded range, which was seen in three separate approaches (Stone and Sunnucks 1993). Firstly, standardised genetic variance (F_{st}) may be interpreted as the proportion of genetic variance due to differences between populations. In the native range, F_{st} was only 0.03 indicating little genetic subdivision. In contrast, F_{st} was 0.19 for invaded continental Europe, and 0.16 in Britain. However, much of the variance in Britain occurred between the two regions of South West Britain and Central/SE/N England which had within region F_{st} of only 0.09 and 0.07 respectively. The distinction between these regions reflects the absence of two alleles in South West Britain, probably as a result of receiving only a few colonists from Central/SE/N England.

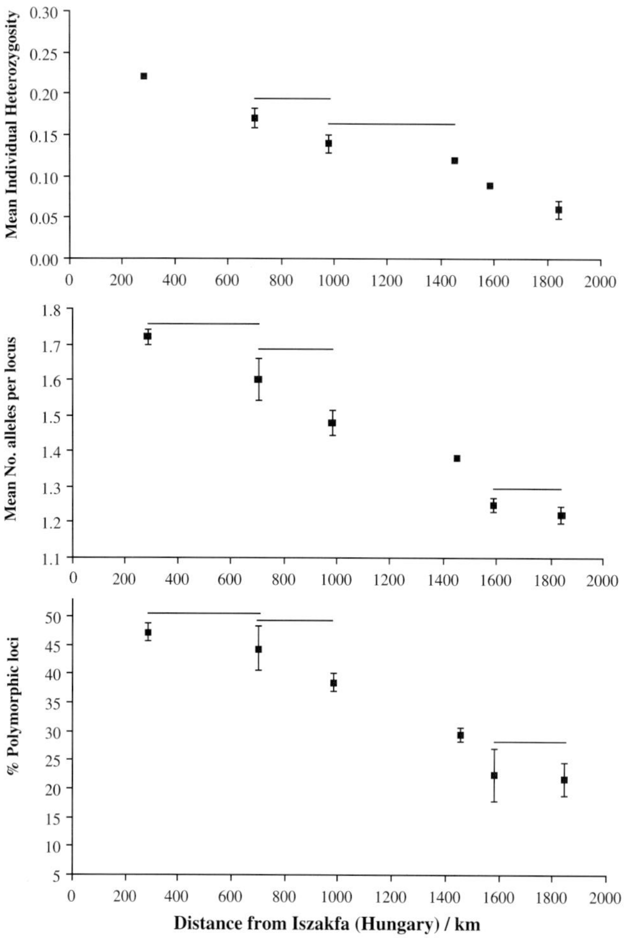

Fig. 2 Three measures of genetic variation (1 standard error) for different geographic regions, plotted against average distance from the native range of *A. quercuscalicis*. The regions are, from left to right: Carpathian basin and Austria (native range) N=15 populations; Italy N =4; continental Europe (invaded range) N =11; Central, SE and N England N =11; SW Britain N =10; Ireland N =5. Mean individual heterozygosity is corrected for expectation under Hardy-Weinberg equilibrium. Means were compared using pairwise Mann-Whitney tests and horizontal lines join means with non-significant differences (p<0.05). (Sark in the British Channel Islands was excluded as one locus was not scored, and Agen in France was excluded because it seems to have been invaded from Italy and is thus not representative of the rest of the European invaded range).

Secondly, genetic isolation in the invaded range was shown by spatial autocorrelation analysis. Over the whole range of *A. quercuscalicis*, neighbouring sites were significantly more similar than by chance, and distant sites were more different. However, within the native range, no allele showed significant spatial structure, indicating high gene flow. In mainland Britain, three out of the five alleles available for analysis showed significantly more similarity than by chance in sites closer to each other. However, there was no significant similarity between populations 230–410 km apart since much of the spatial patterning was probably due to the two subdivisions within Britain, consistent with F_{st} results.

Thirdly, if populations were founded by individuals from a nearby population, then phenetic analysis should group samples on the basis of geographic proximity. There was indeed high concordance between geographic locality and phenetic trees based on similarity of allele frequencies (Fig. 3).

Patterns of linkage disequilibrium (LD)

LD was calculated and N_e estimated for 34 sites with the largest sample sizes. Many estimates of LD had wide confidence limits, probably due to the samll sample size (Bartley *et al.* 1992). Nonetheless, there is important information in the data. Native range populations had generally high estimates of

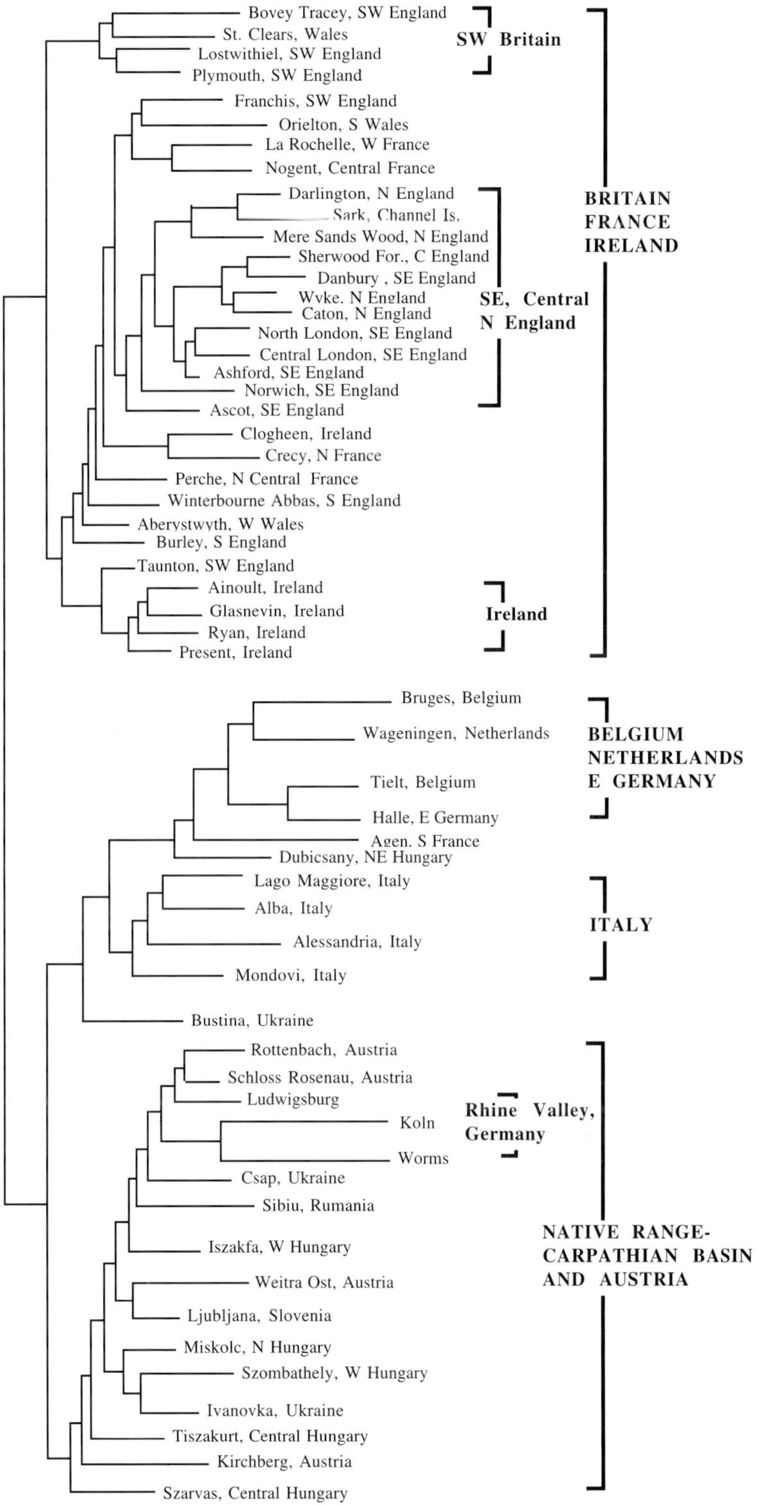

Fig. 3. Unrooted Wagner network of genetic similarity at all loci estimated by Modified Rogers Distance (Swofford and Selander, 1981). Cophenetic correlation = 0.81.

N_e, and only 7.5% of population/locus combinations were in significant LD suggesting that these populations were mostly well-established and panmictic. The situation was similar for two Italian populations. In contrast, many sites in the invaded range had higher LD and lower N_e. This indicates recent founding through very few individuals, or recent local extinction and recolonisation perhaps through low habitat availability or high predator attack. Four populations had especially low N_e and three of these were probably established only recently or were located where Turkey oaks are rare (Halle, E Germany, N_e=3 (1–18, 95% confidence interval); Wageningen, Netherlands, N_e=4 (1–17) and Glasnevin, Ireland, N_e=1 (0–4). Such low estimates of N_e are in agreement with the conclusion from the loss of genetic variation, that founding population sizes in the invaded range were very small (Stone and Sunnucks 1993). It is not clear why Wörms (Rhine Valley, Germany) N_e=2 (1–6) should have high LD, but this site is also unusually lacking in variation for the region (Stone and Sunnucks 1993) which also indicates few founders. Perhaps surprisingly, British and French populations showed comparatively little LD with only 8.4% of locus-pair/populations showing significant disequilibrium. This may reflect different intensities of ecological factors over the range of the species, for example, parasitoid attack on gallwasps is much lower in Britain than in the native range (Schönrogge *et al.* 1995). That study also showed a marked increase (from west to east) in the frequency with which agamic wasps did not emerge from their galls in the first year but underwent prolonged diapause and emerged up to two years later.

Summary of genetic effects of colonisation on A. quercuscalicis

The colonisation of western Europe by *A. quercuscalicis* is one of the best documented natural insect invasions yet studied. Both the timescale over which the invasion has taken place, and its direction, are known (Stone and Sunnucks 1993). All data agree that *A. quercuscalicis* expanded into western Europe from the Carpathian Basin in a stepwise fashion with populations founded by a small number of individuals from a neighbouring area to the east. This simple pattern may reflect the topography of the region since the likely route to the Carpathian Basin from western Europe is through the Danube and Rhine Valleys, which form a very narrow 'corridor' (Fig. 1).

There are some exceptions to the rule of directional stepping-stone colonisation, as expected in areas with few geographic barriers and with available habitat. France, for example, has many areas from which it might be colonised, and more complicated genetic patterns might be expected. There is indeed little spatial resolution in the data with sites spread through the relevant groupings in Fig. 3. In particular, genetic data indicate that Agen in southern France was colonised from Italy whereas northern France was invaded via the Danube-Rhine route (Fig. 1). Also, Halle in E Germany is clearly linked to Belgium/Netherlands rather than the Rhine Valley, indicating a secondary movement in the opposite direction to the major invasion. Data are also consistent with migration to Italy from the Carpathian Basin, although it is also possible that Italy contained a separate smaller glacial refuge population (Fig. 1).

There were profound effects of genetic sampling during the colonisation process. Firstly, this species showed the most dramatic and linear loss of genetic variability yet demonstrated for an invading animal, and colonisation greatly reduced genetic variation relative to the native range. The loss of allelic variation was so severe that Irish samples have only 30% of the alleles and 20% of the heterozygosity seen in the native range. Secondly, genetic subsampling during colonisation resulted in populations with diverse genetic compositions, unlike in the native range. Some alleles were lost in the invaded range (mostly those originally at low or moderate frequency) while others showed high variance in frequency. In addition, allelic associations were changed and significant correlations between alleles (LD) occurred in some new populations. Thirdly, genetic subdivision was greatest in the invaded range. Low gene flow in the invaded range of *A. quercuscalicis* was indicated by spatial allele patterning, F_{st}, and phenetic analysis.

The genetic effects of colonisation on insect populations

Genetic variation in colonising species

Colonising insects may lose much heterozygosity and allelic diversity through founder events and genetic drift (Black *et al.* 1988; Gasperi *et al.* 1991). The loss of genetic variation in *A. quercuscalicis* was extremely high which was probably due to numerous bottlenecks during directional stepping-stone invasion (Stone and Sunnucks 1993). However, such losses of variation are not inevitable. If founding populations are large, grow rapidly, and gene flow is high, there may be little genetic differentiation between native and introduced ranges (Bryant *et al.* 1981; Berlocher 1984; Daly and Gregg 1985; Kambhampati *et al.* 1990). Conversely, if the number of founders is small and colonies grow slowly or are cut off from the source, both founder effect and subsequent genetic drift can result in loss of alleles. These processes reduce heterozygosity less than allelic diversity, unless N_e is very low or small population size is sustained (Nei *et al.* 1975).

Natural selection is another possible cause of genetic patterning. Climatic gradients have been suggested to act on the gallwasps *Diplolepis rosae* and *D. mayri* in northern Europe (Stille 1985a,b) and may affect ecological characteristics such as budburst which can alter oviposition site availability for the galling aphid *Kaltenbachiella japonica* (Komatsu and Akimoto 1995). However, it may be possible to distinguish the effects on allozymes of drift and founder effect from those of selection, by comparison of patterns of several loci (Berlocher 1984; Sokal *et al.* 1987). In the case of *A. quercuscalicis*, selection appears to have played, at most, only a small role in the loss of alleles and in generating changes in allele frequencies and altering association between alleles since it would be necessary to propose very complicated and unusual selection gradients to generate the observed patterns. A selection hypothesis would need to explain why all seven loci showed a trend towards monomorphism, variance in allele frequencies increased westwards, some alleles were lost only locally, and some alleles showed patterning within Britain but not over most of Europe.

The likely fitness effects of losses of genetic variation are the subject of intense debate, and will take a great deal of research to resolve. However, there is clear evidence that inbreeding and loss of variation in insects and other animals can cause serious loss of fitness (Brakefield and Saccheri 1994; Vrijenhoek 1994; Frankham 1995 a, b).

Evolutionary change through loss of variation can be profound. A spectacular example comes from fire ants, *Solenopsis* spp. (Ross and Trager 1990). During colonisation of the USA from South American, *S. invicta* lost only rare allozyme alleles, and genetic structure of populations was similar in the colonised and native areas. However, new populations showed unusual biological characteristics such as having multiple queens, and producing many sterile diploid males. This reflects loss of alleles at highly-variable sex-determining genes (Cook and Crozier, 1995). Native fire ant populations had at least 86 sex-determining alleles, whilst the invading ones had only 10 to 13 (Ross and Trager 1990). In addition to reorganising the social biology of the species, the colonisation bottleneck of *S. invicta* appears to have undermined reproductive isolation from *S. richteri* (Ross and Trager 1990).

It is expected that variance in allele frequencies will be higher among colonising populations where colonisation results in genetic sampling. This was true for *A. quercuscalicis* and has been reported in other species such as the walnut husk fly *Rhagoletis completa*, a species which has rapidly invaded California since the 1920s (Berlocher 1984). If populations pass through very small numbers of individuals, certain combinations of alleles may be over-represented (LD). Highest LD was detected in the invaded range of both *A. quercuscalicis* and *R. completa.*

Very severe genetic sampling, as observed in *A. quercuscalicis*, may result in the generation of unusual and diverse genotypes by stochastic events. Generation of novel genotypes may be an especially helpful characteristic for invading species since they will often be in new or variable environments (Parker *et al.* 1977).

Colonising species are often genetically more uniform at the centre of their range than in peripheral areas, particularly if the latter comprise islands or patches of resource (McCauley 1991). Such patterns were seen in *A. quercuscalicis*, and have been reported for other species (Berlocher 1984). However, it is not a foregone conclusion that even colonising species utilising patchy resources will show genetic subdivision. Several factors operate to determine the genetic structure of colonising or transient populations.

General migration ability of species is very important in determining genetic patterns. In general, species with higher dispersal ability, such as aphids, are less genetically subdivided (Daly 1989). Shufran *et al.* (1991) showed that most of the genetic variation in the Kansas populations of the pest aphid *Schizaphis graminum* was represented by individuals on a single sorghum leaf. Relative lack of genetic structure has been reported in many other groups such as moths (Daly and Gregg 1985) and flies (Eber *et al.* 1992). At the other extreme, Hebert *et al.* (1991) showed that the host-alternating and galling aphid *Melaphis rhois* was organised into small local inbreeding populations.

Variance in individual dispersal distance appears to be important in genetic structure. Long-distance migration may have profound effects on new populations, because immigrants into sparsely populated areas may have a great effect on gene frequencies, and populations in front of the 'advancing wave' of colonisation may also act as sources (Nichols and Hewitt 1994). Empirical demonstrations include studies of *Drosophila pseudoobscura* in desert isolates (Jones *et al.* 1981), and the tephritid fly *Urophora cardui* (Eber and Brandl 1994).

Host use and dispersal ability are likely to be linked. A species with a naturally patchy or transient host is expected to evolve suitable migration abilities, while one which uses stable or evenly-distributed hosts is expected to put resources into other activities (Dixon *et al.* 1993). Selection for dispersal ability has been demonstrated in the spider mite *Tetranychus urticae* (Li and Margolies 1994). Thus the great genetic subdivision of *A. quercuscalicis* in its invaded range, in sharp contrast to the approximate panmixia in the Carpathian Basin, may reflect an interaction between low migration ability and patchiness of Turkey oak in the invaded areas.

Genetic structure in situations of non-equilibrium

Most population genetic theory and data concern situations where populations have reached demographic and genetic equilibria. Although equilibrium may not even be the norm, data and theory on non-equilibrium conditions are generally lacking (Whitlock and McCauley 1990). Non-equilibrium genetic states are seen in colonising species but also in populations subdivided with limited gene flow between compartments (metapopulations). Many studies of insect populations have revealed metapopulation dynamics, and the factors and processes involved are now starting to be understood. One body of data focuses on tephritid flies with similar lifecycles, feeding on thistles (Zwölfer 1988; Seitz and Komma 1984; Eber *et al.* 1992; Eber and Brandl 1994). Seitz and Komma (1984) compared the population genetics of *Urophora cardui*, a monophagous gall-former on the thistle, *Cirsium arvense*, and the polyphagous *Tephritis conura*. Local populations of *U. cardui* are distinct but show only weak genetic patterns. In contrast, *T. conura* is more generally distributed, and has very little genetic structure. Extinction of populations of *U. cardui* through parasitoid attack, and recolonisation with low dispersal ability were proposed as causes for random local differentiation. The wider host range of *T. conura* probably results in more refugia from parasitoids and easier migration giving rise to genetic homogeneity. Eber and Brandl (1994) have extended this work, showing repeated extinction/ recolonisation to be a regular feature of the ecology of *U. cardui*.

Factors affecting genetic and demographic outcomes of extinction/recolonisation have been the subject of theoretical and experimental studies by a group at Vanderbilt University (Whitlock and McCauley 1990; McCauley 1991; Whitlock 1992). Whitlock and McCauley (1990) highlighted the importance of non-equilibrium dynamics in population structure, and predicted how several factors would affect the genetic population structure. These factors include number of founders, probability of common origin of founders, and kin-structured dispersal. Two models of recolonisation have been proposed. In the 'migrant pool model' colonists come at random from the whole metapopulation, while in the 'propagule model' they are from a single source. In Whitlock and McCauley's analyses, the former can reduce or enhance differentiation between populations, the latter can only enhance it. Whitlock (1992) presented tests of these predictions, examining the genetics and ecology of the forked fungus beetle *Bolitotherus cornutus*. The major outcome of this work was a demonstration that populations are not well described by theoretical considerations based on assumptions of genetic equilibrium.

Conclusion

Studies of colonising and subdivided populations constitute extremely important and informative areas of evolutionary and population genetic research. Insects often provide appropriate vehicles for such studies, because of their abundance and diversity of biology and history. Population genetics of non-equilibrium situations is in its infancy compared with analyses at equilibrium. The evolutionary processes in colonising species and those with metapopulation structure will only be understood following further experimental and theoretical investigations.

Acknowledgements

Genetic work was funded by and carried out at the Institute of Zoology, London. Ecological work was carried out at Imperial College London at Silwood Park, Ascot, with support from the Department of the Environment and NERC. We thank Andrea Taylor and Dinah Hales for commenting on the manuscript, and Nick Webb and Thomas Boyce for helpful suggestions as part of the review process.

References

Bartley D, Bagley M, Gall G, Bentley B (1992) Use of linkage disequilibrium data to estimate effective size of hatchery and natural fish populations. *Conserv Biol* **6**:365–375

Berlocher SH (1984) Genetic changes coinciding with the colonisation of California by the walnut huskfly *Rhagoletis completa*. *Evolution* **38**:906–918

Black WC IV, Ferrari JA, Rai KS, Sprenger D (1988) Breeding structure of a colonizing species: *Aedes albopictus* (Skuse) in the United States. *Heredity* **60**:173–181

Brakefield PM, Saccheri IJ (1994) Guidelines in conservation genetics and the use of the population cage experiments with butterflies to investigate the effects of genetic drift and inbreeding. In: Loeschcke V, Tomiuk J, Jain SK (eds). *Conservation Genetics*. Birkhauser Verlag, Basel, Switzerland. pp 105–179

Bryant EH, van Dijk H, van Delden W (1981) Genetic bottleneck in *Musca autumnalis*. *Evolution* **35**:872–881

Bryant EH, Meffert LM (1992) The effect of serial founder-flush cycles on quantitative genetic variation in the housefly. *Heredity* **70**:122–129

Cook JM, Crozier RH (1995) Sex determination and population biology in the Hymenoptera. *Trends Ecol & Evol* **10**:281–286

Daly JC (1989) The use of electrophoretic data in the study of gene flow in the pest species *Heliothis armigera* (Hübner) and *H. punctigera* Wallengren (Lepidoptera: Noctuidae). In: Loxdale HD, den Hollander J (eds). *Electrophoretic studies on agricultural pests*. Clarendon Press, Oxford. pp 115–141

Daly JC, Gregg P (1985) Genetic variation in Heliothis in Australia: species identification and gene flow in two pest species *H. armigera* (Hübner) and *H. punctigera* Wallengren (Lepidoptera: Noctuidae). *Bull Entomol Res* **75**:169–184

Dixon AFG, Horth S, Kindlmann P (1993) Migration in insects: cost and strategies. *J Anim Ecol* **62**:182–190

Endler JA (1992) Genetic heterogeneity and ecology. In: Berry RJ, Crawford TJ, Hewitt GM (eds). *Genes in ecology*, Blackwell Scientific Publications, Oxford. pp 315–334

Eber S, Brandl R (1994) Ecological and genetic spatial patterns of *Urophora cardui* (Diptera: Tephritidae) as evidence for population structure and biogeographical processes. *J Anim Ecol* **63**:187–199

Eber S, Brandl R, Vidal S (1992) Genetic and morphological variation among populations of *Oxyna parietina* (Diptera: Tephritidae) across a European transect. *Can J Zool* **70**:1120–1128

Frankham R (1995a) Inbreeding and extinction: A threshold effect. *Conserv Biol* **9**:792–799

Frankham R (1995b) Conservation genetics. *Annu Rev Genet* **29**:305–327

Gasperi G, Guglielmino CR, Malacrida AR, Milani R (1991) Genetic variability and gene flow in geographical populations of *Ceratitis capitata* (Wied.) (Medfly). *Heredity* **67**:347–356

Hails RS (1994) The population dynamics of the gallwasp *Andricus quercuscalicis*. In: Williams MAJ (ed) *Plant galls: organisms, interactions, populations.* Systematics Association Special Volume no. 49. Clarendon Press, Oxford, pp 391–404

Hebert PD, Finston TL, Foottit R (1991) Patterns of genetic diversity in the sumac gall aphid, *Melaphis rhois. Genome* **34**:757–762

Jones JS, Bryant SH, Lewontin RC, Moore JA, Prout T (1981) Gene flow and the geographical distribution of a molecular polymorphism in *Drosophila pseudoobscura. Genetics* **98**:157–178

Kambhampati S, Black WC, Rai KS, Sprenger D (1990) Temporal variation in genetic structure of a colonising species – *Aedes albopictus* in the United States. *J Hered* **64**:281–287

Komatsu T, and Akimoto S (1995) Genetic differentiation as a result of adaptation to the phenologies of individual host trees in the galling aphid *Kaltenbachiella japonica. Ecol Entomol* **20**:33–42

Li J, Margolies DC (1994) Responses to direct and indirect selection on aerial dispersal in *Tetranychus urticae. Heredity* **72**:10–22

Mayr E (1982) Speciation and macroevolution. *Evolution* **36**:1119–1132

McCauley DE (1991) Genetic consequences of local population extinction and recolonization. *Trends Ecol & Evol* **6**:5–8

Nei M, Maruyama T, Chakraborty R (1975) The bottleneck effect and genetic variability in populations. *Evolution* **29**:1–10

Nichols RA, Hewitt GM (1994) The genetic consequences of long distance dispersal during colonization. *Heredity* **72**:312–317

Parker ED, Selander RK, Hudson RO, Lester LJ (1977) Genetic diversity in colonising parthenogenetic cockroaches. *Evolution* **31**:836–842

Puterka GJ, Black WCI, Steiner WM, Burton RL (1993) Genetic variation and phylogenetic relationships among worldwide collections of the Russian wheat aphid, *Diuraphis noxia* (Mordvilko), inferred from allozyme and RAPD-PCR markers. *Heredity* **70**:604–618

Ross KG, Trager JC (1990) Systematics and population genetics of fire ants (*Solenopsis saevissima* complex) from Argentina. *Evolution* **44**:2113–2134

Schönrogge K, Stone GN, Crawley MJ (1995) Spatial and temporal variation in guild structure: parasitoids and inquilines of *Andricus quercuscalicis* (Hymenoptera, Cynipidae) in its native and alien ranges. *Oikos* **72**:51–60

Seitz A, Komma M (1984) Genetic polymorphism and its ecological background in tephritid populations. In: Woehrmann K, Loeschke V (eds). *Population biology and evolution*. Springer Verlag, Heidelberg, pp 143–158

Shufran KA, Black WC IV, Margolies, DC (1991) DNA fingerprinting to study spatial and temporal distributions of an aphid *Schizaphis graminum* (Homoptera: Aphididae). *Bull Entomol Res* **81**:303–313

Sokal RR, Oden NL, Barker JSF (1987) Spatial structure in *Drosophila buzzatii* populations: simple and two dimensional spatial autocorrelation. *Am Nat* **129**:122–142

Stille B (1985a) Population genetics of the parthenogenetic gall wasp *Diplolepis rosae* (Hymenoptera: Cynipidae). *Genetica* **67**:145–151

Stille B (1985b) Host plant specificity and allozyme variation in the parthenogenetic gall wasp *Diplolepis mayri* and its relatedness to *D. rosae*. *Entomol Gen* **10**:87–96

Stone GN, Sunnucks P (1993) Genetic consequences of an invasion through a patchy environment - the cynipid gallwasp *Andricus quercuscalicis* (Hymenoptera: Cynipidae). *Mol Ecol* **2**:251–268

Sunnucks P, Stone GN, Schönrogge K, and Csóka G (1994) The biogeography and population genetics of the invading gallwasp *Andricus quercuscalicis* (Hymenoptera: Cynipidae) In: Williams MAJ (ed) *Plant galls: organisms, interactions, populations*. Systematics Association Special Volume no. 49. Clarendon Press, Oxford, pp 351–368

Swofford DL, Selander RB (1981) BIOSYS 1: a FORTRAN program for the comprehensive analysis of data in population genetics and systematics. *J Hered* **72**:281–283

Vrijenhoek RC (1994) Genetic diversity and fitness in small populations. In: Loeschcke V, Tomiuk J, Jain SK (eds). *Conservation genetics*. Birkhauser Verlag, Basel Switzerland, pp 37–53

Whitlock M, McCauley DE (1990) Some population genetic consequences of colony formation and extinction: genetic correlations within founding groups. *Evolution* **44**:1717–1724

Whitlock MC (1992) Nonequilibrium population structure in forked fungus beetles: extinction, colonization, and the genetic variance among populations. *Am Nat* **139**:952–970

Zwölfer H (1988) Evolutionary and ecological relationships of the insect fauna of thistles. *Annu Rev Entomol* **33**:103–122

THE GENETIC AND ENVIRONMENTAL COMPONENTS OF HOST DEPENDENT STRATIFICATION OF HERBIVOROUS INSECT POPULATIONS

Paul J. De Barro, Thomas N. Sherratt and Norman Maclean

ABSTRACT

Numerous studies on the biology of various species of herbivorous insect have demonstrated marked variability in fitness among different genotypes with respect to different species of host. However, a firm link between biological variation and population structure at the genetic level has not yet been demonstrated. The evidence for genetic variation in population structure based on host utilisation of herbivorous insects is reviewed. The underlying genetic and environmental causes of this variation and the evidence for trade-offs in host utilisation and conditioning are discussed with particular reference to studies of the grain aphid, *Sitobion avenae* (F.). Molecular techniques such as DNA fingerprinting and RAPD-PCR are used to demonstrate the existence of host based population stratification. Reciprocal host transfer studies were used to demonstrate the roles that trade-offs in host utilisation and conditioning play in creating the host based population stratification found between wheat and wild grasses in southern England.

Key words: Sitobion avenae, population stratification, host utilisation

Frontiers of Population Ecology, R.B. Floyd, A.W. Sheppard and P.J. De Barro (eds), CSIRO Publishing, Melbourne 1996, pp. 497–503

Studies on a number of oligophagous and polyphagous herbivorous insect species have demonstrated the existence of considerable genetic variability in individual performance on different hosts within the species' range (Via 1984a,b; Rausher 1983,1984; reviewed by Fox and Morrow 1981; Diehl and Bush 1984; Futuyma and Peterson 1985). From these and other studies it appears that specialisation within a species is not uncommon and is to some extent genetically based. Service and Lenski (1982) further suggested that this genetic diversity could only be maintained if no genotype was the most fit on all hosts. This has been demonstrated a number of times, for example, *Uroleucon rudbeckia* (Hemiptera: Aphididae) on rudbeckia (*Rudbeckiae lanciniata*) (Service and Lenski 1982) and black pine scale (*Nuculaspis californica* (Hemiptera: Coccidae)) (Edmunds and Alstad 1978). Furthermore, while genetic variation with respect to host utilisation is not uncommon, it does not necessarily result in population stratification, but where it occurs, it suggests that shifting between host species may compromise individual fitness.

The underlying cause of genetic variation with respect to host utilisation is often unclear, but has two possible origins; first, genetic differences among individuals may result in differential host utilisation (Jaenike 1990; Futuyma and May 1993), and second, conditioning may result in the prior host experience of the individual or its mother influencing its ability to utilise the current host (reviewed by Diehl and Bush 1984 and Papaj and Prokopy 1989). Because one explanation is genetic and the other environmental, their implications on the development of plant-herbivore interactions will be different. The genetic explanation is likely to result in host based population stratification that is independent of such events as gene flow (Via 1991a) and a high degree of rigidity in the way in which populations are structured with respect to the utilisation of different potential host species. The environmental (conditioning) explanation requires a severe isolating mechanism, such as the reduction in fitness as a result of a reduced ability to feed on alternative hosts (Maynard Smith 1966; Futuyma and Peterson 1985; Papaj and Prokopy 1988; Prokopy *et al.* 1988, 1989) and, host selection is likely to be more plastic where the spatial utilisation of a particular array of potential hosts reflects the original patterns of colonisation.

Genetic variation in host utilisation has been recorded within species for a number of insect orders. Coleopteran herbivores such as the milkweed beetle (*Tetraopes tetrophthalmus* (Cerambycidae)) on two species of *Asclepias* (Price and Wilson 1976), *Lochmaea capreae* (Chrysomelidae) on birch and willow (Kreslavsky *et al.* 1981), and mountain pine beetle (*Dendroctonus ponderosae* (Scolytidae)) on two species of *Pinus* (Stock and Amman 1980) show stratification in host utilisation. Lepidopteran examples include fall cankerworm (*Alsophila pometaria* (Geometridae)) on a number of tree species (Mitter *et al.* 1979; Schneider 1980; Futuyma *et al.* 1984; Futuyma and Philippi 1987), swallowtail butterflies (*Papilio zelicaon* and *P. oregonius* (Papilionidae)) on two species of Umbelliferae (Thompson *et al.* 1990), and the leafmining ermine moth (*Yponomeuta padellus* (Yponomeutidae)) on a number of sympatric host species (Menken 1981). Similarly, dipteran and hemipteran examples include apple maggot fly (*Rhagoletis pomonella* (Tephritidae)) feeding on apple and hawthorn (Bush 1969; Papaj and Prokopy 1988; Prokopy *et al.* 1988, 1989) and cabbage leafminer (*Liriomyza brassicae* (Agromyzidae)) on three different host genera (Tavormina 1982), *U. caligatum* (Aphididae) on several species of goldenrods (Moran 1981) and pea aphid (*Acyrthosiphum pisum* (Aphididae)) on lucerne and strawberry clover (Via 1991a,b).

While these studies show host based stratification at the genetic level, it is not clear whether the underlying mechanism is genetically controlled or environmentally induced. If stratification is under genetic control some form of trade-off between host fitness and host utilisation would be expected, however, there is little experimental data to support its existence (Shaw 1986; Futuyma and Peterson 1985; Futuyma and Philippi 1987; Fox 1993). The only insect examples are the pea aphid on strawberry clover and lucerne (Via 1991a,b), the hessian fly (*Mayetiola destructor* (Cecidomyiidae)) on wheat (Gallun 1978), the fall armyworm (*Spodoptera frugiperda* (Noctuidae)) on rice and corn (Pashley 1988) and the bean aphid (*Aphis fabae* (Aphididae)) on broad beans and garden nasturtium (Mackenzie 1996). The two spotted mite (*Tetranychus urticae*) also shows clear evidence of fitness trade-offs between lima beans and potato, and lima beans and cucumber (Gould 1979; Fry 1990, 1992).

Similarly, studies which have examined the role of conditioning in host based population stratification are also uncommon. Jaenike (1988) provides several examples where prior host experience has led to population stratification, but perhaps the best example comes from studies of the apple maggot fly on apple and hawthorn (Bush 1969; Prokopy *et al.* 1988,1989) where prior adult experience leads to isolation resulting in host based genetic divergence within a population. In contrast, there are few examples where juvenile choice influences adult host choice (Papaj and Prokopy 1989) and it is generally considered unlikely to lead to genetic divergence within a population. The potential for conditioning is perhaps greatest in hemimetabolous species which as juveniles may already carry in embryonic form, the next and at times subsequent generations.

Perhaps the best known group where this occurs is the aphids. Aphids are parthenogenetic in at least part of their life-cycle and produce live offspring which already contain embryonic offspring. Therefore, in any one mother or nymph there are the potential offspring of the next two generations. As a consequence, these offspring are already experiencing the host on which the mother/nymph is feeding and may therefore be conditioned without ever having fed and so may shift host preference away from unexperienced hosts (reviewed by Klingauf 1987). A newly borne nymph with very limited reserves may be negatively affected by exposure to a new host species. The same outcome would also be expected if the individual was genetically predetermined to do better on certain hosts. Numerous studies have observed clonal variability in response to different hosts (e.g. Weber 1985a,b,c, 1986; McCauley *et al.* 1990a,b, 1991; Via 1991a), but few studies have attempted to determine whether host adaptation in aphids is controlled by genetic (Via 1991a,b), environmental or a combination of components (Mackenzie 1996).

A recent genetic survey of parthenogenetic over-wintering populations (using restriction fragment length polymorphism, RFLP) of the grain aphid (*Sitobion avenae*) on wheat and road-side grasses (De Barro *et al.* 1995a) showed population stratification with respect to host, but not to geographic location, suggesting that stratification was rigid and independent of such events as gene flow and may therefore have had a genetic basis. In addition, De Barro *et al.* (1994) noted that genetic variability of aphids in a wheat field declined over the life of the crop suggesting differential fitness played an important role in shaping the genetic variability of populations of *S. avenae*. However, whether the genetic variation shown by RFLPs was associated with phenotypic characters of ecological/biological importance was not known and as Via (1990) warned may not be present.

While an increasing number of studies of herbivorous insect populations have employed various molecular techniques such as allozyme electrophoresis, DNA fingerprinting and RAPD-PCR to examine population structure, very few have attempted to determine the underlying biological basis for the patterns found. Our study (De Barro *et al.* 1995a) presented clear evidence of population structuring with respect to host utilisation but the factors causing and maintaining the structuring remain unknown, although a reduction in fitness was suspected. To investigate whether there was any variability in clonal fitness, we compared life history traits of *S. avenae* clones collected from wheat and a major component of road-side grasses, cocksfoot (*Dactylis glomerata*) at the end of winter when the population stratification was observed. Furthermore, by comparing aphid fitness in a number of clones over a number of generations on both the host from which they were collected and on the alternative, the roles of environmental (conditioning) and genetic components in creating the observed population stratification were examined (De Barro *et al.* 1995b).

Significant genetic variability was found among clones overwintering on wheat and cocksfoot and while there were no changes in performance (survival to adult, adult weight and fecundity) when clones were moved between the same host over three generations, wheat clones on wheat performed better than cocksfoot clones on cocksfoot. Furthermore, clones varied significantly in performance within each host. When clones from the host of origin were transferred to the alternative host in the second generation, there was generally a significant decline in performance with the greatest reductions being noted in adult weight and fecundity. The subsequent transfer in the third generation back to the host of origin generally resulted in an intermediate performance thus supporting the role of conditioning. In addition, when clones were maintained on the alternative host for two or three generations, the change in performance was maintained suggesting a significant genetic component involving a trade-off in the relative ability to utilise different hosts.

The earlier studies of genetic variation of *S. avenae* and other species did not examine the temporal aspect of the observed stratification. Since variation in the chemistry and morphology of host species is an important cause of herbivore specialisation (Thompson 1988) and can change either seasonally or as the host matures, it is possible that stratification in host utilisation may be transitory. Therefore, the temporal aspect of this population stratification was examined by comparing aphid populations in paired wheat/cocksfoot sites at two locations in Hampshire, southern England during the period April to July. RAPD-PCR, as with DNA fingerprinting, showed that there was significant stratification with respect to host utilisation that was independent of location. However, by July the April stratification had broken down. In addition, the number of clones observed declined over the four months of sampling from 219 in April to 136 in July, supporting similar findings made 12 months earlier using DNA fingerprinting (De Barro *et al.*1994) and again suggesting a role for differential fitness.

There was clear and consistent differentiation between clones found on wheat and cocksfoot during April. This in itself is not surprising given founder effects and genetic drift. What is surprising is that the differences are consistent between locations, which indicates intense and uniform preference for, or selection by the host plant. Furthermore, stratification declines over time. This may indicate that selection is operating in the overwintering population but not in the summer populations, or alternatively, host preference has changed with time. These observations tell us that at certain times of the year populations are stratified with respect to host, but it is not clear whether it is under genetic or environmental control. However, the observation that location was unimportant in April strongly suggests a genetic basis and is supported by the host transfer study using aphids collected from overwintering populations.

The difference in performance of clones, even when tested on the plant species from which they were collected, agrees with earlier studies (e.g. Weber 1985a,b,c, 1986; McCauley *et al.* 1990a,b, 1991; Via 1991a). Variation in performance is not unexpected and a possible explanation is the heterogeneous selection experienced by the clones under field conditions. Nevertheless this indicates (a) that clones consistently performed better on one plant species, usually the species of origin than they did on an alternative host species and (b) the relative fitness of clones varied consistently with host. Observation (a) indicates adaptation of clones to particular host species, while observation (b) provides evidence for a trade-off in performance. The discovery of adaptation in the wheat and cocksfoot clones strongly suggests that the habitats from which they were collected, provides sufficient selection to overcome the confounding effects of gene flow.

The clear demonstration of trade-offs in terms of negative correlations in performance on different hosts is rare, yet of the six studies that have good evidence for trade-offs, three involve aphids (Via 1991a, Mackenzie 1996; this study). This suggests that aphids are exceptional in terms of host based fitness trade-offs.

The observed spatial and temporal genetic structure of *S. avenae* in southern England (De Barro *et al.* 1995b; this study) and the evidence for host specialisation are mutually compatible and supportive. It is clear that the *S. avenae* found on wheat and cocksfoot in March/April differ quite markedly in their genetic constitution and this study supports the idea that one contributory factor is the differential performance of clones.

This study also provides strong evidence that maternal experience can influence offspring performance, at least to a degree (reviewed by Mousseau and Dingle 1991). A number of papers have reported that aphids can become 'acclimated' to certain hosts (Mackenzie 1992, 1996; McCauley *et al.* 1990a, b, 1991) but many of these studies did not identify individual clones. Via (1991b) showed that three generations of rearing *A. pisum* clones on an alternative (non-origin) host crop were not sufficient to reverse the specialisation for their original host (Via 1991b). These data were rather limited, with only two clones and three replicates, but nevertheless, did show a trend towards modification of performance based on prior experience (Via 1994).

The role of host quality in host based population stratification is further supported by the temporal nature of the stratification between *S. avenae* on wheat and cocksfoot. *S. avenae* that overwinter on grasses are forced to feed on leaves, but perform better on and prefer ears (Watt and Dixon 1981). This suggests that host quality selects certain clones which do particularly well on the leaves and that these proliferate over winter at the expense of less well adapted individuals (Thompson 1988). However, once ears are produced (May onwards) the selective pressure is removed allowing stratification to breakdown and the subsequent mixing of individuals from different hosts species.

This study clearly demonstrates that genetic and environmental components can combine to bring about host based population stratification. The fact that aphids appear to be the sole reliable demonstration of trade-offs in host related fitness is curious. Why aphids should demonstrate this form of trade-off so readily and chewing insects so rarely is unknown, but suggests that the answer may lie in the more intimate relationship between host and herbivore found with sucking insects (Farrow and Floyd this volume).

References

Bush GL (1969) Sympatric host race formation and speciation in frugivorous flies of the genus *Rhagoletis* (Diptera, Tephritidae). *Evolution* 23:237–251

De Barro PJ, Sherratt TN, Carvalho GR, Nicol D, Iyengar A, Maclean N (1994). An analysis of secondary spread by putative clones of *Sitobion avenae* within a Hampshire wheat field using the multilocus $(GATA)_4$ probe. *Insect Molecular Biology* 3:253–260

De Barro PJ, Sherratt TN, Carvalho GR, Nicol D, Iyengar A, Maclean N (1995a) The use of the multilocus $(GATA)_4$ probe to investigate geographic and micro-geographic genetic differentiation in two aphid species over southern England. *Molecular Ecology* 4:375–382

De Barro PJ, Sherratt TN, Markovic O, Maclean N (1995b) An investigation of the differential performance of clones of the cereal aphid, *Sitobion avenae* on two host species. *Oecologia* 104:379–385

Diehl SR, Bush GL (1984) An evolutionary and applied perspective of insect biotypes. *Annu Rev Entomol* 29:471–504

Edmunds GF, Alstad DN (1978) Coevolution in insect herbivores and conifers. *Science* 199:941–945

Fox CW (1993) A quantitative analysis of oviposition, preference and larval performance on two hosts in the bruchid beetle, *Callosbruchus maculatus*. *Evolution* 47:166–175

Fox LR, Morrow PA (1981) Specialization: species property or local phenomenon. *Science* 211:887–893

Fry JD (1990) Trade-offs in fitness on different hosts: evidence from a selection experiment with phytophagous mites. *Am Nat* 136:569–580

Fry JD (1992) On the maintenance of genetic variation by disruptive selection among hosts in a phytophagous mite. *Evolution* 41:279–283

Futuyma DJ, Cort RP, van Noordwijk I (1984) Adaptation to host plants in the fall cankerworm (*Alsophila pometaria*) and its bearing on the evolution of host affiliation in phytophagous insects. *Am Nat* 123:287–296

Futuyma DJ, Peterson SC (1985) Genetic variation in the use of resources by insects. *Annu Rev Entomol* **30**:217–238

Futuyma DJ, Philippi TE (1987) Genetic variation and coevolution in response to host plants by *Alsophila pometaria* (Lepidoptera: Geometridae). *Evolution* **41**:269–279

Futuyma DJ, May RM (1993) The coevolution of plant/insect and host/parasite relationships. In Berry RJ, Crawford TC, Hewitt GM (eds) *Genes in ecology*. Blackwell Scientific, Oxford, pp 139–166

Gallun RL (1978) Genetics of biotypes B and C of the hessian fly. *Ann Entomol Soc Am* **71**:481–486

Gould F (1979) Rapid host change evolution in a population of the phytophagous mite *Tetranychus urticae*. *Evolution* **33**:791–802

Jaenike J (1988) Effects of early adult experience on host selection in insects: some experimental and theoretical results. *J Insect Behav* **1**:3–15

Jaenike J (1990) Host specialization in phytophagous insects. *Annu Rev Ecol Syst* **21**:243–273

Klingauf FAC (1987) Host plant finding and acceptance. In: Minks AK, Harrewijn P (eds) *Aphids their biology, natural enemies and control. Vol. 2A*. Elsevier, Amsterdam, pp 209–223

Kreslavsky AG, Mikheev AV, Solomatin, VM, Grizenko VV (1981) Genetic exchange and isolating mechanisms in sympatric races of *Lochmaea capreae* (Coleoptera: Chrysomelidae). *Zool Zh* **60**:62–68

Mackenzie A (1992) The evolutionary significance of host-mediated conditioning. *Antenna* **16**:141–150

Mackenzie A (1996) A trade-off for host plant utilization in the black bean aphid, *Aphis fabae*. *Evolution* **50**:155–162

Maynard Smith J (1966) Sympatric speciation. *Am Nat* **100**:637–650

McCauley Jr GW, Margolies DC, Reese JC (1990a) Feeding behaviour, fecundity and weight of sorghum- and corn-reared greenbugs on corn. *Entomol Exp App* **55**:183–190

McCauley Jr GW, Margolies DC, Reese JC (1990b) Field assessment of greenbug (Homoptera: Aphididae) demography on corn. *Environ Entomol* **21**:1072–1076

McCauley Jr GW, Margolies DC, Collins RD, Reese JC (1991) Rearing history affects demography of greenbugs (Homoptera: Aphididae) on corn and grain sorghum. *Environ Entomol* **19**:949–954

Menken SBJ (1981) Host race and sympatric speciation in small ermine moths, Yponomeutidae. *Entomol Exp App* **30**:280–292

Mitter C, Futuyma DJ, Schneider JC, Hare JD (1979) Genetic variation and host plant relations in a parthenogenetic moth. *Evolution* **33**:777–790

Moran N (1981) Intraspecific variability in herbivore performance and host quality: a field study of *Uroleucon caligatum* (Homoptera: Aphididae) and its *Solidago* hosts (Asteraceae). *Ecol Entomol* **6**:301–306

Mousseau TA, Dingle H (1991) Maternal effects in insect life histories. *Annu Rev Entomol* **36**:511–534

Papaj DR, Prokopy RJ (1988) The effect of prior adult experience on components of habitat preference in the apple maggot fly (*Rhagoletis pomonella*). *Oecologia* **76**:538–543

Papaj DR, Prokopy RJ (1989) Ecological and evolutionary aspects of learning in phytophagous insects. *Ann Rev Entomol* **34**:315–350

Pashley DP (1988) Quantitative genetics, development, and physiological adaptation in host strains of fall armyworm. *Evolution* **42**:93–102

Price PW, Wilson MF (1976) Some consequences for a parasitic herbivore, the milkweed longhorn beetle, *Tetraopes tetrophthalmus*, of a host-plant shift from *Asclepias syriaca* to *A. verticillata*. *Oecologia* **25**:331–340

Prokopy RJ, Diehl SR, Cooley SS. (1988) Behavioural experience for host races is *Rhagoletis pomonella* flies. *Oecologia* **76**:138–147

Prokopy RJ, Cooley SS, Opp B (1989) Prior experience influences fruit residence time of male apple maggot flies. *J Insect Behav* **2**:39–48

Rausher MD (1983) Conditioning and genetic variation as causes of individual variation in the oviposition behaviour of the tortoise beetle, *Deloyala guttata*. *Animal Behaviour* **31**:743–747

Rausher MD (1984) Trade-offs in performance on different hosts: evidence from within- and between-site variation in the tortoise beetle *Deloyala guttata*. *Evolution* **38**:582–595

Schneider JC (1980) The role of parthenogenesis and female aptery in the fall cankerworm *Alsophila pometaria* Harris (Lepidoptera: Geometridae). *Ecology* **61**:1082–1090

Service PM, Lenski RE (1982) Aphid genotypes, plant phenotypes and genetic diversity: a demographic analysis of experimental data. *Evolution* **36**:1276–1282

Shaw RG (1986) Response to density in a wild population of the perennial herb *Salvia lyrata*: Variation among families. *Evolution* **40**:492–505

Stock MW, Amman G.D (1980) Genetic differentiation among mountain pine beetle populations from lodgepole pine and ponderosa pine in northeast Utah. *Ann Entomol Soc Am* **73**:472–478

Tavormina SJ (1982) Sympatric genetic divergence in the leaf-mining insect *Liriomyza brassicae* (Diptera: Agromyzidae). *Evolution* **36**:523–534

Thompson JN (1988) Coevolution and alternative hypotheses on insect/plant interactions. *Ecology* **69**:893–895

Thompson JN, Wehling W, Podolsky R (1990) Evolutionary genetics of host use in swallowtail butterflies. *Nature* **344**:148–150

Via S (1984a) The quantitative genetics of polyphagy in an insect herbivore. I. Genotype-environment interaction in larval performance on different host plants. *Evolution* **38**:881–895

Via S (1984b) The quantitative genetics of polyphagy in an insect herbivore. II. Genetic correlations in larval performance within and among host plants. *Evolution* **38**:896–905

Via S (1990) Ecological genetics and host adaptation in herbivorous insects: the experimental study of evolution in natural and agricultural systems. *Annu Rev Entomol* **35**:421–446

Via S (1991a) The genetic structure of host plant adaptation in a spatial patchwork: demographic variability among reciprocally transplanted pea aphid clones. *Evolution* **45**:827–852

Via S (1991b) Specialized host plant performance of pea aphid clones is not altered by experience. *Ecology* **72**:1420–1427

Via S (1994) Population structure and local adaptation in a clonal herbivore. In: Real LA (ed) *Ecological genetics*. Princeton University Press, pp 58–85

Watt AD, Dixon AFG (1981) The role of cereal growth stages and crowding in the induction of alatae in *Sitobion avenae* and its consequences for population growth. *Ecol Entomol* **6**:441–447

Weber G (1985a) On the ecological genetics of *Sitobion avenae* (F.) (Hemiptera, Aphididae). *J App Entomol* **100**:100–108

Weber G (1985b) On the ecological genetics of *Metopolophium dirhodum* (Walker) (Hemiptera, Aphididae). *J App Entomol* **100**:451–458

Weber G (1985c) Genetic variability in host plant adaptation of the green peach aphid, *Myzus persicae. Entomol Exp App* **38**:49–56

Weber G (1986) Ecological genetics of the host exploitation in the green peach aphid, *Myzus persicae. Entomol Exp App* **40**:161–168

SECTION 5

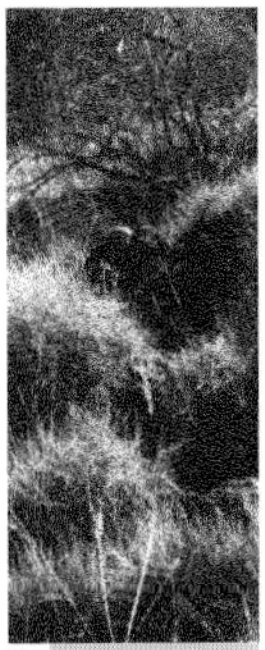

Managing populations

INTRODUCTION

Preceding sections have covered various aspects of population regulation, two-species interactions, spatial processes that effect population dynamics, and molecular ecology. In contrast, the contributions in this final section focus on the application of the theory and tools of population ecology to the management of plant and animal populations.

Steve Morton begins with a discussion of the role of population ecology in land management in Australia. He considers whether population ecology has successfully contributed to the resolution of problems in land management and concludes that while a study of ecology may provide useful scientific background, the resolution lies not with ecological understanding, but with society's use of the knowledge. Three examples are presented where the socio-economic imperatives are not necessarily in line with the ecological ones and several suggestions are made as to how the competing imperatives may be managed by the ecologist. It is indisputable that if scientific research is to be done, then it must be done well, however, Morton also suggests that scientists develop a sense of pragmatism to better judge when proposed research is unlikely to yield an adoption of its results. Finally, population ecologists may need to liaise with social scientists if population ecology and land management are to be reconciled.

The next paper in this section, by David Choquenot and Nick Dexter, is a practical example of the use of mathematical models to describe and predict spatial variation in the abundance of animals. In their study of feral pig populations in rangelands, they used spatial factors to determine whether pigs persisted at a given location and in what density they were likely to occur. Spatial variation in pig abundance and persistence was shown to be influenced by the availability of food and the ability to forage when temperatures were high.

Kent Williams and Laurie Twigg examine the likely effect on rabbit populations of sterilising females with a genetically-modified free-living immunocontraceptive myxoma virus. Here, the Nicholsonian themes of density-dependence and two species interaction are brought into play as Williams and Twigg describe the first year of a study aimed at determining the responses of rabbit populations in eastern and western Australia to various levels of sterility, and quantifying the minimum proportion of female wild rabbits that will need to be sterilised to reduce population growth to zero.

The use of sterile insect release to manage populations of the sheep blowfly, *Lucilia cuprina* is presented by Rod Mahon. Unlike the previous study, there is an added complication in the use of sterile males in that they are less fit than wild males. Mahon describes the interaction between sterile and fertile males and its effect on population size and examines the outcome in terms of the economic cost of a sterile male program. It is concluded that as the competitiveness of the released sterile males is less than that of the wild male, the cost-effectiveness of inundative releases declines as the number of wild flies declines. As a consequence, Mahon presents a series of strategies aimed at reducing the rare male effect and thereby making releases as cost-effective as possible. Interestingly, the proposed strategies suggest that previous sterile male programs against the New World screw-worm fly, while highly successful, may have been more costly than necessary.

The ability to predict the success of biological control of weeds is the subject of the fifth paper in this section. The Nicholsonian theory of density-dependent regulation, competition and two species interactions are used to assess the likely outcome of weed biological control. Mark Lonsdale concludes that while there is a greater need for plant population ecologists to be involved in weed biological control it is equally necessary that plant ecologists utilise the insect population theory first developed by Nicholson sixty years ago.

The use of fire as a means of controlling and containing highly invasive weeds is discussed by Tony Grice and Joel Brown. Fire is shown to have a significant effect on the growth, survival, seed production and seed viability of rubber vine, *Cryptostegia grandiflora* and chinee apple, *Ziziphus mauritiana*. The results suggest that while fire is very effective at controlling rubber vine, its ability to control chinee apple is dependant upon a greater frequency of fires to counter the ability of this species to rapidly regenerate.

In Bill de la Mare's contribution, the sometimes contentious issue of managing marine resources in a sustainable manner is examined. The classical approach to management involves the development of a model that is predictive and adaptive. The problem with this approach is that two key factors, maximum sustainable yield and carrying capacity, can not be reliably estimated and so classical management failed. In its place, de la Mare discusses the development and implementation of the Revised Management Procedure using whaling and krill harvesting as examples. This procedure was developed to enable stable catch limits to be set without endangering the long-term survival of the stock and to ensure that catch sizes were as large as could be sustained by the stock. In addition, de la Mare suggests a process that enables management strategies to be developed and tested through simulations rather than the trial and error of actual implementation.

The final paper, by Mary Myerscough, discusses new insights into classical models of harvesting and stocking. As with the previous paper, the aim is to develop models to predict long term stability and persistence for both conservation and resource management purposes. Again, Nicholsonian theory forms the basis of the approach to simulating the interaction between rates of harvesting and stocking. The management of both predator and prey densities is shown to be essential to achieve stability and persistence of both groups.

LAND MANAGEMENT AND POPULATION ECOLOGY

Stephen R. Morton

ABSTRACT

Land management is an increasingly all-inclusive concept widely discussed in today's scientific circles and in society, but whose breadth would have appeared unfamiliar to A.J. Nicholson and his colleagues. Population ecology in Nicholson's day concentrated on the factors causing population growth and limitation. In many cases, the search for understanding about these issues sprang from applied questions: how could this or that pest be controlled given better knowledge of its population dynamics? Today, we are trying hard not only to grasp fundamentals of the population ecology of many different types of organisms, but also to mesh that knowledge into a demanding framework of sustainable management and the conservation of biodiversity. How successfully are scientists achieving that goal? I ask this question of Australian science with reference to several species constituting either pests or desirable resources: perennial grasses of the pastoral lands; red kangaroos, *Macropus rufus*; plague locusts, *Chortoicetes terminifera*; and Leadbeater's possum, *Gymnobelideus leadbeateri*. The review confirms how uncertain is the transference of ecology into land management.

Key words: sustainable management, population dynamics, conservation, biodiversity

Frontiers of Population Ecology, R.B. Floyd, A.W. Sheppard and P.J. De Barro (eds), CSIRO Publishing, Melbourne, 1996 pp. 509–529

Introduction

Population ecology is the study of the numerical attributes – numbers, sex ratio, rate of increase, and so on – of some biologically meaningful unit of a plant or animal species. There is little dispute about the meaning of population ecology, which was well summed up at the birth of the field by Cole (1957). Plenty of room exists for debate about the appropriate level of detail for a particular investigation, or about the interpretation of population analyses, but most biologists feel comfortable that population ecologists have a fair idea where they are heading.

But let us think for a moment about why we undertake population ecology. There are, broadly speaking, two reasons why we might be interested in analysing the dynamics of plant or animal populations. We might, as ecologists, be interested purely in what makes nature work; or alternatively, we may be land managers or conservation biologists with a commitment to using population analysis to improve human management of natural resources or of pests. Of course, during their working days most population ecologists will swing back and forward across the spectrum characterised by these two poles. My point, though, is that the concept of management intersects inevitably with much of population ecology, and as soon as it does so, the clear picture provided by the definitions of the field becomes a little less precise.

My essay begins with a discussion of how the demands of land management upon scientists have changed in recent times, and proceeds to a brief assessment of some examples of the use of population ecology in management. As will become clear, there are problematic but important signals emerging for scientists from such an overview.

Land management in Australia

The work of A.J. Nicholson (1948, 1954, 1957) in population ecology was primarily aimed at fundamental understanding. His work was not entirely divorced from application, though, because his study organism, the sheep-blowfly *Lucilia cuprina*, had long been a considerable problem for wool-growers. However, it is apparent that the context in which population ecology is applied has changed dramatically since Nicholson's time. It seems to me that the very term 'land management' has come to mean quite different things to us in the 1990s from what it meant in the 1940s and 1950s.

Indeed, it is not at all certain that the term land management would even have been used in Australia 50 years ago. My reading has not been comprehensive, but it suggests that where a phrase similar to this – primarily 'land administration' – was used during the 1800s and in the first 50 or 60 years of this century it meant the allocation of land to a particular use, and involved intimately the legalities of sale, rental, tenure, and boundaries (Powell 1976; Frawley 1994). Almost always the use considered was resource extraction in some sense – agriculture, pastoralism, forestry, or mining. Because of the centrality of land administration to the economy and culture of Australia in this earlier period, the State Departments of Lands were probably among the most powerful governmental agencies. Frawley (1994) described 19th century Australia as an era of exploitative pioneering, but pointed out that after federation the mood was one of national development which, it was argued, should involve the wise use of resources, although the achievement of these aims was frequently limited. Because of this context, scientists of the first 60 years of this century almost universally thought in quite focussed ways about the application of their work as fundamental contributors to the well-being of the country's agricultural and mining industries (Powell 1976; Frawley 1994); and indeed they were. A fine and later example of this orientation is the extraordinary series of reports from 1953 to 1976 in the Land Research Series from CSIRO's Division of Land Research and Regional Survey (e.g. Perry 1962).

Sometime during the 1960s a shift in orientation was set in motion. The history of this change in attitude does not yet appear to have been fully written, as it is so close in time to us (Frawley 1994). The change involved a dramatic broadening of the values attached to the land. The altered attitude first had its effects in the decision to remove grazing stock from part of the high country in the Snowy Mountains in 1958, in debate about sand-mining and oil-drilling on the Great Barrier Reef, and in the dispute over clearing of land for agriculture in the Little Desert of Victoria (Robbin 1993, 1994; Frawley 1994). With the subsequent bitter public dispute over the flooding of Lake Pedder in the early 1970s, and the election of the Whitlam Labor Government in 1972, the issue of land management shot rapidly onto the Australian political stage. In the past 20 years, we have witnessed a virtually continuous debate in the community about the allocation and management of land (Toyne 1994). We have seen gradual moves towards consideration of conservation as a valid and vital land use, towards the building of a representative network of national parks and reserves, towards sustainable land management in agricultural and pastoral lands, and towards considerations of the maintenance of biodiversity. Concurrently, we have seen the rise of Departments of Environment and Land Management, the dramatic development of tourism as an industry, a decline in the influence of Departments of Lands, increasing political influence of 'greens', recognition of Aboriginal land title, and declining effectiveness of traditional agricultural and pastoral lobby groups. Graetz (1995) points out that these trends – and as yet unsuspected ones – will continue to affect our society in profound ways.

Today, the notion that our society should manage its land more effectively is widely acknowledged. For example, Roberts (1993, pp 1) wrote that "land management is the manipulation of all elements constituting the...landscapes on which humans depend...(and that)...land resource management should preserve potential, keep options open, optimise quality of life for humans, conserve biodiversity and recognise equity". In these words he has attempted to express the fact that society has shifted its aim and lifted its sights. We might, as individuals, disagree with how well our society is meeting those aims – and I certainly think we have a very long way to go – but people of A.J. Nicholson's era would be astonished at the speed with which publicly-expressed values are in the process of changing in Australia.

So today we find that land management requires sustainable use of resources, together with the maintenance of populations of native plants and animals. To achieve this aim, ecologists would need to work at various levels, including of course in population ecology. How well are population ecologists equipped to carry out this task, and how well are we dealing with the rapidly changing expectations of society? I want to address these questions by looking briefly at four cases where population ecology has been or is becoming a central part of a land management issue. The selected cases involve a wide array of land-management questions: a natural resource under exploitation, namely perennial grasses of the rangelands; a pest, the plague locust *Chortoicetes terminifera;* pests which may better be regarded as a resource, the kangaroos *Macropus* spp.; and a species at risk of extinction, Leadbeater's possum *Gymnobelideus leadbeateri.*

Native species under exploitation

Soon after European settlement of Australia, people quickly realised that the vast grasslands and woodlands of the interior were suitable for extensive pasturing of stock. The pastoral industry, based on sheep and cattle, was a mainstay of the country's economy in earlier times, and remains significant despite a decline in its importance relative to other industries. The pastoral industry is dependent completely upon grasses growing in response to rainfall; agricultural development and improved pastures are generally not economically viable in the climatically uncertain and low-productivity environment of inland Australia (Pickup and Stafford Smith 1993; Wilcox and Cunningham 1994). The long-term well-being of the industry depends, therefore, on the

Table 1. Population flux over a decade of *Astrebla* spp. on a 30 m^2 plot (summarised from Table 3 of Orr and Evenson 1991; note that two typographical errors in the original versions of lines b and f have been corrected in line with a personal communication from D.M. Orr)

	Ungrazed	Grazed
(a) Number of plants in 1975	75	86
(b) Number of plants in 1986	78	95
(c) Net change	+3	+9
(d) Net reproductive rate	1.0	1.1
(e) Number of plants recruited between 1975 and 1986	95	400
(f) Number of plants lost between 1975 and 1986	92	391
(g) Plants present in 1975 and alive in 1986	56	52
(h) Percent survival of plants in (a)	75	60

maintenance of the resource under exploitation, i.e. the grasses (Hodgkinson 1992). Of most significance here are perennial grasses. They provide the most stable production, being able to respond to rainfall by deploying leaf tissue quickly and also having the capacity to persist into dry periods through acquisition of moisture from extensive root systems (e.g. Westoby 1980).

It became apparent during the first few decades of pastoral occupation that Australian rangelands were easily damaged (see Ratcliffe 1936; Lunney 1994). Stock were held near permanent waters for long dry periods and so degraded the vegetation and the soil. Palatable grasses diminished in abundance, unpalatable species increased, and in some areas almost all vegetation was removed, leading to increased rates of soil erosion. The consequence was frequently a damaged resource and a lower ability to sustain stock. Recognition of these problems led to calls for research to identify the requirements for 'sound management', particularly in the reproduction and demography of perennial plants under grazing (e.g. Perry 1967).

Perry's (1967) call for research into plant population ecology had been anticipated especially in the Mitchell grasslands (*Astrebla* spp.) of Queensland. At approximately yearly intervals from 1945 until 1959, and then again in 1968 and 1970, Williams and Roe (1975) recorded the presence of individually marked tussocks of predominantly *Astrebla lappacea* in transects through a 12 ha exclosure near Cunnamulla, in south-western Queensland. Their extraordinarily detailed results suggested that life-span was as long as 23 years, and that during the 26 years of the study at least 12 cohorts of plants had established following summer rainfalls. Williams and Roe (1975) were unable to analyse convincingly the impact of grazing on Mitchell grass, but they certainly could not detect any effect. Later work by Orr and Evenson (1991) and Orr (1991) analysed populations of *Astrebla* (again mainly *lappacea*) at Blackall, central-western Queensland, and at Julia Creek, north-western Queensland. At each location, individual plants were sampled annually for 10 years, and the data used to calculate population flux and life-spans. These later results confirmed the demographic picture of long life obtained at Cunnamulla, but showed that recruitment was not universally sporadic (Table 1).

The important management question, however, concerns the impact of grazing on the plants. Orr (1980) showed that heavy grazing at Blackall reduced the range of tussock sizes, but did not appear to disadvantage small tussocks or seedlings (Fig. 1). Orr and Evenson (1991) conducted an unreplicated experiment at Blackall which suggested that grazing had a slight negative impact on the survival of adult plants but enhanced the survival of seedlings (Fig. 2). Thus, under grazing there appeared to be an increased number of new plants, but an increased annual mortality of all plants,

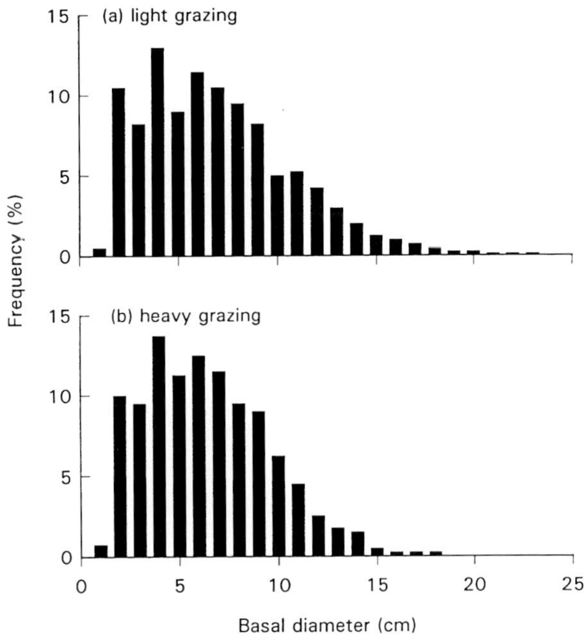

Fig. 1. Frequencies of tussock sizes of *Astrebla* spp. under light (a) and heavy (b) grazing (redrawn from Fig. 1 of Orr 1980).

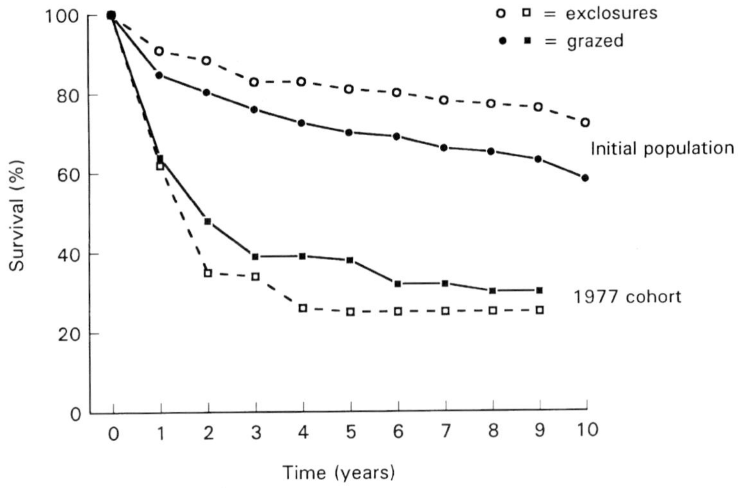

Fig. 2. Impact of grazing on survival of *Astrebla* spp. (redrawn from Fig. 3 of Orr and Evenson 1991). The initial population is shown by circles, and the 1977 seeding cohort by squares; open symbols show exclosures and closed symbols grazed plants.

such that densities were maintained. Although rigorous analyses of the effects of grazing on *Astrebla* are still lacking, Orr's studies, together with the observations of many other workers (e.g. Roe and Allen 1993; Wilcox and Cunningham 1994), suggest that Mitchell grass will persist under most current grazing regimes. The plants seem to establish sufficiently frequently to limit the chance of a transition in vegetational composition occurring (see Westoby *et al.* 1989), despite the often irregular recruitment.

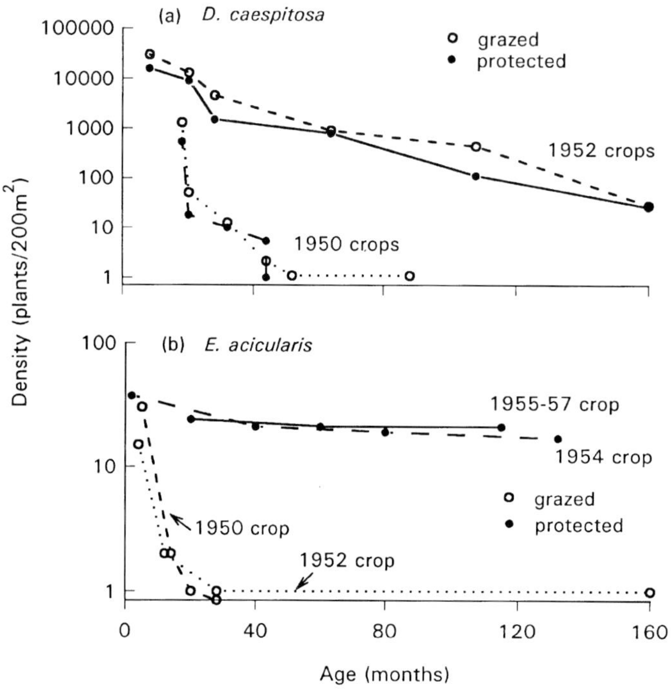

Fig. 3. Survival of crops of (a) *Danthonia caespitosa* and (b) *Enteropogon acicularis* in grazed (open symbols) and protected (closed symbols) grassland (redrawn from Figs 1 and 2 of Williams 1970).

But it is pertinent now to ask whether management of the Mitchell grasslands would have altered if the scientific results had suggested that utilisation of the resource was causing the plants to decline. Would scientific data on this point have changed management? To obtain a view on this question, let us turn to consideration of the demography of grasses in semi-arid New South Wales.

Williams (1970) and Williams and Roe (1975) reported demographic studies of the perennial grasses *Danthonia caespitosa* and *Enteropogon acicularis* near Deniliquin in south-central New South Wales. Plants on quadrats in ungrazed and grazed plots were sampled at irregular intervals between 1949 and 1968, and survival rates of cohorts calculated. Although replication seems inadequate from the present-day perspective, the results suggested that mortality was high in *D. caespitosa* in both ungrazed and grazed situations, and that survivorship was affected by factors other than grazing (Fig. 3a). In contrast, that of *E. acicularis* was high in ungrazed plots but low in grazed ones (Fig. 3b). The former species exhibited a much shorter half-life than the latter. The work was conducted in an area described as a 'disclimax community' resulting from 100 years of heavy grazing by sheep. Williams (1970) noted that *E. acicularis* was the only 'climax' species remaining after the disappearance under grazing of *Cymbopogon*, *Dichanthium*, *Eulalia* and *Themeda*. His comment implied that *E. acicularis* was probably the last remaining species from a suite of plants sensitive to grazing, and that the absent species probably displayed similar population dynamics under grazing. Further work by Hodgkinson (1991, 1992) in semi-arid woodlands near Louth, western New South Wales, demonstrated that seed production of certain species of grasses is dramatically affected by grazing pressures that have little short-term effect on survival of the individuals producing that seed (Fig. 4). Unlike the Mitchell grasslands, in New South Wales and in many other regions grazing has substantially altered the perennial grass composition and consequently reduced the capacity of the pasture to respond to different patterns of rainfall. These vast areas of the rangelands have lost palatable perennial grass cover and can thereby be regarded as degraded.

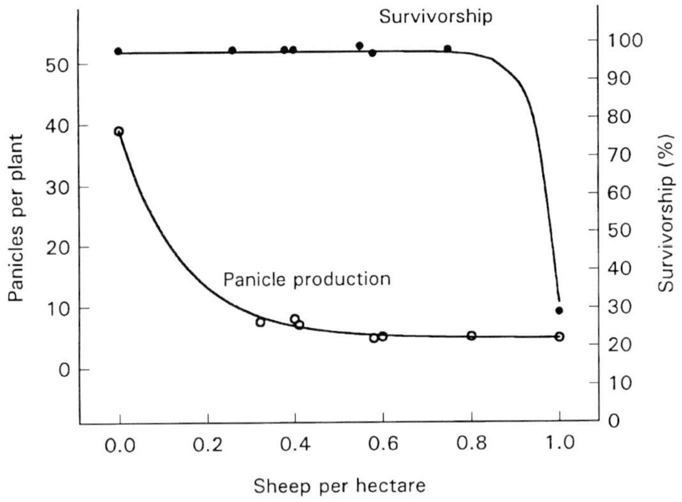

Fig. 4. Survivorship (solid symbols) and panicle production (open symbols) for *Thyridolepis mitchelliana* in relation to stocking intensity of sheep (redrawn from Fig. 4 of Hodgkinson 1992). Data for panicles come from one flowering period, and those for survivorship after 3.5 years of grazing.

Despite the results flowing from research in population ecology, many observers have continued to observe or predict degradation (e.g. Harrington *et al.* 1984; Payne *et al.* 1987; Friedel *et al.* 1990; Bastin *et al.* 1993; Mott and Tothill 1994; Pickard 1994). It seems, therefore, that in many vegetation formations of the Australian rangelands analysis of the demography of grasses has accounted for the degradation of the pasture under grazing but has apparently failed to improve management. Why is this so? One explanation is that species respond differently to grazing in different parts of their ranges (Hodgkinson 1992), and managers of these extensive lands are unable to take account of all the potential interactions. This may well be true, but we must also ask whether the current social and economic circumstances are ever likely to provide for sustainable pastoral enterprises (Fitzhardinge 1994). The essential elements of sustainable grazing – lowered stocking rates and avoidance of risk to pasture during and just after drought – have been known for nearly a century. It is evident that recent scientific opinion about sustainability (Pickup and Stafford Smith 1993) is identical in its principles to that expressed by the Royal Commission of 1901 into the Western Division of New South Wales (see Lunney 1994) and by Ratcliffe (1936). It seems apparent, under the present circumstances, that in the short-term it pays many leaseholders to use their country in such a way as to degrade it in the long-term (Milham 1994). I have reached the conclusion that ecologists should examine carefully their instincts to call for more research into ecological functioning of the rangelands in order to improve pastoral management, and instead should urge Government to examine the social and cultural roots of the continuing crisis of the rangelands. My first example suggests, therefore, that population ecology has not contributed successfully to resolution of a land management problem because that problem's principal cause lies not with ecological understanding but with society's use of that knowledge.

A pest

The Australian plague locust *Chortoicetes terminifera* has been recorded as a plaguing species for 150 years (Key 1938; Andrewartha 1940; Clark 1947; Casimir 1962; Magor 1970). The distribution of the locusts is extremely uneven, but upsurges in their numbers occur from time to time leading to the development of migratory swarms. The locust plagues invade many parts of the agricultural belt of South Australia and New South Wales, and to a lesser extent those of Victoria and Queensland (Fig. 5). Locusts eat live vegetation, and because they can occur at very high densities the amount of material they consume can be great, thus constituting a substantial economic problem. In addition,

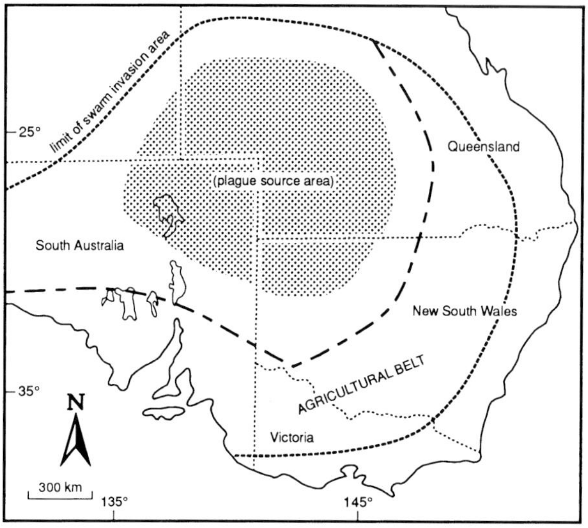

Fig. 5. Distribution of the plague locust (redrawn from Fig. 2 of Wright 1987).

they are highly mobile, particularly when abundant, and so have the capacity to transfer their impact rapidly across large areas. The highly noticeable activities of locusts therefore brought them to the attention of the agriculturally-oriented Australian Governments of the early part of this century, and hence scientific effort has been focussed on them for more than 50 years. The amount and quality of work on this species is truly awe-inspiring, and in order to arrive quickly at my destination I am able only to select a couple of highlights from an immense array of gems.

Plague locusts are a particular problem in the grasslands of central western New South Wales, where their damage tends to be greater because they occur in areas of crop production as well as pastures. It seems widely accepted, though, that the clearing of land for pasture may well be the primary reason why the impact of locusts is so severe. The animals require a mosaic of bare ground for basking, low green vegetation for feeding, and taller tussocky vegetation (generally less than 1 m) for nocturnal shelter. The extensive clearing of the original woodland and shrubland in the late nineteenth century seems to have turned an unfavourable environment into a highly desirable one from the perspective of the locusts (Clark 1950). Nevertheless, once the problem had been created, control was deemed economically imperative.

Population studies at Trangie in central western New South Wales were conducted from 1965 to 1974, initially by Clark (1974) and subsequently by Farrow (1979, 1982). Clark (1974) established that two or three generations per year are produced by locusts in this area during spring and summer. Population dynamics are complicated by the fact that, during winter, diapausing eggs as well as overwintering nymphs are present in the soil (Fig. 6). The timing of the generations is such that eggs are present in the soil during most months of the year, and that active stages (either nymphs or adults) are also present. Locusts take from 6–10 weeks to mature from hatching through to adult reproduction. Clark (1974) showed that rainfall in a 3-week period at the time of hatching was directly related to numbers of adults present in that generation; the rain favours the survival of nymphs by producing a suite of food plants.

Clark (1974) was aware that migration was also a major influence on densities; his own work demonstrated that night flights were major events in this species (Clark 1969,1971). This issue was followed up by Farrow (1979, 1982), who showed that changes in population size were mainly due to migration. He measured densities on his study plots while at the same time collecting flight-trap

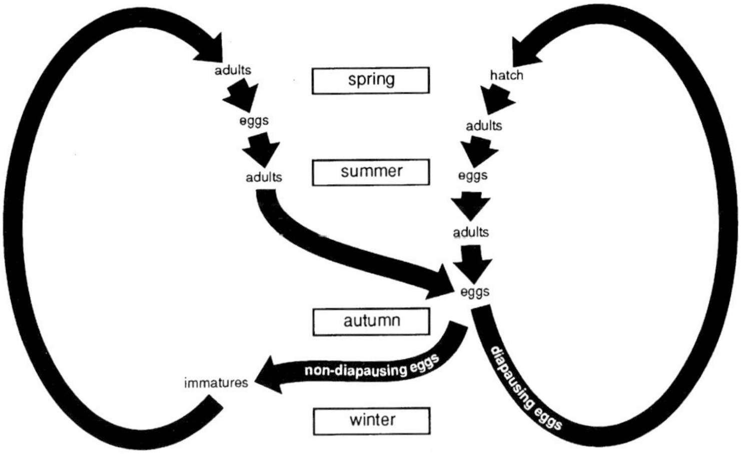

Fig. 6. Modified life cycle of the plague locust (simplified from Fig. 2 of Clark 1974).

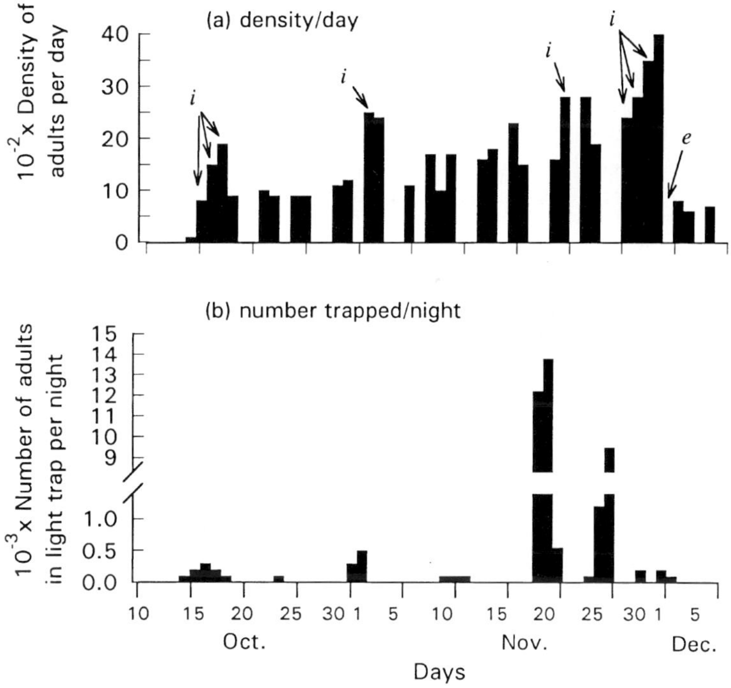

Fig. 7. Dispersal and migration of plague locusts (redrawn from Fig. 3 of Farrow 1979). The upper histogram (a) shows daily population fluctuations, and the lower (b) light-trap catches. Enhanced light-trap catches are coincident with immigration (i) and emigration (e).

samples; the results suggested that both localised dispersal movements and long-range migrations were involved (Fig. 7). Of special prominence were the major population displacements, which Farrow (1975, 1977, 1979) found took place at night primarily when upper-level airflow was from the north. It became evident that locusts were migrating during disturbed weather, over as many as three consecutive nights, and involved immigrants travelling from as far north as 1500 km. Population studies showed that local recruitment was insufficient to maintain populations in central western New South Wales, and that these populations were sustained by immigration from the north (Farrow 1982). It is now clear that big plagues occur when immigrants arrive from the north (Bryceson and Wright 1986; Wright 1987).

Where exactly do the animals come from? Andrewartha (1940) and Casimir (1962) had suggested that the arid interior was the major source of invasions, and they turned out to be correct. Clark *et al.* (1969) established that populations of locusts bred regularly on stony downs and flood plains in the Channel Country of Queensland, a region normally experiencing aridity (i.e. about 200 mm per annum) but occasionally receiving heavy summer rains. Hunter (1989) showed that Mitchell grasses *Astrebla* spp. put on green growth following rainfalls of 20–40 mm, and suggested that areas of these grasses were the primary engine for development of plagues. Simulation models of the development of plagues suggest that five of the six major plagues between 1934 and 1984 originated in the arid interior, when good spring and summer rains allowed two or three consecutive generations of successful breeding (Fig. 8; Wright 1987). Thus, after heavy drought-breaking rains, extensive concentration, multiplication and development of gregarious tendencies occur in the locusts, and mass eruptions of populations invade the agricultural lands far to the south.

Recognition that losses of agricultural production due to depredations of the plague locust are controlled by synoptic events and rainfall in south-western Queensland to a greater extent than anything else led immediately to the possibility of effective control. In 1976 the Australian Plague Locust Commission began operations to monitor, forecast and control populations of locusts. The Commission uses aerially-applied fenitrothion (an organophosphate poison) against swarms of locusts when it is suspected that plagues might develop; in bad years, as much as 130 tonnes of insecticide will be sprayed (Australian Plague Locust Commission 1994). The Commission does an excellent and professional job of controlling plague locusts.

In this case, population ecology eventually led to a most effective management outcome: on behalf of agriculture, a serious pest is able to be controlled. Yet I cannot help wondering how long this management solution might be acceptable. In the Introduction above, the immense attitudinal change to land management occurring in major segments of society was noted. Many people with conservation leanings might not regard the poisoning of locusts (and therefore of many other organisms) as an adequate response to the problem; they might suggest that these activities do not really form part of sustainable land management as it is coming to be known. The Commission is investigating the impact on non-target organisms of the application of fenitrothion (Gordon 1981; Australian Plague Locust Commission 1994). Nevertheless, will resistance in society to the short-term poisoning of large areas of Australia gradually intensify?

Perhaps questions about alternatives will soon gain prominence. It is difficult to see how biological control could be effective, given the fact that the species is a native and is unlikely to be solely susceptible to an introduced pathogen or parasite. Instead, efforts are being made to develop non-persistent biological insecticides (e.g. Herron and Baker 1991), and in particular application of the fungus *Metarhizium* sp. looks promising (Milner and Prior 1994). Bullen (1975) pointed out that serious plagues are infrequent and that the economic damage from locusts was often less than is imagined; he suggested that a strategy of crop insurance to compensate those individual farmers severely affected by a plague would perhaps be more cost-effective than outright control. Later analyses have not universally supported this viewpoint (Wright 1986), but further social change in the coming decades may force more detailed consideration of the full economic and environmental costs and benefits of this technical success story.

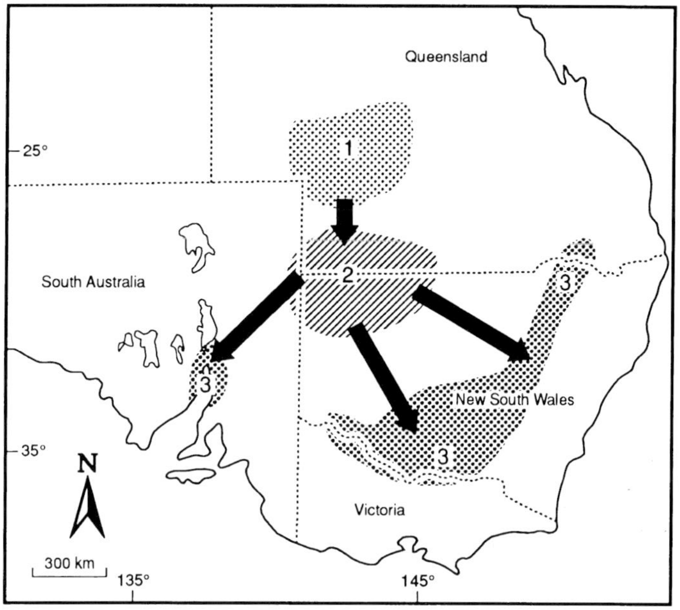

Fig. 8. Sources of swarms of plague locusts in 1979 (redrawn from Fig. 4 of Wright 1987). 1: spring laying; 2: summer breeding; 3: autumn swarms laying in plague area.

A pest, or a resource?

The killing of kangaroos for the sale of skins and for meat began soon after the European settlement of Australia. Commercial exports of hides and (later) meat were extensive, building to levels of nearly 1.5 million skins in about 1965; the main species involved were red kangaroos *Macropus rufus*, eastern grey kangaroos *M. giganteus* and western grey kangaroos *M. fuliginosus*. Nevertheless, the principal reason for the destruction of kangaroos was the perception that they were a pest, especially in the rangelands of eastern Australia. In Queensland alone, over a period of 40 years prior to 1917 some $2 million was paid as Government bounties for 26 million scalps (Poole 1984). Pastoralists in many areas saw their livelihoods threatened by what they perceived as swarms of kangaroos, which have probably increased in abundance in some regions since European settlement due to increased availability of water and altered environmental conditions (Australian National Parks and Wildlife Service 1988). But the pastoralists' demands for control (e.g. Doohan 1971) ran headlong into the growing environmental consciousness of the 1960s and 1970s, and out of that confrontation grew my third case study in population ecology.

Conservationists of the late 1960s claimed that kangaroos were likely to be exterminated, whether it be through hunting for hides and meat or by destruction as pests. Their protestations led to an order by the Commonwealth's Minister for Customs to halt the export of kangaroo products, and to an inquiry by a Parliamentary Select Committee in 1970 into the conservation and exploitation of kangaroos (Commonwealth of Australia 1971). Among many recommendations, the inquiry called – not surprisingly – for research into the ecology and the numbers of kangaroos. The controversy led to subsequent international debate resulting ultimately in a ban on the commercial importation of kangaroo products into the United States. Substantial research effort then ensued. Much of that effort focussed on breeding biology, but I want to look primarily at the survey work which set out to estimate the abundance of the animals.

The immensity of the country occupied by the red and grey kangaroos demanded broad-scale sampling, and fortunately it became apparent at that time that aerial surveys might be practical.

Caughley (1974) quickly recognised the potential for applying to kangaroos the aerial survey techniques for counting large mammals that he had helped develop in Africa; after considerable experimentation and testing (Caughley *et al.* 1976; Caughley 1977, 1979; Caughley and Grigg 1981), his opinion was confirmed that aerial survey was a cost-effective method. Teams of trained observers fly along predetermined transects at fixed height and speed, and record the kangaroos seen. The breadth of the transect is prescribed by marks or streamers attached to the wing-struts of aircraft. Standardisation of speed, height, and transect width between surveys ensures that the counts are repeatable and that population trends can be determined. Correction factors are then developed to adjust the raw counts to actual densities. Such correction factors are likely to be refined continually as more information is gathered (e.g. Bayliss and Giles 1985; Short and Hone 1989). It is widely agreed that virtually all aerial surveys under-estimate actual densities, although there is little doubt that errors in correction factors are conservative. In short, precision is high but accuracy is still a matter for debate.

By the early 1980s precise estimates were being obtained of the densities of kangaroos over the vast inland areas of Western Australia, South Australia, New South Wales and Queensland (Caughley *et al.* 1979; Caughley and Grigg 1981, 1982; Short *et al.* 1984). The first surveys covered 75% of the continent, but subsequent surveys usually excluded areas of the central and western deserts where kangaroo densities had previously been found to be very low. The surveys allowed examination of two aspects of kangaroo population ecology that conservationists had originally claimed placed the species at risk: responses to drought, and lower numbers than suggested by pastoralists.

Aerial survey indeed provided for analysis of the negative effects of drought on populations and on their capacity to recover (e.g. Cairns and Grigg 1993), but there is space only to address the issue of abundance. The surveys provided repeatable estimates of abundance state-by-state, and provided an objective basis for allocation of harvesting quotas by the State management authorities. The summary data for red kangaroos over the past eight years in the three states where the species is most abundant (Fig. 9) show that the animals number in the millions, and that what at first sight looked like substantial commercial kills (5–20% of population estimates) are not appearing to depress abundance. Similar trends are evident with the two species of grey kangaroos, although data for only the eastern are depicted (Fig. 10). The surveys in Queensland have recently switched from the standardised fixed-wing aircraft technique to more localised helicopter searches, and this is a matter of concern for continuity of the long-term monitoring. Nevertheless, the data seem to have been sufficient to satisfy most of the public concern about kangaroo harvesting not only in Australia but also overseas; the United States Government has removed the three species from its List of Endangered and Threatened Wildlife (U.S. Fish and Wildlife Service 1995).

In this case, then, population ecology has provided generally satisfying answers to a deeply-felt public debate. This conclusion is not meant to suggest that all critics have been silenced, for this is not so. Some people feel that no wild animals should be shot, but these appear to be a tiny minority. Many conservationists now accept that the advent of pastoralism, through the control of predators of kangaroos, the expansion of suitable habitat and the provision of more water, has dramatically changed kangaroo populations; therefore, society must take responsibility for kangaroo control (Cameron 1991). However, conservationists are still highly suspicious of the harvesting industry, regarding it as an improper group to control kangaroos. They also remain critical of certain aspects of population monitoring.

But it seems to me that the argument has moved on from a debate between conservationists and those who consider the kangaroo a pest. As knowledge about the parlous state of much of the rangelands has spread, the kangaroo management issue has become inter-twined with the search for solutions to those broader environmental problems (Cameron 1991). Grigg's (1988) suggestion that

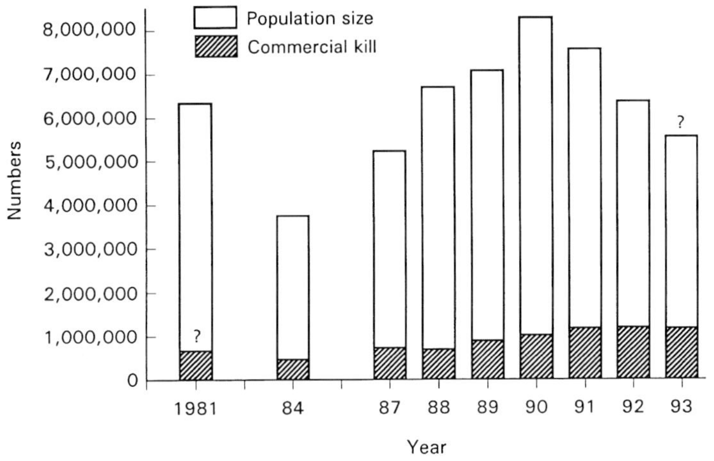

Fig. 9. Estimated population sizes of red kangaroos (open bars), and of commercial quotas for their harvest (hatched portion), in the rangelands of South Australia, New South Wales and Queensland during some of the years between 1981 and 1994 (from Fletcher *et al.* 1990 and from data provided by the Australian Nature Conservation Agency). The ? in 1981 refers to the fact that population estimates were made for kangaroos as a whole; consequently, proportions of red kangaroos were calculated from those observed in later years. In 1993, ? denotes a change in survey technique (to smaller areas of helicopter samples) in Queensland.

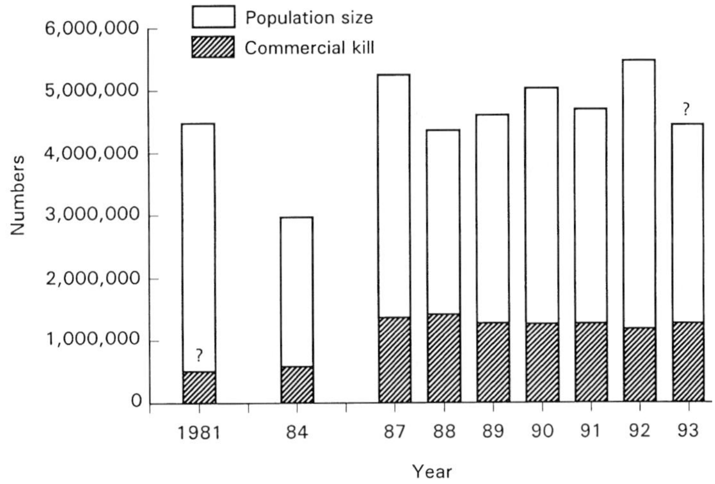

Fig. 10. Estimated population sizes of eastern grey kangaroos (open bars), and of commercial quotas for their harvest (hatched portion), in the rangelands of New South Wales and Queensland during some of the years between 1981 and 1994 (from Fletcher *et al.* 1990 and from data provided by the Australian Nature Conservation Agency). For ?, see legend to Fig. 9.

kangaroos should be 'farmed' instead of sheep continues to reverberate throughout the debate. But farming will only become feasible when there is an economic capacity for enterprises to withstand the decline in harvest during drought and when the value of a kangaroo increases substantially. For increased value to occur, the quotas for harvesting mentioned above would have to be barely sufficient to meet demand. Such a situation is not presently in view – but it might be over the coming years. In addition, it seems likely to me that community opinion will continue to move towards acceptance of harvesting of kangaroos, but away from acceptance of the present environmentally miserable state of much of the sheep-rangelands. It is not yet possible to predict the precise way in which kangaroo management will fit into future management arrangements, although it is likely that population ecology will build on today's framework and thereby continue to play a major role in the eventual solutions.

A conservation problem

Leadbeater's possum *Gymnobelideus leadbeateri* was last seen in 1909 before disappearing until 1961 (Wilkinson 1961). Its reappearance coincided with the gradually developing public interest in conservation matters of the 1960s and 1970s, and many people felt it important not to let the possum slip away a second time. Interest eventually translated into research on the ecology of the possum (Smith 1982), and quickly into realisation that the future of the animal was intimately tied up with use of its habitat for timber production. How has ecology contributed to solutions to this dilemma?

Leadbeater's possum is virtually restricted to forests of mountain ash *Eucalyptus regnans*, alpine ash *E. delegatensis* and shining gum *E. nitens* in the central highlands of Victoria (Lindenmayer *et al.* 1989; Lindenmayer and Possingham 1994). Several studies have indicated that vegetation structure and plant species composition, together with the presence of suitable nest trees, are key features of the habitat requirements of the possum (Smith and Lindenmayer 1988; Lindenmayer *et al.* 1991a). Possums are most likely to occur at sites with a high basal area of *Acacia* spp., the gums of which are an important food source (Smith 1984), and where eucalypt trees support a large quantity of decorticating bark, from which possums can obtain essential invertebrate prey (Lindenmayer *et al.* 1991a; Lindenmayer *et al.* 1994). They are also most likely to be found at sites with numerous hollow-bearing trees; the possums occur as colonies which jointly occupy the hollows. Nest trees appear to be in the late stages of senescence and are typically at least 190 years old (Smith and Lindenmayer 1988; Lindenmayer *et al.* 1991b).

The overall consequence of these habitat requirements is that logging and fire are inimical to the persistence of Leadbeater's possum. Logging operations are likely to create regenerating forest without the necessary nest-trees and foraging habitat. Hence, protection of present old-growth forests is essential for survival of possums. Similarly, wildfire may remove vital habitat characteristics. It kills the acacias that are important for food, and destroys some proportion of the vital nest-trees (Lindenmayer *et al.* 1993). Given the small area occupied by possums, both in geographic terms and in terms of their confinement to relatively tiny patches within that range, it is clear that unless logging can be conducted in a sympathetic way and fire can be managed effectively, then Leadbeater's possum is in trouble.

Lindenmayer and Possingham (1994) used population viability analysis (specifically, ALEX; Possingham *et al.* 1992) to bring these major threads together into coherent management recommendations. They incorporated data on the spatial location and size of potentially suitable habitat patches, combined this with data on the life-history of the species and its density in different types of forest, and simulated the behaviour of metapopulations in present forest blocks over 150 years. These analyses suggested that the probability of extinction of the possums increases with a decrease in patch size (Fig. 11); probability of survival approached 99% in patches greater than 150 ha. This result was not surprising, but Lindenmayer and Possingham (1994) were able to apply these predictions to the forests where possums presently occur. They concluded that extinction was likely

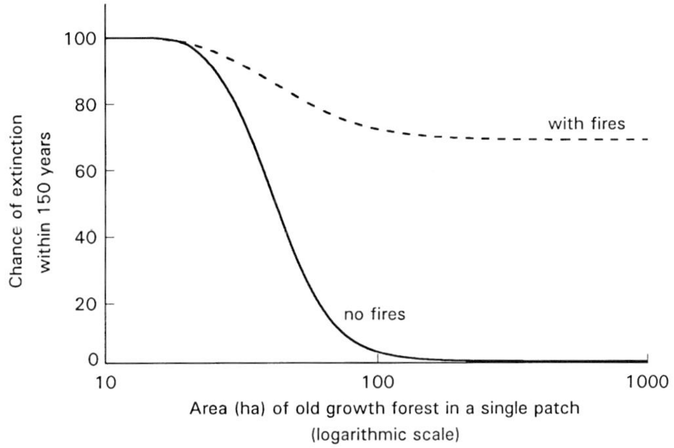

Fig. 11. Probabilities of extinction of Leadbeater's possum in single patches of old growth forest in the absence of wildfire and in its presence, as estimated from population viability analysis (redrawn from Fig. 23 of Lindenmayer and Possingham 1994).

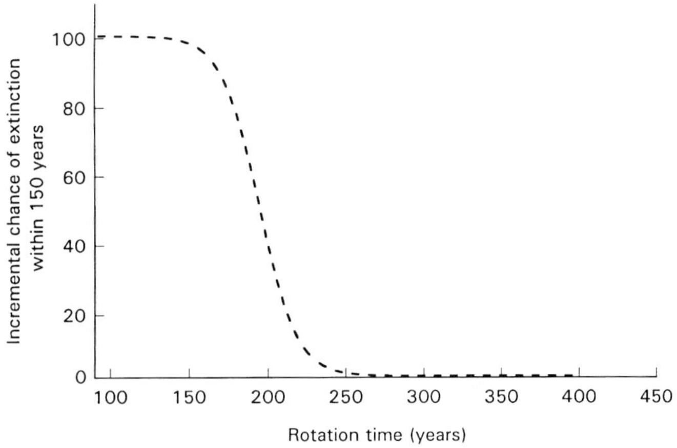

Fig. 12. Probabilities of extinction of Leadbeater's possum in a Victorian forest block, as estimated from population viability analysis (redrawn from Fig. 29 of Lindenmayer and Possingham 1994).

in two forest blocks used for wood production, and that larger patches of habitat were crucial to persistence of the species.

Analysis of the effects of wildfires on the chances of persistence brought out the cumulatively negative impacts of area of habitat and fire-driven habitat change (Fig. 11). Survival of populations is likely to be lowest where fires burn patches of habitat that are relatively small, and where fires are not suppressed for long periods with the result that old nest trees are unable to form. Lindenmayer and Possingham (1994) then predicted the effects of different logging practices on persistence. Because logging alters vegetational structure and competition unfavourably from the point of view of possums, it appears that the only logging regimes that will allow possums some chance of persisting are rotation times greater than 150 years (Fig. 12). Thus, present management strategies appear to be inadequate, and Lindenmayer and Possingham (1994) called for more permanent reservation in order to remove habitat from forestry operations and thereby maximise the probability of Leadbeater's possum surviving.

Lindenmayer and Possingham (1994) interpreted their results cautiously, which is quite proper in view of the fact that population viability analyses are fundamentally untestable (Lacy 1993). They noted areas of the ecology of the montane forests needing more research in order to firm up some of their conclusions, and recommended an adaptive management strategy whereby the population viability analyses are revisited in some years' time when more data become available and when improved conservation strategies are in place.

Will improved management result from this massive investment of time in the field and on a computer? Lindenmayer (pers. comm.) is cautiously optimistic that the outcome will be much better than the present situation. He notes that the profile of Leadbeater's possum in the community is high enough to ensure that the Victorian Government will attempt to institute an improved conservation strategy (Victorian Government 1992), and that new reservations are planned. If he is correct in his optimism, then another case of population ecology successfully being converted into land management will exist. Let us hope that Lindenmayer is right.

Conclusion

The information surveyed here suggests conclusions that I suspect most practising ecologists are aware of, but which most of us suppress in our minds in order not to become overwhelmed. The immediate messages are threefold. First, there is no guarantee that the results of good ecological research will ever find their way into land management. In one of the cases mentioned above (the rangeland grasses), I have serious doubts about the likelihood of uptake of knowledge in the present management context. Secondly, the speed of transition into management may well be governed more by the degree to which the results, and the consequent management recommendations, are at odds with other socio-economic imperatives. If there are powerful forces aligned against a change in management (e.g. the possums), then more and more work is required before change will be instituted. In reverse, if there are no voices speaking on behalf of other organisms and environments affected by the immediate outcomes of the resulting management strategy (e.g. the country poisoned along with locusts), then action may result quickly. Thirdly, the outcome of these conflicting pressures may not be predictable at the outset of the research work. I write this because changes in community opinion and in social currents may be surprisingly rapid, perhaps as a consequence of the results of the scientific work (e.g. the kangaroos), or perhaps not. Subsequently, research may become quickly more or less applicable in response to changing community attitudes. Although I have confined myself to Australian issues, I suspect that we have a lot in common with other countries (e.g. Lindenmayer and Norton 1993).

How is the population ecologist to see his or her way through this welter of conflicting pressures? The answer is "Not easily". Reading the future is notoriously difficult, but we can be certain that the relationship between science and government or the community continues to evolve. Science is no longer considered by government to be outside the political game; scientists increasingly see themselves as being part of the process by which change occurs, rather than being disinterested contributors of information. Nevertheless, it appears to me that politics is still trying to work out how best to incorporate science into itself, just as government and society are still grappling with changing communal perceptions of nature and the environment. But are there some rules of thumb that ecologists might use to judge the likely effectiveness of their research with respect to land management? Let me try to offer some.

The first and undisputable guideline is that good science is clearly essential. I do not wish to imply in my preceding comments that it is acceptable to cut corners just because science is inevitably involved in the political game. The second suggestion is to think hard about land management scenarios before going into a particular research area. Perhaps we would benefit – and our governments too – if scientists asked harder questions about the social and economic circumstances of land management issues before large investment of their time and expertise began. The third recommendation is to avoid becoming part of the management problem by persisting with research, and possibly institutionalising it, beyond the point where it appears likely that application of results will occur. If only it were easy to make the judgement as to when this point is reached! It is evident that unpredictable changes in the social circumstances can improve the chances of uptake substantially and thereby turn a difficult transference task into an easier one. Finally, I suspect that in future ecologists will need to think far more seriously about cooperating with social scientists as a matter of course. We have a few disciplinary barriers to knock down, I believe. Even when a problem does not demand direct involvement with professionals it undoubtedly pays us to stay exceedingly well abreast of the social currents running around the problems we are studying.

Despite the difficulties of meeting all the demands noted in the previous paragraph, I am of the strong belief that ecologists will continue to have a vital role in land management. We have a lot to learn about the expectations of society, and about predicting or taking advantage of the forthcoming changes in attitudes, which will throw up tremendous challenges in the coming decades. But I for one look forward to those challenges.

Acknowledgments

I have gathered a bunch of other people's flowers, and nothing but the thread that binds them is my own. I acknowledge all those people whose hard work and stimulating writings provided the basis for this paper, but especially Roger Farrow, the late Graeme Caughley, Ken Hodgkinson, David Lindenmayer, and the late Owen Williams. Particular thanks are due to Roger Farrow, Gerry Maynes, David Lindenmayer and David Orr for information. I am indebted to Nick Nicholls and David Grice for moral support, to David for preparing most of the figures, and to Nick, David, Roger Farrow and David Lindenmayer for commenting upon a draft of the manuscript. Finally, special thanks to Rob Floyd for his patience.

References

Andrewartha HG (1940) The environment of the Australian plague locust (*Chortoicetes terminifera* Walk.) in South Australia. *Trans R Soc S Aust* **64**:75–94

Australian National Parks and Wildlife Service (1988) Kangaroos in Australia: conservation status and management. *Australian National Parks and Wildlife Service, Occassional Papers* **14**:1–45

Australian Plague Locust Commission (1994) Annual report 1992–93. Department of Primary Industries and Energy, Australian Government Publishing Service, Canberra

Bastin GN, Pickup G, Chewings VH, Pearce G (1993) Land degradation assessment in central Australia using a grazing gradient method. *Rangel J* **15**:190–216

Bayliss P, Giles J (1985) Factors affecting the visibility of kangaroos counted during aerial surveys. *J Wildl Manage* **49**:686–692

Bryceson KP, Wright DE (1986) An analysis of the 1984 locust plague in Australia using multitemporal Landsat multispectral data and a simulation model of locust development. *Agric Ecosyst Environ* **16**:87–102

Bullen FT (1975) Economic effects of locusts in eastern Australia: a report to the Reserve Bank of Australia. Unpublished report. CSIRO Division of Entomology, Canberra

Cairns SC, Grigg GC (1993) Population dynamics of red kangaroos (*Macropus rufus*) in relation to rainfall in the South Australian pastoral zone. *J Appl Ecol* **30**:444–458

Cameron J (1991) Land degradation and the wild harvesting of kangaroos. In: Cameron J, Elix J (eds) *Recovering ground: a case study approach to ecologically sustainable rural land management*. Australian Conservation Foundation, Melbourne, pp 117–133

Casimir M (1962) History of outbreaks of the Australian plague locust, *Chortoicetes terminifera* (Wlk.), between 1933 and and 1959 and analyses of the influence of rainfall in these outbreaks. *Aust J Agric Res* **13**:670–700

Caughley G (1974) Bias in aerial survey. *J Wildl Manage* **38**:921–933

Caughley G (1977) Sampling in aerial survey. *J Wildl Manage* **41**:605–615

Caughley G (1979) Sampling techniques for aerial censuses. *Australian National Parks and Wildlife Service, Special Publications* **1**:9–14

Caughley G, Grigg GC (1981) Surveys of the distribution and density of kangaroos in the pastoral zone of South Australia, and their bearing on the feasibility of aerial survey in large and remote areas. *Aust Wildl Res* **8**:1–11

Caughley G, Grigg GC (1982) Numbers and distribution of kangaroos in the Queensland pastoral zone. *Aust Wildl Res* **9**:365–371

Caughley G, Sinclair RG, Grigg GC (1979) Trend of kangaroo populations in New South Wales, Australia. *J Wildl Manage* **43**:775–777

Caughley G, Sinclair RG, Scott-Kemmis D (1976) Experiments in aerial survey. *J Wildl Manage* **40**:290–300

Clark DP (1969) Night flights of the Australian plague locust *Chortoicetes terminifera* (Walk.), in relation to storms. *Aust J Zool* **17**:329–352

Clark DP (1971) Flights after sunset by the Australian plague locust, *Chortoicetes terminifera* (Walk.), and their significance in dispersal and migration. *Aust J Zool* **19**:159–176

Clark DP (1974) The influence of rainfall on the densities of adult *Chortoicetes terminifera* (Walker) in central western New South Wales, 1969–73. *Aust J Zool* **22**:365–386

Clark DP, Ashall C, Waloff Z, Chinnick L (1969) Field studies on the Australian plague locust (*Chortoicetes terminifera* Walk.) in the 'Channel Country' of Queensland. *Anti-Locust Bull* **44**:1–101

Clark LR (1947) An ecological study of the Australian plague locust (*Chortoicetes terminifera* Walk.) in the Bogan Macquarie outbreak area, N.S.W. CSIRO Bull No. 226, pp 1–71

Clark LR (1950) On the abundance of the Australian plague locust, *Chortoicetes terminifera* (Walker), in relation to trees. *Aust J Agric Res* **1**:64–75

Cole LC (1957) Sketches of general and comparative demography. *Cold Spring Harbor Symp Quant Biol* **23**:1–15

Commonwealth of Australia (1971) Conservation and commercial exploitation of kangaroos: interim report from the House of Representatives Select Committee on wildlife conservation. *Parliamentary Paper* **219**. Parliament of Australia, Canberra

Doohan JJ (1971) The kangaroo as a pest. *Aust Zool* **16**:65–67

Farrow RA (1975) Offshore migration and the collapse of outbreaks of the Australian plague locust (*Chortoicetes terminifera* Walk.) in south-east Australia. *Aust J Zool* **23**:569–595

Farrow RA (1977) Origin and decline of the the 1973 plague locust outbreak in central western New South Wales. *Aust J Zool* **25**:455–489

Farrow RA (1979) Population dynamics of the Australian plague locust, *Chortoicetes terminifera* (Walker), in central western New South Wales. I. Reproduction and migration in relation to weather. *Aust J Zool* **27**:717–745

Farrow RA (1982) Population dynamics of the Australian plague locust, *Chortoicetes terminifera* (Walker), in central western New South Wales. III. Analysis of population processes. *Aust J Zool* **30**:569–579

Fitzhardinge G (1994) An alternative understanding of the relationship between the ecosystem and the social system – implications for land management in semi-arid Australia. *Rangel J* **16**:254–264

Fletcher M, Southwell CJ, Sheppard NW, Caughley G, Grice D, Grigg GC, Beard LA (1990) Kangaroo population trends in the Australian rangelands, 1980–87. *Search* **21**:28–29

Frawley K (1994) Evolving visions: environmental management and nature conservation in Australia. In: Dovers S (ed) *Australian environmental history: essays and cases*. Oxford University Press, Melbourne, pp 55–78

Friedel MH, Foran BD, Stafford Smith DM (1990) Where the creeks run dry or ten feet high: pastoral management in arid Australia. *Proc Ecol Soc Aust* **16**:185–194

Gordon K (1981) The fate of fenitrothion in the environment and its effects on organisms. *Australian Plague Locust Commission, Technical Report* **3**:1–149

Graetz D (1995) The futures of a wide brown land: thriving or surviving. In: Eckersley R, Jeans K (eds) *Challenge to change: Australia in 2020*. CSIRO Publishing, Melbourne, pp 241–268

Grigg GC (1988) Kangaroo harvesting and the conservation of the sheep rangelands. *Aust Zool* **24**:124–128

Harrington GN, Wilson AD, Young MD (eds) (1984) Management of Australia's rangelands. CSIRO, Melbourne

Herron GA, Baker GL (1991) The effect of host stage and temperature on the development of *Hexamermis* sp. (Nematoda: Mermithidae) in the Australian plague locust *Chortoicetes terminifera* (Walker) (Orthoptera: Acrididae). *Nematologica* **37**:213–224

Hodgkinson K (1991) *Identification of critical thresholds for opportunistic management of rangeland vegetation.* Proceedings of the IV International Rangeland Congress, Montpellier, pp127–129

Hodgkinson KC (1992) Elements of grazing strategies for perennial grass management in rangelands. In: Chapman G (ed) *Desertified grasslands: their biology and management.* Linnean Society of London, London, pp 77–94

Hunter DM (1989) The response of Mitchell grasses (*Astrebla* spp.) and button grasses (*Dactyloctenium radulans* (R. Br.)) to rainfall and their importance to the survival of the Australian plague locust, *Chortoicetes terminifera* (Walker), in the arid zone. *Aust J Ecol* **14**:467–471

Key KHL (1938) The regional and seasonal incidence of grasshopper plagues in Australia. CSIRO Bull No. 117, pp 1–87

Lacy RC (1993) VORTEX – a model for use in population viability analysis. *Wildl Res* **20**:45–65

Lindenmayer DB, Boyle S, Burgman MB, McDonald D, Tomkins B (1994) The sugar and nitrogen content of the gums of *Acacia* spp. in the mountain ash and alpine ash forests of central Victoria and its potential implications for exudivorous arboreal marsupials. *Aust J Ecol* **19**:169–177

Lindenmayer DB, Cunningham RB, Donnelly CF, Tanton MT, Nix HA (1993) The abundance and development of cavities in montane ash-type eucalypt trees in the montane forests of the central highlands of Victoria, south-eastern Australia. *For Ecol Manage* **60**:77–104

Lindenmayer DB, Cunningham RB, Tanton MT, Nix HA, Smith AP (1991a) The conservation of arboreal marsupials in the montane ash forests of the central highlands of Victoria, south-east Australia. III. The habitat requirements of Leadbeater's possum, *Gymnobelideus leadbeateri*, and models of the diversity and abundance of arboreal marsupials. *Biol Conserv* **56**:295–315

Lindenmayer DB, Cunningham RB, Tanton MT, Smith AP, Nix HA (1991b) Characteristics of hollow-bearing trees occupied by arboreal marsupials in the montane ash forests of the central highlands of Victoria, south-east Australia. *For Ecol Manage* **40**:289–308

Lindenmayer DB, Possingham HP (1994) The risk of extinction: ranking management options for Leadbeater's possum using population viability analysis. Centre for Resource and Environmental Studies, Canberra

Lindenmayer DB, Norton TW (1993) The conservation of Leadbeater's possum in south-eastern Australia and the northern spotted owl in the Pacific northwest of the U.S.A: management issues, strategies and lessons. *Pac Conserv Biol* **1**:13–19

Lindenmayer DB, Smith AP, Craig SA, Lumsden LF (1989) A survey of the distribution of Leadbeater's possum, *Gymnobelideus leadbeateri* McCoy, in the central highlands of Victoria. *Vic Nat* **106**:174–178

Lunney D (1994) Royal Commission of 1901 on the western lands of New South Wales – an ecologist's summary. In: Lunney D, Hand S, Reed P, Butcher D (eds) *Future of the fauna of western New South Wales.* Surrey Beatty & Sons, Sydney, pp 221–240

Magor JI (1970) Outbreaks of the Australian plague locust (*Chortoicetes terminifera* Walk.) in New South Wales during the period 1937–1962, particularly in relation to rainfall. *Anti-Locust Mem* **11**:1–39

Milham N (1994) An analysis of farmers' incentives to conserve or degrade the land. *J Environ Manage* **40**:51–64

Milner RJ, Prior C (1994) Susceptibility of the Australian plague locust, *Chortoicetes terminifera*, and the wingless grasshopper, *Phaulacridium vittatum*, to the fungi *Metarhizium* spp. *Biol Control* **4**:132–137

Mott JJ, Tothill JC (1994) Degradation of savannah woodlands in Australia. In: Moritz C, Kikkawa J (eds) *Conservation biology in Australia and Oceania.* Surrey Beatty & Sons, Sydney, pp 115–130

Nicholson AJ (1948) Competition for food amongst *Lucilia cuprina* larvae. *Proceedings of the International Entomology Congress* **8**:277–281

Nicholson AJ (1954) An outline of the dynamics of animal populations. *Aust J Zool* **2**:9–65

Nicholson AJ (1957) The self-adjustment of populations to change. *Cold Spring Harbor Symp Quant Biol* **23**:153–172

Orr DM (1980) Effects of sheep grazing *Astrebla* grassland in central western Queensland. I. Effects of grazing pressure and livestock distribution. *Aust J Agric Res* **31**:797–806

Orr DM (1991) Trends in the recruitment of *Astrebla* spp. in relation to seasonal rainfall. *Rangel J* **13**:107–117

Orr DM, Evenson CJ (1991) Effects of sheep grazing *Astrebla* grassland in central western Queensland. I. Dynamics of *Astrebla* spp. under grazing pressure and exclosure between 1975 and 1986. *Rangel J* **13**:36–46

Payne AL, Curry PJ, Spencer GF (1987) An inventory and condition survey of rangelands in the Carnarvon Basin, Western Australia. *Western Australian Department of Agriculture, Technical Bulletin* **73**:1–478

Perry RA (ed) (1962) Lands of the Alice Springs area, Northern Territory, 1956–57. *CSIRO Land Res Ser* **6**:1–280

Perry RA (1967) The need for rangelands research in Australia. *Proc Ecol Soc Aust* **2**:1–14

Pickard J (1994) Land degradation and land conservation in the arid zone of Australia: grazing is the problem and the cure. In: Moritz C, Kikkawa J (eds) *Conservation biology in Australia and Oceania*. Surrey Beatty & Sons, Sydney, pp 115–130

Pickup G, Stafford Smith DM (1993) Problems, prospects and procedures for assessing the sustainability of pastoral land management in arid Australia. *J Biogeogr* **20**:471–487

Poole WE (1984) Management of kangaroo harvesting in Australia (1984). *Australian National Parks and Wildlife Service, Occassional Papers* **9**:1–25

Possingham HP, Davies I, Noble IR, Norton TW (1992) A metapopulation simulation model for assessing the likelihood of plant and animal extinctions. *Math Comp Simul* **33**:367–372

Powell JM (1976) *Environmental management in Australia, 1788–1914*. Oxford University Press, Melbourne

Ratcliffe FN (1936) Soil drift in the arid pastoral areas of South Australia. CSIRO Pamphlet No. 64, pp 1–84

Robbin L (1993) Of desert and watershed: the rise of ecological consciousness in Victoria, Australia. In: Shortland M (ed) *Science and nature: essays in the history of the environmental sciences*. Oxford University Pess, Oxford, pp 115–149

Robbin L (1994) Nature conservation as a national concern: the role of the Australian Academy of Science. *Historical Records of Australian Science* **10**:1–24

Roberts B (1993) *Ground rules: perspectives on land stewardship*. USQ Press, Toowoomba

Roe R, Allen GH (1993) Studies on the Mitchell grass association in south-western Queensland. 3. Pasture and wool production under different rates of stocking and continuous or rotational grazing. *Rangel J* **15**:302–319

Short J, Caughley G, Grice D, Brown B (1984) The distribution and abundance of kangaroos in Western Australia in relation to environment. *Aust Wildl Res* **10**:435–451

Short J, Hone J (1989) Calibrating aerial surveys of kangaroos by comparison with drive counts. *Aust Wildl Res* **15**:277–284

Smith AP (1982) Leadbeater's possum and its management. In: Groves RH, Ride WD (eds) *Species at risk: research in Australia*. Australian Academy of Science, Canberra, pp 129–147

Smith AP (1984) Diet of Leadbeater's possum *Gymnobelideus leadbeateri* (Marsupialia). *Aust Wildl Res* **11**:265–273

Smith AP, Lindenmayer DB (1988) Tree hollow requirements of Leadbeater's possum and other possums and gliders in timber production forests of the Victorian central highlands. *Aust Wildl Res* **15**:347–362

Toyne P (1994) *The reluctant nation: environment, law and politics in Australia*. ABC Books, Sydney

U.S. Fish and Wildlife Service (1995) Endangered and threatened wildlife and plants; removal of three kangaroos from the the List of Endangered and Threatened Wildlife. *US Federal Register* **60**:12887–12906

Victorian Government (1992) *Flora and Fauna Guarantee strategy: conservation of Victoria's biodiversity*. Department of Conservation and Environment, Melbourne

Westoby M (1980) Elements of a theory of vegetation dynamics in arid rangelands. *Isr J Bot* **28**:169–194

Westoby M, Walker B, Noy-Meir I (1989) Opportunistic management for rangelands not at equilibrium. *J Range Manage* **42**:266–274

Wilcox DG, Cunningham GM (1994) Economic and ecological sustainability of current land use in Australial's rangelands. In: Morton SR, Price PC (eds) *R&D for sustainable use and management of Australial's rangelands.* Land and Water Resources Research and Development Corporation Occasional Paper 06/93, Canberra, pp 87–171

Wilkinson HE (1961) The rediscovery of Leadbeater's possum, *Gymnobelideus leadbeateri* McCoy. *Vic Nat* **78**:97–102

Williams OB (1970) Population dynamics of two perennial grasses in Australian semi-arid grassland. *J Ecol* **58**:869–875

Williams OB, Roe R (1975) Management of arid grasslands for sheep: plant demography of six grasses in relation to climate and grazing. *Proc Ecol Soc Aust* **9**:142–156

Wright DE (1986) Economic assessment of actual and potential damage to crops caused by the 1984 locust plague in south-eastern Australia. *J Environ Manage* **23**:293–308

Wright DE (1987) Analysis of the development of major plagues of the Australian plague locust *Chortoicetes terminifera* (Walker) using a simulation model. *Aust J Ecol* **12**:423–437

SPATIAL VARIATION IN FOOD LIMITATION: THE EFFECTS OF FORAGING CONSTRAINTS ON THE DISTRIBUTION AND ABUNDANCE OF FERAL PIGS IN THE RANGELANDS

David Choquenot and Nick Dexter

ABSTRACT

While the distribution and abundance of animals reflects the influence of both spatial and temporal factors, models describing variation in animal abundance have focussed on temporal variation. In contrast, models of variation in individual fitness have emphasised the influence of spatial components of the animal's environment on reproductive success and/or survival. Spatial factors which affect the fate of individual animals should logically have consequences for the populations these individuals constitute. In this paper we examine the role spatial factors play in 1) the probability of species occurrence and persistence at a given location, and 2) the prevailing density of a species where it does occur. Specifically, we examine the influence of spatial variation in the effect of a limiting factor (food availability) on the distribution and abundance of feral pigs inhabiting inland river systems in Australia's rangelands. A habitat-related constraint on the movements of pigs during hot weather appears to affect their foraging efficiency, compounding the rate at which populations decline when food is short wherever habitats offering shade and water are scarce. Stochastic simulation modelling indicates that the spatial distribution of such habitats in this environment ultimately influences the average density of pig populations and whether these populations persist indefinitely, occur periodically or are non-viable (i.e. undergo deterministic extinction).

Key words: habitat, foraging efficiency, stochastic simulation modelling, *Sus scrofa*

Frontiers of Population Ecology, R.B. Floyd, A.W. Sheppard and P.J. De Barro (eds), CSIRO Publishing, Melbourne, 1996 pp. 531–546

Introduction

While the distribution and abundance of animals reflects the influence of both spatial and temporal factors, models which describe variation in animal abundance have focussed more on the influence of temporal factors (Sinclair 1989). In contrast, models of variation in individual fitness have emphasised the influence of spatial components of the animals environment on reproductive success and/or survival (Stephens and Krebs 1986). In particular, optimal foraging theory, which suggests that an animal's habitat preferences reflect the abundance and quality of food available in various habitats, has been used as a point of departure in identifying the consequences of foraging constraints for individual animals. Constraints identified include habitat-related risk of predation (Price 1984; Lima 1985; Lima *et al.* 1986; Hik 1995), intraspecific competition for prime habitat (Morris 1989) and the need to return to a central place to consume prey (Orians and Pearson 1979). Habitat-related factors which affect the fate of individual animals should logically have spatial consequences for the populations these individuals constitute, and hence the mechanisms by which the abundance of these populations is limited and/or regulated (Oksanen and Lundberg 1994). While some conceptual and mathematical models predicting spatial patterns of population abundance have been developed (e.g. Fretwell 1972), empirical studies demonstrating specific mechanisms underlying these patterns are generally lacking (Hassell and May 1985; Levin 1989; Fitzgibbon and Lazarus 1995).

In this paper we argue that the effect of habitat-related foraging constraints on the fecundity and/or mortality of individual animals making up populations lead to spatial variation in the demography of these populations. Further, that this variation will have consequences for both the probability of a species occurring and persisting at a given location, and the prevailing density of a species where it does occur. We evaluate the applicability of this hypothesis to food limited feral pig (*Sus scrofa*) populations inhabiting inland river systems in the semi-arid rangelands of eastern Australia by 1) testing for habitat-related variation in the influence of food availability on the dynamics of these populations, 2) assessing habitat-related environmental constraints on the foraging behaviour of individual pigs, and 3) using simulation models to determine the influence of habitat-related variation in demography on the occurrence, density and persistence of these populations. The generality of the hypothesis to other species is discussed.

Methods and results

Study sites

The data used in this paper come from a three and a half year study of feral pigs on six sites along the Paroo River in northwestern NSW (Choquenot 1994) and a coincident two year study of individual pig behaviour on an area of one of these sites (Dexter 1995) (Fig. 1). The area of the six larger sites is given in Table 1. The behavioural study occurred over a 10400 ha area of the site NP. All sites were selected to span the floodplain of the Paroo River or its major tributary, the Cuttaburra Creek.

Wanaarring, a village on the Paroo to the north of the sites (Fig. 1), receives a mean annual rainfall of 193 mm with slight summer dominance. The reliability of rainfall in a given year and season is low (Table 2). Maximum summer temperatures in the region often exceed 40°C while minimum winter temperatures can fall below 0°C. Evaporation exceeds rainfall in all seasons. The topography of the area is generally flat and hydrology is dominated by the Paroo River and Cuttaburra Creek, which form a series of interconnected channels and anabranches that flow only during flood.

Vegetation was strongly associated with soil type, four major habitats being identifiable:

Shrublands, dominated by hopbush (*Dodonea attenuata*), turpentine (*Eremophila sturtii*) and some other woody shrubs growing on soft red sand of aeolian origin. Stands of taller timber consisting of belah (*Casuarina cristata*), rosewood (*Heterodendrum olefolium*) and occasionally

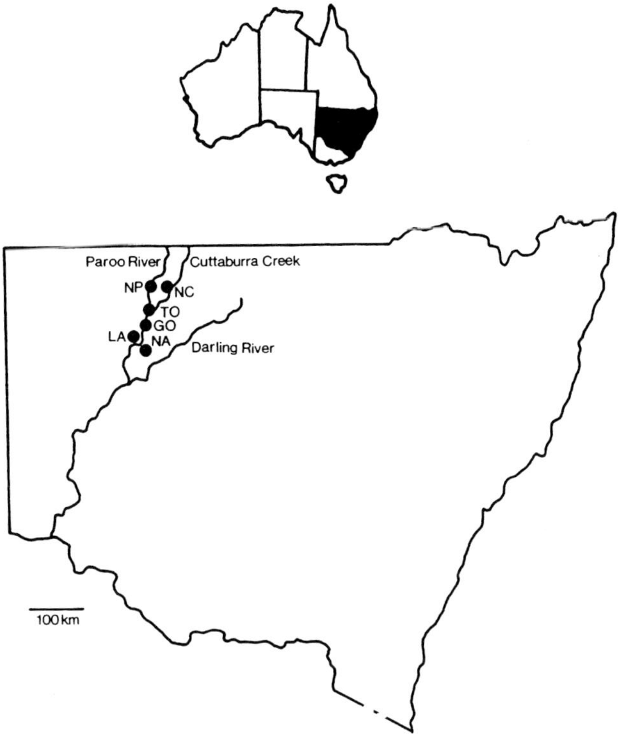

Fig. 1. Location of the six study sites used in the study of feral pig population dynamics (Choquenot 1994). The study of feral pig foraging behaviour (Dexter 1995) was conducted in an area on the larger site NP.

Table 1. Total area, percentage cover of riverine woodland habitat, pig density over the entire site and pig density over riverine woodland habitat on each site prior to experimental reductions in pig density (Choquenot 1994)

Site	Area (ha)	Percent riverine woodland	Pigs/km^2 (overall)	Pigs/km^2 (riverine woodland)
NP	27000	10	3.7	22
NC	21000	7	2.9	21
TO	20000	18	4.9	25
GO	16500	15	4.8	23
LA	16000	14	3.9	21
NA	14000	5	1.8	14

Table 2. Annual and seasonal rainfall (mm) at Wanaaring on the Paroo River, and its reliability. Seasonal rainfall is the mean for three months for the period 1926–1991.

	Summer (Dec–Feb)	Autumn (Mar–May)	Winter (June–Aug)	Spring (Sept–Nov)	Year
Rainfall (mm)	63	52	53	57	193
SD	55	55	24	44	94
CV%	87	106	70	100	49

white cypress (*Callitris* spp.) were distributed through these areas. While shrubland areas did not hold flood waters, they produced substantial growth of annual and perennial grasses (notably *Stipa* spp.) following local rainfall.

Riverine Woodlands, typically dominated by lignum (*Muehlenbeckia florulenta*), interspersed with stands of cooba (*Acacia stenophylla*), growing on the black cracking, self-mulching clays of the Paroo River, Cuttaburra Creek and their associated flood plains. Yapunyah (*Eucalyptus ochropholia*), coolibah (*E. microtheca*) and black box (*E. largiflorens*) occurred largely around the edges of these flood plains, with river red gums (*E. camaldulensis*) and black box in often thick stands along the banks of deeper channels and around waterholes and billabongs. During and immediately following flood, these areas produced a luxuriant growth of grasses (largely *Aristida, Stipa, Panicum, Eragrostis, Sporobolus, Poa* and *Astrebla* spp.), legumes (*Medicago* and *Trigonella* spp.), nardoo (*Marsilea drummondii*) and a variety of forbs.

Woodlands, typically supporting often monospecific stands of canegrass (*Eragrostis australasica* or *Leptochloa digitata*) and less commonly sparse stands of lignum or saltbush on hard, compact mottled red-grey soils more commonly flooded from local rainfall than floodwaters moving down the Paroo. During and following flood, these shallow clay pans produced a flush of grasses (mostly *Aristida, Stipa* and *Poa* spp.) and various forbs including an often lush coverage of wild lettuce (*Lactuca* spp.).

Ephemeral Swamps, which were irregularly flooded areas of red-grey or black soil and were usually ancient creek meanders or terminal basins for small fluvial systems. Ephemeral swamps were generally surrounded by shrublands, were several ha to several km^2 in size and were usually bounded by a narrow strip of trees, mostly black box and bimble box (*E. populnea*). Following flooding, these areas contained a luxuriant but ephemeral growth of annual grasses (primarily *Aristida* spp.) and annual forbs such as *Centipeda cunninghamia, Portulaca interranea* and *Mimulus repens.*

Dingoes (*Canis familiaris*), the only significant non-human predator of pigs in the rangelands (Giles 1980), were excluded from all study sites by a dog-proof fence which runs along the northern and western border of northwestern New South Wales.

Habitat-related variation in population dynamics

As part of a larger experiment to identify factors limiting feral pig abundance in the rangelands (Choquenot 1994), the density of pigs on each of the six study sites was reduced annually to one of three levels, and pig density and pasture biomass monitored quarterly over three and a half years. Pig density was estimated using a combination of helicopter counts and a technique based on proportional take of bait trails (Choquenot 1994). Habitats where pigs were observed during helicopter surveys (corresponding to those described above) were noted and habitat specific correction factors for visibility bias applied to counts (Choquenot 1995). Dry-matter biomass of pasture was estimated using the comparative yield technique (Haydock and Shaw 1975).

Quarterly estimates of pasture biomass and pig density were used to assess the numerical response of pigs to pasture availability. Pigs in the rangelands are primarily herbivorous and display a strong preference for fresh, green pasture (Giles 1980). The exponential rate of increase on an annual basis, estimated from sequential logged density estimates, was regressed on pasture biomass over various time lags. Rate of increase (r) pooled across all sites was best related to pasture biomass lagged one quarter (V). A consonant model of inverted Ivlev form was fitted to the relationship. The model had the form:

$$r = a + c\,(1 - e^{-dV}) \qquad (1)$$

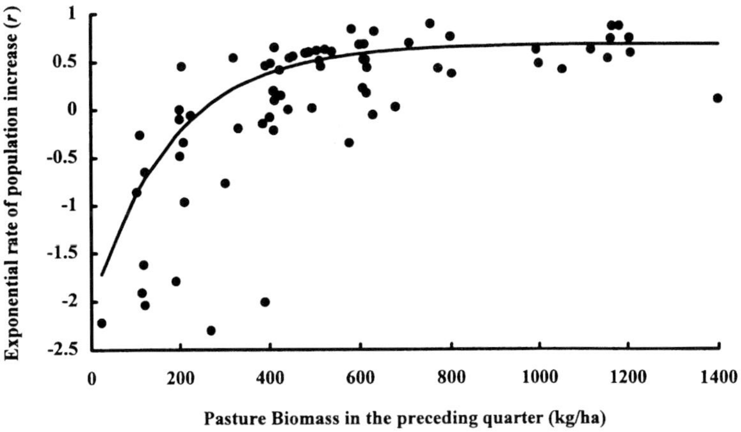

Fig. 2. The numerical response of pigs to variation in pasture biomass (kg/ha) in the preceding quarter. Points are estimates of *r* derived from quarter-to-quarter changes in pig density as a function of pasture biomass lagged one quarter. The line is an inverted Ivlev model fitted by non-linear least squares (Choquenot 1994).

where *a* is the maximum exponential rate of population decrease in the absence of food, *c* is the rate at which *a* is progressively ameliorated with increasing *V* and *d* is a measure of the population's demographic efficiency, quantifying its ability to increase when *V* is low. The maximum or intrinsic rate of population increase (r_m) (Caughley and Birch 1971) is given by *c-a*. Because the least-squares extrapolation used in the fitting procedure typically underestimates the maximum rate of decrease (*a*) in the numerical response (Caughley 1987), this parameter was fixed at the maximum observed rate of decline averaged across all 6 sites ($a = -2.045$ (s.e.=0.178)). Using the entire data set, the fitted model (r^2=0.566, p<0.001) (Fig. 2) was:

$$r = -2.045 + 2.730\,(1 - e^{-0.0055V}) \qquad (2)$$

This model indicates that pig populations increased in abundance when pasture biomass exceeded 251 kg/ha and declined when biomass was below that level. The intrinsic rate of increase on an annual basis (r_m) was (2.730 – 2.045) = 0.685.

Preliminary aerial surveys of each site indicated the density of pigs in riverine woodland habitat was substantially higher than that overall (Table 1), suggesting that the availability of this habitat may have had consequences for the dynamics of pig populations on given study sites. It was hypothesised that any such effect would manifest itself as a systematic relationship between the percentage cover of riverine woodland in an area and the response of the resident pig population to conditions of food shortage. To test this hypothesis, all negative exponential rates of population increase on were regressed on food availability (measured as pasture biomass lagged three months) and percentage cover of riverine woodland, estimated for each site from orthophotographic maps and Landsat imagery (Table 1). Because study sites were placed so that river channels and their associated flood plains ran through the centre of each site, the percentage cover of riverine woodland habitat on each site was directly related to its spatial availability.

The regression indicated that both lagged pasture biomass and the percentage cover of riverine woodlands contributed significantly (α = 0.1) to variation in *r*, a one percent increase in the cover of riverine woodlands increasing *r* by 0.07 over and above variation due to pasture availability (Table 3). This implies that the availability of riverine woodland habitat influenced the effect of food

Table 3. Summary of regression of all negative values of *r* for all study sites on pasture biomass lagged 3 months and the percentage cover of riverine woodlands on each site. The regression was significant ($r^2 = 0.36$, $F = 4.99$, $p = 0.018$)

Parameter	Coefficient	*t*-value	*p*
Intercept	−2.37	−4.55	0.0002
Lagged pasture biomass	0.0002	2.04	0.0559
Percentage cover of riverine woodlands	0.07	1.94	0.0678

shortage on mortality and/or fecundity of feral pigs, such that food shortage resulted in slower average rates of decline in populations inhabiting areas with progressively more of this habitat. To this extent, increasing availability of riverine woodlands appeared to generally mitigate some of the effect of food shortage on rates of change in pig abundance.

Foraging behaviour

In order to investigate whether the relationship between availability of riverine woodland and the effects of food shortage on rates of population decrease was due to some habitat-specific foraging constraint, a detailed study of the foraging behaviour of pigs in this environment was undertaken (Dexter 1995). To assess the effects of environmental variables on habitat selection by pigs, between seven and 22 pigs fitted with radio-collars were tracked during seven sessions between November 1991 and July 1993. Tracking sessions lasted between five and 10 days, during which time the position of pigs was estimated at hourly intervals using triangulation from three or four 12 m telemetry masts. Valid locations were associated with the four habitat classifications described above by overlaying them onto a habitat distribution map, derived from a geometrically corrected SPOT satellite image of the site, ground-truthed for habitat classification (Butson 1992). Percentage cover of the four habitat types within the study area were shrubland (51.4%), riverine woodland (15.1%), woodland (30.2%) and ephemeral swamp (3.3%).

Variation in habitat preference due to food availability and radiant heat load was investigated in order to assess the influence of these factors on foraging behaviour. During each tracking session, the number of locations in each habitat for each pig was expressed as a proportion of the total number of locations for that pig. Because the sum of these proportions for each pig equals one, and hence proportions are composite and non-independent, they were converted to log-ratios (Aitchison 1986) by dividing proportional use of each of three habitats (riverine woodland = U_1, shrubland = U_2, ephemeral swamp = U_3) by the proportional use of the fourth habitat (woodland = U_4), and converting to natural logarithms (Aebischer *et al.* 1993). This gives three log-ratios; $Y_1 = ln(U_1/U_4)$, $Y_2 = ln(U_2/U_4)$, $Y_3 = ln(U_3/U_4)$ from the four proportions of habitat use for each pig in each tracking session. Woodland was used as the denominator in this study, although any of the four habitats could have been used. If a habitat was not used by an individual pig in a given tracking session, a value of 0.001 was entered into analyses. Food availability in each habitat was indexed at each tracking session as the dry matter pasture biomass, measured using the comparative yield technique (Haydock and Shaw 1975). It was assumed that food quality, as far as it potentially influenced foraging behaviour, was similar between habitats. Radiant heat load was indexed for each tracking session as the mean daily maximum temperature over the duration of the tracking session.

The contribution of food availability and radiant heat load to variation in habitat utilisation by pigs was assessed using stepwise multivariate regression, with the log-ratios describing habitat preference of each pig in each tracking session being dependent variables and log transformed pasture biomass (kg/ha) for each habitat (F_1 = woodland, F_2 = riverine woodland, F_3 = shrubland and F_4 = ephemeral swamp) and mean maximum temperature (T) entered as independent variables:

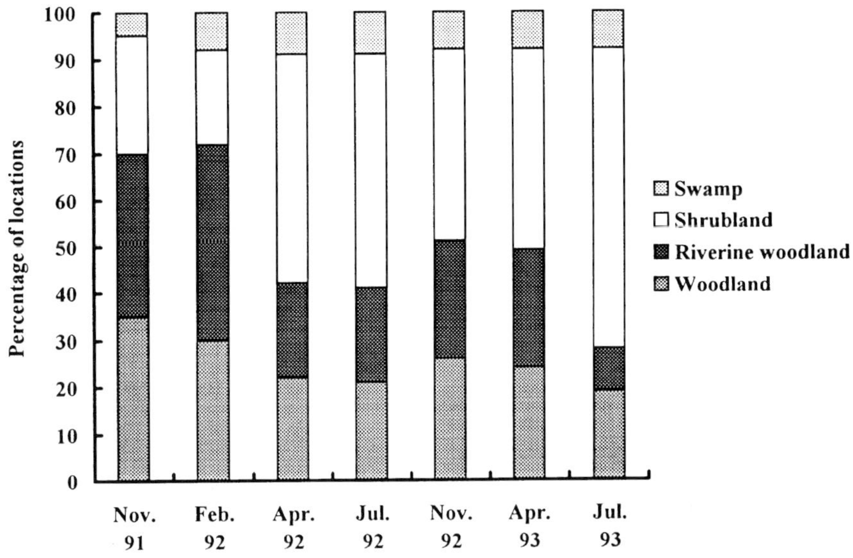

Fig. 3. Change in proportional use of habitats between tracking sessions for all pigs combined.

$$Y_1 Y_2 Y_3 = \alpha + \beta_1 F_1 + \beta_2 F_2 + \beta_3 F_3 + \beta_4 F_4 + \beta_5 T \quad (3)$$

The hypothesis that the regression parameters for each of the independent variables were the same for all dependent variables was tested using Wilk's L criterion, a significant result indicating variables which explain a significant proportion of variance in composite habitat preference.

The mean percentage of the 8,218 valid locations obtained over the seven tracking sessions, in each of the four habitats is shown in Fig. 3. The change in preferred habitat from riverine woodland to shrubland during July (the southern winter) suggests that temperature may have played a part in habitat selection. The multivariate regression analysis of composite habitat selection confirmed this with both food availability in the shrublands and mean maximum daily temperature contributing significantly to overall variation in composite habitat preference (Wilk's $\Lambda = 0.898$, $F_{3,87} = 3.276$, $p = 0.0248$):

$$Y_1 Y_2 Y_3 = \alpha + \beta_1 F_3 + \beta_2 T \quad (4)$$

To investigate how the use of individual habitats was influenced by food availability and maximum daily temperature, four log ratios describing use of each habitat as a proportion of the use of all other habitats were created; $Z_1 = ln[U_1/(U_2+U_3+U_4)]$. $Z_4 = ln[U_4/(U_1+U_2+U_3)]$. Each log ratio was regressed on pasture biomass and temperature as described previously. The significance and sign of regression coefficients indicated the influence of independent variables on utilisation of the numerator habitat in proportion to use of all other habitats. The results of the regressions of Z_1. Z_4 on pasture biomass in the four habitats and mean maximum daily temperature are given in Table 4. Utilisation of woodland increased with decreasing pasture biomass in shrublands. For riverine woodland, utilisation increased with decreasing pasture biomass in shrubland and increasing temperature. Utilisation of shrubland increased with pasture biomass in shrublands and decreasing temperature. For ephemeral swamps, utilisation increased with decreasing temperature.

Overall, these results suggest that pasture availability in shrublands and temperature were the main determinants of habitat choice by pigs. Increasing pasture availability in the shrublands attracted pigs into this habitat, while increasing temperature attracted pigs into riverine woodlands. The

Table 4. Regression equations describing variation in preference for four habitats (Z_1: woodland, Z_2: riverine woodland, Z_3: shrubland, Z_4: ephemeral swamp) with pasture biomass in each habitat and average daily maximum temperature for each tracking session. The only significant independent variables were pasture biomass in shrublands (F_3) and average daily maximum temperature (T)

Regression equation	r^2	F	p
$Z_1 = -1.81 - 0.007F_3$	0.06	5.55	<0.021
$Z_2 = -2.48 - 0.01\ F_3 + 0.045\ T$	0.195	21.80	<0.0001
$Z_3 = 0.762 + 0.006\ F_3 - 0.046\ T$	0.2	11.08	<0.0001
$Z_4 = -1.44 - 0.14T$	0.11	11.39	<0.0011

increasing use of shrublands in response to increasing food availability in this habitat conforms to conventional optimal foraging theory. In contrast, increasing use of riverine woodlands in response to increasing temperature (over and above the influence of food availability), appears to compromise the ability of pigs to forage optimally when temperatures are high by restricting their movements to areas immediately accessible to riverine woodland habitat. As such, the degree to which the need to spend time in riverine woodland habitat compromises the optimality of foraging will depend largely on the spatial extent of riverine woodland in an area; the greater the spatial extent of this habitat, the larger will be the area available to pigs for foraging during hot weather.

Population modelling

An interactive grazing model was constructed to explore the effect of habitat-related variation in rate of decline under conditions of food shortage on the average density and probability of persistence for pig populations in the rangelands. The model was modified from a stochastic red kangaroo-pasture model developed by Caughley (1987). A conceptual diagram of the model is shown in Fig. 4. Models generated 100-year sequences of stochastic rainfall drawn at random from normal distributions with average and variance equal to that for each season according to long-term rainfall records from Wanaaring (Table 2). Pasture grew or died back in response to rainfall according to an empirically derived relationship which accounted for the effects of rainfall on soil moisture, and competition for space between plants in rangelands pastures (Caughley (1987) modified from Robertson (1987)). The relationship has the form:

$$\Delta V = -55.12 - 0.01535V - 0.00056V^2 + 2.5R \tag{5}$$

where DV is the increment of growth or dieback in ungrazed pasture biomass over the subsequent quarter, V is the pasture biomass at the beginning of the quarter and R is rainfall in mm over the quarter. Robertson (1987) found that pasture growth or dieback not accounted for by V and R resulted in a standard deviation around the regression equation equivalent to 52 kg/ha. The pasture growth increment was taken as a random draw from a normal distribution with a mean equal to the solution of equation (5) and a standard deviation of 52.

In all models, red kangaroos were used as a surrogate for all herbivores other than pigs, their numerical and functional responses being taken from Caughley (1987) who modified them after Bayliss (1987) and Short (1987). The numerical and functional responses of red kangaroos to pastures biomass are:

$$r_k = -1.6 + 2\ (1 - e^{-0.007V}) \tag{6}$$

and

$$c_k = 66\ \{1 - e^{(V/34)}\} \tag{7}$$

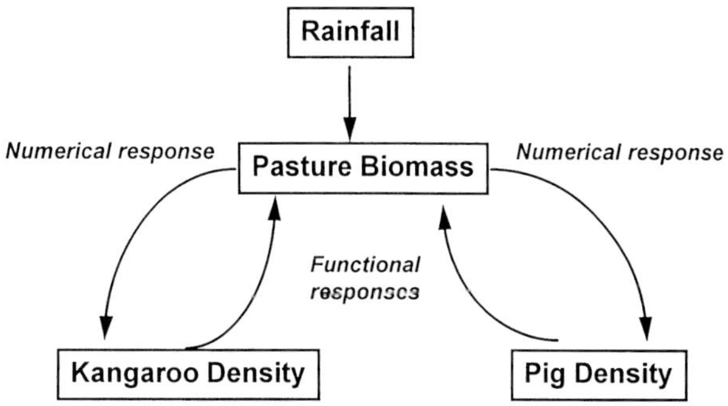

Fig. 4. A conceptual diagram of the interactive model.

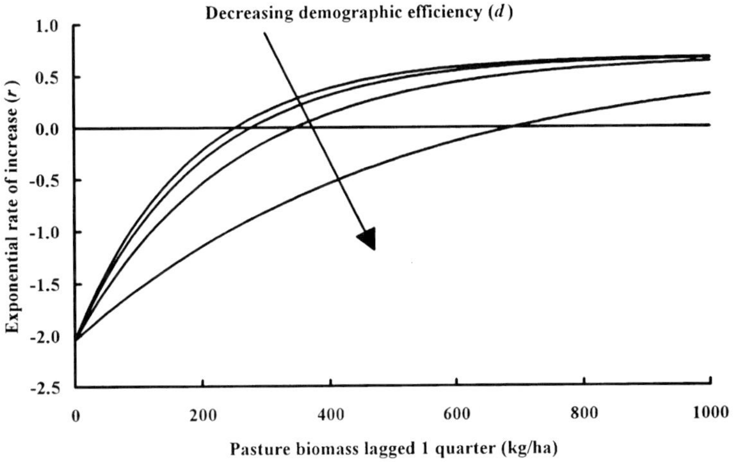

Fig. 5. The effect of increasing the demographic efficiency (*d*) of the numerical response of pigs to pasture biomass (kg/ha) in the preceding quarter.

respectively, where r_k is the annual exponential rate of increase for red kangaroos and c_k is their daily pasture intake rate as a function of metabolic body weight (g/kg$^{0.75}$/day) at given pasture biomass, *V*. Pigs were added to the model through their numerical and functional responses (Choquenot 1994). The numerical response of pigs to variation in pasture biomass (see equation (2)) was described by:

$$r_p = -2.045 + 2.730\ (1 - e^{-dV}) \qquad (8)$$

and the functional response by:

$$c_p = 58\{1 - e^{[(V-92)/302]}\} \qquad (9)$$

As with kangaroos, r_p is the annual exponential rate of increase for pigs and c_p is their daily pasture intake rate, expressed as a function of metabolic body weight (g/kg$^{0.75}$/day) at given pasture biomass, *V*. The parameter *d* in the numerical response of pigs, representing their demographic efficiency, was varied in order to assess the influence of habitat related response of pig populations to periods of food shortage on the density and persistence of the modelled pig population. As the value of *d* increases, the proportion of time a population spends decreasing as opposed to increasing over a given range of pasture availability, increases (Fig. 5). As such, increasing *d* for pigs simulates the

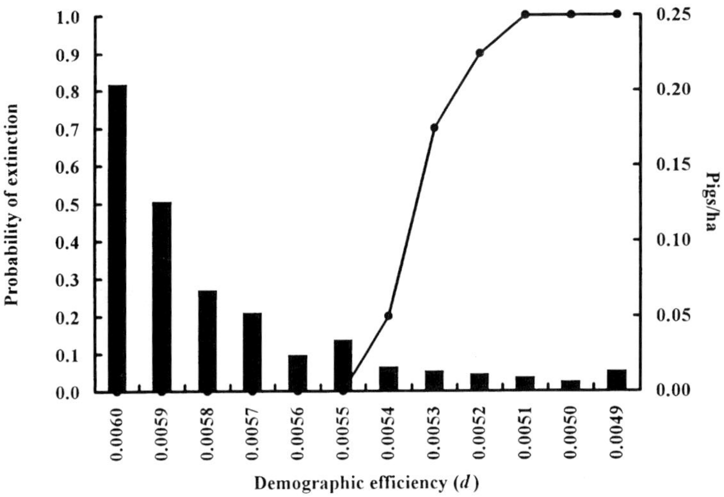

Fig. 6. Variation average pig density (pigs/ha) (bars) and the probability of population extinction (points) as a function of the demographic efficiency (*d*) of the pig population's numerical response, derived from a 100 year simulation model. Pig densities are averages from iterations in which pigs did not go extinct (maximum 10) and extinction probabilities are the proportion of iterations in which pigs went extinct within the 100 years considered.

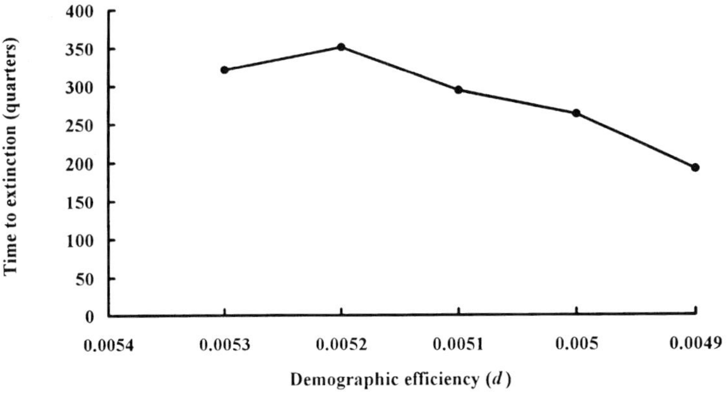

Fig. 7. Variation average time to extinction (quarters) as a function of the demographic efficiency (*d*) of the pig population's numerical response, derived from a 100 year simulation model. Points are averages for all iterations in which pigs went extinct within the 100 years considered.

effect of increasing availability of riverine woodlands, mitigating the effect of food shortage on prevailing *r* by enhancing the efficiency of the numerical response. In order to simplify the model, both kangaroos and pigs were assumed to weigh 35 kg.

Each model was iterated 10 times as *d* for pigs was varied from 0.0060 to 0.0049, over steps of 0.0001. Each iteration had starting values of 0.07 pigs/ha, 0.40 red kangaroos/ha and 250 kg/ha pasture biomass. For each value of *d*, time to extinction or pig density were averaged across iterations where populations did and did not go extinct respectively. The modelled population was considered extinct when pig density fell below 0.0001 pigs/ha. The probability of extinction for each value of *d* was estimated as the proportion of iterations where the pig population went extinct within the 100 years considered.

Figure 6 shows variation in average pig density and the probability of population extinction as a function of *d*. Average pig density declined to minimum viable levels as *d* was varied from 0.0060 to 0.0055, and the probability of population extinction increasing from 0 at $d > 0.0054$ to 1 at $d < 0.0052$. The overlap between average population density and probability of extinction suggests that populations with *d* between 0.0055 and 0.0051 may sometimes persist for over 100 years, but will do so at a low average density. Figure 7 shows variation in average time to extinction (quarters) as a function of *d*, for all iterations leading to extinction. Time to extinction decreased steadily with decreasing *d*.

Discussion

Constraints on foraging behaviour

Because pigs lack sweat glands and other than panting they have no physiological means of cooling themselves (Morrison *et al.* 1967), they are extremely sensitive to high ambient temperatures. For example, domestic pigs die if exposed to ambient temperatures exceeding 38°C (Mount 1968). Because of its potential role as a direct source of mortality, high temperature has been recognised as a major determinant of habitat selection in feral pigs inhabiting hotter environments (van Vuren 1984; Baber and Coblentz 1986). Giles (1980) argued that in order to effectively thermoregulate during periods of hot weather, pigs in Australia's rangelands require regular access to permanent sources of water for drinking and wallowing, and to thick vegetation to avoid prolonged exposure to direct sunlight. Giles (1980) suggested that this requirement restricted the distribution of pigs to riverine woodlands during hot weather, because only this habitat provided both water (in the form of permanent waterholes) and cover (in the form of extensive stands of lignum and taller trees and shrubs). The analysis of foraging behaviour described in this paper confirms Giles' (1980) suggestion, high ambient temperatures being associated with a preference by pigs for riverine woodland habitat. It appears likely that this preference reflects a requirement to constrain movements to this habitat in order to thermoregulate when temperatures are high, and that under such conditions this constraint compromises the ability of pigs to forage in habitats with the highest availability of food. As a consequence, the spatial availability of riverine woodlands in an area appears to limit the effective grazing range of pigs and hence food availability, when temperatures are high.

Belovsky (1984) described a continuum in the potential foraging behaviour of animals which related decisions about how long and where to forage, to a trade-off between nutrient/energy accumulation and risk minimisation. At one end of this continuum, nutrient maximisers attempt to ingest the greatest quantity of some nutrient in the time available for foraging. At the opposite extreme, time minimisers achieve their minimum nutritional requirement in the least amount of foraging time, maximising time available for other fitness enhancing activities. If more efficient nutrient intake incrementally improves survival and reproductive output, a species will be inclined to the nutrient maximiser end of the continuum. If time spent foraging incrementally increases exposure to sources of direct mortality such as predators or adverse weather, a species will tend to the time minimiser end of the continuum. For many species, compromise between nutrient accumulation and risk of exposure to sources of direct mortality results in foraging behaviour which is suboptimal for these factors in isolation, but optimal when their effects are considered together (Sih 1982; Dill 1983; Werner *et al.* 1983; Kotler 1984; Mittlebach 1981; Lima 1985; Lima *et al.* 1986; Stephens and Krebs 1986). These and other studies have considered the consequences of habitat selection, modified by risk of exposure to sources of direct mortality, on the fitness of individuals in wild populations. For example, Skogland (1991) demonstrated that female reindeer dispersed to higher altitudes at parturition to avoid predation by wolves, thereby foregoing more nutritious forage available at lower altitudes. Skogland (1991) argued that disadvantages accruing to females by forsaking better foraging conditions at lower altitudes were more than balanced by the reduction in risk of predation to both the female and her offspring. Skogland (1991) pointed out that for long-lived animals predation generally has a more profound effect on individual fitness than do the incremental effects of better or worse foraging conditions. Hence risk of predation or some other source of direct mortality may influence habitat preference more than optimal foraging decisions alone. In this sense, pigs may optimise their individual fitness by restricting activity to riverine woodland habitat over part of the day during hot weather, despite the fact that this may limit the amount of food to which they have access by constraining 1) the absolute area over which they can forage and/or 2) the types of habitats to which they have access.

Foraging constraints, variation in rates of decrease and population persistence

In this study, the constraint placed on the ability of pigs to forage optimally by the requirement to spend increasing time in riverine woodland habitat as temperature increased, at least partially explains why the abundance of pigs declined more rapidly under conditions of food shortage where the availability of this habitat was low. Figure 8 summarises the links between availability of riverine woodland habitat, the demographic efficiency of the numerical response, average population density and the probability and speed of population extinction in a conceptual model. Declining efficiency in the numerical response of pigs means that populations inhabiting areas with incrementally less riverine woodland habitat occur at progressively lower average densities. At some lower threshold availability of this habitat, the demographic efficiency of pig populations will be low enough that resident pig populations will undergo periodic extinction. As percentage cover of riverine woodlands decreases below this threshold, the probability of extinction increases while time to extinction decreases, suggesting that the demise of pig populations at such locations will be both inevitable and increasingly rapid whenever high temperatures and food shortage coincide. Such areas could be considered non-viable for pigs, extinction being effectively deterministic. Hence, the effect of availability of riverine woodland habitat on pig density and population persistence in the rangelands potentially determines spatial variation in the average density of a population and limits to its geographic distribution at any point in time. In reality, differences in the potential for areas to be recolonised following local extinction will have as much influence on whether a given area contains periodic or non-viable populations of pigs, as does its percentage cover of riverine woodlands. Hence, the persistence of pigs in areas at the edge of their local distribution likely reflects some interaction between proximity to persistent populations (which will effect the rate or probability of recolonisation) and percentage cover of riverine woodland (which effects the probability and speed of extinction following recolonisation).

The associations of habitat, population dynamics and persistence summarised in Fig. 8 suggests that pig populations will only persist in the rangelands along major river systems and their associated flood plains. Most sources of information on the local distribution of pigs in the rangelands confirm this (Hone and Waithman 1979; Giles 1980; Saunders 1988; Wilson *et al.* 1992; Dexter 1995), although Giles (1980) suggested that on occasion individual pigs (primarily large boars) occurred around permanent water sources isolated from floodplain habitat. Similarly, studies of pig populations in other parts of the world confirm a strong association of pigs with areas containing moist, shaded habitats (Barrett 1982; Baber and Coblentz 1986; Gerard *et al.* 1991).

Demographic efficiency and the edge of a population's range

The modelling presented here suggests habitat-related variation in the efficiency of the numerical response influences the local distribution of pigs in the rangelands and variation in this distribution through time. Caughley *et al.* (1988) considered that population ranges which are determined by the availability of a resource that is used consumptively will display a step function in r_m at the edge

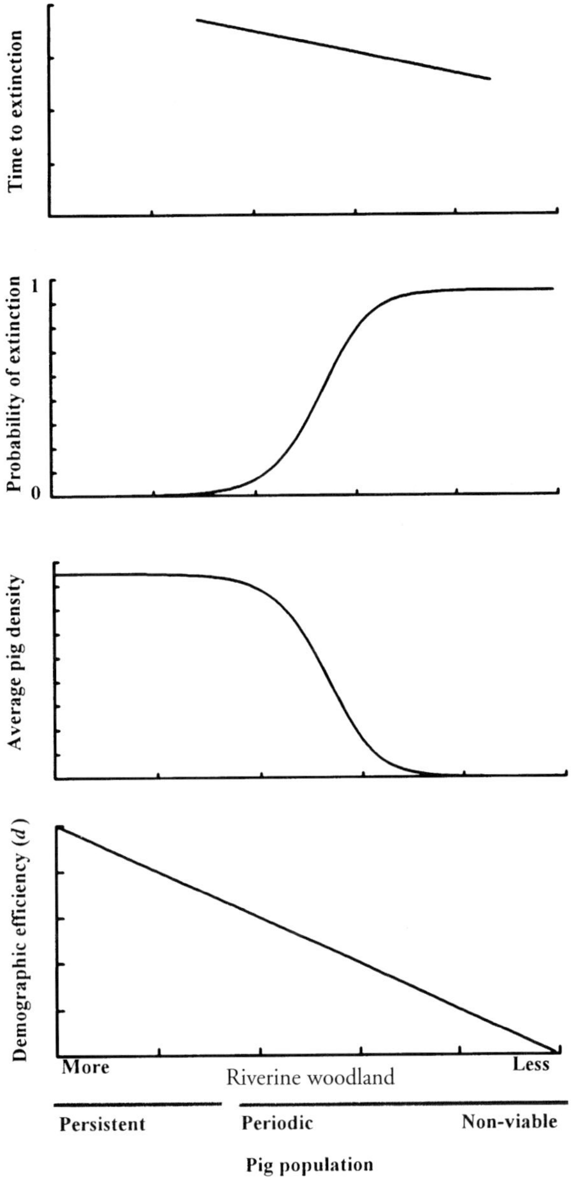

Fig. 8. A conceptual model of how the percentage cover of riverine woodland habitat, through its effect on demographic efficiency (*d*), influences average pig density and probability of population extinction in the rangelands. Consequences for the persistence of pig populations in areas with given coverage of riverine woodland are summarised at the bottom of the figure. Persistent populations never go extinct, while areas with less riverine woodland range from those where pigs occur periodically to those which are non-viable for pigs.

of their range, the population having the same maximum potential rate of growth wherever they and the limiting resource occur together. For the same reason, Caughley *et al.* (1988) argued that the body condition of animals in such populations will also step at the population boundary. In this respect, they suggested a population equivalent of Fretwell's (1972) 'ideal free distribution', which proposes that individual animals will assort amongst habitats in such a way that resource availability leads to equal fitness amongst all individuals in a population. If individuals achieve such an assortment, the populations they make up will have the same r_m and average body condition across a species range, but an average density which reflects spatial variation in resource availability across this range.

However, the prediction that individual fitness, r_m, and average body condition will be constant across a population's range, while density will vary with resource availability, implies that the population is consistently at or close to ecological carrying capacity (Caughley *et al.* 1988; Pulliam 1989). The pig populations modelled here did not maintain any real equilibrium, their density tending to be over- or under-abundant in relation to their food resources at any point in time. As such, the average density and probability of persistence of these populations were determined by the cumulative effects of rates of change in their abundance, which reflected temporal variation in the abundance of limiting resources rather than any constant spatial pattern in their availability. Consequently, any factors which influence the spatial pattern of resource availability (either through direct effects on the resources or on an animal's ability to procure them) have the potential to influence both the density distribution of the population or the ability of a population to permanently or periodically inhabit a given area. In the current study the availability of riverine woodland habitat influenced the accessibility of food resources to pigs during hot weather, which affected average population density, probability of persistence and time to extinction over a given run of seasonal conditions. Hence, the spatial extent of pig populations in the rangelands appears more dependent on factors which determine accessibility of limiting resources, than on any underlying spatial pattern of availability of those resources. To what degree habitat or other constraints on accessibility to limiting resources restrict the range of other species is unknown. However, such constraints may explain why many ostensibly generalist herbivores do not occur wherever forage, water and climatic conditions allow.

Caughley *et al.* (1984) examined variation in exponential rates of increase (r) over 6 years for red kangaroos in far western and central New South Wales, as a function of local rainfall accumulated over the previous 12 months. The two sampling locations represented the geographic centre and eastern edge of the species range in eastern Australia, respectively. Caughley *et al.* (1984) fitted parabolas to relationships between r and rainfall for the 2 locations. These parabolas indicated that maximum rate of increase (r_m) was similar for kangaroos in the centre and at the edge of their range (0.40), but an index of the hypothetical maximum rate of decline (estimated from the intercept of fitted parabolas which predict r for each population if no rain fell over the previous 12 months) was 39% lower at the edge (intercept = –2.8) than at the centre of the species range (intercept = –1.7). If this index provides a reasonable relative measure of the propensity of these populations to decline under conditions of food shortage, demographic processes leading to lower red kangaroo abundance at locations progressively further east in New South Wales may be qualitatively similar to those which lead to lower densities of pigs at locations progressively further from river systems in the rangelands. That is, demographic mechanisms which determine average population density and the probability of population persistence may have more to do with factors which affect the accessibility of limiting resources than with any underlying spatial pattern to the abundance of those resources. Similarly, Jarman (1974) and Sinclair (1985) demonstrated that predation risk influenced the foraging behaviour of ungulate species in eastern Africa. Although demographic consequences of modified grazing behaviour were not measured, Sinclair (1985) provided evidence that such behaviour increased interspecific competition between ungulates, and that this may have affected species' distributions and consequently ungulate community composition. Unequivocal tests of the role of constraints on foraging behaviour in determining species or population ranges requires evidence of a direct link from the constraint, through foraging behaviour to demographic processes which affect the populations density and/or probability of persistence in given areas. The study described in this paper provides such a test for feral pig populations in the rangelands.

Acknowledgments

This paper is based on the results of two related research projects, to which many individuals and organisations contributed. Financial and operational support for this work came from the Wildlife Exotic Disease Preparedness Program of the Bureau of Resource Sciences, the Australian Wool Research & Development Corporation and the Wildlife Monitoring Program of the Australian Nature Conservation Agency. Brian Lukins, Barry Kay, Dave Mula, Russel Pizel and Phil Boglio assisted with various aspects of the work. Jim Hone, Judy Caughley, Glen Saunders, Chris Dickman, Peter Jarman, Wendy Ruscoe and 2 anonymous referees improved earlier drafts of the manuscript.

References

Aebischer NJ, Robertson PA, Kenward RE (1993) Compositional analysis of habitat utilisation from animal radiotelemetry data. *Ecology* **74**:1313–1325

Aitchison J (1986) *The statistical analysis of compositional data.* Chapman & Hall, London

Baber DW, Coblentz BE (1986) Density, home range, habitat use and reproduction in feral pigs on Santa Catalina Island. *J Mammal* **67**:512–525

Barrett RH (1982) Habitat preferences of feral hogs, deer, and cattle on a Sierra foothill range. *J Range Manage* **35**:342–346

Bayliss P (1987) Kangaroo dynamics. In: Caughley G, Shepherd N, Short J (eds) *Kangaroos: their ecology and management in the sheep rangelands of Australia.* Cambridge University Press, Cambridge, pp 119–134

Belovsky GE (1984) Herbivore optimal foraging: a comparative test of three models. *Am Nat* **124**:97–115

Butson B (1992) A geographical approach to the ecology of pigs in northwestern New South Wales. B.Sc (Hons) thesis, University New South Wales, Canberra

Caughley G (1987) Ecological relationships. In: Caughley G, Shepherd N, Short J (eds) *Kangaroos: their ecology and management in the sheep rangelands of Australia.* Cambridge University Press, Cambridge, pp 159–187

Caughley G, Birch LC (1971) Rate of increase. *J Wildl Manage* **35**:658–663

Caughley G, Grice D, Barker R, Brown B (1988) The edge of the range. *J Anim Ecol* **57**:771–785

Caughley J, Bayliss P, Giles J (1984) Trends in kangaroo numbers in western New South Wales and their relation to rainfall. *Aust Wildl Res* **11**:415–422

Choquenot D (1994) *The dynamics of feral pig populations in the semi-arid rangelands of eastern Australia.* Ph.D thesis, University Sydney, Sydney

Choquenot D (1995) Habitat related visibility bias in helicopter counts of feral pigs in Australia's semi-aird rangelands. *Wildl Res* **22**:569–578

Dexter N (1995) *The behaviour of feral pigs in northwestern New South Wales and its implications for the epidemiology of foot-and-mouth disease.* Ph.D thesis, University New England, Armidale

Dill LM (1983) Adaptive flexibility in the foraging behaviour of fishes. *Can J Fish Aquat Sci* **40**:398–408

Fitzgibbon CD, Lazarus J (1995) Anti-predator behaviour of Serengeti ungulates: individual differences and population consequences. In: Sinclair ARE, Arcese P (eds) *Serengeti II: research, management and conservation of an ecosystem.* University of Chicago Press, Chicago, pp 274–296

Fretwell SD (1972) *Populations in a seasonal environment.* Princeton University Press, Princeton

Gerard JF, Cargnelutti B, Spitz F, Valet G, Sardin T (1991) Habitat use of wild boar in a French agroecosystem from late winter to early summer. *Acta Therio* **36**:119–129

Giles JR (1980) *Ecology of feral pigs in New South Wales.* Ph.D thesis, University Sydney, Sydney

Hassell MP, May RM (1985) From individual behaviour to population dynamics. In: Sibly RM, Smith RH (eds) *Behavioural ecology: ecological consequences of adaptive behaviour.* Blackwell, Oxford, pp 3–32

Haydock KP, Shaw NH (1975) The comparative yield technique for estimating dry matter yields of pasture. *Aust J Exp Agr Anim Husb* **15**:663–670

Hik DS (1995) Does risk of predation influence population dynamics? Evidence from the cyclic decline of snowshoe hares. *Aust Wildl Res* **22**:115–129

Hone J, Waithman J (1979) Feral pigs are spreading. *Agric Gaz NSW* **90**:12–13
Jarman PJ (1974) The social organisation of antelope in relation to their ecology. *Behav* **48**:215–266
Kotler BP (1984) Risk of predation and the structure of desert rodent communities. *Ecology* **65**:689–701
Levin SA (1989) Challenges in the development of a theory of community and ecosystem structure and function. In: Roughgarden J, May RM, Levin SA (eds) *Perspectives in ecological theory*. Princeton University Press, Princeton, pp 242–255
Lima SL (1985) Maximizing feeding efficiency and minimizing time exposed to predators: a trade-off in the black-capped chickadee. *Oecologia* **66**:60–67
Lima SL, Valone TJ, Caraco T (1986) Foraging efficiency-predation risk trade-off in the grey squirrel. *Anim Behav* **33**:155–165
Mittlebach GG (1981) Foraging efficiency and body size: a study of optimal diet and habitat used by bluegills. *Ecology* **62**:1370–1386
Morris DW (1989) Density-dependent habitat selection: testing the theory with fitness data. *Evol Ecol* **3**:80–94
Morrison SR, Bond TE, Heitman H (1967) Skin and lung moisture loss from swine. *Trans Am Soc Agric Eng* **10**:691–696
Mount LE (1968) Adaptation of swine. In: Hafez ESL (ed) *Adaptations of domestic animals*. Lee & Febiger, Philadelphia, pp 277–291
Orians GH, Pearson NE (1979) On the theory of central place foraging. In: Mitchell DJ, Stairs GR (eds) *Analysis of ecological systems*. Ohio State University Press, Columbus, Ohio, pp 154–177
Oksanen L, Lundberg P (1994) Optimization of reproductive effort and foraging time in mammals: the influence of resource level and predation risk. *Evol Ecol* **9**:45–56
Price MV (1984) Microhabitat use in rodent communities: predator avoidance or foraging economies? *Neth J Zool* **34**:63–80
Pulliam HR (1989) Individual behaviour and the procurement of essential resources. In: Roughgarden J, May RM, Levin SA (eds) *Perspectives in ecological theory*. Princeton University Press, Princeton, pp 25–38
Robertson G (1987) Plant dynamics. In: Caughley G, Shepherd N, Short J (eds) *Kangaroos: their ecology and management in the sheep rangelands of Australia*. Cambridge University Press, Cambridge, pp 50–68
Saunders GR (1988) The ecology and management of feral pigs in New South Wales. M.Sc thesis, Macquarie University, Sydney
Short J (1987) Factors affecting food intake of rangelands herbivores. In: Caughley G, Shepherd N, Short J (eds) *Kangaroos: their ecology and management in the sheep rangelands of Australia*. Cambridge University Press, Cambridge, pp 84–99
Sih A (1982) Optimal patch use: variation in selective pressure for efficient foraging. *Am Nat* **120**:666–685
Sinclair ARE (1985) Does interspecific competition or predation shape the African ungulate community? *J Anim Ecol* **54**:899–918
Sinclair ARE (1989) Population regulation in animals. In: Cherret JM (ed) *Ecological concepts: the contribution of ecology to an understanding of the natural world*. Blackwell Scientific Publications, Oxford, pp 197–241
Skogland T (1991) Ungulate foraging strategies: optimization for avoiding predation or competition for limiting resources? *Trans 18th Inter Union Game Biol*, 1987, Krakow, pp 161–167
Stephens PW, Krebs JR (1986) *Foraging behaviour*. Princeton University Press, Princeton
van Vuren D (1984) Diurnal activity and habitat use by feral pigs on Santa Cruz Island, California. *Calif Fish Game* **70**:140–144
Werner EE, Gilliam JF, Hall DJ, Mittlebach GG (1983) An experimental test of the effects of predation risk on habitat use by fish. *Ecology* **64**:1540–1548
Wilson G, Dexter N, OBrien P, Bomford M (1992) *Pest animals in Australia*. Bureau of Rural Resources & Kangaroo Press, Canberra

Responses of Wild Rabbit Populations to Imposed Sterility

C. Kent Williams and Laurie E. Twigg

ABSTRACT

The efficacy of reducing fertility in wild rabbit populations as a means of reducing rabbit abundance is being examined in two similar experiments in south-eastern and south-western Australia, each on 12 discrete populations. In the first year, 0%, 40%, 60% or 80% of all females in the respective populations were sterilised randomly by surgical ligation of fallopian tubes. In the second year the female recruits were sterilised at the same levels.

Results from the first year of sterility treatments indicate that density-dependent responses in survival of juveniles sustained rabbit abundance and compensated for reduced production of young, and environmental factors constrained population abundance. Young were produced in direct proportion to the level of fertility in the populations, indicating that fertile females did not respond to the sterility of other females or the presence of reduced numbers of young; the rate of production of young per fertile female was similar among all sterility treatments. The populations with higher levels of sterility recruited higher proportions of the fewer young produced. A reduction in mortality of juveniles compensated for the reduced production of young caused by sterilisation.

At the end of the first breeding season, the populations with higher levels of sterility contained fewer rabbits, but abundances became similar among treatment levels after the period of summer mortality of young. There was a near-significant trend for reduced numbers of young recruited into the 80% sterility treatment; the reduced mortality of young in these populations did not compensate fully for the imposed sterility. The survival, loss and immigration of adult rabbits were unaffected by the level of sterility in the treated populations.

Within populations, the sterilised females seemed to survive longer than the fertile females in the first year. However, any responses in survival and abundance of adult rabbits to sterility are likely to be more protracted than for kittens and are yet to be documented. The experiment needs to proceed for another two years before it will be possible to assess fertility control as a means of reducing abundance of wild rabbits.

Key words: population dynamics, recruitment, immunocontraception, myxomatosis

Frontiers of Population Ecology, R.B. Floyd, A.W. Sheppard and P.J. De Barro (eds), CSIRO Publishing, Melbourne, 1996 pp. 547–560

Introduction

The European wild rabbit continues to affect Australian agriculture and native biota in spite of great expenditure and effort on rabbit control for more than a century. In Australia currently we use mechanical, chemical and biological methods in attempts to contain rabbit numbers below levels perceived to cause significant damage to pastures or crops, or below levels that prevent regeneration of palatable native flora in conservation areas (Williams *et al.* 1995). Although these methods are very effective when applied thoughtfully and diligently, their high cost and the continuing serious impact of rabbits over extensive areas of southern Australia motivate the ongoing search for cheap, effective and extensive means of controlling rabbits (Williams *et al.* 1995). Biological means particularly are sought because the initial devastating effect of myxomatosis met these three criteria and gave us high expectations. Current research concentrates on two methods of biological control: the potential of Rabbit Calicivirus Disease which has reduced wild rabbit populations by about half in Spain (Cooke 1994), and genetic modification of the myxoma virus to cause infertility in rabbits (Tyndale-Biscoe 1994a,b).

The concept of managing pest populations by controlling fertility is attractive for humanitarian reasons and because of species-specificity and safety. For example, the sterile male technique safely eradicated large populations of the New World screw-worm fly (Spradbury 1994). The sterile male technique is not suitable for some insect pests and may not work on mammals because the females may mate several times in a season and possibly with several partners, and costs of obtaining and sterilising sufficient male mammals would be prohibitive. Previous attempts to control mammalian fertility have concentrated on chemical or hormonal disruption of reproductive cycles, but for wild populations there are problems of dosage and repeated administration to individuals and problems of hormonal modification of behaviour (Bomford 1990).

Controlling the fertility of the wild European rabbit in Australia seems highly desirable, but it is not known whether the mating system, social structure, biology and population dynamics of the rabbit are suitable for such methods (see Caughley *et al.* 1992). The present study investigates the suitability of fertility control for limiting the abundance of wild rabbits.

The present research on rabbits focuses on immunological contraception with its intended extensive dissemination using a genetically-modified free-living immunocontraceptive myxoma virus (GMIV) that would transmit between rabbits in the normal way, commonly via biting insects such as mosquitoes and rabbit fleas, with little or no human aid (Tyndale-Biscoe 1994a,b). The challenge is to develop the concept to a reality that is effective, safe and cheap.

Previous research has identified or measured some of the biological processes and their rates or levels in the complex of rabbit, insect vectors and virus, but others are not known and experimentation is needed. One of these unknowns is the demographic responses of rabbit populations to sterility. The relationship between the proportion of rabbits sterilised and the number of young recruited into the breeding population may not be linear. These responses will affect the threshold level of sterility in the rabbit population required to reduce the rate of increase or cause decline in rabbit abundance. This threshold level will indicate the minimum prevalence of myxomatosis in the wild rabbit populations required for the GMIV to limit wild rabbit abundance.

We report on progress in the first year of the experiments that assess responses of rabbit populations in eastern and western Australia to various levels of sterility, and quantify the minimum proportion of female wild rabbits that must be sterilised to reduce the rate of increase to zero. We substituted surgical sterilisation, ligation of the female's fallopian tubes, for the sterilising effect of the proposed GMIV. This technique prevents conception among predetermined proportions of females in the populations without interfering with hormones and reproductive behaviour.

Methods

The experiments were replicated in western and eastern Australia in order to increase the generality and power of the conclusions. They were designed to be as similar as possible, addressing the same problem but in contrasting habitats. In south-western Australia the Mediterranean climate is more strongly seasonal, the rabbits are distributed along roadside verges bordering pastures where they shelter in the dense scrub and infrequently use burrows which are necessary for breeding. Rabbit-proof fencing was erected to separate the study populations spaced along the roadside verges. In south-eastern Australia the climate is also Mediterranean but rainfall is more evenly distributed through the year. Here the rabbits commonly use warrens which usually are distributed on hills of pasture with few trees or shrubs. The study sites are hilly and contain irregularly shaped clusters of warrens. The sites are separated by areas devoid of warrens and burrows or from which rabbits have been cleared by warren-ripping and fumigation.

These differences between east and west required different trapping strategies and methods to achieve the same objectives. Nevertheless, both employ cage traps assisted by strategic patterns of fencing, and both methods trap all rabbits in the populations during the late summer-autumn trapping session. Rabbits when first trapped are given numbered eartags to identify individuals permanently. Rabbits are trapped late in the winter-spring breeding seasons to assess the production of young, and in late summer and early autumn to quantify recruitment of surviving young into the breeding populations.

The experimental design and protocol are very similar in the east and west. Each include twelve study populations in three replicates of four treatments of surgical sterilisation of female rabbits in proportions of 0%, 40%, 60% and 80%. The treatment levels are complemented by randomly assigned sham operations so that 80% of all females caught receive surgery. Treatments were assigned to sites randomly in the east and, in the west, randomly stratified by pasture biomass, rabbit densities and farm practices. Individual females are assigned randomly to surgical treatments. Most surgery is undertaken in late summer-autumn when myxomatosis normally has passed through the populations, the summer high mortality rate of young has declined, most surviving rabbits are mature, and breeding has not begun. Overall, few immigrants are encountered. Untagged rabbits trapped after March, but estimated from body weight to have been born before 31 March are deemed to be immigrants and enter the random treatment protocol immediately, receiving surgery if prescribed.

We are determining any responses to the levels of sterility by monitoring the populations of rabbits, their burdens of fleas and prevalence of myxomatosis. The sites were trapped initially in September 1992, before treatments began, as a prelude to regular monitoring. The monitoring of eastern sites comprises enumeration by trapping and marking individuals over four days per month on each site during February and March (late summer-autumn), July (winter-early breeding season) and November (spring-late breeding season). The western sites were trapped during January to March (late summer-autumn), May (early breeding season) and (September-October) late breeding season. Trapping data and observations of tagged and untagged rabbits indicate that virtually all rabbits present were trapped during the trapping sessions.

Analyses of covariance were used to determine differences among the sterility level treatments. This corrects for different initial numbers of rabbits on the respective sites. The two covariates used were the numbers of rabbits on the sites at the start of treatments in February and March 1993 (east and west), representing the late summer-autumn adult population, and the numbers of rabbits present in the following May (west) or July (east), representing the adult breeding population before young were present. Therefore for eastern rabbits the trapping data for February-March 1993 and July 1993 were used as pre-treatment enumerations of the adult populations on the study sites in the

Table 1a. Mean numbers of eastern rabbits, adjusted for covariates, on the three replicate sites per treatment before and after implementing sterility treatments, compared among sterility treatments by analysis of covariance. The covariates are the prior numbers of total rabbits on sites during February–March 1993 and adults only in July 1993. The study sites excluding the buffer zones each occupy an area ranging from 1.16 to 3.99 ha.

Trap month Covariates Sterility level	Sept 92 None	Feb–Mar 93 None	July 93 Feb–Mar 93	Nov 93 Feb–Mar & July 93	Feb–Mar 94 Feb-Mar & July 93
0%	49	86	45	82	40
40%	40	57	32	54	50
60%	68	86	50	62	41
80%	69	53	49	45	35
Probability	0.504	0.719	0.202	0.068	0.545

Table 1b. Mean numbers of western rabbits, adjusted for covariates, on the three replicate sites per treatment before and after implementing sterility treatments, compared among sterility treatments by analysis of covariance. The covariates are the prior numbers of total rabbits on sites during January–March 1993 and adults only in May 1993. The study sites each occupy about 16–23 ha on a scrub-pasture interface of 320–420 m.

Trap month Covariates Sterility level	Sep–Nov 92 None	Jan–Mar 93 None	May 93 Jan–Mar 93	Nov 93 Jan–Mar & May 93	Jan–Mar 94 Jan-Mar & May 93
0%	27	55	29	60	87
40%	23	59	38	52	60
60%	27	40	32	60	77
80%	30	52	32	26	62
Probability	0.944	0.706	0.295	0.190	0.495

non-breeding and breeding seasons respectively, before demographic responses to sterility were possible. These pre-treatment numbers were similar among treatments (Table 1a). The equivalent covariates for the west were the enumerations for March 1993 and May 1993 (Table 1b). In calculations of proportions of young rabbits recruited or lost, the data for one eastern site were culled to prevent inequitable weighting to the very few young produced there.

We report on some aspects of rabbit demography for the first year since treatments began in February and March 1993. We illustrate the population processes with results from the eastern and western studies, and indicate, where comparable, similarities and differences in the outcome.

Results

Rabbit abundance

The numbers of rabbits trapped in each population during the year are shown in Fig. 1 for the eastern and western experiments. Rabbit numbers at the start of the trials, before treatments were implemented, were more variable within sterility treatments in the east than in the west. The changing abundances during the year reflect the concentrations of breeding and mortality. There was little or no breeding in the late summer, January to March 1993. The rabbits bred during the following winter and spring, and the high numbers in November included many young rabbits. Then numbers declined as breeding diminished and the numbers of young rabbits declined in

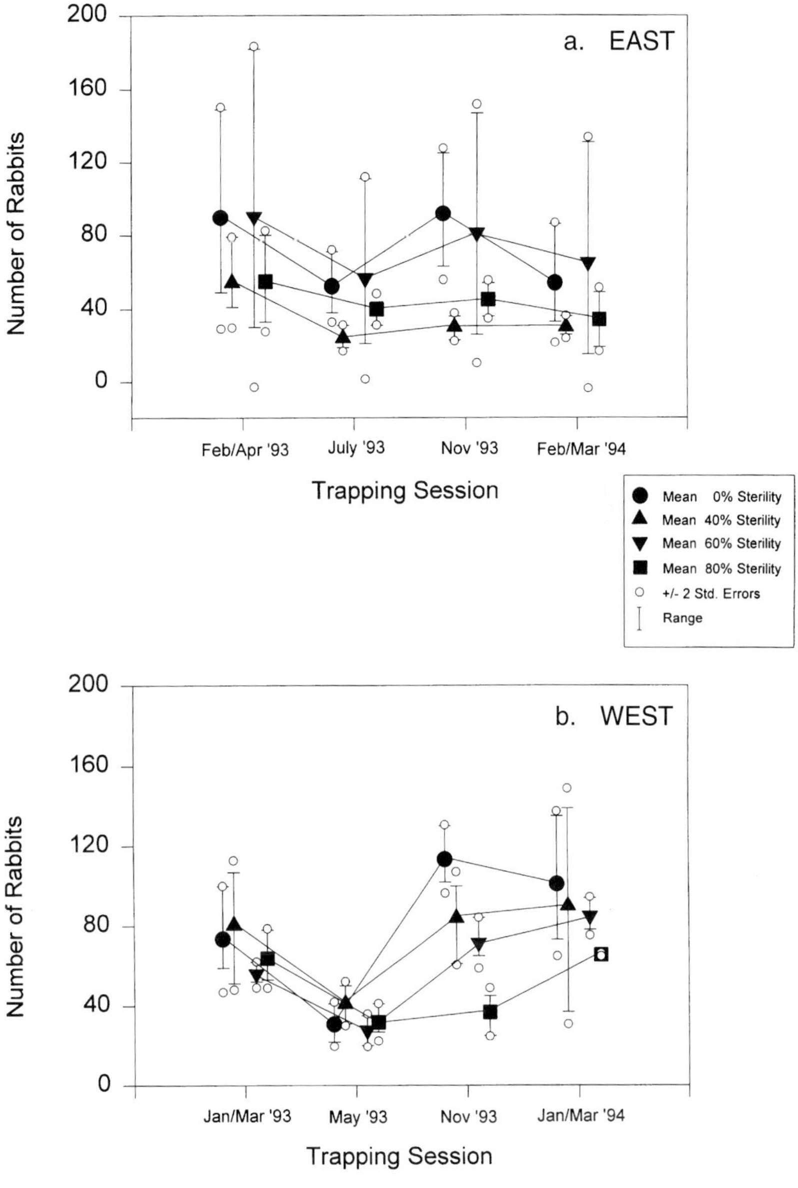

Fig. 1. Numbers of rabbits trapped on 12 eastern and 12 western sterility sites.

summer when conditions were dry and myxomatosis was prevalent. Predation also would have contributed to the loss of young. Over the year, from summer to summer, rabbit numbers tended to decline in the eastern trial and they were stable or tended to increase in the west.

Effects of population sterility on rabbit abundance

The numbers of rabbits trapped in trapping sessions are compared among sterility treatments by analyses of variance and covariance in Tables 1a and 1b. The population numbers shown are mean adjusted numbers for the three replicates of each treatment. There was no significant difference among sterility treatments for all enumerations before treatments were implemented, at the time of implementation, and in the early breeding assessment. The first occasion when sterility could affect population enumerations was late in the breeding season, November 1993 when many young were present. Then rabbit numbers had increased on the 0%, 40% and 60% sterility sites, but not on the 80% sterility sites. Enumerations indicated near-significant differences among treatments (east p = 0.068, west p = 0.190) and fewest rabbits in the 80% sterility treatment. After the summer non-breeding period, January-March 1994, rabbit abundance again became similar among treatments (east p = 0.545, west p = 0.495). The influence of sterility on rabbit numbers disappeared during summer.

Therefore, both experiments demonstrated that rabbit abundance at the end of the breeding season tended to reflect the imposed sterility treatments, and this disparity among treatments had disappeared by the end of summer. The analysis below seeks the sources of those responses.

Effects of population sterility on numbers of adult rabbits

The sterility treatments had no significant effect on the numbers of adults in the populations in the first year (east $p \geq 0.221$, west $p \geq 0.201$). Survival of adults over summer, estimated as the numbers of adults that were trapped during the 1993 breeding season and trapped again in January-March 1994, were similar among treatments (east p = 0.125, west p = 0.570). Losses of adults, those rabbits trapped during the 1993 breeding season but not trapped in January-March 1994, also were similar among treatments (east p = 0.478, west p = 0.297). There were few immigrant adult rabbits on the eastern sites during summer, and these were similar among treatments (east p = 0.111); the migration of adult rabbits on the western sites is yet to be examined. Therefore, abundance of adult rabbits did not respond to sterility in the first year; the dynamics of the progeny must have caused the observed population responses to sterility.

Effects of population sterility on production of young rabbits

The numbers of young produced differed significantly among sterility treatments (Table 2, east p = 0.059, west p = 0.002). The numbers produced correlated inversely with the level of sterility imposed (Fig. 2, east $r = -0.552$, $p<0.10$, west $r = -0.889$, $p<0.001$). The production of young was reduced proportionally on the sites treated with higher levels of sterility.

Table 2. Adjusted mean numbers of young rabbits produced and recruited, compared among sterility treatments by analysis of covariance, for eastern and western rabbit populations. Young produced were the numbers of untagged rabbits weighing less than 1000g (east) and 1200g (west) trapped on sites in July or November 1993 in the east, and in the 1993 breeding season in the west. Young recruited were trapped in the breeding season and retrapped in January or February to March 1994. The means are adjusted by the covariates, the numbers of adult rabbits on the sites in late summer–autumn and early breeding season before the sterility treatments were effective.

	East		West	
Sterility level	Produced	Recruited	Produced	Recruited
0%	47	8	94	24
40%	21	9	62	16
60%	28	10	57	24
80%	16	4	24	10
Probability	0.059	0.355	0.002	0.074

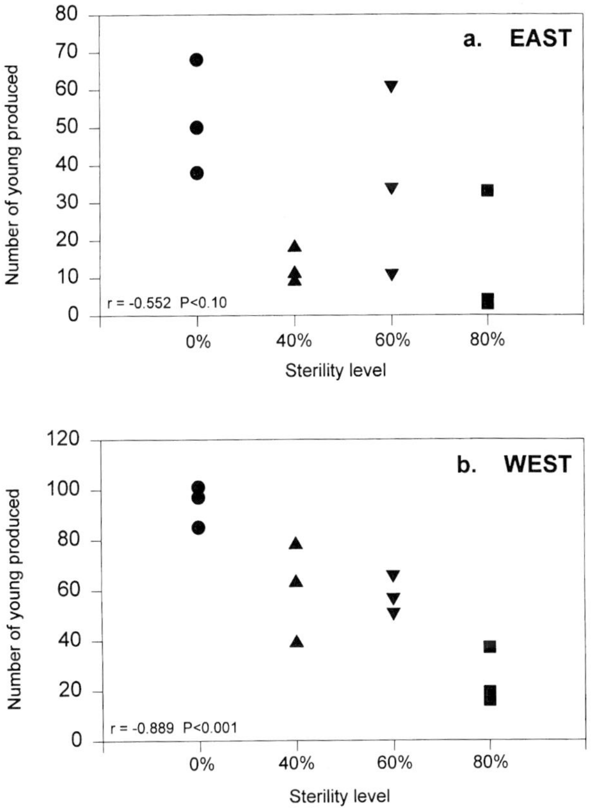

Fig. 2. Numbers of young produced on eastern and western study sites where females in the populations were sterilised in four different proportions.

This inverse correlation of production of young with sterility level suggests that the females which were not sterilised did not respond to the direct consequences of the sterility treatments, namely the failure of sterilised females to breed and the lack of their young on the sites during the breeding season. The fertile females did not seem to compensate by producing more (or fewer) young. This inference is supported by the comparison of the numbers of young trapped relative to the numbers of unsterilised adult females trapped on the sites during the breeding season; the numbers of young per adult fertile female were similar among sterility treatments (east $p = 0.687$, west $p = 0.592$). The sterility treatments did not induce reproductive compensation.

Effect of population sterility on recruitment of young rabbits

The numbers of young surviving the summer approximates the numbers of young recruited into the breeding stratum of the population. Survival of young through the summer was estimated for all rabbits weighing less than 1000g (east) and 1200g (west) when trapped during the 1993 breeding season, by the method used above for adults.

The comparison of recruitment of young among sterility treatments followed similar trends in the eastern and western experiments although the statistical outcomes tended to differ (east $p = 0.355$, west $p = 0.074$). Both experiments showed a trend for similar numbers of young recruited in the 0%, 40% and 60% sterility treatments and fewer young recruited in the 80% sterility treatment (Table 2). The unadjusted numbers did not correlate significantly with sterility level (Fig. 3a,b, east

Figure 3a, b. Number of young recruited relative to the level of sterility imposed on eastern / western sterility sites

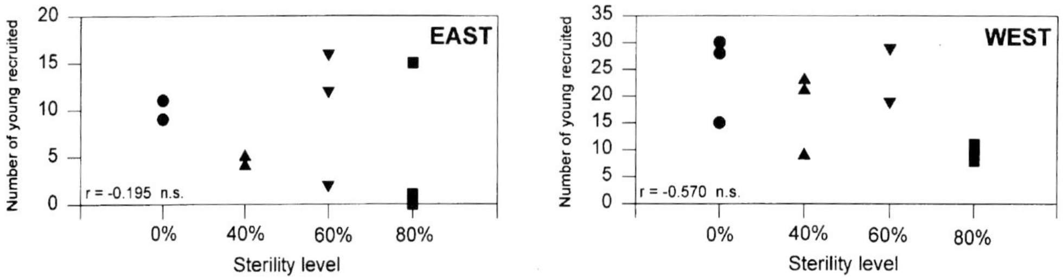

Figure 3c, d. Number of young recruited relative to the number produced on eastern / western sterility sites

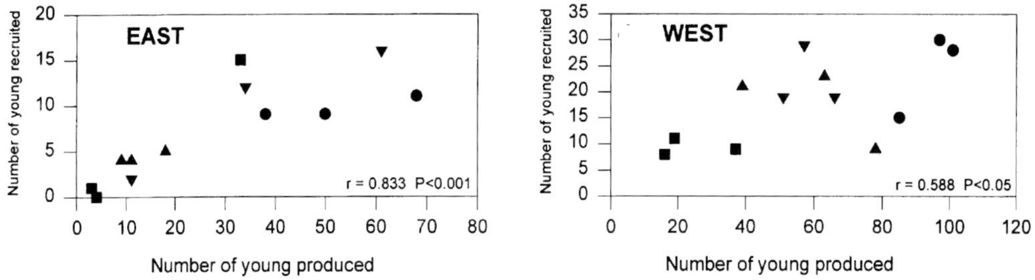

Figure 3e, f. Proportion of young recruited relative to sterility level imposed on eastern / western sterility sites

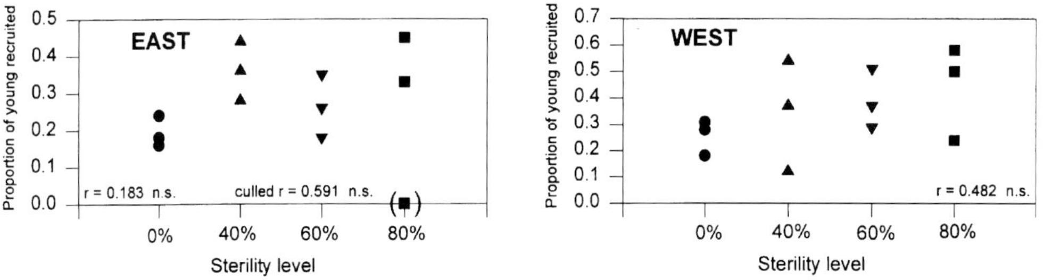

Figure 3g, h. Proportion of young recruited relative to the number produced on eastern / western sterility sites

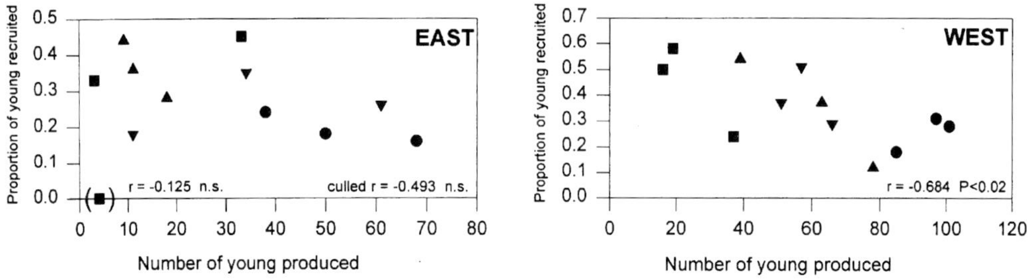

Fig. 3. Number (a–d) and proportion (e–h) of young recruited relative to the level of sterility imposed (a, b, e and f) and the number produced (c, d, g and h) on eastern and western sites.

Figure 4a, b. Number of young lost relative to the level of sterility imposed on eastern / western sterility sites

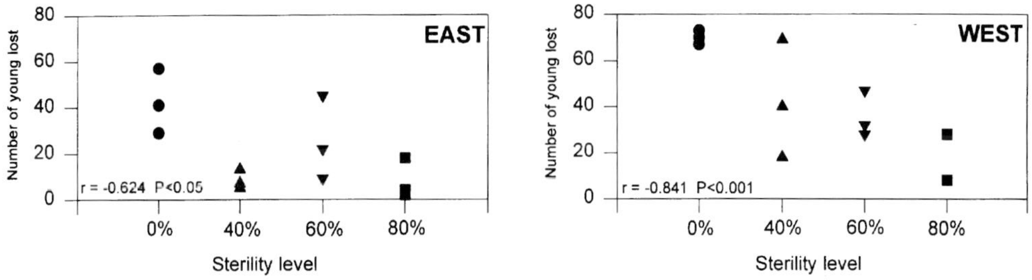

Figure 4c, d. Number of young lost relative to the number produced on eastern / western sterility sites

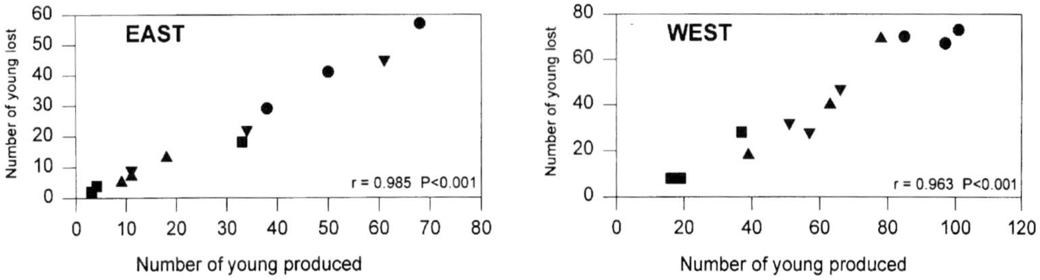

Figure 4e, f. Proportion of young lost relative to sterility level imposed on eastern / western sterility sites

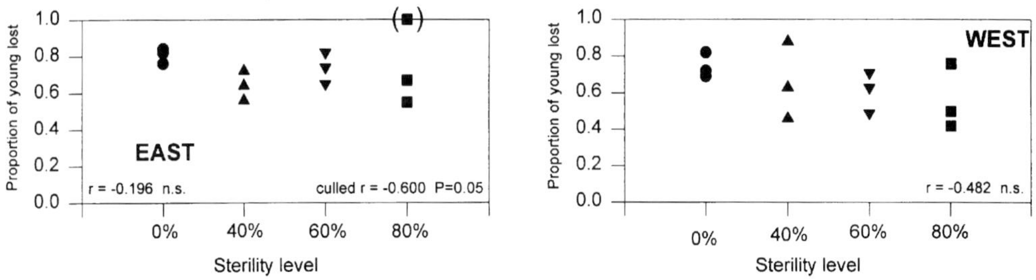

Figure 4g, h. Proportion of young lost relative to the number produced on eastern / western sterility sites

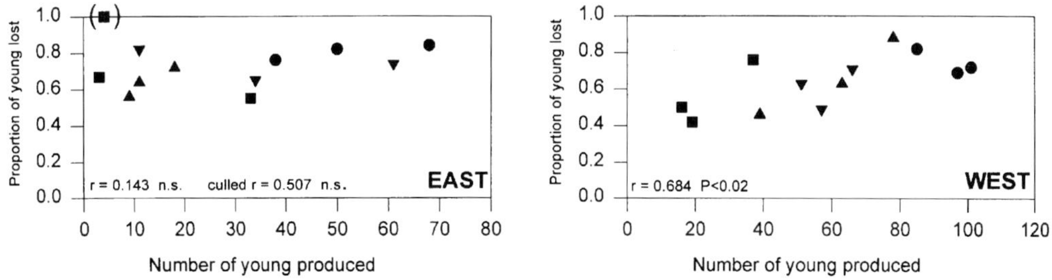

Fig. 4. Number (a–d) and proportion (e–h) of young lost relative to the level of sterility imposed (a, b, e and f) and the number produced (c, d, g and h) on eastern and western sites.

$r = -0.195$, n.s., west $r = -0.570$, $p<0.10$). The numbers of young recruited correlated positively with the numbers of young produced (Fig. 3c,d, east $r = 0.833$, $p<0.001$, west $r = 0.588$, $p<0.05$).

Recruitment was expressed also as a proportion of the numbers of young produced, using the actual numbers of young trapped. The proportion recruited did not correlate significantly with the level of sterility imposed (Fig. 3e,f, east culled $r = 0.591$, $p<0.10$, west $r = 0.482$, $p>0.10$). Nevertheless, both experiments suggest a non-significant trend for a higher proportion of young recruited in the higher sterility treatments.

The processes that tended to correlate the proportions of young recruited and the imposed level of sterility probably resulted from the numbers of young produced rather than from any direct effects of female sterility. The proportions of young recruited plotted against the numbers of young produced (Fig. 3g,h) showed that smaller proportions of young were recruited where more young were produced (east culled $r = -0.493$, n.s., west $r = -0.684$ $p<0.02$). The proportions of young rabbits recruited were related inversely to their cumulative abundance on the sterility sites. However, the actual number recruited is the factor of demographic and ecological value; this was measured and equates to the product of the proportion recruited and the number produced (see Table 2 for the mean numbers recruited, adjusted for the initial sizes of the populations).

Effect of population sterility on loss of young rabbits

The numbers of young rabbits not recruited, those lost to the sites by death or emigration, differed significantly among sterility treatments (east $p = 0.046$, west $p = 0.002$). The unadjusted numbers lost correlated inversely with the level of sterility imposed (Fig. 4a,b, east $r = -0.624$, $p<0.05$, west $r = -0.841$, $p<0.001$); fewer young were lost from the sites treated with higher levels of sterility. The numbers lost correlated positively with the numbers of young produced (Fig. 4c,d, east $r = 0.985$, $p<0.001$, west $r = 0.963$, $p<0.001$).

The proportions of young rabbits not recruited correlated inversely with the level of sterility imposed (Fig. 4e,f, east culled $r = -0.600$, $p = 0.05$, west $r = -0.482$, n.s.). The proportions of young not recruited were smaller in the higher sterility treatments. The proportions of young not recruited correlated positively with the number of young produced (Fig. 4g,h, east culled $r = 0.507$, n.s., west $r = 0.684$, $p<0.02$).

Effects of sterilisation

Myxomatosis transmission

The proportions of rabbits with antibodies to myxoma virus were high in the eastern populations, the means for the sterility treatments ranging from 63% to 79% in February-March 1994. The prevalence in the western populations were even higher, the mean prevalence among the sterility treatments ranging from 86% to 99% in January-March and October-December 1993. The sterility treatments did not differ significantly in prevalence of myxoma antibodies (east $P = 0.850$, west $p = 0.164$). Therefore the transmission efficiencies of vectors, including the European rabbit flea, were similar among the sterility treatments before and during the first year of the experiment.

Body weight

Body weights of female rabbits may reflect potential survival and future demographic responses in rabbit populations to surgical sterility. We compared body weights in samples of the various sterility classes of rabbit across eastern and western sites, regardless of treatment levels. Mean weights were calculated for a sample of rabbits that were trapped in late summer-autumn 1993 and retrapped one year later.

In the east the loss of body weight during summer was least in the sterilised females. The mean values declined from 1651g to 1613g between late spring and early autumn, whereas the body weights of the sham-operated females, excluding those pregnant, declined from 1640g to 1539g and the unoperated non-pregnant females declined in weight from 1650g to 1550g. The pattern was similar in the west. The sterile females declined in weight from 1913g in spring to 1756g in autumn, whereas the fertile females declined from 1884g in spring to 1697g in autumn. In spring the weights of sterile and fertile females were similar ($p = 0.373$), whereas in the following autumn they differed ($p = 0.004$). The benefit of sterility to body weight was not evident after the breeding season and stress of lactation but it became apparent after the additional stresses of summer.

Longevity

The proportions of rabbits that survived one year after sterilisation was implemented were determined for unoperated females, sterilised females, sham-operated females and males, to identify any effects of surgical sterilisation on longevity as detected by recapture. The proportion of classes of eastern rabbits retrapped after one year was 26% for the unoperated females, 26% for the sham-operated females, and 39% for tubally-ligated females (scaled deviance $p<0.05$, generalised linear model, logit regression and binomial errors), and 20% for the males. In the west also the sterilised females survived better than the fertile females. The proportion of sterilised females surviving the summer was 59% compared to 42 % of the sham-operated females, 39% of the untreated females (scaled deviance $p<0.01$), and 46% of the males. Survival was similar for males and unoperated females in both eastern and western rabbits (scaled deviance $0.20< p<0.30$). This common pattern may reflect a lower propensity of sterilised females to move from the sites, or an increased apparent longevity. The effects on body weight noted above suggest that sterility probably extended longevity. These survival levels are within the range recorded for various Australian climatic regions, and exceed the average level, 17%, recorded in the same eastern area, Canberra, between 1966 and 1974 (Gilbert *et al.* 1987) when rabbit densities were higher than during this study, and exceed the average level of 20% recorded for Chidlow in south-western Australia between 1971 and 1976 (Gilbert *et al.* 1987).

Discussion

By manipulating fertility and thus varying the numbers of young produced, we have identified experimentally, a demographic process that many long-term field studies have described for populations of rabbits in diverse parts of southern Australia and New Zealand (Gilbert *et al.* 1987). In this first year of these experiments we observed that the wild rabbit populations produced many more young than the numbers of adults that were lost to the populations, and many of these young also were lost. The production of more young than are normally incorporated into the breeding stratum of the population is the basis of the compensation in survival of young during summer for the reduced production of young that the experimental surgical sterilisation caused. The limited numbers recruited suggests that environmental limitations on population size resulted in variable replacement of adult losses by recruitment, and loss of the surplus young by death or emigration.

The message for immunocontraception as a means of controlling wild rabbit populations is that we would need to overcome the rabbit's reproductive insurance by which populations persist. That is, we would need to prevent the production of the surplus young before the population size would decline. In both the eastern and western studies the required level of sterility of females was about 80% in 1993–4. The similar outcomes of the two experiments support our confidence in this conclusion.

The demographies of the eastern and western rabbit populations in contrasting habitats responded similarly in the first year of experimentally imposed sterility. All demographic responses were observed in the young of the populations. The numbers of adult rabbits did not respond to sterility, although responses in subsequent years seem likely because of apparently enhanced survival of sterilised females. The reproductive output of individual unsterilised females remained unchanged by the treatments and did not compensate for the sterility imposed on other females. Thus the production of young correlated inversely with the level of sterility imposed.

Greater survival of young rabbits compensated for the reduced production of young. The mortality of young declined inversely with the level of female sterility, and higher proportions of the young produced survived where sterility was higher and fewer young were produced. However, where 80% of the females were sterilised the production of young tended to be insufficient and the populations tended to decline. Therefore, the compensatory increase in survival seemed near its effective limit at about 80% of females sterilised, in this first year of treatment.

The compensation by survival and mortality of young during summer for reduced production of young in winter-spring suggests that the rabbit populations produced more young than the environment could support. We are unable to identify at this stage the factors causing this mortality, although predation and myxomatosis are likely major contributors. Nevertheless, we can infer some characteristics of the mortality and survival. The correlation of mortality with numbers of young produced, and the lack of response in adult mortality, suggest that the young rabbits competed for survival with their own age class, not with the adult rabbits. Thereby the young rabbits were more susceptible than the adults to the agents causing the density-dependent mortality. Both predation and myxomatosis behave in this way, but so do other mortality factors such as coccidiosis and shortage of nutritious herbage.

The means by which the environment limited the abundance of rabbits in these areas during summer is likely to be complex and we have little information. Emigration and death are difficult to distinguish. The causes of death sometimes can be recognised but they cannot be apportioned, and in many cases the causes interact. Foxes, cats and birds of prey take rabbits from the sites. Myxomatosis is common and must contribute to many deaths and predispose animals to predation. In summer the vegetation usually dries and is depleted by stock and rabbits. Crowding in warrens and drying of vegetation would contribute to predation losses, to depletion of herbage and the resulting increased exposure to predators through loss of cover and extension of feeding excursions. Crowding would contribute also to increased transmission of myxomatosis and coccidiosis, and possibly to other ways of increasing mortality. These processes may be involved, to varying degrees, in changing the mortality of young rabbits in proportion to their abundance in the warrens. In general terms, production of more young rabbits than the environment could support seemed to cause the density-dependent mortality of young and compensation for the imposed sterility.

Most of the young produced were lost, but the numbers recruited correlated with the numbers produced. Also the proportion recruited of the numbers produced tended to correlate inversely with the numbers of young produced. These outcomes imply that the recruitment of young was limited by environmental constraints. This limit to recruitment in the east, where population numbers declined, was about half that in the west where the populations tended to increase; annual variation in seasons is also involved in setting the limits to recruitment. In this first year those adjustments caused a decline in the rabbit populations in the eastern study and an increase in the west. Irrespective of these different trends, the experiments yielded similar outcomes, indicating a robust conclusion.

The experiment implies some conditions for the introduction of a GMIV based on the myxoma virus.

- Between 60% and 80% of females would need to be infected with the GMIV for it to reduce population numbers. These levels were achieved by field strains in both eastern and western experiments.
- If the GMIV is about equally transmissible as the local field strains of myxoma, a prolonged high level of effort in disseminating the GMIV will be required in order to achieve those levels of infection in competition with the local field strains.
- A lower level of effort would be needed if the GMIV were more transmissible than the local field strain. Least effort would be needed if competition were avoided, such as in the absence of field strains. This prospect seems slim because myxomatosis was present on one or more of the study sites in all seasons.
- The best prospect of minimising effort in dissemination seems to be to use a highly transmissible strain of virus and to time the release effort just prior to the most common time of spread of field strains.

The indication in the data that sterile females may survive longer than fertile females suggests that sterile females may increase in proportion in the population progressively to an equilibrium level. The higher proportion of sterile females will tend to reduce population productivity and reduce the proportion of females that the GMIV will need to infect. Conversely, the higher proportion of sterile females may reduce numbers of infective rabbits, reduce numbers of susceptible rabbits, possibly reduce flea numbers, and reduce transmission of the GMIV. These interactions are complex and perhaps might be clarified by mathematical models based on experimental data.

The demographic outcome of the first year of these experiments represents only the initial events of a response that will take several years to attain some stability, particularly any responses of adult rabbits. The experiment is being conducted in two different environments, but it will need also to continue for several years through variation in weather for its outcome to apply generally. Until the experiments have progressed further we cannot judge the eventual responses of wild rabbit populations to the imposition of sterility. The first year has revealed the first datum of a process of change that would occur over generations of rabbits.

Acknowledgments

These studies in western and eastern Australia involve the efforts of many people. In Western Australia the main contributors were: Gary Gray, Tim Lowe, Gary Martin, Suzanne McLean, Sandy Griffin, Cath O'Reilly, Dave Robinson and Leonie Monks. In eastern Australia the main contributors were: Chris Davey, Bob Moore, John Bray, Louise Silvers, Justin Stanger, Lyn Hinds, Roger Pech, Jenny Grigg, John Wright, John Libke, and Don Wood. Bob Moore prepared the figures. We thank Ian Parer, Roger Pech and Charley Krebs for constructive criticism of the manuscript.

We are grateful for the assistance and cooperation of the Yass Rural Lands Protection Board, the ACT Department of the Environment, Agriculture and Land Care, and the landholders in the west and east on whose properties the studies are being conducted.

This research was funded by the Co-operative Research Centre for Biological Control of Vertebrate Pest Populations and by the wool growers of Australia through the International Wool Secretariat.

References

Bomford M (1990) A role for fertility control in wildlife management? Bulletin No. 7. Bureau of Rural Resources, Department of Primary Industries and Energy, Canberra, 1990

Caughley G, Pech R, Grice D (1992) Effect of fertility control on a population's productivity. *Wildl Res* **19**:623–627

Cooke BD (1994) RHD in wild European rabbits. In: Munro RK, Williams RT (eds) *Rabbit haemorrhagic disease: Issues in assessment for biological control*. Bureau of Resource Sciences, Department of Primary Industries and Energy, Canberra, pp 148–151

Gilbert N, Myers K, Cooke BD, Dunsmore JD, Fullagar PJ, Gibb JA, King DR, Parer I, Wheeler SH, Wood DH (1987) Comparative dynamics of Australasian rabbit populations. *Aust Wildl Res* **14**:491–503

Spradbury JP (1994) Screw-worm fly: a tale of two species. *Agric Zool Rev* **6**:1–62

Tyndale-Biscoe CH (1994a) Virus-vectored immunocontraception of feral mammals. *Reprod Fertil Dev* **6**:281–287

Tyndale-Biscoe CH (1994b) The CRC for Biological Control of Vertebrate Pest Populations: fertility control for wildlife conservation. *Pac Conserv Biol* **1**:163–169

Williams CK, Parer I, Coman BJ, Burley J, Braysher ML (1995) *Managing vertebrate pests: rabbits*. Bureau of Resource Sciences and CSIRO Division of Wildlife and Ecology, Canberra, pp 284

Frequency dependent competitiveness and the sterile insect release method

Rod J. Mahon

ABSTRACT

The competitiveness (*C*) of released, mass-reared Australian sheep blowfly, *Lucilia cuprina* has been shown to vary as their frequency in the mating population changes with *C* declining as the frequency of released males increases. While the male *L. cuprina* studied carried a Y-autosome translocation, reanalysis of published data on genetic control programs suggest that this effect may be exhibited by other mass reared insects, and is not restricted to this particular *Lucilia* strain. Simple stochastic models are employed to determine the consequences of this phenomenon to the efficacy of the sterile insect release method (SIRM). When *C* is frequency dependent, the overall cost of SIRM programs to eradicate target populations is increased as 1) it is necessary to release for a longer period, and 2) additional sterile males are required. This effect is especially evident when large numbers of sterile insects are released, causing high released: wild ratios, which is commonly the approach employed. However, by monitoring the wild population throughout the course of the release program and adjusting release numbers accordingly, acceptable competitiveness levels may be maintained. The financial benefits of such a strategy depend on the relative cost of rearing and releasing flies, but considerable savings can be expected for typical SIRM programs.

Key words: genetic control model, frequency dependence

Frontiers of Population Ecology, R.B. Floyd, A.W. Sheppard and P.J. De Barro (eds), CSIRO Publishing, Melbourne, 1996 pp. 561–572

Genetic control and SIRM

In 1955 Kniplings proposal to release sterilised male insects to suppress a natural population (Knipling 1955) heralded a radically new form of insect control. Before the advent of genetical methods, control relied on the exclusion of insects from a particular environment, or modification of the environment by various means. Traditionally, chemicals, pathogens or natural enemies are introduced into the pests environment in order to reduce pest survival or their reproductive success.

Irradiation of pupal stages of insects to introduce dominant lethal chromosomal mutations was found to be an ideal method to sterilise the New World screw-worm, *Cochliomyia hominivorax* (Bushland and Hopkins 1951) and subsequently a great variety of insects. Alternative forms of imparting genetic death to natural populations have also been developed, for example, the use of small doses of radiation to produce strains of insects that carry chromosomal rearrangements. Serebrovsky (1940) proposed to make use of individuals carrying homozygous translocations that were capable of breeding with individuals carrying similar genomes, but not with individuals with normal chromosomes. Serebrovsky's proposed manipulation of segments of chromosomes was the first of many such proposals that involve the modification of the chromosomal structure of a pest. Arguably, the most sophisticated of these that have been field tested are the compound chromosomes produced in the Australian sheep blowfly, *Lucilia cuprina* (Foster *et al.* 1976) and the T-Y-autosome translocations linked to recessive lethals, also in *L. cuprina* (Whitten *et al.* 1976). In addition to chromosomal manipulation, exploitation of the phenomena of hybrid sterility, cytoplasmic incompatibility and conditional lethals have been proposed as 'genetic control' techniques.

Competitiveness

All genetic control proposals (including the Sterile Insect Release Methods (SIRM) as a specific type of genetic control) currently under consideration involve releases of mass reared insects that pass on their deleterious factors through mating with their wild counterparts. Genetic death may be invoked at mating, as with SIRM, or it may occur over a number of generations, for example, the T-Y autosome female killing system (Whitten *et al.* 1976). The ability of released males to survive and rendezvous with receptive females when in competition with wild males, is critical for the success of genetic control methods. The ability of released males to mate with wild females has been termed competitiveness (C) (Haisch 1970; Fried 1971), and can be expressed as

$$C = \left(\frac{\textit{matings by released males}}{\textit{matings by wild males}} \right) \times \left(\frac{\textit{wild males in mating population}}{\textit{released males in mating population}} \right)$$

Wild males are assumed to have a C value of 1.0.

Most field studies of competitiveness of mass-reared insects have shown that released males fail to mate as often as expected from their frequency in the male population and C is therefore less than 1.0. Examples include the mosquito, *Culex pipiens fatigans* (Grover *et al.* 1976), the melon fly, *Dacus cucurbitae* (Iwahashi *et al.* 1983), a species of tsetse, *Glossina palpalis gambiensis* (Rogers and Randolph 1985), and the Mediterranean fruit fly, *Ceratitis capitata* (Wong *et al.* 1986; Wong *et al.* 1992). The competitiveness of strains of the Australian sheep blowfly, *Lucilia cuprina* carrying chromosomal rearrangements have also been examined. Mass-reared males carrying T-Y autosome translocations were found to have a competitiveness of approximately 0.3 (Mahon 1983; Foster *et al.* 1985a) however strains carrying compound chromosomes were less fit (Foster *et al.* 1985a) with C estimates averaging 0.2 in small trials (Mahon 1983).

Frequency dependent competitiveness in *Lucilia cuprina*

A field trial of the ability of a T-Y-autosome translocation strain of *L. cuprina* to suppress a wild population was recently conducted on Flinders Island, Tasmania (R. Mahon unpublished). During this trial, the competitiveness of released males was unexpectedly found to vary spatially and over time. Mating efficiency (or competitiveness) of released males was strongly correlated to their frequency relative to that of wild males in the mating population and declined as the released : wild ratio increased. This phenomenon of frequency dependent mating success of released males was independent of fly density and unaffected by either weather conditions or the relative distribution of released and wild flies (R. Mahon unpublished).

In the light of the findings from the Tasmanian trial, R. Mahon and K. Wardhaugh (unpublished) re-examined data from previous field trials of Y-autosome translocation strains of *L. cuprina.* Their analysis of published data (Fig. 1a) showed evidence of frequency dependent mating success among males released in the Wee Jasper experiment (Foster *et al.* 1985b). In addition, preliminary analysis of data from a trial conducted on an island in South Australia also yielded evidence of the phenomenon (R. Mahon and K. Wardhaugh, unpublished).

Rare Male Effects in other species

Frequency dependent mating success of released *L. cuprina* is an example of the rare male effect (RME), or rare male advantage, first described in laboratory *Drosophila melanogaster* (Petit 1951). RME has been found on many occasions in laboratory experiments, largely within species of the genus *Drosophila* (see reviews by Knoppien 1985; Partridge 1988), but there are few reports of the effect from field studies. RME's have been reported in nature among morphs of ladybirds in Britain (Muggleton 1979) and less convincingly (Knoppien 1985) among milkweed beetles (Eanes *et al.* 1977).

Variation of *C* of mass-reared and released insects has been reported on two occasions. Krafsur *et al.* (1987) citing Krafsur (unpublished), suggested that density dependent or frequency dependent matings may occur during releases of sterile New World screw-worm, *Cochliomyia hominivorax* but did not present any evidence. A more substantive report was made by Rogers and Randolph (1985) who showed that *C* was inversely correlated to the numbers of sterile males released during SIRM trials of the tsetse fly *G. palpalis gambiensis* conducted by Cuisance *et al.* (1978). Rogers and Randolph concluded that the released males were influenced by a density-dependent effect.

R. Mahon and K. Wardhaugh (unpublished) examined the genetic control literature in order to determine if rare male effects are restricted to *L. cuprina* or are more widespread among released insects. Three data sets were extracted from the literature in order to test for the presence of frequency dependence of *C.* Two of these studies, the tsetse study of Cuisance *et al.* (1978) and studies involving the Mediterranean fruit fly, *C. capitata* (Wong *et al.* 1986; Wong *et al.* 1992) provided no evidence of a rare male effect on competitiveness. However, in the third, that of the melon fly *D. cucurbitae* (Iwahashi *et al.* 1983), there was a strong indication of frequency dependent mating success (see Fig. 1b) (R. Mahon and K. Wardhaugh, unpublished).

Implications of RME for SIRM

Hitherto, all models of genetic control, regardless of whether they utilise SIRM or are based on chromosomally altered organisms, have assumed that *C* was a constant. This has led to the belief that inundative releases that produce high release : wild ratios offer the best control strategy. However, in circumstances where *C* is inversely proportional to the frequency of released males, as appears the case in *L. cuprina* and *D. cucurbitae*, inundative releases are likely to become increasingly cost-ineffective as pest numbers decline. Accordingly, there is a need to develop alternative release

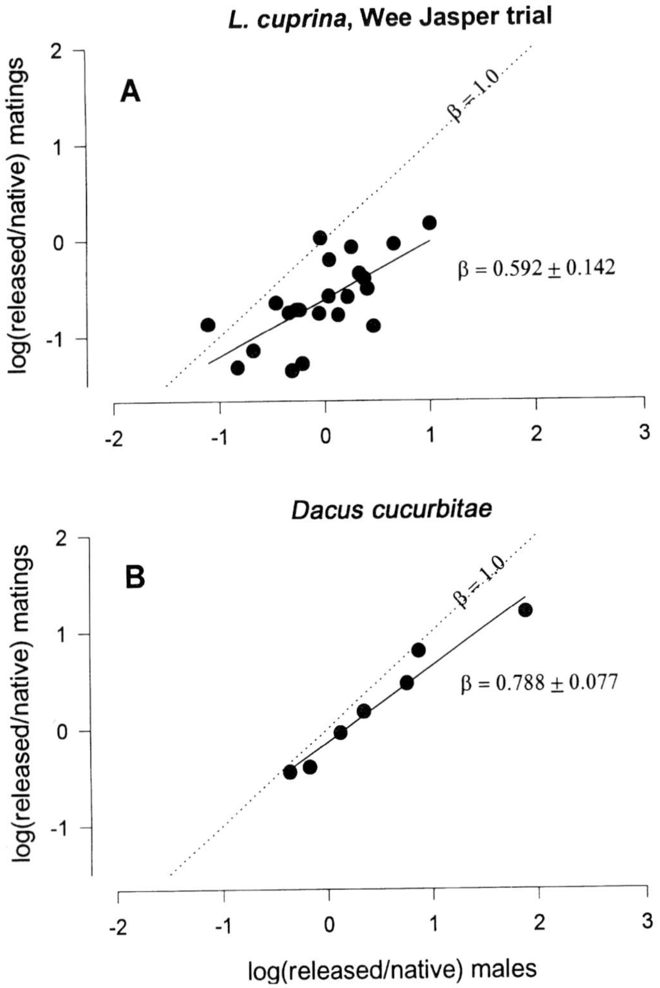

Fig. 1. Tests for frequency dependent mating success. Ayala (1972) proposed that frequency dependent mating success can be measured by the regression of the logarithm of the ratios of mated individuals on the logarithm of the ratios at which the individuals are present in the mating population. For any value of *C*, if the slope (β) of the regression deviates significantly from unity, frequency dependent mating success is indicated. A – Ayala's test on data from a field trial at Wee Jasper (Foster *et al.* 1985b) of *L. cuprina* T-Y autosome translocation males. B – Ayala's test on data from field tests of competitiveness of released *D. cucurbitae* during an eradication program on the island of Kume, (Iwahashi *et al.* 1983).

strategies that compensate for such effects. As a first step, simple stochastic models have been constructed to mimic the impact of frequency dependent mating in a field population of *L. cuprina.*

Modeling constraints

In these analyses, the models of genetic control examined were all discrete-generation models of a hypothetical insect population consisting of 1000 males and 1000 females. Released males introduced each generation were fully sterile and their survival and distribution mimicked that of the wild males. Generation time was constant, and females were assumed to mate only once. Other assumptions in the model were that the target population was isolated, (i.e. no migration) and that it was uniformly distributed. Density dependent regulation of the target population was excluded. Eradication was claimed when the number of females in a generation was reduced to below one.

The dynamics of the female population is represented by:

$$F_{g+1} = R\left[F_g\left(\frac{M_g}{M_g + M_s \times C}\right)\right]$$

Where F_g and F_{g+1} are the number of wild females in generation g and g+1 respectively, R is the intrinsic rate of increase applied to that generation, M_g is the number of wild (fertile) males in generation g, M_s the number of sterile males released in that generation and C is the competitiveness of released males. In models where C is allowed to vary, the value of C is that derived (R. Mahon and K. Wardhaugh, unpublished) from field data gathered during the first year of releases of a Y-autosome translocation strain of *L. cuprina* on Flinders Island, Tasmania, and is given by the expression:-

$$Log(C) = -0.304 - 0.600\log\left(\frac{Released\ \ Males}{Wild\ \ Males}\right) \qquad (1)$$

Model 1: *C*, release rates and rates of increase constant

The first model (Table 1) is presented in order to provide a baseline for subsequent comparisons, and examines the classical situation where C and release rates are constant. Table 1 illustrates that 'eradication' is achieved in fewer generations when, a) the competitiveness of released males is high, b) released : wild ratios are high and c), the rate of increase of the wild population is low. This simple model is at the core of all SIRM models, however more complex models e.g. Geier 1969; Berryman *et al.* 1973; Itô *et al.* 1989 remove one or more of the constraints of the simple model, e.g. density dependent regulation of the target population, multiple matings by females etc.

Model 1A (Table 2) is a specific case of the first model. In the situation illustrated, the rate of increase is fixed at unity, the initial release to wild ratio is 5 : 1 and a constant number of sterile males is introduced each generation. Competitiveness of released males is fixed at 0.19 which is the fitted value (from equation 1) for the competitiveness of released *L. cuprina* at a 5:1 release to wild ratio. Under these conditions, it took 5 generations to achieve eradication and 25 000 sterile males were required (Table 2). The cost of different release strategies (columns 5 to 11 in Table 2) will be considered later.

Model 2: Release rates constant, *C* frequency dependent

In the next model (model 2, Table 2), the rate of increase is set at 1.0 and a constant number of sterile males is introduced each generation, varying from 90 000 to 1000 depending on the initial release: wild ratio required. Competitiveness is now permitted to vary according to the frequency dependent function described in equation 1. It varies among the initial populations in response to the selected starting released : wild male ratios (90:1, 60:1...1:1), and also changes between generations as the number of wild males in the mating population declines. As in the previous model, releases are made at a constant rate each generation until eradication is achieved. Under these conditions, the number of generations required to eradicate vary from 4 (requiring 360 000 sterile males) when the initial release to wild ratio is 90:1, to 9 (requiring only 9000 males) when the initial

Table 1. Model 1. The number of generations required to eradicate a population consisting of 1000 males and 1000 females under various combinations of (1) initial release : wild ratios (R:W), (2) competitiveness (C) and, (3) rates of increase of 1× to 20×.

		Rate of Increase										
R:W	C	1	2	3	4	5	6	7	8	9	10	20
90	0.1	3	3	3	4	4	5	5	6	7	*	*
90	0.2	2	3	3	3	3	3	4	4	4	4	*
90	0.3	2	2	3	3	3	3	3	3	3	4	5
90	0.4	2	2	3	3	3	3	3	3	3	3	4
90	0.5	2	2	2	3	3	3	3	3	3	3	4
90	0.6	2	2	2	2	3	3	3	3	3	3	4
90	0.7	2	2	2	2	2	2	3	3	3	3	3
90	0.8	2	2	2	2	2	3	3	3	3	3	3
90	0.9	2	2	2	2	2	2	3	3	3	3	3
90	1	2	2	2	2	2	2	2	3	3	3	3
9	0.1	5	*	*	*	*	*	*	*	*	*	*
9	0.2	4	6	*	*	*	*	*	*	*	*	*
9	0.3	3	4	6	*	*	*	*	*	*	*	*
9	0.4	3	4	5	7	*	*	*	*	*	*	*
9	0.5	3	4	4	5	7	*	*	*	*	*	*
9	0.6	3	3	4	5	6	8	*	*	*	*	*
9	0.7	3	3	4	4	5	6	8	*	*	*	*
9	0.8	3	3	4	4	5	5	6	9	*	*	*
9	0.9	3	3	4	4	4	5	6	7	10	*	*
9	1	3	3	3	4	4	5	5	6	7	*	*
1	0.1	15	*	*	*	*	*	*	*	*	*	*
1	0.2	10	*	*	*	*	*	*	*	*	*	*
1	0.3	8	*	*	*	*	*	*	*	*	*	*
1	0.4	7	*	*	*	*	*	*	*	*	*	*
1	0.5	6	*	*	*	*	*	*	*	*	*	*
1	0.6	5	*	*	*	*	*	*	*	*	*	*
1	0.7	5	*	*	*	*	*	*	*	*	*	*
1	0.8	5	*	*	*	*	*	*	*	*	*	*
1	0.9	5	*	*	*	*	*	*	*	*	*	*
1	1	5	*	*	*	*	*	*	*	*	*	*

* Eradication not achieved by generation 18

ratio was 1:1. A comparison of the results for the 5 :1 released to wild ratio in model 2 and model 1A, illustrates the effect of frequency dependence of *C* on the suppressive effect of releases. While in the first generation the competitiveness values are the same in both models (0.189), by allowing *C* to vary between generations in model 2, in subsequent generations the release to wild ratio progressively increases until it reaches 5000 to 1 in the final generation. Thus in the 7 generations required to eradicate, competitiveness values are 0.189, 0.127, 0.078, 0.043, 0.020, 0.008 and 0.003 respectively. As a consequence of this decline, two additional releases of 5000 sterile males are required to achieve eradication. Similar increases in the number of generations that must be treated occur for all initial release : wild male ratios (data not shown).

Table 2. The number of generations required, the number of males required and 'costs' of genetical control programs to eradicate a target population of 1000 males and 1000 females under 3 different models: Costs for the eradication of *L. cuprina* are in AU$. Costs for the more general models (columns 6 to 11) are in arbitrary units (see text for explanation).

Model 1A – constant competitiveness of 0.19, and constant release rates;

Model 2 – Competitiveness varies according to equation 1 (see text);

Model 3 – specific release to wild ratios are maintained by varying the number of released males, thereby maintaining a constant competitiveness for each R:W.

Model conditions		Eradication requirements			Eradication costs. The number of sterile males that can be reared for the cost of one release #					
Initial R:W	C	Gens	Males x1000	Cost (AU$) (*L. cuprina*)	100000	20000	10000	1000	100	10
Model 1A										
5	0.19	5	25	11.60	5.25	6.25	7.5	30	255	2505
Model 2										
90	varies	4	360	69.20	7.6	22	40	364	3604	36004
60	varies	4	240	47.60	6.4	16	28	244	2404	24004
30	varies	5	150	32.50	6.5	12.5	20	155	1505	15005
20	varies	5	100	23.50	6	10	15	105	1005	10005
10	varies	6	60	17.40	6.6	9	12	66	606	6006
5	varies	7	35	14.00	7.35	8.75	10.5	42	357	3507
1	varies	9	9	11.52	9.09	9.45	9.9	18	99	909
Model 3										
90	0.03	6	119.8	28.16	7.2	12.0	18.0	126.0	1204	11985
60	0.04	7	83.5	2.72	7.8	11.2	15.3	90.5	841	8353
30	0.06	7	45.4	15.87	7.5	9.3	11.5	52.4	460	4546
20	0.08	8	32.1	14.57	8.3	9.6	11.2	40.1	329	3218
10	0.12	10	18.0	14.24	10.2	10.9	11.8	28.0	190	1812
5	0.19	no erad	10.3							
1	0.50	no erad	3.0							

See text for details of rearing : release costs

Model 3: Release rates variable, *C* frequency dependent

In model 2 it was seen that as the number of wild males declines and especially in the latter generations, released males become particularly ineffectual. Therefore, perhaps it would be appropriate to limit the decline of competitiveness by releasing fewer flies as the suppression process proceeds. Model 3 (Table 2) examines that approach by maintaining the initial release : wild ratios (90:1, 60:1...to 1:1) throughout all generations by manipulating release numbers. Under this scenario, competitiveness is determined by the initial released : wild male ratios and remains the same until eradication is achieved. Introducing this factor into the model reduces the efficacy of SIRM as the numbers of generations required to achieve eradication increases for a given initial release to wild ratio (Table 2), and furthermore, where the lower released to wild ratios are employed, (5:1 and 1:1) eradication does not occur. Under the conditions applied in this model, if the release to wild ratio multiplied by *C* is less than one, eradication is not possible. However, where eradication is achieved,

Table 3. Model 4. The number of generations (Gens) and the numbers of sterile males (×1000) required to eradicate a population of 1000 males and 1000 females when the release to wild ratios (90:1 to 1:1) are held constant each generation, except when expected release numbers (based on the number of wild females in the previous generation and the release : wild ratio) falls below a certain number (either none, 1000 100 or 10). If so, the pre-set minimum number is released.

	Minimum release number							
	None		10		100		1000	
R:W	Gens	No. males	Gens	No. males	Gens	No. males	Gens	No. males
90	6	119.8	6	119.8	6	119.8	6	121.3
60	7	83.5	7	83.5	7	83.5	6	84.9
30	7	45.4	7	45.4	7	45.5	6	46.8
20	8	32.1	8	32.1	8	32.2	7	34.5
10	10	18.0	10	18.0	9	18.2	7	20.4
5	*	*	12	10.3	10	10.5	8	13.9
1	*	*	16	3.0	14	3.5	9	9

* Eradication not achieved by generation 18

(released : wild ratios 10 : 1 or greater, Table 2) despite the increase in the number of generations required to eradicate, the numbers of sterile males required is reduced by as much as 2–3 fold.

Model 4: Release rates variable, *C* frequency dependent, minimum release rates

In model 3, very few flies are released during the final generations. This is particularly relevant in those situations where the initial (and sustained) release to wild ratio was low, (10:1, 5:1 and 1:1). Model 4 is essentially the same as the previous model with an additional condition that prevents release numbers falling below a set number, (1000, 100 or 10). Under these conditions, (Table 3) the lower initial released to wild ratios (5:1 and 1:1) that previously failed to eradicate now achieve eradication, even when the minimum is set at 10 males. Furthermore, comparison of the number of generations and the number of flies required for eradication between models 3 and 4 (Table 3) reveals that under certain conditions, eradication is achieved a generation or more earlier with only marginal increases in the requirement for sterile males. Under such conditions, setting minimum release rates are likely to result in a reduction of the total cost of an eradication program.

Economic considerations

Once the capital investment required for a mass-rearing facility has been made, the major costs of a genetic control program are rearing expenses and the cost incurred to release the genetically modified males. Rearing costs for insects vary greatly depending on the ingredients used and labour input required. Larvae of *L. cuprina* are reared on animal protein (meat-meal) cotton linters and water and the cost of rearing is estimated to be between \$1.10 and \$1.40 per 1000 males depending on the size of the facility (J.A. Stone, pers. comm.), with the cost declining as the capacity of the facility increases. The ingredients used to rear the New World screw-worm, *C. hominivorax* are more expensive, and include spray dried blood, egg powder, milk powder and gelling agent (Taylor *et al.* 1991). This species is perhaps three times more expensive to rear than *L. cuprina*. The peculiar biology of tsetse fly must make rearing costs for this fly even more expensive while at the other end of the scale, rearing costs for species of fruit fly are likely to be considerably less.

The cost associated with the placement of genetically modified males of different species into the target population is perhaps less variable than rearing costs if aerial releases are employed. Insects are light and if chilled occupy little space, therefore aircraft costs while releasing insects would largely be determined by the distance flown over the release 'grid' rather than flights to and from the loading point. Consequently release costs would be largely independent of the number of insects released and could be considered constant for each generation. There would of course be some costs that would depend on the number of individuals released such as packaging, aircraft positioning etc, however these will be ignored in this analysis. Again using *L. cuprina* as a yardstick, aerial releases along lanes spaced 4 km apart should give adequate coverage, given the dispersal ability of this species (Vogt and Morton 1991). Two releases would adequately cover the mating period and each aerial release would cost approximately AU\$0.22 per km^2 treated (J.A. Stone pers. comm.).

Vogt and Morton (1991) provide two estimates of the density of female *L. cuprina,* one of 4394 and the other 645 per km^2. At the average value of these estimates (2,519 per km^2) the area occupied by an initial population of 1000 females would be 0.4 km^2. Therefore the two releases to treat this area each generation would cost $\$0.22 \times 2 \times 0.4 = \$0.18$.

In a contrived situation with discrete generations of *L. cuprina,* if the cost to rear 1000 *L. cuprina* is \$1.1 and to release males in the target area \$0.18, we can calculate total eradication costs for the various models. These costs are presented in column 5 of Table 2. Employing the strategy of adjusting release rates to maintain a constant competitiveness (model 3) for a SIRM program against this species, enables costs to be reduced for all initial released : wild ratios. The savings are less marked when initial release to wild rates are low (for example from \$17.40 (model 2) to \$14.24 (model 3) when a release to wild ratio of 10:1 is employed (table 2)), but become more significant at higher initial released to wild ratios.

Columns 6–12 present eradication costs in a different format. The values in the table are essentially the sum of rearing plus release costs however to accommodate a range of rearing and release costs, each column represents a unique ratio of the two values ie, cost to rear 1000 males / cost to release for one generation over the area initially occupied by 1000 females. Comparison of the overall 'cost' of alternative models is only appropriate within columns. Thus in column 6, rearing cost is low compared to releasing costs (the cost of rearing 100 000 males is the same as that to release for one generation), whereas in column 12 rearing costs are high relative to release (only 10 sterile males can be reared for the cost of a release). Comparison of 'costs' between model 2 and 3 show that if release costs do not exceed the cost of rearing 20 000 males, adjustment of release rates to maintain a constant *C* will generally result in reductions in overall cost. Savings increase as release costs increase relative to rearing costs. Under the most favourable conditions, this stratagem would reduce costs by 1/3 of those in which a constant release rate is employed.

Summary and conclusions

The models developed in this study indicate that the phenomenon of frequency dependent mating success of released males will make genetic control more expensive. Two strategies have been proposed to reduce this increase. Modification of release rates as the wild population declines tends to ameliorate the increase. However, to make use of this strategy (which of the two, offer the greatest savings) accurate knowledge of the released : wild ratio throughout the program is required. Monitoring costs will therefore be an additional economic burden, but given the savings that may be achieved through knowledge of changes in numbers of the target population, monitoring costs are likely to be trivial. Under certain conditions, defining a minimum economic release rate may also assist in the reduction of program costs.

The models also indicate that minimal costs to eradicate are most frequently achieved at low initial released to wild ratios. This suggests that while the 'overkill' eradication strategy (Krafsur *et al.* 1987) adopted to eradicate the New World screw-worm fly has been remarkably successful, it may also have been more costly than necessary. Similarly, the high release to wild ratios used in SIRM against Mediterranean fruit fly, for example (Wong *et al.* 1986; Wong et al 1992) may also be inefficient. However, the simulations of SIRM for the control of *D. cucurbitae* (Itô and Koyama 1982) and of female killing genetic control systems for *L. cuprina* (Foster 1991) both show that populations regulated by density dependent factors are difficult to eradicate at low release : wild ratios. Thus the choice of release strategy employed against a pest population should be determined on the basis of detailed ecological data.

The costing system employed in this study is perhaps overly simplistic, as program expenses beyond rearing and release costs were ignored. For example, unless detection of the pest is perfect, it would be necessary to maintain releases for several generations after eradication is achieved. Capital equipment, overheads and administrative costs would also be factors in the overall cost of genetical control programs, and these would be partially dependent on the numbers of insects reared. However, despite the simplicity of the costing system, it is proposed that the models presented would be appropriate to maximise costs : benefits when the competitiveness of released insects are found to be frequency dependent. Modifications required would include information on the cost structures (rearing, release, overheads) for a given program and the specific relationship between *C* and the frequency of released males. Exploitation of techniques that will reduce pest numbers prior to the implementation of SIRM may also prove economically sound.

On the evidence of the successful application of SIRM to eradicate the New World screw-worm, proposals to employ genetic methods to control insect pests proliferated during the 1970's. The environmentally clean and evolutionary sound technique has great appeal. However, nearly 40 years after Kniplings proposal, the number of successful programs is limited. The problems and impediments to more widespread adoption of the technique have been numerous, but many of the problems can be attributed to the expense of implementation. Frequently, to be economically viable, eradication must be achieved, thereby enabling the initial high cost to be recovered over a long period. The suggestion that competitiveness may be frequency dependent, and that this effect will cause a deterioration of cost to benefits ratios, will not assist proponents of genetical control techniques. Clearly, where genetical control is under consideration, detailed studies of the field competitiveness of released insects is warranted.

Acknowledgments

The author would like to thank Dr Keith Wardhaugh, CSIRO Division of Entomology for helpful suggestions regarding the manuscript and to John Stone, of Jasam Consulting, who kindly provided estimates of the cost of rearing and releasing *Lucilia cuprina.*

References

Ayala FJ (1972) Frequency-dependent mating advantage in *Drosophila. Behav Genet* **2**:85–91

Berryman AA, Bogyo TP, Dickmann LC (1973) Computer simulation of population reduction by release of sterile insects: II The effects of dynamic survival and multiple mating. In: *Proceedings of the panel on computer models and application of the sterile-male technique.* International Atomic Energy Agency Vienna, pp 31–43

Bushland RC, Hopkins DE (1951) Experiments with screwworm flies sterilised by X-rays. *J Econ Entomol* **44**: 725–731

Cuisance D, Politzar H, Taze Y, Sellin E, Bourdoiseau G, Fevrier J, Koch K (1978) Presentation succincte des travaux sur l'experimentation de la methode du male sterile contre *Glossina palpalis gambiensis* a Bobo-Dioulasso (Haute-Volta) au cours de L'annee 1978. *Centre de Recherches sur les Trypanosomoses Animales Bobo-Dioulasso, Haute- Volta,* pp 1–46

Eanes WF, Gaffney PM, Koen RK, Simon CM (1977) A study of sexual selection in natural populations of the milkweed beetle *Tetraopes tetraopthalmus* In: Christiansen FB, Fenchel TM (eds) *Lecture notes in biomathmatics 19. Measuring selection in natural populations.* Springer, Berlin, pp 49–64

Foster GG (1991) Simulation of genetic control. Homozygous-viable pericentric inversion in field-female killing systems. *Theor Appl Genet* **82**:368–378

Foster GG, Madern RH, Helman RA (1985a) Field trial of a compound chromosome strain for genetic control of the sheep blowfly *Lucilia cuprina. Theor Appl Genet* **70**:13–21

Foster GG, Vogt WG, Woodburn TL (1985b) Genetic analysis of field trials of sex linked translocation strains for genetic control of the Australian sheep blowfly, *Lucilia cuprina* (Wiedemann). *Aust J Biol Sci* **38**:275–293

Foster GG, Whitten MJ, Konovalov C (1976) The synthesis of compound autosomes in the Australian sheep blowfly, *Lucilia cuprina. Can J Genet Cytol* **18**:169–177

Fried M (1971) Determination of sterile insect competitiveness. *J Econ Entomol* **64**:869–872

Geier PW (1969) Demographic models of population response to sterile release procedures for pest control. In: *Insect ecology and the sterile-male technique.* International Atomic Energy Agency, Vienna, pp 33–41

Grover KK, Suguna SG, Uppal DK, Singh KRP, Ansari MA, Curtis CF, Singh D, Sharma VP, Panicker KN (1976) Field experiments on the competitiveness of males carrying genetic control systems for *Aedes aegypti. Entomol Exp Appl* **20**:8–18

Haisch A (1970) *Some observations on decreased vitality of irradiated Mediterranean fruit fly. Sterile male technique for control of fruit flies.* International Atomic Energy Agency, Vienna, pp 71–75

Itô Y, Koyama J (1982) Eradication of the melon fly: Role of population ecology in the successful implementation of the sterile insect release method. *Prot Ecol* **4**:1–28

Itô Y, Miyai S, Hamada R (1989) Modelling systems in relation to control strategies. In: Robinson AS, Hooper G (eds) *Fruit flies, their biology, natural enemies and control.* Elsevier, Amsterdam, pp 267–279

Iwahashi O, Itô Y, Shiyomi M (1983) A field evaluation of the sexual competitiveness of sterile melon flies, *Dacus (Zeugodacus) cucurbitae. Ecol Entomol* **8**:43–48

Knipling EF (1955) Possibilities of insect control or eradication through the use of sexually sterile males. *J Econ Entomol* **48**:495–462

Knoppien P (1985) Rare male mating advantage: A review. *Biol Rev Camb Philos Soc* **60**:81–117

Krafsur ES, Whitten CJ, Novy JE (1987) Screwworm eradication in north and central America. *Parasitol Today* **3**:131–137

Mahon RJ (1983) The competitiveness of strains of potential use in the genetic control of the sheep blowfly. In: *Sheep blowfly and flystrike in sheep.* Department of Agriculture, New South Wales, Sydney, pp 268–273

Muggleton J (1979) Non-random mating in wild populations of polymorphic *Adalia bipunctata. Heredity* **42**:57–65

Partridge L (1988) The rare male effect: what is its evolutionary significance? *Philos Trans R Soc London B Biol Sci* **319**:525–539

Petit C (1951) Le rôle de l'isolement sexuel dans l'évolution des populations de *Drosophila melanogaster. Bull Biol Fr Belg* **85**:392–418

Rogers D J, Randolph SE (1985) Population ecology of tsetse. *Annu Rev Entomol* **30**:197–216

Serebrovsky AS (1940) On the possibility of a new method for the control of insect pests. Originally published in 1940 in *Zoologicheskii Zhurnal* 19:618–630. English translation in *Sterile-male technique for eradication of harmful insects.* International Atomic Energy Agency, Vienna 1969, pp 123–127

Taylor DB, Bruce JC, Garcia R (1991) Gelled diet for screwworm (Diptera: Calliphoridae) mass production. *J Econ Entmol* **84**:927–935

Vogt WG, Morton R (1991) Estimation of population size and survival of sheep blowfly, *Lucilia cuprina,* in the field from serial recoveries of marked flies affected by weather dispersal and age-dependent trappability. *Res Popul Ecol* **33**:141–163

Whitten MJ, Foster GG, Vogt WG, Kitching RL, Woodburn TL, Konovalov C (1976) Current status of genetic control of the Australian sheep blowfly *Lucilia cuprina* (Wiedemann) (Diptera: Calliphoridae) *Proc XV Int Cong Entomol*, pp 129–139

Wong TTY, Kobayashi RM, McInnis DO (1986) Mediterranean fruit fly (Diptera: Tephritidae): Methods of assessing the effectiveness of sterile insect releases. *J Econ Entomol* 79:1501–1506

Wong TTY, Ramadan MM, Herr JC, McInnis DO (1992) Suppression of a Mediterranean fruit fly (Diptera: Tephritidae) population with concurrent parasitoid and sterile fly releases in Kula, Maui, Hawaii. *J Econ Entomol* 85:1671–1681

THE BALANCE OF WEED POPULATIONS

W.M. Lonsdale

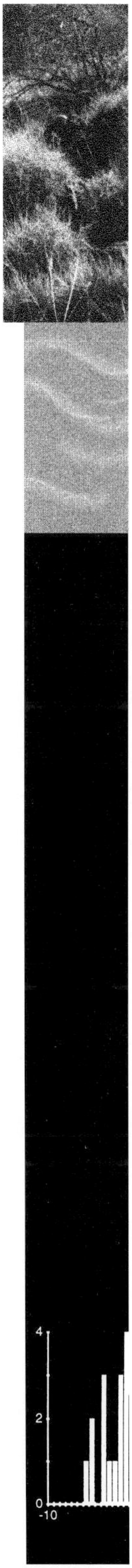

ABSTRACT

The Nicholsonian themes of density-dependence, competition, and host-parasitoid interactions, which have shaped animal population ecology, also loom large for population ecologists studying the biological control of weeds. Indeed, density-dependence and competition are perhaps more easily investigated in plants than in insects. Examples are given of situations where disturbance has allowed weeds to flourish, and where strong density-dependence has then regulated the weed population at its new higher abundance. Furthermore, some of the insights gained from the theory of arthropod host-parasitoid interactions can enhance our understanding of success and failure in the biological control of weeds. It is argued however that it will always be extremely difficult to make reliable predictions of success in biological control of weeds. This is because critical aspects of the interaction between the population of the weed and that of the control agent can only be known from field measurements following the release of the agent. What theory can do is to indicate the kind of performance our control agents will have to give in order to control the weed population.

Key words: biological control, plant population ecology, density-dependence, competition

Frontiers of Population Ecology, R.B. Floyd, A.W. Sheppard and P.J. De Barro (eds), CSIRO Publishing, Melbourne, 1996 pp. 573–588

Introduction

The *balance* of weed populations? The title of this paper, chosen for its resonance with those of the classic papers of Nicholson (1933) and Nicholson and Bailey(1935), will strain the indulgence of Australian readers at least. Although it is difficult to be precise about weed abundances, we know that the number of introduced species that has become naturalised in Australia has increased at the rate of four to six species per year since European settlement (Groves 1986), with no signs of deceleration. The floras of Australian ecosystems now include on average 18% exotic species (W.M. Lonsdale, unpublished), with some sites as high as 60%. It seems that the abundance of weeds is on the increase.

Nevertheless, we commonly see weed species reach stable equilibria in terms of biomass *per unit area*, even as their range and overall abundance continues to increase. So it *is* reasonable to talk about balance and how it is maintained in weed populations (see 'Density-dependence and competition' below). Moreover, it is in disturbing the massive stability of these equilibria that the difficulties of weed control lie. For example, the prickly shrub *Mimosa pigra* L. grows as dense, practically pure stands, thousands of hectares in area across Northern Australia. The tough mature shrubs at densities around 10^0 m^{-2} are backed up by seed banks at densities in the order of 10^4 m^{-2} (Lonsdale *et al.* 1988). Disturbing this stable equilibrium, to return the ecosystem to native vegetation, requires great expense of time and money – roughly \$500 ha^{-1} over 3 or more years, to kill the mature stands, clear them, and deal with the recruiting seedlings.

Some of the most successful work in theoretical population ecology has emerged from the studies of insect host-parasitoid interactions begun by Nicholson. Populations of higher plants, however, have more complicated biologies than such systems, and present theoreticians with a much greater challenge (Rees and Lawton 1993). Firstly, simply counting plants is rarely sufficient to characterise a population because plants of a similar age can vary 50 000-fold in size (Harper 1977). Secondly, recruitment can be highly unpredictable or episodic (e.g. Wellington and Noble 1985). Thirdly, plant communities are often not at equilibrium but are in transition (Sprugel 1991). Fourthly, because plants in general must colonise new sites by the dispersal of propagules, the process takes generations, so the currently colonised ground that we are studying may in fact be far less suitable for the species than the unoccupied ground that we are ignoring (Crawley 1990).

Nevertheless, plant population models derived from those developed for animals have proved remarkably successful at describing the behaviour of natural populations (e.g. Caughley and Lawton 1981; Watkinson 1985a; Watkinson *et al.* 1989; Lonsdale *et al.* 1995). What follows is an attempt to explore the application of the Nicholsonian themes of density-dependence, competition, and host-parasitoid interactions to the biological control of weeds, by reference to theory and models.

Density-dependence and competition in weeds

Despite arguably giving rise to much of contemporary ecology, the seemingly irreconcilable stances on the role of density-dependence in natural populations taken by Nicholson (1957) on the one hand, and Andrewartha and Birch (1954) on the other, now appear more like semantic arguments (Clark *et al.* 1967; Hassell 1986). Nicholson clearly recognised the importance of density-independent factors in determining abundance (e.g. Nicholson 1933). At the same time, Andrewartha felt that weather could act as a form of density-dependence, by killing the proportion of the population inhabiting less favourable situations (Davidson and Andrewartha 1948). For animals, Strong's idea of 'density vagueness' might best represent the modern synthesis of the extreme positions. Here, populations exhibit high variance in demographic performance at intermediate densities, and density distinctly affects performance only at high densities (Strong 1986). It is a logical necessity that there is some density-dependent regulation but this is hard to demonstrate at normal densities in animal populations. Populations cannot persist by a juggling act, balancing their rate of increase against a pot-pourri of unpredictable environmental factors (Hassell 1986). Indeed, more recent reanalyses of the thrips data of Andrewartha and Birch (Smith 1961; Varley *et al.* 1973; Schaffer and Kot 1986) have demonstrated strong evidence for population regulation.

For plants, it is far easier to manipulate density and show such regulatory phenomena as density-dependent fecundity. It is hard to imagine an intense debate in plant ecology on the detection of density-dependence (cf. Wolda and Dennis 1993; Dennis and Taper 1994; Holyoak and Lawton 1993; Hanski *et al.* 1993; Wolda *et al.* 1994). Indeed, some of the best documented generalisations in plant ecology come from studies of populations where competition is clearly occurring (White 1980). In plant ecology, it is not the occurrence of density-dependence, but its precise manner of operating, that is the subject of discussion (e.g. the case of density-dependent mortality, or self-thinning; Weller 1987a,b; Lonsdale 1990).

Population cycles in plants and animals

Nicholson's concepts of scramble and contest competition (Nicholson 1954) underlie a recent debate on the difference between the population ecology of plants and animals. Rees and Crawley (1989) have argued that cyclical behaviour is much more common in animals than in plants. However, the number of demonstrated cases of animal population cycles in the field is in fact rather small, e.g. lemmings (*Lemmus* and *Dicrostonyx* spp.), voles (*Microtus*; Krebs and Myers 1974) and certain other boreal vertebrates (see Sinclair, this volume), and insect pests such as the larch bud moth, *Zeiraphera diniana* (Anderson 1981) and the Colorado potato beetle, *Leptinotarsa decemlineata* (May 1981a).

The number of plant species for which cycles have been demonstrated is still smaller – perhaps only one (the annual plant *Erophila verna*; Symonides *et al.* 1986; Rees and Crawley 1991). For both plants and animals, the detection of cyclical behaviour requires observation over many generations (see Moran 1952, 1953; Poole 1978), and there are few organisms for which such studies have been done (Silvertown 1991). Attractive idea though it may seem, then, the evidence for this fundamental difference between plants and animals is lacking.

Even if we take this difference for read, there is disagreement on the reasons for it. Rees and Crawley (1989) believe the answer lies in the modularity of plants, which develop through the repeated iteration of similar units (Harper 1981). This contrasts with animals, which have a unitary structure with a determinate development process. Thus, plants (particularly annuals) do not have significant reproductive thresholds, whereas animals do (Rees and Crawley 1989). Consequently, plants would not exhibit the full range of dynamic behaviour, because, they argue, cyclic dynamics and chaos are unlikely without reproductive thresholds (Rogers 1986; Lomnicki 1988).

Silvertown (1991), by contrast, argues that the key difference between plants and animals is that plants are competing for light, leading to asymmetric competition (contest; Nicholson 1954), while symmetric (scramble) competition is the generality for animals. Symmetric competition, combined with the existence of a reproductive threshold (which does not necessarily follow from the modularity of plants – Silvertown 1991), is destabilising (Readshaw and Cuff 1980) and leads to population cycles.

We can probe these questions from a different angle by further reference to insect population theory. Annual plant populations with discrete generations can be modelled by an equation of the form

$$N_{t+1} = \lambda N_t \left(1 + aN_t\right)^{-b} \qquad (1)$$

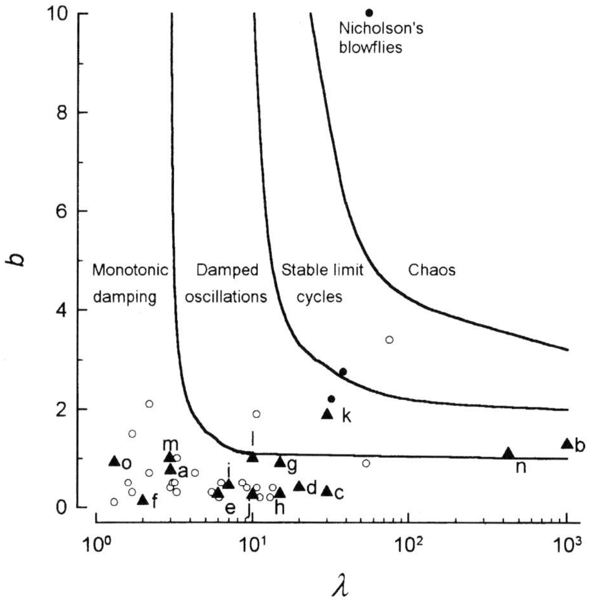

Fig. 1. Stability boundaries between the density-dependent parameter, *b*, and the population growth rate, λ, in Eq. 1 (redrawn from Hassell *et al.* 1976). Insect populations (see Hassell *et al.* 1976 for sources) are shown as circles (hollow are field, and filled are laboratory populations), while plants are shown as filled triangles. The plants are (a) *Vulpia fasciculata; Salicornia europaea* in (b) low and (c) high marsh, *Cakile edentula* in (d) seaward, (e) middle and (f) landward sides of a dune, (g) *Rhinanthus angustifolius,* (h) *Floerkea proserpinacoides,* (i) *Polygonum confertiflorum,* (j) *Diamorpha smallii,* (k) *Androsace septentrionalis,* (a-k measured from graphs of Watkinson 1985b), *Sida acuta** (l), a weed at a research station in northern Australia (Lonsdale *et al.* 1995), (m) *S. acuta* in a cattle paddock (Lonsdale *et al.* 1995), (n) *Hyptis suaveolens* (W.M. Lonsdale and G. Flanagan, unpublished), and (o) *Sorghum intrans** in good patches (Watkinson *et al.* 1989). Reasonably accurate values of λ are shown only for species marked *. The rest are gross overestimates, being the seed output per plant.

which Watkinson (1980) developed from the plant yield-density equation of Bleasdale and Nelder (1960). It is a difference equation relating the population in generation $t + 1$, N_{t+1}, to that in generation t, N_t; λ is the number of offspring per plant or per insect produced under ideal conditions, and a and b are constants defining the density-dependent feedback term. It is a widely used formulation in plant population ecology (see e.g. Crawley and Rees, this volume), but is identical to that proposed by Hassell (1975) for single species populations of insects. From a Nicholsonian point of view, b is the defining term for the nature of the competition, which ranges from scramble ($b \to \infty$) to contest ($b \to 1$). Hassell *et al.* (1976) have explored the stability of this equation as λ and b vary (Fig. 1). Values for some plant populations are superimposed on the graph, which also shows the insect populations (including Nicholson's blowflies) plotted by Hassell *et al.* (1976). Notice that most of the plants have values of b close to 1 – this seems the generality for plants (Watkinson 1980). Consequently, they have to reach very high values of λ even to get into the region of damped oscillations. One might expect that weedy species in Australia, the home of many rampant exotic plants with few natural enemies, might provide a few candidates with high λ values. However, even such vigorous tropical weeds as *Sida acuta* and *Senna obtusifolia* in northern Australia do not make it into the critical region of the graph (Fig. 1). Three plant species (*Hyptis suaveolens,* a tropical annual shrub, *Androsace septentrionalis*, a dune annual herb, and low-marsh populations of *Salicornia europaea*, a salt marsh annual) appear to be pre-disposed to damped oscillations (Fig. 1), but these are as yet undemonstrated. Note also that the values of λ shown for the plants are mostly gross overestimates, being the maximal seed output, with no allowance for mortality from seed to adulthood. This is of course usually prodigious. Lastly, it is important to note that most of the animal populations are also firmly entrenched in the damping region of the graph. Laboratory animal populations such as Nicholson's blowflies tend to exhibit cyclic or chaotic behaviour (e.g. Nicholson 1954 and Fig. 1). Natural populations, by contrast, tend to have a stable equilibrium point (Hassell *et al.* 1976; May 1981a; Harwood and Rohani, this volume).

Thus, it is not necessarily true that animal populations show cyclical behaviour and plants do not. The evidence is not strong for either, and theory (Eq. 1) says that stable behaviour should

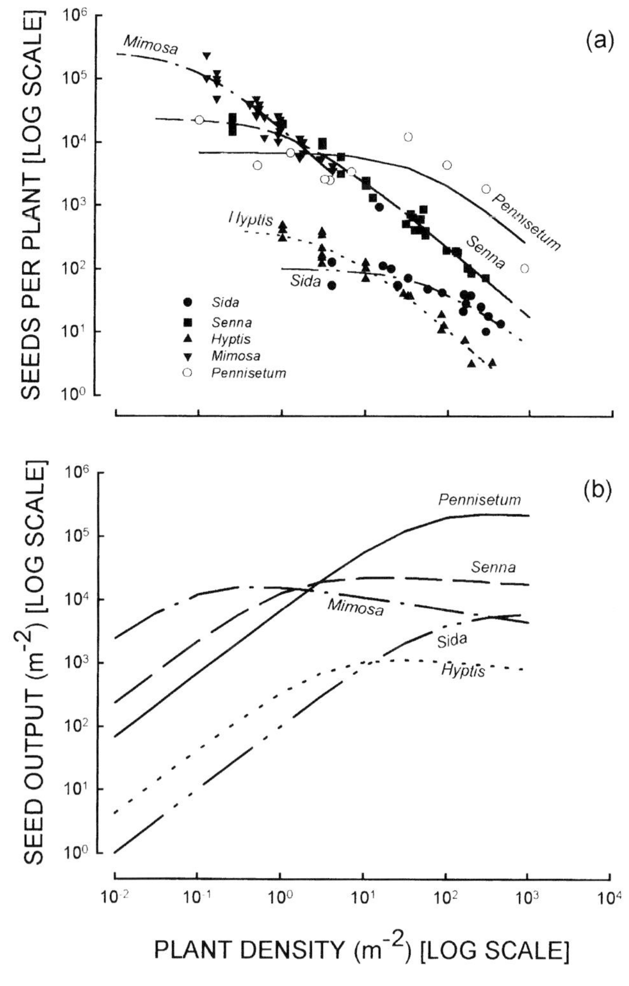

Fig. 2. The relationship between (a) fecundity and density, and (b) population seed output and density, for five weeds of tropical Australia (see Table 1). The curves in (a) are Eq. 2 fitted to the data (i.e. seeds per plant), while those in (b) are the same curves, but with seed output per plant converted to a seeds per unit area basis.

preponderate for both plants and animals. Furthermore, this is only true for single, non-interacting populations, which of course are not to be found in the real world. When there are two or more interacting species, whether plant or animal, they are much more likely to exhibit dynamic behaviour (May 1981a). It is clear from population theory that time-lags, competition between plants, the interaction between a pollinator and its host, plant-herbivore interactions (May 1974, 1981b), and seasonal changes, could all result in plant population cycles.

Density-dependent fecundity and biological control

The weeds in Fig. 2a are all exotic weeds naturalised in the Top End of the Northern Territory. All have responded to disturbance of the native vegetation caused by exotic ungulates (see e.g. Lonsdale 1993), and produce dense stands of near-monospecific vegetation. While their general abundance in Northern Australia may still be increasing, strong density-dependent fecundity regulates the density of all of them. All show negative relationships between seed output and density, that are reasonably well fitted by

$$F = f(1 + aN)^{-b} \tag{2}$$

Table 1. Demographic characteristics of five weeds naturalised in tropical Australia, and the consequences for biological control. The parameters *f*, *a* and *b* are from Eq. 2. The nature of competition in Nicholsonian terms is defined by *b* (for scramble, $b \rightarrow \infty$; for contest, $b \rightarrow 1$). Q is the ratio N_{min}/N_{max} where N_{min} is the critical density gauged from Fig. 2b, at which population seed output starts to decline with density, and N_{max} is the peak density seen in the field. A biological control agent would have to reduce flowering plant density to a fraction Q of the peak value before there would be any impact on overall seed output. Of course, whether this reduction in seed output has any impact on the next generation depends on whether the plant population is seed limited.

Species	Life form	f (seeds m^{-2})	a (m^{-2})	b	Q
Sida acuta	Malvaceous annual shrub	10^2	0.016	1.0	0.98
Senna obtusifolia	Leguminous annual shrub	2.4×10^4	0.8	1.1	0.01
Hyptis suaveolens	Annual herb	4.3×10^2	0.25	1.1	0.03
Mimosa pigra	Leguminous perennial shrub	2.8×10^5	10.0	1.2	0.07
Pennisetum polystachion	Perennial grass	6.9×10^3	0.02	1.1	0.11

an equation related to Eq. 1 (Watkinson 1980), where *F* is the seed output per plant, and *f* is the maximum seed output of the plant under conditions where competition is not occurring (see Table 1). As density declines, the number of seeds per plant increases to take up the slack, until the curve flattens off as the individual plants approach their maximal seed output.

Three of the weeds shown (*M.pigra*, *H. suaveolens*, and *S. acuta*) are the subject of biological control research by the CSIRO Division of Entomology. It is important, for control purposes, to find out from these graphs how far the control agents would have to reduce the flowering plants below peak densities to have an impact on seed output of the population as a whole (see Fig. 2b and *Q* in Table 1). One species, *S. acuta*, declines in total seed output at the first sign of a fall in flowering plant density ($Q = 0.98$). The remainder, though, require a reduction in peak plant density of 90–99 % to even begin to reduce seed output. Moreover, seed availability rarely limits plant populations (Andersen 1989; Crawley 1990). For the species in Table 1, a depression in seed output will only affect plant density in the next generation if the populations are seed limited. These species are probably quite typical of the problems faced in biological control of weeds, and they show why tackling weeds with seed-feeding insects is unlikely to provide the sole answer (see also Lonsdale 1993). It is possible however that seed feeders will reduce rates of spread, since, in the dispersal phase, populations are more likely to be seed-limited (see Crawley and Rees, this volume). In this case, the degree of reduction will depend on the shape of the seed dispersal curve and the relationship between mortality and distance (Lonsdale 1993). Later in the paper (see 'Host-parasitoid interactions') I will be looking at the population dynamics of seed-feeding beetles, and the uphill battle they face in having any effect on *M. pigra*.

For one of the weeds discussed above, *S. acuta*, we have gone along way towards achieving successful biological control. *S. acuta* is a tough malvaceous weed of overgrazed pastures, a shrub that dies back each year during the tropical dry season, and recolonises at the start of the following wet season through dense seedling recruitment. The weed is native to Mexico and Central America. From its native habitat, the Division of Entomology introduced the foliage feeding beetle *Calligrapha pantherina* (Chrysomelidae) for biological control (Forno *et al.* 1992). The beetle has been successful in controlling the weed in the Darwin region. We investigated the nature of this control in an insecticidal exclusion experiment (Lonsdale *et al.* 1995). The beetle, we found, reduced *S. acuta* seed output by an order of magnitude, from 8000 m^{-2} to 730 m^{-2}. We then used a model similar to that in Eq. 1, but incorporating known levels of density-independent mortality besides density-dependent fecundity (Watkinson *et al.* 1989), to predict what effect this reduction would have on

Fig. 3. Frequency distributions of plant population density of the tropical annual weed *Sida acuta* (a) before and (b) one year after defoliation by chrysomelids that reduced seed output by an order of magnitude. Hollow bars are populations measured in the field, hatched bars are predicted distributions from a population model (redrawn from Lonsdale *et al.* 1995).

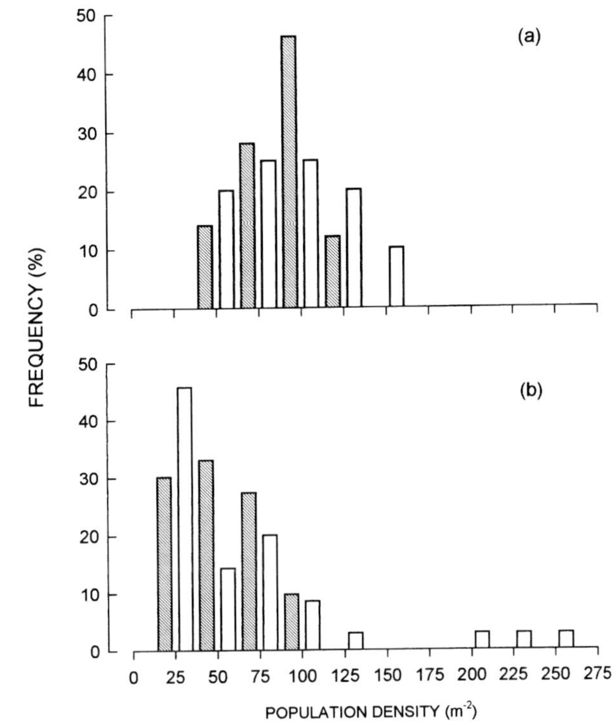

flowering plant density in the next generation. The model accurately predicted, not only the fall in density, but also the change in frequency distribution of densities (Lonsdale *et al.* 1995, Fig. 3). This was encouraging, but we still lack a good understanding of the dynamics of the insect population. Without it we cannot model the interaction as it develops over the years. What is interesting about this type of model is the interaction between density-dependent and independent factors, and the way the stochastic variation in the environment is built into what is a deterministic model through the random variation (between limits observed in the field) of the density-independent mortality (Fig. 4).

Insecticidal exclusion experiments do not always work. In testing the effects of a stem boring moth, *Neurostrota gunniella*, on *M. pigra*, we found that a systemic insecticide dramatically reduced the insect's density, but caused a suppression of seed production by excluding native insect pollinators. Despite a measurable effect on plant growth rate, any effect of our insect on fitness had been masked. In such circumstances, we need a different approach. Underwood (1994), working in the field of environmental impact assessment, has advanced the idea of using what he terms asymmetrical experiments. They are an extension of what are termed BACI (Before-After-Control-Impact) designs, developed to detect impacts of environmental disturbance while avoiding as far as possible the pseudoreplication involved in BACI designs. Applied in biological control of weeds, asymmetrical experiments would involve a release site at which the performance of the plant population is measured before and after the release, with several control sites, at which similar before and after measurements are taken, for comparison. A significant interaction between treatment (control vs. release) and time (before vs. after release) would indicate a significant impact of the agent on the weed population (cf. Underwood 1994). They may well prove the way forward for assessing

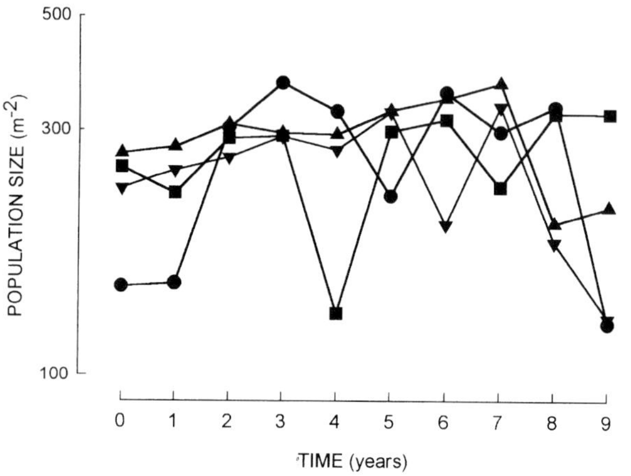

Fig. 4. Time course for population density in four typical model replicates of *Sida acuta* (redrawn from Lonsdale *et al.* 1995). A normal run of the model would have 1000 replications (see Watkinson *et al.* 1989).

the impact of weed biological control agents where the use of insecticides to provide controls is undesirable or impractical.

Host-parasitoid interactions and biological control of weeds

Nicholson's work on host-parasitoid interactions (Nicholson 1933; Nicholson and Bailey 1935) has given rise to some of the most successful exercises in theoretical population biology (see Murdoch and Nesbit, this volume). Some possibilities and limitations in applying arthropod predator-prey theory to our understanding of success and failure in biological control of weeds follow.

Predicting the outcome of control programs

A considerable amount of effort and expense in biological control goes into screening the control agents – typically three scientist-years per species – to ensure specificity to the target weed. It would obviously improve efficiency if predictions could narrow down the field of candidates to agents that will control the weed. In considering whether it will ever be possible to make such predictions accurately, a useful place to start is the modelling exercises carried out by Caughley and Lawton (1981) and Watkinson (1986). These were carried out on cases of successful biological control of weeds, the former on the *Opuntia/Cactoblastis* data (Dodd 1940; Monro 1967), and the latter on *Carduus/Rhinocyllus* (Harris 1984). Both papers adapted the same model, originally developed from the classic Lotka-Volterra equations to describe arthropod predator-prey interactions (May 1981b). As used by Caughley and Lawton (1981) and Watkinson (1986), it incorporates logistic plant growth, a functional response of the insect to plant density, exponential herbivore population growth and intense competition between the larvae at high densities. The equations are

$$dV/dt = r_1 V(1 - V/K) - cH[V/(V + D)] \tag{3}$$

for the plant population dynamics, and

$$dH/dt = r_2 H(1 - JH/V) \tag{4}$$

for the dynamics of the control agent. The parameters for the model (Caughley and Lawton 1981) are as follows: V = the plant population density or biomass, t = time, r_1 = the intrinsic rate of increase of the plants, K = the maximum ungrazed plant density or biomass, c = the maximum rate of food intake per herbivore, D = the density or biomass of the plant at which the herbivore functional response saturates, H = the density of the biocontrol agent, r_2 = the intrinsic rate of increase of the biocontrol agent, and J = a proportionality constant related to the number of plants needed to sustain a herbivore at equilibrium.

To give the model a topical flavour for this Symposium, I have adapted it here to describe one of the proudest achievements of the CSIRO Division of Entomology. This is the famous case of the floating fern *Salvinia molesta*, controlled by the introduced weevil *Cyrtobagous salviniae* at Lake Moondarra in northern Queensland. Basic data were presented by Room *et al.* (1981, 1984), and Room (1990). In attempting to recreate the time course of the weed's abundance, I did not attempt to model abundance before Spring 1980 because climate was up to then driving the changes in abundance (Room *et al.* 1981). At the start of the period in question (October 1980 to August 1981), V = 5900 *t* (from Fig. 3 in Room *et al.* 1981), while r_1 = 0.22, as measured from the growth trend for the growing season before the introduction of the weevil. K was set at 19000 *t* from October to December, then was increased to 40 000 *t*. This was the projected autumn peak given the area of the lake in 1981, following heavy rains in December and January (Room *et al.* 1981). Parameters c and D cannot be estimated. However, c can be set (perhaps unintuitively) large to reflect the structural damage done by the weevils, which far outweighs the small amount of ingested material, while D can be set small to reflect its great searching efficiency. H was calculated at the start by increasing the number of weevils introduced in June 1980 at r_2 until October. This gave 1.2×10^6 weevils, while r_2 itself (1.68 month^{-1}) came from Table 1 of Room (1990), for weevils living on *S. molesta* containing 2% dry weight of nitrogen, which is the concentration in Lake Moondarra (Room *et al.* 1984). The order of magnitude of J was estimated from the fact that 6×10^6 weevils were living in about 500 *t* of salvinia in May 1981 (Room *et al.* 1981), i.e. 10^{-5} or thereabouts (this is an underestimate because the weevils and the weed were not at equilibrium at this stage).

The actual time course of salvinia and weevil abundance gleaned from the data was best represented with c set at 0.00006*t* month^{-1} weevil^{-1}, J at 0.00011*t* and D set at 2. The model then does a reasonable job of describing what happened in Lake Moondarra (Fig. 5). The most notable divergence between theory and reality is in the weevil population's size in April (Fig. 5a), which Room *et al.* (1981) estimated at 10^8, where the model gives 6×10^7. However, the weevil population estimate is very elastic, being based on only two quite small samples, and with a difference between them of 30% (see Room *et al.* 1981).

Some interesting discoveries emerged from my attempts to fit the model to the data. Firstly, the additional release of 1500 weevils in January 1981, following the first 1500 released in June 1980, was of no significance, according to the model. By January, the model weevil population would have been in the tens of millions (Fig. 5a) – this was necessary to give a population of about 100 million by April. The addition of 1500 more weevils at this stage, then, would have been of little consequence. Secondly, the actual equilibrium population size for the weed in the presence of the weevil is less than 1*t* (Room *et al.* 1981), very much lower than the value in the model, which was about $10^2 t$. Thus, the model is not a complete picture of the dynamics of the weevil and the fern. Perhaps we must invoke host-plant refugia or interference between weevils to explain this strong stable depression in weed biomass (see discussion of q values below).

The main finding from this modelling exercise is that, though a few parameters can describe the interaction between the weed and the biocontrol agent reasonably, it will always be extremely difficult to make reliable predictions of success in biological control before the release of a control agent. A multitude of factors affect the outcome of control programs, including plant life history, genetics of the natural enemy and the environmental conditions (Chaboudez and Sheppard 1992). The model above is probably the bare minimum required to describe such an interaction (Caughley and Lawton 1981), but, effective though it is, its parameters would have to be measured in the field *following* release of the control agent, and may be quite site-specific (Crawley 1983; Room 1990).

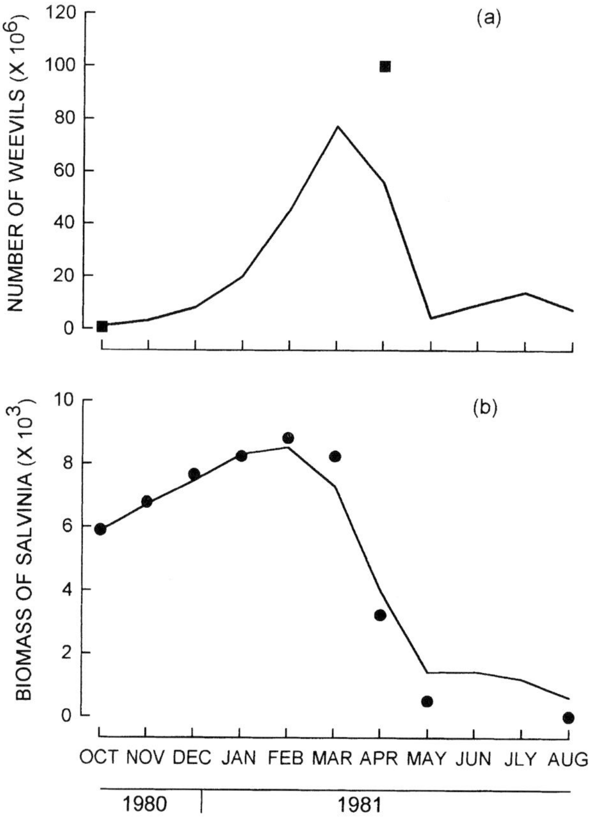

Fig. 5. A typical predator-prey type model (Eqns 3 and 4) adapted for use in biological control of weeds by Caughley and Lawton (1981), and here applied to the famous case of the control of the floating fern *Salvinia molesta* by the weevil *Cyrtobagous salviniae* at Lake Moondarra in Queensland (Room *et al.* 1981). (a) the weevil population, modelled (—) and actual (■), and (b) the weed population, modelled (—) and actual (●). See text for further details.

In the native range, other organisms, e.g. parasites of the weevil, may be involved in the interaction (see Sheppard and Woodburn, this volume), so measurements of the above parameters in the native range would not translate to the much simpler system in the introduced range. It is obvious, for example, that the values of r_1 and r_2 used in Fig. 5 could be very different in different systems. Indeed, Room (1990) suggests that the ratio of r_1 to r_2 is 1.5 to 2.5, whereas the ratio of the values that worked in my simulation was 0.13. Note also that the spectacular success of the weevil at many sites around the tropics has not been repeated in Kakadu National Park, where cyclical behaviour seems to be occurring (Julien and Storrs 1994). Predator-prey theory tells us that cyclical behaviour will occur in situations where $r_1 / r_2 = 2$ provided the carrying capacity for the prey, *K*, is much larger than the equilibrium prey density in the presence of the predator (e.g. Fig. 5.2 in May 1981b). With *S. molesta* and *C. salviniae*, this latter ratio can be in the order of 10^4:10^0 (Room *et al.* 1981). It may be, therefore, that the system meets the theoretical conditions for cyclical behaviour in Kakadu, but not in Lake Moondarra.

This leads to another important quantity to emerge from predator-prey theory, *q* (Beddington *et al.* 1978), that may lend some insights into success in biological control of weeds. This (*q*) is the ratio of host equilibrium density in the presence of the predator or parasitoid (N_c) to that in its absence, the host's carrying capacity (*K*). Beddington *et al.* (1978) compiled a variety of field and laboratory examples of *q* for arthropod predators and their prey; these are redrawn in Fig. 6 together with values of *q* obtained from cases of successful weed biological control in the literature (particularly Caughley

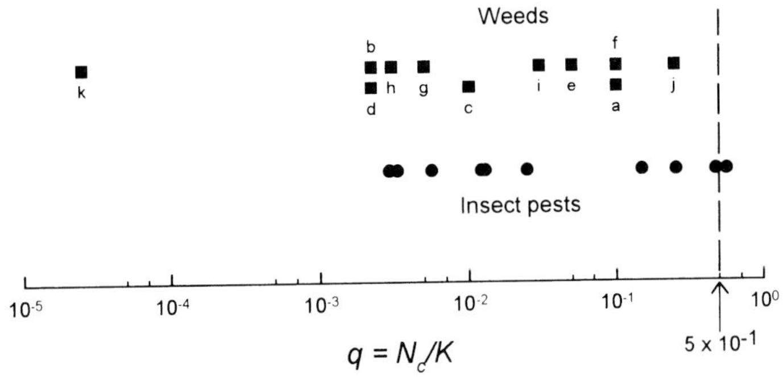

Fig. 6. Values of *q* for (●) insect host-parasitoid interactions (redrawn from Beddington *et al.* 1978), and (■) cases of successful biological control of weeds. The weed *q*-values are very crude estimates, which serve only to demonstrate that they are well below the critical value of 0.5. Where various values are given for a particular weed, the lowest is presented. See 'Host-parasitoid interactions' for further details. The weed species (and their control agents) are (a) *Alternanthera philoxeroides* (*Agasicles hygrophila*), (b) *Carduus thoermeri* (*Rhinocyllus conicus*), (c) *Eichhornia crassipes* (*Neochetina* spp.), (d) *Hydrilla verticillata* (a–d from Julien 1992), (e) *Carduus nutans* (*Rhinocyllus conicus*), (Harris 1984), (f) *Opuntia* spp. (*Cactoblastis cactorum*), (g) *Hypericum perforatum* (*Chrysolina quadrigemina*), (h) *Senecio jacobaea* (*Tyria jacobaeae*), (i) *Chondrilla juncea* (*Puccinia chondrillina*), (j) *Melampyrum lineare* (*Atlanticus testaceous*) (f–k from Table 7.3 of Caughley and Lawton 1981), and (k) *Salvinia molesta* (*Cyrtobagous salviniae*) (Room *et al.* 1981).

and Lawton 1981; Julien 1992). Note that there is considerable overlap for insects and weeds, and all the weeds lie below $q = 0.5$, within the range 0.3 to 0.00003.

Theoretically, values of *q* below 0.5 can *only* be maintained stably in two situations. There must either be (a) interference between the predators/herbivores or (b) spatial heterogeneity for the host in the form of refugia from the predators/herbivores (Beddington *et al.* 1978, Caughley and Lawton 1981, Hassell 1981). Interference may result for example when the herbivore lays eggs in a clumped distribution, which results in some larvae eating themselves out of house and home, while some plants remain unscathed (e.g. *Opuntia* and *Cactoblastis*; Caughley and Lawton 1981). The control of the weed *Hypericum perforatum* by *Chrysolina* spp. beetles in the United States provides an interesting example of the role of spatial heterogeneity (Huffaker 1964), where the beetles' reluctance to oviposit in the shade has provided a stabilising refugium for the weed. To maintain a strong *stable* depression in weed density, then, we need the control agents to compete strongly at high densities, or we need refugia for the weed.

The picture that emerges from all this is that, while we may not be able to predict precisely which agent is going to work, we can use theory to see what broad attributes an agent must possess, to have a measurable, sustainable impact on the weed. However, it should be noted that this kind of analysis of biological control systems, wherein it is assumed that natural enemies control the pests by reducing their density to a new low, but non-zero, stable equilibrium, is coming under increasing challenge (Murdoch *et al.* 1985; McEvoy *et al.* 1993). Many successful control systems are in fact characterised by local extinctions and reinvasions, and many such systems appear to lack the mechanisms that yield stability in such models (McEvoy *et al.* 1993). We should always take models with a grain of salt. As Crawley and Rees (present volume) have pointed out: "models define what is possible (given certain conditions and assumptions). It is the job of observation and experiment to separate the actual from the possible".

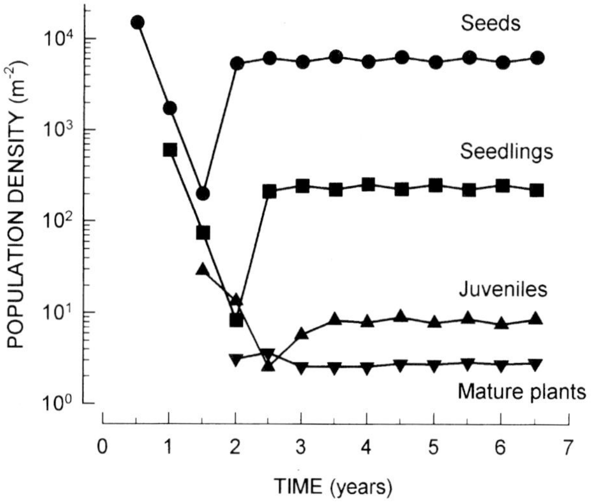

Fig. 7. Typical output from a new matrix model (W.M. Lonsdale and G.S. Farrell, unpublished) for the tropical woody weed *Mimosa pigra*, showing the recovery of a stand from the seed bank after the canopy has been killed. (Seedling height is ≤ 100 cm; 100 cm < juvenile height ≤ 2 m; 2 m < mature plant height ≤ 6 m).

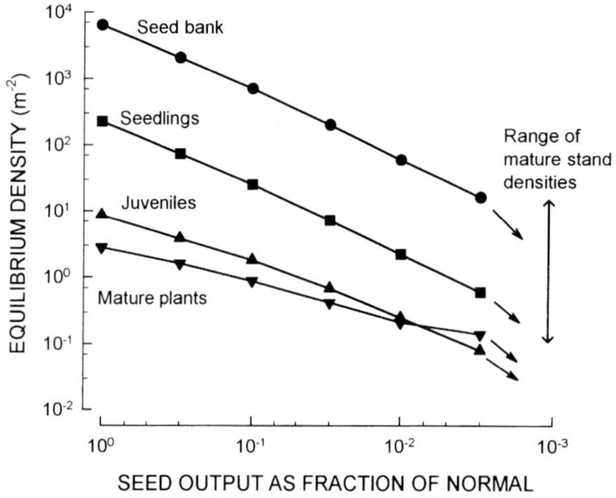

Fig. 8. A preliminary sensitivity analysis of the mimosa model (W.M. Lonsdale and G.S. Farrell, unpublished) showing the effect on the different components of a stand of reducing seed output to various fractions of the normal level. Note that the seed bank declines almost in proportion as the seed output, while the mature plants decline in density about one third as fast. As seed output is reduced to a thousandth of normal, all size classes decline to extinction. However, even strong depressions in seed output do not reduce mature stand density below what is currently seen in closed stands.

Models can give us an idea of what kind of performance our control agent must give, to be effective. We are in the process of developing a matrix model for *Mimosa pigra* (W.M. Lonsdale and G.S. Farrell, unpublished) that will allow us to do just this. Unusually for a matrix model, it incorporates density-dependent fecundity, survival, and growth. Figure 7 shows some basic output from the model, describing a situation where the canopy plants have been killed, but the massive seed bank in the order of 10^4 seeds m^{-2} has not been affected. The graph shows that a dense thicket will have developed within two years, and the thicket will be at equilibrium again by about 4 years, if nothing is done to control emergent seedlings.

What would be the effect of reducing seed output by varying amounts on the equilibrium density of the different size classes? The model suggests that reducing seed output greatly affects the seed bank, but an order of magnitude change reduces mature plant density only by a third (Fig. 8). When seed output is reduced to about one thousandth of the normal amount, all size classes head to extinction. However, even down to about 0.3% of normal seed output, mature stand density is still within the range seen in normal thickets. The mature plants are plastic enough in size to maintain a closed canopy down to 0.1 plant m^{-2} or so. Clearly, seed output has to be reduced dramatically to have any effect on the thickets.

This reduction may be achieved by attacking the plant, or, less likely, by seed predators. Two species of bruchid beetle, *Acanthoscelides puniceus* and *A. quadridentatus*, were introduced early in the biological control program against *M. pigra*, but neither has ever achieved a high percentage infestation of the seeds. A knowledge of the pattern of seed production in this weed, and some understanding of host-parasitoid interactions, explains why. The seed output is highly seasonal, so that during six months of the year there is little or no seed fall. Suddenly, within the space of a few weeks at the start of the wet season, the seed population explodes (Lonsdale 1988). It does so at a value of r_1 in the order of 0.77 $week^{-1}$, while the bruchids lag behind at only r_2 = 0.22 $week^{-1}$ (G. Flanagan, unpublished). Like parasitic wasps, the bruchids rely upon their prey, the seeds, to complete their life cycle, one larva developing within each infested seed. (Adult bruchids feed on pollen). Thus, the bruchids constantly lag behind the seed population. At best, only about 10% of the annual seed production would be destroyed (W.M. Lonsdale *et al.*, unpublished). This situation would be improved if other control agents reduce the total seed output, or if the bruchids were able to survive in large numbers during the annual dearth of seeds and flowers.

Conclusions

Briese (1993) has argued convincingly that plant ecologists should have more input into the biological control of weeds, rather than, as often happens, studies of the plant being carried out by entomologists working outside their area of specialisation. There is a small paradox here, however: our central need is for plant *population* ecology (a view probably shared by Briese 1993), carried out by specialists in that field, but this field in turn owes much to, and can still gain from, advances made by entomologists. The body of insect population theory that started with Nicholson sixty years ago, and has continued with the work of Hassell and his colleagues up to the present, providing a rich store of insights for workers on the biological control of weeds. It has yet to be sufficiently exploited.

Acknowledgments

I thank A.N. Andersen, D.T. Briese, J. Cullen, G.S. Farrell, I.W. Forno, P.M. Room, A.W. Sheppard and C.S. Smith for helpful comments on the manuscript.

References

Andersen AN (1989) How important is seed predation to recruitment in stable populations of long-lived perennials? *Oecologia* 81:310–315

Anderson RM (1981) Population ecology of infectious disease agents. In: May RM (ed) *Theoretical ecology. 2nd ed.* Blackwells, Oxford

Andrewartha HG, Birch LC (1954) *The Distribution and abundance of animals.* University of Chicago Press, Chicago

Beddington JR, Free CA, Lawton JH (1978) Characteristics of successful natural enemies in models of biological control of insect pests. *Nature* 273:513–519

Bleasdale JKA, Nelder JA (1960) Plant population and crop yield. *Nature* 188:342

Briese DT (1993) The contribution of plant biology and ecology to the biological control of weeds. In: Swarbrick J (ed) *Proc 10th Aust/14th Asian-Pacific Weed Conf.* Council of Australian Weed Science Societies, Brisbane

Caughley G, Lawton JH (1981) Plant-herbivore systems. In: May RM (ed) *Theoretical ecology. 2nd ed.* Blackwells, Oxford

Chaboudez P, Sheppard AW (1992) Are particular weeds more amenable to biological control? – a reanalysis of mode of reproduction and life history. In: Delfosse ES Scott RR (ed) *Proc Eighth Int Symp Biol Contr Weeds*. CSIRO, Melbourne

Clark LR, Geier PW, Hughes RD, Morris RF (1967) *The Ecology of Insect Populations in Theory and Practice*. Methuen, London

Crawley MJ (1983) *Studies in Ecology 10: Herbivory*. Blackwells, Oxford

Crawley MJ (1990) The population dynamics of plants. *Philos Trans R Soc Lond B Biol Sci* **330**:125–140

Davidson J, Andrewartha HG (1948) The influence of rainfall, evaporation and atmospheric temperature fluctuations in the size of a natural population of *Thrips imaginis* (Thysanoptera). *J Anim Ecol* **17**:200–222

Dennis B, Taper ML (1994) Density dependence in time series observations of natural populations: estimation and testing. *Ecol Monogr* **64**:205–224

Dodd AP (1940) *The biological campaign against prickly-pear*. Government Printer, Brisbane

Forno IW, Kassulke RC, Harley KLS (1992) Host specificity and aspects of the biology of *Calligrapha pantherina* (Col: Chrysomelidae), a biological control agent of *Sida acuta* (Malvaceae) and *S. rhombifolia* in Australia. *Entomophaga* **37**:409–417

Groves RH (1986) Plant invasions of Australia: an overview. In: Groves RH, Burdon JJ (eds) *Ecology of biological invasions*. CUP, Cambridge UK

Hanski I, Woiwod I, Perry J (1993) Density dependence, population persistence, and largely futile arguments. *Oecologia* **95**:595–598

Harper JL (1977) *Population biology of plants*. Academic Press, London

Harper JL (1981) The concept of population in modular organisms. In: May RM (ed) *Theoretical ecology*. 2nd ed. Blackwells, Oxford

Harris P (1984) *Carduus nutans* L., nodding thistle and *C. acanthoides* L., plumeless thistle (Compositae). In: Kellener JS, Hulme MA (eds) *Biological control programmes against insects and weeds in Canada* 1969–1980. CAB International, London

Hassell MP (1975) Density-dependence in single-species populations. *J Anim Ecol* **44**:283–295

Hassell MP (1981) Arthropod predator-prey systems. In: May RM (ed) *Theoretical ecology. 2nd ed.* Blackwells, Oxford

Hassell MP (1986) Detecting density dependence. *Trends Ecol & Evol* **1**:90–93

Hassell MP, Lawton JH, May RM (1976) Patterns of dynamical behaviour in single-species populations. *J Anim Ecol* **45**:471–486

Holyoak M, Lawton JH (1993) Comments arising from a paper by Wolda and Dennis: using and interpreting the results of tests for density-dependence. *Oecologia* **95**:592–594

Huffaker CB (1964) Fundamentals of biological weed control. In: DeBach P, Schlinger EI (eds) *Biological control of insect pests and weeds*. Chapman and Hall, London

Julien MH (1992) *Biological control of weeds: a world catalogue of agents and their target weeds*. 3rd edn. CAB International and ACIAR, Wallingford, UK

Julien M, Storrs M (1994) *Control of Salvinia molesta in Kakadu National Park*. Unpubl. report to the Australian Nature Conservation Agency. CSIRO Division of Entomology, Canberra

Krebs CJ, Myers JH (1974) Population cycles in small mammals. *Adv Ecol Res* **8**:267–399

Lomnicki A (1988) *Population ecology of individuals*. Princeton University Press, Princeton, USA

Lonsdale WM (1988) Litterfall in an Australian population of *Mimosa pigra*, an invasive tropical shrub. *J Trop Ecol* **4**:381–392

Lonsdale WM (1990) The self-thinning rule: dead or alive? *Ecology* **71**:1373–1388

Lonsdale WM (1993) Rates of spread of an invading species: *Mimosa pigra* in northern Australia. *J Appl Ecol* **81**:513–521

Lonsdale WM, Harley KLS, Gillett JD (1988) Seed bank dynamics in *Mimosa pigra*, an invasive tropical shrub. *J Appl Ecol* **25**:963–976

Lonsdale WM, Farrell GS, Wilson CG (1995) Biological control of a tropical weed: a population model and experiment for *Sida acuta*. *J Appl Ecol* **32**:391–399

May RM (1974) Biological populations with non-overlapping generations:stable points, stable cycles and chaos. *Science* **186**:645–647
May RM (1981a) Models for single populations. In: May RM (ed) *Theoretical ecology.* 2nd ed. Blackwells, Oxford
May RM (1981b) Models for two interacting populations. In: May RM (ed) *Theoretical ecology.* 2nd ed. Blackwell Scientific Publications, Oxford
McEvoy PB, Rudd NT, Cox CS, Huso M (1993) Disturbance, competition, and herbivory effects on ragwort *Senecio jacobaea* populations. *Ecol Monogr* **63**:55–75
Monro J (1967) The exploitation and conservation of resources by populations of insects. *J Anim Ecol* **36**:531–547
Moran PAP (1952) The statistical analysis of gamebird records. *J Anim Ecol* **21**:154–158
Moran PAP (1953) The statistical analysis of the Canadian lynx cycle. *Aust J Zool* **1**:163–173
Murdoch WW, Chesson J, Chesson PL (1985) Biological control in theory and practice. *Am Nat* **125**:344–366
Nicholson AJ (1933) The balance of animal populations. *J Anim Ecol* **2**:131–178
Nicholson AJ (1954) An outline of the dynamics of animal populations. *Aust J Zool* **2**:9–65
Nicholson AJ (1957) The self-adjustment of populations to change. *Cold Spring Harbor Symp Quant Biol* **22**:153–172
Nicholson AJ, Bailey VA (1935) The balance of animal populations. *Proc Zool Soc Lond* **3**:551–598
Poole RW (1978) *An introduction to quantitative ecology.* McGraw Hill, New York
Readshaw JL, Cuff WR (1980) A model of Nicholson's blowfly cycles and its relevance to predation theory. *J Anim Ecol* **49**:1005–1010
Rees M, Crawley MJ (1989) Growth, reproduction and population dynamics. *Funct Ecol* **3**:645–653
Rees M, Crawley MJ (1991) Do plant populations cycle? *Funct Ecol* **5**:580–582
Rees M, Lawton JH (1993) What can models tell us? In: Fowden L, Mansfield T, Stoddart J (eds) *Plant adaptation to environmental stress.* Chapman and Hall, London
Rogers AR (1986) Population dynamics under exploitation competition. *J Theor Biol* **119**:363–368
Room PM (1990) Ecology of a simple plant-herbivore system: biological control of *Salvinia. Trends Ecol & Evol* **5**:74–79
Room PM, Harley KLS, Forno IW, Sands DPA (1981) Successful biological control of the floating weed salvinia. *Nature* **294**:78–80
Room PM, Forno IW, Taylor MFJ (1984) Establishment in Australia of two insects for biological control of the floating weed *Salvinia molesta. Bull Entomol Res* **74**:505–516
Schaffer WM, Kot M (1986) Chaos in ecological systems: the coals that Newcastle forgot. *Trends Ecol & Evol* **1**:58–63
Silvertown JS (1991) Modularity, reproductive thresholds and plant population dynamics. *Funct Ecol* **5**:577–580
Smith FE (1961) Density dependence in the Australian thrips. *Ecology* **42**:403–407
Sprugel DG (1991) Disturbance, equilibrium, and environmental variability: What is 'natural' vegetation in a changing environment? *Biol Conserv* **58**:1–18
Strong DR (1986) Density-vague population change. *Trends Ecol & Evol* **1**:39–42
Symonides E, Silvertown J, Andreasen V (1986) Population cycles caused by overcompensating density-dependence in an annual plant. *Oecologia* **71**:156–158
Underwood AJ (1994) On beyond BACI: sampling designs that might reliably detect environmental disturbances. *Ecol Appl* **4**:3–15
Varley GC, Gradwell, GR, Hassell, MP (1973) *Insect population ecology.* Blackwells, Oxford.
Watkinson AR (1980) Density-dependence in single species populations of plants. *J Theor Biol* **83**:345–357
Watkinson AR (1985a) On the abundance of plants along an environmental gradient. *J Ecol* **73**:569–578
Watkinson AR (1985b) Plant responses to crowding. In: White J (ed) *Studies on plant demography: a Festschrift for John L. Harper.* Academic Press, London
Watkinson AR (1986) Plant population dynamics. In: Crawley MJ (ed) *Plant ecology.* Blackwells, Oxford

Watkinson AR, Lonsdale WM, Andrew MH (1989) Modelling the population dynamics of an annual plant: *Sorghum intrans* in the wet-dry tropics. *J Ecol* **77**:162–181

Weller DE (1987a) A reevaluation of the −3/2 power rule of plant self-thinning. *Ecol Monogr* **57**:23–43

Weller DE (1987b) Self-thinning exponent correlated with allometric measures of plant geometry. *Ecology* **68**:813–821

Wellington AB, Noble IR (1985) Post-fire recruitment and mortality in a population of the mallee *Eucalyptus incrassata* in semi-arid, south-eastern Australia. *J Ecol* **73**:645–656

White J (1980) Demographic factors in populations of plants. In: Solbrig OT (ed) *Demography and dynamics of plant populations*. Blackwells, Oxford

Wolda H, Dennis B (1993) Density-dependence tests, are they? *Oecologia* **95**:581–591

Wolda H, Dennis B, Taper ML (1994) Density dependence tests, and largely futile comments: answers to Holyoak and Lawton (1993) and Hanski, Woiwod and Perry (1993). *Oecologia* **98**:229–234

THE POPULATION ECOLOGY OF THE INVASIVE TROPICAL SHRUBS *CRYPTOSTEGIA GRANDIFLORA* AND *ZIZIPHUS MAURITIANA* IN RELATION TO FIRE

A.C. Grice and J.R. Brown

ABSTRACT

Several exotic shrubs are significant invaders of the tropical woodlands of northern Australia. They cause problems for pastoral industries and are likely to induce significant change in woodland communities. Fire, which has been an important factor in the evolution and ecology of the tropical woodlands, has potential as a control agent for these shrubs by altering population structure and rates of increase and spread. It affects several stages in the life cycles of two important species, *Cryptostegia grandiflora* and *Ziziphus mauritiana*. Fire can reduce the viability of recently dispersed seeds; the survival rates of seedlings, juvenile and established plants; and the seed output of established plants. While fire may kill a large proportion of *C. grandiflora*, even small *Z. mauritiana* resprout vigorously following a late dry season fire. These effects suggest that fire may play a role in the control and containment of at least some exotic shrubs of the tropical woodlands, in a way analogous to that advocated for native shrubs of the southern semi-arid woodlands.

Key words: invasive shrubs, tropical woodlands

Frontiers of Population Ecology, R.B. Floyd, A.W. Sheppard and P.J. De Barro (eds), CSIRO Publishing, Melbourne, 1996 pp. 589–597

Introduction

The vegetation of much of northern Australia is characterised by a grassy understorey beneath an open woodland that is commonly dominated by *Eucalyptus* spp. (Mott and Tothill 1984). Large areas lack a prominent shrub stratum. However, several exotic shrub species are invasive in these communities and can have major effects on the structure and composition of the vegetation. This has consequences for the pastoral industry, the prominent land user in the region, and for the conservation of tropical woodland ecosystems (Humphries *et al.* 1991). It is, therefore, important to understand the nature of these invasions and develop measures to counter them.

Fire has probably been significant in keeping the tropical woodlands open and grass-dominated. Changes in fire regimes in recent decades may have facilitated colonisation by exotic shrubs. We are using these ideas as a basis for testing the potential for using fire in control and containment strategies for two species, *Cryptostegia grandiflora* (Roxb.) R. Br. (Asclepiadaceae, rubber vine) and *Ziziphus mauritiana* Lam. (Rhamnaceae, chinee apple or Indian jujube).

C. grandiflora is native to Madagascar and was introduced to Australia in the late 1800s. It is now widespread in north eastern Australia where its range extends from the Gulf of Carpentaria and the southern half of Cape York Peninsula to Brisbane (Humphries *et al.* 1991). The species favours riparian habitats where it grows as a climber that can dominate the vegetation by smothering the canopies of even tall trees. Elsewhere, it commonly occurs as a free standing, often multi-stemmed shrub. *C. grandiflora* produces a typical anemochorous propagule.

Z. mauritiana is a spiny shrub or small tree that is native to southern Asia and eastern Africa. It was introduced to Australia sometime in the 1800s, being widely planted around early northern settlements. It now occurs as numerous, relatively isolated populations, especially in Queensland, but also in the Northern Territory and Western Australia (Anderson 1993) and often forms dense thickets. The fruit of *Z. mauritiana* is a drupe that, in north east Queensland is dispersed by wallabies (e.g. *Macropus agilis*), domestic cattle, feral pigs and perhaps some birds, though many remain beneath the canopy of the parent plant, especially where cattle are denied access. In north eastern Australia, both *Z. mauritiana* and *C. grandiflora* are at least partly deciduous in the dry season.

Fire could influence any of several stages in the life cycles of these shrubs and so alter population structures and rates of increase and spread. We present early results to test the hypotheses that fire can reduce the populations of recently dispersed seeds, the survival rates of juvenile and established plants, and their reproductive output.

Methods

Study site

The study site was located on Lansdown Research Station, 40km south of Townsville in north east Queensland, where 80% of the average annual rainfall (861mm) falls between December and March (Cook and Russell 1983). The vegetation is a highly modified woodland that includes *C. grandiflora* and *Z. mauritiana.* Cattle were excluded from May 1993 to facilitate the accumulation of herbaceous material as fuel for prescribed fire. The herbaceous vegetation of the site was dominated by the introduced *Urochloa mosambicensis* and *Stylosanthes* spp. and a mixture of native grasses (e.g. *Heteropogon contortus, Bothriochloa decipiens, Themeda triandra*). The site was divided into nine plots, each approximately 1.3 ha. Six randomly chosen plots were burnt on 31 August 1994 during the dry season. Fuel loads ranged from 3200–5200 kg/ha, fire speeds varied between 0.4 and 2.6 m/sec and fireline intensities were calculated to be 2200–18300 kW/m.

Fates of established plants

In April 1993, 60 *C. grandiflora* and 90 *Z. mauritiana* were tagged in each plot and their heights recorded. Plants were selected from three height classes [<100 cm (small), 100–199 cm (medium) and ≥200 cm (large)]. Heights were again recorded in April 1994. At the time of the fire, a large proportion of plants of both species had low foliage cover levels due to dry season deciduousness.

Responses to fire were assessed by revisiting tagged plants. Within one week of the fire, cover of live foliage on each shrub was recorded by rating plants on a scale of 0–100 (0 = no foliage, 100 = plant in full leaf). Thereafter, similar assessments were made monthly. Mean foliage covers for burnt and unburnt plants in each size class were calculated using data for plants with cover values greater than zero. In December 1994, each burnt plant was classified as to whether its regrowth was from basal shoots, from buds in the canopy or both.

Reproduction of C. grandiflora

The effect of fire on reproductive performance of *C. grandiflora* was assessed by recording reproductive status in October, November and December 1994. Plants were classed as reproductive if they bore flower buds, open flowers or fruit. Counts of mature pods per plant were made in December 1994. Neither burnt nor unburnt *Z. mauritiana* initiated reproductive activity between August and December 1994.

Effects on seed

At the time of the fire, mature fruits were present on both *C. grandiflora* and *Z. mauritiana*. Mature fruits of *Z. mauritiana* had also accumulated on the soil surface beneath parent plants and seeds were being released from the pods of *C. grandiflora*. The day after the fire, naturally fallen fruits of *Z. mauritiana* were collected from the soil surface in burnt (two locations) and unburnt plots, and individual pods of *C. grandiflora* were harvested from shrubs on burnt and unburnt plots. Because of small seed size and their scattered distribution it was not possible to collect recently released seed of *C. grandiflora* from the soil surface. Instead, prior to the fire, mature seed of *C. grandiflora* were harvested and placed on the soil surface at two locations in plots that would be burnt and a third location in a plot that would remain unburnt.

Viability of seeds was tested using the method of Grice and Brown (1994). Seeds of *C. grandiflora* were placed in Petri dishes in a dark growth cabinet at 25°C. Samples from each soil surface location (2 × 100 seeds per location) and from individual pods (2 × 50 seeds per pod) were tested separately. Cumulative germination after 10 days was taken as a measure of viability. Seeds of *Z. mauritiana* were sown in pots in a glasshouse following removal of the woody endocarp. For each location, twenty five seeds were sown in each of three pots. Cumulative germination after 17 days was taken as a measure of viability.

Results

Fates of established plants

By one week after the fire, all *C. grandiflora* on burnt plots were leafless. At the same time, 56% of *C. grandiflora* on unburnt plots were leafless. Defoliation of burnt *Z. mauritiana* was also severe with only one large plant retaining any live leaves one week after the fire. At the same time, 9.5% of *Z. mauritiana* on unburnt plots were leafless.

The proportions of plants bearing live foliage changed greatly over the six-month post-fire period and the pattern of change differed markedly between the two species. Only 4% of small and 16% of medium-sized burnt *C. grandiflora* bore any live leaves even six months after the fire, though the proportion of unburnt plants bearing live leaves was over 80% by November 1994 (Fig. 1). Greater

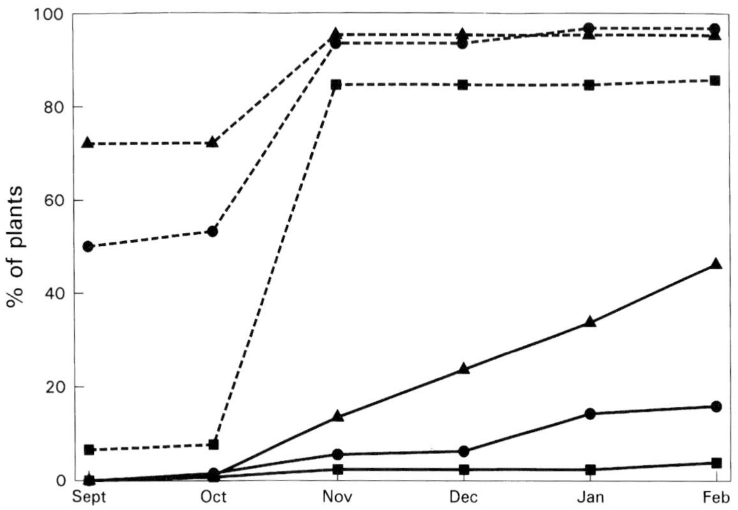

Fig. 1. Percentage of burnt (solid line) and unburnt (broken line), small (square), medium (circle) and large (triangle) *C. grandiflora* with live leaf at monthly intervals after fire.

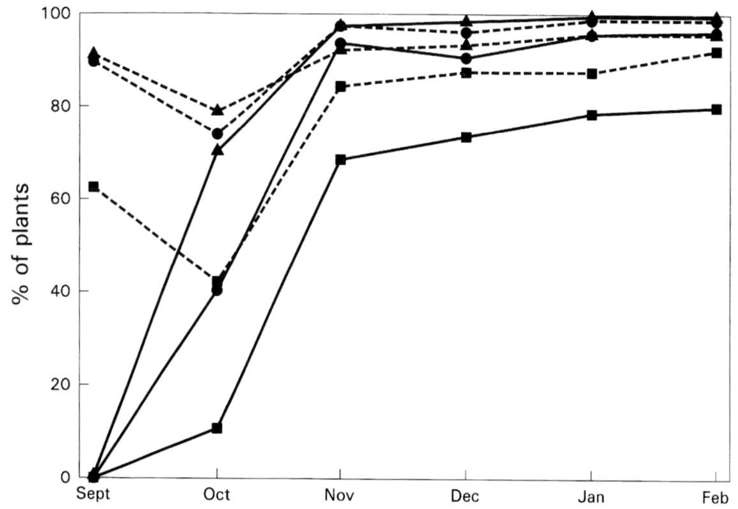

Fig. 2. Percentage of burnt (solid line) and unburnt (broken line), small (square), medium (circle) and large (triangle) *Z. mauritiana* with live leaf at monthly intervals after fire.

proportions of large burnt *C. grandiflora* showed post-fire recovery with 46% bearing live leaves by February 1995 (Fig. 1). For all three height classes, there were significant differences between burnt and unburnt plants bearing live foliage in February (χ^2 = 151.47, $p < 0.001$ for small plants, χ^2 = 111.25, $p < 0.001$ for medium plants; χ^2 = 29.92, $p < 0.001$ for large plants).

After the fire, the proportion of burnt *Z. mauritiana* bearing live leaves rapidly approached that of unburnt plants (Fig. 2). By October 1994, similar proportions of burnt (70%) and unburnt (79%) large *Z. mauritiana* bore live leaves (χ^2 = 2.29, $p > 0.05$). For medium-sized plants, differences were significant in October (χ^2 = 23.69, $p < 0.001$) but not in December (χ^2 = 2.26, $p > 0.05$) or February

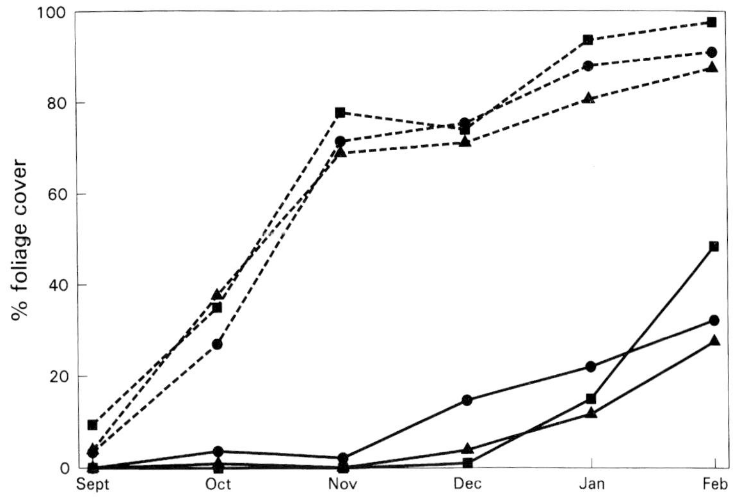

Fig. 3. Mean foliage cover of burnt (solid line) and unburnt (broken line), small (square), medium (circle) and large (triangle) *C. grandiflora* at monthly intervals after fire (based only on data for plants with cover > 0).

(χ^2 = 1.10, $p > 0.05$). For small plants, the proportion of plants bearing live foliage remained significantly greater for unburnt plants (χ^2 = 26.61, $p < 0.001$ in October; χ^2 = 4.81, $p < 0.05$ in February).

Mean foliage cover (based only on data for plants with cover greater than zero) increased with the proportion of plants exhibiting post-fire recovery but cover of burnt *C. grandiflora* remained below that of their unburnt counterparts (Fig. 3). Mean foliage cover of large burnt *C. grandiflora* were less than 15%, well below those on unburnt plots (80%) (Fig. 3).

Mean foliage cover on foliated unburnt *Z. mauritiana* of all height classes fell between the fire and November 1994. Thereafter, they remained steady or increased slightly. By contrast, cover of foliated plants that had been burnt increased, in some cases markedly, so that by December, mean foliage cover of burnt plants was similar to or greater than that of unburnt plants in each height class (Fig. 4).

The position of regrowth on burnt *Z. mauritiana* varied with the height class of the plant. Most regrowth on plants that were under 300 cm high was from basal shoots. Less than 1% of plants under 150 cm high produced new shoots from the canopy. 95% of plants over 300 cm high produced new shoots from buds in the canopy but many of these produced basal shoots as well (Fig. 5). 7% of burnt *C. grandiflora* produced basal shoots and 4% produced new shoots in the canopy by December 1994.

Reproduction of C. grandiflora

Many *C. grandiflora* had maturing pods at the time of the fire and many of these remained on the shrubs and released seeds after the fire but they could be distinguished from pods produced after the fire. No plants on burnt plots produced any new pods in September to December 1994 but unburnt plants that were over 1m high did produce pods in this period. 67% of unburnt large plants produced pods by December at an average of 6.7 (± 1.4 s.e.) pods per plant.

Effects on seed viability

Viability of seeds of *C. grandiflora* was reduced by the fire. Seeds recovered from pods burned whilst on the shrubs had germination percentages significantly lower than those of seeds removed from

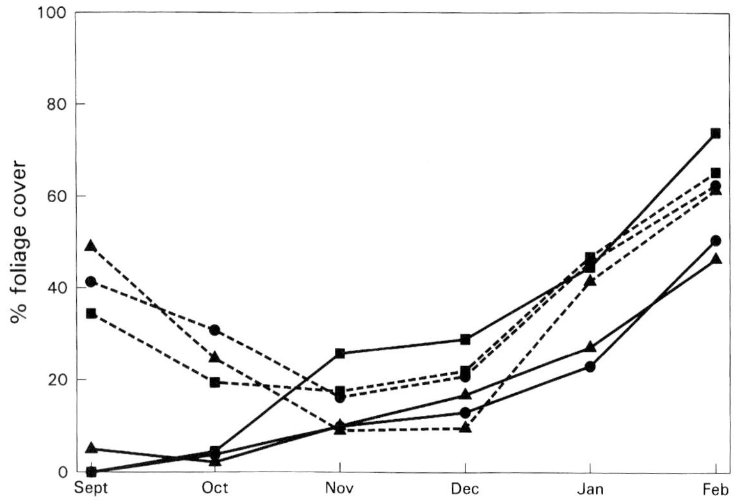

Fig. 4. Mean foliage cover of burnt (solid line) and unburnt (broken line), small (square), medium (circle) and large (triangle) *Z. mauritiana* at monthly intervals after fire (based only on data for plants with cover > 0).

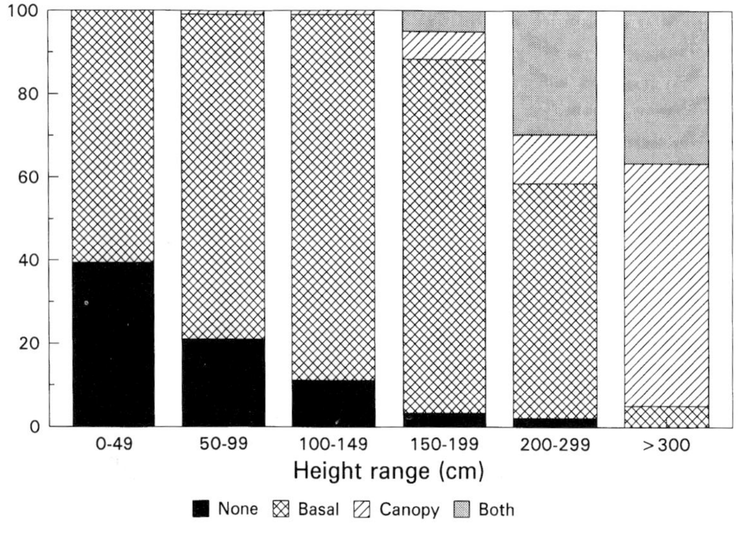

Fig. 5. Percentage of plants exhibiting no regrowth, basal regrowth, canopy regrowth and both basal and canopy regrowth in six height classes of burnt *C. grandiflora* in December 1994.

unburnt pods ($\chi^2 = 307.44$, $p < 0.001$) (Table 1). Seeds of *C. grandiflora* placed on the soil surface in burnt plots had germination percentages significantly lower than seeds placed on the soil in unburnt plots ($\chi^2 = 159.38$, $p < 0.001$). Germination of naturally fallen seed of *Z. mauritiana* collected from the soil surface was also significantly reduced by fire (burnt vs unburnt: $\chi^2 = 71.87$, $p < 0.001$) (Table 1). For seeds of both species collected from burnt plots, there was significant variation between samples from different locations.

Table 1. Mean percentage germination of seeds collected from burnt and unburnt plots one day after fire. (a) *C. grandiflora* seeds removed from pods; (b) naked seed of *C. grandiflora* placed on soil surface; (c) naturally fallen fruits of *Z. mauritiana*.

Species	Seed source	Germination (%)
(a) *C. grandiflora*	unburnt pods	100
	burnt pod 1	0
	burnt pod 2	19
	burnt pod 3	2
(b) *C. grandiflora*	unburnt seeds	97
	burnt, location 1	17
	burnt, location 2	35
(c) *Z. mauritiana*	unburnt plots	57
	burnt plot, location A	5
	burnt plot, location B	0

Discussion

These data show that fire can be an important factor in the population ecology of both *C. grandiflora* and *Z. mauritiana*, influencing the growth and survival of established plants, the reproductive output of plants that survive fire, and the viability of seeds both before and after dispersal.

Six months after fire, when a majority of unburnt plants of both species had produced their new season's foliage, most burnt *C. grandiflora* showed no signs of recovery. This effect was greatest in plants under 200 cm high where results suggest the possibility of up to 90% mortality, though leaflessness six months after fire does not necessarily indicate mortality. In contrast with *C. grandiflora*, most *Z. mauritiana* of all sizes had shown some degree of recovery within two months of the fire and the maximum mortality rate attributable to fire is only 20%, and that only for plants less than 50 cm high.

Differences in foliage cover between burnt and unburnt *C. grandiflora* indicate that fire can reduce the vigour of plants that survive the fire. New leaves are usually produced in the early part of the wet season but this was delayed by fire in the late dry season. Regrowth of *C. grandiflora* occurred mainly from basal shoots. *Z. mauritiana* again contrasts with this picture with the fire stimulating vigorous growth at a time when unburnt plants were still dormant. In smaller *Z. mauritiana*, regrowth was mostly from new shoots that arose from just above or just below ground level. In large plants, new shoots arose from buds in the canopy or epicormic buds on the larger stems.

Fire curtailed seed production in *C. grandiflora.* This species can flower and seed at least twice in a year. During this study, a reproductive episode occurred in unburnt plants soon after vegetative growth commenced in September. The average number of pods produced by unburnt plants over 200 cm high equates to about 1800 seeds per plant or more than 48 000 seeds per ha (A.C. Grice unpublished). No pod production occurred in burnt plants. Moreover, fire reduced the viability of seed that was on the soil surface or in pods borne on the shrubs during the fire. Fire also reduced the viability of seed of *Z. mauritiana* that was on the soil surface. Variation between samples collected from different locations are probably attributable to differences in fire intensity, residence time or other aspects of fire behaviour.

These effects make fire a potentially useful tool in the management of populations of these important invasive shrubs, particularly as they frequently occur where land use is extensive and the need is for inexpensive broadacre control options. However, while a single fire may significantly reduce a population of *C. grandiflora*, the control of *Z. mauritiana* would require multiple fires or integration of fire with other techniques.

C. grandiflora is wind dispersed and it is likely that seeds can spread considerable distances from parent plants. Plants establishing from these seeds could form populations of widely scattered individuals that are difficult to detect and expensive to control by conventional techniques but may be readily controlled by fire, particularly if burnt when they are small. A containment strategy for the species could utilise the impact of fire on the survival of small, relatively isolated plants and the reproductive output of parent populations. Fire could also be used to manage heavier infestations through its effects on reproductive output and survival of plants of a wide range of sizes.

Information available so far suggests fire is likely to be less valuable in the management of *Z. mauritiana*. Individual plants are capable of rapid and vigorous resprouting after fire. However, high defoliation can be obtained and this would make regrowth more accessible for control by conventional techniques in the manner advocated for native shrubs in Australian semi-arid woodlands (Noble *et al.* 1992).

The impacts of fire on the population ecology of both *C. grandiflora* and *Z. mauritiana* have analogues in the semi-arid woodlands of southern Australia. In those woodlands, a range of native woody species proliferated as fire regimes were altered with the advent of pastoralism. Prior to European settlement, periods of high rainfall led to both large scale germination of shrubs and accumulation of grass fuel. The fires that this made possible killed most shrub seedlings. Livestock grazing both removed grass fuel and reduced competition faced by shrub seedlings and so increased the probability of shrub establishment. Consequently, formerly open woodlands were transformed to dense shrublands (Hodgkinson and Harrington 1985).

The shrubs of the southern semi-arid woodlands differ from one another in their responses to fire in the same way as *C. grandiflora* and *Z. mauritiana*. Some species (e.g. *Eremophila mitchellii*, *E. sturtii*) resprout vigorously while others (e.g. *Acacia aneura*, *Callitris glaucophylla*, *Dodonaea* spp.) suffer high (80–95%) mortality (Hodgkinson and Harrington 1985). In spite of these differences, however, fire is recommended as a means of controlling shrubs overall because of a general susceptibility of seedlings. The same may apply to exotic shrub weeds of the tropical woodlands though *Z. mauritiana* only 30 days old are capable of resprouting from the base (J.R. Brown and A.C. Grice unpublished).

Further work on the role of fire in the population ecology of invasive tropical shrubs should examine the longer term prospects for recovery of shrubs that have been burned, the impacts of fires in different seasons and the effects of consecutive fires.

Acknowledgments Lindsay Whiteman contributed substantially to data collection and management. Bill Beyer, Brendan Ebner, Jeff Corfield, Janice Jackson, Richard Walton and Lindsay Whiteman assisted with fire management. Jim Noble provided useful comments on this manuscript.

References

Anderson E (1993) *Plants of central Queensland.* Queensland Department of Primary Industries, Brisbane

Cook SJ, Russell JS (1983) The climate of seven CSIRO field stations in northern Australia. *Aust CSIRO Trop Crops Pastures Tech Pap* 25:1–38

Grice AC, Brown JR (1994) Invasive shrubs in the tropical woodlands: seed and seedling ecology in the development of control strategies. *8th Biennial Conference, Australian Rangeland Society*, pp 253–254

Hodgkinson KC, Harrington GN (1985) The case for prescribed burning to control shrubs in eastern semi-arid woodlands. *Aust Rangel J* 7:64–74

Humphries SE, Groves RH, Mitchell DS (1991) *Plant invasions of Australian ecosystems: a status review and management directions*. Report to the Australian National Parks and Wildlife Service, Endangered Species Program and CSIRO, Kowari 2, ANPWS, Canberra

Mott JJ, Tothill JC (1984) Tropical and subtropical woodlands. In: Harrington GN, Wilson AD, Young MD (eds) *Management of Australia's rangelands*. CSIRO, Melbourne, pp 255–269

Noble JC, Grice AC, MacLeod ND, Müller WJ (1992) Integration of prescribed fire and sub-lethal chemical defoliation for controlling shrub populations in Australian semi-arid woodlands. *Proceedings of the First International Weed Control Conference*, Melbourne, pp 362–364

Some recent developments in the management of marine living resources

W.K. de la Mare

ABSTRACT

Some recent developments in the management of marine living resources are reviewed to examine the changing relationship between ecological knowledge, including uncertainty, and the formulation of scientific advice on catch limits and other management measures. Two case studies are used to illustrate these developments. The first is the case of whaling where the Scientific Committee of the International Whaling Commission has developed a revised management procedure. Computer simulation modelling was extensively used to develop a procedure which was robust to the many uncertainties which had previously undermined scientific assessments in whaling management. The principles underlying this development are discussed, along with the methodology, and an outline of the way in which the procedure works, despite the inability to resolve major biological uncertainties. The second case study is for krill in the Antarctic for which the Scientific Committee for the Conservation of Antarctic Marine Living Resources has calculated precautionary catch limits for the South Atlantic and Western Indian Ocean sectors of the Southern Ocean. This development also relies heavily on computer simulations. The implications of these approaches are discussed, particularly in the interaction between resource management, computer modelling and ecological knowledge.

Key words: simulation modelling, whales, krill, carrying capacity, maximum sustainable yield

Frontiers of Population Ecology, R.B. Floyd, A.W. Sheppard and P.J. De Barro (eds), CSIRO Publishing, Melbourne, 1996 pp. 599–616

Introduction

A scientific approach has been a central tenet of the management of marine living resources since the early years of this century. That management has frequently failed cannot be disputed. It is of course not accurate to lay the blame for the failings of fisheries management solely on the science. Gross failures of management occurred at the political and institutional level, and these have overshadowed the weaknesses in the scientific approach to fisheries management.

In more recent times the political and institutional failings have begun to be corrected. One consistent theme emerging from these corrections is the reaffirmation of scientific approaches to management as the common denominator on which institutions, politicians and managers can agree. Given this renewed reliance on science we should ask what sort of science is required for the task. Was science let off the hook by failures at the institutional level? What would have happened if managers had invariably followed scientific advice? Recent developments have begun to re-evaluate the science of living resource management to see what works, and to decide where we go from here.

Some recent negative comments have received wide publicity, which, while highlighting well known problems to many closely involved in living resource management, obviously came as news to the public at large (Ludwig *et al.* 1993; Holmes 1994). These comments have correctly underlined the lack of scientific consensus on interpretation of data and the development of scientific advice as a key factor in management failure. So another key question is how to develop a scientific consensus in formulating management advice, particularly in cases of continuing and intractable uncertainty.

The classical approach to living resource management

The classical scientific approach to living resource management was predictive and adaptive (Walters and Hilborn 1976). Predictive, in that studies of the natural history and demography of exploited animal populations, such as growth and mortality rates, were used to predict the rate of fishing mortality which would satisfy management objectives such as Maximum Sustainable Yield (*MSY*). Adaptive, in that the determining the response of the exploited population to fishing was attempted by examining data collected from the fishery. These data were used both to estimate the yield directly from simple production models (Fig. 1), which specify the yield as a function of population depletion, and to refine predictions based on natural history and demography. These are the approaches correctly criticised by Ludwig et al (1993).

Case studies in marine living resource management

Two case studies from the Antarctic illustrate emerging approaches to marine living resource management that avoid many of the problems of the classical approach. They cover pretty much the extremes of the range of living resource management, from whales, typifying populations with low yields and sluggish population dynamics, and krill, a relatively short-lived species with highly variable recruitment. The whale case is important for revealing why the classical approach is not sufficient.

Case study 1: The IWC and whales

Antarctic whaling is one of the most notorious examples of mismanagement of marine living resources. However, most of the damage was done before any serious attempt at management was undertaken. The International Whaling Commission (IWC) did attempt in the 1960's and 70's to put whale management on a sound scientific foundation. This culminated in what became known as the New Management Procedure (NMP), which was a pinnacle of the classical approach. The NMP (Fig. 1) was based on a typical surplus production model. Catch limits were to be set at 90% of *MSY* (that is, a 10% safety factor was allowed for), which, given the assumed production model,

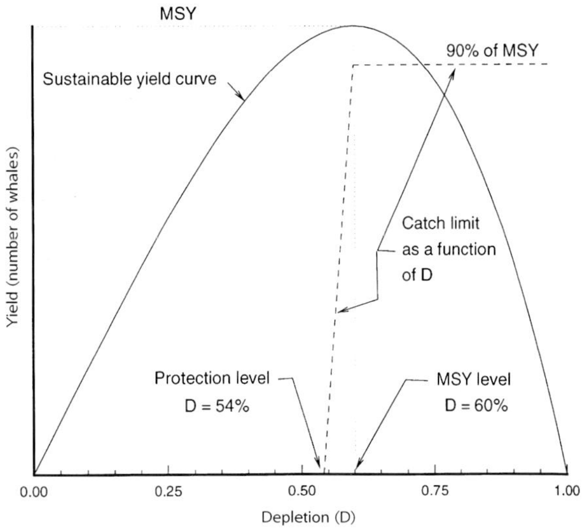

Fig. 1. The International Whaling Commission's old management procedure showing its integral relationship with a classic production model. The production model gives the sustainable yield as a function relative depletion *D*, where *D* is the ratio of the exploited population size to the carrying capacity, *K*. A catch limit for a stock was set at 90% of the estimate of *MSY* for the stock so long as it was estimated to be at or above the population level where the sustainable yield would be a maximum (known as *MSYL*, assumed to be at *D* = 0.6*K*) Stocks were protected if *D* was estimated to be less than 90% of *MSYL*. Catch limits were proportionately reduced for stocks where *D* was estimated to lie in the range *D* = 90% to 100% of *MSYL*.

would in theory would lead to an exploited population stabilising at ~74% of the unexploited abundance (carrying capacity, *K*). Stocks more than 10% below the equilibrium population level at which the sustainable yield is maximum (known as the *MSY* level, *MSYL*) were to be protected. Since *MSYL* was assumed to be at 60% of *K*, the protection level was 54% of *K*. However, when the NMP was negotiated, it was realised that there were some stocks for which there was no idea of how depleted they were. The way around this difficulty was to include a 'stable stock' provision which allowed catches to continue at the average level if there was no positive evidence of stock decline.

What went wrong with classical management

To apply the NMP required estimates of *MSY* and *K*. Estimating *K* in turn required the designation of a 'stock', that is, a self-contained population with negligible rates of immigration/emigration (Holden and Raitt 1974). The problem was that *MSY* and *K* could not be estimated reliably.

One approach to estimating *MSY* was to calculate the parameters of the production model by considering detailed demographic parameters such as pregnancy rates, ages at sexual maturity and the rate of natural mortality. The approach fails if any one of the required parameters cannot be estimated to the required degree of precision. Table 1 gives a brief summary of the types of information used in attempts to apply the NMP.

MSY is nearly directly proportional to the natural mortality rate *M*. Natural mortality can be estimated from age data collected during the early stages of exploitation, before fishing mortality affects the age structure. However, samples from the commercial catches are not random samples from a population, and therefore these give biased estimates (de la Mare 1985). The other demographic parameter estimates were also biased by the effects of non-random sampling, and some, such as juvenile survival rate were unobservable from the data. The net result was that *MSY* could not be estimated by this approach to within the 10% 'safety factor' built in to the NMP.

A second approach to estimating the parameters of the production model was based on the response of the populations to exploitation. This involved fitting a population model to time trends in Catch per Unit Effort data (CPUE), which was assumed to be an index proportional to population size.

Table 1. List of parameters needed to apply the IWC's 'New Management Procedure' (adopted in 1976)

Parameter	Comment
MSY level	No reliable estimates
MSY rate	Predicted from the following parameters
Age dependent natural mortality rates	and making assumptions on how they
Survival of neonates	change relative to population size:
Pregnancy rate	
Age at first parturition	
Population size	
Stock identity	No reliable method known
Stock size	
Stock depletion	

However, models of CPUE data later showed that it would not be proportional to abundance, in large part because the commercial whaling fleets did not, and could not, randomly sample CPUE (Beddington 1979; Cooke 1985). Moreover CPUE data were highly variable from year to year, and this tended to mask trends in the data (de la Mare 1984). These problems contributed to the failure of both the estimation of population depletion and *MSY* and also the implementation of the 'stable stock' provision of the NMP.

The Scientific Committee of the IWC became unable to give the Commission unambiguous advice on catch limits, and it was widely agreed that the NMP was flawed. At the time the NMP was designed, scientists were confident that they could reliably estimate the quantities set out in Table 1. However, as attempts were made to apply the procedure, and the methods of estimation were more closely scrutinised, it became obvious that this confidence was misplaced. The parameter estimation methods produced estimates with poor precision or with substantial bias, and often both. One of the NMP's major flaws was that it had no provision for taking uncertainty in the parameter estimates into account. Simulation studies of its properties showed that, even when applied with correct assumptions to fill the many gaps, it did not work well. It gave catch limits which fluctuated wildly from year to year, and led to a substantial probability that stocks would be over-exploited (de la Mare 1986).

These shortcomings are not unique to the IWC and the NMP, they are common to many fisheries management methods, including some of those which have been adopted by the Commission for the Conservation of Antarctic Marine Living Resources (CCAMLR).

Nonetheless, the NMP had one feature which paved the way for the next steps in living resource management. This was the incorporation of decision rules which specified what management action was to follow given certain assessments of the state of the stocks. That the NMP failed because these decision rules were incomplete should not detract from the fact that the incorporation of specific decision rules was a major advance.

How could it be it fixed

In 1993, the Scientific Committee recommended to the IWC a Revised Management Procedure (RMP). How was this developed, and how does it work? Does it fix the problems of the NMP?

The RMP was developed by going back to basics, considering a complete model of the management system, as shown in Fig. 2. Thus the development of the RMP required consideration of the

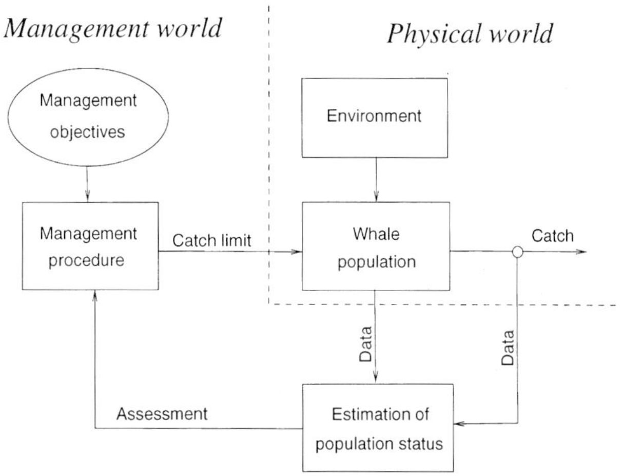

Fig. 2. Schematic overview of a living resource management system. A management procedure can ideally be regarded as a set of rules which designate catch limits or other management measures as both a function of a set of management objectives and assessments of the state of the exploited populations. This definition is chosen to emphasise that catch limits or other measures will change if the management objectives themselves are changed. The state of the exploited population depends on both the effects of exploitation, and variations arising from changes in the environment.

objectives the management procedure is meant to achieve, and the information necessary and sufficient to achieve the objectives.

Management Objectives

The management objectives were set by the Commission:

- Stable catch limits – desirable for the orderly development of the whaling industry
- Acceptable risk that a stock not be depleted (at a certain level of probability) below some chosen level (e.g. a fraction of its carrying capacity), so that the risk of extinction of the stock is not seriously increased by exploitation
- Make possible the highest continuing yield from the stock

Although these are rather vague, they were sufficient for the development process to begin.

Information available for management

No major breakthroughs were made in studying the demography of whales stocks since the collapse of the NMP. In fact, uncertainty over *MSY* and *MSYL* has increased (IWC 1994a). In contrast, over the 1980's considerable progress was made in estimating abundance using sightings surveys, based on line transect theory (Buckland *et al.* 1993). This methodology overcame the serious weakness of CPUE in that surveys were designed to estimate the density of whales throughout the area surveyed, not just places with commercial concentrations. Importantly, it gives absolute estimates, not an index. The other data which, in principle, could be collected reliably was the numbers of whales caught. The Committee decided to base the RMP only on those data in which it could rely, and these were the absolute abundance data from sightings surveys and the numbers of whales caught.

Table 2. Single stock simulation tests used in the development the RMP (based on IWC 1992). The approach was to choose parameter values which bounded the range, or changed in ways more extreme than would be regarded as likely. More than one ecological factor may be the proximate cause of the variation in the parameter used in setting up the simulation. For example, a change in carrying capacity (*K*) may arise through the effects of climate change or through inter-species competition, but the consequences of either can be captured in a single test. All trials were conducted for cases where abundance surveys carried out every 5, 7 and 10 years (unless otherwise stated).

Different population depletions at start of RMP: 5, 20, 30, 40, 60 and 99% of *K*

Different *MSY* rates (MSYR): 1, 2.5, 4 and 7% (sustainable yield as a percentage of the equilibrium population size at *MSYL*)

Bias in abundance estimates:
Surveys overestimate abundance by 50% (P_{est}/P_{true} = 1.5)
Surveys underestimate abundance by 50% (P_{est}/P_{true} = 0.5)
Linear trend in survey estimate bias, P_{est}/P_{true} increases from 0.5 to 1.0 over 99 years
Linear trend in survey estimate bias, P_{est}/P_{true} declines from 1.5 to 1.0 over 99 years

Historic catch pattern:
30 years of historic catch data (base case)
Catches followed by 25 years of protection before RMP applied
Historic catch data have been under-reported by 50% (subsequently tested with 10% and 0% of historic catch reported)

Variations in population model:
Age structured population model, age at sexual maturity = age at recruitment = 7 years (base case) and age at sexual maturity = age at recruitment = 10 years
Time lag of 25 years in density dependent response
Different values of *MSYL* = 40%, 60% and 80% of *K*
Linear trend in carrying capacity (*K*) - doubling or halving over the 100 year management period
Sinusoidal trend in *K* - minimum at years 0 and 100, maximum at year 50
Sinusoidal trend in *K* - maximum at years 0 and 100, minimum at year 50
MSYR changes with time, rectangular cycles with 66 year period, a) 1% - 4% - 1% and b) 4% - 1% - 4%
MSYR changes with time, rectangular cycles with 28 year period, a) 1% - 4% - 1% and b) 4% - 1% - 4%
Piecewise linear stock recruitment relationship with *MSYL* = 40% of *K* and *MSYL* = 80% of *K*
MSYR and *K* show a simultaneous linear decline by 50% over the 100 year period

Random parameters:
MSYR = *U*[0.1%, 5%]; depletion = *U*[1% to 99%] (of *K*); catch history *U*[15,40](years)

Irregular episodic events (e.g. epidemics):
50% of the population die in any given year with probability 0.02.
As above, but with survey bias = 150%

Survey frequency:
Survey every 5 years (base case)
Survey every 2, 10 and 20 years
Only one survey at the first implementation of the RMP
Survey interval random between 1 and 9 years
Surveys cease after 20 years

Random error in surveys:
Combined Poisson and lognormal distribution with coefficient of variation (*CV*) ~ 0.4
Lognormal distributions with *CV* = 0.2 and *CV* = 0.4
Gamma distribution with *CV* = 0.4

The Committee proceeded by means of computer simulations; the real world bounded by the dotted line in Fig. 2 was replaced by simulated whale populations. The data available for management were generated from appropriate statistical models. The aim was to develop a management procedure which was robust to known or suspected sources of uncertainty. Unlike the testing of the NMP, which was only carried out on real stocks, the developers were able to make their mistakes quickly, and in ways which did not matter.

A wide range of simulated whale populations were used, with different values of *MSY* and *MSYL*, different population dynamics and so on, so that the major factors about which the Committee were uncertain were all put to the test (see Table 2). The aim was not to develop a single model for the real world but a range of models which covered the range of uncertainty in our understanding of whale population dynamics and the factors which might affect them. Environmental trends were simulated, with whale population abundance either increasing or declining over time. In some models half the whales died from time to time in epidemics. Problems with the data were also simulated, for example historic catch data was under-reported and bias was added to the population abundance estimates. A fractional factorial 'experiment' was carried out to determine whether there were important interactions between the various factors (there were none; the different factors were basically additive in their effects on the performance criteria).

The development process became a kind of competition, with different groups of developers devising procedures and comparing results on a common set of performance criteria (see Table 3). The competition was beneficial, leading to a convergent evolution of procedures so that they all gave rather similar results, even though several quite different approaches to management procedures were included. The final versions all worked very much better than the first attempts, and the diversity of approaches gives us some assurance that the results obtained are probably close to being as good as they could be. Interestingly, no procedure developed on purely theoretical grounds met the objectives as well as those developed heuristically. The total development time was about six years.

In the end a procedure developed by Dr. Justin Cooke was adopted as the basic method for calculating catch limits (a full specification is given in IWC 1994b). The procedure uses a very simple population model.

$$P_{t+1} = P_t - C_t + 1.4184\mu\left(1 - \left(\frac{P_t}{K}\right)^2\right)$$

where P_t is the population in year t, C_t is catch in year t, μ is a productivity parameter, and K is the carrying capacity. There are no demographic parameters such as natural mortality, pregnancy rate, ages at sexual maturity, in fact virtually no biology at all. Catch limits are calculated from a simple control law (Fig. 3).

Table 3. Performance criteria used in evaluating the IWC's revised management procedure (based on IWC 1992). These statistics were calculated from either 400 or 100 replications of the management procedures for each of the trials set out in Table 2.

Quantity	Statistics
Total catch over 100 years	Mean, Median, 5th and 95th percentiles
Final population size (in year 100)*	Median, 5th and 95th percentiles
Estimate of continuing yield at 100 yrs	Median, 5th and 95th percentiles
Lowest population point at any time*	Median, 5th and 95th percentiles
Variability in catches	Annual average variation in catch limit

*Rescaled to *K* in cases where *K* varies over time

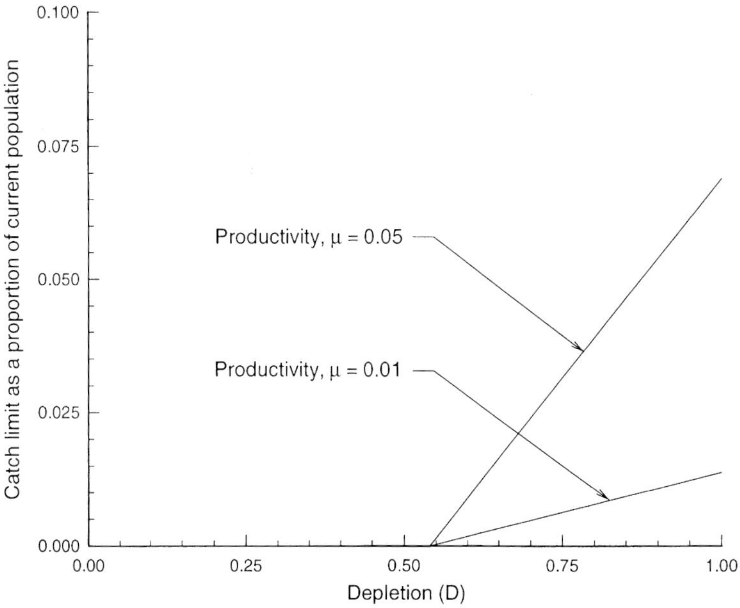

Fig. 3. In the IWC's revised management procedure, the nominal catch limits which contribute to the 'posterior' distribution of catch limits are calculated from a simple control law. The nominal catch limits are a function of both the relative population size *D* (as a fraction of *K*), and the value of the 'productivity' parameter μ, which is not a single fixed parameter but which has a 'prior' probability distribution corresponding to *MSY* rates in the range 0 to 5% of the exploitable population size. *D* also has a statistical distribution which depends, among other things, on the estimates of stock abundance from surveys and the record of past catches.

The procedure differs from the NMP in a fundamental way because uncertainty about the population dynamics of whales is taken into account by calculating catch limits in a probabilistic way, using a modified form of Bayes' theorem. Before applying the model to the data it is assumed that: *MSY* rate can be anywhere between 0% and 5%, depletion (P_t / K) can to be anywhere between 0% and 100%, and the population estimates can be biased anywhere between 0% and 167% of the true value.

In the first instance, instead of a single catch limit, a probability distribution of catch limits is calculated which takes into account the prior uncertainty in productivity, depletion and bias, and the additional information about these parameters provided by the available absolute abundance estimates. A large number of population trajectories from the model are calculated systematically over the assumed ranges for the productivity, depletion and bias parameters (see Fig. 4). The catch limit from each trajectory is assigned a weight based on the prior probabilities and the likelihood of the trajectory given the abundance data. The actual catch limit is 41st percentile of this posterior probability distribution (see Fig. 5), which was found to give a reasonable compromise between the three components of the management objectives. A brief example is given in Table 4 of the performance of the procedure in relation to conservation of a stock. The procedure has the property that catch limits are dependent on the precision of the abundance estimates. Imprecise abundance estimates result in lower catch limits than those obtained with more precise estimates.

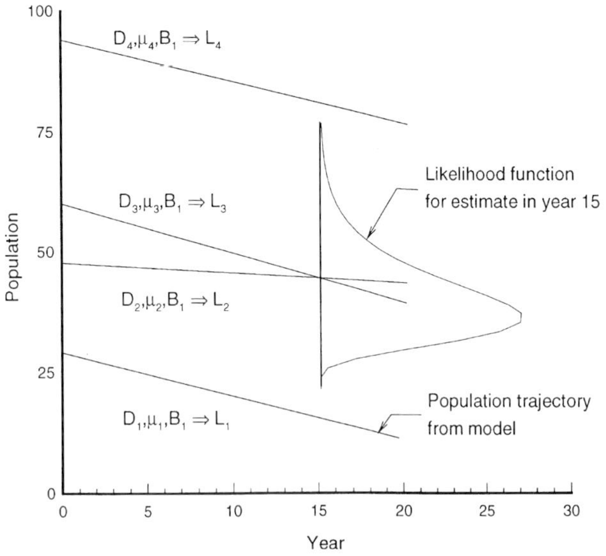

Fig. 4. Catch limits (L) in the RMP are calculated in a probabalistic way which takes into account 'prior' probabilities for the productivity parameter μ, the depletion D and the bias in abundance estimates B. The statistical likelihood of the population abundance is combined with the prior probability using a modified form of Bayes' theorem. Population trajectories are systematically calculated from the simple population model with the recorded catches (see text) to give a large range of population trajectories, each with an associated prior probability derived from the prior probabilities for the parameters μ, B and D. Four such trajectories are shown on the figure. Each trajectory leads to a catch limit from the application of the control law (Fig. 2), and that catch limit has the same prior probability as its associated trajectory. The 'posterior' probability for each catch limit L is proportional to the product of the prior probability of L multiplied by the likelihood of the associated population trajectory in relation to the population abundance estimates. However, the abundance estimate likelihood is down-weighted by a factor of 16, to reduce the influence of sample information over prior information. In the examples shown, L_1 and L_4 would have low posterior probability densities because they have model trajectories which have low likelihoods according to the abundance estimate. L_2 and L_3 have the same likelihood from the abundance estimate, and so these two catch limits would have similar posterior probability densities.

The stock identity problem

The catch limit procedure only solved half the management problem. It was confirmed early on that the most critical problem for the success or failure of the procedure was to do with stock identity. Like most management bodies, the IWC set catch limits by geographic areas, under the assumption that each area contained a unit stock (i.e. negligible mixing between areas). However, no reliable methodology has yet been developed for drawing up such areas to ensure that that assumption could be met in practice. Whales are highly migratory, and there is evidence that different components of populations migrate to different places at different times, and migration patterns can vary from year to year. The following schematic examples illustrate how failures to identify adequate stock boundaries can lead to failure of a management procedure.

Stock identity problem 1: coastal whaling; designating a too large management area

Figure 6 sets out one example of how failing to correctly identify stocks can lead to over-exploitation. The management area contains more than one stock, a catch limit is set appropriate for the total whale abundance in the management area, but the catches all come from a local stock inhabiting coastal waters. Thus the coastal stock will be over-exploited and collapse, even though the catch would be sustainable if it were spread over the three stocks.

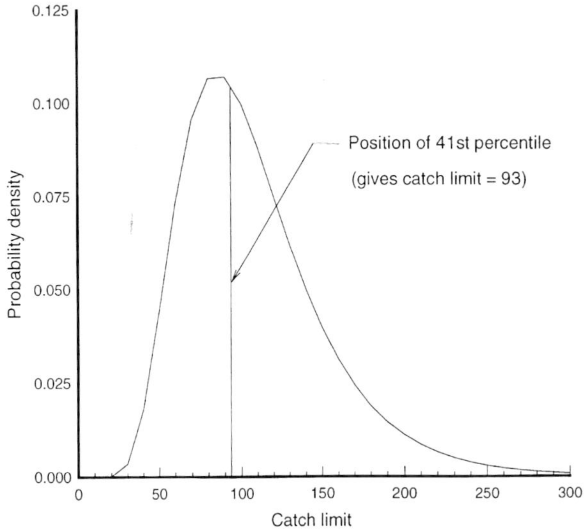

Fig. 5. Application of the catch-limit algorithm (Fig. 4) leads to a 'posterior' probability distribution for catch limits. The actual catch limit is taken as the 41st percentile of this distribution. This was determined from the result of the simulation trials as giving a reasonable compromise between the IWC's objectives for maximising the yield, minimising the risk to the populations and maintaining stability in catch limits.

Table 4. Some performance statistics for the RMP related to conservation. The procedure is applied for 100 years with abundance estimates every five years with a *CV* of 0.4. The initial state of the population is either previously unexploited, or depleted to either 60% of *K* or 30% of *K*. The conservation objective is met because their is a low probability of a population which is initially above 0.5 becoming reduced to below this level, and populations which start below this level there is a low probability that they would be depleted any further. In other words the recovery of depleted stocks has a very low probability of being inhibited by the setting of substantial catch limits.

Case 1: True MSY = 1%			
Depletion at time of first application	100%	60%	30%
5 percentile of lowest population size	62.0%	57.9%	30.0%
Case 2: True MSY = 1% and abundance, estimates biased up by 1.5			
Depletion at time of first application	100%	60%	30%
5 percentile of lowest population size	48.1%	46.6%	30.0%

Stock identity problem 2: pelagic whaling, fleets take catches just inside boundaries to minimise distance steamed

Figure 7 shows an example where catch limits are set appropriately for three stocks but for operational reasons (to minimise the distance steamed) the whaling fleet concentrates on a smaller area than that occupied by all three stocks, and the three catch limits are inadvertently taken from a single stock. This operational pattern of catches being taken near the management area boundaries was not unusual in the Antarctic (e.g. Kasamatsu and Shimadzu 1986).

Stock identity problem 3: variability in whale distribution

In some areas, there have been a number of repeat surveys, which have shown that the total variability in abundance estimates is considerably greater than the sampling variability from the individual surveys. This constitutes evidence that whales distribute themselves differently from year

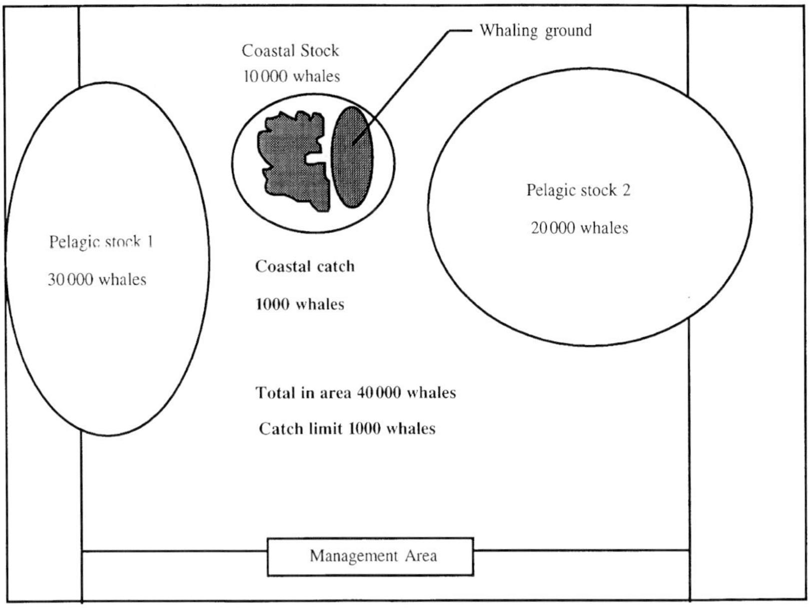

Fig. 6. Failure to correctly designate management areas can lead to failure to achieve management objectives. In this case, whaling is confined to a small coastal whaling ground, but the whales in the large management area, which includes the coastal ground, have been assumed to comprise a single large stock. The total number in the management area has been used in calculating the catch limits. If, as is shown here, there is an isolated coastal stock, the catch rate of 10% from the coastal stock will lead to its over-exploitation.

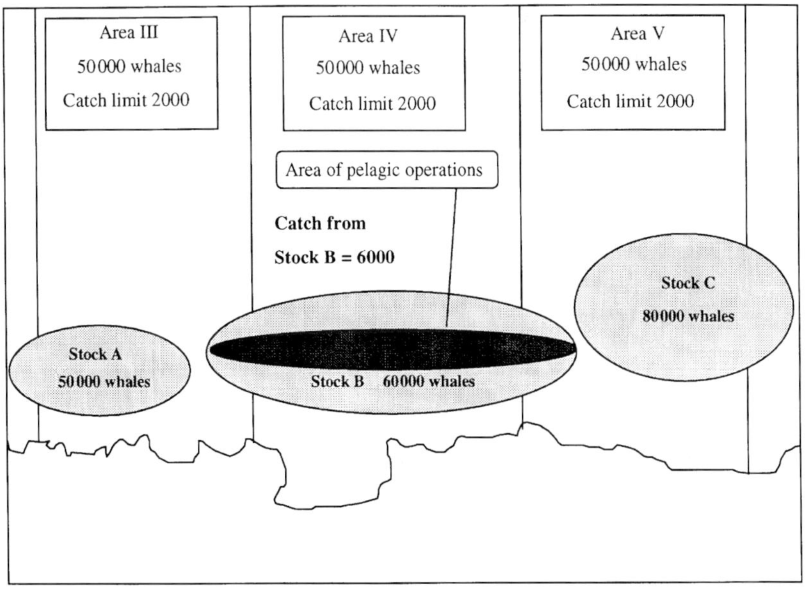

Fig. 7. Another way in which failure to correctly designate management areas can lead to management failure is shown. In this case of a pelagic whaling operation takes catches in three management areas, but for operational efficiency (in order to minimise the distance travelled by the fleet) only operates in a small part of two of them. This may result in all three areas catch limits in fact being taken out of only one stock.

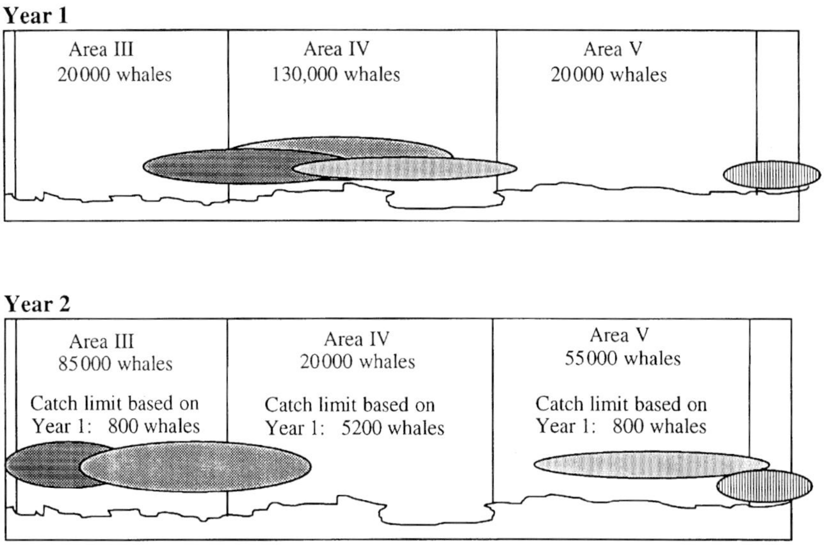

Fig. 8. Another form of management problem was found when whale behaviour leads to variability in the distribution of whales from a range of stocks from year to year, particularly when these overlap in some years and not others. In the case shown, in the year of the survey, there was a substantial overlap of three stocks, and accordingly, a high abundance estimate would be obtained. The corresponding high catch limit would be too large for the residual stock in subsequent years. The variability in distribution contributes to uncertainty in abundance estimates. The RMP sets lower catch limits when the uncertainty in abundance estimation is high, which helps to solve this particular stock-identity problem.

to year, possibly depending on factors such as spatial variability in food abundance. If this causes stocks to overlap in some years but not others, catch limits set by area may be too high on average for the stock normally found there (see Fig. 8).

Stock identity solution

The general solution to all the stock identity problems so far identified is to break management areas into sub-areas, and apply a separate catch limit to each. This ensures that catching is more evenly distributed, and that local catches are related to local abundance. This tends to be somewhat conservative, but areas can be re-defined as more is learned about population identity and distribution. Figure 9 shows the consequences of this procedure in the pelagic whaling case shown in Fig. 7.

The role of ecological knowledge

At first sight there is very little ecological knowledge incorporated in the IWC's RMP. However, this is not a correct impression. While the RMP itself does not use biological parameters explicitly, the process whereby it was developed incorporated both all that we knew about the relevant biological systems, plus all the insights we had on uncertain features about whale biology and behaviour.

In general, a management procedure for marine living resources need not be explicitly based on estimates of biological parameters. However, we need to ensure that our full extent of biological knowledge and key areas of our ignorance are taken into account in developing procedures. In evaluating the performance of prospective procedures we have the opportunity to determine which ecological features are critical to the success of management. If further research is needed, we will

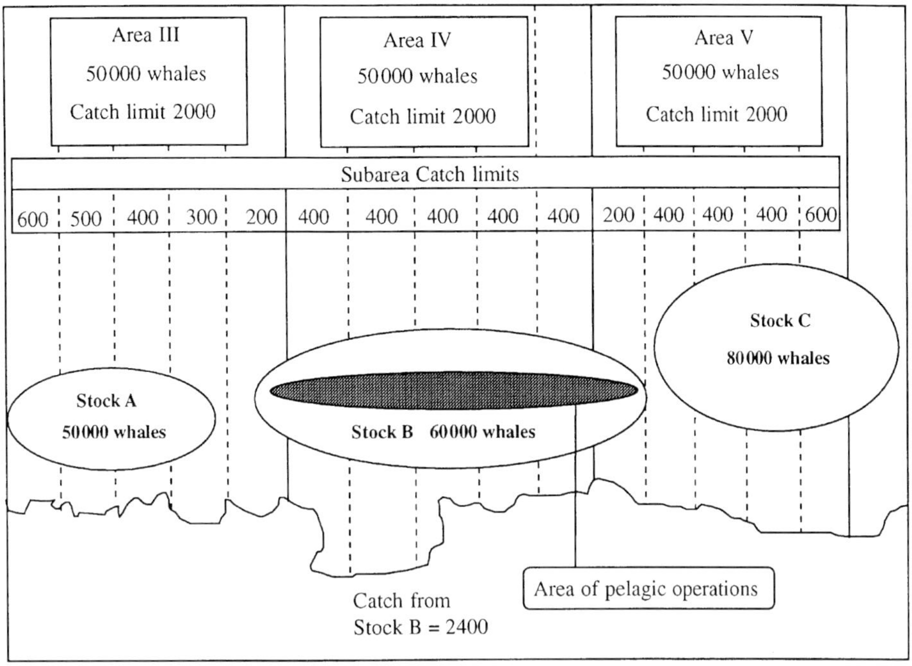

Fig. 9. All classes of the stock-identity problem are minimised if the catch limits are set for small subareas. The figure shows how this applies to the pelagic stock identity problem shown in Fig. 7. Because each subarea limit is related to the local abundance in that subarea, the total catch taken out of the Area IV stock is now directly related to the size of that stock. The fleet has to operate over the whole of each management area it is to take the whole of the catch limit which applies to it.

know where it should be directed. If it is impractical to dispel a critical uncertainty, we need to modify the management procedure until it is no longer critical. If new knowledge becomes available, the management procedure is re-tested to see if it requires adjustment. Such knowledge may be not only enhanced ecological knowledge, but would also include any declines of certainty about our earlier knowledge, or the revelation of new sources of uncertainty.

The RMP also specifies the steps to follow when new ecological knowledge or uncertainties are to be taken into account. For example, should we some day determine that there is a negligible probability for any of the baleen whales that *MSY* rate is less than say 2%, instead of 1%, then the RMP stipulates that the procedure is adjusted by carrying out new simulation trials which incorporate the revised ecological knowledge. The actual alterations to the procedure could occur in either adjustments to the prior distribution on productivity, or in the percentile used in determining the catch limit from the 'posterior' catch limit distribution, or both simultaneously. However, the fact that the currently used prior probability on *MSY* of 0% to 5% roughly corresponds to our current beliefs about the likely yields from whale stocks should not be over-interpreted as being the mechanism whereby the procedure would be adjusted if a better prior probability distribution for *MSY* rate becomes available. Of course this is an option, but not the only one. It is in reformulating the trials that the new information will be taken into account, and the adjustments to the procedure will be determined by the performance on trials, not by ad hoc adjustments to the elemental components of procedure, no matter how reasonable they may appear.

Case Study 2: CCAMLR and krill

CCAMLR is often held up as an example of the future of fisheries management because it embodies an ecosystem management concept. Article II of the CCAMLR convention requires not only that fisheries be managed so as to ensure the conservation of target species, but also in a way such that fisheries on prey species allows for the conservation of their predators. For example, the management of fisheries for krill should take into account the food requirements of predators such as penguins, seals and whales. This led to a great deal of research activity on a wide range of species being justified on the grounds that it would be useful to CCAMLR. But can such claims be sustained when there has yet to be any systematic analysis of what in principle is required to meet such objectives, nor any expression of those objectives in terms which admit a scientific interpretation?

The concept of evaluating management procedures by simulation as a means for developing CCAMLR's management approach was put before the Commission in 1986 (de la Mare 1987). For a few years a special Working Group met to develop approaches to conservation. Unfortunately, it became clear that there was no consensus on the idea that management procedures should be developed in a coherent way using prospective evaluation. CCAMLR has tended to follow an ad hoc approach, picking elements of a management policy, usually on the advice of the Scientific Committee, to solve specific problems, usually after they have occurred. Since 1990, the Scientific Committee for the Conservation of Antarctic Marine Living Resources (SC-CAMLR) has begun to move more systematically in developing management measures for particular fisheries, and computer simulation is beginning to be used at least as a tool for management under uncertainty.

The area where computer simulation methods have played a major part is in the development of precautionary catch limits for krill. These are calculated probabilistically using Monte Carlo integration. A simple population model of a krill population, which includes random variability in recruitment, is run hundreds of times with values for growth, mortality and abundance drawn at random from suitable statistical distributions. This approach allows us incorporate both natural variability in the krill population and uncertainty in the parameter estimates. The method follows the basic approach developed by Beddington and Cooke (1983) and was applied to krill by Butterworth *et al.* (1992).

The simulation model is used to calculate a distribution of population sizes both in the absence of fishing and at various fishing mortalities. These distributions are used to determine the proportion (γ) of an estimate of the unexploited biomass (B_0; from a hydroacoustic survey) that can be caught each year. CCAMLR has developed the following three part decision rule for determining the value of γ:

i. choose γ_1, so that the probability of the spawning biomass dropping below 20% of its pre-exploitation median level over a 20-year harvesting period is 10%,

ii. choose γ_2, so that the median krill escapement in the spawning biomass over a 20-year period is 75% of the pre-exploitation median level and

iii. select the lower of γ_1 and γ_2 as the level of γ for calculation of krill yield.

Each of the many simulations starts with a biomass of krill drawn from a statistical distribution which reflects the properties of the biomass survey estimates. The biomass is divided into a number of age classes. In each simulation year the biomass is recalculated by adding an amount for annual growth and deducting an amount corresponding to natural mortality. The biomass of each year's recruits is added and the effects of a constant annual catch of γB_0 is deducted. Variability in the simulated population biomass in each year arises because the recruitment to the population in each year is drawn from a statistical distribution which reproduces the statistical properties of the estimates of proportional recruitment (obtained from the length compositions collected during krill

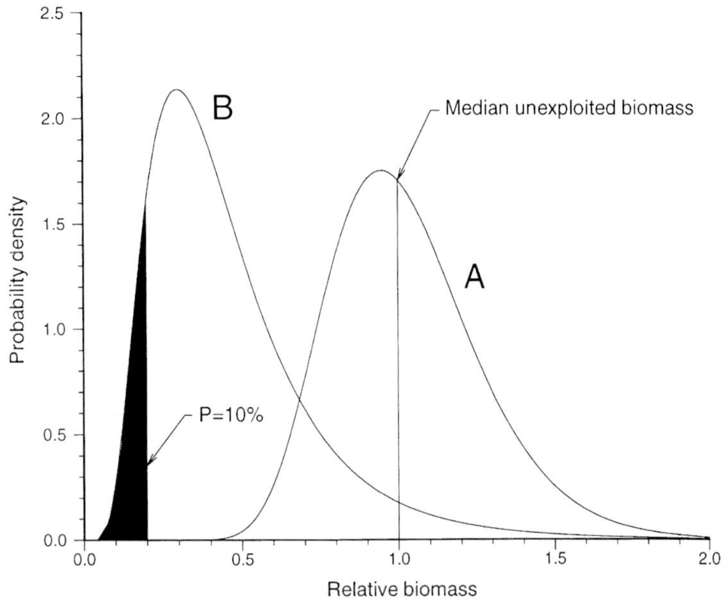

Fig. 10. Precautionary catch limits for krill are calculated using a population model to derive statistical distributions of the effects of fishing on krill abundance. These distributions take into account uncertainty about krill population dynamics by 'Monte Carlo' integration. The krill model is applied with demographic parameters drawn from statistical distributions which reflect our uncertainty about them. Further variability in the outcomes arises from the variable nature of recruitment typical in relatively short lived animals such as krill. The distribution marked *A* is the distribution of unexploited biomass from the model, which takes into account both the effects of variable recruitment, and the uncertainty in the biomass estimate. The distribution *B* is the statistical distribution of lowest population biomasses over twenty years of simulation. The 10th percentile of this distribution is used in one of the criteria for selecting an exploitation rate γ for setting krill precautionary catch limits.

estimates of proportional recruitment (obtained from the length compositions collected during krill surveys (see de la Mare 1994a, b).

A value for γ is selected as the value which give a statistical distribution of the outcomes of the many simulations which meet selected criteria. The model allows for uncertainty in estimates of key demographic parameters such as growth and mortality, by drawing values for each parameter from appropriate statistical distributions for each repetition of the model.

The model is run with $\gamma = 0$ (i.e. no catches) to produce the distribution of unexploited spawning stock biomass (shown in Fig. 10 as distribution A). This distribution determines the median unexploited spawning stock biomass. When γ is greater than zero, the simulated biomass is reduced by the effects of fishing.

The first criterion, or decision rule, requires the value of γ which leads to a 10% probability of the spawning biomass dropping below 20% of its pre-exploitation median level over a 20-year harvesting period. Applying this criterion requires the examination of the statistical distribution of the lowest population size (in terms of spawning biomass) in any year over the twenty years of each simulation, collected over hundreds of replicates. This distribution, for a given value of γ, is shown in Fig. 10 as distribution B. The probability of attaining a lowest spawning stock biomass less than

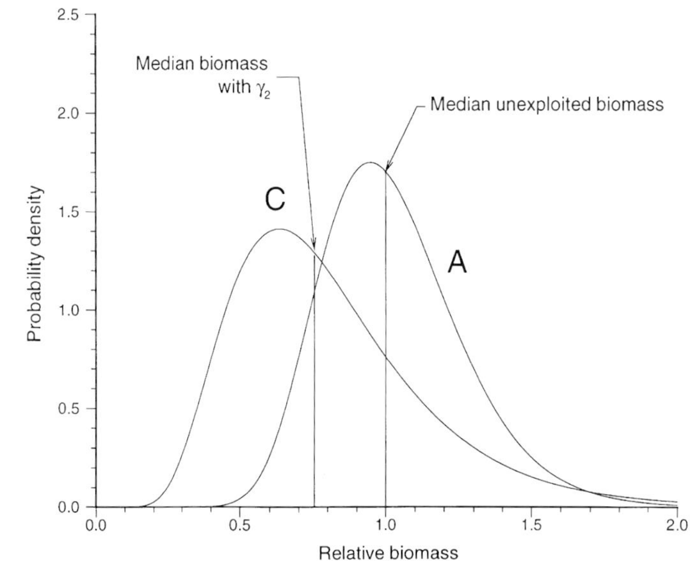

Fig. 11. A second criterion for setting krill precautionary catch limits is derived from the statistical distribution of krill abundance at the end of twenty years of exploitation. This is shown as distribution *C*. The exploitation rate is determined by that which results in the median of *C* being at 0.75 of the median of *A*, where *A* is the statistical distribution for the unexploited biomass.

20% of the pre-exploitation biomass is estimated from the relative frequency of this event over the set of replications, for a range of values of γ. The required value of γ is that which has this relative frequency at 10%.

This first part of the decision rule was aimed at meeting the requirement for stable recruitment in the krill stock by not allowing the spawning biomass to drop to very low levels below which the chance for successful recruitment may be impaired. This particular decision rule, however, is derived from a single-species approach. The second part of the decision rule was derived as a first attempt to give some explicit effect to the requirements under Article II to limit the effects on predators of fisheries for their prey.

The second part of the rule also leads to a value of γ, which is determined by the statistical distribution of the spawning stock biomass at the end of the twenty year period used in each simulation. The criteria embodied in this part of the rule are illustrated in Fig. 11. As before, *A* is the distribution of unexploited spawning stock biomass. *C* is the distribution of stock sizes after twenty years of exploitation corresponding to a given γ. The selected value of γ_2 is that which results in *C* having a median which is 75% of the median of *A*.

The values of γ_1 and γ_2 will usually be different, and so the third part of the decision rule chooses one of the two values. Whether γ_1 or γ_2 is the greater depends largely on the degree of variability in recruitment and the variance of the estimate of unexploited biomass B_0. γ_1 and γ_2 can be designated as the 'recruitment criterion' and the 'predator criterion' respectively. The lower of the two values is chosen because it means that the criterion corresponding to that part of the decision rule is just attained, and the criterion corresponding to the higher value of γ will be exceeded. Choosing the higher γ would automatically lead to a failure to fulfil one or other of the two criteria.

The levels used in the two criteria are somewhat arbitrary and they will need to be revised from time to time. The recruitment criterion will need to be revised to take into account any information which becomes available on the relationship between stock and recruitment. Revising the predator criterion depends on better information on the functional relationship between abundance of prey and recruitment in predator populations. The 75% level is chosen as the mid-point between taking no account of predators (i.e. treating krill as a single species fishery), and providing complete protection for predators (i.e. no krill fishery). CCAMLR has begun to develop models to explore the possible form of these functional relationships. However, it will take considerable time to acquire the information needed to provide advice on revised values for either the recruitment or the predator criterion levels. Nonetheless, CCAMLR has taken the first steps in expressing some of its objectives for managing krill fisheries in scientifically interpretable terms.

In this case some of the key uncertainties about the population dynamics have been taken into account in determining catch limits based on limiting the risks of adverse outcomes against two criteria. The missing step for the this analysis is that no prospective evaluation has yet been done to determine how well the analysis holds up in more complex cases which apply due to uncertain stock identity and the transport of krill between management areas due to ocean currents.

A paradigm for developing management procedures

The whale case study provides a good general model of how to develop a management procedure for marine living resources. Of course the details of the management procedures for fish or squids or krill or whales will all be different. Nonetheless, the same basic development paradigm is applicable. The most important element of this paradigm is to evaluate possible management procedures to see if they can work in principle, using simulation or other methods. To carry out such evaluations we need to specify the objectives for management in a scientifically interpretable way and consider the kinds of observations about the exploited population which can be reliably obtained. We need also to develop models for both the resource and the patterns of operation of the fisheries.

The utility of various types of information needs to be evaluated in the context in which it will be used. Although a method may seem to give disappointing accuracy or precision, this does not necessarily mean that it will fail to work in practice. Conversely, a method which gives precise estimates of some parameter may not be sufficient for achieving the management objectives. This can only be found out by some form or analysis or testing, and the best time to carry out testing is before trying it out on real fisheries.

The IWC experience has demonstrated that it is feasible to develop scientific consensus on ranges of parameters and the kinds of behaviours that species and ecosystems may exhibit, even in a body notorious for the depth and breadth of its disagreements. Evaluating possible management procedures does not require scientists to agree that any given effect may or may not happen, only that it cannot be ruled out. This is clearly a more tractable consensus than getting scientists to agree that an observed change in a fish stock is due to over-fishing, and that a given management action will rectify the problem.

By developing management procedures based on decision rules, decisions about management are made prospectively, that is, agreements about what to do when certain situations arise are made in advance. If the management procedure has been shown to be sound, there may even be a reasonable chance that pre-agreed decision rules will be followed. Obtaining a consensus to make painful adjustments only after the need for them has become apparent has proved to be a major problem for many fishery management bodies, and has resulted in many cases of fisheries collapse, or the maintenance of stocks at unproductive levels.

We now have powerful new scientific tools to use in developing methods for the management of marine living resources. It is essential that we use these tools if we are to have scientifically based management which will achieve ecologically sustainable utilisation and conservation of Marine Living Resources. CCAMLR, for example, has set itself ambitious goals in attempting an ecosystem approach to management. Attempting to reach such goals by trial and error on managing real fisheries will be both a trial and an error. Making our mistakes quickly by simulation can spare us from making them in reality.

References

Beddington JR (1979) On some problems of estimating population abundance from catch data. *Rep Int Whaling Comm* **29**:149–154

Beddington JR, Cooke JG (1983) The potential yield of fish stocks. *FAO Fish Biol Tech Pap* **242**:1–47

Buckland ST, Anderson DR, Burnham KP, Laake JL (1993) *Distance sampling: estimating abundance of biological populations*. Chapman and Hall, London, pp 446

Butterworth DS, Punt AE, Basson M (1992) *A simple approach for calculating the potential yield of krill from biomass survey results*. Selected Scientific Papers 1991. CCAMLR, Hobart, pp 191–206

Cooke JG (1985) On the relationship between catch per unit effort and whale abundance. *Rep Int Whaling Comm* **35**:511–520

de la Mare WK (1984) On the power of catch per unit effort series to detect declines in whale stocks. *Rep Int Whaling Comm* **34**:655–664

de la Mare WK (1985) On the estimation of mortality rates from whale age data, with particular reference to the minke whale (*Balaenoptera acutorostrata*) in the Southern hemisphere. *Rep Int Whaling Comm* **35**:239–250

de la Mare WK (1986) Simulation studies on management procedures. *Rep Int Whaling Comm* **36**:429–450

de la Mare WK (1987) *Some principles for fisheries regulation from an ecosystem perspective*. Selected Scientific Papers, 1986 CCAMLR, Hobart, pp 323–340

de la Mare WK (1994a) Estimating krill recruitment and its variability. *CCAMLR Science* **1**:55–70

de la Mare WK (1994b) Modelling krill recruitment. *CCAMLR Science* **1**:49–54

Holden MJ, Raitt DFS (1974) Manual of fisheries science. Part 2. Methods of resource investigation and their application. *FAO Fish Biol Tech Pap* **115**:214

Holmes R (1994) Biologists sort the lessons of fisheries collapse. *Science* **264**:1252–1253

International Whaling Commission (1992) Report of the fourth comprehensive assessment workshop on management procedures. *Rep. Int Whaling Commn* **42**:305–321

International Whaling Commission (1994a) Report of the Working Group on *MSY* rates. *Rep Int Whaling Comm* **44**:181–189

International Whaling Commission (1994b) The revised management (RMP) for baleen whales. *Rep Int Whaling Comm* **44**:145–152

Kasamatsu F, Shimadzu Y (1986) Operating pattern of Antarctic minke whaling by the Japanese expedition in the 1984/85 season. *Rep Int Whaling Comm* **36**:205–206

Ludwig D, Hilborn R, Walters C (1993) Uncertainty, resource exploitation, and conservation: lessons from history. *Science* **260**:17–36

Walters CJ, Hilborn R (1976) Adaptive control of fishing systems. *J Fish Res Board Can* **33**:145–159

Bifurcations, Structural Stability and Persistent Populations; New Insights into Classical Models with Harvesting and Stocking

Mary R. Myerscough

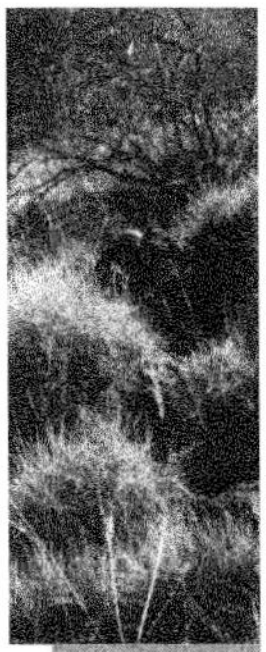

ABSTRACT

The construction and use of any mathematical or theoretical model in population biology requires assumptions about the populations which are being modelled. Changing the assumptions slightly or including extra assumptions may dramatically alter either the quantitative or qualitative outcomes of the model or both. Thus it is important to know whether a model is structurally stable; that is, how much its outcomes are likely to change with minor changes in formulation.

This question is explored for a group of classical predator-prey models of the MacArthur-Rosenzweig type. In particular it is shown how including harvesting or stocking of either predator or prey or both in the model can make the model behave in more complex ways than in the unharvested case. This behaviour can be hard to predict intuitively even in these very simple and well-known models. The effect of using different predator response functions in harvested predator-prey equations is also examined.

Key words: predator-prey models, MacArthur-Rosenzweig equations, oscillations, predator response function

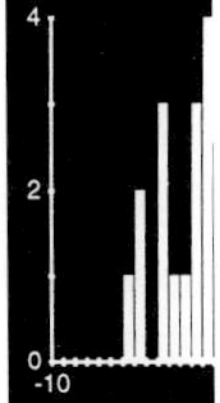

Frontiers of Population Ecology, R.B. Floyd, A.W. Sheppard and P.J. De Barro (eds), CSIRO Publishing, Melbourne, 1996 pp. 617–630

What is structural stability and why is it important?

Mathematical models in ecology are useful in at least two different ways; to give a general theoretical framework and to make predictions, often using data from a particular ecological system. (This statement of course ignores the modelling inherent in statistical analyses of data which is an important area but fundamentally different from the type of model considered here.) The logistic growth equation, for example, is a useful description of how resource-limited populations grow and has become an important idea in the discussion and measurement of the growth and reproduction of individual species. Species are often classified, for example, as r-strategists or K-strategists depending on their survival strategies when under stress. For a model to be helpful, either as a predictive tool or as theory, it must be reasonably robust and bear some relation to the system under study (although it is unlikely to encapsulate every feature observed in a real system).

The behaviour of a model may change when the statement of the model is altered. Such alterations may or may not involve changes in the underlying assumptions of the model. Changes in model behaviour can come about either through bifurcations when a parameter in the model is altered or through a change in the mathematical formula that underlies the model. For example, in the classical non-dimensional discrete logistic model,

$$N_{t+1} = r\,N_t\,(1 - N_t)$$

if r is between one and three the population settles to a steady state in N_t. Once r becomes greater than 3.0 and less than about 3.8 the solution becomes oscillatory. The period of the oscillation will depend on the exact value of r. When r is greater than about 3.8 the model population becomes chaotic. This qualitative change in the model outcome as the parameter r changes is an example of a bifurcation. The mathematical formulation of the model might be changed to give

$$N_{t+1} = r\,(N_t)^2\,(1 - N_t)$$

This mathematical formulation is slightly different from the logistic model. This model assumes that the growth of the modelled population is slower at low population densities than for the logistic growth model. Its behaviour, however, is similar to the logistic model with a steady state which bifurcates to oscillatory solutions and thence to chaos as r increases. The values of r where the bifurcations occur will differ but the behaviour will be qualitatively similar. Thus the model can be said to be structurally stable under such a change because there is no qualitative change in model behaviour.

A simple example of a structurally unstable model is the Lotka-Volterra predator-prey equations for two species. These are

$$\frac{dx}{dt} = r\,x - p\,x\,y$$

$$\frac{dy}{dt} = -s\,y + q\,x\,y$$

where x represents the prey population, y the predator population and r, s, p and q are parameters in the model. It is assumed that, in the absence of the predator, the prey population grows at a rate proportional to the number of prey, that the rate of predation is proportional to both predator and prey populations and that the predator dies at a rate proportional to predator population in the absence of prey. The model predicts that both prey and predator populations will oscillate at a fixed period with the amplitude of the oscillation depending on the initial populations. This is illustrated in Fig. 1. If these equations are changed slightly by making the prey population resource-limited, *viz*

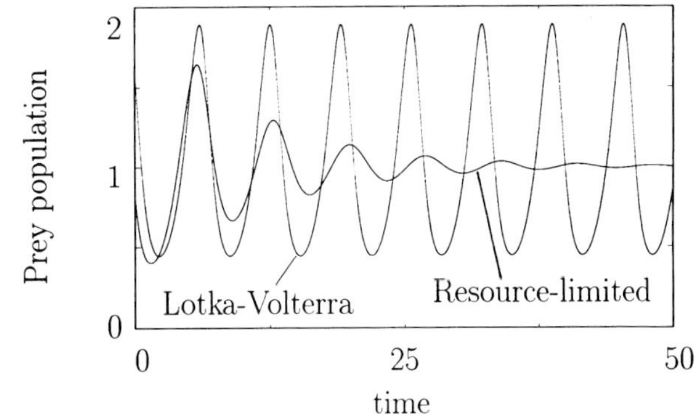

Fig. 1. Simulations of prey populations for Lotka-Volterra and resource-limited Lotka-Volterra models. The conventional Lotka-Volterra model predicts that the populations undergo persistent oscillations. The resource-limited model which requires only an extra term in the equations, predicts that the populations will tend to a steady state via damped oscillations. In this simulation the value of K in the resource-limited model was set to 4.

$$\frac{dx}{dt} = r\,x(1 - x/K) - p\,x\,y$$

$$\frac{dy}{dt} = -s\,y + q\,x\,y$$

then the populations will settle to a steady state via ever decreasing oscillations when K is sufficiently large (Fig. 1). Of course, if K is very large, the behaviour of the model populations will initially be very similar to the Lotka-Volterra model but eventually the two models will diverge as the resource-limited model tends to steady state equilibrium populations. Thus the Lotka-Volterra model is structurally unstable because a small change in its formulation qualitatively alters the model's outcome. It may be useful in making short-term predictions but its instability makes questionable its use as a theoretical framework.

Thus a model is structurally stable if a small change in its mathematical formulation leaves the behaviour of the model substantially unchanged. A model may, of course, be structurally stable to one change in its formulation but not to another. Structural instability differs from bifurcation which is a qualitative change which occurs as parameters in the model are altered.

Models continue to be used to construct and test theory. For example, the stabilising effects of aggregation and patchiness in host parasitoid systems have been discussed in a series of papers using continuous time deterministic models (Ives 1992; Godfray and Pacala 1992; Murdoch *et al.* 1992). As much as anything these papers show how apparently small alterations in the assumptions of a model and hence its formulation can significantly change the model's predicted outcome and lead to different conclusions as to what is happening in the field.

Data is not always helpful in formulating a model or choosing parameters to fit a certain field situation. Morris (1990) shows that even the type of statistical test used to fit data can alter the outcome of a discrete time model. Further, it may not be clear from data what model is most suitable in a given situation. Morris shows that two different discrete time models can fit data equally well but give different predicted outcomes. Thus modellers need to look carefully at the assumptions behind their models' formulations (Berryman 1992). If it is not clear what formulation is most appropriate from knowledge about the system being modelled, then several models should be analysed before making any predictions or proposing hard and fast theory about the underlying ecology.

In the rest of the paper we consider the behaviour and changes in behaviour as a classical predator prey model is harvested or stocked. These are explored using both mathematical analyses and simulations. The results are presented in diagrams which summarise possible qualitative outcomes of the model. These diagrams allow comparison between slightly different models and hence we can assess the likely effect of a particular structural change in a model.

A classical predator-prey model

In this section we consider the classical MacArthur-Rosenzweig equations (Rosenzweig and MacArthur 1963),

$$\frac{dH}{d\tau} = rH\left(1 - \frac{H}{K}\right) - aPF(H) - R \qquad (1)$$

$$\frac{dP}{d\tau} = -sP + bPF(H) - S$$

where H represents the prey population, P the predator population, τ the time and K, a, b, r and s are parameters in the model. The first term on the right hand side of each equation ($rH(1-H/K)-sP$) models the intrinsic growth of each species in the absence of the other. The prey population is resource-limited and the predators, in the absence of prey, decrease at a rate proportional to predator population. The second term on the right hand side of the first equation represents the death of prey due to predation. Predation is assumed to be proportional to the number of predators and a nonlinear function, $F(H)$ of prey density. This is the predator response function. To begin, we will consider a simple 'Holling Type 2' (Holling 1959) predator response function, $H/(c+H)$ where c is a constant relating to handling time, the time a predator takes to hunt or digest the prey before it is ready to hunt its next prey. The growth of the predator population is given by the second term in the second equation. It is proportional to the predation rate. The constant harvesting rates of prey and predator are R and S respectively. If R or S are negative these terms can represent stocking or some other type of input of individuals into the system. In fact R and S need not represent harvesting or stocking but can represent any movement of predator or prey into or out of the system at a constant rate. This might include immigration or death due to disease for example. For brevity, however, we will continue to refer to them as harvesting rates.

The equations can be rescaled for convenience to reduce the number of parameters. To do this we set

$$x = \frac{H}{K} \qquad t = s\tau \qquad y = \frac{a}{rK}P$$

$$\gamma = \frac{r}{s} \qquad \mu = \frac{b}{s} \qquad \alpha = \frac{c}{K} \qquad \rho = \frac{R}{rK} \qquad \phi = \frac{aS}{brK}$$

and the equations then become

$$\frac{dx}{dt} = \gamma(x(1-x) - f(x)y - \rho) \qquad (2)$$

$$\frac{dy}{dt} = y(-1 + \mu f(x)) - \phi$$

where x is the scaled prey population, y the scaled predator population and t is the scaled time variable. The parameter γ relates to the linear growth of prey. The parameter μ determines the predator's efficiency in converting prey to biomass; a high μ value reflects a high predator birth rate or long life expectancy or both. The rescaled harvesting rates are ρ and ϕ for prey and predator respectively. The rescaled predator response function is $f(x) = x/(x+\alpha)$.

We will examine the effects on the model ecosystem of harvesting and stocking just one species to begin with and then explore what happens if both predator and prey are harvested or stocked. The changes that occur as ρ and ϕ are changed are essentially bifurcations. We next look at the effects on the changes of population behaviour due to changes in harvesting rates when the formulation of the predator response function is altered. That is, we look at changes in the way the changes in model outcome occur.

The behaviour of model populations can sometimes be predicted, without solving Eq. 2 explicitly, by examining the possible equilibrium states of the populations. A suite of mathematical techniques exists to do this (see, for example, Edelstein-Keshet 1988; Gray and Roberts 1988; and for an application of these methods to an ecological system, Myerscough *et al.* 1992). Such analysis can be helpful, but does not always give the complete picture. For example, the model populations may have a stable steady state where both predator and prey coexist but this steady state may not be reached from some initial populations. Thus, if population evolution is being examined it is important to look not only at the equilibrium states but also at their 'basins of attraction', that is, what initial populations will evolve to the equilibria. Brauer and Soudack (1979a,b, 1981a,b) have calculated basins of attraction for some harvested predator-prey models.

In the next section we take a naive approach in investigating predictions that Eq. 2 make of the effects of harvesting and stocking. We then show how analysis can help make sense of the naive approach and give a sense of the underlying structure of the model.

Simulations and summaries

To appreciate the effects of harvesting or stocking we must first look at the behaviour of the model for the unharvested system. This unharvested model predicts that the predator and prey populations will either approach a steady equilibrium population or will tend to stable oscillations, depending on the values of the parameters μ and α. A theorem, Kolmogorov's theorem, states that, for these equations, these are the only two possible types of qualitative behaviour (see, for example, May 1973; Edelstein-Keshet 1988). In practice, when the populations are undergoing oscillations, the populations of predator and prey can become so small that they are effectively zero. Oscillations occur for high values of α the nondimensional handling time or μ, which gives a measure of the predator reproduction and death rates (Myerscough *et al.* 1996). When α and μ are low the predator becomes extinct. For intermediate values both predator and prey coexist at a steady state. The populations will tend either to the steady state or to oscillations when these exist, regardless of the initial populations. Thus the unharvested system has, in theory at least, an inherent stability which ensures that the prey will never become extinct. When equilibria with both predator and prey exist then the predator will also persist.

We first consider what happens if the predator remains unharvested while the prey is harvested or stocked. Figure 2 shows simulations when $\alpha = 0.1$, $\mu = 1.25$ and $\gamma = 5$ for various values of ρ. The predator is not harvested or stocked, so $\phi = 0$. When $\rho = 0$ the populations oscillate at equilibrium. Increasing ρ to 0.01 sends both populations to extinction. Stocking the model system by making ρ negative leads to a reduction in the amplitude of the oscillations and an increase in the average predator population. At about $\rho = -0.16$ the oscillations die out and the populations tend to a steady state. Thus including harvesting of prey destabilises both predator and prey populations and leads to extinctions; including stocking of prey stabilises the system.

In the rest of this section we consider cases where predator harvesting or stocking occurs with or without prey harvesting. We now consider the case where the prey is unharvested. Some simulations for this case are presented in Fig. 3. We use the same set of parameters as above. If neither predator nor prey is harvested then the populations oscillate in a similar ways to the population simulations shown in Fig. 2a. Increasing ϕ, that is, allowing the predator to be harvested, causes the equilibrium oscillations to decrease in amplitude and period. These look similar to those in Fig. 2c. The

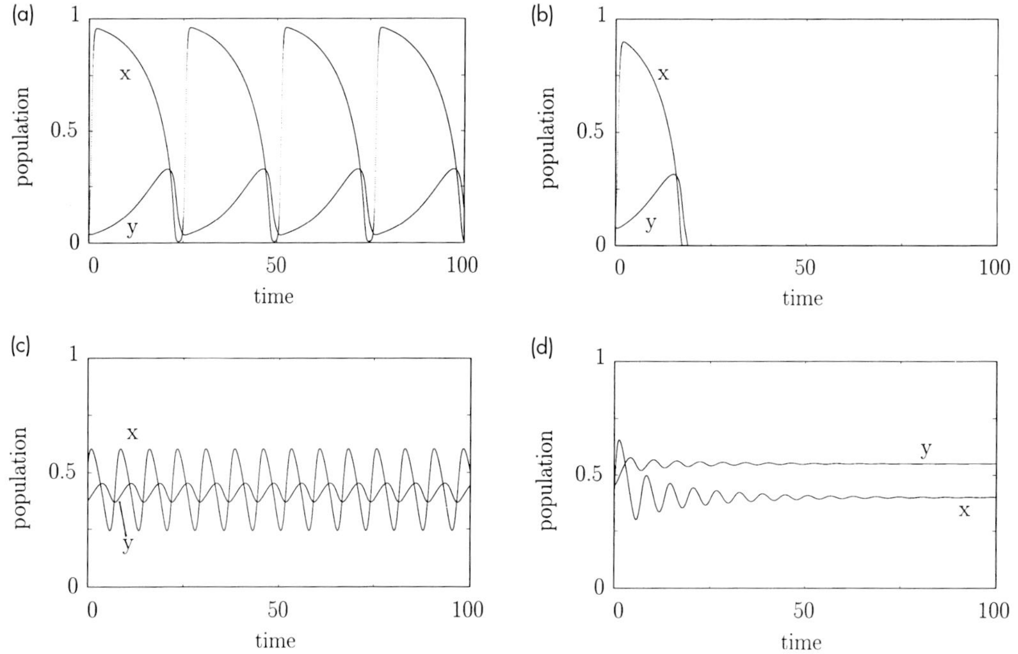

Fig. 2. Simulations showing the predicted behaviour of the model prey population x and predator population y as a function of time when the predator is unharvested, that is $\phi = 0$, for various levels of prey harvesting and stocking. The values of the model parameters are $\mu = 1.25$, $\gamma = 5$ and $\alpha = 0.1$. Diagram (a) shows that populations oscillate when both predator and prey are unharvested, i.e. $\rho = \phi = 0$. When the prey is harvested ($\rho = 0.01$ here) both predator and prey go to extinction as shown in (b). If the prey is stocked at a rate given by $\rho = -0.1$ the populations oscillate with a smaller amplitude and faster period, shown in (c), than the unstocked case. If the prey is stocked at a faster rate, such as $\rho = -0.2$ shown in (d), then the oscillations damp out to leave the populations at a steady state.

populations may not, however, settle to these oscillations even though the oscillations are stable or attracting. If the predator population is too high or too low then either the predator only or both the predator and prey will go to extinction. As ϕ continues to increase the oscillations die out and settle to a steady state where both predator and prey coexist. Whether the model populations tend to this steady state as they do in Fig. 3a or become extinct as in Fig. 3b will depend on whether the initial populations are in the basin of attraction of the steady state. For yet higher values of ϕ the predator becomes extinct and the prey tends to its equilibrium population in the absence of predator.

If the predator is stocked, that is, $\phi > 0$, then the oscillations persist until ϕ is about -0.09. The populations evolve to these stocking oscillations from any initial populations. The oscillations, however, pass through regions where the prey population is very low; in practice the prey may become extinct and the oscillations not persist. When stocking rates are higher the populations tend to a stable steady state with prey populations very low or effectively zero. At these values stocking is merely compensating for a high death rate as the predators cannot reproduce because of lack of prey due to over-predation. In summary, the stable oscillations in populations when $\phi = \rho = 0$ can be reduced to stable steady state populations by either harvesting or stocking predator at an appropriate rate. Over-harvesting or over-stocking can both lead to extinctions. The populations will reach the steady state in the harvested regime if the initial populations lie in the basin of attraction; any initial populations will reach steady state in the stocking regime.

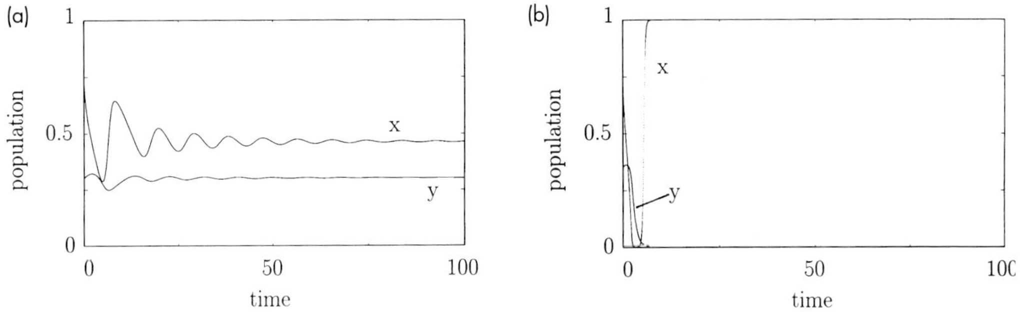

Fig. 3. Simulations of prey population x and predator population y when the prey is unharvested ($\rho = 0$) but the predator is harvested or stocked. The model parameters are the same as in Fig. 2. The predator is harvested at a rate $\phi = 0.085$. In (a) the initial populations are $x = 0.75$ and $y = 0.3$ and the populations tend to a stable equilibrium state via damped oscillations. In (b), where the initial conditions, $x = 0.75$ and $y = 0.35$, are outside the basin of attraction, the predator goes to extinction and the prey goes to $x = 1$, its equilibrium population in the absence of predators.

The nature of the equilibrium states and the type of behaviour predicted by the model can be summarised graphically by using mathematical analysis and simulations together. Figure 4 shows how the outcome of model changes for different values of ρ and predator life-expectancy or birthrate μ when the predator is unharvested. Figure 5 shows the same thing for different values of predator harvesting ϕ and μ when the prey is unharvested. It is apparent from this diagram that the range of harvesting rates for which oscillatory populations occur is very small. When the predator's birth rate is comparatively high, that is above about $\mu = 2$, then there is no steady state in the harvested regime and one or both populations will go to extinction. For intermediate values of μ there is a range of predator harvesting rates where there is no equilibrium state with both predator and prey coexisting when prey is unharvested. This range lies between two sets of values of ϕ where equilibrium states exist. This means that there may be some regimes where *increasing* harvesting could, in principle, prevent extinction. This steady state for the higher range of ϕ, however, has a very small basin of attraction. That is, populations tend to the steady state and persist only for a very small range of initial populations. This steady state occurs for a very small range of ϕ and so it is unlikely that such a phenomenon would ever be observed.

The advantage of diagrams such as Figs. 4 and 5 is that they allow changes in the equilibrium behaviour of model equations to be easily seen. Figure 6 shows such a diagram for $\rho = 0.05$. The most obvious and significant change between this picture and Fig. 5 is the loss of the region of oscillation for $\phi < 0$ when the predator is being stocked. Harvesting the prey destabilises the system and introducing predator harvesting or stocking will not restabilise the model populations. The differences between Fig. 5 where $\rho = 0$ and Fig. 6 where the prey harvesting rate ρ is not zero, could be regarded as the result of a structural change. Not only does the model outcome change as ϕ changes when $\rho > 0$ but the sequence of possible changes as ϕ is increased or decreased also alters from the sequence when $\rho = 0$.

When the prey is stocked, that is $\rho < 0$, the type of behaviour that the model shows is similar to that when $\rho = 0$. This behaviour is summarised in the diagram in Fig. 7. When $\phi = 0$, however, for values of μ above about 2.2, the populations tend to a steady state where predator and prey coexist. Below this value of μ the populations tend to stable oscillations when $\phi = 0$. The populations will tend to this steady state and these oscillations irrespective of the initial populations. As for the case $\rho = 0$ the populations will oscillate at equilibrium as ϕ increases and then, as ϕ continues to increase, the predator is driven to extinction and the prey tends to a steady state population. Stocking prey tends to stabilise the system and allow higher levels of predator harvesting before the predator is driven to extinction.

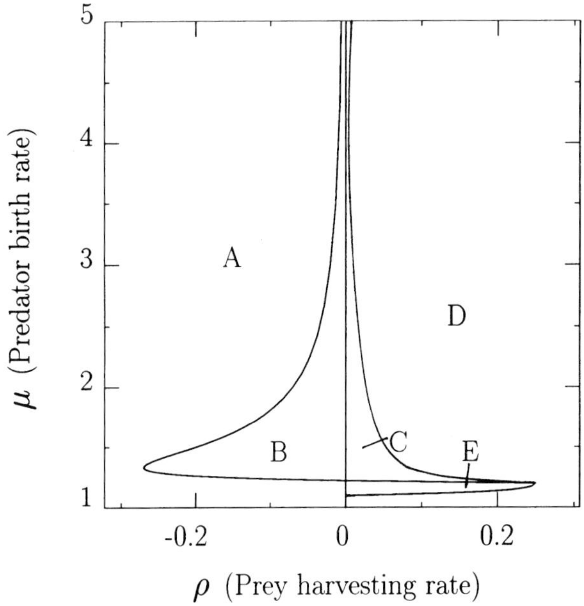

Fig. 4. This diagram summarises the different predicted behaviours of the model for different values of prey harvesting rate ρ and predator birth rate or life expectancy m when the predator is unharvested. The lines divide the diagram into regions where the model displays qualitatively different behaviours. In region A, a stable steady state exists which all initial populations will tend towards. Region B, stable oscillations exist which all initial populations will tend to. In region C stable oscillations exist within a finite basin of attraction. Some initial populations will evolve to these oscillations. From other initial populations either one of both populations will go extinct. Region D, either the predator only or both the predator and prey will go to extinction Region E, either one or both populations will go extinct or they will both go to a steady state which lies in a basin of attraction, depending on initial conditions. Two other types of behaviour are distinguished in this system. They are not shown on this diagram but they occur in other situations. On the diagrams where they occur they are denoted by F where the predator goes extinct and the prey tends to $x = 1$, its population in the absence of predator and G where both predator and prey become extinct. The values of the model parameters which are used in this diagram are $\gamma = 5$ and $\alpha = 0.1$.

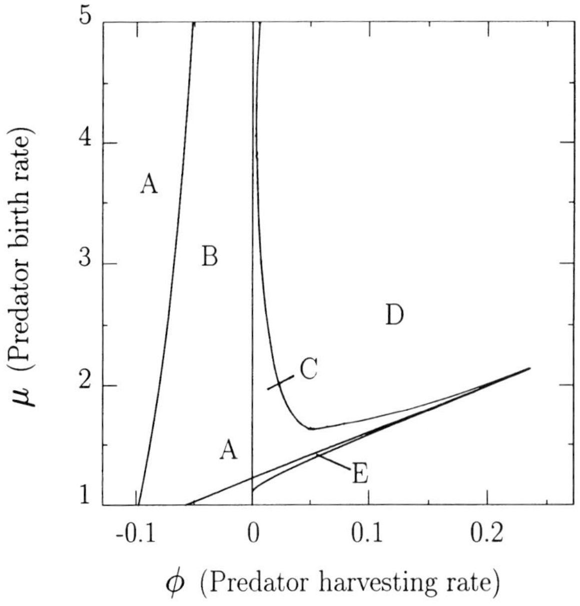

Fig. 5. Summary of model predictions as a function of predator harvesting rate ϕ and predator birth rate μ when prey is unharvested and unstocked ($\rho = 0$). The letters A–E denote regions of different behaviour. Their meanings are given in the caption to Fig. 4. The values of α and γ are as for Fig. 4.

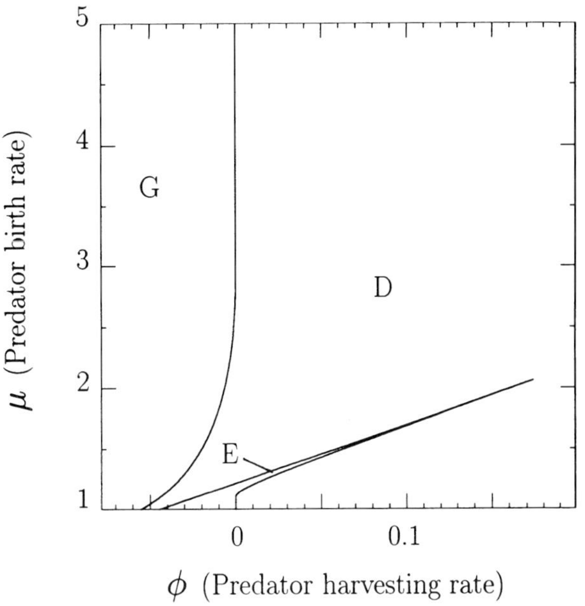

Fig. 6. Summary of model predictions as a function of predator harvesting rate ϕ and predator birth rate μ when prey is harvested (ρ = 0.05). Refer to the caption of Fig. 4 for the letters D, E and G. The parameter values are as for Fig. 4. Harvesting prey reduces the range of possible behaviours. There is a very small region where oscillations exist along the boundary between region D and E but this is not shown.

Harvesting with a Type 3 predator response function

Diagrams such as Figs. 4, 5, 6 and 7 show at a glance the type of behaviour expected from a model in any particular parameter regime. They also can provide some indication of how changing a model's formulation (as opposed to just changing its parameter values) can change its outcome. In this section we examine how the predicted results from the model change if the predator response function (PRF) is changed. We have been using the Holling Type 2 response function, $f(x) = x / (x + \alpha)$. This function models predator behaviour when the predator takes a finite amount of time to catch, consume or digest the prey and when the predator has no alternative prey. Instead of this Type 2 response we look at the model with a Type 3 response function. The unscaled form of this function is $F(H) = H^2 / (H^2 + d)$. After rescaling the function becomes $f(x) = x^2 / (x^2 + \beta)$ where $\beta = d / K^2$. This form of PRF is sigmoid and is illustrated in Fig. 8. It models predator behaviour where the predators seek alternative prey when the modelled prey population is low (Holling 1959). The model continues to assume, however, that consumption of the modelled prey is vital for the predator to reproduce. The two PRFs are shown in Fig. 8 for α = 0.1 and β = 0.04. This value of β was chosen as it produced a function curve comparable to the Type 2 curve with α = 0.1 used in simulations in the previous section.

Simulations were performed with μ = 1.25, β = 0.04 and γ = 5. The results of these simulations and analysis are summarised in Figs. 9, 10, 11 and 12. No population oscillations exist when both predator and prey are unharvested (Fig. 9). This is in contrast to the model with the Type 2 PRF where there are always oscillations when $\rho = \phi = 0$. For the unharvested system with the Type 3 PRF the populations always reach a steady state where both predator and prey coexist. This agrees with the received wisdom that Type 3 PRFs are stabilising. It is possible, however to find some parameter ranges where a model with a Type 2 PRF predicts more stable behaviour than a Type 3 PRF model (Myerscough *et al.* 1996). The effects of harvesting or stocking the prey only are summarised in Fig. 9. This should be compared to Fig. 4 which is the equivalent diagram for the Type 2 PRF. The Type 3 PRF model predicts that when the prey is stocked the populations tend to a steady state where both coexist. Harvesting only the prey destabilises the system. As ρ increases, the equilibrium

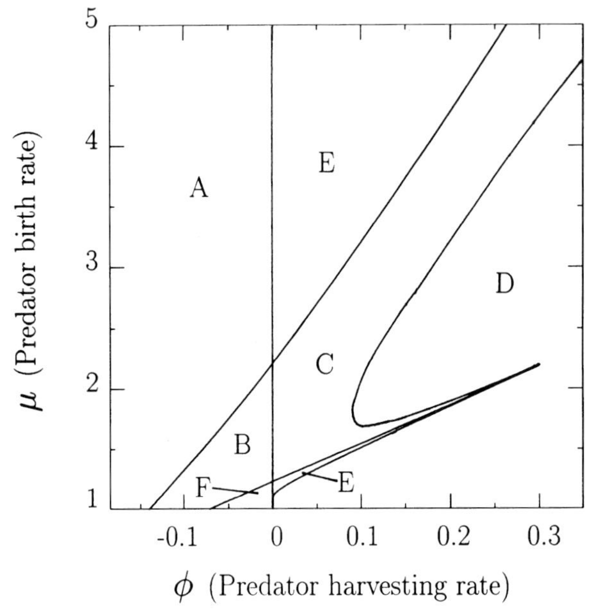

Fig. 7. Summary of model predictions as a function of predator harvesting rate ϕ and predator birth rate μ when prey is stocked (ρ = –0.05). Refer to the caption of Fig. 4 for the letters D, E and G. Stocking prey increases the range of values of ϕ and μ for which the model predicts either a stable equilibrium or stable oscillations. The values of *a* and *g* are as for Fig. 4.

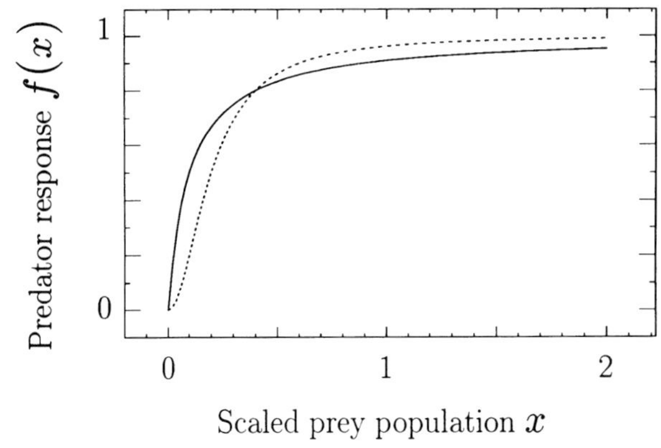

Fig. 8. Holling Type 2 and Type 3 predator response functions. The solid line is the Type 2 function given by $x/(x + \alpha)$ where α = 0.1. The dotted line is the Type 3 function given by $x^2/(x^2 + \beta)$ where β = 0.04. These values of *a* and *b* were chosen so that the functions are roughly similar.

populations cease to be steady and become oscillatory. The populations will only tend to these oscillations provided that the initial predator population is not too high and the initial prey population is not too low. Otherwise the prey become extinct, followed by the predator. As ρ becomes even higher there is no equilibrium where predator and prey coexist and either the predator or both predator and prey become extinct. This behaviour is essentially similar to that of the model with the Type 2 PRF.

The model behaviour when the predator is harvested or stocked is summarised in Figs. 10, 11 and 12. When only the predator is harvested and ρ = 0, the behaviour of the model with the Type 3 PRF, illustrated in Fig. 10, is similar to that of the model with the Type 2 PRF whose behaviour is summarised in Fig. 5. The only major difference is that oscillations only occur when μ is greater than about 1.5. The same is true when the prey is stocked as well as the predator being harvested or stocked; except for the lack of oscillations at low μ the type of changes which occur in the Type 3

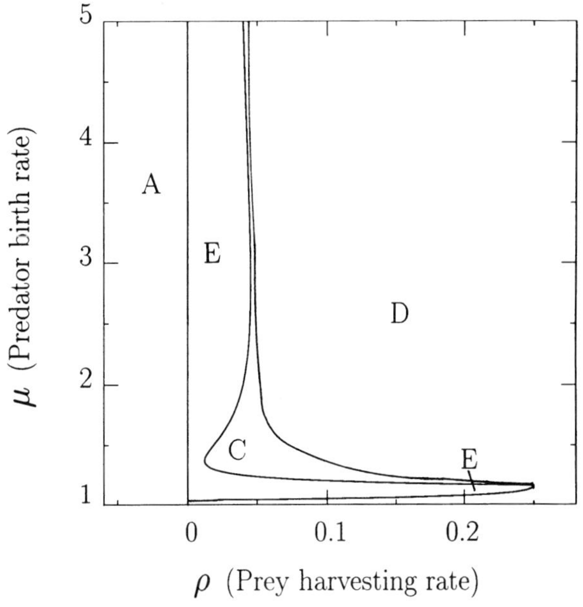

Fig. 9. Summary diagram of the model predictions when the Type 3 PRF is used. Here β = 0.04 and γ = 5. The diagram shows the predicted qualitative behaviour of the model for a range of prey harvesting rates ρ and predator birth rates μ when the predator is unharvested ϕ = 0. The meanings of the letters A–E are given in the caption of Fig. 4. This diagram should be compared to Fig. 4, the comparable diagram for the Type 2 PRF model. Note that the horizontal scales of Figs 4 and 9 are different.

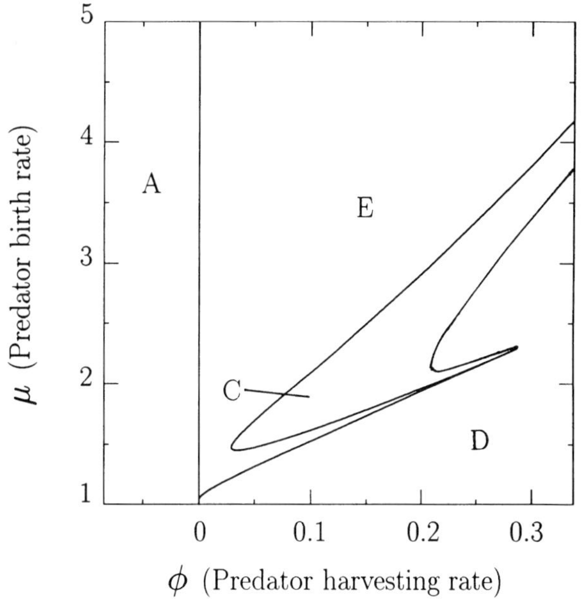

Fig. 10. Summary diagram of the model behaviour as a function of predator harvesting ϕ and predator birth rate μ when the Type 3 PRF is used and the prey is unharvested (ρ = 0). Parameter values are as for Fig. 9. The meanings of letters A, C, D and E which denote the different regions are given in the caption of Fig. 4. This diagram should be compared with Fig. 5 in the Type 2 PRF model. Note that the horizontal scale is different in Fig. 5.

PRF model (Fig. 11) are the same for that of the Type 2 PRF model (Fig. 7), although the exact values of ϕ where these changes occur are different.

The type of behaviour seen in simulations using the Type 3 PRF model with low levels of prey harvesting as predator stocking rates change can, however, be quite different from that of the Type 2 PRF model in similar circumstances. A series of simulations for the Type 3 PRF model with μ = 1.25 and ρ = 0.05 as stocking rates increase shows that when ϕ = 0 the populations tend to

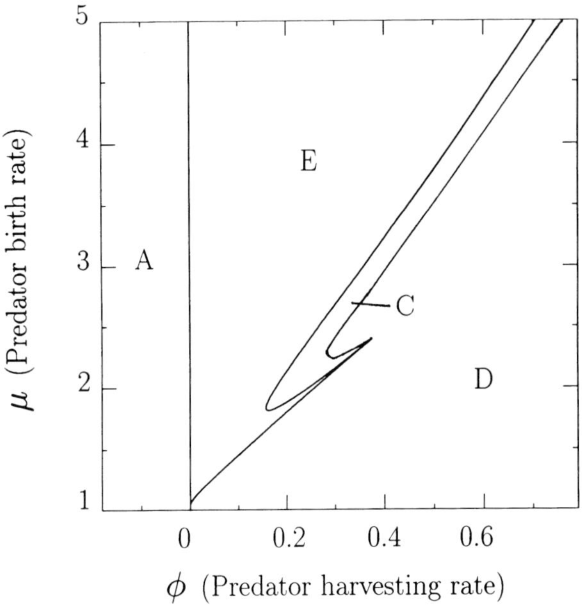

Fig. 11. Summary of model predictions as a function of predator harvesting rate ϕ and predator birth rate m when prey is stocked ($\rho = -0.05$) and the Type 3 PRF is used. Parameter values are as in Fig. 9 and the key to the letters denoting the different regions is given in the caption to Fig. 4. This diagram is comparable to Fig. 7 which gives the model behaviour for the Type 2 PRF (and has a different horizontal scale from Fig. 11).

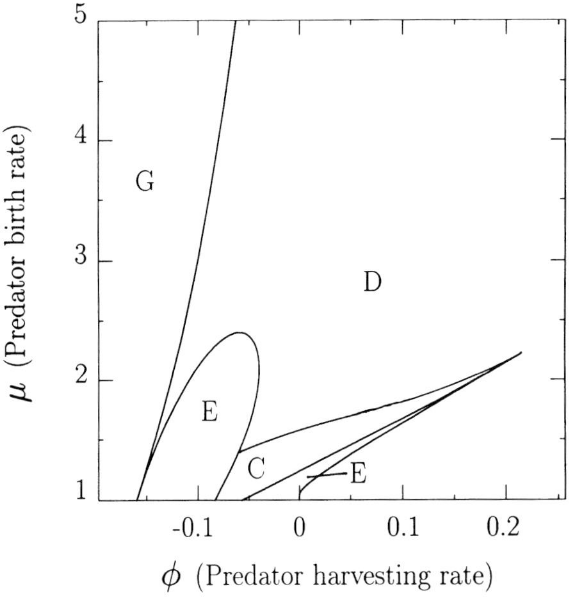

Fig. 12. Summary diagrams of Type 3 PRF model behaviour as a function of predator harvesting rate and predator birth rate when prey is harvested at the rate $\rho = 0.05$. Parameter values are as for Fig. 9 and the key to the labels C,D,E and G is given in the caption to Fig. 4. The most significant feature of this diagram, when compared to other summary diagrams, particularly Fig. 8, the equivalent diagram for the Type 2 PRF model, is the region E when the predator is stocked i.e., $\phi < 0$. In this region a steady state exists but it lies in a finite basin of attraction and is not accessible from all initial populations.

stable oscillations. When ϕ is about -0.07 these oscillations disappear and the populations either go to extinction or they tend to a stable, coexisting steady state. For the populations to tend to the steady state the initial populations must be close to the steady state as there is a distinct basin of attraction around this steady state; just how close will depend on the value of the predator stocking rate ϕ. If the populations are perturbed from the steady state to outside the basin of attraction then they will go to extinction. Mathematically this effect is due to the presence of an unstable limit cycle in the xy phase plane. As the stocking rate continues to increase this steady state disappears altogether and the populations always go to extinction, regardless of their initial values. These extinctions of both species under a regime of predator stocking are not seen in the Type 2 PRF model. The parameter range where this behaviour occurs is shown in the diagram in Fig. 12. Comparing this figure to Fig. 6 it is immediately clear that the behaviour of the model with a Type 3 PRF as ϕ and μ change when $\rho = 0.05$ is significantly different from that of the Type 2 model. Thus substituting a Type 3 PRF introduces a structural change in the model as it introduces an entirely new type of behaviour in the model populations. This new behaviour is not seen for higher values of prey harvesting such as $\rho = 0.1$. In these cases the Type 3 PRF model is similar to the Type 2 PRF model at the same value of ρ. Thus the model is structurally unstable to a change in the PRF from a Type 2 function to a Type 3 function. This instability is only manifest for some values of ϕ and ρ.

Discussion and conclusions

Three themes run through the results of this paper: the changes wrought in model behaviour when parameters are changed, the changes which occur when the model is changed and the importance of initial conditions where the steady state or equilibrium oscillation has a finite basin of attraction.

Generally speaking, harvesting prey or predator or both destabilises the system. For prey harvesting in the Type 2 PRF model this is particularly apparent in Fig. 4 and in comparing Fig. 5 with Fig. 6. However, for low values of prey birth rate μ there is always a small range of predator harvesting rates where a steady state exists and is attainable from some initial populations, even if the prey is harvested. This is a somewhat counter-intuitive result and is only apparent from analysing the model. The precise effect of predator harvesting or stocking depends on the rate of prey harvesting or stocking (Figs. 5, 6, and 7). Increasing predator harvesting tends to drive the predator to extinction while increasing predator stocking drives the prey to extinction. These are expected results of the model. However, changing the underlying rate of prey harvesting alters the way that predator harvesting affects the system. For example, when prey is neither harvested nor stocked then stocking predator causes population oscillations for a range of stocking rates near zero (Fig. 5). If, on the other hand, the prey is also being stocked then stocking the predator will not produce oscillations (Fig. 7).

Altering the model so that there is a Type 3 PRF instead of a Type 2 PRF in the model changes the way in which parameter change affects model predictions. For example, the Type 3 PRF model never shows oscillations when either predator or prey is being stocked (Figs. 9 and 10), two cases where there are stocking oscillations in the Type 2 PRF model (Fig. 4 and 5). The Type 3 PRF model does, however, admit the possibility of oscillations when the prey is harvested while the predator is stocked (Fig. 12). The Type 2 model in this case has no sizeable range of oscillations (Fig. 6). In summary, changing the PRF does lead to a structural change in the model. The Type 3 PRF model is generally more stable than the Type 2 PRF in that it has a greater range of harvesting rates where steady states exist. There are however a few surprises so that a modeller should be a little wary and think carefully about theoretical considerations before choosing the PRF. (See Berryman (1992) for a discussion of theoretical considerations in model building.) Data may fit equally well to both a Type 2 or a Type 3 PRF (see, for example, Hagen and Mann 1992). Analysis can be extremely helpful in pointing up the differences in the way that different models behave.

It is clear from the results in this paper that there are some harvesting rate ranges where initial populations affect the final outcome and some ranges where initial conditions are unimportant. This indicates the dangers of considering only steady states and behaviour near steady states (stability of steady states in particular) when assessing models or using them predictively. This strongly suggests that the wise modeller should beware and take care to use not only analysis but also simulations!

Acknowledgments

I thank W.L. Hogarth for useful discussions, Paul Wormell for assistance with computers and text formatting and J.M. Myerscough for practical assistance during the writing of this paper. This work was supported in part by an Australian Research Council small grant.

References

Berryman AA (1992) On choosing models for describing and analysing ecological time series. *Ecology* **73**:694–698

Brauer F, Soudack AC (1979a) Stability regions and transition phenomena for harvested predator-prey systems. *J Math Biol* **7**:319–337

Brauer F, Soudack AC (1979b) Stability regions in predator-prey systems with constant-rate prey harvesting. *J Math Biol* **8**:55–71

Brauer F, Soudack AC (1981a) Constant-rate stocking of predator-prey systems. *J Math Biol* **11**:1–14

Brauer F, Soudack AC (1981b) Coexistence properties of some predator-prey systems under constant rate harvesting and stocking. *J Math Biol* **12**:101–114

Edelstein-Keshet L (1988) *Mathematical models in biology*. Random House, New York

Godfray HCJ, Pacala SW (1992) Aggregation and the population dynamics of parasitoids and predators. *Am Nat* **140**:30–40

Gray BF, Roberts MJ (1988) A method for the complete qualitative analysis of two coupled ordinary differential equations dependent of three parameters. *Proc R Soc Lond Ser A* **416**:361–389

Hagen NT, Mann KH (1992) Functional response of the predators American lobster *Homarus americanus* (Milne-Edwards) and Atlantic wolffish *Anarhichas lupus* (L.) to increasing numbers of the green sea urchin *Strongylocentrotus droebachiensis* (Müller). *J Exp Biol Ecol* **159**:89–112

Holling CS (1959) The components of predation as revealed by a study of small-mammal predation of the European pine sawfly. *Can Entomol* **91**:293–320

Ives AR (1992) Continuous-time models of host-parasitoid interactions. *Am Nat* **140**:1–29

May RM (1973) *Stability and complexity in model ecosystems*. Princeton University Press, Princeton

Morris WF (1990) Problems in detecting chaotic behavior in natural populations by fitting simple discrete models. *Ecology* **71**:1849–1862

Murdoch WM, Briggs CJ, Nisbet RM, Gurney WSC, Stewart-Oaten A (1992) Aggregation and stability in metapopulation models. *Am Nat* **140**:41–58

Myerscough MR, Gray BF, Hogarth WL, Norbury J (1992) An analysis of an ordinary differential equation model for a two species predator-prey system with harvesting and stocking. *J Math Biol* **30**:389–411

Myerscough MR, Darwen MJ, Hogarth WL (1996) Stability, persistence and structural stability in a classical predator-prey model. *Ecol Modell* **89**:31–42

Rosenzweig ML, MacArthur RH (1963) Graphical representation and stability conditions of predator-prey interactions. *Am Nat* **97**:209–223

CONTRIBUTORS

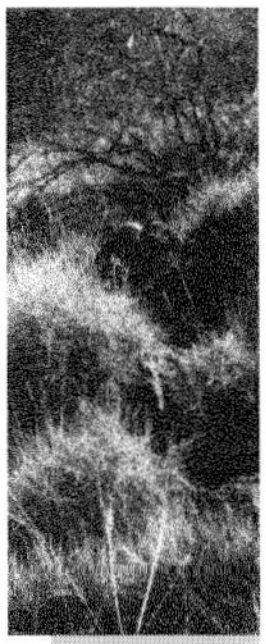

Anderson, Marti J., School of Biological Sciences, Institute of Marine Ecology, University of Sydney, NSW 2006, Australia

Ballard, J. William O., CSIRO Division of Entomology, PO Box 1700, ACT 2601, Australia. Current address: The Field Museum, Roosevelt Rd at Lake Shore Drive, Chicago, IL 60605, USA

Barlow, Nigel D., Biological Control Group, AgResearch, Canterbury Agriculture and Science Centre, PO Box 60, Lincoln, Canterbury New Zealand

Begon, Michael, Population Biology Research Group, Department of Environmental and Evolutionary Biology, University of Liverpool, Liverpool L69 3BX, UK

Boutin, Stan, Department of Biological Sciences, University of Alberta, Edmonton, Alberta T6G 2E9, Canada

Bowers, Roger G., Population Biology Research Group, Department of Environmental and Evolutionary Biology, University of Liverpool, Liverpool L69 3BX, UK

Brown, Joel R., CSIRO Division of Tropical Crops and Pastures, Private Mail Bag, PO Aitkenvale, QLD 4814, Australia

Burdon, Jeremy J., Centre for Plant Biodiversity Research, CSIRO Division of Plant Industry, GPO Box 1600, Canberra, ACT 2601, Australia

Cappuccino Naomi, Department of Zoology, University of Texas at Austin, Austin, TX 78712, USA

Chesson, Peter, Ecosystem Dynamics Group, Research School of Biological Sciences, Australian National University, ACT 0200, Australia

Choquenot, David, Vertebrate Pest Research Unit, NSW Agriculture, Agricultural Research and Veterinary Centre, Forest Rd, Orange, NSW 2800, Australia

Cockburn, Andrew, Evolutionary Ecology Group, Division of Botany and Zoology, Australian National University, Canberra, ACT 0200, Australia

Crawley, Mick J., Department of Biology, Imperial College, Silwood Park, Ascot SL5 7PY, UK

Daly, Joanne C., CSIRO Division of Entomology, GPO Box 1700, Canberra, ACT 2601, Australia

De Barro, Paul J., Department of Biology, University of Southampton, Bassett Crescent East, Southampton SO9 3TU, UK. Current address: CSIRO Division of Entomology, GPO Box 1700, Canberra, ACT 2601, Australia

de la Mare, W.K. (Bill), Australian Antarctic Division, Channel Highway, Kingston, Tasmania 7050, Australia

Dexter, Nick, Department of Ecosystem Management, University of New England, Armidale, NSW 2351, Australia

Dixon, A.F.G. (Tony), School of Biological Sciences, University of East Anglia, Norwich NR4 7TJ, UK

Drake, V. Alistair, Department of Physics, University College, University of New South Wales, Australian Defence Force Academy, Canberra, ACT 2600, Australia

Efford, Murray, G., Manaaki Whenua – Landcare Research, Private Bag 1930, Dunedin, New Zealand

Farrow, Roger A., CSIRO Division of Entomology, GPO Box 1700, Canberra, ACT 2601, Australia

Fitter, Alastair H., Department of Biology, University of York, York YO1 5DD, UK

Fleming, Richard A., Canadian Forest Service, Forest Pest Management Institute, PO Box 490, Saulte Ste. Marie, Ontario P6A 5M7, Canada

Floyd, Robert B., CSIRO Division of Entomology, GPO Box 1700, Canberra, ACT 2601, Australia

Fox, David R., CSIRO Biometrics Unit, Private Bag, Wembley, WA 6014, Australia

Galway, Nora J., CSIRO Division of Entomology, PO Box 1700, ACT 2601, Australia

Gatehouse, A. Gavin, School of Biological Sciences, University of Wales, Bangor, Gwynedd LL57 2UW, UK

Grice, A.C. (Tony), CSIRO Division of Tropical Crops and Pastures, Private Mail Bag, PO Aitkenvale, QLD 4814, Australia

Harrison, Susan P., Division of Environmental Studies, University of California at Davis, Davis, CA 95616, USA

Harwood, John, Natural Environment Research Council, Sea Mammal Research Unit, High Cross, Madingley Rd, Cambridge CB3 0ET, UK

Kindlmann, Pavel, Faculty of Biological Sciences, University of South Bohemia, Branisovska 31, CS37005 Ceske Budejovice, Czech Republic

Kingsland, Sharon E., Department of History of Science, Medicine and Technology, Johns Hopkins University, Baltimore, Maryland 21218, USA

Knell, Robert J., Population Biology Research Group, Department of Environmental and Evolutionary Biology, University of Liverpool, PO Box 147, Liverpool L69 3BX, UK. Current address: Department of Zoology, University of the Witwatersrand, Private Bag 3, WITS 2050, South Africa

Krebs, Charles J., Department of Zoology, University of British Columbia, 6270 University Blvd, Vancouver V6T 1Z4, Canada

Lavery, Shane, Queensland DPI Agricultural Biotechnology Centre and Department of Zoology, University of Queensland, QLD 4072, Australia

Lonsdale, W. Mark, CSIRO Division of Entomology, PMB 44, Winnellie, NT, 0821, Australia. Current address: CSIRO Biological Control Unit, Campus International de Baillarguet, 34982 Montpellier-sur-lez CEDEX, France

Mackerras, Ian M., Deceased

Maclean, Norman, Department of Biology, University of Southampton, Bassett Crescent East, Southampton SO9 3TU, UK

Mahon, Rod J., CSIRO Division of Entomology, c/- Institute Haiwan, PO Box 520, 86009 Kluang, Johor D.T. 6009, Malaysia

McCarthy, Michael A., School of Forestry, University of Melbourne, Parkville, Victoria 3052, Australia

McGarvey, Rick, SARDI/SAASC, PO Box 120, Henley Beach, SA 5022, Australia

Moritz, Craig, Department of Zoology and Centre for Conservation Biology, University of Queensland, QLD 4072, Australia

Morton, Stephen R., CSIRO Division of Wildlife and Ecology, P.O. Box 84, Lyneham, ACT 2602, Australia

Murdoch W.W. (Bill), Department of Biological Sciences, University of California, Santa Barbara, CA 93106, USA

Myerscough, Mary R., School of Mathematics and Statistics, University of Sydney, NSW 2006, Australia

Newsham, Kevin K., ITE, Monks Wood, Abbots Ripton, Huntingdon PE17 2LS, UK

Nisbet, Roger M., Department of Biological Sciences, University of California, Santa Barbara, CA 93106, USA

Possingham, Hugh P., Department of Environmental Science and Management, University of Adelaide, Roseworthy Campus, Roseworthy, SA 5371, Australia

Rees, Mark, Department of Biology, Imperial College, Silwood Park, Ascot SL5 7PY, UK

Ridsdill-Smith, T. James, CSIRO Division of Entomology, Centre for Mediterranean Agricultural Research, Private Bag, Wembley, WA 6014, Australia

Ritchie, Mark E., Ecology Center and Department of Fisheries and Wildlife, Utah State University, Logan, UT 84322-5210, USA

Rohani, Pejman, Natural Environment Research Council, Sea Mammal Research Unit, High Cross, Madingley Rd, Cambridge CB3 0ET, UK

Sait, Steven M., Population Biology Research Group, Department of Environmental and Evolutionary Biology, University of Liverpool, Liverpool L69 3BX, UK

Sequeira, Richard, School of Biological Sciences, University of East Anglia, Norwich NR4 7TJ, UK

Sheppard, Andrew W., CSIRO Division of Entomology, GPO Box 1700, Canberra, ACT 2601, Australia

Sherratt, Thomas N., Department of Biology, University of Southampton, Bassett Crescent East, Southampton SO9 3TU, UK. Current address: Department of Biological Sciences, University of Durham, Science Laboratories, South Road, Durham DH1 3LE, UK

Sinclair, A.R.E. (Tony), Department of Zoology, University of British Columbia, Vancouver, V6T 1Z4, Canada

Stone, Graham N., NERC Centre for Population Biology and Department of Biology, Imperial College at Silwood Park, Ascot SL5 7PY, UK

Sunnucks, Paul, Conservation Genetics Group, Institute of Zoology, Regent's Park, London NW1 4RY, UK. Current address: School of Biological Sciences, Macquarie University, NSW 2109, Australia

Taylor, Andrew D., Department of Zoology, University of Hawaii, 2538 The Mall, Edmondson 152, Honolulu, HI 96822, USA

Thompson, David J., Population Biology Research Group, Department of Environmental and Evolutionary Biology, University of Liverpool, Liverpool L69 3BX, UK

Twigg, Laurie E., Agriculture Protection Board, Bougainvillea Ave, Forrestfield, WA 6058, Australia

Underwood, A.J. (Tony), School of Biological Sciences, Institute of Marine Ecology, Zoology Building A08, University of Sydney, NSW 2006, Australia

Vet, Louise E.M., Department of Entomology, Wageningen Agricultural University, PO Box 8031, NL-6700 EH Wageningen, The Netherlands

Walker, Paul, A., CSIRO Division of Wildlife and Ecology, PO Box 84, Lyneham, ACT 2601, Australia

Watkinson, Andrew R., School of Biological Sciences, University of East Anglia, Norwich NR4 7TJ, UK

Wellings, Paul W., CSIRO Division of Entomology, GPO Box 1700, Canberra, ACT 2601, Australia

Williams, C. Kent, CSIRO Division of Wildlife and Ecology, PO Box 84, Lyneham, ACT 2602, Australia

Woodburn, Timothy, CSIRO Division of Entomology, GPO Box 1700, Canberra, ACT 2601, Australia

Wratten, Stephen D., Department of Entomology and Animal Ecology, PO Box 84, Lincoln University, Canterbury, New Zealand

INDEX

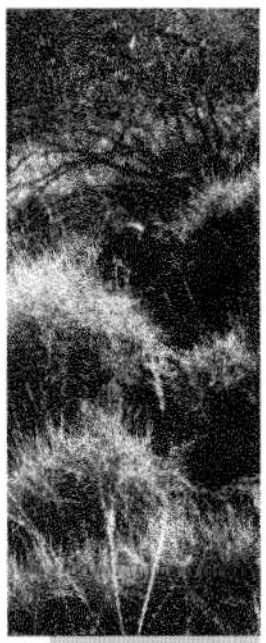